ISBN 978-0-265-93590-3
PIBN 10912314

FAMILIAR

LECTURES ON BOTANY,

PRACTICAL, ELEMENTARY, AND PHYSIOLOGICAL,

WITH

- A NEW AND FULL DESCRIPTION

OF

THE PLANTS OF THE UNITED STATES,
AND CULTIVATED EXOTICS, &c.

FOR THE USE OF

SEMINARIES, PRIVATE STUDENTS, AND PRACTICAL
BOTANISTS.

BY MRS. ALMIRA H. LINCOLN—Now MRS. LINCOLN PHELPS,
PRINCIPAL OF THE PATAPSCO FEMALE INSTITUTE OF MARYLAND.
Author of The Fireside Friend, A Series of works on Botany, Chemistry, and Natural
Philosophy, &c.

NEW EDITION, REVISED AND ENLARGED; .
ILLUSTRATED BY MANY ADDITIONAL ENGRAVINGS.

NEW YORK:
PUBLISHED BY HUNTINGTON AND SAVAGE,
216 PEARL STREET.
1846.

PREFACE.

This work was prepared, originally, with the view of being used as a text book in the class-room, and by private students, teaching in a simple and inductive manner the Science of Analytical Botany, as also Vegetable Physiology. It did not profess to contain a sufficient number of descriptions of genera and species to furnish a complete manual for the Botanist in collecting and labelling plants; those which were described were chiefly the more common, such as the student would be most likely to meet with in his botanical excursions, or could readily be collected for illustrations before classes, and for teaching the mode of analysing and classifying.

The extensive circulation of this work has encouraged the Author and publishers to incur new labor and expense to adapt it more fully to the demands of the public. These demands, according to the testimony of teachers in various sections of the country, are for a greater number of generic and specific descriptions of plants. We have, therefore, added extensively to the catalogue of *Southern* and *Western* plants, as also to that of more northern latitudes. So that the book will now contain descriptions of most of the plants of the United States, and cultivated exotics. We except such of the Cryptogamia and Grasses as are too obscure in their characteristics for the attention of the general student; as also some *new Species*, which appear to have been separated from their proper and established relations, in order to gratify the vanity of imaginary discoverers, or to enable them to compliment their friends by giving their names to the *supposed* new Species.

With the *Flora* of *Northern, Southern,* and *Western* plants now presented to the public, in connexion with the Familiar Lectures on Botany, we hope to have rendered our work such as will fully answer public expectation.

Patapsco Female Institute,
(Ellicott's, near Baltimore, Maryland,)
March 1, 1845.

TO TEACHERS.

THE author indulges the hope that this book will not only afford assistance but gratification to Teachers, in the pursuance of the severe and often *ennuyant* duties of their profession. It is hoped that it may serve to interest and quicken the dull intellects of some pupils, to arrest the fugitive attention of others, and to relax the minds of the over-studious, by leading them all into paths strewed with flowers, and teaching them that these beautiful creations of Almighty Power are designed, not merely to delight by their fragrance, color, and form, but to illustrate the most logical divisions of Science, the deepest principles of Physiology, and the goodness of God.

The best time for commencing botanical studies seems to be that of the opening of flowers in the spring; though, where circumstances render it convenient to begin in winter, assistance is offered by engravings. The arrangement of subjects might be altered, in pursuing the study *without the aid of natural flowers*. The *Second* part, which treats of the various organs of plants, the formation of buds, and other subjects connected with vegetable physiology ; the *Fourth* part, which gives the history of the science, with the distinctions in the kingdoms of nature, might be studied to advantage, before attending much to the principles of classification, which are mostly illustrated in the First and Third parts.

On the first meeting of a botanical class, after some explanation as to the nature of the study they are about to commence, each member should be presented with a flower for analysis. The flower selected should be a simple one, exhibiting in a conspicuous manner the different organs of fructification; the lily and tulip are both very proper for this purpose. The names of the different parts of the flower should then be explained, and each pupil directed to dissect and examine the flower. After noticing the parts of fructification, the pupils will be prepared to understand the principles on which the artificial classes are founded, and to trace the plant to its proper class, order, &c. At each step, they should be required to examine their flowers, and to answer simultaneously the questions proposed; as, how many stamens has your flower? Suppose it to be a lily, they answer *six*. They are then told it is of the sixth class. How many pistils? They answer *one*—they are told it is of the first order. They should then be directed to take their books and turn to the sixth class, first order, to find the genus. In each step in the comparison they should be questioned as above described, until, having seen in what respects their plant agrees with each general division, and differs from each genus under the section in which it is found, they ascertain its generic name. They should

be taught in the same manner to trace out its species: they will perceive
.at each step some new circumstance of resemblance or difference, until
they come to a species, the description of which answers to the plant
under consideration.

Technical terms should be explained as the pupil proceeds. The advantage in this kind of explanation, over that of any abstract idea, is, that
it is manifested to the senses of the pupils by the object before them. If
a teacher attempt to define the words *reason*, *will*, &c., or any other abstract terms, there is danger that the pupil may, from misunderstanding
the language used in the explanation, obtain but a very confused and imperfect idea of the definition; and, indeed, what two authors or philosophers give to abstract terms the same definition? Though mankind do
not, in the purely mental operations, exhibit an entire uniformity, yet, in
their external senses, they seldom disagree. A flower which appears to
one person to be composed of six petals, with corolla bell-form, and of a
yellow color, is seen to be so by another. Pupils who find it difficult to
understand their other studies (which in early youth are often too abstract),
are usually delighted with this method of analyzing plants; they feel that
they understand the whole process by which they have brought out the
result, and perhaps, for the first time, enjoy the pleasure of clear ideas
upon a scientific subject.

It is necessary, before the meeting of the class, to have a suitable number of plants collected, so that all may have specimens. In examining
the pupils as they proceed in their study, each one, besides reciting a lesson, should be required to give an analysis of one or more plants; sometimes the whole class having similar flowers; sometimes giving to each
pupil permission to bring any plant she chooses. This, also, at public
examinations, is a satisfactory method of testing their knowledge of the
subject. With respect to those portions of the work to which their attention should most particularly be paid, it must be left to the judgment of
the teacher. Whatever relates to modes of classification, and makes part
of a system, should be noted; many remarks, illustrations, and quotations,
are designed merely for reading, without being considered as important
matter for recitation.

The name of the Natural Order is connected with the name of each
genus under the head of Descriptions of Species; indeed, the subject of
the *natural affinities* of plants is kept in view through the whole work,
although the artificial system is considered as the groundwork of botanical knowledge. The origin of the generic name is also given, as far as
this could be ascertained with any degree of certainty.

The analysis at the bottom of each page, is designed rather to suggest
the leading subjects, than as a form of questions; for every experienced
teacher must perceive the importance of varying his mode of questioning.

CONTENTS

OF THE LECTURES AND APPENDIX.

PART III.—CLASSIFICATION.

PART IV.

PART V.—APPENDIX.

BOTANICAL names of plants are formed according to the analogies of the ancient languages, chiefly the Latin. Some of the most common terminations of names of Genera and Species, are in *a, um, us,* and *is;* for example, the generig names GERARDIA, TRIFOLIUM, PRUNUS, and IRIS; and the specific names, *virginicum,* *candidum, blandus,* and *officinalis.* A great proportion of botanical names terminate in *a,* in which case the word has the sound of *a* in *father,* as Rosa, Viola, &c.

The letter *e* at the end of a word is always sounded; for example, *Anemone,* pronounced *anem''o-ne.*

The *e* is long before *s,* when it ends a word, as *Bicor''nes,* pronounced *Bicor-nees.*

In words that end in *ides,* the *i* is long, as in *Hesper''ides.*

The vowels *ae* and *oe,* are often used as diphthongs, and then have the sound of *e,* as *Hepatica,* pronounced *Hepat'ice,* and *Di-œcia,* pronounced *Di-e-cia.*

C and *g,* as in English, are soft before *e, i,* and *y,* and hard before *a, o,* and *u.* The soft sound of *c* is like *s,* the hard sound like *k.* The soft sound of *g,* is like *i,* the hard sound like *g,* in the word *gave;* thus *Algæ* is pronounced *Al''je.* *Musci* is pronounced *Mus''ci.*

The letters *ch* are hard like *k,* as in *Orchis,* pronounced *Or'-kis.*

Accent and Quantity.

The marks over the Generic and Specific names, in the Description of Genera and Species, have reference not only to the *syllable which is to be accented,* but to the *quantity of the vowel* in the accented syllable, as either *long* or *short.*

Those syllables over which the *single mark* is placed, have the *vowel pronounced long,* as in *Fra-ga'-ria;* those over which the *double mark* is placed, have the vowel *short,* as in *He-pat''i-ca;* in the latter case, the stress of voice is thrown upon the consonant; the two marks may, therefore, be considered as indicating that the consonant, as well as the vowel, is accented.

Words of *two syllables* always have the *accent on the first;* if the syllable *end with a vowel, it is long,* as in *Cro'-cus;* if it *end with a consonant, it is short,* as in *Cac''-tus.*

Figures, and other Characters.

The figures at the right hand of the name of the Genus, in the Description of Species, refer to the Class and Order of the Plant in the Artificial System; the word following the figures, and included in a parenthesis, designates the natural order the plant. (For the characteristics of these orders, see Appendix, from page 27 to 32.)

The following characters denote the duration of the plant :—

⊕ Annual—♂ Biennial—♃ Perennial—♄ Woody.

Colour of Corollas.

r. red, p. purple, g. green, b. blue, w. white, y. yellow. The union of any two or more of these characters, denotes that the different colours are united.

Ex. stands for exotic.

S. stands for south, referring to a region south of the Middle States.

Time of Flowering.

Mar. March, Ap. April, M. May, J. June, Ju. July, Au. August, S. September, Oc. October, Nov. November.

Localities.

Can. Canada, N. E. New England, Car. Carolina, *height,* i. and m. inches, f. and ft. feet.

INTRODUCTION.

LECTURE I.

IMPORTANCE OF SYSTEM.—ADVANTAGES TO BE DERIVED FROM THE STUDY OF
BOTANY.

THE universe consists of *matter* and *mind*. By the faculties of mind with which God has endowed us, we are able to examine into the properties of the material objects by which we are surrounded.

If we had no sciences, nature would present exactly the same phenomena as at present. The heavenly bodies would move with equal regularity, and preserve the same relative situations, although no system of Astronomy had been formed. The laws of gravity and of motion, would operate in the same manner as at present, if we had no such science as Natural Philosophy. The affinities of substances for each other were the same, before the science of Chemistry existed, as they are now. It is an important truth, and one which cannot be too much impressed upon the mind in all scientific investigations, that no systems of man can change the laws and operations of Nature; though by systems, we are enabled to gain a knowledge of these laws and relations.

The Deity has not only placed before us an almost infinite variety of objects, but has given to our minds the power of reducing them into classes, so as to form beautiful and regular systems, by which we can comprehend, under a few terms, the vast number of individual things, which would, otherwise, present to our bewildered minds a confused and indiscriminate mass. This power of the mind, so important in classification, is that of *discovering resemblances*. We perceive two objects, we have an idea of their resemblance, and we give a common name to both; other similar objects are then referred to the same class or receive the same name. A child sees a flower which he is told is a rose; he sees another resembling it, and nature teaches him to call that also a rose. On this operation of the mind depends the power of forming classes or of generalizing.

Some relations or resemblances are seen at the first glance; others are not discovered until after close examination and reflection; but the most perfect classification is not always founded upon the most obvious resemblances. A person ignorant of Botany, on beholding the profusion of flowers which adorn the face of nature, would discover general resemblances, and perhaps form in his mind, some order of arrangement; but the system of Botany now in use, neglecting the most conspicuous parts of the flower, is founded upon the observation of small parts of it, which a common observer might not notice.

System is necessary in every science. It not only assists in the acquisition of knowledge, but enables us to retain what is thus acquired; and, by the laws of association, to call forth what is treasured up in the storehouse of the mind. System is important not only in the grave and elevated departments of science, but is essential in the most common concerns and operations of ordinary life. In conducting any kind of business, and in the arrangement of household

By the faculties of mind we examine the properties of matter—Human science cannot alter the laws of nature—Power of the mind to form classes—Classification not always founded upon the most striking resemblances, as in Botany—Importance of system.

2

concerns, it is indispensable to the success of the one, and to the comfort of those interested in the other. The very logical and systematic arrangement which prevails in Botanical science, has, without doubt, a tendency to induce in the mind the habit and love of order; which, when once established, will operate even in the minutest concerns. Whoever traces this system through its various connexions, by a gradual progress from individual plants to general classes, until the whole vegetable world seems brought into one point of view, and then descends in the same methodical manner, from generals to particulars, must acquire a habit of arrangement, and a perception of order, which is the true practical logic.

The study of Botany seems peculiarly adapted to females; the objects of its investigation are beautiful and delicate; its pursuits, leading to exercise in the open air, are conducive to health and cheerfulness. It is not a sedentary study which can be acquired in the library, but the objects of the science are scattered over the surface of the earth, along the banks of the winding brooks, on the borders of precipices, the sides of mountains, and the depths of the forest.

A knowledge of Botany is necessary to the medical profession. Our Almighty Benefactor, in bestowing upon us the vegetable tribes, has not only provided a source of refined enjoyment in the contemplation of their beautiful forms and colours; in their fragrance, by which, in their peculiar language, they seem to hold secret communion with our minds; He has not only given them for our food and clothing, but with kind, parental care, has, in them, provided powers to counteract and remove the diseases to which mankind are subject. For many ages plants were the only medicines known, or used; but modern discoveries in Chemistry, by forming compounds of previously existing elements, have, in some degree, superseded their use. Although the science of medicine has received much additional light from Chemistry, it may perhaps in modern days have occupied the attention of medical men too exclusively; inducing them to toil in their laboratories to form those combinations which nature has done, much more perfectly, in the plants which they pass unheeded. It is probable that the medicinal productions of the animal and mineral kingdoms, bear but a small proportion to those of the vegetable.

When our forefathers came to this country, they found the natives in possession of much medical knowledge of plants. Having no remedies prepared by scientific skill, the Indians were led, by necessity, to the use of those which nature offered them: and, by experience and observation, they had arrived at many valuable conclusions as to the qualities of plants. Their mode of life, leading them to penetrate the shades of the forest, and to climb the mountain precipices, naturally associated them much with the vegetable world. The Indian woman, the patient sharer in these excursions, was led to look for such plants as she might use for the diseases of her family. Each new and curious plant, though not viewed by her with the eye of a botanist, was regarded with scrutinizing attention; the colour, taste, and smell, were carefully remarked, as indications of its properties. But the discoveries and observations of the Indians have perished with themselves; having had no system for the classification or description of plants, nor any written language by which such a system might have been conveyed to others, no other vestige remains than uncertain tradition, of their knowledge of the medicinal qualities of plants.

The study of nature, in any of her forms, is highly interesting and useful. But the *heavenly bodies* are far distant from us;—and were they within our reach, are too mighty for us to grasp, our feeble minds seem overwhelmed in the contemplation of their immensity. *Animals*, though affording the most striking marks of designing wisdom, cannot be dissected and examined without painful emotions. But the *vegetable world* offers a boundless field of inquiry, which may be explored with the most pure and delightful emotions. Here the Almighty seems to manifest himself to us, with less of that dazzling sublimity which it is almost painful to behold in His more magnificent creations; and it would seem, that accommodating the vegetable world to our capacities of observation, He had especially designed it for our investigation and amusement, as well as our sustenance and comfort.

The study of Botany naturally leads to greater love and reverence for the Deity. We would not affirm, that it does in reality always produce this effect; for, unhappily, there are some minds which, though quick to perceive the beauties of nature, seem blindly to overlook Him who spread them forth. They can admire the gifts, while they forget the giver. But those who feel in their hearts a love to God, and who see in the natural world the workings of His power, can look abroad, and adopting the language of a christian poet, exclaim,

"My father made them all."

Division of the Lectures.

Having endeavoured to convince you that the study you are about to commence, is recommended by its own intrinsic utility, and especially by its tendency to strengthen the understanding and improve the heart, we will now present you with the arrangement which we propose to follow.

We will divide our course of study into *Four Parts*, viz.:

PART I. Will be chiefly devoted to teaching the *Analysis* of *Plants*, or lessons in *Practical Botany*.

PART II. We shall here consider the various organs of the plant, beginning with the root and ascending to the flower; this part will include what is usually termed *Elementary Botany ;* it will also contain remarks upon the uses of the various organs of plants, the nature of vegetable substances, and other circumstances connected with *Vegetable Physiology*.

PART III. In this part we shall consider the different *systems of Botany*. We shall examine some of the most important *Natural families ;* and then proceed to give a detailed view of the *Linnæan System ;* remarking upon some of the most interesting genera and natural families found under each class and order.

PART IV. In this part we shall consider the Progressive appearance of Flowers during the season of blossoming; their various phenomena produced by the different states of the atmosphere, light, &c.; and their geographical distribution. After giving a History of the progress of botanical science, we shall, in a general view of Nature, consider the distinction between organized and unorganized matter, with their analogies and contrasts.

Heavenly bodies—Animals—Study of the vegetable world—The study of Botany tends to piety.—Division of the subject into *four parts*—I. Practical Botany—II. Vegetable Physiology—III. Systematic Botany—IV. Various phenomena of Plants, History of Botany, and General Views of Nature.

PART I.

LECTURE II.

THE Universe, as composed of *mind* and *matter*, gives rise to various sciences. The SUPREME BEING we believe to be *immaterial*, or *pure mind.*

The knowledge of *mind* may be considered under *two general heads.*

1. THEOLOGY,* or that science which comprehends our views of the Deity, and our duties to Him.

2. PHILOSOPHY OF THE HUMAN MIND, or *metaphysics,*† which is the science that investigates the mind of man, and analyzes and arranges its faculties.

The knowledge of *matter*, which is included under the general term, *Physics*, may be considered under *three general heads.*

1. NATURAL PHILOSOPHY, which considers the effects of bodies acting upon each other by their mechanical powers; as their weight and motion.

2. CHEMISTRY, in which the properties and mutual action of the elementary atoms of bodies are investigated.

3. NATURAL HISTORY, which considers the external forms and characters of objects, and arranges them in classes.

NATURAL HISTORY is divided into *three branches.*

1. ZOOLOGY,‡ which treats of animals.

2. BOTANY, which treats of plants.

3. MINERALOGY, which treats of the unorganized masses of the globe; as stones, earths, &c. GEOLOGY, which treats of minerals as they exist in masses, forming rocks, is a branch of mineralogy.

Having thus presented you with this general view of the natural sciences, we will now proceed to that department which is to be the object of your present study.

Departments in Botany.

BOTANY§ treats of the vegetable kingdom, including every thing which grows, *having root, stem, leaf, or flower.* This science comprehends the knowledge of the methodical arrangement of plants, of their structure, and whatever has relation to the vegetable kingdom. The study of plants may be considered under two general heads.

1st. The classification of plants by means of comparing their different organs, is termed *Systematic Botany.*

2d. The knowledge of the relations and uses of the various parts of plants with respect to each other, is termed *Physiological Botany.* This department includes Vegetable Anatomy.

* From the Greek *Theos*, God, and *logos*, a discourse.
† From *meta*, beyond, and *phusis*, nature. This term originated with Aristotle, who, considering the study of the intellectual world as beyond that of the material world, or physics, called it *meta ta phusis.*
‡ From *zoe*, life, and *logos*, a discourse.
§ From the Greek, *botane*, an herb.

Sytematic Botany is divided into the *artificial* and *natural methods*. The artificial method is founded upon different circumstances of two organs of the plant, called the *pistils* and *stamens*. Linnæus, of Sweden, discovered that these organs are common to all plants, and essential to their existence. Taking advantage of this fact, he founded divisions, called classes and orders, upon their *number*, *situation*, and *proportion*. By this system, plants which are unlike in their general appearance, but agree in certain particulars of their stamens and pistils, are brought together; thus in a dictionary, words of different signification are placed together from. the mere circumstance of agreement in their initial letters.

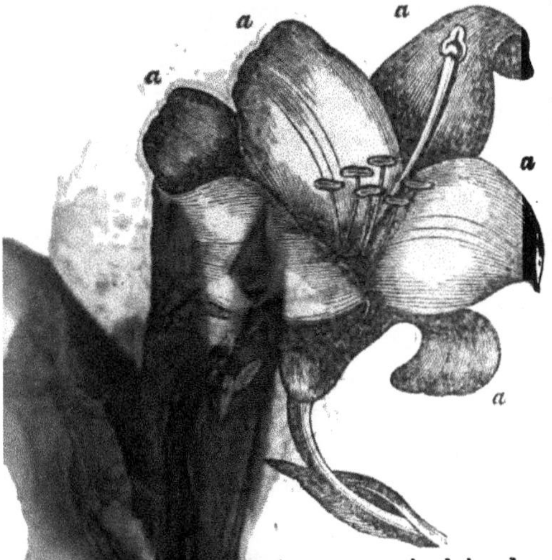

Before you can learn the principles on which the classification of plants depends, it is necessary that you should become acquainted with the parts of a flower;—you have here the representation of a white Lily. (*See fig.* 1.) At first this flower up in a green bud, by degrees it changes its colour, and into a blossom.

Explanation of the parts of a flower as seen in the Lily.

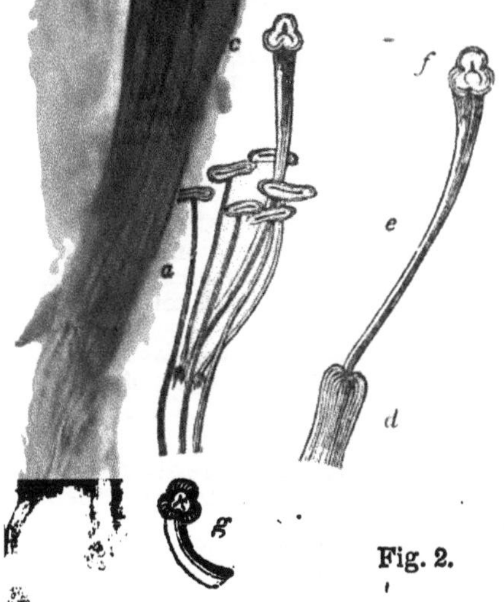

The envelope is called the *corolla* from *corona*, a crown.

The pieces which compose the corolla are called *petals*. (Fig. 1. *a.*)

The six thread-like organs within the corolla are called stamens; each stamen consists of a *filament*, (Fig. 2. *a*,) and an *anther* (*b*.) The anther contains the *pollen*, a fine powder, which serves to give life to the young seed. When the flower comes to maturity, the anthers burst and scatter the pollen. In the centre of the flower is the *pistil*, (Fig. 2. *c ;*) this consists of the germ, (*d*,) the *style*, (*e*,) and the stigma, (*f.*) The germ contains the young seeds, called *ovules ;* these are contained in one or more cavities, called *cells*. The end of the stem which supports the organs of the flower, and which in some plants is very broad, is called the *receptacle*, (Fig. 2. g.)

Fig. 2.

Artificial Method—The flower enveloped in the bud—Corolla—Petals—Stamens. Parts of a stamen—Pistil—Parts of the pistil—Receptacle.
2*

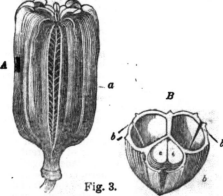

Fig. 3.

When the seed is ripe, the germ is then called the *pericarp*, from the Greek *peri.* around, and *karpos* fruit. Pericarps are of different kinds; that of the lily is called a *capsule*, (signifying casket,) (Fig. 3. *A*;) it is of a dry membraneous texture, and when ripe opens by the separation of pieces, called *valves.* In the capsule at *A*, is seen a longitudinal opening, with fibres connecting the valves as appears in a mature state. (Fig. 3. *B*) represents the capsule as if cut transversely to show its three cells (*b ;*) each cell contains two triangular seeds (*c.*)

The lily, although a beautiful flower, is deficient in one organ, which is common to the greater part of flowers ; this is the *calyx*, or cup, which is usually green, and surrounds the lower part of the corolla, as in the Pink.

When the calyx consists of several parts, these are called *sepals* and sometimes leaves of the calyx.

The organs we have now considered, are as follows:

Calyx—the cup, surrounding the corolla, the parts are called sepals.
Corolla—the blossom, the parts are petals.
Stamens—next within the corolla, the parts are the anther, pollen, and filament.
Pistil—central organ, the parts are the germ, style, and stigma.
Receptacle—which supports the other parts of the flower.
Besides these, there are in the mature plant, the
Pericarp—containing the seed.
Seed—rudiment of a new plant.

· Botanical Analysis.

Although the examination of the different organs of the flower may properly be called *analysis*, because it is the observation of constituent parts singly ;—yet when the botanist speaks of *analyzing plants*, he understands an examination of their organs with reference to determining their place in some botanical system.

We will now proceed to the analysis of some plants, that we may thus introduce the pupil to what we believe the best system of botanical arrangement for popular use.

Division of Plants into Classes, &c.

According to the system we shall adopt, all plants are divided into *twenty-one classes.* Each class is divided into Orders, the Orders into Genera,* and the Genera into Species.

The name of the *genus* may be compared to the family name; that of the *species*, to the individual or christian name; for example : the

---

* The plural of *genus*, a family or tribe.

Pericarp—Describe that of the lily—Calyx—Sepals—Enumerate the parts of a flower—What is meant by analyzing plants ?—Classes—Orders—Genera—Species.

Rosę family contains many different species; as *Rosa alba*, the white rose, *Rosa damascena*, the damask rose, &c. The specific or individual name in Botany, is placed after the family name, as *Rosa alba*, which is rose white, instead of white rose : this circumstance is probably owing to the use of Latin terms; as in that language the adjective is generally placed after the noun, instead of before it, as in English.

LECTURE III.

METHOD OF ANALYZING PLANTS.—ANALYSIS OF THE PINK, LILY, ROSE, AND POPPY.

WHEN you begin to analyze plants, you will meet with many new terms. It will be necessary in these cases, to resort to the vocabulary of botanical words ;* by the observation of plants, connected with definitions, you will soon become familiar with the technical terms of Botany.

We will now proceed to analyze a flower in order to ascertain its botanical name. We will commence with the Pink, as you are provided with a drawing which you can examine if you have no natural flower.†

Analysis of the Pink.

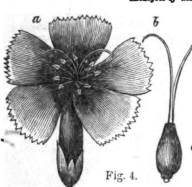

The first step, is to find the class. We will suppose this flower to belong to one of the *first ten classes ;* in this case, all you have to do is, to ascertain the *number of stamens,* as by this circumstance, these classes are arranged.

Because there are *ten* stamens, (Fig. 4. *a,*) the Pink is in the 10th class, the name of which is, *Decandria.* The second step is to find the *order.*

Fig. 4.

In the first 12 classes, the orders depend on the number of pistils ;—these you must count; —because you find *two,* (Fig. 4. *b,*) you know your flower belongs to the 2d order ;—the name of which is Digynia.

You must now turn to the " Description of the Genera of plants ;"‡ find class 10th, order 2d. The third step is to ascertain the genus of your plant; for this purpose, you must compare it with each *genus,* until you find it described.

* This is placed in the latter part of this volume.
† In analyzing a natural flower, it is necessary to separate the parts; first, if there is a calyx, remove it carefully, then take off the corolla, or if it is monopetalous, cut it open with a knife. A microscope is necessary if the organs are very small.
‡ See Table of Contents.

1st. 'HYDRANGEA. *Calyx 5 toothed, superior ;'*—your calyx is 5 toothed, (see the notches around the top of it, Fig. 5, *a*,) but it is not *superior*, that is, it does not stand upon the germ. You must go to the next genus.

2d. 'SAXIFRAGA. *Calyx 5 parted, half superior,'*—but your calyx is not half superior, or partly above the germ. You must go to the next genus.

3d. 'SAPONARIA. *Calyx inferior*, 1 *leafed, tubular, 5 toothed,'*—so far the description agrees with the Pink; next, '*calyx without scales.'* In this particular, your flower, the calyx of which has scales, (Fig. 5. *b*,) does not correspond with the description ;—therefore you must look further.

* Fig. 5.

4th. 'DIANTHUS. *Calyx inferior, cylindrical,* 1 *leafed, with* 4 *or* 8 *scales at the base ; petals 5,* (Fig. 4, *a*,) *with claws (long and slender at the base ;)—capsule cylindrical, celled, dehiscent (gaping.'*) Fig. 5 at *c*, represents the ripe capsule of the pink opening at the top by the parting of its valves ;—at *d*, it appears cut crosswise, and showing that it has but one cell, and many seeds. Fig. 4, at *c*, represents the capsule, as seen in the germ, when the pink is in blossom. Your flower agreeing with every particular in the description of the last-mentioned genus, you may be certain that you have found the generic or family name of the Pink, which is DIANTHUS.

But there are several species in this genus; you wish to know to which the Pink belongs; and this process constitutes a fourth step in your analysis.

Turn to the *Description of Species of Plants,** and look for Dian thus. Now compare the description of each species, with a Pink having the leaves and stem before you; 1st. '*Armeria, flowers aggregate,'* (in a thick cluster ;) this does not agree; you must look further.

2d. '*Barbatus, flowers fascicled,'* (crowded together,) but your flower grows singly on each stalk.

3d. '*Caryophyllus, flowers solitary, scales of the calyx sub-rhomboid,* (somewhat diamond-shape ;) *very short, petals crenate,* (scolloped on the edge,) *beardless,'* (without any hair or down.)

The Pink answers to this description. It is also added that the 'leaves are *linear*,' which signifies long and of nearly equal width; '*subulate*,' which signifies pointed at the end, like a shoemaker's awl; '*channelled*,' which signifies furrowed.

You have now found the botanical name of your plant to be DI-ANTHUS *Caryophyllus ;* and that it belongs to

         Class 10th, DECANDRIA.   Order 2d, DIGYNIA.

In this way it should be labelled for an *herbarium* or collection of dried plants.

You will remember, that in this process, *four* distinct steps have been taken; first, to find the *class ;* second, the *order ;* third, the *genus ,* and fourth, the *species.*

You can now proceed with the analysis of any plant which belongs

* See Table of Contents.

Fourth step in the analysis of a plant.

to the first ten classes, in the same manner as you have done with the Pink; as all these classes depend upon the number of stamens.

Analysis of the Lily.

In analyzing the Lily, you can refer to Figures 1st, 2d, and 3d; —you will find this flower belonging to the 6th class, the name of which is HEXANDRIA; and to the 1st order, MONOGYNIA. (In the description of Genera, see Class 6th, Order 1st.) This order, containing many genera, is divided into several *sections*.

1st Section contains flowers, ' *with a calyx and corolla*.'

The Lily has no calyx, therefore you will not find it in this section.

2d Section. ' *Flowers issuing from a spatha*.'

The Lily has no spatha ' or sheath at its base,' therefore it is not in this section.

3d Section. ' *Flowers with a single, corolla-like perianth*.'

The Lily has such a corolla-like envelope, therefore you may expect to find it described under this section. You can proceed, as in the Pink, to compare each genus with your flower, till you find one which corresponds with the Lily.

' HEMEROCALLIS. ' *Corolla six parted.* This shows that the corolla is all of one piece, with six divisions in the border. The Lily has six petals, therefore you need look no farther in this genus.

LILIUM.' Now compare each particular in this description with your flower, (looking out the terms in the vocabulary,) and you will find an agreement in every respect.

In the description of a genus, nothing is usually said about any part of the plant, except the different organs of *the flower*; in the species, the distinctions are chiefly drawn from different circumstances of the *leaves, stems, &c.*

The flowers of two plants may agree in the organs of fructification, while the leaves, stalks, and branches, are very unlike; in this case, the plants are considered as belonging to different species of the same genus.

Thus, the shape of the leaves, the manner in which they grow on the stem, its height, with the number of flowers growing upon it, the manner in which they grow, whether erect or nodding, these, and other circumstances, distinguish the different species. The colour, a quality of the flower usually the most striking, is, in botany, little regarded; while many other particulars, which might at first have been scarcely noticed, except by botanists, are considered as important.

In the 11th class, Icosandria, and the 12th class, Polyandria, we are to remark, not only the number of stamens which is always more than ten; but the *manner in which they are inserted*, or the part of the flower on which they are situated. If, in pulling off the corolla, the stamens remain upon the calyx, the plant belongs to the 11th class; but if the corolla and calyx may be both removed, and the stamens still remain on the receptacle, the plant is of the 12th class.

It is said that no poisonous plant has the stamens growing on the calyx; it is in the 11th class that we find many of our most delicious fruits, as the Apple, Pear, &c.

Analysis of the Rose.

The rose, on account of its beauty, is one of the most conspicuous flowers in the 11th class; it is considered as one of the most inter-

esting of the vegetable race, and is often, dignified with the title of
"queen of flowers."

You will perceive, on examining the Rose, that its numerous stamens are attached to the calyx. A more perfect idea of their situation may be obtained by removing the petals, and cutting the calyx
longitudinally. Therefore, because it has more than ten stamens
growing upon the calyx, it belongs to the 11th class, Icosandria.
The pistils being more than ten, it is of the 13th order, Polygynia. It
belongs to the genus Rosa.

The shape of the calyx is '*urnform;*' the calyx is '*inferior,*' or
below the germ; it is '*five cleft,*' or has five divisions around the
border; '*it is fleshy,*' or thick, '*contracted* towards the top;' '*petals
5,*' (this is always the case with a rose in its natural state, unassisted
by cultivation;) '*seeds numerous, bristly, fixed to the sides of the ca
lyx within.*'

There is no seed vessel, or proper pericarp to the rose; but the
calyx swells and becomes a dry, red berry, containing many seeds.

The genus Rosa contains many species, distinguished one from
another, by the different shape of the ger   roughness of the stems, the presence or absence   of the
leaves, and the manner in which the flo   ks,
whether solitary, crowded together in pa
ther they are erect, or drooping.

The Moss rose, (ROSA *muscosa,*) is distin
resembling moss, which cover the stems of
are a collection of glands containing a resinous

The apple blossom appears like a little rose;
thick and pulpy, and at length constitutes that par
call the fruit, though strictly speaking, the see
On examining an apple, you may notice, at the e
the five divisions of the calyx.

Analysis of the Poppy.

The Poppy affords a good illustration of the 12th cla
dria; here are numerous stamens, always *more than ten,*
more than a hundred, growing upon the receptacle; the Po
but one pistil, and therefore belongs to the first order, Monogy
the genus is PAPAVER. The Poppy has a '*calyx of two leaves or*
pals,' but these fall off as soon as the blossom expands, and are the
fore called '*caducous;*' the corolla (except when double) '*is f*
petalled;' it has no style, but the stigma is set upon the germ, an
therefore said to be *sessile.*

The germ is large and somewhat oblong, the stigma is flat a
radiated. The pericarp is one-celled, or without divisions, it ope
at the top, by pores, when the seeds are ripe. The species of Papaver which is cultivated in gardens, is the *somniferum,* which name
signifies to produce sleep. It is often called Opium Poppy.

The analysis of even one or two flowers, cannot fail of suggesting
thoughts of the beauty of a system which so curiously identifies the
different plants, described by botanists, and points to each individual of the vegetable family the place it must occupy. Even one hour
spent by a person in following a plant from class to order, and from
order to genus, until its name and specific character were ascertained, would be of great value, should this be all of botany he was
ever to learn.

Why is it in the 11th class?—why the 13th order?—Generic characters of the Rose
—Circumstances which distinguish the different species of the genus Rosa—Apple
blossom and fruit—Analysis of the Poppy—The analysis of one or two flowers useful

In the commencement of a new science, however, it is not to be expected that every idea, or principle of arrangement, will seem perfectly clear, as such may often relate to other principles not yet explained. In architecture, we know it would be impossible to form a clear idea of the use or beauty of a particular part of an edifice, until it was considered in its relation to the whole. The beginner in any branch of scientific knowledge, is not like one travelling a straight road, where every step is so much ground actually gained; but the views which he takes are like the faint sketches of a painter, which gradually brighten, and grow more definite as he advances.

An idea was formerly entertained, that students must learn perfectly, every thing as they proceed; but this appears to be founded upon a wrong view both of the nature of the mind, and of the sciences. The memory may be so disciplined as to retain a multitude of words, but words are only valuable as instruments of conveying knowledge to the mind; and if, after a careful attention to a subject, something in your lessons may appear obscure, you must not be discouraged; the confusion may arise from want of clearness in an author's style, or the subject may be connected with something which is to follow, therefore, you should patiently proceed, with the hope and expectation that difficulties will gradually disappear.

We shall not at present give any more examples of analyzing plants. With even the little practice you have now had, you can analyze flowers of any of the first thirteen classes; but it is necessary for you to know before proceeding farther, that the two circumstances of the *number* and *insertion* of the stamens, are not all that are considered in the arrangement of the classes;—this was not sooner observed, that your minds might not be confused with too many new ideas.

You are now prepared to comprehend the general features of the Linnæan system, and to study the whole of the classes and orders in a connected view. Before proceeding to this, it seems necessary that you should have some knowledge of Greek and Latin numerals. In our next lecture we shall commence by this necessary preparation, and shall then explain the characters of the classes and orders, and illustrate the same by drawings. Sensible objects are of great assistance to the mind, by enabling it to form definite ideas of the meaning of words. In abstract studies we cannot have such aid; and in order to comprehend instructions given upon them, it is necessary that the definitions of words should be well understood. Many persons are satisfied with a general notion of the meaning of abstract terms; thus, they speak of ' a *sensation* of pity,' when they mean an *emotion*. A more critical knowledge of the meaning of words, would enable them to perceive, that *sensation* is a term appropriated to that state of the mind which immediately follows the presence of an external object; it depends on the connexion between the body and the mind. The mind, separated from all the organs of sense, could have no *sensations ;* but it could have *emotions*, for they are feelings which the mind has, independently of the senses.

The great advantage of pursuing studies which relate to material objects, is, as we have before remarked, in being able to illustrate principles, and define terms by a reference to those objects themselves, or to delineations of them.

Remarks respecting the commencement of a new science—Words of use only as instruments—Assistance which the mind derives from sensible objects—Example of using terms indefinitely.

LECTURE IV.

WE shall now present you with a list of Latin and Greek numerals; these it is necessary to commit to memory, in order that you may understand the names given to the classes and orders. It is not in Botany alone that a knowledge of these numerals will be useful to you; many words in our common language are compounded with them; as, *uniform*, from *unus*, one, and *forma*, form;—*octagon*, from *octo*, eight, and *gonia*, an angle, *hexagon*, *pentagon*, &c.

NUMERALS.

Latin.	Numbers.	Greek.	Latin.	Numbers.	Greek.
Unus,	1.	Monos, single.	Duodecem,	12.	Dodeka.
Bis,	2.	Dis,—twice.	Tredecem,	13.	Dekatreis.
Tres,	3.	Treis.	Quatuordecem,	14.	Dekatettares.
Quatuor,	4.	Tettares.	Quindecem,	15.	Dekapente.
Quinque,	5.	Pente.	Sexdecem,	16.	Dekaex.
Sex,	6.	Hex.	Septendecem,	17.	Dekaepta.
Septem,	7.	Hepta.	Octodecem,	18.	Dekaōkto.
Octo,	8.	Okto.	Novemdecem,	19.	Dekaennea.
Novem,	9.	Ennea.	Viginti,	20.	Eikosi.
Decem,	10.	Deka.	Multus,	Many.	Polus.
Undecem,	11.	Endeka.			

The Classes of Linnæus.

In the first place, all plants are arranged in two grand divisions, *Phenogamous*, when the stamens and pistils are visible, and *Cryptogamous*, when the stamens and pistils are too small to be visible, by the naked eye. The former division includes 20 classes, the latter only the 21st.

The classes are founded upon distinctions observed in the STAMENS. All known plants are divided into *twenty-one classes.*

The first twelve classes are named by prefixing Greek numerals to ANDRIA, which signifies *stamen.*

CLASSES.

	Names.	Definitions.
	1. MON-ANDRIA,	*One Stamen.*
	2. DI-ANDRIA,	*Two Stamens.*
	3. TRI-ANDRIA,	*Three Stamens.*
	4. TETR-ANDRIA,	*Four Stamens.*
Number of Stamens.	5. PENT-ANDRIA,	*Five Stamens.*
	6. HEX-ANDRIA,	*Six Stamens.*
	7. HEPT-ANDRIA,	*Seven Stamens.*
	8. OCT-ANDRIA,	*Eight Stamens.*
	9. ENNE-ANDRIA,	*Nine Stamens.*
	10. DEC-ANDRIA,	*Ten Stamens.*

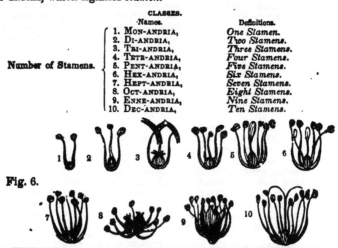

Fig. 6.

Words compounded with Latin and Greek numerals—Latin numerals—Greek numerals—Two grand divisions of plants—Classes, on what founded?—how many?—first twelve, how named?

| Number of Stamens, and their position, relative to the Calyx and Receptacle. | 11. ICOS-ANDRIA,* (Eikosi.) 20 | Over ten Stamens inserted on the Calyx. |
| | 12. POLY-ANDRIA, (Polus.) many. | Over ten Stamens inserted on the Receptacle. |

Fig. 7.

The two following classes are named by prefixing Greek numerals to DYNAMIA, which signifies *power* or *length*.

| Number and relative length of Stamens. | 13. DI-DYNAMIA, | Two Stamens longer or more powerful than the other two. |
| | 14. TETRA-DYNAMIA, | Four Stamens longer or more powerful than the other two. |

Fig. 8.

The two following classes are named by prefixing Greek numerals to the word ADELPHIA, which signifies *brotherhood*.

| Connexion of Stamens either by filaments or anthers. | 15. MON-ADELPHIA, | Stamens united by their filaments in one set or brotherhood. |
| | 16. DIA-DELPHIA, | Two brotherhoods. |

The next class is named by prefixing SYN, signifying *together*, to GENESIA, which signifies *growing up*.

| | 17. SYN-GENESIA, | Five united anthers, flowers compound. |

Fig. 9.

The next class is named by an abbreviation of the word GYNIA, which signifies pistil, prefixed to ANDRIA, showing that the stamens and pistils are united.

| | 18. GYN-ANDRIA, | Stamens growing out of the Pistil. |

The two following classes are named by prefixing numerals to œCIA, which signifies a house.

| Position of Stamens relative to the Pistil. | 19. MON-œCIA, | Stamens and Pistils on separate corollas upon the same plant, or in one house. |
| | 20. DI-œCIA, | Stamens and Pistils in separate corollas upon different plants, or in two houses. |

* The name of this class does not now designate its character, since the number of stamens is often more or less than twenty.

Classes which depend on the number of stamens—those which depend on number and position—number and relative length—What classes depend on the connexion of the stamens?—Explain the signification of their names—What classes depend on the position of the stamens?—What does Gynandria signify?—Monœcia?—Diœcia?

3

Fig. 10.

The name of the 21st class is a compound of two Greek words,
CRYPTO and GAMIA, signifying a concealed union.

Natural Families. } 21. CRYPTO-GAMIA, { *Stamens and Pistils invisible, or too
small to be seen with the naked eye.*

Fig. 11.

Lichens.   Mushrooms.   Ferns.   Mosses.

The number of classes as arranged by Linnæus, was twenty-four.
Two of them, Poly-adelphia, (many brotherhoods,) which was the
eighteenth class; and Poly-gamia, (many unions,) the twenty-third
class, are now, by many botanists,* rejected as unnecessary. The
eleventh class, Dodecandria, which included plants whose flowers
contain from twelve to twenty stamens, has been more recently
omitted. The plants which were included in these three classes have
been distributed among the other classes.

The Orders of Linnæus.

The orders of the first twelve classes are founded upon the number of PISTILS.

The orders are named by prefixing Greek numerals to the word
GYNIA, signifying *pistil.*

ORDERS.

	Names.	No. of pistils.
	1. MONO-GYNIA,	1.
	2. DI-GYNIA,	2.
	3. TRI-GYNIA,	3.
	4. TETRA-GYNIA,	4.
Orders found in	5. PENTA-GYNIA,	5.
the first twelve	6. HEXA-GYNIA,	6. *this order seldom found.*
classes.	7. HEPTA-GYNIA,	7. *this still more unusual.*
	8. OCTO-GYNIA,	8. *very rare.*
	9. ENNEA-GYNIA,	9. *very rare.*
	10. DECA-GYNIA,	10.
	13. POLY-GYNIA, *over ten pistils.*	

The classes vary as to the number of orders which they contain.
The orders of the 13th class, Didynamia, are but two.

1. GYMNOSPERMIA. From GYMNOS, signifying naked, and SPERMIA,
Seeds usually four, lying in   signifying seed, implying that the seeds are not
the calyx.   covered by a seed vessel.

2. ANGIOSPERMIA. From ANGIO, signifying bag or sack, added to
Seeds numerous in a capsule. SPERMIA, implying that the seeds are covered.

_____

* A few writers still retain the 24 classes of Linnæus;—but in the works of Eaton,
Torrey, Beck, and Nuttall, only 21 are adopted.
_____

What does Cryptogamia signify?—Classes omitted—Orders of the first twelve
classes, on what founded?—How are the orders named?—Orders of the class Didynamia.

The orders of the 14th class, Tetradynamia, are two, both distinguished by the form of the fruit.

1. SILICULOSA. Fruit, a *silicula*, or *roundish* pod.
2. SILIQUOSA. Fruit, a *siliqua*, or *long* pod.

The orders of the 15th class, Monadelphia, and of the 16th class, Diadelphia, are founded on the number of stamens, that is, on the characters of the first twelve classes, and they have the same names, as Monandria, &c.

The 17th class, Syngenesia, has its five orders distinguished by different circumstances of the florets, as:

1. EQUALIS. Stamens and pistils *equal*, or in proportion; that is, each floret has a *stamen*, a *pistil*, and one seed. Such florets are called *perfect*.
2. SUPERFLUA. Florets of the disk perfect, of the ray containing *only pistils*, which without stamens are *superfluous*.
3. FRUSTRANEA. Florets of the disk perfect, of the ray neutral, or without the stamen or pistil; therefore *frustrated*, or useless.
4. NECESSARIA. Florets of the disk staminate, of the ray pistillate; the latter being *necessary* to the perfection of the fruit.
5. SEGREGATA. Florets separated from each other by partial calyxes, or each floret having a perianth.

The orders of the 18th class, Gynandria, of the 19th class, Monœcia, and the 20th class, Diœcia, like those of the 15th and 16th classes, depend on the number of stamens.

The orders of the 21st class, Cryptogamia, constitute six natural families.

1. FILICES,—includes all *Ferns*, having the fruit on the leaves.
2. MUSCI,—Mosses.
3. HEPATICAE,—Liverworts, or succulent mosses.
4. ALGAE,—Sea-weeds, and frog spittle.
5. LICHENES,—Lichens, found growing on the bark of old trees, old wood, &c.
6. FUNGI,—Mushrooms, mould, blight, &c.

Note.—No confusion is produced in taking the character of some classes, for orders in other classes; for example : if you have a flower with ten stamens, *united by their filaments into one set*, you know by the definition of the classes that it belongs to the *class* Monadelphia ; you can then, because it has ten stamens, place it in the *order* Decandria.

LECTURE V.

METHOD OF ANALYZING PLANTS BY A SERIES OF COMPARISONS—GENERAL RE-
MARKS UPON PLANTS—METHOD OF PRESERVING PLANTS FOR AN HERBARIUM
—POISONOUS PLANTS, AND THOSE WHICH ARE NOT POISONOUS.

THE dissection of a plant is, properly, analysis; the meaning of the term being a separation : but when we speak of analyzing plants, we mean something more than examining each part of the flower ; this is, indeed, the first step in the process; but by analysis, we learn the Class, Order, Genus, and Species of the plant. A person engaged in ascertaining the name of a plant, may be said to be upon a *Botan-*

Of Tetradynamia—Of the classes Monadelphia and Diadelphia—Of the class Syngenesia—Of the classes Gynandria, Monœcia, and Diœcia—Of the class Cryptogamia—Meaning of the word analysis—How used in Botany

ical Journey, and the plant being his Directory; if he can read the botanical characters impressed on it by the hand of Nature, he will, by following system, soon arrive at his journey's end.[*]

Let us suppose, then, we have before us a plant in blossom, of whose *name* and *properties* we are ignorant.—The *name* must be first ascertained, and this can only be done with certainty by the Linnæan system.

In the first place we have *two comparisons* to make.

1st. Whether the *Stamens* and *Pistils* are VISIBLE.

2d. Whether they are INVISIBLE.

If the Stamens and Pistils are not visible, we have already arrived at the class, which is CRYPTOGAMIA.

If, however, the *Stamens* and *Pistils* are *visible*, we have now two *comparisons* to make.

1st. Whether the flowers have *stamens and pistils* on the same corolla.

2d. Whether the *Stamens and Pistils* are placed *on different corollas*.

If the Stamens and Pistils are on different flowers, we then shall find our plant either in the class *Diœcia* or *Monœcia;* according as the Stamens and Pistils are on different flowers, proceeding from the same root, or from different roots.

But if our plant has the Stamens and Pistils both enclosed in the *same corolla*, we must next examine,

1st. Whether the *Anthers* are *separate*, or,

2d. Whether the *Anthers* are *united*.

If we find *five anthers united* around the *pistil*, we have found the class of our plant; it is SYNGENESIA.

If the *Anthers* are separate, we must proceed to a *fourth stage*, and see,

1st. Whether the *filaments* are *separate*, or,

2d. Whether the *filaments* are *united* with each *other*, or,

3d. Whether the *filaments* are *united* to the *pistil*.

If the latter circumstance is ascertained, we need search no farther; our plant is in the class GYNANDRIA.

If the flower has not the filaments united to the pistil, we must ascertain if the filaments are united with each other; if they are so, and in two parcels or sets, the flower is in the class DIADELPHIA, but,

If in *one* parcel or *set*, it is in the class MONADELPHIA.

But if the *filaments* are *separate*, we must next examine,

1st. Whether these are *similar* in length, or,

2d. Whether they are of different lengths.

(Of different *lengths*, those only which have *four* or *six stamens* are to be regarded.)

If we find our flower has *six stamens, four long* and *two short*, we need go no farther, this is the class TETRADYNAMIA.

If the flower has four stamens, two long, and *two short*, it is in the class DYDYNAMIA.

If our flower comes under none of the foregoing heads, we must then count the number of stamens; if these amount to *more* than ten, we must then consider their *insertion*, as,

[*] Thornton.

What two comparisons to be first made in analyzing a plant—When the stamens and pistils are enclosed in the same corolla, what is next to be considered?—When the anthers are separate, what must be done?—If the filaments are separate, what must be observed?—If the flower has not stamens of unequal length, what is to be observed?

1st. Whether inserted *on the calyx or corolla,* or,
2d. Whether inserted *on the Receptacle.*

If we find the Stamens inserted on the *Receptacle,* the flower is in the class POLYANDRIA; but if on the *Calyx* or *Corolla,* it is in ICOSANDRIA.

If our flower has less than twenty stamens, with none of the peculiarities above mentioned, of *connexion, position,* or *length,* we have only to count the number of stamens, in order to be certain of the class; if there are *ten* stamens, it is in DECANDRIA; and so on through the nine remaining classes. This is the true analytical process; but when we put plants together to form a species, and species together to form a genus, and genera together to form an order, and orders together to form a class, we then proceed by Synthesis, which means putting together.

General Facts relating to Vegetables.

Plants are furnished with pores, by which they imbibe nourishment from surrounding bodies. The part which fixes the plant in the earth, and absorbs from it the juices necessary to vegetation, is the *root ;* this organ is never wanting.

The *stem* proceeds from the root; sometimes it creeps upon the earth, or remains concealed in its bosom; but generally, the stem ascends either by its own strength, or, as in the case of vines, by supporting itself upon some other body. The divisions of the stem are its *branches ;* the divisions of the branches are its *boughs.* When the vegetable has no stem, the flower and fruit grow from the tops of the root; but when the stem exists, that or its branches bear the leaves, flowers, and fruits. *Herbs* have generally soft, watery stems, of short duration, which bear flowers once, and then die.

Trees and *shrubs* have solid and woody stems; they live and bear flowers many years.

Small bodies of a round or conical form, consisting of thin scales, lying closely compacted together, appear every year upon the stems, the boughs, and the branches of trees. They contain the *germs* of the productions of the following years, and secure them from the severity of the seasons. These germs, and the scales which cover them, are called *buds.* The buds of the trees and shrubs of equinoctial countries, have few scales, as they are less needed for protection against inclemencies of weather.

Leaves, like flowers, proceed from buds; the former are the lungs of vegetables; they absorb water and carbonic acid from the atmosphere, decompose them by the action of rays of light, and exhale or give out oxygen gas.

Vegetables, like animals, produce others of their kind, and thus perpetuate the works of creation. The organs essential to the perfection of plants, are the stamens and pistils. Those plants in which the stamens and pistils are manifest, are called *Phenogamous ;* where these are rather suspected than demonstrated to exist, they are called *Cryptogamous.* The presence of a stamen and pistil only constitutes a *perfect* flower; but in general, these organs are surrounded with an inner envelope, called the *corolla,* and an outer one, called the *calyx.* When there is but one envelope, as in the tulip, this is often called by the more general term of *perianth,* which signifies, surrounding the flower. Persons ignorant of botany, give exclusively

---

When is the flower in one of the first ten classes?—Difference between analysis and synthesis—Stem—Branches—Boughs—Herbs—Trees and Shrubs—Buds—Leaves - Phenogamous and Cryptogamous plants.

the name of *flower* to these envelopes, which are often remarkable for the brilliancy of their colours, the elegance of their forms and the fragrance of their perfumes.

Method of preserving Plants, and of preparing an Herbarium.

Plants collected for analysis, may be preserved fresh many days, in a close tin box, by occasionally sprinkling them with water; they may also be preserved by placing their stems in water, but not as well by the latter, as the former method. While attending to the science of Botany, you should keep specimens of all the plants you can procure. An herbarium neatly arranged is beautiful, and may be rendered highly useful, by affording an opportunity to compare many species together, and it likewise serves to fix in the mind the characters of plants. It is a good method in collecting plants for an herbarium, to have a port-folio, or a book in which they may be placed before the parts begin to wilt. Specimens should be placed between the leaves of paper, either newspaper or any other kind which is of a loose texture, and will easily absorb the moisture of the plants; a board with a weight upon it should then be placed upon the paper containing them; the plants should be taken out frequently at first; as often as once or twice a day, and the paper dried, or the plants placed between other dry sheets of paper. Small plants may be dried between the leaves of a book. Plants differ in the length of time required for drying as they are more or less juicy; some dry in a few days, others not sooner than two or three weeks. When the specimens are dry, and a sufficient number collected to commence an herbarium, a book should be procured, composed of blank paper, (white paper gives the plants a more showy appearance.) A quarto size is more convenient than a folio. Upon the first page of each leaf should be fastened one or more of the dried specimens, either with glue or by means of cutting through the paper, and raising up loops under which the stems may be placed. By the sides of the plants should be written the *class, order, generic,* and *specific name;* also the place where found, and the season of the year. The colours of plants frequently change in drying; the blue, pale red, and white, often turn black, or lose their colour; yellow, scarlet, violet, and green, are more durable. An herbarium should be carefully guarded against moisture and insects; as a security against the latter, the plants may be brushed over with corrosive-sublimate.

Botanical Excursions.

As a healthful and agreeable exercise, we would recommend frequent botanical excursions; you will experience more pleasure from the science, by seeing the flowers in their own homes; a dry grove of woods, the borders of little streams, the meadows, the pastures, and even the waysides, will afford you constant subjects for botanical observations. To the hardier sex, who can climb mountains, and penetrate marshes, many strange and interesting plants will present themselves, which cannot be found except in their peculiar situations; of these you must be content to obtain specimens, without seeing them in their native wilds. You will, no doubt, easily obtain such specimens, for there is, usually, among the cultivators of natural science, a generosity in affording assistance, and imparting to others the treasures which nature lavishes upon those who have a taste to enjoy them.

Poisonous Plants, and those which are not Poisonous.

In collecting flowers, you should be cautious with respect to *poisonous* plants. Such as have five stamens and *one pistil*, with a corolla of a dull lurid colour, and a disagreeable smell, are usually poisonous; the Thorn apple (*stramonium*) and the Tobacco are examples. The Umbelliferous plants, which grow in *wet* places, have usually a nauseous smell: such plants *are poisonous*, as the water hemlock. Umbelliferous plants which grow in dry places, usually have an aromatic smell, and *are not poisonous*, as Caraway and Fennel.

Plants with *Labiate* corollas, and containing their seeds in capsules, are often poisonous, as the Foxglove; (Digitalis;) also, such as contain a *milky juice*, unless they are compound flowers. Such plants as have horned or hooded nectaries, as the Columbine and Monk's-hood, are mostly poisonous.

Among plants which are seldom poisonous, are the compound flowers, as the Dandelion and Boneset; such as have labiate corollas, with seeds lying naked in the calyx, are seldom or never poisonous; the Mint and Thyme are examples of such plants. The *Papilionaceous* flowers, as the pea and bean; the *Cruciform*, as the radish and mustard, are seldom found to be poisonous. Such plants as have their stamens standing on the calyx, as the rose and apple, are never poisonous; neither the grass-like plants with glume calyxes, as Wheat, Rye, and Orchard-grass, (Dactylis.)

Proper Flowers for Analysis.

In selecting flowers for analysis, you must never take double ones; the stamens (and in many cases the pistils also) change to petals by cultivation, therefore you cannot know by a double flower, how many stamens or pistils belong to it in its natural state. Botanists seem to view as a kind of sacrilege, the changes made by culture, in the natural characters of plants; they call double flowers, and variegated ones, produced by a mixture of different species, *monsters* and *deformities*. These are harsh expressions to be applied to Roses and Carnations, which our taste must lead us to admire, as intrinsically beautiful, although their relative beauty, as subservient to scientific illustration, is certainly destroyed by the labour of the florist. The love of native wild flowers is no doubt greatly heightened by the habit of seeking them out, and observing them in their peculiar situations. A Botanist, at the discovery of some lowly plant, growing by the side of a brook, or almost concealed in the cleft of a rock, will often experience more vivid delight than could be produced by a view of the most splendid exotic. Botanical pursuits render us interested in every vegetable production: even such as we before looked upon as useless, present attractions as objects of scientific investigation, and become associated with the pleasing recollections, arising from the gratification of our love of knowledge. A peculiar interest is given to conversation by an acquaintance with any of the natural sciences; and when females shall have more generally obtained access to these delightful sources of pure enjoyment, we may hope that scandal, which oftener proceeds from a want of better subjects, than from malevolence of disposition, shall cease to be regarded as a characteristic of the sex. It is important to the cause of science, that it should become *fashionable ;* and as one means of effecting this, the parlours of those ladies, who have advantages for intellectual improvement, should more frequently exhibit specimens of their own scientific taste. The fashionable *et ceteras* of scrap books, engravings, and albums, do not reflect upon their possessors any great degree of credit. To paste pictures, or pieces of prose or poetry, into a book; or to collect in an album the wit and good sense of others, are not proofs of one's own acquirements; and the possession of elegant and curious engravings, indicates a full purse, rather than a well stored mind; but *herbariums* and books of *impressions of plants,** drawings, &c. show the taste and knowledge of those who execute them.

It is unfortunately too much the case, that female ingenuity, (especially in the case of young ladies after leaving school,) is in a great degree directed to trivial objects, which have no reference either to utility, or to moral and intellectual improvement. But a taste for scientific pursuits once acquired, a lady will feel that she has no time for engagements, which neither tend to the good of others, nor to make herself wiser or better.

---

* *Manner of taking impressions of leaves.*—Hold oiled paper over the smoke of a lamp until it becomes darkened; to this paper, apply the leaf, having previously warmed it between the hands, that it may be pliant. Place the lower surface of the leaf upon the blackened paper, that the numerous veins which run through its extent, and which are so prominent on this side, may receive from the paper a portion of the smoke. Press the leaf upon the paper, by placing upon it some thin paper, and rubbing the fingers gently over it, so that every part of the leaf may come in contact with the sooted oil-paper. Then remove the leaf, and place the sooted side upon clean white paper, pressing it gently as before; upon removing the leaf, the paper will present a delicate and perfect outline, together with an accurate exhibition of the veins which extend in every direction through it, more correct and beautiful than the finest drawing.

Female ingenuity too often directed to trivial objects.

PART II.

LECTURE VI

THE exercises which constitute the principal part of our previous course of lectures, are chiefly designed to assist you in practical botany. It is not expected that you are to be the passive receivers of instruction, but that you are to compare with real objects, the descriptions which are presented; by doing this faithfully, you will find your minds gradually strengthened, and more competent to compare and judge in abstract studies, where the subjects of investigation are in the mind only, and cannot, like the plants, be looked at with the eyes, and handled with the hands.

All our thoughts, by means of the senses, are originally derived from external objects. Suppose an infant to exist, who could neither hear, see, taste, smell, nor feel; all the embryos of thought and emotion might exist within it; it might have a soul capable of as high attainments as are within the reach of any created beings; but this soul, while thus imprisoned, could gather no ideas; the beauty of reflected light, constituting all the variety of colouring; the harmony of sounds, the fragrant odours of flowers, the various flavours, which are derived from our sense of taste, the ideas of soft, smooth, or hard; all must for ever remain unknown to the soul confined to a body having no means of communication with the world around it. The soul, in its relation to external objects, may be compared to the embryo plant, which, imprisoned within the seed, would for ever remain inert, were no means provided for its escape from this confinement, and no communication opened between it and the air, the light, and vivifying influence of the earth.

Since our first ideas are derived from external nature, is it not a rational conclusion that we should add to this original stock of knowledge, by a continued observation of objects addressed to our senses? After the years of infancy are past, and we begin to study *books*, should we, neglecting sensible objects, seek only to gain ideas from the learned; or, in other words, should we, in the pursuit of human sciences, overlook the works of God?

Having now enabled you to understand the method of analyzing plants, we shall proceed to consider more fully the different organs of plants, with the uses of each, in the vegetable economy.

In plants, as well as animals, each part or organ is intimately connected with the whole; and the vegetable, as well as the animal being, depends for its existence on certain laws of organization.

We shall consider the vegetable organs under two classes; the first, including such organs as *promote the growth* of the plant, as the root, leaves, &c.; the second, such as *perfect the seed*, and thus provide for the reproduction of the species, called organs of fructification.

Study of external objects strengthens the mind—Abstract studies facilitated by acquaintance with the natural sciences—Our first ideas gained by the senses—Analogy between the soul and the embryo plant—We should not confine our attention exclusively to books—Vegetable, as well as animal existence, depends on certain laws of organization—Two kinds of organs of vegetables.

Of the Root.

The *root* (radix) is that part of the vegetable which enters the earth, and extends in a direction contrary to the growth of the stem; it supports the plant in an upright position, and at the same time gives nourishment to every part of it. There are exceptions to the general fact, of a root being fixed in the ground; some plants, as the pond-lily, grow in water, and are called *aquatic*, (from *aqua*, water,) some, like the mistletoe, have no root, but fix themselves upon other plants, and derive sustenance from them; such are called *parasites*.*

The Root consists of two parts, the *Caudex, or main body* of the Root, and the *Radicle, or fibres;* these are capillary tubes, which absorb the nourishment that is conveyed to other parts of the plant. This nourishment ascending through the stem, experiences in the leaves and green parts of the plant, an important change, effected, in part, through the agency of air and light; and a portion of it, through a different set of vessels, flows back, in what is called the returning sap, or *cambium.*

Between the Caudex and stem is a point, called the *neck, or root stock;* any injury to this part is followed by the death of the plant.

Duration of Roots.

Roots, with respect to duration, are *annual, biennial*, or *perennial.*

Annual Roots—are such as live but one year. They come from the seed in the spring, and die in autumn, including such as are raised from the seed every year; as peas, beans, cucumbers, &c.

Biennial Roots—are such as live two years. They do not produce any flowers the first season, the next summer they blossom, the seeds mature, and the roots die. The roots of cabbages are often, after the first season, preserved in cellars during the winter. In the spring they are set out in gardens, and produce flowers; the petals of which, in time, fall off, and the germ grows into a pod which contains the seed. The root having performed this office, then dies, and no process can restore it to life; the flowering is thought to exhaust the vital energy or living principle. The onion, beet, and carrot, are biennial plants.

Perennial Roots—are those whose existence is prolonged a number of years to an indefinite period; as the asparagus, geranium, and rose; also trees and shrubs. Climate and cultivation affect the duration of the roots of vegetables. Many perennial plants become annual by transplanting them into cold climates: the garden nasturtion, originally a perennial shrub in South America, has become in our latitude an annual plant.

Forms of Roots.

There are many varieties in the forms of roots; the most important are the *branching, fibrous, spindle, creeping, granulated, tuberous*, and *bulbous.*

1st. *Branching root*, (Fig. 12.) This is the most common kind; it consists of numerous ramifications, resembling in appearance the

---

* The word *parasite*, from the Greek *para*, with, and *sitos*, corn, was first applied to those who had the care of the corn used in religious ceremonies, and were allowed a share of the sacrifice; afterward it was applied to those who depended on the great, and earned their welcome by flattery; by analogy, the term is now applied to plants which live upon others.

Definition of the root—Aquatic roots—Parasites—Division of the root—Annual roots—Biennial—Perennial roots—Classification of roots as founded upon their forms—Branching root.

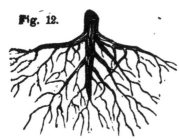

Fig. 12.

branches of a tree; some of these branches penetrate to a great depth in the earth, and others creep almost horizontally near its surface. Experiments have been made, which show, that branches by being buried in the soil may become roots; and roots, by being elevated in the atmosphere become branches covered with foliage. We often see the upturned roots of trees, throwing out leaves. Branching roots terminate in fibres or *radicles;* these are in reality the proper roots, as they imbibe, through pores, the nourishment which the plant derives from the earth. Nature furnishes his nourishment in the moisture, and various salts, which are contained in the soil.

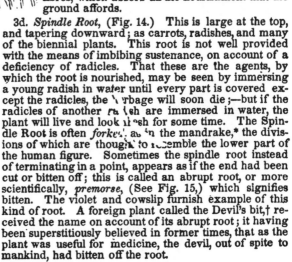

Fig. 13.

2d. *Fibrous Root,* (Fig. 13.) This consists of a collection of thread-like parts; as in many kinds of grasses, and most annual plants. The fibres usually grow directly from the bottom of the stem, as may easily be seen by pulling up a handful of the most common grass. The fact that grass of various kinds will live and flourish in a soil too dry and barren to produce other vegetation, is owing to the abundance of the fibres, which absorb all the nourishment that the ground affords.

Fig. 14.

3d. *Spindle Root,* (Fig. 14.) This is large at the top, and tapering downward; as carrots, radishes, and many of the biennial plants. This root is not well provided with the means of imbibing sustenance, on account of a deficiency of radicles. That these are the agents, by which the root is nourished, may be seen by immersing a young radish in water until every part is covered except the radicles, the herbage will soon die;—but if the radicles of another radish are immersed in water, the plant will live and look fresh for some time. The Spindle Root is often *forked,* as in the mandrake,* the divisions of which are thought to resemble the lower part of the human figure. Sometimes the spindle root instead of terminating in a point, appears as if the end had been cut or bitten off; this is called an abrupt root, or more scientifically, *premorse,* (See Fig. 15,) which signifies bitten. The violet and cowslip furnish example of this kind of root. A foreign plant called the Devil's bit,† received the name on account of its abrupt root; it having been superstitiously believed in former times, that as the plant was useful for medicine, the devil, out of spite to mankind, had bitten off the root.

* *Atropa mandragora.* The word mandrake is said to be derived from the German *Mandragen,* resembling man.
† *Scabiosa succisa,* or a kind of Scabious.

Fibrous roots—Spindle root—Importance of radicles—Forked spindle root—Premorse root.

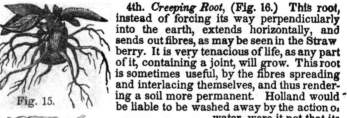

Fig. 15.

4th. *Creeping Root*, (Fig. 16.) This root, instead of forcing its way perpendicularly into the earth, extends horizontally, and sends out fibres, as may be seen in the Straw berry. It is very tenacious of life, as any part of it, containing a joint, will grow. This root is sometimes useful, by the fibres spreading and interlacing themselves, and thus rendering a soil more permanent. Holland would be liable to be washed away by the action o, water, were it not that its coasts are bound together by these creeping plants. This root will grow in sandy, light soils, which scarcely produce any other vegetation.

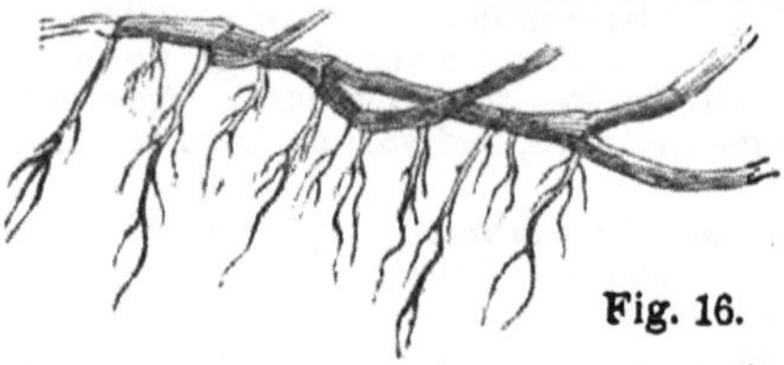

Fig. 16.

5th. *Granulated Root*, (Fig. 17.) This consists of little bulbs or *tubers*, strung together by a thread-like radicle; this form approaches to that of some varieties of the tuberous.

Fig. 17.

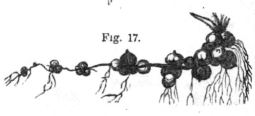

Fig. 18.

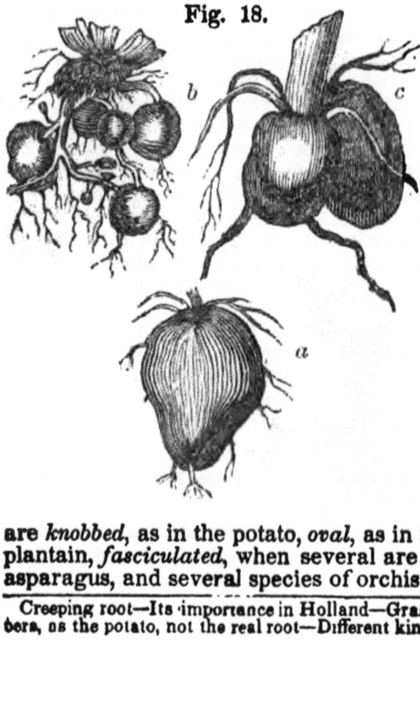

6th. *Tuberous Root*. This kind of root is hard, solid, and fleshy; it consists of one knob or tuber; as in the potato, *a;* or of many such, connected by strings or filaments, as in the artichoke, *b*. These tubers are reservoirs of moisture, nourishment, and vital energy The potato is in reality but an excrescence, proceeding from the real root; and it is a singular fact that this nutritious substance is the product of a plant whose fruit (often termed potato balls) is poisonous. The root of some of the orchis plants, (Fig. 18. *c*.) consists of two tubers, resembling the two lobes into which a bean may be divided. Tuberous roots are *knobbed*, as in the potato, *oval*, as in the orchis, *abrupt*, as in the plantain, *fasciculated*, when several are bundled together, as in the asparagus, and several species of orchis.

Creeping root—Its importance in Holland—Granulated root—Tuberous root—Tubers, as the potato, not the real root—Different kinds of tuberous roots.

Fig. 19.

Fig. 19, at *a*, shows a root of the Ophris, one of the orchis tribe of plants. It is composed of a mass oi crowded tubers. It is called a *grumose* root. At *b*, is a *fasciculated* tuberous root, as in the asphodel. At *c*, the tubers are suspended from an upright body or caudex, as in the root of the *Spiræa filipendula.*

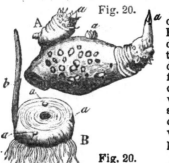

Fig. 20.

Fig. 20.

Roots sometimes produce a kind of bud, or little bulb, called by the French botanists, *turion.* It appears doubtful whether this, and indeed the bulb, should be considered under the head of roots or buds. The figure at A shows a tuberous root crowded with turions, some of which, *a, a,* are in a germinating state. At B, is a bulbous root (crocus) showing the turions at *a, a,* while at *b,* appears one which is partially developed.

7th. *Bulbous root,* a fleshy root, of a bulbous or globular form. It seems designed to enclose and protect the future plant against cold and wet. Bulbous plants belong chiefly to the great division of Monocotyledons, or those whose seeds have but one cotyledon ; they produce some of the earliest flowers of spring, and afford the most beautiful ornaments of the garden. Among them are the Hyacinth, the Crown Imperial, the Lily, and the Tulip, with a great variety of other splendid and interesting flowers. The use of the bulb being to preserve the young plant from the effect of cold, we see the bountiful agency of providence in the number of bulbous plants in cold countries.

Bulbs seem to be analogous to buds, and in some plants grow like them upon stems or branches ; as in the tiger-lily and tree-onion ; in the latter, the bulbs or onions grow upon the stalks in clusters of four or five, continuing to enlarge, until their weight brings them to the ground, where they take root. This is a *viviparous* plant, or one which produces its offspring alive ; such plants as produce seeds, or such animals as produce their offspring from eggs, are called *oviparous.* Bulbs are *solid,* as in the turnip, (Fig. 21, *a,*) *tunicated,* or coated, as in the onion *b,* and *scaly,* as in the white lily *c.*

Explain Fig. 19—Explain Fig. 20—Bulbous root—Use of the bulb—Analogous to buds—Viviparous and oviparous plants.

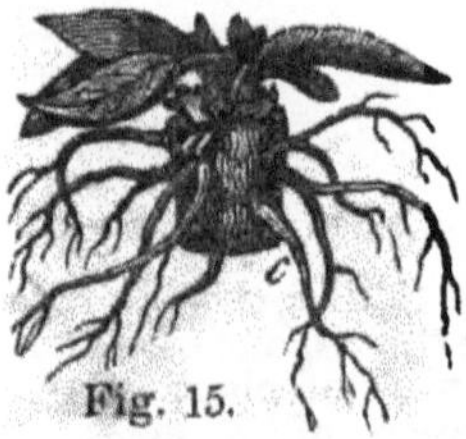

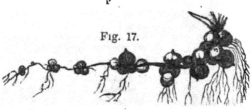

Fig. 15.

4th. *Creeping Root*, (Fig. 16.) This root, instead of forcing its way perpendicularly into the earth, extends horizontally, and sends out fibres, as may be seen in the Straw berry. It is very tenacious of life, as any part of it, containing a joint, will grow. This root is sometimes useful, by the fibres spreading and interlacing themselves, and thus rendering a soil more permanent. Holland would be liable to be washed away by the action o₁ water, were it not that its coasts are bound together by these creeping plants. This root will grow in sandy, light soils, which scarcely produce any other vegetation.

Fig. 16.

Fig. 17.

5th. *Granulated Root*, (Fig. 17.) This consists of little bulbs or *tubers*, strung together by a thread-like radicle; this form approaches to that of some varieties of the tuberous.

Fig. 18.

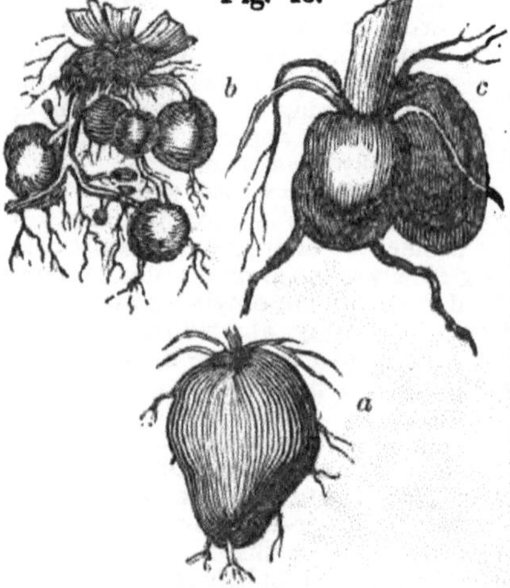

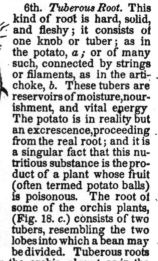

6th. *Tuberous Root.* This kind of root is hard, solid, and fleshy; it consists of one knob or tuber; as in the potato, *a;* or of many such, connected by strings or filaments, as in the artichoke, *b.* These tubers are reservoirs of moisture, nourishment, and vital energy The potato is in reality but an excrescence, proceeding from the real root; and it is a singular fact that this nutritious substance is the product of a plant whose fruit (often termed potato balls) is poisonous. The root of some of the orchis plants, (Fig. 18. *c.*) consists of two tubers, resembling the two lobes into which a bean may be divided. Tuberous roots are *knobbed*, as in the potato, *oval*, as in the orchis, *abrupt*, as in the plantain, *fasciculated*, when several are bundled together, as in the asparagus, and several species of orchis.

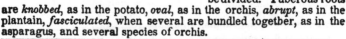

Creeping root—Its importance in Holland—Granulated root—Tuberous root—Tubers, as the potato, not the real root—Different kinds of tuberous roots.

Fig. 19.

Fig. 19, at *a*, shows a root of the Ophris, one of the orchis tribe of plants. It is composed of a mass oɪ crowded tubers. It is called a *grumose* root. At *b*, is a *fasciculated* tuberous root, as in the asphodel. At *c*, the tubers are suspended from an upright body or caudex, as in the root of the *Spiræa filipendula*.

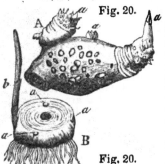

Fig. 20.

Fig. 20.

Roots sometimes produce a kind of bud, or little bulb, called by the French botanists, *turion*. It appears doubtful whether this, and indeed the bulb, should be considered under the head of roots or buds. The figure at A shows a tuberous root crowded with turions, some of which, *a, a,* are in a germinating state. At B, is a bulbous root (crocus) showing the turions at *a, a,* while at *b*, appears one which is partially developed.

7th. *Bulbous root*, a fleshy root, of a bulbous or globular form. It seems designed to enclose and protect the future plant against cold and wet. Bulbous plants belong chiefly to the great division of Monocotyledons, or those whose seeds have but one cotyledon; they produce some of the earliest flowers of spring, and afford the most beautiful ornaments of the garden. Among them are the Hyacinth, the Crown Imperial, the Lily, and the Tulip, with a great variety of other splendid and interesting flowers. The use of the bulb being to preserve the young plant from the effect of cold, we see the bountiful agency of providence in the number of bulbous plants in cold countries.

Bulbs seem to be analogous to buds, and in some plants grow like them upon stems or branches; as in the tiger-lily and tree-onion; in the latter, the bulbs or onions grow upon the stalks in clusters of four or five, continuing to enlarge, until their weight brings them to the ground, where they take root. This is a *viviparous* plant, or one which produces its offspring alive; such plants as produce seeds, or such animals as produce their offspring from eggs, are called *oviparous*. Bulbs are *solid*, as in the turnip, (Fig. 21, *a*,) *tunicated*, or coated, as in the onion *b*, and *scaly*, as in the white lily *c*.

Fig. 21

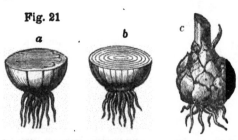

Some bulbs die after the blossoming of the plant, and new ones are formed from the base or sides of the original bulb, which, in their turn, produce plants. This is the fact with respect to the orchis tribe; in which every year one bulb or tuber dies, and the other throws out a new stem, (see Fig. 19, *c*;) by this means it changes its position, though slowly, since it takes but one very short step each year.

Gardeners take up their bulbous roots as often as once in two or three years. In some plants the new bulbs are formed *beside* the old ones; thus they become crowded, and produce inferior flowers. Many kinds, as the tulip and the narcissus, form the new bulbs *under* the old ones, and these become at length too deep in the earth; while the new bulbs of the crocus and gladiolus, and some others, grow *above* the old ones, and on account of being too near the surface, are liable to be injured by frosts and drought.

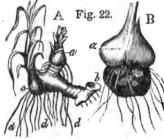

Fig. 22 shows at A, a root of Solomon's seal, (*Convallaria ;*) *a*, *a*, are the young bulbs of the plant; *b* marks the spot from which the decayed stalk of the former year has fallen; *d, d* are the fibres or true root of the plant.

At B, is a root of the Ixia, or Blackberry Lily; *a* shows the young bulb formed above the parent one, which is withering in consequence of imparting its vigour to its offspring.

The bulbous root might more properly be termed the *bulbiferous* or bulb-bearing root, since all that is truly a root is the fibrous part.

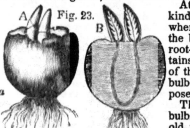

At A, Fig. 23, is a root of this kind; *a* shows the disk or surface where the fibres are attached to the base of the bulb; this is the root-stalk. The bulb above it contains the leaves, stems, and flowers of the plant. B shows the same bulb cut vertically, in order to expose the embryo plant.

The production by means of bulbs, is only a continuation of the old plant, while by means of the seed, a new plant is brought forth This is an important distinction; and it is observed that in process of time, a plant continued by means of reproduction, whether by

Different forms of bulbous roots—Difference in the production of plants by means of bulbs and seeds—Reasons for taking up bulbous plants—Explain Fig. 22—Explain Fig. 23—Difference between the continuation of plants by bulbs, &c. and by raising from the seed.

bulbs, grafting, or any other manner, ultimately dwindles and degenerates as if worn out with old age, and it becomes necessary to renew its vigour by producing a young plant from the seed. This is the case with the potato, for the farmer often finds his stôck degenerated, and is obliged to provide himself with new roots produced from the seed.

The specific character of plants is sometimes taken from the root, and in some cases the specific name; as SOLANUM *tuberosum*, the potato, and RANUNCULUS *bulbosus*, the bulbous ranunculus. The tuberous and bulbous roots distinguish those species from all others of the families Solanum and Ranunculus.

The forms of roots are so various, that it is impossible to give names to all; even in the same species of plants, the root presents many varieties of form. In the potato, for example, we see some roots round, and of an even surface, others long and oval, and some very knobbed and irregular; but yet amidst all this variety there is a prevailing uniformity, and we can usually at one glance distinguish a potato, by its form, from all other vegetables. It might, at first, have appeared as if there could be little interesting in the consideration of roots, which are destitute of that symmetry of parts and liveliness of colouring, which is exhibited in other organs of the plant. We find, on casting a rapid glance over the face of the earth, that all this variety in the form of roots is not without its peculiar use. Mountains being exposed to winds, we find them covered with plants which have branching roots with strong and woody fibres. These fastening themselves into the clefts of rocks, take firm hold, and the trees they support, seem undauntedly to brave the violence of storms and tempests. Spindle roots abound in rich, soft grounds, which they can easily penetrate. Damp and loose soils are rendered fit for the use of man, by being bound together by creeping and fibrous roots. We find here, as in every part of nature, proofs of a wise Creator, who makes naught

"In vain, or not for admirable ends."

We have now described those roots which grow by being fixed in the earth. But besides these, there are plants which are not fixed, but float about in the water; some grow upon other plants, and some seem to derive sustenance from air alone.

Of the first kind, or *aquatic roots*, is the Lemna or duckmeat, which grows in stagnant water, having thread-like roots, not confined to any fixed place. The water star-grass,* previous to its blossoming, floats about, and is nourished by its suspended fibres; after flowering, it sinks to the bottom, its roots become fixed, and its seeds ripen. These seeds germinating, a new race of plants appear, which rise to the surface of the water, blossom, and sink to the earth, producing in turn their successors. Some of the Cryptogamous plants, particularly of the genus Fucus, exist in a wandering manner, often forming islands of considerable size. In the Gulf of Florida, the *Fucus natans* is very abundant; this, by voyagers, is often called gulf-weed, and is sometimes found in masses extending many miles, and,

"Sailing on ocean's foam,
Where'er the surge may sweep, the tempests breath prevail."

How strikingly analogous this poor weed to many a human being, blown about on the ocean of life, by every breath of passion or ca-

* *Callitriche aquatica.*

Specific character and name taken from the roots—Roots of the same species sometimes vary in form—Utility in the variety of form in roots—Aquatic roots.

price! Who would not rather, like the mountain oak, meet the storms of life firmly rooted in virtuous principles, than to be floated along even by the breath of pleasure, without end or aim, forgetful of the past, and careless of the future? To the virtuous, afflictions serve but to strengthen them in goodness; so,

> "Yonder oaks! superior to the power
> Of all the warring winds of heaven do rise,
> And from the stormy promontory tower;
> While each assailing blast increase of strength supplies."

We find some roots growing on other plants, and appearing to derive sustenance from their juices. These are called *parasites*; this term is often applied to persons who are willing to live in dependance upon others; and so despicable does this trait of character appear, that we almost conceive it a kind of meanness, even for a plant to live without elaborating its own food. Parasitic plants are common in tropical regions; sometimes many kinds are found upon the same tree, presenting a curious variety of foliage. In our climate, except in the Cryptogamous family, as lichens, mosses, &c. we have but few genera of these plants.* The Dodder and Mistletoe are celebrated parasitic plants.

Some plants grow without roots; these are called *air plants:* they are furnished with leaves or stems which seem to *inhale*, but not to *exhale* fluids; their substance is usually fleshy and juicy; some of them flourish in the most dry and sandy places, exposed to a burning sun; as the *Stapelia*, sometimes called the *vegetable camel*. The *Epidendrum* grows and blossoms for years, suspended from the ceiling of a room, and nourished only by air.

Many roots, as the rhubarb, wild-turnip, blood-root, &c. possess important medicinal properties. The growth of the root is most rapid in autumn; at this season, the sun being less powerful, and the air more charged with moisture, the juices condense in the lower part of the plant, and nourish it, but as the season becomes cold, vegetation is checked; the winter is the best time to collect roots for medicinal purposes, because their peculiar virtues are then most concentrated

---

LECTURE VII.

OF THE STEM.

THE stem is the body of a plant, whether it be a tree like the oak, a shrub like the lilac, or an herb like the poppy; its use is to sustain the branches, leaves, and flowers, and to serve as an organ of communication between them and the root, conducting from the latter to the former, the animal and vegetable substances, salts, and earthy matter, which the radicles, by their mouths, suck up for the nourishment of the plant. The influence of light and air is, through the medium of the stem, conveyed from the leaves to the root.

* In the vicinity of Troy, I have seen a very beautiful species of the *Pterospora*, growing upon a branch of the whortleberry. Its colour was a bright crimson, which contrasted finely with the white flowers and green leaves of the plant on which it grew.

Parasitic plants—Air plants—Proper time to collect roots for medicinal purposes—Stem, its use.

'If a plant be watered by any coloured liquid, the stem will, in time, show that this fluid has ascended into it. There is also in the stem a set of vessels to carry downward the juices, which have passed through peculiar processes in the leaves of the plant.

But of the circulation of fluids in the vegetable substance we shall speak more particularly hereafter. Our present object is, to describe the external appearance of the vegetable organs, and not their internal structure; or, in other words, it is the *anatomy* and not the *physiology* of plants, which we are now attempting to explain.

The different kinds of stems have been divided into seven classes, as follows—

*Caulis,** or proper stem, *Culm, Scape, Peduncle, Petiole, Frond,* and *Stipe.*

1st. *Caulis,* or proper stem, is such as is seen in forest trees, in shrubs, and in most annual plants. The caulis is either simple, as in the White lily; or branching, as in the Geranium. The branching is the more common form. You have here (Fig. 24) the representation of a *caulis,* or proper stem (*a ;*) a *peduncle,* or flower stalk (*b ;*) and a *petiole,* or leaf stalk (*c.*)

Fig. 24.

2d. *Culm,* or straw, (Fig. 25,) is the kind of stem which you see in grasses and rushes. The culm is either *without knots,* as in the Bulrush, *jointed* or *knotted,* as in Indian corn, *geniculated,* or bent like an elbow, as in some of the grasses. Those culms which are bent, are also knotted, though they may be knotted without being bent. The Bamboo, Sugar Cane, and various species of Reeds, have stems of the culm kind; some of them, particularly the Bamboo, are known to attain the height of forty feet.

Fig. 25.

Fig. 26.

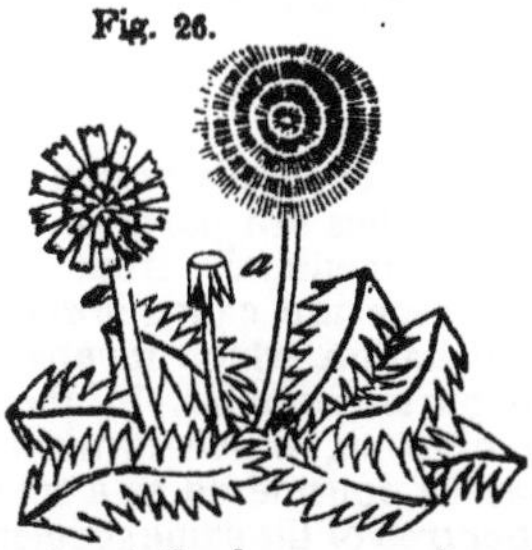

3d. *Scape,* (Fig. 26, *a, a,*) a stalk springing from the root, which bears the flower and fruit, but not the leaves: as the Dandelion, the Cowslip, and the Lily of the Valley. Plants with scapes are sometimes called stemless plants; in this case, the scape would be considered as a peduncle proceeding from the root.

4th. *Peduncle,* or flower stalk, is but a subdivision of the caulis or stem; (See Fig. 24, *b ;*) it bears the flower and fruit, but not the leaves; when the peduncle is divided, each subdivision is called a pedicel. In determining the species of plants, we often consider the length of the peduncle, compared with the flower; as, whether it is longer or shorter. When there is no peduncle or flower stalk, the flowers are said to be *sessile.*

* This kind of stem is by the French called *tige;* the *i* should be sounded like *e,* the *g* soft like *j,* as in *teje.* The word Caulis is from the Greek *Kaulos,* a stem.

5th. *Petiole*, or leaf stalk, is a kind of stem, like a fulcrum, supporting the leaf, as the *peduncle* supports the flower; it is usually green, and appears to be a part of the leaf itself. The peticle of many plants is somewhat in the form of a cylinder; but the upper surface is rather flattened, the under surface convex. You will find this remark useful, in distinguishing the foot-stalks of compound leaves from young branches, with which they are sometimes confounded. In most cases, the leaves and flowers are supported by distinct foot-stalks, but sometimes the foot-stalk supports both the leaf and flower. The Petiole is often compared with the leaf, as the peduncle is with the flower, as to its relative length, in the different species.

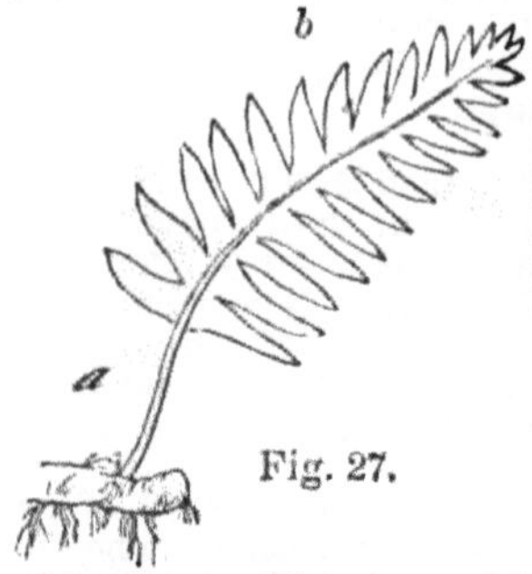

6th. *Frond.* (Fig. 27.) The term frond, belongs entirely to Cryptogamous plants. This term however is applied to the leaf rather than the stem; in this sketch of the fern, the leafy part, *b*, is the frond; this bears the flower and fruit. Linnæus considered the leaves of palm-trees as fronds; we shall hereafter remark upon the different internal structure of their stems from those of the oak and other plants which are termed *cauline*, because their stem is a *caulis*. Plants with fronds are monocotyledonous.

Fig. 27.

7th. *Stipe.* The stem of the fern (Fig. 27, *a*,) is called a stipe. By observations of geologists it is ascertained that *stiped* plants were created before *cauline* ones; petrifactions of the former being found in the lower formations of the earth, while no remains of cauline plants are ever found there. The stalk of a fungus or mushroom is called a stipe. The term is also applied to the slender thread, which in many of the compound flowers, elevates the hairy crown with which the seeds are furnished, and connects it with the seed. Thus, in a seed of the Dandelion, which is here represented, the column (Fig. 28, *a*,) standing on the seed (*b*,) and elevating the down (*c*,) is the stipe.

Fig. 28.

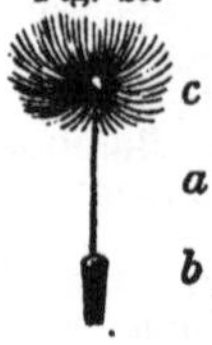

Here is a mushroom with the cap (Fig. 29, *d*,) elevated on its stipe (*e*.)

Branches. The stem is either *simple*, or *divided* into branches. The branches are parts of the plant which proceed immediately from the trunk; the division of these are called *branchlets;* a diminutive appellation, which means a little branch. These parts resemble, in their formation, the trunk or stem, which furnishes them; the branch may be considered as a tree, implanted upon another tree of the same species. Branches sometimes grow without any apparent order in their arrangement; sometimes they are *opposite;* sometimes *alternate;* and sometimes, as in the pine, they form a series of rings around the trunk. Some branches

Fig. 29.

Peduncle—Petiole—Frond—Which part of the fern is its frond?—Which the stipe? —Difference between stiped and cauline plants—Which first formed?—Different applications of the term stipe—Stipe of a dandelion seed—Stipe of a mushroom— Branches—Branchlets—Various appearances of branches.

are *erect*, as in the poplar, others *pendent*, as in the willow, and some, as in the oak, form nearly a right angle with the trunk These various circumstances constitute distinctive characters in plants, a knowledge of which is very necessary to the painter. Of all our forest trees, perhaps none, in the disposition of its branches, presents a more beautiful and graceful aspect than the elm.

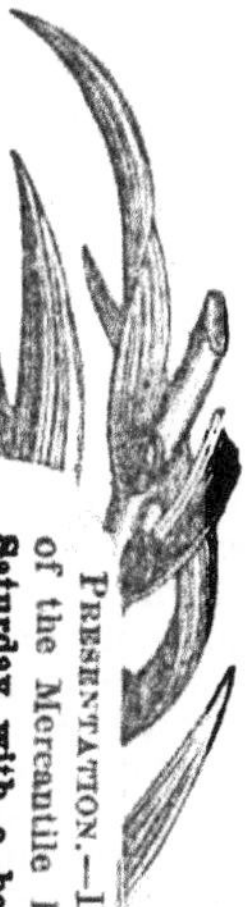

The branches of trees, as they grow older, usually form a more open angle with the trunk than at first We often see branches form a very acute angle, but a the tree advances in age, the angles enlarge mor and more, until the branch becomes pendent.

Some stems are remarkable for bearing little *bulbs*, called *bulbilles*, in the axils of their leaves. These, like the bulbous root, contain within them the germ of a new plant. The LILIUM *bulbiferum*, or tiger-lily, is of this description. (Fig. 30.) The bulbs are of a red-brown colour, about the size of a large gooseberry. They begin, soon after they are formed, to detach themselves from the plant, and falling upon the ground, shoot out fibres and take root. This splendid flower may thus be rapidly increased.

A remarkable phenomenon is described by travellers, as being exhibited by the stems of the Banyan tree of India, *Ficus Indicus;* these stems throw out fibres, which descend and take root in the earth. In process of time, they become large trees; and thus from one primitive root, is formed a little forest. This tree is called by various names; as the Indian-God-tree, the arched-Fig-tree, &c. The Hindoos plant it eir temples, and in many cases, the tree itself serves them emple. Milton speaks of this tree, as the one from which and Eve obtained leaves to form themselves garments; he was not the fig-tree renowned for fruit, but

'ig. 30.

PRESENTATION.—Lieut. Pinckney S. Cone, of the Mercantile Battery, was presented on Saturday with a handsome sword, belt and sash, and a pair of Smith & Wesson's revolvers, by his friends. The affair was a pleasant one, and the gifts well deserved.

"Such as at this day to Indians known
In Malabar or Decan, spreads her arms,
Branching so broad and long, that in the ground
The bended twigs take root, and daughters grow
About the mother tree, a pillar'd shade
High over-arched, and echoing walks between."

Ficus Indicus.

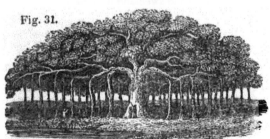

Fig. 31.

You have here, a representation of this wonderful tree, which is said to be capable of giving shelter to several thousand persons.

All the varieties of stems, which we have now considered, may be included under two divisions; 1st, such as grow *externally*, having their wood arranged in concentric layers;

Branches alter in their angles as they grow older—Bulb-bearing stems—Rooting stems.

the others being in the centre of the trunk, and the newest forming the outer layer. This kind of stem may be seen in the oak and other forest trees in our climate, and also in most of our common herbaceous plants; these spring from seeds with two cotyledons, and are called *dicotyledonous.*

2d. Stems which grow *internally*, as palms and grasses: here the wood, instead of circling around the first formed substance, is pushed outwards by the development of new fibres in the centre; this kind of stem belongs to plants whose seeds have but one cotyledon, and are therefore called *monocotyledonous.**

LECTURE VIII.

OF BUDS.

MOST leaves and flowers proceed from scaly coverings called buds. The scales envelop each other closely; the exterior ones being dry and hard, the interior moist, and covered with down; they are also furnished with a kind of resin or balsam, which prevents the embryo from being injured by too much moisture. Buds have been known to lie for years in water, without injury to the germ within.

The sap is the great fountain of vegetable life; by its agency new buds are yearly formed to replace the leaves and flowers destroyed by the severity of winter. Branches also originate from buds. Linnæus supposed that buds spring from the pith, this being found necessary to their formation and growth. The bud is a protuberance formed by the swelling of the germ; and as, for this purpose, the agency of an additional quantity of sap is needed, we see the bud appearing at the axils of leaves, or the extremities of branches and stems, where there is an accumulation of this fluid. If you plant a slip of Geranium, you will observe that it either sprouts from the axil of a leaf, or from knots in the stem, which answer the same purpose as the leaf, by slightly interrupting the circulation of the juices, and thus affording an accumulation of sap necessary for the production of a new shoot.

Some botanists distinguish the different periods of the bud as follows: first, the point in the plant which gives rise to the bud, is called the *eye;* when this begins to swell and become apparent, it is termed the *button;* and when it begins to unfold, the *bud.*†

Herbs and shrubs have buds, but these usually grow and unfold themselves in the same season, and are destitute of scales; while the buds of trees are not perfected in less than two seasons, and, in some cases, they require years for their full development. You have, no doubt, observed in the spring, the rapid growth of the leaves and branches of trees; and perhaps, have also noticed, that as summer advances, the progress of vegetation seems almost suspended. But nature, instead of resting in her operations, is now busy in providing for the next year; she is turning the vital energies of the plants to

---

* These two kinds of stem have by some French botanists been called *exogenous* and *endogenous:* these words are derived from the Greek; the first signifying to *grow* externally, the second, to *grow* internally.

† These terms in French, are *l'œil,* the eye, *bouton,* the button, and *bourgeons,* the bud.

Dicotyledonous stems—Monocotyledonous stems—Description of buds—Agency of sap—The eye, button, and bud—Herbs and shrubs destitute of scaly buds.

the formation of buds. Those little embryo plants, so nicely wrapped up in downy scales as to be able to bear the coldness of winter, in the ensuing spring will come forth from their snug retreats, and taking the places of the leaves which had withered in autumn, delight us with new verdure and beauty.

The poet Cowper, in the following lines on the formation of buds, shows us the improvement which the pious make, in observing the phenomena of nature.

> "When all this uniform uncoloured scene,
> Shall be dismantled of its fleecy load,
> And flush into variety again,
> From dearth to plenty, and from death to life,
> Is Nature's progress, when she lectures man
> In heavenly truth; evincing, as she makes,
> The grand transition, that there lives and works
> A soul in all things, and that soul is GOD.
> HE sets the bright procession on its way,
> And marshals all the order of the year;
> HE marks the bounds which winter may not pass,
> And blunts his pointed fury; in its *case,*
> *Russet and rude, folds up the tender germ,*
> Uninjured, with inimitable art;
> And ere one flowery season fades and dies,
> Designs the blooming wonders of the next."

Some French botanists,[*] have explained the formation of the scaly covering of buds in a manner somewhat different from the generally received opinion. They suppose, that in the latter part of summer, the *eye* is formed, and that the young shoot forces its way through the bark, but the young leaves which would put forth, becoming chilled by the ungenial atmosphere of the coming winter, contract and harden, and at length form scales; and that these scales afterward protect the new leaves, which, urged by the same vegetable instinct, are, in their turn, seeking to emerge into light and air. If we admit this explanation with respect to the formation of scales, it seems not difficult to account for the covering of varnish, which defends the embryo leaves and flowers from moisture. When the leaf becomes a scale, it then absorbs from the sap but a portion of what was destined for its use, and the remaining sap may be converted into the resinous substance, or varnish. With respect to the downy coat upon the inside of the scales, this may be seen in the rudiments of the leaves, if examined before the bud is developed. These hypotheses do not, in any degree, derogate from the wisdom of Him, who, "with art inimitable, folds up the tender germ;" for whether He acts by secondary causes, or "speaks, and it is done," design is alike apparent in all his works.

The term bud, in common language, extends to the rudiments of all plants, whether with scales or without, which originate upon other living plants. Buds with scales are chiefly confined to the trees of cold countries. In the northern part of the United States, there are few trees which can endure the cold weather, without this security. In Sweden, it is said, there is but one shrub† destitute of buds, and this, from the peculiarity of its situation, is always protected from the inclemencies of weather.

* De Candolle, and others.
† A species of *Rhamnus*, which grows under trees, in marshy forests.

---

considering this subject, you cannot but have been impressed with a sense of the goodness of that great Being who watches with unceasing care over his vast creation. To observe the progress of life, whether in the vegetable or animal kingdom, is highly interesting to an investigating mind. Man may plant and water, but God alone giveth the increase.

A bud lives, an infant lives; both are destined to grow, and to pass through physical changes: but the bud, although active with a principle of life, knows not its own existence; while the infant becomes conscious of its own powers and faculties, capable of loving those who have contributed to its well being, and especially of adoring the great Author of its existence.

It is delightful, while gratifying our natural love of knowledge, by inquiring into the economy of nature, to be thus met at every step, with new proofs of the goodness and wisdom of the Author of Nature, particularly as manifested towards the human race. To discover the character of the Deity, should indeed be the end and aim of all knowledge; and should an occasional digression from our subject retard your progress in botanical investigations, the loss would be slight, compared to the gain of one pious and devout aspiration.

When we become so deeply engaged in philosophical speculations, as to forget Him whose works we study, we have wandered from the path of true knowledge. It was not thus that Newton studied the laws of matter, or Locke and Watts the laws of mind, or Paley the animal and vegetable physiology; these great and good men, made their rich treasures of knowledge subservient to one great design, that of learning the character of God, and their duty to him, and of instructing their fellow-men in these sublime and important truths.

LECTURE IX.

OF LEAVES.

You all know what is meant by the leaf of a vegetable; but were you called on to give a definition of the term *leaf*, you might find it more difficult than at first you would imagine. Young persons are often disconcerted, when asked by their teachers to explain some word of which they have an idea, and yet find themselves unable to give a definition; but although the pupil may be surprised at this fact, it is not unaccountable to those who know, that it is not always easy to convey our conceptions to the minds of others. To give correct definitions of terms, is one of the greatest difficulties in science.

The manner in which different persons describe objects, varies with the degree of knowledge possessed respecting their properties. For example; in attempting to describe *common salt*, if a person knew nothing more of it than his unassisted senses had informed him, he would speak of its colour, taste, and other obvious properties. One familiar with the principles of chemistry, would first speak of the *materials* which compose salt; he would describe it as

---

Comparison between a bud and an infant. —The goodness of God particularly manifested towards the human race—Philosophical speculations should not lead us to forget the Author of nature—Difficulty in giving correct definitions—Descriptions of objects vary with our knowledge of their properties—Example: common salt.

a compound substance, consisting of chlorine and sodium. In the first definition, given without any reference to scientific principles, there is nothing so definite as to afford a certain mark of distinction between salt and other substances; in the chemical definition, we have a test for salt, in a knowledge of its composition, which distinguishes it from all other substances.

In botanical definitions, we do not include the constituent elements of the vegetable substance; this belongs to the department of chemistry, but we consider the external forms and uses of the various parts of the plant.

The leaf is an expansion of the fibres of the bark, connected by a substance, called the *cellular tissue;* the whole is covered with a green coat, or skin, called the *cuticle.* Leaves are furnished with pores called *stomas,* for exhaling and inhaling gases. They present to the air a more extended surface than all the other vegetable organs, and are of great importance by imbibing suitable nourishment, and throwing off such gases as would be useless or injurious to the plant.

We have seen how the bud is formed, and by what wise means the principle of life which it contains, is protected through the cold and dampness of winter. In the spring, when the sun, having recrossed the equator, is advancing towards our hemisphere, the vegetable world, quickened by its influence, begins to awaken from a dormant state; the buds expand, and bursting their envelopes, the new branches, bearing leaves and flowers, come forth.

The manner in which the leaf lies wrapped up in the scales of the bud, is called *Foliation;* this presents an interesting study, and is said to be sufficiently various, in different families of plants, to afford a mark of distinction between them.

Fig. 37.

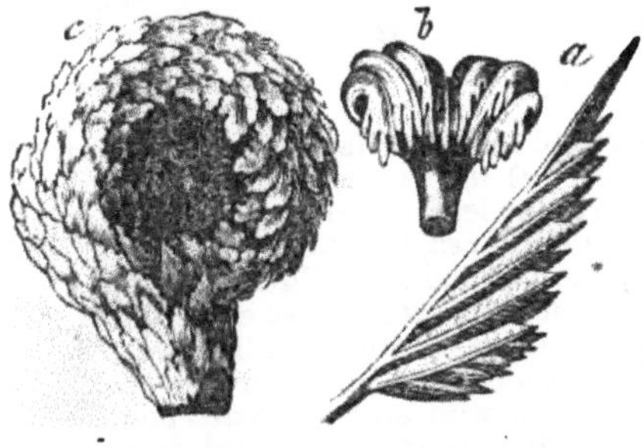

Figure 37, at *a,* shows a young leaf of the currant; this is *folded.* At *b,* is a young leaf of the Aconitum, (monk's-hood;) this is *inflected.* At *c,* is the young leaf of a fern, (*aspidium,*) this is *circinate,* or rolled from the summit towards the base.

Some plants are destitute of leaves; they are then called *Aphyllous,* from the Greek, *a,* to want, *phyllon,* a leaf.

In determining the species of plants, the leaves are much regarded. Specific names are often given from some circumstance of the leaf; the HEPATICA *triloba* is that species of the Hepatica, which has leaves with three divisions, called *lobes.* The VIOLA *rotundifolia,* is a species of violet with round leaves.

A knowledge of the various appearances presented by leaves, is of great importance to the botanical student; in order to become acquainted with these, much practice in the analysis of plants is necessary. Engravings will assist you in understanding the definitions, but you must chiefly consult nature.

Definition of the leaf—Utility of leaves to the whole plant—The period at which leaves appear—Foliation—Aphyllous plants—Leaves furnish specific characters.

Leaves considered with regard to the manner in which they succeed each other in different stages of the plant.

1. *Seminal*, leaves which come up with the plant when it first appears above the surface of the earth; as in the garden bean; these leaves are only the cotyledons, or lobes of the seed, which, after nourishing the young plant, decay.

2. *Primordial*, leaves growing immediately after the seminal leaves, and resembling them in position, form, and size. The primordial leaf, according to the fanciful idea of a French botanist, is a sketch which nature makes before the perfection of her work.

3. *Characteristic*, leaves which are found in the mature state of the plant; or according to the idea above advanced, nature, in them, perfects her design.

It is not always, however, that this process, with regard to change of leaves, takes place; as in many cases, the proper, or characteristic leaf, is the only one which appears.

Form of Leaves.

The *form* of the leaf is expressed by various terms borrowed from the names of different objects; as *palmate*, hand-shaped; *digitate*, from digitus, the finger, &c. We will illustrate some of the most common forms of simple leaves, leaving you to consult the vocabulary for many terms, which it would be too tedious to attempt to define in the body of this work.

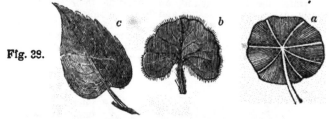

Fig. 38.

Orbicular, or the round leaf; the Nasturtion affords an example of this kind, (See Fig. 38, *a;*) this is also *peltate*, having its petiole inserted into the centre of the leaf, and thus resembling a shield.

Reniform, (from the Latin *ren*, the kidney,) or as it is sometimes called *kidney-form;* the Ground-ivy (*Glechoma*) has a leaf of this kind, (See Fig. 38, *b ;*) it is *crenate*, or has a margin with scalloped divisions; *ciliate*, being fringed with hairs, like eyelashes.

Cordate, (from the Latin *cor*, the heart,) or *heart-shaped*. Fig. 38, *c*, represents a cordate leaf with an *acuminated* point, that is, acute and turned to one side; the margin is *serrated*, or notched like the teeth of a saw; this kind of leaf may be seen in the *Aster cordifolium*, or aster with a heart-shaped leaf.

Fig. 39.

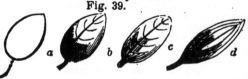

Ovate, *obovate*, *oval ;* these are terms derived from the Latin *ovum*, an egg; suppose the figure at 39, *a*, to represent an egg; you observe that one end is broader than the other; now, if to this broad end you add a petiole, prolonging it into a mid-rib with some lateral divisions, you have, as at *b*, the representation of an *ovate* leaf. If the petiole were placed at the narrowest end, it would be an *obovate* leaf. An *oval* leaf (*c*,) is when both the ends are of equal breadth. When the length is much greater than the breadth, the leaf is said to be *elliptical*, as at *d*.

Fig. 40.

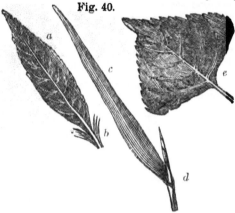

Lanceolate : this kind of leaf may be seen in the peach-tree; it is represented at Fig. 40, *a ;* this is *acuminate*, with a *serrulated* or slightly notched margin ; at *b*, may be seen the cleft *stipules* or appendages of the leaf.

Linear, as the grasses and Indian corn ; Fig. 40, *c*, represents a leaf of this kind ; it is *sheathing* or encloses the stem by its base, as may be seen at *d*.

Deltoid, from the Greek letter, *delta* Δ ; this kind of leaf is represented at *e*, Fig. 40; the Lombardy poplar affords an example of the same.

Fig. 41.

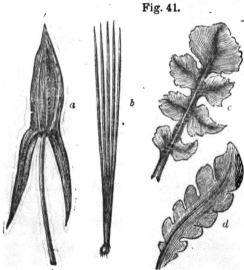

Sagittate (from *sagitta* an arrow,) or arrow-shaped leaf; this is represented at *a*, Fig. 41 ; the *Sagittaria*, an aquatic plant, affords an example of this leaf.

Acerose, or needle-shaped ; this is represented at *b*, Fig. 41. Leaves of this kind are mostly clustered together, as in the pine ; they are *subulate*, or pointed like a shoemaker's awl; they are *rigid* and *evergreen.*

Trees with acerose leaves, are usually natives of mountainous or northern regions; any other kind of leaves would, in these situations, be overpowered by the weight of snow, or the violence of tempests; but these admit the snow and wind through

Elliptical—Lanceolate—Linear—Sagittate—Acerose.

their interstices. Their many points and edges, presented even to a gentle breeze, produce a deep solemn murmur in the forest; and when the storm is abroad and the tempest high,

> "The loud wind through the forest wakes,
> With sound like ocean's roaring, wild and deep,
> And in yon gloomy pines strange music makes."—

Burns, in describing such a scene, says; "this is my best season for devotion: my mind is wrapt up in a kind of enthusiasm to Him, who 'walks on the wings of the wind.'"

Pinnatifid, may be seen at Fig. 41, *d;* leaves of this form are sometimes finely divided, like the teeth of a comb; they are then said to be *pectinate.*

Lyrate, differs from pinnatifid in having its terminating segment broader and more circular. (See Fig. 41, *c.*)

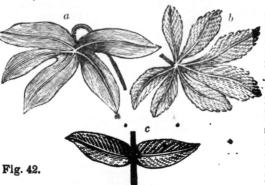

Palmate, or hand shaped, (Fig. 42, *a;*) one species of the passion flower (*Passiflora cærulea*) affords a good example of this kind of leaf. The oblong segments, like fingers, arise from a space near the petiole, which may be considered as resembling the palm of the hand.

Fig. 42.

Digitate, or fingered leaf (Fig. 42, *b,*) differs from the palmate in having no space resembling the palm of a hand; but several distinct leafets arise immediately from the petiole, as may be seen in the Horse Chestnut.

Connate, (Fig. 42, *c;*) the bases of opposite leaves are united so as to appear one entire leaf.

Fig. 43.

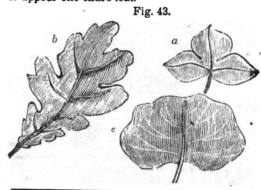

Lobed, when leaves are deeply indented at their margins, they are said to be lobed, and according to the number of these indentations, they are said to be *three lobed, four lobed,* &c. Fig. 43, *a,* represents a three lobed leaf, as may be seen in the *Hepatica triloba.*

Pinnatifid—Lyrate—Palmate—Digitate—Connate—Lobed.

Sinuate, from the Latin *sinus*, a bay; this term is applied to leaves which have their margins indented with deep roundish divisions, as the leaf at *b*, Fig. 43.

Fig. 44.

Emarginate, denotes a slighter indentation, as the leaf at *c*, Fig. 43.

Flabelliform, or fan-shaped, (from *flabellum*, a fan;) this form of the leaf is seen in some of the palms. In China they are used for fans, and sold to foreign merchants for the same purpose. Fig. 44 is a representation of the dwarf fan palm.

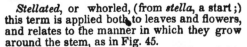

Stellated, or whorled, (from *stella*, a start;) this term is applied both to leaves and flowers, and relates to the manner in which they grow around the stem, as in Fig. 45.

Tubular: there are many varieties of this kind; the leaf of the onion is a complete tube. The Sarracenia or side-saddle flower has the sides of its leaf united, forming a cup which is found filled with liquid, supposed to be a secretion from the vessels of the plant. In some countries of the torrid zone is the wild pine,

Fig. 45.

(*Tillandsia*,) the leaves of which are hollowed out at their base, so as to be capable of containing more than a pint of fluid. A traveller says, "by making an incision into the base of this leaf, and collecting in our hats the water which it contained, we could obtain a sufficient supply for the relief of the most intense thirst." This water is not a secretion from the plant, but is deposited during the rainy season.

c   Fig. 46.   *a*
b

The pitcher-plant (*Nepenthes distillatoria*, Fig. 46,) affords a most singular, tubular appendage, to its lanceolate leaf; beyond the apex of the leaf *a*, the mid-rib extends in the form of a tendril; at the extremity of this tendril is the cylindrical cup or pitcher *b*, about six inches in length and one and a half in diameter; it is furnished with a lid, *c*, which opens and shuts with changes in the atmosphere. The cup is usually found filled with pure water, supposed to be a secretion from the plant. Insects, which creep into it are drowned in the liquid, except a small species of shrimp, which lives by feeding on the

Sinuate—Emarginate—Flabelliform—Stellated—Tubular.
5*

rest. The pitcher-plant is a native of Ceylon, where it is called monkey-cup, on account of its being frequented by these animals for the purpose of quenching their thirst.

Compound Leaves.—When several leafets grow on one petiole, the whole is termed a compound leaf, as in the rose. .

Fig. 47.

Pinnate ; Fig. 47, *a,* represents the petiole or principal leaf stalk bearing leafets arranged opposite to each other ; these may be either petioled or sessile. *b, b,* represent the stipules, the whole taken together forms one compound pinnate leaf. The term pinnate is from the Latin *pinna,* a wing or pinion.

Binate ; when two leafets only spring from the petiole, as in Fig. 47, *c.*

Fig. 48.

Ternate ; when three leafets arise from the petiole, as Fig. 48, *a. Biternate* is a second division of threes, as Fig. 48, *b. Triternate* is a third division of threes, as Fig. 48, *c.*

Decompound, when a pinnate leaf is again divided, or has its leaves twice compound, as Fig. 49, *a.* At *b,* is a representation of *tri-compound* leaves.

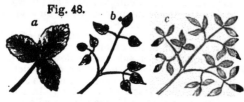

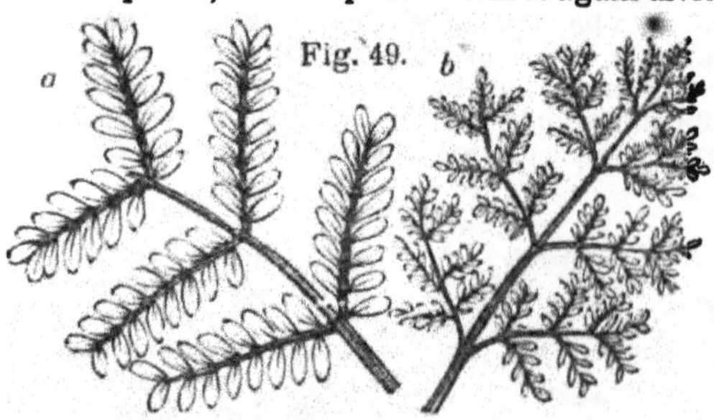

Fig. 49.

We shall now add some miscellaneous examples of various kinds of leaves for the examination of the pupil.

Fig. 50 at *a,* is a leaf of the *Ilex aquifolium,* (holly ;) it is *oval* and *dentate,* with *spinescent* teeth.

b, is a leaf of the *Malva crispa,* (mallows ;) it is *seven-lobed, crisped* or irregularly platted, and finely *crenulate.*

c, is a leaf of the *Hydrocotyle tridentata ;* it is *cuneiform, dentate* at the summit.

d, is a leaf of the *Corchorus japonicus ;* it is *oval-acuminate,* doubly *denticulate.*

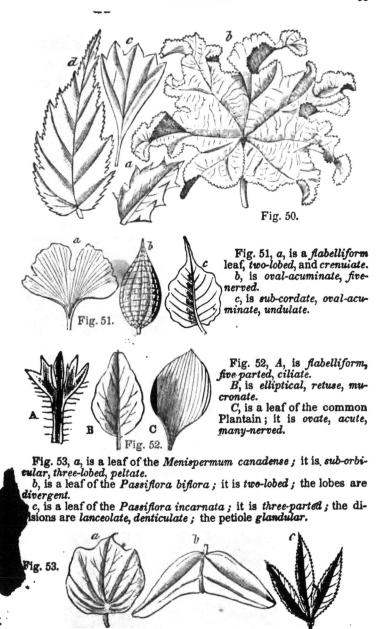

Fig. 50.

Fig. 51, *a*, is a *flabelliform* leaf, *two-lobed*, and *crenulate*.
b, is *oval-acuminate*, *five-nerved*.
c, is *sub-cordate, oval-acuminate, undulate*.

Fig. 51.

Fig. 52, A, is *flabelliform, five-parted, ciliate*.
B, is *elliptical, retuse, mucronate*.
C, is a leaf of the common Plantain; it is *ovate, acute, many-nerved*.

Fig. 52.

Fig. 53, *a*, is a leaf of the *Menispermum canadense*; it is, *sub-orbicular, three-lobed, peltate*.
b, is a leaf of the *Passiflora biflora*; it is *two-lobed*; the lobes are *divergent*.
c, is a leaf of the *Passiflora incarnata*; it is *three-parted*; the divisions are *lanceolate, denticulate*; the petiole *glandular*.

Fig. 53.

Explain Fig. 51—Fig. 52—Fig. 53.

Fig. 54, *a*, is *seven-lobed, denticulate, peltate.*

b, is a leaf of thē *Passiflora serrata;* it is *seven-lobed;* the divisions are *lanceolate, denticulate, veined, glandular.*

c, is a leaf of the *Alchemilla hybrida*, it is *nine-lobed, denticulate, plicate.*

Fig. 54.

Fig. 55.

Fig. 55, *a*, is a leaf of the *Jatropha multifida;* it is *many-parted;* the divisions are *pinnatifid.*

b, is a leaf of the *Helleborus niger;* the leafets are *sub-petioled,* mostly *acuminate, denticulate, veined.*

Fig. 56, *a*, is a leaf of the *Pæonia officinalis,* (Peony;) it is *three-parted, decompound.*

b, is a leaf of the *Geranium pratense;* it is *seven-parted, laciniate.*

c, is a leaf of the *Leontodon taraxacum,* (dandelion;) it is *runcinate*

Fig. 56.

Fig. 57, *a*, is a *trifoliate* leaf; the leafets are *ob-cordate, entire.*

b, is *digitate, five-leaved;* the leafets are *lanceolate, denticulate.*

c, has the petioles *stipuled* and *articulated;* the leafets are *oval* and *acuminate.*

Fig. 57.

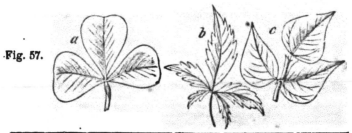

Fig. 58.

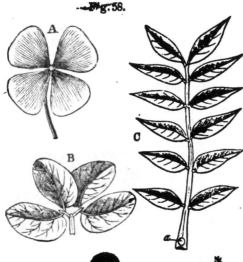

Fig. 58, *A*, is *four-leaved;* the leafets are *cuneiform, very entire.*

B, is a *mimosa* leaf; it is twice *binate.*

C, is thrice *binate, articulate.*

Fig. 59.

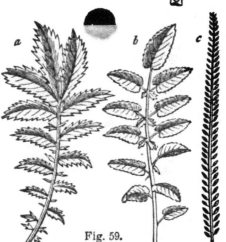

Fig. 59, *a*, is *interruptedly pinnate.*

b, is *unequally pinnate;* the leafets are *stipuled.*

c, *pinnate;* the *rachis* large and *compressed.*

Fig. 60.

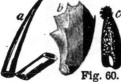

Fig. 60, at *a*, is *cylindrical*, and *fistulous*, as in the onion.

b, is a *fleshy* leaf, *deltoid* and *dentate.*

c, a leaf which is *sub-ovate*, and *bearded* at the summit.

NOTE.—It is recommended to the pupil to practise drawing the various leaves which are given for examples; and to collect as many specimens of leaves as possible.

Explain Fig. 58—Fig. 59—Fig. 60.

Leaves with respect to Magnitude. ━━

Leaves vary in size, from the small leaves of some of the forest trees of our climate; to the spreading Palms and Bananas of the torrid zone. As we approach the torrid zone, the leaves increase in magnitude ; we can, however, scarcely credit the reports of travellers, who say, that the Talipot-tree, in the Island of Ceylon, produces leaves of such size, that twenty persons may be sheltered by one single leaf. Although this account may be exaggerated, there is no doubt of the fact, that the leaves of the torrid zone are of a wonderful size ; and that whole families, in those regions, can make their habitations under the branches of trees. Here we see the care of a kind Providence, which, in countries parched the greater part of the year by a vertical sun, has formed such refreshing shelters. Mungo Park, in his travels in Africa, remarks upon the many important uses of palm-leaves ; serving as covering to cottages, baskets for holding fruit, and umbrellas for defence against rain or sun. These leaves answer as a substitute for paper, and were so used by the eastern nations. Many suppose that the scriptures of the Old Testament were originally committed to palm-leaves.

The magnitude of leaves often bears no proportion to the size of the plants to which they belong. The oak, and other forest-trees, bear leaves, which appear very diminutive, when compared with those of the cabbage, or burdock.

Leaves, with respect to *Duration*, are,

Caducous, such as fall before the end of summer ;

Deciduous, falling at the commencement of winter ; this is the case with the leaves of most plants, as far as 30° or 40° from the equator ;

Persistent, or permanent, remaining on the stem and branches amidst the changes of temperature ; as the leaves of the pine and box ;

Evergreen, preserving their greenness through the year ; as the fir-tree and pine, and generally all cone-bearing and resinous trees ; these change their leaves annually, but the young leaves appearing before the old ones decay, the plant is always green.

In our climate, the leaves are mostly deciduous, returning in autumn to their original dust, and enriching the soil from which they had derived their nourishment. In the regions of the torrid zone, the leaves are mostly persistent and evergreen ; they seldom fade or decay in less than six years ; but the same trees, removed to our climate, sometimes become annual plants, losing their foliage every year. The passion-flower is an evergreen in a more southern climate.

Leaves with respect to Colour.

Leaves have not that brilliancy of colour which is seen in the corolla or blossom ; but the beauty of the corolla, like most other external beauty, has only a transient existence ; while the less showy leaf remains fresh and verdant after the flower has withered away.

The substance of leaves is so constituted as to absorb all the rays of light except *green ;* this colour is of all others best adapted to the extreme sensibility of our organs of sight. Thus, in evident accommodation to our sense of vision, the ordinary dress of nature is of the only colour upon which our eyes, for any length of time, can rest without pain.

But although green is almost the only colour which leaves reflect, the variety of its shades is almost innumerable.

Palm-leaves—Leaves not corresponding in magnitude to the size of the plant—Duration—Colour of leaves—Different shades in the colour of leaves.

"No tree in all the grove but has its charms,
Though each its *hue* peculiar ; *paler* some,
And of a *wannish gray ;* the willow such,
And poplar, that with silver lines his leaf ;
And ash far stretching his umbrageous arm ;
Of *deeper green* the elm ; and deeper still,
Lord of the woods, the long surviving oak."*

The contrast between their shades, in forests, where different families of trees are grouped together, has a fine effect, when observed at such a distance as to give a view of the whole as forming one mass.

A small quantity of iron, united to oxygen in the vegetable substance, and acted upon by rays of light, is said to give rise to the various colours of plants.† If this theory is correct, the different shades of colour in plants, must be owing to the different proportion in which the iron and oxygen are combined.

To quote the words of a celebrated chemist "When Nature takes her pencil, iron is the colouring she uses."

LECTURE X.

ANATOMY AND PHYSIOLOGY OF LEAVES—THEIR USE IN THE VEGETABLE SYSTEM—APPENDAGES TO PLANTS.

Leaves are compared to the lungs of animals; they are organs for *respiring, perspiring,* and *absorbing.* When leaves are wanting, as in the Prickly Pear, (*Cactus,*) the green surface of the stem appears to perform their office. If you will observe a dead leaf which has for some time been exposed to the action of the atmosphere, you may see its *skeleton,* or *frame-work ;* this consists of various fibres minutely subdivided, which originate from the petiole. This skeleton of the leaf may be examined to advantage, after boiling the leaves slightly, or rubbing them in water ; the *cuticle,* or skin, easily separates, and the pulp, or cellular texture, may then be washed out from between the meshes of the veined net-work; thus, the most minute cords of the different vessels become perceptible, with their various divisions and subdivisions; these form what is called the *vascular* system. (See Fig. 61.)

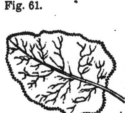

Fig. 61.

Though in external appearance, the organs which compose the *vascular* system of plants, are analogous to the bones which constitute the foundation of the animal system, yet they are rather considered as performing the office of veins and arteries. They are found to be

* Cowper
† This idea coincides with the supposition, that the green colour of leaves is changed to brown by the loss of an acid principle; that the petals of flowers change from purple to red by an increase of acid. The base of this acid is *oxygen.*

What is the cause of these different shades of colour?—the use of leaves in the vegetable economy—Skeleton of the leaf—Vascular system.

tubular; in some cases, this is ascertained by the naked eye; in others, it may be beautifully illustrated by immersing the fibres of the leaf in some coloured liquid; on taking them out, they are found to contain internally a portion of the liquid; this experiment proves them to be *transparent*, as well as *tubular*.

The covering of this frame of the leaf is the *cuticle*, and a pulpy substance, called the *parenchyma*, or *cellular* texture. Some leaves contain much more of this than others, of course they are more pulpy and juicy; it is found, as its name *cellular* would denote, to consist of a mass of little cells, various in size in different leaves; in some, with the most powerful magnifiers, the cells are scarcely perceptible; in others, they may be seen with the naked eye. These cells are of important use in the secretion and communication of substances through the leaf; and may thus be considered as a kind of gland, having a communication with the vascular system.

The covering of the leaf, or the *cuticle*,* guards the vascular and cellular system from injury, and is the medium by which the leaf performs the important functions of absorbing nourishment, and throwing off such substances as are useless or hurtful. The cuticle is sometimes covered with downy, or hairy glands, which seem to afford security against changes of weather; such plants are capable of enduring a greater degree of heat than others. In some cases, the cuticle is covered with a transparent varnish, which preserves the plant from injury by too much moisture, and adds to the beauty of the leaves. The trees of Abyssinia and some other countries, which are subject to long rains, and continued moisture, are thus shielded from the injurious effects of the weather.

When the surface of the cellular tissue is more ample than the vascular net-work, the leaf is *rugose*, as seen at Fig. 62, *a ;* where, for every swelling of the upper surface of the leaf, there is a correspondent depression of the under surface; the sage has a leaf of this kind. When the net-work exists, but the meshes are destitute of cellular tissue, the leaf presents the appearance of lattice-work, and is said to be *cancellated ;* the leaves of an aquatic plant of Madagascar (*Hydrogeton fenestralis*, Fig. 62, *b*,) are of this kind. Another example of this leaf is seen in the *Claudea elegans*, a species of marine Algæ, found in New Holland, (Fig. 62, *c ;*) the veins are parallel to the sides, and cross the nerves.

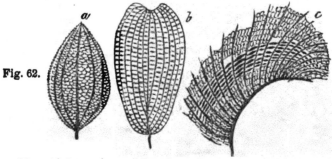

Fig. 62.

* The cuticle is sometimes called *epidermis*, from *epi*, around, and *derma*, skin; the true skin being not the outer covering, but a cellular substance beneath: thus, the thin skin upon the back of the hand, which so easily becomes rough, is the cuticle, or epidermis, (sometimes called the scarf-skin,) while the real skin is below.

How ascertained to be tubular and transparent—Cellular texture—Cuticle—Important office of the leaf—What is a rugose leaf?—What is a cancellated leaf?—Explain Fig. 62.

'These two are the only plants known which have cancel ated
leaves.

Some of the uses of Leaves.

Leaves perform a very important office, in sheltering and protecting the flowers and fruit; the fact of their inhaling or absorbing air
is thought to have been proved, by placing a plant under an exhausted receiver, permitting the *leaves only* to receive the influence of air ;
the plant remained thrifty in this situation for a length of time ; but
as soon as the whole plant was placed under the receiver, it withered and died.*

The upper surface of leaves is usually of a deeper green, and supposed to perform a more important part in respiration, than the under surface.   The upper surface also repels moisture ; you may perceive upon a cabbage-leaf after a shower, or heavy dew, that the
moisture is collected in drops, but has no appearance of being absorbed by the leaf. It has been found that the leaves of plants, laid
with their surface upon water, wither almost as soon as if exposed
to the air ; although the leaves of the same plants, placed with their
under surfaces upon water, retain their freshness for some days. But
few among the vegetable tribes are destitute either of leaves, or
green stems, which answer as a substitute.   The *Monotropa,* or Indian pipe, is of pure white, resembling wax-work.   Mushrooms are
also destitute of any green herbage. It is not known in what manner
the deficiency of leaves is made up to these vegetables.

The period in which any species of plant unfolds its leaves, is
termed *Frondescence.* Linnæus paid much attention to this subject ;
he stated, as the result of his investigations, that the opening of the
leaf-buds of the Birch-tree, was the most proper time for the sowing
of barley.   The Indians of our country had an opinion, that the best
time for planting Indian corn was when the leaves of the White Oak
first made their appearance ; or according to their expression, are
of the size of a squirrel's ears.

One of the most remarkable phenomena of leaves, is their *irritability,* or power of contraction upon coming in contact with other
substances. Compound leaves possess this property in the greatest
degree ; as the sensitive plant, (MIMOSA *sensitiva,*) and the American
sensitive plant, (CASSIA *nictitans ;*) these plants, when the hand is
brought near them, seem agitated as if with fear ; but as plants are
destitute of intelligence, we must attribute this phenomenon to some
physical cause ; perhaps the warmth of the hand, which produces the
contractions and dilatations of the leaves.

The *effect of light* upon leaves is very apparent, plants being almost uniformly found to present their upper surfaces to the side on
which the greatest quantity of light is to be found. It has already
been observed, that plants throw off oxygen gas ; but for this purpose they require the agency of light.

Carbonic acid gas is the food of plants ; this consists of carbon and
oxygen, and is decomposed by the agency of light ; the carbon becomes incorporated with the vegetable, forming the basis of its substance, while the oxygen is exhaled, or thrown off into the atmosphere.

Many plants close their leaves at a certain period of the day, and

* I give this experiment on the authority of Barton; but although the respiration of
leaves seems not to be doubted, this experiment may not be thought a fair one; for it
would seem very difficult, to place a plant under a receiver, with the leaves exposed to
the air, without, at the same time, admitting any air into the receiver.

Few plants are destitute of leaves—Frondescence—Irritability—Effect of light—
What effect has light upon the carbonic acid gas imbibed by plants ?

open them at another; almost every garden contains some plants in which this phenomenon may be observed; it is particularly remark able in the sensitive plant, and the tamarind-tree. This folding up of the leaves at particular periods, has been termed *the sleep* of plants; a celebrated botanist,* remarks, "this may be as useful to the vegetable constitution, as real sleep to the animal." Linnæus was led to observe the appearance of plants in the night, from a circumstance which occurred in raising the Lotus plant; he found one morning some very thrifty flowers, but on looking for them at night, they were no longer visible. This excited his attention, and he began to watch their unfolding. He was thus led to investigate the appearance of other plants at the same time, and to observe their different manner of sleeping. He found, as darkness approached, that some folded their leaves together, others threw them back upon their petioles, or closed their corollas, thus exhibiting a variety of interesting phenomena. This state of relaxation and repose seems to depend on the absence of light; with the first rays of the morning sun, the leaves recommence their chemical labours by drawing in oxygen, the fibres of the roots begin to imbibe sustenance from the earth, and the whole vegetable machinery is again set in motion. It is not solar light alone which seems capable of producing its effect on plants; this has been proved by the following experiment. A botanist placed the sensitive plant in a dark cave, and at midnight lighted it up with lamps; the leaves which were folded up, suddenly expanded; and when, at midday, the lights were extinguished, they again as suddenly closed.

Falling of the Leaf.

The period at which leaves fall is termed the *Defoliation*† of the plant. The "fall of the leaf" may be referred to two causes; the *death of the* leaf, and the *vital action of the parts to which it is attached.* If a whole tree be killed by lightning, or any sudden cause, the leaves will adhere to the dead branches, because the latter have not the energy to cast them off. The development of buds, the hardening of the bark, and the formation of wood, accelerate the fall of the leaf. Heat, drought, frosts, wind, and storms, are all agents in their destruction.

About the middle of autumn, the leaves of the Sumac and Grapevine begin to look red, those of the Walnut, brown, those of the Honeysuckle, blue, and those of the Poplar, yellow; but all sooner or later take that uniform and sad hue, called the *dead-leaf* colour. The rich autumnal scenery of American forests is regarded by the European traveller with astonishment and delight, as far exceeding any thing of the kind which the old world presents. Painters, who have attempted to imitate the splendid hues of our forests, have, by foreigners, been accused of exaggeration; but no gorgeous colouring of art can exceed the bright scarlet, the deep crimson, the rich yellow, and the dark brown, which these scenes present.

After what you have now learned of the anatomy and physiology of leaves, you will probably be induced to pay attention to them in their different stages; from their situation in the bud, to their full growth and perfection; you will feel a new interest in their change of colour, now that you understand something of the philosophy of this change;—even the dry skeletons of leaves, which the blasts of autumn strew around you, may not only afford a direct moral lesson,

* Sir J. E. Smith.
† From *de*, signifying to deprive of, and *folium*, leaf.

as emblematical of your own mortality; but, in examining their stiuc-
ture, you may be led to admire and adore the power which formed them

Appendages to Plants.

Plants have a set of organs, the uses of which are less apparent
than those we have been considering; but we should not infer, because the design for which they have been formed, is in some measure concealed from us, that they were made for no purpose, or exist
by mere accident; let us rather, with humility, acknowledge that this
blindness must be owing to the limited nature of our own faculties.
It would be impious for us to imagine, that all the works of God
which we cannot comprehend are useless.

The organs to which we now refer are called by the general name
of appendages; they are the following: *Stipules, Prickles, Thorns,
Glands, Stings, Scales, Tendrils, Pubescence,* and *Bracts.*

Fig. 63.

1st. *Stipules* are membranous or leafy scales, usually
in pairs, at, or near the base
of the leaf, or petiole. The
stipules furnish characters
used in botanical distinctions.
They are various in their forms
and situations, are found in
most plants, though sometimes wanting. In the garden violet, VIOLA
tricolor, (Fig. 63, *a, a,*) the stipules are of that form called *lyrate-
vinnatifid,* while the true leaf (*b*) is oblong and crenate. The most
natural situation of the stipules is in pairs, one on each side of the
base of the foot-stalk, as in the sweet pea; some stipules fall off
almost as soon as the leaves are expanded, but, in general, they remain as long as the leaves.

2d. *Prickles* arise from the bark; they are straight, hooked, or
forked. They are usually found upon the stem, as in the rose; but
in some cases, they cover the petiole, as in the raspberry; in others,
they are found upon the leaf or the calyx, and in some instances,
upon the berry; as in the gooseberry.

3d. *Thorns* are distinguished from prickles, by growing from the
woody part of the plant, while the prickle proceeds only from the
bark. On stripping the bark from the rose-bush, the prickles will
come away with it; but let the same experiment be made with a
thorn-bush, and although the bark may be separated, the thorn will
still remain projecting from the wood.

Fig. 64.

In this draw-
ing, you will
observe the
thorn, (*a,*) to
remain on the
stem, while the
bark (*b*) has
been peeled off. In the prickle (*c*) the whole appears separated from
the plant. The thorns, in some plants, have been known to disappear by cultivation. The great Linnæus imagined, that in such
cases, the trees were divested of their natural ferocity, and became
tame. We may smile at such a fanciful idea, but should remember
that great men have their weaknesses; and that when persons become enthusiasts in any science, they are in danger of tracing analogies or resemblances, which exist in their own minds, rather than

Different kinds of appendages—Stipules—Prickles—Thorns—Thorns in some cases
made to disappear.

in nature. A more rational opinion is given by another botanist, viz.—that thorns are in reality bulbs, which a more favourable situation converts into luxuriant branches. But in many cases, they do not disappear even under circumstances the most favourable to vegetation. Thorns have been compared to the horns of animals.

4th. *Glands* are roundish, minute appendages, sometimes called tumours or swellings; they contain a liquid secretion which is supposed to give to many plants their fragrance. They are sometimes attached to the base of the leaf, sometimes they occur in the substance of leaves; as in the lemon and myrtle, causing them to appear dotted when held to the light. . They are found on the petioles of the passion-flower, and between the teeth and divisions of the leaves of many plants.

5th. *Stings* are hair-like substances, causing pain by an acrid liquor, which is discharged upon their being compressed; they are hollow, slender, and pointed, as in the nettle.

6th. *Scales* are substances, in some respect resembling the coarse scales of a fish; they are often green, sometimes coloured, and are found upon all parts of vegetables, as upon the roots of bulbous plants, and upon the stems and branches of other plants. They are *imbricated* upon the calyxes of most of the compound flowers. You have seen in buds, how important the scales are, in protecting the embryo plant during the winter. Scale-like calyxes surround the flowers of grasses, under the name of *glumes.* Scales. envelop and sustain the stamens and fruit of the pine, oak, chestnut, &c.

Fig. 65.

7th. *Tendrils,* or claspers, are thread-like appendages, by which weak stems attach themselves to other bodies for support; they usually rise from the branches, in some cases from the leaf, and rarely from the leaf-stalk or flower-stalk. You have here the representation of a tendril. Tendrils are very important and characteristic appendages to many plants. In the trumpet-flower and ivy, the tendrils serve for roots, planting themselves into the bark of trees, or in the walls of buildings. In the cucumber and some other plants, tendrils serve both for sustenance and shade. Many of the papilionaceous, or pea-blossom plants, have twining tendrils, which wind to the right, and back again. Among vegetables which have tendrils, has been discovered that property, which some have called, the instinctive intelligence of plants. A poetical botanist represents the tendrils of the gourd and cucumber, as, " creeping away in disgust from the fatty fibres of the neighbouring olive." The manner in which tendrils stretch themselves forward to grasp some substances, while they shrink from others, is indeed astonishing; but instead of imagining that they have a preference for some, and a dislike for other objects, it is more philosophical to conclude that these effects arise from physical causes, which do not the less exist because we can not discover them. It has been ascertained by experiments, that the tendrils of the vine, and some other plants, recede from the light, and seek opaque bodies. The fact with respect to leaves is directly the reverse of this, for they turn themselves round to seek the light.

Some plants creep by their tendrils to a very great height even to the tops of the loftiest trees, and seem to cease ascending, only because they can find nothing higher to climb. One of our most beautiful climbing plants is the CLEMATIS *virginica*, or virgin's bower, which has flowers of a brilliant whiteness. Its pericarps, richly fringed, are very conspicuous in autumn, hanging in festoons from the branches of trees, by the sides of brooks and rivers.

8th. *Pubescence* includes the down, hairs, woolliness, or silkiness of plants. The pubescence of plants varies in different soils, and with different modes of cultivation. The species in some genera of plants are distinguished by the direction of the hairs. The microscope is often necessary in determining with precision, the existence and direction of the pubescence. It has been suggested that these appendages may be for similar purposes as the fur, hair, and bristles of animals, viz. to defend the plants from cold, and injuries from other causes.

Fig. 66.

9th. *The Bract*, or floral leaf, is situated among, or near the flowers, and is different from the leaves of the plant. You may, in Fig. 66, observe the difference between the real leaves (*b, b,*) and the bract (*a;*) the former being *cordate* and *crenate*, the latter *lanceolate* and *entire.*

In some plants, as in several species of sage, the transition from leaves to bracts is so gradual, as to render it difficult to distinguish between them, and a considerable part of the foliage is composed of the bracts. In the crown-imperial, the stem is terminated by a number of large and conspicuous bracts These appendages are sometimes mistaken for the calyx. Bracts are *green* or *coloured, deciduous* or *persistent.* The orchis tribe have green leaf-bracts. No plants of the class *Tetradynamia* have bracts.

We have, in regular order, considered the first of the two classes of vegetable organs, viz.: such as tend to the support and growth of the plant, including *root, stem, leaf,* and *appendages;* we shall next examine the class of organs whose chief use appears to be that of bringing forward the fruit

LECTURE XI.

CALYX.

WE are now to consider the second division of vegetable organs, viz.: such as serve for the reproduction of the plant, called organs of fructification. Their names were considered when commencing the analysis of flowers; but we are now to examine them with more minute attention, and to remark upon their different uses in the vegetable economy.

You are no doubt pleased to have arrived at that part of the plant, which is the ornament of the vegetable kingdom. Flowers are delightful to every lover of nature; a bouquet, or even the simplest blossom, presented by a friend, interests the heart. How many pleasant thoughts are awakened by the fresh and perfumed incense which ascends from flowers!—their odour has been poetically termed, the language by which they hold communion with our minds. Females are usually fond of flowers; but until recently, the greater number have only viewed them as beautiful objects, delighting the senses by their odour and fragrance, without being aware that they, lovely as they seemed, might be rendered doubly interesting, by a scientific knowledge of the relations and uses of their various parts. Even at the present period, there are those who spend years in cultivating plants, ignorant of their botanical characters, when a few hours study might unfold to them the beautiful arrangement of Linnæus, and open to their mental vision a world of wonders.

Although every part of a plant offers an interesting subject for study, the beauty of the blossom seems, by association, to heighten the pleasure of scientific research. Flowers are indeed lovely, but like youth and beauty they are fading and transient; they are, however, destined for a higher object than a short-lived admiration; for, to them is assigned the important office of producing and nourishing the fruit; like them should the young improve the bloom of life, so that when youth and beauty shall fade away, their minds may exhibit that fruit, which it is the business of youth to nurture and mature.

Calyx.

The calyx is frequently wanting; as in the lily and tulip. The corolla is also wanting in many plants; as, in most of the forest trees, which, to a careful observer, may seem to produce no flower; but the presence of a stamen and pistil, is in botany considered as constituting a *perfect flower.* These two organs are essential to the perfection of the fruit; and when a flower is destitute either of stamens or pistils, it is termed *imperfect.* A flower is said to be *incomplete* when any of the seven organs of fructification are wanting.

The word calyx is derived from the Greek, and literally signifies a cup; it is the outer cover of the corolla, and usually green; when not green, it is said to be *coloured.* This organ is an expansion of the bark of the flower-stalk, as appears from its colour and texture. The calyx usually envelops the corolla, previous to its expansion, and afterward remains below or around its base. Sometimes the calyx consists of one leaf or sepal only, it is then called *monosepalous;* when it consists of several distinct leaves, it is called *polysepalous;* when one calyx is surrounded by another, it is, *double;* when one calyx belongs to many flowers, it is *common.*

In the calyx are three parts, very distinct in calyxes which are long and cylindric; these are, 1st, the *tube* which rises from the base; 2d, the *throat,* above the tube; and 3d, the *mouth,* or the upper and expanded part; the tube of the calyx is *cylindric* in the pink, and *prismatic* in the stramonium.

The position of the calyx with respect to the germ offers an important mark of distinction between different genera, and also between different natural families of plants. The calyx is said to be *superior* when it is situated on the summit of the germ, as in the apple; it is *inferior,* when situated below the germ, as in the pink. In many plants the calyx is neither superior nor inferior, but is situated around the germ.

Flowers delightful—Many who cultivate them ignorant of their botanical characters—Flowers analogous to youth—Calyx, sometimes wanting—Description of the calyx—Parts of the calyx—Position with respect to the germ.

When the calyx drops off before the flower fully expands, it is called *caducous;* the petals of the poppy are, at first, enclosed in a calyx of two large green leaves, but these fall off before the flower is full blown. When the calyx withers and drops off with the corolla, it is called *deciduous.* In many plants it remains until the fruit is matured; it is then called *persistent* Upon a pea-pod, for example, the calyx may be seen as perfect as it was in the blossom. On examining an apple or pear, the dried leaves of the calyx may be seen on the top of the fruit; this shows that the calyx was superior, as well as persistent.

According to the divisions of Linnæus, there are seven kinds of calyxes; viz.

Perianth, Involucrum, Ament, Spatha, Glume, Calyptra, Volva.

Perianth. This term is derived from the two Greek words, *peri,* around, and *anthos,* flower. This is the only real calyx or cup, as the term cup does not properly apply to the other kinds. A good example of the perianth calyx is presented in the rose, where it is *urn-form,* with divisions at the top resembling small leaves. In the pink, the perianth is long and tubular, having the border *dentate* or toothed. The holly-hock, hibiscus, and many other plants, have a double perianth. The term perianth is often used when a flower has but one envelope, as in the tulip; and more especially in cases where it is difficult to determine whether this envelope should be called a corolla or calyx.

Involucrum. This term is derived from the Latin, *involvo,* to wrap up; this kind of calyx is usually found at the base of an umbel, as in the carrot. It is said to be *universal,* when it belongs equally to the whole of an aggregate flower; and *partial,*[*] when it encloses one floret which, with others, constitutes a compound or aggregate flower. The term involucrum is also applied to the membranous covering in the fructification of ferns.

Ament or *catkin,*[†] is a kind of calyx, by some classed as a mode of inflorescence; it consists of many chaffy scales, ranged along a thread-like stalk or receptacle; each scale protects one or more of the stamens or pistils, the whole forming one aggregate flower. The ament is common to forest trees, as the oak and chestnut; and is also found upon the willow and poplar. In some trees, the staminate flowers are enclosed in an ament, and the pistillate in a perianth.

Fig. 67.

Spatha, or sheath. It is that kind of calyx which first encloses the flower, and when this expands, bursts lengthwise and often appears at some distance below it. The wild turnip, or Arum, furnishes an example of this kind of calyx, enclosing a kind of inflorescence called a *spadix,* (Fig. 67. *a.*) From the peculiar appearance of the spadix as it stands up surrounded by the spatha, it is sometimes called *Jack-in-the-box.* The spatha is common in many of our cultivated exotics, as in the Narcissus, where it appears brownish and withered, after the full expansion of the flower. You see here it re-

* See Fig. 128, *a, a.* † See Fig. 91.

presentation (Fig. 67) of the Spatha of the Arum (b,) and of the Narcissus, (c.) In the Egyptian Lily, the spatha is white and permanent, and the stamens and pistils grow upon different parts of the spadix. Palms have a spadix which is branched, and often bears a great quantity of fruit.

Glume, is from the Latin word *gluma*, a husk. This is the calyx of the grasses, and grass-like plants. In the oat and wheat it forms

the *chaff*, a part which is thrown away as worthless. In the oat, (Fig. 68,) the glume calyx is composed of two pieces called valves; in some kinds of grain of but one, in others

Fig. 68.

of more than two valves. To the glume belongs the *awn* or beard. The corolla of grasses is husky, like the calyx, and is sometimes considered as a part of it. Some botanists consider that there is in the grasses, neither calyx nor corolla, and that these scales are only membranous bracts.

Calyptra. This term is derived from the Greek, and signifies a veil. It is the cap, or hood, of pistillate mosses, resembling in form and position, the extinguisher of a candle.*

Volva, the ring, or wrapper of the fungus plants. It first encloses the head of the Fungus, afterward bursts and contracts, remaining on the stems, or at the root.†

We have now considered the different kinds of calyx. We find that this organ is not essential, since it is wanting in some plants, but its presence adds to the completeness of the flower; in some cases it is the most showy part; as in the Lady's-ear-drop, where it is of a bright scarlet-colour, and in the Egyptian Lily, where it is pure white.

The calyx is of use in protecting the other parts of the flower before they expand, and afterward supporting them in their proper position. Pinks, having petals with long and slender claws, which would droop or break without support, have a calyx. Tulips having firm petals, and each one resting upon a broad strong basis, are able to support themselves, and they have no calyx. In some plants, the calyx serves as a seed-vessel; as in the order *Gymnospermia*, of the class *Didynamia*, where there are four seeds lying in the bottom of the calyx.

---

* See Fig. 153, a.   † See Fig. 157, d.

LECTURE XII. '

COROLLA.

THE term *Corolla*, or corol, is derived from the Latin, *corona*, a crown or chaplet. As the calyx is formed by a continuation of the fibres of the outer bark, the corolla is a continuation of the inner coat of the same. The texture of the corolla is delicate, soft, watery, and coloured. It exhales carbonic acid gas, but not oxygen, neither in the dark, nor when acted upon by light. The cuticle, or outward covering of the corolla, is of an extremely fine texture. The rich and variegated colours of flowers, are owing to the delicate organization of the corolla; and to this cause the transient duration of this organ may also be attributed.

The corolla exhibits every variety of colour except black; florists present us with what they term black roses, and we see others which approach this colour, yet none are perfectly black; the darkest being but a very deep shade of purple. Corollas are white, yellow, blue, violet, &c.; in some, different colours are disposed, and blended; in others, they meet abruptly, without any intermediate teint. The colour of the corolla, in the same species, often varies without any assignable cause. This fact is strikingly illustrated in the *Four o'clock*, (MIRABILIS;) the flowers of which are sometimes of pale yellow, sometimes bright crimson, and often richly variegated. These varieties are the result of circumstances not under the control of man; the florist watches these changes, and, as far as possible, avails himself of them in the production of new beauties in the vegetable kingdom.

The corolla, before blossoming, is folded in the calyx, as the leaves are within the scales of the leaf-bud, and the whole is then called the flower-bud. In most cases, the calyx and corolla are so distinctly marked, that it is perfectly easy to distinguish them. The colour usually constitutes a very striking mark of difference; the calyx being ordinarily green, and the corolla of a more lively hue. But the colour is not always a criterion, for in some cases the calyx is beautifully coloured. In the FUSCHIA, (*Lady's ear-drop*,) the calyx is of a bright scarlet; you might at first think it to be the corolla; but if you remove the searlet coat, you may see, wrapped around the eight stamens, a purple covering; on taking off each piece carefully, you will find four petals,* as distinct as the petals of a rose; you will then perceive that the outer covering must be the calyx.

Linnæus made the following distinction between the corolla and the calyx; viz. that the corolla has its petals alternate with the stamens, and the calyx has its leafets arranged opposite to them. This rule is not found to be invariable; it has led some botanists to call that the corolla which others have named the calyx. It seems that we must come to the conclusion that nature has not placed any absolute limits between these two organs.

The corolla sometimes falls off soon after the flowering, as in the poppy; it is then said to be *caducous;* sometimes it fades and withers upon the stalk, as in the blue-bell; it is then said to be *marescent,* or withering.

* Some botanists call these nectaries, but this seems to be making an unnecessary confusion in terms; for they have as much the appearance of petals, as those of a rose or pink.

Each simple part, of which the corolla is composed, is called a *petal.* A flower with petals is said to be *petalous;* without petals, *apetalous.* The petals are said to be *definite,* when their number is not more than twenty, *indefinite,* when they exceed that number.

If the corolla is formed of one single piece, or petal, it is *monopetalous;* if of more than one, it is *polypetalous.* You may sometimes find a difficulty in determining whether a corolla is composed of one piece or more; for monopetalous flowers often have deep divisions, extending almost to the base of the corolla; but they must be divided at the base; that is, be in separate pieces, in order to be considered *polypetalous.* The parts into which a corolla naturally falls, may be considered as so many petals.

Monopetalous corollas, (see Fig. 70,) consist of the *tube, throat,* and *limb.* The *tube* is the lower part, having more or less the form of a tunnel. The *throat* is the entrance into the tube; it is either open, or closed by scales or hairs. The *limb* is the upper border of the corolla.

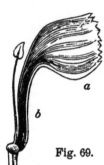

Polypetalous corollas consist of several petals. Each petal consists of two parts, the *lamina,* and *claw.*

The *lamina,* (Fig. 69, *a,*) is the upper, and usually the thinner part of the petal; its margin is sometimes *entire,* or without divisions, as in the rose; sometimes notched, or crenate, as in the pink. The lamina corresponds to the limb of monopetalous corollas.

The *claw,* (Fig. 69, *b,*) is the lower part of the petal, and inserted upon the receptacle; it is sometimes very short, as in the rose; in the petal of the pink, as seen at Fig. 69, it is long and slender. The claw is analogous to the tube of monopetalous corollas.

Fig. 69.

The corolla is *superior* when inserted above the germ, *inferior,* when below. It is *regular,* when each division corresponds to the other. The rose and pink have regular corollas. When the parts do not correspond with each other, a corolla is *irregular;* as in the pea and the labiate flowers.

Different forms of Monopetalous Corollas.

Fig. 70.

Monopetalous corollas may, according to their forms, be divided as follows:

1st. *Bell-form,* (campanulate, from *campanula,* a little bell; here the tube is not very distinct, as the corolla gradually spreads from the base; as in the blue-bell, hair-bell, &c. At Fig. 70, is the representation of a bell-form corolla; it is monopetalous; the limb, *a,* is five-parted; calyx, *b,* five-parted; corolla superior. The blue-bell of the gardens offers a fine illustration of this kind of corolla.

Parts of the corolla—Polypetalous corollas, how divided?—Forms of monopetalous corollas—Polypetalous—Corolla, superior—Inferior—Regular—Irregular—Bell form.

Fig. 71. 2d. *Funnel-form*, (*infundibuliformis*, from *infundibulum*, a funnel;)  having a tubular base, and a border opening in the form of a funnel, as the Morning-glory, Fig. 71.

3d. *Cup-shaped*, (*Cyathiformis*, from *cyathus*, a drinking-cup;) differing from funnel-shaped, in having its tube, and border, less spreading; and from bell-form, in not having its tube appear as if scooped at out the base, Fig. 72.

Fig. 72.

4th. *Salver-form*, (*hypocrateriformis*, from the Greek *krater*, an ancient drinking glass called a *salver;*) this has a flat, spreading border, proceeding from the top of a tube, Fig. 73.

Fig. 73.

Fig. 74.

5th. *Wheel-form*, (*rotate*, from *rota*, a wheel;) having a short border without any tube or with a very short one, Fig. 74.

This kind of corolla may be seen in the mullein.

6th. *Labiate*, (from *labia*, lips;) consists of two parts, resembling the lips of a horse, or other animal. Labiate corollas are said to be *personate*,* having the throat closed, or *ringent*,† with the throat open. You have a labiate corolla of the ringent kind, at Fig. 75. The term labiate is also applied to a calyx of two lips. *Bi-labiate* is sometimes used in the same sense as labiate.

Fig. 75.

Different forms of Polypetalous Corollas.

1st. *Cruciform*, (from *crux*, a cross;) consisting of four petals of equal size, spreading out in the form of a cross, as the radish, cabbage, &c. Fig. 76.

Fig. 76.

2d. *Caryophyllous*, having five single petals, each terminating in a long claw, enclosed in a tubular calyx, as the pink, Fig. 77.

Fig. 77.

3d. *Liliaceous*, a corolla with six petals, spreading gradually from the base, so as to exhibit a bell-form appearance, as in the tulip and lily.

4th. *Rosaceous*, a corolla formed of roundish spreading petals, without claws, or with very short ones, as the rose and apple.

---

* From *persona*, a mask.
† From *ringo*, to grin, or gape.

Funnel-form—Cup-shaped—-Salver-form—Wheel-form.

Labiate corollas, how divided ?—Forms of polypetalous corollas—Cruciform—Caryophyllous—Liliaceous—Rosaceous.

5th. *Papilionaceous*, a flower with a banner, two wings, and a keel; the name is derived from the word *papilio*, a butterfly, on account of a supposed resemblance in form, as the pea-blossom, Fig. 78.

If a corolla is not, in form, like any of those we have descɪ bed, it is said to be *anomalous*.

Fig. 78.

Odour of Flowers.

The odour of flowers has its origin in the volatile oils. elaborated by the corolla; its production results from causes both external and internal, but, in both cases, equally beyond our observation. Temperature renders the odour of flowers more or less sensible; if the heat is powerful, it dissipates the volatile oils more rapidly than they are renewed: if the heat is very feeble, the volatile oils remain concentrated in the little cells where they were elaborated; under these circumstances the flowers appear to possess but little odour. But if the heat is neither too great nor too little, the volatile oils exhale without being dissipated, forming a perfumed atmosphere around the flowers.

You perceive the reason, that when you walk in a flower garden in the morning or evening, the flowers seem more fragrant than in the middle of the day. The air being more charged with humidity, is another cause of an increase of fragrance at those times; as the moisture, by penetrating the delicate tissue of the corollas, expels the volatile oils. There are some exceptions to the laws just stated; for some flowers are only odorous during the night, and others during the day. Some flowers exhale fetid odours, which attract such insects as are usually nourished by putrid animal substances. Many flowers exhale sweet odours; but, however odours may differ, in the sensations which they produce, it is certain, that powerful ones have a stupifying, narcotic effect upon the nerves, and that it is dangerous to respire, for any great length of time, even the most agreeable of them, in a concentrated state.

One important office of the corolla, is to secure those delicate and important organs which it encloses, the stamens and pistils, from all external injury, and to favour their development. After the germ is fertilized by the influence of the pollen, the corolla fades away, and either falls off or remains withered upon the stalk; the juices which nourished it then go to the germ, to assist in its growth, and enable it to become a perfect fruit.

Another use of the corolla seems to be, to furnish a resting-place for insects in search of honey.

The corolla is supposed by Darwin to answer the same purpose to the stamen and pistils, as the lungs in the animal system; each petal being furnished with an artery which conveys the vegetable blood to its extremities, exposing it to the light and air under a delicate moist membrane; this vegetable blood, according to his theory, is then collected and returned in correspondent veins, for the sustenance of the anthers and stigmas, and for the purpose of secreting honey.

Saint Pierre* thinks the corolla is intended to collect the rays of the sun, and to reflect them upon the stamens and pistils which are placed in the centre or focus.

After all our inquiries into the uses of the corolla, we are obliged to acknowledge that it appears less important in the economy of vegetation, than many less showy organs. It seems chiefly designed to beautify and enliven creation by the variety and elegance of its forms, the brilliancy of its colouring, and the sweetness of its perfume.

MMMM   *Nectary.*

In many flowers there is an organ called the *nectary*, which secretes a peculiar fluid, the honey of the plant; this fluid constitutes the principal food of bees and various other species of insects.

Linnæus considered the nectary as a separate organ from the corolla; and every part of the flower which was neither stamen,

Fig. 79. pistil, calyx, nor corolla, he called a nectary; but he undoubtedly applied the term too extensively and vaguely. The nectary is not to be confined to any particular part of the flower. Sometimes it is a mere *cavity*, as in the lily. The crown imperial, Fig. 79, exhibits in the claw of each of its petals a nectary of this kind; each one being filled with a sweet liquid, the secretion of the flower. If these drops are removed, others immediately take their place. The six nectariferous glands at the base of the corolla are represented in the figure; the petals are supposed to be cut in order to show the base of the flower.

In the Ranunculus, (Butter-cup,) the nectary is a production of the corolla in the form of a *scale;* in the violet, a process of the same, in the form of a *horn* or *spur*. In the Columbine, (Aquilegia,) the nectary is a separate organ from the petals, in the form of a *horn*. In the monkshood, one of the petals being concave, conceals the nectaries; they are therefore said to be *hooded*.

In monopetalous corollas, the tube is supposed to answer the purpose of a nectary in secreting honey. In the honeysuckle, we find at the bottom of the tube a nectariferous liquid; yet there is no appearance of any gland or organ, by which it could have been secreted, unless we suppose the tube to have performed this office.

With respect to the purpose for which honey is secreted by the nectary and other parts of the flower, there seems, among authors, to be a difference of opinion. Darwin supposes this to be the food with which the stamens and pistils are nourished, or the unripe seeds perfected. Smith asserts, that the only use of honey, with respect to the plant, is to tempt insects, which, in procuring it, scatter the dust of the anthers, and fertilize the flower, and even carry the pollen from the barren to the fertile blossoms; this is particularly the case

* This ingenious author remarks, that man seems the only animal sensible to the sweet impressions made by the colour and odour of plants upon the senses; but we think he has asserted too much. Do not the brute creation seem to enjoy, by the sense of smelling, the freshness of the verdant fields? But man is very apt to say, "See all things for my use."

in the fig-tree. Although in the case of plants whose stamens and
pistils are on separate flowers, we see this advantage arising from
the fact of insects being attracted by the honey, yet the greater
number of plants do not need any assistance in conveying pollen to
the stigmas.. Some imagine that honey contributes to the perfection
of the stamens: but plants that do not appear to secrete honey, have
perfect stamens. One thing, however, is certain with respect to this
fluid, that without detriment to the plant, it yields to the industrious
bee the material for the manufacture of honey, a luxury highly
valued from the most ancient times. Virgil knew that bees made
honey from the juices which they gathered from flowers; and we
indeed, on this subject, know but little more than he has beautifully
expressed in his pastorals.

Although we are ever discovering something new and wonderful
in the economy of nature; and, in some cases, seem permitted to
search into the hidden mysteries of her great Author, yet in our researches we are continually made sensible of the limited nature of
our own faculties; and a still, small voice, seems to whisper to man,
in the proudest triumphs of his reason, "Hitherto shalt thou go, but
no farther."

LECTURE XIII.

STAMENS AND PISTILS.

ALTHOUGH the calyx and the corolla may be wanting, the stamens
and pistils are indispensable to the perfection of the fruit. They are
in most plants enclosed by the same envelope, or stand on the same
receptacle; in the class Monœcia they are on different flowers which
spring from one common root; and in Diœcia, they are on different
flowers, springing from different roots. Yet, however distant the
stamens and pistils may be, nature has provided ways by which the
pollen from the staminate flowers may be conveyed to the pistillate,
and there assist in perfecting the seed. That you may the better
understand this curious process, and the organs by means of which
it is carried on, we will examine each one separately.

Stamens.

Stamens are thread-like parts which are exterior with respect to
the pistil, and interior to the corolla. They exhibit a variety of positions with respect to the pistil. These positions seldom vary in
the same family, and they have therefore been taken by the celebrated Jussieu as one of the fundamental distinctions in his classification, called the "Natural method." If the stamens are inserted
upon the pistil, as in umbelliferous plants, they are said to be *epigynous* (from *epi*, upon, and *gynia*, pistil;) if the stamens are inserted
under the germ, as in cruciform plants, they are said to be *hypogynous* (from *hypo*, under, and *gynia*, pistil;) when the stamens are inserted upon the calyx, and thus stand *around* the germ, as in the rosaceous plants, they are said to be *perigynous*, (from *peri*, around, and
gynia, pistil.)

When a corolla is monopetalous, the number of the stamens is,
usually, either equal, double, or half that of the divisions of the corolla
he stamens in such flowers never exceed twenty.

Reflections—Stamens and pistils necessary to the perfection of the fruit—Definitions of the stamen—Positions with respect to the pistil—Divisions of monopetalous
corollas usually in proportion to the number of stamens.

In polypetalous corollas, the number of stamens is so much greater. When they equal the divisions of the cord usually alternate with these divisions. When the number of is double the divisions of the corolla, half of the stamens are placed in the intervals of the divisions, and the remaining half before each lobe of the corolla, corresponding to the intervals in the divisions of the calyx. If any of the stamens are barren or without anthers, they will be found to be those which are placed before the lobes of the corolla.

In commencing the analysis of flowers according to the Linnæan system, you learned that the *number* of stamens, their *position, relative length*, and *connexion*, taken either singly or in combination, afford certain and distinctive marks for purposes of classification.

In the first place we find the stamens differing in *number*, in different plants; some plants have but one, some two, and so on till we come to ten; when they have more than ten, we find the number in the same plant varies, and therefore we cannot depend on this circumstance for further classification.

We then resort to *position*, and consider whether the stamens are inserted upon the calyx or the receptacle, thus furnishing an eleventh and a twelfth class.

Inequality in the length of stamens, when they are either four or six, furnishes us with a thirteenth and fourteenth class.

The *connexion or union of stamens* gives us the fifteenth class, where the filaments of the stamens are united in one set; the sixteenth class, where they are in two sets; the seventeenth, where the anthers of the stamens are united.

The three remaining classes of phenogamous plants are distinguished by the *position of the stamens with respect to the pistils*. In the eighteenth class the stamens stand on the pistil; in the nineteenth the stamens and pistils are on separate flowers on the same plant; in the twentieth they are on separate plants; and in the twenty-first they are invisible.

Parts of the Stamen.—The *Filament*, is so called from *filum*, a thread. Filaments vary in their form; some are long and slender, as in the pink; others are short and thick, as in the tulip. They are usually smooth, but in the mullein they are bearded; in the spider-wort (*Tradescantia*) they are covered with down. In most cases a filament supports but one anther, but sometimes it is forked and bears two or more; in some instances, many filaments have but one anther. When the filaments are enclosed in the tube of the corolla they are said to be *inserted*, when they extend out of it, *exserted*. In some cases the filament is wanting, and the anther is sessile, or immediately attached to the coralla.

In double flowers, the stamens, which seem to be intimately connected with the parts of the corolla, are changed to petals. This is the effect of cultivation, which, by affording the stamens excess of nourishment, causes them to swell out, and thus assume the form of petals. In some double flowers almost every trace of the stamens disappears; in others, it is easy to perceive the metamorphosis which they have undergone, as they retain something of their original forms. In double flowers the anthers usually disappear, which shows that the filaments have absorbed all the nourishment. In

double roses some stamens appear entirely changed, others retaining something of their form, and others are still perfect. When all
the stamens disappear, no perfect fruit is produced. On account of
this degeneration of the stamens, cultivated flowers are not usually
so good for botanical analysis as wild ones. The single flower exhibits the number of parts which nature has given to it. The rose
in its native state has but five petals.

Anther, is a little knob or box usually situated on the summit of
the filament; it has cells or cavities which contain a powder called
the pollen; this is yellow, and very conspicuous in the lily and
tulip. You have here the representation (Fig. 80) of a stamen with

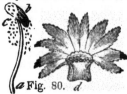 its filament a, its anther b, and the dis-
charging pollen c. In many flowers the
filament is wanting; the anthers are then
said to be sessile; that is, placed immediately upon the corolla, as at d, which
represents a flower cut open, showing
its stamens growing sessile in the
throat.

a Fig. 80. d

Fig. 81.

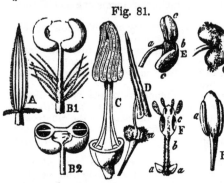

The figure at A, re-
presents a magnified
stamen,* with a lance-
olate anther, denticu-
late at the sides, with
two hairy appendages;
filament short.

At B 1, is a magni-
fied stamen,† with the
filament bearded at the
base; the anther is two-
lobed, reniform. B 2
shows the two cells in
each lobe, which is cut
horizontally.

At C,‡ the three fila-
ments are distinct at the base, and connected at the upper part; anthers, adnate, linear, twisting.

At D,§ the anther is sagittate, the filament bent, and glandular in
the middle, (at a.)

At E, is a stamen of the Thyme, (family of the Labiatæ;) the
lobes of the anthers c, are divergent; a, is the filament, b, the con-
nective of the anthers.

At F, is a stamen of the Laurus; a, cordate, pedicelled glands; b,
pubescent filament; c, anther opening by four valves, throwing out
pollen.

At G, is a stamen of the genus Lavendula; the anthers are reniform, cilicate, opening transversely, lobes confluent at the summit,
divergent at the base.

At H, a stamen of the genus Begonia; the filament is enlarged at
the summit; the two lobes of the anther a, a, adnate at the sides,
parallel distant.

* Of the Cerinthe major, (family of the Boragineæ.)
† Of the Tradescantia virginica.
‡ Of the Cucumber family.
§ Of the Linden family.

Pistils.

In the centre of the flower stands the pistil, an organ essential to the continued existence of the plant. Like the stamens, the pistils vary in number in different plants, some having but one, and others hundreds. Linnæus founded the orders of his first twelve classes on the number of these organs. When they are more than ten, he did not rely upon their number, which in this case is found to vary in individuals of the same genus.

The pistil consists of three parts, the *germ, style,* and *stigma.* It may be compared to a pillar; the germ, (Fig. 82, *a,*) corresponding to the base; the style (*b,*) to the shaft; and the stigma (*e,*) to the capital.

Fig. 82.

The figure at (*g*) represents the pistil of the poppy, the germ or base is very large; you will perceive that the style is wanting, and the stigma is *sessile,* or placed immediately on the germ. The style is not an essential part, but the stigma and germ are never wanting; so that these two parts, as in the poppy, often constitute a pistil.

Germ. The germ, or *ovary,* contains the rudiments of the fruit, or (*ovules,*) yet in an embryo state. A distinction is to be made between the germ here spoken of, and the germ of the bud.* This germ in the flower, is the future fruit, though in passing to maturity it undergoes a great change. You would scarcely believe that the pumpkin was once but the germ of a small yellow flower. The germ is said to be *superior,* when placed above the calyx, as in the strawberry; *inferior,* when below it, as in the apple. The figure of the germ is roundish in some plants, cordate and angled in others; but its various forms can better be learned by observation than description.

Style. This, like the filament, is sometimes wanting; when present, it proceeds from the germ, and bears the stigma on its summit. It is usually long and slender, of a cylindrical form, consisting of bundles of fibres, which transmit to the germ, from the stigma, the fertilizing pollen.

Stigma. This word signifies perfecting. The stigma is the top of the pistil, and always present; if the style be wanting, it is placed upon the germ, and said to be *sessile,* as in the tulip. The stigma is various in size and form; sometimes it is a round head; sometimes hollow and gaping, more especially when the flower is in its highest perfection; it is generally downy, and always more or less moist, with a peculiar, viscid fluid.

You have, in the following page, a representation of the pistils of several different genera of plants, most of which are magni-

* In strict scientific language, the base of the pistil is the *ovary,* and the germ of the bud is the *gemma.*

Pistil, situation and number—Orders founded upon the pistil—Parts of the pistil—Germ—Style—Stigma.

7*

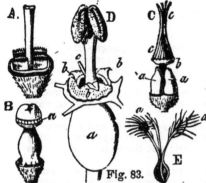

fied. Fig. 83, **A**, shows the pistil of the Cynoglossum. The style is cylindric; stigma depressed or flattened at the top. Four ovaries or rudiments of seeds.

B, shows the pistil of the Tournefortii. The stigma is hemispherical, sub-sessile, surrounded with a glandular hood, *a*. **C**, shows the pistil of the Helitropium: *a*, four ovaries, two of which only are visible in the cut; *b*, a short style; *c*, a conical, four-parted stigma.

Fig. 83.

D, shows a pistil of the genus Cucumis; *a*, is the ovary adhering to the calyx; *b*, three abortive stamens; *c*, cylindric style; *d*, three-lobed stigmas.

E, pistil of the Rumex genus; *a, a*, plumose stigmas.

Use of the Stamens and Pistils.

In a former part of our Lectures, it was observed that the stamens and pistils were necessary to the perfection of the fruit; we will now explain to you the manner in which they conduce to this important object; as you are now acquainted with the different organs and their names, you are prepared to understand the explanation.

The pollen, which, in most flowers, is a kind of farina, or yellow -dust, is thrown out by the bursting of the anther, which takes place in a certain stage of the flower. The pollen is very curiously formed; although appearing like little particles of dust, upon examining it with a microscope it is found to be composed of innumerable organized corpuscles.* These little bodies, though usually yellow, are sometimes white, red, blue, &c. In order to observe them well, it is necessary to put them upon water; the moisture, by swelling them, renders their true form perceptible. They are *oblong* in the Umbelliferous plants, *globular* in the Syngenesious, and *triangular* in some others. In some their surface is smooth, in others armed with little points. They are connected together by minute threads, as in the honeysuckle, &c. These particles of pollen thus placed upon water, swell with the moisture until they burst; a liquid matter is then thrown out, and, expanding upon the surface of the water, appears like a light cloud.

Fig. 84.

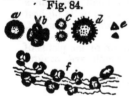

The figure represents the pollen of several different kinds of plants as seen under a magnifier, when placed upon water. At *a*, is a grain of pollen of one of the Mallows-like plants, it is globular, hispid. At *b*, the grain of the pollen is four-lobed. This belongs to the Orchis family. At *c*, is the pollen of the Aster. At *d*, is the pollen of the Hibiscus, globular, muricated. At *e*, is the pollen of the Nasturtium; angular. At *f*, is the pollen of the honeysuckle.

* Little bodies or particles of matter.

If you have paid attention to what has been said respecting the pollen, you perceive that wonders exist in nature, which are entirely unperceived by a careless observer. Who could have imagined that the yellow dust seen upon the lily or tulip, and scarcely visible upon many other flowers, exhibited appearances so interesting? It is in part to show you the almost unlimited extent of the field of observation, in the works of nature, that we have dwelt upon this subject.

Another purpose, and one more connected with our present design, in calling your attention to this subject, is to show the use of the pollen in the vegetable economy. You have seen the effect of moisture upon the pollen; you will recollect that the stigma was said to be imbued with a liquid substance, and that the anther, when ripe, throws out the pollen by the spontaneous opening of its lids or valves; the pollen coming in contact with the moist stigma, each little sack explodes, and the subtle penetrating substance which it contains, being absorbed by the stigma, passes through minute pores into the germ..

In the germ are seeds formed, but these seeds require the agency of the pollen to bring them to the perfection necessary for producing their species. You perceive now why the stamens and pistils are so essential to the perfection of a plant. Nature does not form a beautiful flower, and then leave it to perish without any provision for a future plant; but in every vegetable provides for the renewal of the same.

The real use of stamens and pistils was long a subject of dispute among philosophers, till Linnæus explained it beyond a possibility of doubt; these organs have from the most remote antiquity been considered of great importance in perfecting the fruit. The Date palm, which was cultivated by the ancients, bears stamens and pistils on separate trees; the Greeks discovered, that in order to have good fruit it was necessary to plant the two kinds of trees near together, and that without this assistance, the dates had no kernel, and were not good for food.

In the East, at the present day, those who cultivate palms select trees with pistillate flowers, as these alone bear fruit. When the plant is in blossom, the peasants gather branches of the wild palm-trees, with staminate flowers, and strew the pollen over their cultivated trees.

Pistillate flowers are called *fertile*, staminate, *infertile* flowers.

As moisture causes the pollen to explode, rains and heavy dews are sometimes injurious to plants; the farmer fears wet weather while his corn is in blossom. Nature has kindly ordered that most flowers should either fold their petals together, or hang down their heads when the sun does not shine; thus protecting the pollen from injury.

The fertilization of the fig is said to be accomplished by insects. In this singular plant, the fruit encloses the flower; it is, at first, a hollow receptacle, lined with many flowers, seldom both stamens and pistils in the same fig. This receptacle has a small opening at the summit. The seeds are fertilized by certain little flies, fluttering from one fig to the other, and thus carrying the pollen from the staminate to the pistillate flowers.

Although the fertilization of plants, where the stamens and pistils are on separate flowers, depends a little upon chance, the favourable chances are so numerous, that it is hardly possible, in the order of nature, that a pistillate plant should remain unfertilized. The particles of the pollen are light and abundant, and the butterflies, honey-bees, and other insects, transport them from flower to flower.

The winds also assist in executing the designs of nature.

The pollen of the Pines and Firs, moved by winds, may be seen rising like a cloud above the forests; the particles being disseminated, fall upon the pistillate flowers, and rolling within their scaly envelopes, fertilize the germs.

A curious fact is stated by an Italian writer, viz. that in places about forty miles distant, grew two palm-trees, the one without stamens, the other without pistils; neither of them bore seed for many years; but in process of time, they grew so tall as to tower above all the objects near them. The wind, thus meeting with no obstruction, wafted the pollen to the pistillate flowers, which, to the astonishment of all, began to produce fruit.

The number of plants in which the pistils and stamens are on different flowers, is few, compared to those which have these important organs enclosed within the same corolla; as in our herbaceous plants, and the trees of hot countries, whose leaves being always present, might impede the passage of the pollen from other trees. On the contrary, the trees of cold climates have generally the stamens and pistils on separate flowers, blossoming before the leaves come forth, and in a windy season of the year. Those which blossom later, as the oak, are either peculiarly frequented by insects, or like the numerous kinds of firs, have leaves so little in the way, and pollen so excessively abundant, that it can scarcely fail of gaining access to the pistillate flower.

In all cases the pollen and stigma are in perfection at the same time; in those flowers where the stamens and pistils are together, and of an equal length, some are drooping and some erect, but where the stamens are longer than the pistil, the flower is usually erect; where they are shorter, the flower is pendent; nature thus provides for the fertilization of the germ by the fall of the farina upon the stigma.

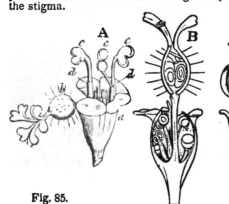

Fig. 85, at *A*, represents a flower of the genus Euphorbia.* It is monœcious; in the centre of the perianth, *a*, is the infertile flower, consisting of several double stamens, *c c*, upon jointed filaments, *dd*. *b*, is the fertile flower, with a petal-like stigma. At *B*, is the same flower before blossoming; it is represented as cut

Fig. 85.

* Euphorbia *illyrica*.—Mirbel.

Facts stated by an Italian writer—Trees of hot countries have mostly stamens and pistils on the same corolla—Trees of cold countries have the stamens and pistils on separate flowers—Methods by which the objects of nature are accomplished—Explain Fig. 85.

vertically, in order to show its internal structure at this period. The Figure at *C*, shows the same flower after its fertilization. Before the maturity of the blossom, the pistil was above the stamens, as seen at *B*. At the expansion of the perianth it was below the stamens, as at *A*, *b*;—resuming its erect position, we see the pistil at *C*, its germ having become a fruit filled with swelling seeds.

Fig. 86.

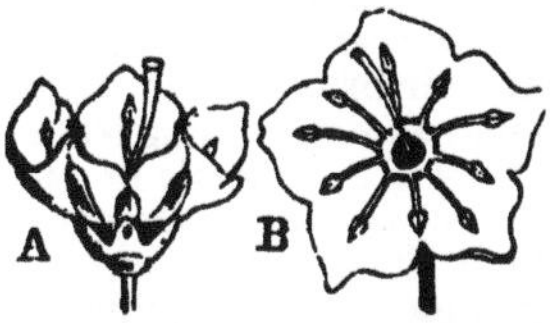

In the Laurel (Kalmia) the ten stamens are confined by their anthers in ten cavities of the five-parted, monopetalous corolla. When the flower is in a state of maturity, the anthers suddenly spring from their confinement, and scatter their pollen upon the stigma. Fig. 86, at A, represents the flower as it appears before its perfect expansion; at B, it is seen as it appears after that period.

Interesting as is the subject of the various means, contrived by Providence, for the continuation of the vegetable tribes, the limits of our work will not permit us to extend our inquiries in this department of our science. But if there are any who hold Botany to be a trifling science, let them examine into the grand principles which it develops, unfolding to the view of man the workings of Creative wisdom in one vast domain of nature. Not that we presume to say this wisdom is yet fully understood; the greatest Botanist, in the midst of his discoveries, must experience a feeling of humiliation at his own ignorance of nature. Facts that when discovered seem so simple, that we wonder a child should not have discovered them, have eluded the research of great men;—and at this moment philosophers are groping for truths, which in due time will be elicited and incorporated into the elements of science to be learned and understood by children.

LECTURE XIV.

INFLORESCENCE—RECEPTACLE.

Having given our particular attention to the important uses of the stamens and pistils, we shall now proceed to consider the various ways in which flowers grow upon their stalks; this is called their *inflorescence*, or mode of flowering.

Inflorescence.

We are now to consider the corolla or flower under three aspects:

With respect to the organs which it contains.

The branches which support it.

The flowers which are near it, or which grow on the same peduncle.

1st. *The corolla with respect to the organs which it contains.*

The corolla, when it is monopetalous, supports the stamens; the number of which in this case always corresponds to the number of divisions of the limb of the corolla. When the corolla is polypetalous, the stamens are inserted upon the calyx or upon the receptacle;

Explain Fig. 86—Inflorescence—Flower considered under three aspects—What is said of the corolla with respect to the organs which it contains?

their number is then usually double the number of petals; as in the pink, which has ten stamens and five petals. When inserted *beneath* the germ or base of the pistil, the corolla is said to be *hypo-gynous,* (underneath the style, or inferior ;) as in the *stramonium.* When it is inserted into the calyx and *surrounds* the germ, as in the currant, it is said to be *peri-gynous,* (around the style, or enveloping it.) When the corolla is inserted *upon* the germ, as in the trumpet-honey-suckle, it is said to be *epi-gynous,* (upon the germ, or superior.)

2d. *The corolla with respect to the branches which support it.*

The disposition of flowers upon their branches is analogous to that of leaves ; thus, flowers are either *radical,* coming from the root, or *cauline,* coming from the stem; they are *peduncle* or *sessile, solitary, scattered,* or *opposite, alternate* or *axillary.* Sometimes they are *unilateral,* growing on one side of the branch ; and sometimes fixed equally upon all parts of the peduncle, and pointing in different directions.

3d. *The corolla with respect to the flowers which surround it, or which grow on the same peduncle.*

The different modes of division of the common peduncle, into lesser peduncles or supports, cause a great difference in the appearance and situation of flowers, and exhibit a variety of forms of inflorescence. The green part which comes from the stem and supports the flower, is called the *peduncle ;* sometimes it is called the foot-stalk of the flower or fruit. The divisions of the peduncle are called *pedicels.*

When the plant is one-flowered, the flower is usually inserted at the end of the stem; the peduncle in that case is scarcely distinct from the stem.

The most common kinds of inflorescence are as follows:

Fig. 87.

1st. *Whorl,* (Fig. 87,) an assemblage of flowers surrounding the stem, or its branches, constitutes a whorl, or ring; this is seen in mint and many of the labiate plants. Flowers which grow in this manner, are said to be *verticillate,* from the Latin *verto,* to turn. Leaves surrounding the stem in a similar manner, are said to be *stellate,* or star-like.

2d. *Raceme*, (Fig. 88, *a*,) consists of numerous flowers, each on its own stalk, and all arranged on one common peduncle, as in the locust and currant.

3d. *Panicle*, (Fig. 88, *b*,) bears the flowers in a kind of loose, subdivided bunch or cluster, without any regular order; as in the oat, and some other grasses. A panicle contracted into a compact, somewhat ovate form, as in the lilac, is called a *thyrse*, as a bunch of grapes.

Fig. 88.

4th. *Spike*, (Fig. 89, *a*,) this is an assemblage of flowers arising from the sides of a common stem; the flowers are sessile or with very short peduncles; as the grasses and mullein. A spike is generally erect. The lowest flowers usually blossom and fade before the upper ones expand. When the flowers in a spike are crowded very close, an *ear* is formed, as in *Indian corn*.

5th. *Umbel*, (Fig. 89, *b*,) consists of several flower-stalks, of nearly equal length, spreading out from a common centre, like the rays of an umbrella, bearing flowers on their summits; as fennel and carrot.

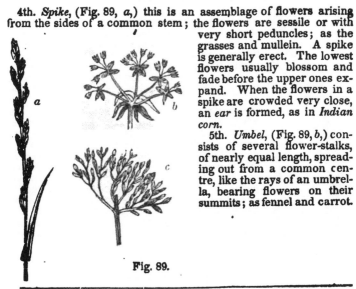

Fig. 89.

6th. *Cyme,* (Fig. 90, *c,*) resembles an umbel in having its common stalks all spring from one centre, but differs in having those stalks irregularly subdivided; as the snowball and elder.

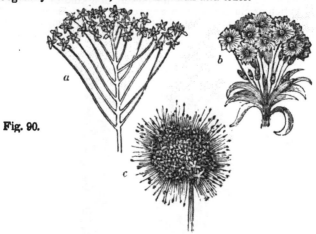

Fig. 90.

7th. *Corymb,* (Fig. 90, *a,*) or false umbel; when the peduncles rise from different heights above the main stem, but the lower ones being longer, they form nearly a level or convex top; as the yarrow.

8th. *Fascicle,* (Fig. 90, *b,*) flowers on little stalks variously inserted and subdivided, collected into a close bundle, nearly level at the top; as the sweet-william; it resembles a corymb, but the flowers are more densely clustered.

9th. *Head,* (Fig. 90, *c,*) or tuft, has sessile flowers heaped together in a globular form; as in the clover, and button bush, (*cephalanthus.*)

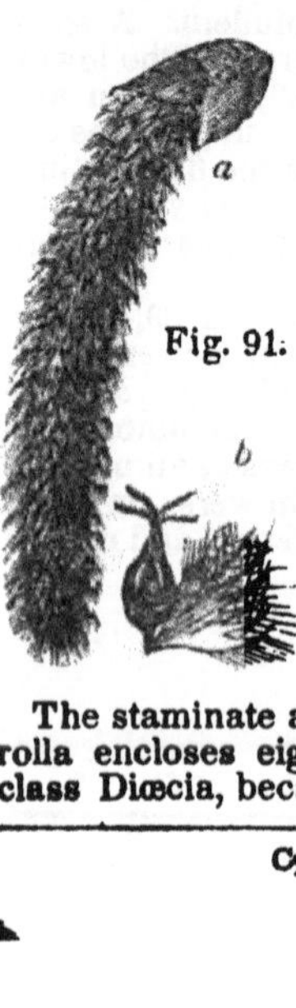

10th. *Ament* or catkin, is an assemblage of flowers, composed of scales and stamens, or pistils arranged along a common thread-like receptacle, as in the chestnut and willow; this, though described under the divisions of the calyx, is only a mode of inflorescence. The scales of the ament are properly the calyxes; the whole aggregate, including scales, stamens or pistils, and filiform *receptacle,* constitutes the ament. At Fig. 91, is the representation of the ament of the poplar, containing pistillate flowers; this is oblong, loosely imbricated, and cylindrical; the calyx is a flat scale, with deep-fringed partings. At *b,* is a representation of the fertile or pistillate flower; the calyx or bract is a little below the corolla, which is cup-shaped, of one petal, and crowned with an egg-shaped, pointed germ; the germ is superior, and bears four, (sometimes eight) stigmas.

Fig. 91.

The staminate ament resembles the pistillate, except that its corolla encloses eight stamens, but no pistil. The poplar is in the class Diœcia, because the pistillate and staminate flowers are on different trees; and of the order Octandria, because its barren flowers have eight stamens.

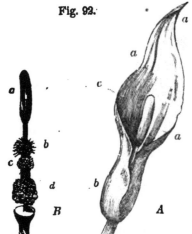

Fig. 92.

11th; *Spadix*, is an assemblage of flowers growing upon a common receptacle, and surrounded by a spatha or sheath. At Fig. 92, A, *a*, is a representation of the blossom of the wild turnip, (*arum ;*) *a* represents the spatha, which is erect, sheathing, oblong, convolute at the base, *b;* and it is compressed above and below the middle; *c*, represents the spadix, which, from its club-shaped appearance, is called *clavi-form*, (from *clava*, a club.)

At *B*, is the spadix divested of the spatha; *a*, is the claviform summit; *b*, a ring of filaments without anthers; *c*, a ring of sessile anthers; *d*, a dense ring of pistillate flowers with sessile stigmas; each germ produces a one-celled, globular berry. This plant is of the class Monœcia, because its staminate and pistillate flowers are separate, but yet grow on the same plant; it is in the order Polyandria, because its stamens are numerous.

Receptacle.

The receptacle is the extremity of the peduncle, it is also called the *clinanthe*,* from *kline*, bed, and *anthos*, flower; at first it supports the flower, and afterward the fruit. As this is its only use, it may properly be considered in connexion with the organs of fructification. In simple flowers, as the tulip, the receptacle is scarcely to be distinguished from the peduncle, but in compound flowers it is expanded, and furnishes a support for the flowers and fruit. Receptacles are of various kinds; as,

1st. *Proper*, which supports but one flower, as in the violet and lily.

2d. *Common*, which supports many florets, the assemblage of which forms an aggregate or compound flower, as in the sunflower and dandelion. The common receptacle presents a great variety of forms; as *concave, convex, flat, conical,* or *spherical.* In the fig it is concave, and constitutes the fruit. As to its surface, the receptacle is *punctate*, as in the daisy; *hairy*, as in the thistle; *naked*, as in the dandelion; *chaffy*, as in the chamomile; it is pulpy in the strawberry, and dry in most plants.

3d. *Rachis*, is the filiform receptacle which connects the florets in a spike, as in a head of wheat.

Our examination of the flower is now completed. We shall, in our next lecture, proceed to consider the change which takes place, after the bloom and beauty of the plant have faded. We shall find that organs, at first scarcely perceptible, begin to develop

* Sometimes *torus*, from the Latin, signifying bed.

themselves, until the character of the fruit is fully exhibited. So in the heart of youth, the germs of virtue or vice may, for a while, be apparently dormant and inactive, but growing more vigorous and powerful, they at length unfold themselves, and reveal either a character matured into what is lovely and desirable, or marked with qualities of a disagreeable and deleterious nature.

LECTURE XV.

THE FRUIT—PERICARP—PARTS OF THE PERICARP—LINNÆUS'S CLASSIFICATION OF FRUITS—MIRBEL'S CLASSIFICATION OF FRUITS.

The Fruit.

THE fruit is composed of two principal parts, the *pericarp* and *seed.* The term pericarp is derived from *peri* around, and *karpos* seed or fruit; it signifies surrounding the seed. All that in any fruit which is not the seed belongs to the pericarp.

Let us now inquire into the progress of the fruit from its first appearance in the germ to its mature state. When you analyze a flower, you often find it necessary to ascertain the number of cells contained in the germ. In making this examination, what appearance does the interior of the germ present, when exposed by cutting it horizontally? You see there minute bodies of a pale green colour, and an apparently homogeneous nature: each of these is called an *ovule,*[*] and their outer covering, an *ovary.* These ovules, before the fertilization of the germ by the pollen, are scarcely perceptible; after this period, and the fading of the corolla, the ovules increase in size, and the embryo and other parts which constitute the seed become manifest. The ovary enlarges with the growth of the ovules; the use of this covering is not confined to the mere protection of the seeds from injury, but it is furnished with glands, which secrete such juices as are necessary for the growth and development of the ovules. As the ovary becomes more mature, it takes the name of *pericarp.* Pericarps in their growth become either woody or pulpy; the latter absorb oxygen gas and throw off carbonic acid; saccharine juices are elaborated in their cellular integument. In another stage, the pulpy substance passes through a slight fermentation, the organization is disturbed, the juices sour, the pulp decomposes, and putrefaction ensues. Such is the change which you may see in pulpy fruits during their progress towards maturity and subsequent decay.

Parts of the Pericarp.

The germ being fertilized, the parts of the flower which are not necessary for the growth of the fruit, usually fade, and either fall off or wither away. The pericarp and seed continue to enlarge until they arrive at perfection. Every kind of fruit[†] you can behold has been once but the germ of a flower. The size of fruit is not usually proportioned to that of the vegetable which produced it. The pumpkin and gourd grow upon slender herbaceous plants, while the large oak produces but an acorn.

---

[*] From *ovum,* an egg.
[†] The term fruit, in common language, is limited to pulpy fruits which are proper fo. food; but in a botanical sense, the fruit includes the seeds and pericarps of all vegetables.

Fruit, the two principal parts—Derivation and signification of the word pericarp—Ovules—Ovary—Use of the ovary—Its name in a mature state—Pulpy pericarps—Germ—Size of the fruit not in proportion to the plant that produces it.

In some fruits the pericarp seems to consist of three parts—

1st. The *epicarp*,[*] the skin of the fruit, or membranous part which surrounds it, and which is a kind of epidermis;

2d. The *sarcocarp*,[†] a part more or less fleshy, corky or coriaceous, often scarcely perceptible, and covered by the epicarp;

3d. The *endocarp*,[‡] an internal membrane of the fruit, which lines the cavity, and by its folds forms the partitions and cells.

In the peach, for example, the skin is the epicarp; the pulpy, cellular substance which absorbs the juices of the fruit is the sarcocarp: the shell which encloses the kernel, deprived of moisture, and rendered dry and tough, is the endocarp. The endocarp is also called the *putamen*.

In most fruits the pericarp consists of the following parts:

1st. *Valves* or *external pieces,* which form the sides of the seed vessels. If a pericarp is formed of but one, it is *univalved;* the chestnut is of this kind. A pericarp with two valves is said to be *bivalved*, as a pea-pod. The pericarp of the violet is *trivalved;* that of the stramonium *quadrivalved*. Most valves separate easily when the fruit is ripe; this separation is known by the term *dehiscence.*

2d. *Sutures* or *seams*, are lines which show the union of valves; at these seams the valves separate in the mature stage of the plant; they are very distinct in the pea-pod, which has two sutures.

3d. *Partitions* or *dissepiments*, are internal membranes which divide the pericarp into different cells; these are *longitudinal* when they extend from the base to the summit of the pericarp; they are *transverse* when they extend from one side to the other.

4th. *Column* or *Columella*, the axis of the fruit; this is the central point of union of the partitions of the seed vessels; it may be seen distinctly in the core of an apple.

5th. *Cells*, are divisions made by the dissepiments, and contain the seeds; their number is seldom variable in the same genus of plants, and therefore serves as an important generic distinction.

6th. *Receptacle* of the fruit, is that part of the pericap to which the seed remains attached until its perfect maturity; this organ, by means of connecting fibres, conveys to the seed, for its nourishment, juices elaborated by the pericarp.

Some plants are destitute of a pericarp, as in the labiate flowers, compound flowers, and grasses; in these cases the seeds lie in the bottom of the calyx, which performs the office of a pericarp.

Linnæus's Division of Pericarps.

Linnæus made a division of fruits into nine classes, viz.: *Capsule, Silique, Legume, Follicle, Drupe, Nut, Pome, Berry, and Strobilum.*

1st. CAPSULE, a little chest or casket; this is a hollow pericarp which opens spontaneously by pores, as the poppy, or by valves, as the pink. The internal divisions of the capsule are called *cells;* these are the chambers appropriated for the reception of the seeds; according to the number of these cells, the capsule is *one-celled, two-celled,* &c. The membranes by which the capsule is divided into cells are called *dissepiments*, or partitions; these partitions are either parallel to the valves or contrary. The columella is the central pillar in a capsule; and is the part which connects the several internal partitions with the seed. It takes its rise from the recep-

---

* From *epi*, upon, and *karpos*, fruit. † From *sarx*, flesh, and *kàrpos*, fruit. ‡ From *endo*, within, and *karpos*, fruit.

---

tacle, and has the seed fixed to it, all around. In one-celled capsules this is wanting. (For the capsule, see Fig. 94.)

2d. SILIQUE or *Siliqua*, is a two-valved pericarp or pod, with the seeds attached alternately to its opposite edge, as mustard and radish. The proper silique is two-celled, being furnished with a partition which runs the whole length of this kind of pericarp; there are some exceptions to this, as in the celandine. *Silicle*, (*silicula*, a little pod,) is distinguished by being shorter than the silique, as in the pepper-grass. This difference in the form of the silique and silicle, is the foundation of the distinction in the orders of the class *Tetradynamia*.

3d. LEGUME is a pericarp of two valves, with the seeds attached only to one *suture*, or seam, as the pea. In this circumstance it differs from the silique, which has its seeds affixed to both sutures. The word *pod* is used in common language for both these species of pericarp. Plants which produce the legume, are called *leguminous*. The greater number of these plants are in the class *Diadelphia*. The tamarind is a legume filled with pulp, in which the seeds are lodged.

4th. FOLLICLE is a one-valved pericarp, which opens longitudinally on one side, having its seed loose within it; that is, not bound to the suture. We have examples of this in the dog's-bane, (*Apocynum*,) which has a double follicle, and in the milk-weed, (*Asclepias*.)

5th. DRUPE, (Fig. 101,) a stone fruit, is a kind of pericarp which has no valve, and contains a nut or stone, within which there is a kernel. The drupe is mostly a moist, succulent fruit, as in the plum, cherry, and peach. The nut or stone within the drupe, is a kind of woody cup, commonly containing a single kernel, called the *nucleus;* the hard shell, thus enveloping the kernel, is called the *putamen;* the stone of a cherry or peach, may furnish an example.

6th. NUT, is a seed covered with a shell resembling the capsule in some respects, and the drupe in others; as the walnut, chestnut, &c.

7th. POME, (Fig. 102,) is a pulpy pericarp without valves, but containing a membranous capsule, with a number of cells which contain the seeds. This species of pericarp has no external opening or valve. The apple, pear, quince, gourd, the cucumber and melon, furnish us with examples of this kind of pericarp. With respect to form, the pome is *oblong, ovate, globular*, &c., the form of fruits being much varied by climate and soil. Every child knows that apples are not uniform in their size or figure; with respect to the number of cells also, the apple is variable.

8th. BERRY, (Fig. 104,) is a succulent, pulpy pericarp, without valves, and containing naked seeds, or seeds with no other covering than the pulp which surrounds them; the seeds in the berry are sometimes dispersed promiscuously through the pulpy substance, but are more generally placed upon receptacles within the pulp. A compound berry consists of several single berries, each containing a seed united together; as in the mulberry, (Fig. 108.) Each of the separate parts is called an *acinus*, or grain. The orange and lemon are berries with a thick coat. There are some kinds of berries, usually so called, that, according to the botanical definition of a berry, seem scarce entitled to the name; for the pulp is not properly a part of the fruit, but originates from some other organ. In the mulberry, the calyx becomes coloured and very juicy, surrounded by seeds like a real berry. What is commonly called the berry in the strawberry, is but a pulpy receptacle studded with naked seeds. In the fig, the whole fruit is a juicy calyx, or common receptacle, containing in its cavity innumerable florets, each of which has a proper calyx of its own, which becoming pulpy invests the seed, (Fig. 107.) The paper mulberry of China is an intermediate genus between the mulberry and fig, resembling a fig laid open, but without any pulp in the common receptacle.

9th. STROBILUM, (Fig. 105,) is a catkin or ament hardened and enlarged into a seed vessel, as in the pine ; this is called an aggregate, or compound pericarp. In the most perfect examples of this kind of fruit, the seeds are closely enveloped by the scales, as by a capsule. The Strobilum is of various forms, as *conical, oblong, round,* or *ovate.* The intelligent student will now perceive how much instruction be derived from the study of the various kinds of fruits. And, though the rich gifts of God in this department of nature may be taken of by the creatures of his bounty, with the relish which he y enables us to enjoy, still we cannot but feel, that in the enjoy arising from the philosophical contemplation of these His works, there is an exercise of higher and nobler faculties. The external sense is "of the earth, earthy," the mental enjoyment may be shared with us by angels. The blessedness of heaven, we have reason to believe, will in part consist in studying and admiring the wisdom of God, as displayed in the works of his hand.

MIRBEL'S CLASSIFICATION OF FRUITS, OR PERICARPS.

The following classification of fruits, by one of the most eminent botanists of the age, is given for the more advanced pupil. It is not introduced as being a part of the elements of Botany. The teacher will do well, therefore, to pass over the remainder of this lecture, leaving the pupil to read it at leisure, or to study it in the course of a reviewing lesson.

Mirbel has divided the fruits of all phenogamous plants into two classes ; 1st, *Gymnocarpes,* which include all such as are not masked or covered by any organ, which conceals their true character. 2d, *Angiocarpes,* which include all fruits covered by any organ, which disguises them from observation.

CLASS I.   GYMNOCARPES.
Fruits not covered.

ORDER 1st. CARCERULARES, (from *carcer,* a prison,) simple fruits, without valves, and which never open spontaneously. This order includes the fruits of *syngenesious plants,* of the *grasses,* &c.

Fig. 93.

*Cypsela,** (from *Kupselion,* a coffer.) The pericarp is one-celled, one-seeded, adhering; the seed is erect, with the radicle pointing to the hilum ; it is *monocephalous,* and crowned by the border of the calyx, prolonged in scales, in ridges, or an egret. Figure 93 represents a pericarp of this genus ; it is of the *syngenesious* family ; the pericarp (*a*) is turbinate, (shaped like a top;) its surface is pubescent and furrowed ;

* This is the *achenium,* or *acine,* of some writers.

---

it is *indehiscent*, (not opening when ripe;) *monospermous*, (having one seed;) the egret (*c*) is *sessile* and *plumose*, and the embryo is *dicotyledonous* and fleshy. At *b*, is the same pericarp, cut longitudinally, and exposing an inner half of one of the cotyledons. In this genus are the pericarps of the Dandelion, the Oyster-plant, Lettuce, &c.

*Cerion ;** in this genus the embryo is situated upon the side of the perisperm; cotyledon one, large and fleshy. The germ is clothed with a *pileole ;†* the radicles are contained in *coleorhizes.* The fruit of Indian-corn wheat, of the grasses and rice, are found here.

Carcerula ;‡ the characters of this genus are variable; it includes all fruits of the order *Carcerulares,* which do not come within the two preceding genera; the buckwheat, elm, and rhubarb, are examples.

Fig. 94.

ORDER 2d. CAPSULARES, simple fruits, having capsules which open when in a mature state; they have their origin from a single ovary, free, or adhering to the calyx; they have valves, and consequently sutures, and open by the separation of the valves.

Capsule. You see here, (Fig. 94,) a capsular fruit; it is the seed of the martagon-lily, (*Lilium martagon ;*) *a*, represents the capsule open, as it appears in a mature state; *b*, the same cut transversely, showing the seeds. All capsular fruits which do not belong to the other genera in this order, are here included. They are monocephalous, as in the lily, or polycephalous, as in Nigella; they do not adhere to the calyx, and have one or many cells.

Legume, is an irregular, bivalve, elongated pericarp; it is monocephalous, free, the two valves joined by two sutures, an upper and lower; it contains seeds in one cell, a placenta along the lower suture. The embryo has two cotyledons, and a radicle bordering on the hilum. The legume is sabre-form in the bean; cylindric in the Cassia, compound in the pea, and articulated in Hedysarum, where it is called a loment.

Fig. 95.

Fig. 95, *a*, represents the fruit of the *Astragalus ;* it is swollen; the cell is longitudinal; *b* is the same legume cut transversely in order to show the two cells.

Silique, a bivalved pericarp, peculiar to the *Cruciferæ,* having its seeds attached to both the upper and lower valves. The silique is divided by a longitudinal partition, formed by the dilated placenta, and bearing the seeds.

* The same as *caryopsis.*
† For an illustration of these terms, see plate 115, with its explanation, or the vocabulary.
‡ This includes what some call the *utricle,* others the *scleranthus,* or *samara.*

Fig. 96.

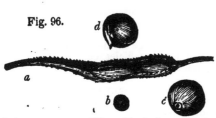

Fig. 96, *a*, represents a *silique*, the fruit of the SINAPIS *alba*, (white mustard;) this is said to be *rostrate*, terminating like a bird's beak. *b*, represents a globular seed; *c*, the same magnified; *d*, shows the seed dividing, and the embryo making its appearance. The *silicula* is a variety of the same genus.

Fig. 97.

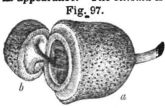

Pyxides, (from *puxis*, a box;) it has two valves, an upper and lower, the latter is attached to the receptacle, while the former opens like the lid of a box. This genus may be illustrated by the fruit of the genus *Lecythis*, (Fig. 97;) *a*, represents the lower valve, *b*, the upper valve or lid of the pericarp. To this genus belong the fruit of the Anagalis, Hyosciamus, and Gomphrena globosa, or bachelor's button.

ORDER 3d. DIERESILLA, (from *diæresis*, division,) contains simple fruits, which divide into many carpels ranged symmetrically round a central axis. These carpels are formed by the adhering valves of the pericarp, which in the maturity of the fruit separates, and the carpels appear like so many little nuts; as in the seed of the nasturtion, which easily falls into parts.

Cremocarp, (from *kremao*, to suspend, and *karpos*, fruit;) this kind of fruit derives its origin from an ovary surmounted with two styles, and often crowned by the limb of the calyx. It has two cells, and two seeds. It divides itself into two seeds, suspended by their summit to a slender central axis, usually two-forked. Each seed contains a depending embryo, clothed with a membranous and adhering tegmen, and having a horny perisperm. The embryo is very small, and has two cotyledons. The coriander is a spherical cremocarp; the caraway is ellipsoid. The seeds of the carrot and parsley and other umbelliferous plants belong to this genus.

Regmate, (from *regma*, opening with noise,) containing many seeds which are enclosed by two valves opening by an elastic movement, as Euphorbia.

Fig. 98.

The cut represents a pericarp of the Euphorbia; it consists of four carpels;—in the ripe fruit, the panextern or outer covering is thrown off by an elastic movement of the valves; *a*, represents the entire fruit, and *b*, the same cut transversely, showing four seeds.

Dieresil,* a variable genus, containing such fruits in the order as do not properly come under the two other divisions, as the nasturtion, geranium, hollyhock, &c.

* The *samara* of Gærtner.

Pyxides—Order Dieresilia—Genus Cremocarp—Regmate—Dieresil.

ORDER 4th. ETAIRIONNAIR, (from *etairoi*, associates,) contains compound fruits, proceeding from ovaries, bearing the styles; this order contains two genera.

Double Follicle, as in the milk-weed, (*ascle pias*,) having two follicles, each formed of one valve, folded lengthwise, and adhering at its edges.

Etairon*, having many seeds ranged round the imaginary axis of the flower, as the ranunculus and anemone.

Here is the fruit, (Fig. 99,) of the *Aconitum*, (monk's-hood,) which belongs to this order; it is composed of three pods united in one compound fruit; *a*, shows one of the valves in a dehiscent state; *b*, represents a seed cut longitudinally.

Fig. 99.

The *Clematis* is a caudate etairon, the *Pæonia* is divergent and dehiscent.

ORDER 5th. CENOBIONNAIR, (from *koinobion*, a community,) compound fruits without valves or sutures, proceeding from ovaries without any adhering styles; this order contains but one genus.

Cenobion,† includes fruit of the labiate plants and some others. Figure 100, represents the pericarp of the genus *Gomphia*; it is composed of five *companions*, *a*, as Mirbel calls each of the one-celled divisions which stand around an ovoid germ, destitute of any style; *b*, represents one of these divisions cut vertically; it contains one seed.

Fig. 100.

ORDER 6th. DRUPACES, simple, succulent fruits, containing a nut. This order has but one genus.

Drupe, this pericarp is composed of a woody or bony panintern,‡ called the nut, and of a panextern,‡ sometimes dry and membranous, at others fleshy or pulpy; this character is peculiar to this fruit. It may be regular or irregular, monocephalous or polycephalous, adhering to the calyx or free. The cherry has a pulpy panextern, the peach fleshy, the walnut woody. The AMYGDALIS *persica*, Fig. 101, *a*, is a succulent drupe, of a roundish form, and furrowed on the side; the nut of this drupe is an ellipsoid, one-celled and one-seeded,

Fig. 101.

* The *syncarp* of Richard.
† Called by De Candolle, *Sarcobase* and *Microbase*.
‡ The panextern includes what is sometimes called epicarp and sarcocarp, the panintern is the same as the endocarp.

Order Etairionnair—Double Follicle—Etairon—Describe the fruit of the Aconitum—Order Cenobionnair—Cenobion—Order Drupaces—Drupe.

b, represents the peach deprived of one half of its pulpy exterior, or panextern, and exposing the nut or panintern; *c*, represents the nut divested of one of its valves, and showing the seed *d*.

ORDER 7th. BACCATI, (from *bacca*, a berry,) simple, succulent fruits, containing many separate seeds. The genera in this order are the following:

Pyridion,* (from *perideo*, to lie around;) this is a regular fruit, crowned with the adhering calyx. The pericarp is fleshy, and has several cells, each of which contains one or more seeds; the embryo has two cotyledons, which are large and fleshy. This genus contains the apple and pear. The apple, (*Malus communis*,) Fig. 102, has a round, fleshy pericarp, crowned with the calyx; the seeds are enclosed in five carpels, or cells, ranged around in the axis of the fruit; the cells are composed of membranaceous valves. The seeds are tunicated, or coated; *a*, represents an entire pyridion; *b*, the same cut vertically; and *c*, the same transversely.†

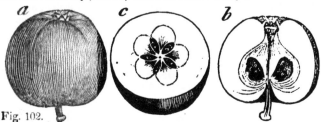

Fig. 102.

Pepo, (from the Latin *pepo*, a melon;) this is a regular monoce phalous fruit with a radiating placenta, containing many seeds; the panextern is solid and dry; the panintern is pulpy. The watermelon is globular, and the cucumber oblong. Fig. 103, represents the cUCUMIS *anguria*, sometimes called prickly cucumber; *a*, is the entire *pepo*, which is *spinous*, three-celled, and many-seeded. The cells and seeds are shown by the same fruit cut transversely, as at *b*; *c*, represents a seed, this is tunicated and dicotyledonous; *d*, the same cut vertically.

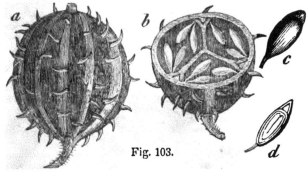

Fig. 103.

* Called *Pome*, by Linnæus.
† A singular fact is observable in the fruit of the apple: when cut in slices transversely, it exhibits in its substance an exact representation of the five petals which existed in the flower; I have never, in any botanical work. met with a notice of this phenomenon, and know not on what physiological principles it can be explained.

Bacca, contains all the fruits of this order not found in the other genera. The pericarp of the currant, whortleberry, orange, bar-

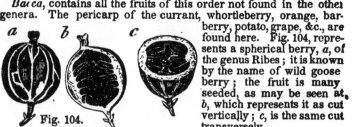

berry, potato, grape, &c., are found here. Fig. 104, represents a spherical berry, *a*, of the genus Ribes; it is known by the name of wild goose berry; the fruit is many seeded, as may be seen at, *b*, which represents it as cut vertically; *c*, is the same cut transversely.

Fig. 104.

CLASS II. ANGIOCARPES.

Fruits which are covered by a bract or foliaceous envelope.

This class is divided into five genera, as follows:

1st. *Strobilum* or *cone*, a collection of carcerular fruits concealed by scales, formed of bracts or peduncles, whose union produces a globular or conical body, as the juniper, pine, &c. Fig. 105, represents the fruit of the pine, which is composed of woody, close, and indehiscent cupules. The glands are membranous, one-celled, and one-seeded; *a*, is an entire *strobilum*; *b*, is the same, cut vertically the placenta, extending lengthwise through the fruit, is large. The pine-apple, *Bromelia*, is of this genus of fruits.

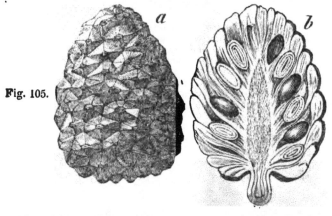

Fig. 105.

2d. *Calybion*,* (from *kalubion*, a little cabin;) fruits of this genus are composed of a cupule or cup of variable forms, and of carcercu-lars enveloped entirely, or in part, by the cupule. The carcerculars of calybions are called glands. The gland of the oak is partly con-cealed in its cupule, that of the beech entirely concealed, and also of the yew, (*Taxus*;) in the latter are two cupules, one enclosing the other; the exterior one is succulent, and of an orange red; the interior, which is hard and woody, encloses the fruit.

* This includes what some writers call the *gland* and the *nut*.

Bacca—Enumerate the orders in the class Gymnocarpes, with the genera of each-Describe the class Angiocarpes—Strobilum—Calybion.

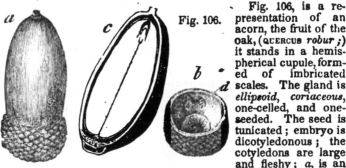

Fig. 106.

Fig. 106, is a representation of an acorn, the fruit of the oak, (QUERCUS *robur*;) it stands in a hemispherical cupule, formed of imbricated scales. The gland is *ellipsoid, coriaceous,* one-celled, and one-seeded. The seed is tunicated; embryo is dicotyledonous; the cotyledons are large and fleshy; *a*, is an entire *calybion*; *b*, the cupule, *d*, two abortive glands; *c*, the gland cut vertically, showing the embryo near its apex.

Fig. 107.

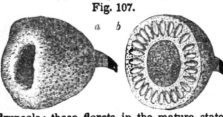

Sycone, (from *sucon,* a fig.) This is a genus of fruits formed by the enlargement of the clinanthe or receptacle, into a hollow fleshy substance, covered within by numerous florets, each of which contains a drupeole; these florets in the mature state of the fruit disappear, leaving only seeds imbedded in the cellular substance of the pericarp. The cavity within becomes gradually filled by the increase of cellular tissue, until, as in the fig, it entirely disappears. Fig. 107, *a*, represents a sycone, the fruit of the *Ambora*, which belongs to the fig tribe of plants; this remains open at its summit, and is more woody in its texture than the common fig, (*Ficus carica.*) *b*, represents the fruit, cut transversely, with the seeds circularly arranged within the sarcocarp.

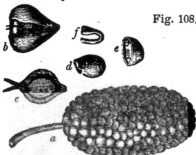

Fig. 108.

4th. *Sorose,* (from *soros,* a collection;) this genus contains many fruits united in a spike, or catkin, and covered with succulent floral envelopes, as the mulberry. Fig. 108, *a*, represents the fruit of the MORUS *rubra,* (red mulberry,) which is an example of the genus *sorose;* it is of an oblong form; each little drupe is surrounded by a succulent pericarp; the nut is one-seeded; *b*, represents a detached perianth, containing a *drupeole;* *c*, drupeole; *d*, a nut; *e*, the same cut transversely; *f*, the embryo.

Describe an acorn—What is a Sycone?—Describe the Sorose—What does Fig. 108 present?

Synopsis of Mirbel's Orders and Genera of Pericarps.

CLASS I.

Fruit naked, GYMNOCARPES.

ORDER 1. CARCERULARES, simple fruits, remaining closed.

Genera, {
1. Cypsela,
2. Cerion,
3. Carcerula.
}

ORDER 2. CAPSULARES, simple fruits, which open at maturity.

Genera, {
1. Capsule,
2. Legume,
3. Silique and Sillicle,
4. Pyxides.
}

ORDER 3. DIERESILIA, simple fruits, which divide into many parts when ripe.

Genera, {
1. Cremocarp,
2. Regmate,
3. Dieresil.
}

ORDER 4. ETAIRONNAIR, compound fruits, proceeding from a germ to which the style adheres.

Genera, {
1. Double Follicle,
2. Etairon.
}

ORDER 5. CENOBIONNAIR, compound fruits, proceeding from a germ not bearing the style.

Genera, { 1. Cenobium.

ORDER 6. DRUPACES, simple and succulent fruits, contained in a nut.

Genera, { 1. Drupe.

ORDER 7. BACCATI, simple, succulent fruits, containing many separate seeds.

Genera, {
1. Pyridion,
2. Pepo,
3. Bacca.
}

CLASS II.

Covered fruits, ANGIOCARPES.

Genera, {
1. Calybion,
2. Strobilum,
3. Sycone,
4. Sorose.
}

LECTURE XV.

THE SEED—SYNOPSIS OF THE EXTERNAL ORGANS OF PLANTS.

THE seed may be considered as that link in the chain of vegetable existence which connects the old and new plant; were this destroyed, were nature to fail in her operation of perfecting the seed, what a change would the earth soon exhibit! One year would sweep away the whole tribe of annual plants; beautiful flowers, medicinal herbs, and our most important grains for the sustenance of man and beast, would vanish for ever. Another year would take from us many of our most useful garden vegetables, and greatly reduce the number of our ornamental plants. Year after year the perennials would vanish, until the earth would present but one vast scene of vegetable ruin. The ancient pines and venerable oaks, instead of the smiling aspect of ever-renovating nature which they now witness, would stand alone in solitary grandeur, the mournful remains of a once

---

Repeat the Synopsis of Mirbel's classification—What is the seed? its form—What would be the appearance of the earth, if plants should cease to produce perfect seed?

beautiful and fertile world! And why, my young friends, are we never filled with alarm, lest the provisions of nature should fail? It is because we know that a Being, unchangeable in purpose, and omnipotent in means, directs the course of physical events, and He has promised that while the earth remaineth, "seed-time and harvest shall not cease."

We have seen, in the progress of our inquiries, that while the present plan is diffusing around it beauty and fragrance, administering to the necessities and luxuries of man, the watchful care of that Being who never slumbers nor sleeps, is, by a slow but certain progress, perfecting that part which is destined to continue the species, and which. "is the sole end and aim of all the organs of fructification."*

The seed is the ovule in a mature state; it is that internal part of the fruit which envelops the complete rudiment of a new plant, similar to that from which it received its existence. Seeds are various in their form; the mustard is globular; some species of beans are oblong; the cocoa-nut is ovoid; the buckwheat is angular, &c.

The seed consists of three principal parts, viz.: the *eye*, *husk*, and *kernel.*

1st. The *Eye*, or *hilum*, is the scar formed by the separation of the funicle, a membrane or thread, which connected the seed with the pericarp, and conveyed to the former the necessary nourishment. This connecting membrane is usually very short; but in the *magnolia* and some other plants it is several inches in length. When the seed is fully ripe, the connexion between it and the pericarp

Fig. 109.

ceases by the withering and separation of the funicle, leaving upon the outer surface of the seed the mark of its insertion. This scar, called the eye, is very conspicuous in the bean, which also exhibits the pore through which the nourishment was conveyed to the internal parts of the seed. That part of the seed which contains the eye is called the *base;* the part opposite is called the *apex.*

Fig. 109 represents the garden bean; it is an oblong, tunicated seed; between its two thick cotyledons, at *a*, may be seen the *hilum* or eye.

2d. The *Husk* is the outer coat of the seed, which, on boiling, becomes separate; as in peas, beans, Indian corn, &c.; this skin is also called the *spermoderm*, from the Greek *sperma*, signifying seed, and *derma*, skin. The *spermoderm* or skin of the seed, consists of three coats, analogous to the three divisions of the pericarp; the externa. *skin*, called the *testa* or *cuticle*, corresponds to the *epicarp;* the cellular tissue, called *mesosperm*, corresponds to the *sarcocarp;* and the internal skin, or *endosperm*, corresponds to the *endocarp*, or inside skin of the pericarp.† The husk surrounds the kernel, and is essential, as the kernel, which was originally a fluid, could not have been formed without its presence.

3d. The *Kernel* includes all that is contained within the husk or *spermoderm;* it is also called the *nucleus* or almond of the seed.

* Linnæus.
† These three divisions may not always seem distinct, as in some cases, the *mesosperm* is scarcely to be separated from the cuticle.

The kernel is usually composed of the *albumen, cotyledon,* and *embryo.*

The *A.oumen* is that part of the kernel which invests the cotyledons or lobes, and is thought to afford the same support to the germinating emoryo, that the white of an egg does to a chicken. Both in respect to hardness and colour, the albumen, in many seeds, greatly resembles the white of a boiled egg. It is not considered an essential part of the seed, because it is sometimes wanting ; but when present, it supports and defends the embryo while imprisoned in the seed, and serves for nutriment when it begins to germinate. It has no connexion with the embryo, and is always so distinct as to be easily detached from it. Albumen makes up the chief part of some seeds, as the grasses, corn, &c.; in the nutmeg, which has very small cotyledons, it is remarkable for its variegated appearance and aromatic quality. It chiefly abounds in plants which are furnished with but one cotyledon.

Fig. 110.

Fig. 110 represents the cotyledons of the bean, as divested of the husk ; *a*, represents the cotyledons ; *b* and *c*, the embryo ; *d*, shows the petioles or stems of the cotyledons.

Cotyledons, (from a Greek word, *kotule,* a cavity,) are the thick, fleshy lobes of seeds, which contain the embryo. In beans they grow out of the ground in the form of two large leaves. They are the first visible leaves in all seeds, often fleshy and spongy, of a succulent and nourishing substance, which serves for the food of the embryo at the moment of its germinating. Nature seems to have provided the cotyledons to nourish the plant in its tender infancy. After seeing their young charge sufficiently vigorous to sustain life without their assistance, the cotyledons in most plants wither and die. Their number varies in different plants, and there are some plants which have none.

Acotyledons, are those plants which have no cotyledons in their seeds; such as the *cryptogamous* plants, *mosses,* &c.

Mono-cotyledons, are such as have but one cotyledon or lobe in the seed ; as the *grasses,* the *liliaceous* plants, &c.

Di-cotyledons, are such plants as have two cotyledons; they include the greatest proportion of vegetables ; as the *leguminous,* the *syngenesious,* &c.

Poly-cotyledons, are those plants the seeds of which have more than two lobes; the number of these is small; the *hemlock* and the *pine* are examples.

The number of cotyledons seldom varies in the same family of plants ; it nas therefore been assumed by some botanists as the basis of classification; but there are difficulties attending a method wholly dependant on these organs. In order to be certain as to their number, it is necessary to examine the seed in a germinating state ; this is often difficult. The natural method of Jussieu is in part founded upon the number of cotyledons.

The *Embryo* is the most important part of the seed ; all other parts seem but subservient to this, which is the point from whence the life and organization of the future plant originate. In most dicotyledonous seeds, as the bean, orange, and apple, the embryo may be plainly discovered. Its internal structure, before it begins to vegetate, is very simple, consisting of a uniform substance, enclosed in its appropriate bark or skin. When the vital principle is excited to action, vessels are formed and parts developed which were before invisible. The embryo is usually central and enclosed by the cotyledons; sometimes it is no more than a mere point or dot, and in some cases, altogether invisible to the naked eye.

The embryo consists of the *plume* and *radicle.*

The *Plume*, or plumula, which is the ascending part, unfolds itself into herbage.

The *Radicle*, or descending part, unfolds itself into roots. At Fig. 111 appears the embryo in a germinating state; *a*, represents the radicle, *b*, the plume, *c*, the funicle, by means of which the plant is still connected to the cotyledons, and receives from them its nourishment.

To use the words of an ancient botanist, "the embryo continues imprisoned within its seed, and remains in a profound sleep, until awakened by germination, it meets the light and air, to grow into a plant, similar to its parent."

Fig. 111.

"Lo! on each seed, within its slender rind,
Life's golden threads in endless circles wind;
Maze within maze the lucid webs are roll'd,
And as they burst, the living flame unfold.
The pulpy acorn, ere it swells, contains
The oak's vast branches in its milky veins,
Each ravell'd bud, fine film, and fibre-line,
Traced with nice pencil on the small design.
The young Narcissus, in its bulb compressed,
Cradles a second nestling on its breast;
In whose fine arms a younger embryo lies,
Folds its thin leaves, and shuts its floret-eyes;
Grain within grain, successive harvests dwell,
And boundless forests slumber in a shell."*

There are various appendages which may or may not be present without injury to the structure of the seed.

Aigrette, or *egret*, sometimes called *pappus*, is a kind of feathery crown with which many of the compound flowers are furnished, evidently for the purpose of disseminating the seed to a considerable distance, by means of winds; as the dandelion. It includes all that remains on the top of the seed after the corolla is removed.

Stipe, is a thread connecting the egret with the seed. The egret is said to be *sessile*, when it has no stipe, *simple* when it consists

* These lines, which so beautifully set forth the manner in which the embryo is contained within the seed or bulb, are not strictly philosophical, as to the fact of the future generations lying enfolded, the one within the other; it is true, that we may in many seeds, by the help of a microscope, discern the form of the future plant, but we cannot believe that this is the miniature image of another plant, which contains another, and so on through successive generations; for the fact is established, that a seed does not produce a plant without being fertilized by the pollen. We may say that a seed contains within itself the *elements* of future generations; but not their *images*, except that of the immediate plant which is to issue from the perfected seed.

What are the parts of the embryo?—Plume—Radicle—What is the egret?—Stipe?

of a bundle of hairs without branches, *plumose* when each hair has other little hairs arranged along its sides, like the beards on a feather.

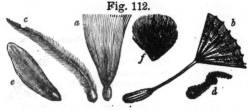
Fig. 112.

In Fig. 112, *a*, represents the *capillary*, or hair-like egret; *b*, is a pedicelled egret; *c* and *d*, show the style remaining, and forming a *plumose train*, as in the virgin's bower and Geum; *e*, a wing, as may be seen in the fir; *f*, a sessile egret.

General Remarks upon Seeds.

The number of seeds in plants is variable; some have but one; some, like the umbelliferous plants, have two; some have four. The number varies from these to thousands. A stalk of Indian corn is said to have produced, in one season, two thousand seeds. A sunflower four thousand. A capsule of the poppy has been found to contain eight thousand seeds. It has been calculated that a single thistle seed will produce, at the first crop, twenty-four thousand, and at the second crop, at this rate, five hundred and seventy-six millions. In the same species of plants the number of seeds is often found to vary. The apple, and many other fruits, might be given as examples.

Seeds, according as they vary in size, have been divided into four kinds; *large*, from the size of a walnut to that of the cocoa-nut; *middle* size, neither larger than a hazel nut, nor smaller than a millet seed; *small*, between the size of the seeds of a poppy and a bell-flower; *minute*, like dust or powder, as in the ferns and mosses.

When a pericarp separates itself from the parent plant, or when the valves of the fruit open, the fruit has ceased to vegetate; like the leaves at the end of autumn, it has lost its vital principle, and becomes subject to the laws which govern inorganized matter.

The maturity of the seed marks the close of the life of annual plants, and the suspension of vegetation in woody and perennial ones. Nature, in favouring by various means the dispersion of these seeds, presents phenomena worthy of our admiration, and these means are as varied as the species of seeds which are spread over the surface of the earth.

The air, winds, rivers, seas, and animals, transport seeds and disperse them in every direction. Those which are provided with feathery crowns, or egrets, as the dandelion and thistle, or with wings, as the maple and ash, are raised into the air and even carried across the seas. Linnæus asserted that the ERIGERON *canadense* was introduced into Europe from America, by seeds wafted across the Atlantic Ocean. "The seeds," says he, "embark upon the rivers which descend from the highest mountains of Lapland, and arrive at the middle of the plains, and the coasts of the seas. The ocean has thrown, even upon the coasts of Norway, the nuts of the mahogany, and the fruit of the cocoanut-tree, borne on its waves from the far distant, tropical regions; and this wonderful voyage has been performed without injury to the vital energy of the seeds."

Number of the seeds variable—Size variable—Separation of the pericarp from the plant—What is denoted by the maturity of the seed?—Dispersion of seeds, how effected?—Seeds carried by water.

Some fruits, endowed with elasticity, throw their seeds to a considerable distance. In the oat, and in the greater number of ferns, this elasticity is in the calyx. In the *Impatiens*, wild cucumber, and many other plants, it resides in the capsule. The pericarp of the IMPATIENS* upon being touched, when the seeds are ripe, suddenly folds itself in a spiral form, and, by means of its elastic property, throws out its seeds.

Animals perform their part in this economy of nature. Squirrels carry nuts into holes in the earth. The Indians had a tradition, that these animals planted all the timber of the country. Animals also contribute to the distribution of seeds by conveying them in their wool, fur, or feathers.

Although distance, chains of mountains, rivers, and even seas, do not present obstacles sufficient to prevent the dispersion of plants, *climate* forms an eternal barrier which they cannot pass. It is not unlikely, that in future times the greater part of vegetable tribes which grow between the same *parallels of latitude*, may be common to the countries lying between them; this may be the result of the industry of man, aided by the efficient means which nature takes to promote the same object in the dissemination of seeds; but no human power can ever cause to grow within the polar circles, the vegetables of the tropics, or those of the poles at the equator. Nature is here stronger than art. That something may be done to promote the growth of tropical plants in our climate is true, but how different are they with us, from the same species in their own genial climate;—we toil and watch for years to nurture an orange or lemon tree, which after all is stinted in its growth, while in its own native home the same plant would have grown spontaneously in luxuriant beauty.

The diffusion of seeds completes the circles of vegetation, and closes the scene of vegetable life. The shrubs and trees are despoiled of their foliage, the withered herbs decompose, and restore to the earth the elements which they have drawn from its bosom. The earth, stripped of its beauty, seems sinking into old age;—but, although the processes of nature may have been unseen and unmarked by man, innumerable germs have been formed, which wait but the favourable warmth to decorate with new brilliancy this terrestrial scene.

So fruitful is nature, that a surface a thousand times more extended than that of our globe, would not be sufficient for the vegetables which the seeds of one single year would produce, if all should be developed; but great quantities are eaten by men and animals, or left to perish in unfavourable situations. Some are carried into the clefts of rocks, or buried beneath the ruins of vegetables; here, protected from the cold, they remain inactive during the winter season, and germinate as soon as the early warmth of spring is felt. Then the pious botanist, beholding the vegetable species with which the earth begins to be clothed, and seeing successively all the types or representations of past generations of plants, admires the power of the Author of nature, and the immutability of His laws.

In concluding our examination of the external organs of plants, we will give a synoposis of the principal ones, with their subdivisions, as heretofore explained.

* The IMPATIENS of the garden is sometimes called *Ladies'-slipper*, sometimes *Balsamine*.

Elasticity of some fruits—Agency of animals—Effect of climate upon the dispersion of plants—Circle of vegetation completed—Concluding remarks.
9*

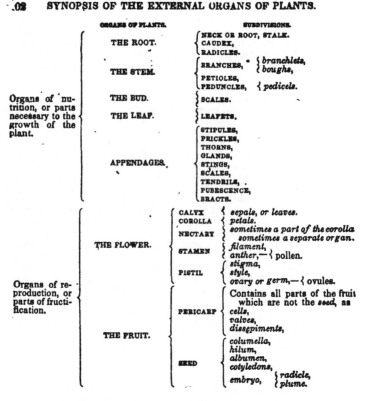

ORGANS OF PLANTS.   SUBDIVISIONS.

Organs of nutrition, or parts necessary to the growth of the plant.

- THE ROOT. { NECK OR ROOT, STALK. CAUDEX, RADICLES.
- THE STEM. { BRANCHES, { *branchlets, boughs,* PETIOLES, PEDUNCLES, { *pedicels.*
- THE BUD. { SCALES.
- THE LEAF. { LEAFETS.
- APPENDAGES. { STIPULES, PRICKLES, THORNS, GLANDS, STINGS, SCALES, TENDRILS, PUBESCENCE, BRACTS.

Organs of reproduction, or parts of fructification.

- THE FLOWER.
  - CALYX { *sepals, or leaves.*
  - COROLLA { *petals.*
  - NECTARY { *sometimes a part of the corolla sometimes a separate organ.*
  - STAMEN { *filament, anther,*— { *pollen.*
  - PISTIL { *stigma, style, ovary or germ,*— { *ovules.*
- THE FRUIT.
  - PERICARP { Contains all parts of the fruit which are not the *seed,* as *cells, valves, dissepiments,*
  - SEED { *columella, hilum, albumen, cotyledons, embryo,* { *radicle, plume.*

LECTURE XVII.

PHYSIOLOGICAL VIEWS—GERMINATION OF THE SEED.

WE have traced the various organs of the plant, through their successive stages of development, from the root to the bud, leaf, and flower, and from the flower to the fruit and seed. We have seen, in imagination, the vegetable world fading under a change of temperature, the "sear and yellow leaf" becoming a prey to the autumnal blasts, and even the fruits themselves exhibiting a mass of decayed matter. Were this appearance of decay and death now presented to us for the first time, how gloomy would be the prospect! How little should we expect the return of life, and beauty, and fragrance! No power short of Omnipotence, could effect this; it is indeed a miracle! But we are so accustomed to these changes, that, "seeing, we perceive not;" we think not of the mighty Being who produces them; we call them the *operations* of *nature;* but what is

---

Enumerate the organs of nutrition—Of reproduction—What are the parts of the root ?—The Stem—Bud—Leaf—Different kinds of Appendages—Divisions of the calyx—Corolla—Nectary—Stamens—Pistil—What are the parts of the fruit ?—What are the parts of the pericarp ?—Parts of the seed—Of the Embryo—What remarks commence this lecture?

nature, or the *laws* of *nature*, other than manifestations of Almighty power?

The word *nature*, in its original sense, signifies *born*, or *produced ;* --let us then look on nature as a created thing, and beware of yielding that homage to the creature which is due to the Creator. The skeptic may talk with seeming rapture of the beauties of nature, but cold and insensible must be that heart, which, from the contemplation of the earth around, and the heavens above, soars not to Him,

" The mighty Power from whom these wonders are."

How impressively is the reanimation of the vegetable world urged by St. Paul, as an argument to prove the *resurrection from the dead !* The same power, which from a dry, and apparently dead seed, can bring forth a fresh and beautiful plant; can assuredly, from the ruins of our mortal frame, produce a new and glorious body, and unite it to the immortal spirit by ties never to be separated.

Leaving the external appearances of the plant, we are now to enter the inner temple of nature, and to examine into those wonderful operations by which vegetable life is called into action and sustained.

Germination. The process of the shooting forth of the seed is termed *germination.* The principle of life contained in the seed does not usually become active, until the seed is placed in circumstances favourable to vegetation. When committed to the bosom of the earth, its various parts soon begin to dilate, by absorbing moisture. Chemical action then commences ; *oxygen* from the air unites to the *carbon* of the seed, and carries it off in the form of *carbonic acid* gas. As the carbon of the cotyledons, by this process, continues to diminish, and oxygen is produced in excess, a sweet sugar-like substance is formed; this being conveyed to the embryo, it is by its new nourishment kindled into active life ; from this period, we may date the existence of the *young plant.*

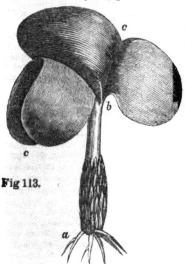

Fig 113.

Bursting through the coats which surrounded it, and which are already enfeebled by their loss of carbon, the embryo emerges from its prison ; the *radicle* shoots downward, and the *plume* rises upward. We then say, the seed has come up, or sprouted. Fig. 113 represents a young dicotyledonous plant, with its radicle, *a*, developed ; its plume, *b*, is yet scarcely perceptible ; its cotyledons, *c*, appear in the form of large, succulent seed-leaves.

The radicle, or descending part, is usually the first to break through the coats of the seeds ; it commences its journey downward, to seek in the soil nourishment for the future plant, and to fix it firmly in the earth. It always takes a downward course, in whatever situation

Meaning of the word nature—Feelings which should be excited by created objects--St. Paul's argument for the resurrection—Describe the process of germination—Describe Fig. 113—Which part of the embryo first escapes from its integuments ?

the seed may have been placed in the ground. A botanist once planted in a pot, six acorns, with the points of their embryos upward. At the end of two months, upon removing the earth, he found that all the radicles had made an *angle*, in order to reach downward. It is supposed that if the root met with no obstruction in going downward, it would always be perfectly straight.

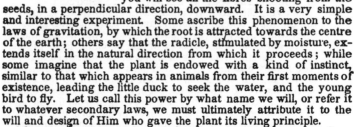

Fig. 114 is a representation of a germinating seed of the *Mirabilis*, (four o'clock;) it will be seen that the radicle, p, has made nearly a right angle in turning-downward ; the plume is not developed. .

Fig. 114.

If you put cotton into a tumbler of water and place upon it some seeds of rye or wheat, you will see all the fibres shooting from the seeds, in a perpendicular direction, downward. It is a very simple and interesting experiment. Some ascribe this phenomenon to the laws of gravitation, by which the root is attracted towards the centre of the earth; others say that the radicle, stimulated by moisture, extends itself in the natural direction from which it proceeds; while some imagine that the plant is endowed with a kind of instinct, similar to that which appears in animals from their first moments of existence, leading the little duck to seek the water, and the young bird to fly. Let us call this power by what name we will, or refer it to whatever secondary laws, we must ultimately attribute it to the will and design of Him who gave the plant its living principle.

After the young root has made some progress, the cotyledons swell, and rising out of the ground, form two green leaves, called *seed-leaves*. When the plume develops its leaves, these seed-leaves being no longer needed wither and decay.

You will recollect that the embryo or germ is composed of two principal parts, the radicle and plume. The radicle, we have just seen, extends itself downward. Soon after this part of the germ has begun its downward course, the plume, (so called from its resembling a little feather,) rises upwards, and soon becomes a tuft of young leaves, with which the stem, if there be one, ascends.

"Some rye being planted in a good soil, at the end of the second day its radicle was discernible. At the end of twenty-four hours the embryo had escaped from its integument. On the second day the fibres of the root had augmented, but the leaves had not appeared. On the fourth day the first leaf began to appear above the ground, at which time the colour was red. . On the fifth day, it had grown to the length of an inch, and its colour was now green, and on the sixth day the second leaf had appeared."[*]

Rye belongs to that class of plants whose seeds have but one cotyledon, and this never rises above the ground to form a seed-leaf. Seeds with but one cotyledon are chiefly composed of albumen, which performs the same office of nourishing the embryo during its germination, as the cotyledons of dicotyledonous plants. In some monocotyledons is perceived under the albumen, a part called *vitellus*, or the yolk; this, like the albumen, is entirely converted into nourishment for the young plant ; it may be seen in the seeds of grasses, and is conspicuous in the Indian corn.

* Sumner.

Describe the experiment with acorns.—Describe Fig. 114—Causes assigned for the downward course of the radicle—Seed leaves—Plume—Experiment with rye—Seeds with one cotyledon—Vitellus.

Fig. 115.

Fig. 115 represents a young monocotyle-donous plant; at *a*, is the cotyledon; at *b*, is the second leaf, which, in the example just given of the rye, appeared on the sixth day; at *c*, is the primordial leaf,* which, at first, en-velops and conceals the other leaves; at *d*, are the several branches of the root, bearing their radicles, and at their base enveloped by a peculiar covering; *e*,† through which the ex-tremities have forced their way.

Earth, though not absolutely essential to germination, is useful, as affording to the vegetable egg a favourable situation, where it may receive the influence of the various agents, which are to perform their offices in the development of its parts. It seems, too, not improbable that some of the constituent elements of earth may be absorbed by the germinating plant and converted into nour-ishment. It is, however, sufficiently apparent that plants may vegetate without earth. The parasite grows upon the bark of other plants; many seeds vegetate in water, and some will grow if moistened and placed on cotton, or any other supporting substance.

Air, is essential to vegetation; under an ex-hausted receiver a seed will not germinate. although possessing every other requisite. Seeds that become imbedded deeply in the ground, do not vegetate, unless accidentally ploughed up, or exposed to the contact of the atmosphere. Acorns supposed to have lain for centuries, have germinated as soon as raised sufficiently near the surface of the earth to receive the influence of air.

You will recollect that in the process of ger-mination, *oxygen gas* unites with the carbon of the seed, and carries it off in the form of carbonic acid. Air furnishes that important agent, oxygen, which is the first moving prin-ciple of vitality.

Carbon constitutes the greater part of the substance of seeds; and this principle, being in its nature opposed to putrefaction, prevents seeds from rotting, previous to their being sown. Some seeds having an abundance of carbon, are capable of being preserved for ages; while others, in which this element exists but in a small proportion, require to be sown almost as soon as ripe; and such as are still more deficient in carbon lose their vital prin ciple before separating from the pericarp.

You can now understand that oxygen is important to germination on account of its agency in removing the carbon which holds the living principle of the seed in bondage.

* Called by Mirbel, the *pileole.*   † The *coleorhize.*

The *absence of light* is favourable to the germination of seeds for light acts upon plants in such a manner as to take away oxygen by the decomposition of carbonic acid gas, and to deposite carbon, now this is just the reverse of the process required in germination, where the carbon must be evolved and the oxygen in excess.

A certain degree of *heat* is necessary to germination. Seeds planted in winter, will remain in a torpid state; but as soon as the warmth of spring is felt, the embryo emerges into life. By increasing heat, the vegetating process of seeds may be hastened; thus the same seed, which with a moderate degree of heat would germinate in nine hours, may be brought to this state in six hours, by an increase of temperature. Too great heat destroys the vital principle; thus corn which has been roasted cannot be made to vegetate. The process of malting consists in submitting some kind of grain, (barley is most commonly used,) to a process which causes an incipient state of germination; this is done by moistening the grain, and exposing it to a suitable degree of warmth; as soon as germination commences, the process is stopped by increasing the heat. The taste of the grain is then found to have become sweetish. The term *malt* is given to grain which has been submitted to this process. When mixed with water it forms a sweet liquor; and the fermentation of this liquor produces beer.

There is a great difference in plants as to their *term of germinating;* some seeds begin to vegetate before they are separated from the pericarp.*

In the greater number of vegetables, however, there is no vegetation until after the opening of the pericarp and the fall of the seed. The time at which different species of seeds, after being committed to the earth, begin to vegetate, varies from one day to some years. The seeds of grasses, and the grain-like plants, as rye, wheat, corn, &c., germinate within two days. The cruciform plants, such as radish and mustard, the leguminous, as the pea and bean, require a little more time. The peach, walnut, and peony, remain in the earth a year, before they vegetate.

All kinds of plants germinate sooner, if they are sown immediately after being separated from their pericarps. Many vegetables preserve their vital principle for years; some lose it as soon as they are detached from their pericarps. This is said to be the case with respect to coffee and tea. The seeds of some of the grasses, as wheat, &c. are said to retain their vital principle even for centuries. It is asserted that mosses, kept for near two hundred years in the herbariums of botanists, have revived by being soaked in water. An American writer† says, that "seeds, if imbedded in stone or dry earth, and removed from the influence of air or moisture, might be made to retain their vegetative quality or principle of life for a thousand years." But he very rationally adds, "life is a property which we do not understand; yet life, however feeble and obscure, is always life, and between it and death, there is a distance as great as existence and non-existence."

* In the month of January, on observing the seeds of a very juicy apple, which had been kept in a warm cellar, I saw that they were swollen, and the outward coat had burst; examining one seed, by removing the tegument and separating the cotyledons, I saw, by the help of a microscope, the embryo as if in a germinating state; the radicle was like a little beak; in the upper part or plume was plainly to be seen the tuft of leaves and the stem.

† E. Barton.

The subjects upon which, in this lecture, we have been engaged, properly come under the head of vegetable physiology, a department of botany highly interesting, but too complicated in its nature to be, to any great extent, presented to the mind of the youthful investigator. The physician finds in the vegetable organization striking analogies to the internal structure of the animal frame; to him the language of physiological botany is familiar, because it is borrowed from his own science. On the other hand, the botanical student, in learning the names and offices of the various internal organs of plants, is making no inconsiderable improvement in the knowledge of the animal economy, and stupid must be that mind which is not, by the consideration of the one, led to reflect upon the organization of the other.✛

LECTURE XVIII.

PHYSIOLOGICAL VIEWS—SOLID AND FLUID PARTS OF VEGETABLES.

The careless observer of nature may consider the trunk of a tree, or a stem of an herb, as very simple in its structure, presenting little more than a homogeneous mass; but the botanical philosopher looks with a far different eye upon the vegetable being. He has learned that plants, like animals, are formed of vessels of different kinds, variously fitted to carry on the operations of imbibing nourishment, of making a chemical analysis of the same, and of appropriating to themselves such elements as are necessary to promote their health and vigour, and of rejecting such as are useless. In short, that they have parts which are analogous to skin, bones, flesh, and blood: that they are living, organized beings, composed of solid and fluid parts; and, like animals, the subjects of life and death.

Plants differ from animals in being destitute of the organs of sense. They can neither see, hear, taste, smell, nor touch. Some vegetables, however, seem to have a kind of sensibility like that derived from the organs of touch; they tremble and shrink back upon coming in contact with other substances; some turn themselves round to the sun, as if enjoying its rays. There is a mystery in these circumstances which we cannot penetrate; it is not yet fully known at what point in the scale of existence animal life ends, and vegetable life commences. Some beings, like the sponge and corals, seem almost destitute of any kind of sensation, and yet they are ranked among animal substances. The subject of the distinctions and analogies between plants and animals, we shall consider more extensively hereafter.

Solid parts of Vegetables.

We shall now treat of the solid portions of the vegetable organization; these are all composed of a *membranous substance*, which exists in every part of the plant, forming by its various modifications, the different textures which the plant exhibits. This mem-

Vegetable Physiology—Its language borrowed from animal physiology—Different aspects of vegetables to the careless observer and the philosopher—Difficult to determine where vegetable life commences—Solid parts of plants.

branous substance appears chiefly under two elementary forms. viz. 1st, that of *cellular texture;* 2d, *vascular texture.*

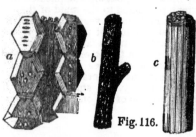

Fig. 116.

1st. *Cellular texture,* (Fig. 116, *a ;*) this, according to the opinion of Mirbel, is composed of a mass of little hexagonal cells, resembling honey-comb. Another writer* compares the appearance of the cellular texture to the froth of fermenting liquor : he considers that each cell is disconnected with the others; while Mirbel believes that the divisions of the membrane, which forms these cells, are common to contiguous cells. The cellular system in animals contains the fat ; in vegetables it is generally filled with resinous, oily, or saccharine juices ; in some cases the cells contain air only. They are usually marked by small dots, (as at *a,* Fig. 116;) these are supposed to be apertures, through which fluids are transmitted from one cell to another.

The cellular texture composes most of the pith, parenchyma, and cotyledons of almost all vegetables. It is abundant in tuberous roots, pulpy and fleshy fruits, and the stems of grasses, and constitutes the principal part of mushrooms, and other cryptogamous plants. In the bark of plants the cellular texture is situated under the cuticle ; it is filled with a juice which varies in colour in different species of plants, but is most commonly green ; it gives its colour to the bark, as the same texture under the human cuticle gives colour to the skin. The green colour of leaves is caused by the cellular texture, which is enclosed on both sides by the cuticle. In the pith of young plants, the cells are filled with watery fluids, but in older plants they are empty, or only filled with air. The petals of flowers owe their beautiful hues to the presence of cellular texture, filled with juices, which refract and reflect the rays of light, in a peculiar manner.

Vascular† texture, consists of tubes, which, like the vessels of the animal frame, are capable of transmitting fluids. These tubes are open at both ends, and are protected by a coating of cellular integument ; their sides are thick and almost opaque. These vessels extend throughout the whole plant, distributing air and other fluids necessary to vegetation. The vascular system of plants presents a variety in form, and also with respect to the functions which the different parts perform.

Some are *entire* vessels, or without any perforation, (Fig. 116, *c ;*) these convey the proper juices of the plant, and generally contain oils and resinous juices.

Porous vessels exhibit many perforations, (Fig. 116, *b ;*) they often separate and again unite, changing at length into cellular integument.

---

* Dutrochet.
† The term *vascular* is derived from the Latin word *vasculum,* a little vessel.

Fig. 117.

Spiral vessels are so called from their form, which
resembles that of a screw,
(Fig. 117, *a ;*) they are
sometimes termed *trachea*,
from a supposed analogy to
the trachea of insects, or
their organs for breathing.
These vessels are formed
of a thread-like fibre turned spirally from right to left.

Annular vessels, (so called from the Latin *annulus*,
a ring,) are so perforated
as to make the tube appear
to be composed of rings,
(Fig. 117, *b.*)

Moniliform vessels (from *monile*, a necklace) resemble, in external
appearance, a string of beads, (Fig. 117, *c ;*) these serve to connect
large vessels, and to convey sap from one set to another.

Mosses, fungi, and lichens, have no vascular system, but their
tissue is all of the cellular kind. The solid substance of plants is all
composed of some varieties of the two kinds of membranes we have
now descibed. Roots and stems are made up of vascular fibres;
these may easily be split longitudinally, as the vessels in this case
are only separated, and the cellular texture easily yields; but in
severing the roots and stems horizontally, greater resistance is to be
overcome, since the tubes are to be cut across.

Vegetables, like animals, have a *system of glands*, or internal vessels, which are made subservient to the purpose of producing changes
in the fluids of the plants;—thus the sap is converted into the proper
juices; and from the same soil and nourishment plants of very different properties are produced.

Mirbel, by the aid of the microscope, succeeded in discovering a
system of glands in the pores or cells, and on the borders of the spiral vessels. There are also external glands, which appear manifest
to the naked eye; as the nectaries of flowers, which secrete or manufacture honey; and the stings of plants, which secrete an acrid substance, which, by penetrating the skin, causes a painful sensation.

Fluid Parts of Vegetables.

The different fluids which are exhibited in the vegetable body may
be considered under three general divisions: 1st, the *sap*, or ascending fluid; 2d, the *cambium*, or descending fluid; 3d, the *proper juices*.

The *sap* is a limpid, inodorous liquid, the elements of which are
imbibed from the earth by pores in the radicles of the root. Every
one knows, that if the earth around the roots of plants is destitute of
moisture, they soon die. Water holding in solution various substances, such as earths, salts, animal and vegetable matter, is absorbed by the radicles; by some unknown process, they convert this fluid
matter into sap, and then, by means of vessels which form what is
called the *sap-wood*, or *alburnum*, this sap ascends through the stems
to the branches; passing through the woody part of the petioles, and
those minute branches of the petiole which form the ribs and veins

Spiral vessels—*Annular*—*Moniliform*—All the solid substance of plants composed of some of these vessels—The use of glands—Glands discovered by Mirbel—
External glands—Three kinds of fluids—What is the sap, and how formed ?—What
the use of the sap-wood ?
10

of the leaf, it enters into the vessels and cells which extend through-out its substance.

The ascending sap is always in circulation, but its energy varies with the season, and the age of the plant. Heat has an important in-fluence in quickening the ascent of the sap; yet, during a dry and hot season, it often appears to ascend but slowly. This is because the absorption of fluids from the earth is checked by the dryness of the soil. The plant, by a little stretch of the imagination, may be con-sidered as thirsty, and thus man may seem not only provident, but humane, in administering to its roots refreshing draughts of water. Even the leaves, at such a period, seem too impatient to wait for supplies by means of the connecting sap-vessels; for if water is sprinkled upon them, they fail not to use their own power of absorp-tion, and upon such an application, may be seen to revive almost in-stantaneously.

When the moisture of the earth coincides with elevation of tem-perature, the sap ascends with the greatest rapidity; this is the case in spring. It is at this period, that incisions are made into the wood of maple-trees, in order to procure sap for the manufacture of sugar. The sap may at this time be seen flowing almost in a stream. It has been thought that the circulation of sap was wholly suspended during winter; this, however, seems not to be the case; for we may observe during this season a gradual development of some parts of the plant; we see many plants preserving the freshness and verdure of their foliage, and mosses putting forth their flowers We must then believe, that the sap is in perpetual motion, suscepti-ble of being accelerated or retarded by changes of temperature, and humidity, or dryness of the earth. The development of buds must be attributed to the ascension and redundancy of the sap, which di-lates and nourishes their parts. In spring, when the ascent of the sap is accelerated, the buds enlarge rapidly, and their complete de-velopment is soon perfected.

The vascular texture appears by its tubes and channels to afford great facilities for the ascension of the sap. In imperfect plants, such as mushrooms and lichens, which are wholly composed of cellular texture, it is not known that there is any ascent of sap, but they seem to be nourished by fluids absorbed from the air.

The question naturally arises, by what force is the sap made to ascend, contrary to the laws of gravitation? Some have asserted, that this phenomenon was owing to the *contraction* and *dilatation* of the air, and of the juices of the plant; others have referred it to the action of *heat;* these two propositions, however, amount to the same thing, since *heat* is the cause of the contraction and dilatation referred to. Some ascribe the ascent of the sap to the *irritability* of the vessels, and the *energy* of *vital power.*

The latter is but a vague and unsatisfactory explanation, since we know neither the cause of this irritability, nor in what this vital power consists. There is no doubt but the ascent of the sap is, in a degree, owing to capillary attraction, assisted by heat. You will recollect that the vessels containing this fluid, were described as very small tubes, no larger than a hair, and, in most cases, much smaller, since few are visible to the naked eye. Those who understand something of Natural Philosophy, know that capillary tubes have the property

What effect has drought upon the plant?—What two circumstances cause the rapid ascent of the sap?—Why are incisions made in maple-trees in the *spring,* rather than at any other period?—Perpetual motion of sap—Cause of development of buds—Vas-cular texture unlike the cellular in affording facilities for the ascension of sap—Ex planations of the pauses of the ascent of the sap.

of raising liquids against the laws of gravitation, and with a force proportional to their smallness of diameter;—this law seems to explain, in some degree, the phenomenon we are considering.

But it is necessary for us now to trace the progress of the sap, after it has ascended to the leaves and extremities of the plant. A considerable portion of it is, by pores in the leaf, exhaled in the form of almost pure water, while the particles of various kinds, which the sap held in solution, are deposited within the substance of the leaf. This process is sometimes termed the *perspiration* of plants; it is visible in some grass-like plants, particularly upon the leaves of Indian corn. If these are examined before sunrise, the perspiration appears in the form of a drop at the extremity of the leaf; the ribs the leaf unite at this point, and a minute aperture furnished for the ge of the fluid, may be discovered.

sap which remains, after the exhalation by means of the is supposed to consist of about one third of that originally ed by the root; this remainder possesses all the nutritive which had, before, been divided through the whole of the t this period, an important change in its nature takes place, ne which has its analogy in the animal economy.

e have compared the sap to the blood of animals, but it is, in ty, more like the animal substance, *chyle*, which is a milk-like separated by digestion, from the food taken into the stomach. erable part of this chyle is converted into blood, which st into the arteries and then into the veins, are by the lat-d to the heart; the heart, by its contractions, sends the lungs. At each *inspiration* of the breath, oxygen from eric air is absorbed by the lungs; here uniting with the e blood, it forms carbonic gas, which is thrown off at tion of the breath. Thus the carbon, which, in the an-is accumulated by feeding on vegetables, and which e diminished, is carried off; it is said that a person in enty-four hours, expires almost one pound of carbon, of charcoal!

now return to the sap in the leaves of plants, and see hether a change takes place, analogous to that in the animal sys-em. We will consider the sap as bearing a resemblance to the animal chyle, and the leaves to the animal lungs. These vegetable lungs are furnished with pores, by which they, too, inhale gases; ut here our comparison fails, since, instead of oxygen, the plant les *carbonic acid ;* this it decomposes, and converting to its own he *carbon*, which is an important element of vegetable com-s, it exhales the *oxygen* necessary for the support of animal Light, however, is necessary for this process of respiration in ant; deprived of this agent, vegetables absorb instead of giving off oxygen.

The carbon which is deposited in the sap, in order to be fitted for the nourishment of the plant, seems to require the further agency of oxygen, to convert it into carbonic acid; this is effected by means of the oxygen, which, during the night, is absorbed by the leaves. At the appearance of light, carbonic acid is again decomposed and oxygen evolved. Besides the oxygen which the plant separates from the carbonic acid inhaled by its leaves, it is undoubtedly fur-

Exhalation of sap—Perspiration of plants—What is the nature of the sap which re-mains after exhalation?—Sap compared to animal chyle—Formation of carbonic gas —In what respect does the comparison between the respiration of plants and animals fail?—What is needed in order to fit the carbon for the nourishment of the plant?

.iished with this gas by the decomposition of water* and other substances which are absorbed by the root.

The *Cambium* is the sap elaborated by the chemical process car-
r:ed on in the leaves, and rendered fit for the nourishment of the
plant.

In tracing the descent of the cambium or returning sap, we shall
not find it passing through the same vessels by which it ascended ;
it is chiefly conveyed by a system of vessels between the liber or inner layer of the bark, and the alburnum or young wood ; here it
contributes both to the formation of an outward layer of new wood
and an inward layer of new bark ; extending also from the extremity of the roots, to the upper extremity of the plant, it furnishes
materials for the formation of new buds and radicles.

If a ring is cut through the bark of a tree, the cambium will be
arrested in its course, and accumulating around the upper edge of
the bark, will cause a ridge or an *annular* protuberance. This vegetable blood being thus prevented from having access to the lower
part of the plant, the roots cease to grow, the sap ascends but feebly,
and in two or three years the tree dies. If the incision is not made
too deep, the wound will soon heal by the union of the disconnected
bark, and the circulation of the cambium proceeds as before. This
experiment proves the importance of this fluid to the existence of
the plant.

The *Proper Juices of Vegetables.* This division comprehends
all the fluids furnished by the plant except the sap, and cambium ;
as oils, gums, &c. These are the product of the cambium, as, in
the animal system, tears are secreted from blood. The secretions,
carried on by the vegetable glands from the cambium, are of two
kinds ; 1st, such as are destined to remain in the plant, as milk, resins, gums, essential and fixed oils ; 2d, such as are destined to be
conveyed out of the plant ; these consist chiefly of vapours and
gases exhaled from flowers, and may, perhaps, more properly be
called *excretions* than *secretions.*

LECTURE XIX.

PHYSIOLOGICAL VIEWS—BARK, WOOD, AND PITH—GROWTH OF A DICOTYLEDO-
NOUS PLANT—GROWTH OF A MONOCOTYLEDONOUS PLANT.

WE have exhibited to your view the minute discoveries made by
the help of the microscope in the solid parts of the vegetable substances ; we have also noticed those important fluids, the circulation of which appears to constitute the life, and produce the growth
of plants. We have now to consider the solid parts already described, as composing the body of the vegetable, and collected
under the three forms of *Bark, Wood,* and *Pith.*

Bark. The bark consists of the *epidermis, cellular integument,* and
cortex.

1st. *Epidermis†* is the skin of the membrane which extends over

---

* Water consists of *oxygen* in union with hydrogen.
† The word *eperdimis* is from *epi,* upon, and *derma,* the skin.

Cambium, or descending sap—How conveyed—Importance of this fluid—What is
the effect of cutting a ring through the bark of a tree ?—What are the proper juices of
vegetables ?—Of what three parts is the body of the vegetable composed ?—Divisions
of the bark—Describe the *epidermis.*

the surface of every vegetable. It is also called the *cuticle*, a name which anatomists have given to the external covering of the animal body. There is a striking analogy between animal and vegetable cuticle or skin. In the animal it varies in thickness from the delicate film which covers the eye, to the thick skin of the hand or foot, the coarser covering of the ox, or the hard shell of the tortoise. In the vegetable, it is exquisitely delicate, as in the covering of a rose leaf, or hard and coarse, as in the rugged coats of the elm and oak. In the birch you may see the cuticle or outer bark peeling off in circular pieces; it seems not to be endowed with the vital principle, and in this respect differs from all other parts of the plant. The cuticle serves for protection from external injuries, and regulates the proportion of absorption and perspiration through its pores. It is transparent as well as porous, so as to admit to the cellular integument the free access of light and air, while it excludes every substance which would be injurious.

It is to the cuticle of wheat, oat, rye, and some of the grasses, that we are indebted for straw and Leghorn hats. In their manufacture the cellular texture is scraped away, so that nothing remains but the cuticle. It has been ascertained that the outer bark of many of the grasses contains silex, or flint;—in the scouring rush, (*Equisetum*,) the quantity of silex is such, that housekeepers find it an excellent substitute for sand, in scouring wood or metals. A peculiar property of the cuticle is, that it is not subject to the same changes as the other parts of bodies; it is, of all substances found upon animal or vegetable matter, the most indestructible. The cuticle is sometimes, like the skin of animals, clothed with wool or down, and it then becomes an important security against the effects of heat and cold. The leaf of the mullein has its cuticle covered with a kind of wool; the pericarp of the peach has a downy cuticle.

2d. *Cellular Texture*, is situated beneath the epidermis or outer skin of the bark; it is filled with a resinous substance, which is usually green in young plants. This cellular layer possesses glands, which, when submitted to the action of light, carry on the process of decomposing carbonic acid gas, by retaining the carbon and evolving the oxygen gas. The cellular integument envelops branches, as well as trunks of trees, and herbaceous stems; it extends into roots, but there it neither retains its green colour, nor decomposes carbonic acid gas. It is the seat of colour, and in this respect analogous to the *cutis*, or true skin of animals, which is the substance situated under the cuticle, and is black in the Negro, red in the Indian, and pale in the American. In the leaves of vegetables, the cellular integument occupies the spaces comprised between the nerves, and is of a green colour; in flowers and fruits it is of various colours. The cellular substance of some aquatic plants is filled with air; in the pine, sumach, &c., it is filled with the proper juices of the plant. This herbaceous envelope of the trunks of trees, after a time dries, appearing on the surface in the form of a cuticle, and often cleaves off. It is renewed internally from the cambium.

The petals of flowers are almost entirely composed of cellular texture, the cells of which are filled with juices fitted to refract and reflect the rays of light, so as to produce the brilliant and delicate teints which constitute so great a portion of their beauty. The fuci,

Uses of the eperdimis, or cuticle—Cellular texture—Glands of the cellular integument—Cellular integument in roots—The seat of colour—Cellular integument in leaves, &c.—In aquatic plants—How renewed in the trunks of trees—Found in the petals of flowers, &c.

10*

a species of sea-weed, and some other succulent plants, appear to be altogether composed of cellular texture.

3d. *Cortex.* Immediately under the cellular integument, we find the true bark, which, in plants that are only one year old, consists of one simple layer; but in trunks of older trees, it consists of as many layers as the tree has numbered years. The cortex is formed of bundles of longitudinal fibres called cortical vessels. The peculiar virtues or qualities of plants chiefly reside in the bark. Here we find the resin of the fir, the astringent principle of the oak, and the aromatic oil of the cinnamon.

The inner layer of the bark is called the *liber ;* it is here only, that the essential, vital functions, are carried on; this integument is so called from *liber,* a book, on account of its fine and thin plates, which are thought to bear some resemblance to the leaves of a book. This substance, by its development, produces new roots, branches, leaves, flowers, and fruits. It is composed of a kind of net-work, which has been compared to cloth; the elongated fibres representing the warp, and the cellular texture the filling up. It has been observed that the cambium descends between the liber and the wood, and that a layer of new liber, and of new wood, are every year formed from that liquid; as the new layer of bark is formed, the old one is pushed outward, and at length, losing its vital principle, it becomes a lifeless crust. The natives of Otaheite manufacture garments from the liber of the paper mulberry. The liber of flax is, by a more refined process, converted into fine linen. This part of the bark is important to the life of vegetables; the outer bark may be peeled off without injury to them, but the destruction of the liber is generally fatal.

The operation of *girdling* trees, which is often practised in new countries, consists in making, with an axe, one or more complete circles through the outer bark and the liber of the trunk. Trees seldom survive this operation, especially if it be performed early in the spring, before the first flow of the sap from the root towards the extremities.

During the repose of vegetation, that part of the liber most recently organized, and which of course retains its vital power, remains inactive between the wood and the outer layers of the bark, until the warmth of spring causes the ascent of the sap. After promoting the development of buds, and the growth of new wood and bark, the liber hardens and loses its vital energy, like that of the preceding year.

Fig. 118, at *A,* represents a young dicotyledonous stem, cut transversely; the inner circle surrounds the *pith ;* the wood extends to the *bark,* which at *a* appears darkly shaded.

At *B,* is a section of the same stem magnified; *a b,* is the bark, *b i,* the wood, and *i k,* the pith.

The divisions of the bark may be seen as follows; *a c,* represents the *cuticle,* or the dry, disorganized part; at *c d,* is the *cellular integument ;* at *d b,* is the *cortex,* the extreme part of which, at *b,* is the ,iber.

Wood. The wood (*lignum*) consists of two parts, *alburnum* or sap-wood, and *perfect wood.*

The *alburnum* is so called from *albus,* white, on account of the paleness of its colour. This is the newly formed wood, and consti

---

What is said of the cortex?—Liber—Annually renewed—Girdling—What ultimately becomes of the liber?—Describe a dicotyledonous or exogenous stem—Of how many parts does the wood consist?—Alburnum

tutes the outer part of the woody substance of the plant. It is at first soft and tender, and in this state appears to be active with the principle of life. As the liber is formed annually from the cambium or descending sap, new layers of alburnum are supposed to have the same origin, and to be formed during the same intervals of time. Most of the sap ascends through the alburnum, though some passes through the perfect wood. The sap which nourishes the buds, passes through the centre of the stem, and from thence is conveyed in appropriate vessels to the buds.

Fig. 118.

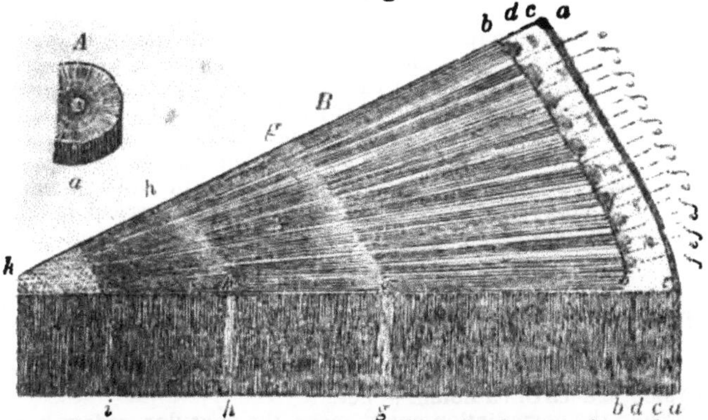

The *perfect wood*, is sometimes called the *heart;* its colour is usually darker than that of the sap-wood, and its texture is firmer and more compact; it is also more durable for timber. It is formed by the gradual concentration and hardening of the alburnum. The wood constitutes the greater part of the bulk of trees and shrubs; when cut across, it is found to consist of numerous concentric layers. It is supposed that one of these circular layers is formed every year. To prove that the wood is deposited externally from the cambium, pieces of metal have been introduced under the bark of trees that were growing, and the wounds carefully bound up; after some years, on cutting them across, as many layers of new wood have been found on the outside of the metal, as years had elapsed since its insertion.

The strength and hardness of wood, is owing to woody fibres extending longitudinally; these fibres are chiefly of vascular texture, and contain sap, and the various secreted juices; some contain only air.

For illustration of the formation of wood, see Fig. 118, *B*, which represents a section of a woody stem of three years' growth; *i h*, next the pith, is a layer of the first year's growth, and the hardest part of the wood; *h g*, is a layer of the second year's growth; and *g b*, of the third; the last is the sap-wood recently formed from the cambium.

Pith. The pith (see Fig. 118, *B*, *k i*) is situated in the centre of the trunk and branches of plants, and is a soft, spongy substance, analogous to the marrow of animals. It is composed of cellular texture. The cells, which are very large in the elder and some other

Perfect wood—How has it been proved that wood is deposited externally?—Strength and hardness of wood—Illustrate the formation of wood by a reference to Fig. 118—Pith.

plants, are filled with fluids when young, but in old branches, the fluids disappear, and the cells are filled with air. In general herbs and shrubs have a greater proportion of pith than trees. It is also more abundant in young than old vegetables; it extends from the root to the summit of the trunk or stem of the plant.

The *medullary** rays are lines which diverge from the pith towards the circumference; they are fibrous textures interwoven in the wood, the alburnum, and the different layers of the bark. The new buds seem to originate from the points at which they terminate. The pith has been compared to the spinal marrow in animals; it appears to be an important part of the vegetable substance, though its offices are perhaps less understood than those of the other parts. The letter *e*, Fig. 118, represents the medullary rays as proceeding from the pith and terminating in the cellular integument.

You are not to expect that every stem or branch of a dicotyledonous plant will present all the various parts which we have described as constituting the vegetable body; neither when they exist are they always distinct, for they often pass into each other in such a manner as render it difficult to define their boundaries. Many species of plants, have no distinct layers of bark, and in many others there is such a similarity between the alburnum and perfect wood, as to render it difficult to distinguish them.

Growth of a Dicotyledonous Plant.

Let us now review the most important circumstance in the growth of a woody plant. Before germination, the substance of the plume or ascending part of the embryo, exhibits a delicate and regular cellular texture; where the liber and medullary rays are to be formed, traces of cambium appear

When the germination commences, the vascular system begins to organize around the pith, and the medullary rays to form; the extremities of these rays exhibit cellular texture, which is soon converted into libers. (See *f*, Fig. 118, which shows the extremities of the medullary rays, and the points where the liber is formed.) While this change is taking place, the cambium, which may be considered a fluid cellular mass, flowing between the bark and the wood, hardens into a new layer of liber, and a new layer of alburnum—the latter is at length changed to this; each year a new layer succeeds, and thus the growth of the vegetable goes on until death completes its term of existence.

Each layer of wood is generally the product of one year's growth; but it is only near the base of the trunk, that the number of layers of wood is a criterion of the age of the tree; for in trees where one hundred layers may be counted near the base, no more than one can be found at the extremity of the branches. These layers, then, do not extend through the length of the tree; but while the base exhibits all the layers which have been formed, the extremity of the branches contains under the bark only the continuation of an annual layer.

The age of branches may be determined by the number of layers of wood at the base of each branch.

We will now consider the manner in which the tree increases in

---

* So called from *medulla*, marrow, a name often given to the pith.

Medullary rays—Pith, to what compared?—Various parts not always distinct in different plants—Appearance of a dicotyledonous plant before germination, or while in embryo—Change at the commencement of germination—Process in the formation of perfect wood—Number of layers of wood near the base of the trunk, a criterion of the age of a tree—How may the age of branches be determined?

height. A seed germinates; the plume rises; the cambium, in developing, gradually becomes less capable of extension; at length, when it is converted into wood, its circulation ceases. The layer of wood then exhibits the form of an elongated cone; at the summit of the cone a bud is formed, from which a new shoot issues; a new layer of alburnum organizes upon the surface of the cone; this, in turn, becomes perfect wood, covering the layer first formed; and thus the tree goes on increasing in height and in diameter. The terminal bud is formed each successive year. After a hundred years of vegetation, a hundred cones might be found boxed within each other in the manner first described; the spaces comprised between the summits of the cones would show the succession and elongation of the annual shoots.

As the wood is formed by the conversion of cambium into alburnum, so from the same liquid the inner layers of bark are formed to renew the waste occasioned by the destruction of the epidermis. While the wood is growing externally, that is, at an increasing distance from the centre, the bark is forming internally, and the new layers are pressing outward.

Growth of Monocotyledonous Plants.

The growth of trunks, as hitherto considered, has relation only to *woody* plants; but between plants which grow from seeds with one cotyledon, and such as grow from seeds with two cotyledons, there is a great difference as to the mode of organization and growth.

The first kind of plants are called *monocotyledonous ;* the second *dicotyledonous.* Their stems, on account of their different modes of growth, have been distinguished into *endogenous*, signifying to grow inwardly; and *exogenous*, signifying to grow outwardly. The discovery of the different modes of growth in these two great divisions of plants, is of recent origin, and constitutes an important era in vegetable physiology.

The stems of *monocotyledonous* or *endogenous* plants have seldom a bark distinct from the other texture; they have no liber, or alburnum disposed in concentric layers; they have no medullary rays; and their pith, instead of being confined to the centre of the stem, extends almost to the circumference.

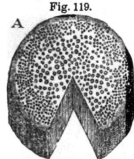

Fig. 119.

The wood is divided into fibres running longitudinally through the stem, (see Fig. 119, where the dots represent the fibres;) each of these fibres seems to vegetate separately; they are ranged around a central support, and are so disposed that the oldest are crowded outwardly by the development of new fibres in the centre of the stem; this pressure causes the external layers to be very close and compact. This mode of increase, little favourable to growth in diameter, produces long and straight stems, nearly uniform in size throughout their whole extent; as the palms and sugar-canes of the tropics, and the Indian corn of our climate. Most of these plants present us with roots of the fibrous kind.

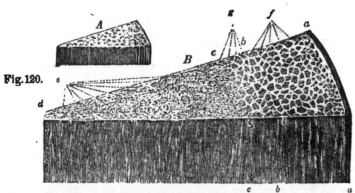

Fig. 120.

Fig. 120, at *A*, represents a section of the stipe or stem of a palm-tree; at *B*, is the same magnified; *a, b*, shows a part of the stipe in which the woody fibres are most dense and hard; *b, c*, shows the fibres less numerous, less compact, and less hard; *c, d*, includes the woody fibres, tender and scattered; the orifices of tubes which have disappeared are seen at *c, a*. In the part *c, d*, the cellular tissue occupies a greater space than at *c, b*, and much more than at *b, a*, where the woody fibre, or vascular texture, predominates. The fibres at *e*, are of new formation; at *f*, they are older, and at *g*, still more ancient; thus the development of the wood in this plant proceeds inversely to that of dicotyledonous plants.

Endogenous plants continue to increase in height, long after they cease to grow in diameter; the stem is gradually extended upward by new terminal shoots, which are formed annually.

The epidermis is formed of the foot-stalks of leaves, which annually sprout from the rim of a new layer of wood; the leaves falling in autumn, their foot-stalks become indurated, and remain upon the outer surface of the plant.

We have now taken a brief view of the most important facts and principles which constitute the science of vegetable anatomy and physiology. Although the vegetable structure is much less complicated than the animal, there are many analogies between them; and many parts of the former have been named, and various phenomena explained, by a reference to names and principles common to animal anatomy and physiology. You cannot therefore expect, at the first glance, to comprehend explanations which presuppose some knowledge of those intricate subjects. By attention to the vegetable structure, you will, doubtless, be induced to think more upon the wonderful mechanism of your own material frames; upon the analogy, and yet infinite difference, between yourselves and the lilies of the field.

In considering these things, we are led to exclaim, in the language of the Psalmist, "Oh Lord, how manifold are thy works, in wisdom hast thou made them all!" The human body is nourished by the same elements as the grass which perisheth; the flowers have a much more refined corporeal substance than you, but how much more precious are you in the sight of the Almighty!

Do you ask, why you are of more value "than the lilies of the field," or even than "many sparrows?" It is the very principle

What is Fig. 120 designed to illustrate?—How is the Epidermis formed?—Reflections on the analogies between the vegetable and animal substances.

within you which enables you to make this inquiry, that renders you thus precious;—it is your soul that raises you above the inanimate and brute creation. Though the body is sister to the worm and weed, the soul may aspire to the fellowship of angels. Oh, then, let me entreat you, suffer not your chief thought to be given to the decoration of the perishable part, the mere temporary dwelling-place of the immortal mind! but seek to prepare this mind for admission into "the glorious company of the spirits of the just made perfect."

LECTURE XX.

PHYSIOLOGICAL VIEWS—CHEMICAL COMPOSITION OF PLANTS—PROXIMATE PRINCIPLES—CHEMICAL ANALYSIS OF THE SAP.

WE have, according to our method of arrangement, considered the anatomy of the vegetable in connexion with its physiology: that is, when treating upon each particular organ, we have remarked upon its uses in the *life* and *growth* of the whole plant. We have treated of the germination of the seed, the minute vessels which constitute the vegetable fabric, with the fluids which circulate through these vessels; we have considered them as constituting, in various ways, three essential parts of woody plants, the bark, wood, and pith. We have inquired into the manner in which these separate parts are formed, and observed the great distinction in the growth of the stems of monocotyledonous and dicotyledonous plants.

Yet, although we have attempted to show how plants *grow*, it is no easy thing to explain how they *live*. The great principle which operates in organic life, appears not to have been laid open to the eye of man. But by a careful observation of facts, we can learn all that it is important for us to know, in order to cultivate plants successfully; their habits, food, and the causes of their diseases and death.

The physician who spends a long and laborious life in the study of the human frame, can give only the result of his *observation*. He finds a certain article efficacious in the relief of a particular disease; but he knows not *why* this should be so; or if he is able to give some reasons, he is ultimately arrested in his speculations by a barrier which he cannot pass. Thus he knows that soda or pearlash corrects acidity in the stomach; ask the reason of this, and he tells you that these are alkalies, substances which neutralize acids, and thus render them harmless; inquire still further, why alkalies do thus affect acids, and the physician is as ignorant as yourselves.

Before closing our view of the vegetable structure, we will, by the aid of chemistry, examine the elements which compose it.

The growth of vegetables, and the increase of their weight, show that they imbibe some external substances, which are incorporated into their own substance. This constitutes *nutrition*, and distinguishes living substances from dead matter. A stone does not receive nourishment, although it may increase by an external accumulation of matter. "Vegetable substances, analyzed by a chemical process, have been found to contain *carbon, oxygen, hydrogen*, and sometimes *nitrogen, sulphur, silex*, the *oxide of iron, soda, magnesia*, and *chalk*."* These different substances are by the root, stems, and leaves of the plant, derived from the earth, air, and water.

* Mirbel, "Elemens de Botanique."

Recapitulation—A difference between the knowledge of facts, and of their causes—Substances which compose plants.

Proximate Principles.

Vegetation produces chemical combinations, which are distinguished by the name of *proximate principles.* Although the proximate principles of plants are very numerous, but few of them are well known; they are the result of the action of the vital forces of plants, and are, therefore, important subjects of investigation to those who pursue the study of physiological botany to any great extent. *Carbon, oxygen, hydrogen,* and *nitrogen,* are the most important of the *ultimate* elements of plants, and the constituent parts of their proximate principles. These principles may be divided into two classes.

I. Those principles which are composed of *carbon, hydrogen,* and *oxygen,* without any nitrogen.

II. Such as contain, besides the substances belonging to the other class, some *nitrogen.* There are few of this class.

The FIRST CLASS of proximate principles is divided into *three orders.*

1st. Principles which have more *oxygen than sufficient to form water.*

2d. Principles in which oxygen and hydrogen exist *in the exact proportion to form water.*

3d. Principles where *hydrogen is in excess.*

The 1st order includes *vegetable acids;* as,

Acetic acid, or pure vinegar; this is generally produced by fermentation from wine, cider, and some other liquids; it is also found in a pure state in the Campeachy wood, and the sap of the elm.

Malic acid may be extracted from green apples and the barberry.

Oxalic acid is found in several species of *sorrel,* belonging to the genera *Oxalis* and *Rumex.*

Tartaric acid is obtained from the tamarind and the cranberry; this acid, combined with potash, forms what is commonly called cream of tartar.

Citric acid is found in the lemon; it is mixed with the malic acid in the gooseberry, the cherry, and the strawberry.

Quinic acid is obtained from the Peruvian Bark, (*Cinchona.*)

Gallic acid is obtained from the oak, and the sumach; it is highly astringent.

Benzoic acid is found in the Laurus *benzoin,* and in the Vanilla this is highly aromatic; it is thought to give the agreeable odour common to balms.

Prussic acid; this acid gives out a strong odour like bitter almonds; it is an active poison; it is obtained from peach-meats and blossoms, from bitter almonds, &c.

The 2d order includes *gum, sugar,* &c.

The Gums. Of these there are many kinds; they have neither taste nor smell; dissolved in water, they form a mucilage more or less thick. The principal gums are,

Gum Arabic, which flows from the plant MIMOSA *nilotiça ;**

Common Gums, such as issue from the peach-tree, the cherry-tree, and many others.

Sugar is a substance which dissolves in water, and has a sweet taste; it is obtained from the sugar-cane, the sugar-maple, from the stalks of Indian-corn, pumpkins, beets, and sweet apples. All vegetables which have a sweet taste, may be made to yield sugar.

* By some writers called ACACIA *Arabica.*

Proximate Principles—What are the most important ultimate elements of plants?—Proximate principles divided into two classes—First class divided into three orders—First order—Second order—Third order.

The 3d order includes *oils, wax, resins, &c.*

Oils. These are fluid and combustible substances, which do not unite with water. They are divided into *Fixed* and *Volatile.* The fixed oils are thick, and have little odour.

The oil of sweet almonds, and *olive oil,* grow thick and opaque by being exposed to the air.

The *Oil of Flaxseed,* called linseed oil, and some other oils, dry without losing their transparency; it is this quality which renders linseed oil so valuable to painters.

The *Volatile oils* are distinguished from the fixed oils by their aromatic odours, and their tendency to fly off, from which circumstance the term *volatile* is derived. Among these oils are those of the orange, lavender, rose, jasmine, peppermint, and wintergreen. They are sometimes greatly reduced by being mixed with alcohol, and are then called *essences.* The volatile oils may be found in a great variety of plants, particularly those of the Labiate family.

The *Aroma* or *aromatic property,* consists chiefly of the odours which are exhaled from plants, containing volatile oil; to this oil is owing the aromatic odour of the ginger plant, of the myrtle, rose, and other sweet-scented plants. Aromatic plants are much more common in hot, than cold countries; most of aromatic spices are found in the equatorial regions.

Wax is found on the surface of the fruit of the bay-berry, (MYRICA *cerifera.*) Beeswax, though an animal production, is made by the bees from the pollen of plants.

Camphor has much analogy with the volatile oils; it is an extract from the LAURUS *camphora,* or camphor-tree of Japan.

RESIN exudes from the pine, and some other trees; it is dry, insoluble in water, but soluble in alcohol, and very inflammable. The people in new countries often use, as a substitute for lamps, pine knots, which abounding in resin, burn with a bright flame. The difference between resin and the volatile oils, appears to consist in the action of oxygen upon the former; for the oil in absorbing oxygen from the air, passes into the resinous state.

Resins mixed with volatile oils form *balsams;* they are thick, odorous, and inflammable substances, as, the balsam copaiva, and the balsam of *Tolu.*

These resins are sometimes mixed with gums, they are then called *gum-resins;* of this kind are gamboge, assafœtida, guaiacum, aloes, an extract from the ALOE *perfoliata.* These gum-resins in flowing from vegetables are sometimes white and liquid like milk, but they usually become brown and hard by exposure to the air.

Indian rubber,[*] or as it is sometimes called, gum elastic, is the product of a South American tree, (SIPHONIA *elastica,*) an East Indian plant, (the URCEOLA *elastica,*) and some other trees in the equatorial regions; by exposure to the air the gum hardens, becomes brown, and takes the appearance of leather; it can neither be dissolved by water nor alcohol. The juice of the milk-weed is said to be similar to that of the plants from which the Indian rubber is obtained.[†]

[*] Caoutchouc.
[†] Mr. H. Eaton, (late professor of Chemistry at Transylvania University, Kent.) informed me that he prepared a small quantity of the juice of the milk weed, (Asclepias,) in such a manner that it could not be distinguished from the imported Indian rubber, either in external appearance, or in its properties.

What substances belong to the third order of the first class of proximate principles? Describe the different vegetable oils—What causes the aroma of plants?—Wax—Camphor—Resins—Indian rubber.

11

The green principle. It is to this principle that all the green parts exposed to light, owe their colour; it undergoes changes in the different states of the plant, in autumn becoming brown or yellow. Davy attributes the change of colour to the formation of an acid. Every one knows that a drop of sour wine, lemon juice, or any other acid, will change green to a brown, or yellowish colour.

The *second class* of proximate principles consists of substances which, like the first class, are formed of carbon, hydrogen, and oxygen; but to these is added *nitrogen*. We here find;

Opium, a narcotic principle, extracted from the poppy. It is soluble in alcohol, slightly in water.

Hematine, the colouring principle from the Campeachy wood.

Indigo, a colouring substance, obtained from several species of *Indigofera,* or indigo plant.

Gluten, is extracted from the cotyledons of the seeds of *leguminous* plants, as peas, beans; and from the albumen of wheat, rye, &c. It is obtained by separation from the starch. Flour owes much of its nourishing properties to gluten, which, in some respects, is analogous to animal principles, being, like them, subject to putrefaction.

Jelly, is the thickened juice of. succulent fruits; as currants, quinces, and apples; it is soluble in hot water, though scarcely so in cold; when heated, it loses its jelly-like form, which is that of a coagulated mass, susceptible of a tremulous motion; by too long boiling, the juice loses this property, which gives to jelly its peculiar appearance. Many colouring principles have never been separated from the substances to which they are united; as those of saffron, logwood, &c.

It has already been suggested, that the red colour of fruits arises from the combination of an acid, with a blue colouring principle. Every beginner in chemistry knows that the effect of mixing an acid with an infusion of blue violets, or any vegetable blue, is to give a red tinge, varying in shade from a purple red to a brilliant scarlet, in proportion to the quantity of acid. It has, upon the same principle, been supposed that the purple, red, and blue colouring of the petals of flowers, is owing to different proportions of acid; this may explain the change of colour which appears in some flowers, which pass from blue to red, as the changeable hydrangea. This change may be attributed to increase of acid,* combining with the blue colouring principle. Some red flowers become blue; they are in this case supposed to have parted with some portion of the acid, which was united with their colouring principle.

Chemical composition of the Sap.

The sap is a transparent, colourless fluid, imbibed by the vegetable from the earth and air; or more properly, from the water existing in them, which holds in solution *oxygen, hydrogen, carbon, nitrogen, earths, mineral-salts,* and *animal* and *vegetable matter.* We might suppose, that being derived from the same source, the sap in all vegetables would be alike, but it is never obtained pure; it is more or less mingled with the *proximate principles,* or proper juices, and thus differs in different species of vegetables; water, however constitutes the principal part in all.

Sap of the *elm* (Ulmus *campestris*) has by analysis been found to

---

* Iron is supposed to be combined with the oxygen of the acid.

What is said of the green principle?—What new element is found in the second class of proximate principles?—What substances are found in this class?—Cause of the red colour of fruit—Of the various hues of the petals of flowers—Sap of the elm.

contain water, volatile matter, acetate of potash, carbonate of lime, vegetable matter, sulphate of potash.

Sap of the beech, (FAGUS *sylvatica,*) contains water, acetate of lime, with excess of acid, acetate of potash, gallic acid, tannin. mucous extract, acetate of alumine.

Sap of the Horse-chestnut, (ÆSCULUS *hippocastanum,*) contains water, extractive mucous matter, nitre, acetate of potash, and carbonate of lime.*

These few examples of the decomposition of vegetable principles show how wide a field is open to the chemist, in the study of vegetable elements.

It may seem wonderful, that of so few elementary substances, such a great variety should exist in the taste, smell, colour, consistence, medicinal and nutritious qualities of vegetable combinations; is it not equally wonderful that with the nine digits and the cipher, we may make such varied combinations of numbers; or with our twenty-six letters of the alphabet, form every variety of composition? Thus, *by various combinations of a few simple principles, are formed all vegetable and animal productions.*

The presence of nitrogen was formerly considered as a test of animal substance, and the want of it of a vegetable substance, but it is now ascertained that animal substances may exist without nitrogen, and that this principle is contained in several vegetables.

The *elements of the compounds being the same,* the question naturally arises, what causes the *great diversity in the properties?.* Two causes may be assigned for this; viz. 1st. The *different proportions in which the elements are combined.* 2d. The *various modes of their combination.*

In vinegar and sugar, the one substance a liquid, and of a sour taste, the other solid and sweet, are found the same elements in *different proportions* and *differently combined.* In gum, starch, and sugar, the *elements are the same,* the *proportion nearly the same,* but *they are combined differently.*

When we know by chemical analysis the combinations which exist in inorganized bodies, we can, by putting the same together, often form similar substances; but we cannot thus form organized bodies; for to these belongs a living principle, which it is not in the power of man to bestow. It is said, that Rousseau, skeptical in science as in religion, declared he would not believe in the correctness of the analysis of vegetable or animal substances, until he should see a young animal, or a thrifty plant, spring into existence from the retort of the chemist. But the power to *create,* the Almighty has not delegated to man; neither is it to be supposed that any future discoveries in science will ever confer it upon him. To study the compound nature of substances, to classify, arrange, and by various combinations to beautify the world of matter, to cultivate the faculties of mind, until stronger and brighter the mental vision sees facts and principles before invisible; these are the high privileges bestowed on man;—but to *add one new particle to matter, or one new faculty to the mind, is beyond the power of the whole human race.*

* These results of the analysis of sap are extracted from Vauquelin.

Sap of the beech—Of the horse-chestnut—All vegetable and animal productions composed of a few simple principles—Illustration—What two causes assigned for the different properties of compounds formed from the same elements?—Organized bodies not produced by the skill of man

PART III.

CLASSIFICATION.

LECTURE XXI

METHOD OF TOURNEFORT—SYSTEM OF LINNÆUS—NATURAL METHODS—METH-
OD OF JUSSIEU—COMPARISON BETWEEN THE CLASSIFICATIONS OF TOURNE-
FORT, LINNÆUS, AND JUSSIEU.

LET us now imagine the whole vegetable kingdom, comprising innumerable millions of individual plants, to be spread out before a
botanist. Could he, in the course of the longest life, number each
blade of grass, each little moss, each shrub, or even each tree? If
he could not even count them, much less could he give to each one a
separate *name* and *description*. But he does not need to name them
separately, because that nature has arranged them into *sorts* or
kinds.

Were you sent into the fields to gather flowers of a similar kind,
you would need no book to direct you to put into one parcel, all the
red clover blossoms, and into another, the *white clover ;* while the
dandelions would form another group. These all constitute different *species*. Nature would also teach you that the red and white clover, although differing from each other in some particulars, yet bear
a strong resemblance. By placing them together you form a *genus*,
and to this *genus* you refer all the *different kinds* or *species* of clover.
When you see the common red, damask, and cinnamon roses, you
perceive they all have such strong marks of resemblance as to en-
title them to be placed together in one genus. But yet you know
that the seed of a damask rose would never produce a red rose.
One species of plants can never produce another species.

The whole number of *species* of plants, which have been named
and described, including many which have been recently discovered in New Holland, and about the Cape of Good Hope, is said to
be 56,000.*

If species of plants were described without any regular order, we
could derive neither pleasure nor advantage from the study of practical botany. When we wished to find the name of a plant, we
should be obliged to turn over the leaves of our books without any
rule to guide us in the search.

The necessity of some kind of system was so apparent, that many
attempts for the methodical arrangement of plants were made before the time of Linnæus; but his system was so superior to all others, that it was no sooner published to the world, than it was adopted by the universal consent of all men of science.

Previous to this time, Tournefort, a native of France, had published an ingenious method of arrangement, beautiful by its simplicity, but imperfect, on account of the vagueness of its application
The characters of his classes were founded upon the *absence, presence,* and *form* of the corolla. Tournefort made *twenty-two* classes ;
these he subdivided into sections or orders.

* As recently reported by the Baron Humboldt, to the French National Institute.

Nature arranges plants into kinds or sorts—Examples—Number of species of
plants—Necessity of order in description—Attempts at arrangement made before the
time of Linnæus—Tournefort's classes, on what founded?

Synopsis of the Method of Tournefort.

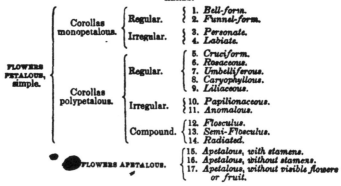

HERBS.

FLOWERS PETALOUS, simple.	Corollas monopetalous.	Regular.	1. *Bell-form.* 2. *Funnel-form.*
		Irregular.	3. *Personate.* 4. *Labiate.*
	Corollas polypetalous.	Regular.	5. *Cruciform.* 6. *Rosaceous.* 7. *Umbelliferous.* 8. *Caryophyllous.* 9. *Liliaceous.*
		Irregular.	10. *Papilionaceous.* 11. *Anomalous.*
		Compound.	12. *Flosculus.* 13. *Semi-Flosculus.* 14. *Radiated.*
FLOWERS APETALOUS.			15. *Apetalous, with stamens.* 16. *Apetalous, without stamens.* 17. *Apetalous, without visible flowers or fruit.*

TREES.

FLOWERS APETALOUS.			18. *Trees apetalous.* 19. *Trees amentaceous.*
FLOWERS PETALOUS.	Corollas monopetalous.		20. *Trees with monapetalous flowers.* 21. *Trees with rosaceous flowers.*
	Corollas polypetalous.		22. *Trees with papilionaceous flowers.*

After having derived from the corolla the distinctions of classes, Tournefort subdivided these into orders, or as he called them, sections. These orders were founded upon observation of the *pistil, calyx, fruit,* &c.

The first step in this classification, or the separation of shrubs and trees, was wrong. The distinction between a small tree and shrub, cannot be accurately settled. Two circumstances were, by Tournefort, relied on, as a foundation for this distinction; first, that shrubs do not form buds for the future year; and secondly, the difference in size of trees and shrubs. With respect to the formation of buds, the distinction is not found to be invariable, as some shrubs do form buds, and some trees do not. As to size, the variation, even in the same species, is such, in different soils and situations, that it cannot be admitted as a mark of distinction.

Different species, even in the same genus, sometimes differ in their stems; some being woody and others herbaceous. Neither is the form of the corolla to be depended on; even in the most natural families of plants, we find flowers of different forms, as in different species in the natural order Solaneæ, where the mullein is wheel-form, the tobacco funnel-form, and the atropa bell-form.

System of Linnæus.

We shall not now attempt to give a full view of the system of Linnæus, as we are hereafter to consider it in detail. We introduce it here merely to compare it with other modes of classification. The removing of plants which are nearly allied in their natural character, to different classes, by means of any artificial principle of classification, ought as far as possible to be avoided; and although the system of Linnæus, as you will find, when we compare it with natural fami-

Synopsis of Tournefort's method—Orders—Defects in Tournefort's classification —Difficulty of determining between trees and shrubs—System of Linnæus not entirely perfect.

11*

lies, is not wholly free from this confusion, it is much more so than any other, which has been invented.

Although we do not now receive the method of Tournefort for practical uses, a knowledge of it may extend your views of botanical science. When we accustom ourselves to take but one view of a subject, we are in danger of acquiring a contracted mode of thought. We are not to suppose that the system of Linnæus is perfect; but may well imagine that men of science will arise, who shall discover principles now hidden, and look back upon what they will call the very imperfect state of our sciences. We should rejoice that the human race is thus destined to a degree of improvement beyond our highest powers of calculation. "What should we think of a savage," says an elegant writer,* "if, in the pride of his ignorance, he was to conceive his own thoughts and feelings to be the noblest of which the human intellect is capable? And perhaps even the mind of a Newton, is but the mind of such a savage, compared to what man is hereafter to become."

The system† of Linnæus has, in its principal features, been laid before you.‡ This system not only includes within it all known plants, but is founded on such principles as must comprehend within it whatever plants may yet be discovered. Its author believed that no plant was destitute of stamens and pistils; but at the same time, that there were species in which these organs were so small, so obscure, of such a singular formation, as to render it difficult, and sometimes impossible, to be certain of their existence, except by the principle of analogy. Therefore, he made the two grand divisions of plants, *Phenogamous*, such as have stamens and pistils *visible*, and *Cryptogamous*, stamens and pistils *invisible*.§

You must not forget, that species, genus, order, and class, are mere abstract terms, denoting certain distinctions which would equally have existed, although we had never observed them, or given them names.

An *Individual* is an organized being, complete in its parts, distinct and separate from all other beings. An oak, a rose, and a moss, are each of them individuals of the vegetable kingdom.

A *Species* includes such individuals as agree in certain circumstances of the roots, stems, leaves, and inflorescence. We have no reason to suppose that any new species, either of animals or vegetables, have been produced since the creation. We sometimes see *varieties* in plants made by cultivation; the stamens and pistils from excess of nourishment, expanding into petals. Varieties are also occasioned by strewing the pollen from one species, upon the stigma of another; but such plants do not produce perfect seed, and therefore cannot reproduce themselves. Colour, taste, and size, are not considered as marks of specific difference.

* Dr. Thomas Brown.
† *System* differs from *method* in having but one single primitive character, and in founding its principal divisions upon the consideration of only one single organ or principle. Linnæus founded his system upon the consideration of the *stamens* as more or less numerous, upon their proportion, connexion, and their absence. Newton founded his system of Natural Philosophy upon *attraction*. The *vital principle* is the foundation of all systems of Physiology. *Method* is not confined to the consideration of one character; it employs all such as are conspicuous and invariable.
‡ See Part I, page 24.
§ Mirbel believes there are some plants absolutely destitute of stamens and pistils; these he calls *agamous*.

Advantages of taking different views of a subject—Human mind destined to progressive improvement—What is the difference between system and method? See note—What is said of the system of Linnæus—The terms species, genus, &c.—What is an individual?—What is a species?

A *Genus* comprehends one or more species, grouped together on account of some resemblance in situation, proportion, and connexion of the organs which constitute the flower. Any one species of a genus may be regarded as a type or example of the others; we may easily refer species which we have not studied to their proper genus, by a knowledge of any one species of that genus. Some genera appear to be distinctly marked by nature; the various species of the rose form a beautiful genus which is known to all, although every one might not be able to describe it to others in such a manner as to be understood; it is chiefly distinguished by its urn-shaped and fringed calyx.

The generic names of plants are derived from various circumstances; in some cases from a peculiarity in the form or colour of the corolla, or some property of the plant; and some are named from distinguished persons. Thus Iris (flag) is named from Iris the rainbow, on account of its various shades of colour. Digitalis (foxglove) is named from *digitus*, a finger, on account of the shape of its corolla, which is like the finger of a glove. Convallaria (lily of the valley) is named from a Latin word *convallis*, signifying, in the valley. Teucrium (germander) is named in honour of Teucer, a Trojan prince. The English name, germander, is supposed to have originated from the word Scamander, the name of a river of ancient Troy. The name of the great Linnæus is commemorated in a beautiful and modest flower, called the *Linnæa borealis*.[*]

Specific names are *adjectives;* generic names are *nouns*. The specific name sometimes indicates the number of leaves, as ORCHIS *bifolia*, (bifolia, signifies two leaves;) or the colour of the corolla, as VIOLA *tri-colour*, (three-coloured violet;) or the form of the root, as SOLANUM *tuberosum*, (with a tuberous root.) Specific names are often derived from the names of persons; thus a species of Origanum is named *tournefortii*, after its discoverer Tournefort.

The system of Linnæus may be illustrated by the following comparison;—as,

Individual persons compose		Families,
Families	"	Towns,
Towns	"	Counties,
Counties	"	States.
Individual plants compose		Species,
Species	"	Genera,
Genera	"	Orders,
Orders	"	Classes.

Thus, as individual persons are the real existences which make up a state; so are individual plants the real existences which compose classes; the words town and county, genus and order, being general terms used to designate certain circumstances of these men and plants.

Natural Families.

After having analyzed a number of plants, you will begin to observe a striking resemblance in many genera, and your own minds will suggest the propriety of arranging them into groups, without any reference to the artificial class or order where they may have

[*] *Borealis*, signifying *northern*, has reference to the situation of the country which gave birth to Linnæus. The Linnæa borealis is not uncommon in New England, and has been found on an island in the Hudson, near Troy.

What is a genus?—A knowledge of one species enables us to recognise all other species of the same genus—Derivations of generic names—Iris—Digitalis—Teucrium—Linnæa borealis—Specific names—Natural families.

been placed. We thus form *natural families*. If the whole vegetable kingdom could thus be distributed into natural tribes, we should need no artificial system. But after selecting a few families, which exhibit striking marks of resemblance, we find genera whose relation to other genera seems doubtful or obscure, and at length find a vast number of plants which seem to have few natural affinities with any other.

Among resemblances which gives rise to natural families, are,

    1st, resemblance in seeds,

    2d, in pericarps, or the envelopes of seeds,

    3d, in stamens and pistils,

    4th, in corollas and calyxes,

    5th, in the modes of infloresence, or the manner in which the flowers grow together upon the stalks,

    6th, in leaves,

    7th, in roots and stems.

In order to form a correct idea of the *natural methods* of classification, it is necessary to observe many plants, and the most constant characters of their organs. To find the place of plants in the artificial classes and orders, it is only necessary to observe the distinctions of the stamens and pistils.

The physician is chiefly conversant with the natural characters of plants, especially with such as are connected by medicinal qualities; he considers one group as *narcotics ;* another as *tonics ;* another as *stimulants*, &c.

The natural method depends for its utility, much upon the artificial system, which enables the student to ascertain the name of a plant, and thus learn its place among the natural orders. For example; suppose that a person meets with the plant commonly called stramonium, and wishes to know its character; by the Linnean System, he soon learns its botanical name, *Datura ;* and this genus he finds belongs to the natural order, *Solaneæ*, characterized by qualities of an active and deleterious nature, as the Tobacco, Foxglove, &c.

The experienced botanist is not always obliged to refer to the artificial system for the natural character of an unknown plant. Being familiar with the characteristics of the different families, he can often determine at once by the habit or general appearance of the plant that it belongs to the lily tribe (*Liliaceæ*,) to the mallows tribe (*Malvaceæ*,) to the wild turnip tribe (*Aroideæ*,) or to other of the conspicuous and well-defined natural orders or families.

To Linnæus belongs the honour of having first suggested the arrangement of plants into natural orders.

He published in 1738 what he modestly termed "Fragments of a natural method," consisting of 58 orders, founded upon the resemblance of plants in their habits, general appearance, or medicinal qualities.

The most popular Natural method is that of Jussieu, a botanist of Paris, improved by De Candolle of Geneva.* The characters employed in this method, are,

    1. *The structure of the Seed, with respect to cotyledons.* A plant

* Professor Lindley of England, has recently published a work on the natural system, which is deservedly popular.

Resemblances which give rise to them—Physicians interested in the natural method—Connexion between the natural and artificial methods—Experienced botanists know plants by their habits—Natural method of Linnæus—Method of Jussieu—What are the characters employed in Jussieu's method ?—How is the structure of the seed considered ?

having no cotyledon is called, *A-cotyledonous*, with *one*, *Mono-cotyle-donous*, and with *two*, *Di-cotyledonous*.

2. *Insertion of the Stamens*. The *stamens* are *above* the germ, *under* the germ, or *around* the germ; in the 1st case, they are *Epi-gynous*, 2d, *Hypo-gynous*, 3d, *Peri-gynous*.

3. *Absence and presence of the Corolla*. *A-petalous*, corolla wanting, *Mono-petalous*, corolla of one piece, *Poly-petalous*, many petals.

4. *Union, or separation of Stamens and Pistils*. *Mono-clinious*, stamens and pistils on the same corollas, *Di-clinous*, stamens and pistils on different corollas.

5. Union or separation of *anthers*. Anthers *distinct*, or anthers *combined*.

Synopsis of Jussieu's Method.

ACOTYLEDONS, - - - - - - - - -				CLASS 1.
MONOCOTYLEDONS,		Stamens *hypogynous*,		2.
		" *perigynous*,		3.
		" *epigynous*,		4.
DICOTYLEDONS.	*apetalous*.	Stamens *epigynous*,		5.
		" *perigynous*,		6.
		" *hypogynous*,		7.
		Corolla *hypogynpus*,		8.
	monopetalous.	" *perigynous*,		9.
		" *epigynous*,	anthers combined,	10.
			anthers distinct,	11.
	polypetalous.	Stamens *epigynous*,		12.
		" *hypogynous*,		13.
		" *perigynous*,		14.
	diclinous.			15.

These classes were at first formed of 100 orders; under the present modifications of Jussieu's method they have been multiplied, by establishing new orders from genera which seemed not to belong to any of the former established orders.

The *acotyledons* include the cryptogamous plants of Linnæus. They are also called *cellulares*, from their being formed of cellular tissue without a vascular system. These are by some botanists called *flowerless plants;*[*] their leaves are destitute of veins. They have no seeds with cotyledons, but are reproduced from a powder-like substance, exhibiting nothing of the parts which constitute the seeds in the other divisions of the vegetable kingdom.

The *monocotyledons*, which consist principally of grasses, palms, and liliaceous plants, are *endogenous* as regards the structure of their stems and branches;—the veins in their leaves, instead of being *reticulate*, or spreading out in various directions like a net, are straight and parallel. This division consists of two large groups;—1st, plants whose flowers have petals, called *Petalloidæ*, as the iris and lily; the calyx and corolla being in three or six divisions;—2d, where, instead of a proper calyx and corolla, the stamens and pistils are surrounded with glume-like bracts; these are called *Glumaceæ;* as in the grasses.

The *dicotyledons* include all the phenogamous plants, except those which belong to the monocotyledonous division. These are *vascular*

* It was long asserted by botanists, that every plant had a flower, although it might be invisible; but the term flowerless is now adopted by many for the cryptogamous family.

in their structure, *exogenous* in their mode of growth, and their leaves are distinguished by branching, reticulate veins.

Comparison of the Methods of Tournefort, Linnæus, and Jussieu.

We have now presented the pupil with the outlines of three modes of classification, exhibiting the plant under a variety of aspects, calculated to give general and extended views of the subject, and at the same time impress the mind with a few important distinctions.

Tournefort dwells chiefly on different aspects and circumstances of the *corolla ;*—Linnæus, of the *stamens* and *pistils ;*—Jussieu, of the *cotyledons* and *insertion* of the stamens.

Of the comparative merits of these methods, we would observe, that Tournefort's cannot be relied on, because the forms of corollas are often indefinite, and vary into each other; that of Jussieu appears too abstract to be used independently of the aid of some more simple method ;—the *number of cotyledons*, though a definite and important character, cannot, in many cases, be determined without the slow process of waiting for the seeds to germinate ;—the *insertion* of stamens and of the corolla often appears doubtful, even to the experienced botanist. Much as this method has been admired, it is but little used; while, on the contrary, that of Linnæus has, for more than half a century, been regarded as the key to botanical knowledge.

The characters used in his system are very apparent; and as it refers to the *number* of parts, rather than to their *forms* or *insertion*, it offers to the mind something positive, which is not found either in the method of Tournefort, or that of Jussieu. Between a corolla bell-form, or funnel-form, there are many intermediate forms, which may be as much like one as the other. The insertion *over* the germ, or. *under* the germ, are distinct, but the insertion *around* the germ sometimes blends with one, sometimes with the other mode. But between one or two *stamens*, or one or two *pistils*, there is no intermediate step, or gradual blending of distinctions, which leaves the student in doubt whether the case before him belongs to the one, or the other.

LECTURE XXII.

CHARACTERS USED IN CLASSIFICATION.

LINNÆUS, in his "Philosophy of Botany," established three kinds of characters to be used in the description of plants.

1st. *Factitious* (or *made*.) That which is, by agreement, taken as a mark of distinction; thus, certain circumstances with respect to stamens and pistils are fixed upon for distinguishing classes and orders. Although nature has formed these organs, the arrangement of plants by their means is an invention of man, or *artificial*.

2d. *Essential Character*. That which forms a peculiar character of one genus, and distinguishes it from all other genera.

3d. *Natural Character*. This is difficult to define, though it is that which is understood by all; it is the general aspect and appearance of the plant, which enables all persons to make a kind of arrangement of plants in their own minds, although they would find it very difficult to explain their reasons for this classification to others. It will appear, from this definition of natural characters, that in some respects, the method of Jussieu is no less artificial than that of Linnæus, since it depends upon particulars which can only be learned and understood by the aid of science; and we must admit that the genera which its orders exhibit, are often as unlike, in habit and properties, as are those which compose the classes of Linnæus.

It is by their natural characters, that persons who have never, perhaps, heard of such a science as zoology or the classification of animals, are enabled to distinguish ferocious beasts from domestic and gentle animals; they see a sheep or cow without any terror, although that individual one they may never have seen before; for nature teaches them to consider that as resembling other sheep and cows, which they know to be inoffensive. This natural character teaches savages to distinguish among the many plants of the forest, those which may administer to their wants, and those which would be injurious.

Even the lower grades of animals have this faculty of selecting by natural characters, nutritious substances, and avoiding noxious ones; thus we see the apparently unconscious brutes luxuriating in the rich pastures prepared for them by a benevolent Creator, and cautiously passing by the poisonous weed, directed by an instinct given them by this same Almighty Benefactor.

A *natural family* is composed of several genera of plants which have some common marks of resemblance, and its name is usually founded upon this general character; as *Labiate* and *Cruciform*, which are derived from the form of the corollas; *Umbellate* and *Corymbiferous*, from the infloresence; *Leguminous*, from the nature of the fruit. In many cases the family takes its name from a conspicuous genus belonging to it; as the *Rosaceæ*, or rose-like plants; *Papaveraceæ*, or poppy tribe, from *Papaver*, the poppy.

Natural families or orders resemble artificial orders in being composed of genera, but the principles on which these are brought together differ widely in the two cases.

In the truly natural families, the classification is such as persons who have never studied botany, might make; thus, dill, fennel, caraway, &c., belong to the Umbellate family, on account of the form in which the little stalks, bearing the flower, and afterward the seed, branch out from one common centre, like the sticks of an umbrella; this general resemblance being observable by all, it seems very natural to class such plants together.

But in the artificial orders, genera which may be very unlike in other respects, are brought together, from the single circumstance of plants having the same number of stamens and pistils. Thus, in the first order of the eighth class, we have the tulip and the bulrush, the lily of the valley and the sweet flag. In the second order of the fifth class, we have the beet and the elm. You will at once perceive the striking disparity between these plants, and that an arrangement, which thus brings them together, is properly called an artificial method.

Many families of plants possess a marked resemblance in form

Why is the method of Jussieu no less artificial than that of Linnæus?—Animals distinguished by natural characters—Savages distinguish plants by these characters—Animals capable of discerning these natural characters—What gives name to a natural family of plants?—In what respect do natural families resemble artificial orders?—How do they differ?—Why may natural families be formed without a knowledge of botany?—Genera in the artificial orders brought together by having the same number of stamens and pistils.

and qualities, and appear evidently as distinct tribes.   If the whole
of the vegetable kingdom could thus be distributed into natural
classes, the study of Botany would be much simplified ; but it has
already been remarked, that there are many plants which cannot be
thus arranged, and no principle has yet been discovered for systematic arrangement which bears any comparison to the Artificial System.   This system may be compared to a *dictionary ;* though by its
use we do not at first find the name for which we seek, and then
learn its definition, as we do in dictionaries of terms ; but we first
learn some of the *characters* of a plant, and with these as our guide,
we proceed to find the name.   Having ascertained the botanical
name, we can easily find to what natural family a plant belongs,
and thus learn its habits, medicinal use, and other important particulars.   The natural method may be considered as the *grammar*
of botany ; for between this, and the artificial system, the same relation exists, as between the grammar and dictionary of a language ; it would be idle to attempt to decide on their comparative
merits, since both are essential to science.

As the subject of classification is so important to a knowledge of
botanical science, we will now consider the general principles on
which it depends.

Rules.

1st. *All botanical classification results from an examination and
comparison of plants.*

2d. *Every organic distinction which establishes between individuals
any resemblance, or any difference, is a character ; that is, a sign by
which they may be known and distinguished.*

3d. *The presence of an organ, its different modification and its absence, are so many characters.*

4th. *The presence of an organ furnishes positive characters, its absence negative characters.*

Positive characters offering means of comparison, show the resemblances and differences which exist between individuals ; those
plants in which these characters present but slight differences should
be collected in groups ; those in which these characters differ more
sensibly, should be separated ; here we follow strictly the laws of
the mind.   But *negative* characters, as they allow no comparison,
can only be employed to separate individuals, and never to bring
them together.

When we say that plants have seeds with one or two cotyledons ;
that they have monopetalous or polypetalous flowers, and are provided with stamens and pistils, we point out particulars where visible
and striking resemblances may be observed ; these *characters*, then,
are positive, since they are founded on something real.

When we say that some plants are destitute of cotyledon, corolla,
stamens or pistils, we do not establish any real basis for the foundation of a comparison.   If we wish to separate plants with monopetalous corollas, from such as have polypetalous corollas, this single
character establishes, at once, the *difference*, which exists between
the two groups, and the *resemblance*, which exists between individuals of each group.   Thus positive characters possess a great advantage over negative ones ; the latter should never be employed

Artificial system of arrangement compared to a dictionary—First learn the characters, then the name—The natural method considered as the grammar of botany—
Mention the first four rules which are given for classification—Positive and negative
characters—Give illustrations of these characters, with their uses—Advantage of positive characters over negative.

when the former can be used; and in proportion as positive characters can be substituted for negative, the science of botany will be perfected.

Positive characters can only be founded upon evident *facts*, and never upon a *presumption of the existence of facts*, derived from analogy. For it is contrary to true philosophy, to suffer hypothetical reasoning to usurp the place of direct observation of facts.

5th. *Positive characters are constant or inconstant.* All seeds produced by plants of the same species have the same structure; all plants which grow from these seeds produce other seeds, similar to those from which they have had their origin; of course the characters derived from the structure of these seeds are *constant.* But among these plants some are large and others small; some may have white corollas, some red, or blue; some are more fragrant than others; of course, *size, colour*, and *odour*, offer *inconstant* characters.

6th. *All real science in Botany must rest upon constant characters;* therefore, these characters are much more important than the others.

7th. *Constant characters may be isolated or coexistent.* The petals of the RANUNCULUS *acris*, (butter-cup,) have a nectary in the form of a scale; this character, although constant, is *isolated,* for it is not necessarily connected with any other characteristic trait. The calyx of the *campanula rotundifolia*, (blue-bell,) adheres to the germ; the germ must of necessity be simple, or without divisions, and the corolla and stamens attached to the interior of the calyx. The character of the adherence, of the calyx to the germ, brings in its train several other characteristics; it is then *coexistent;* and is more important than the isolated character.

8th. *Two orders of characters are derived from the two great divisions of vegetable organs; those of vegetation and reproduction. The characters of reproduction are numerous and often coexistent; one character serving as an index to many others.*

It is seldom that plants which resemble each other in their characters of reproduction, differ much in their characters of vegetation. For example; all plants which have four *didynamous** stamens, attached to a monopetalous, labiate corolla, and four seeds lying uncovered in a monophyllous calyx, have an angular stem and opposite leaves. On the contrary, it frequently happens that plants which resemble each other by the characters of vegetation, differ by those of reproduction. Labiate and caryophyllous plants agree in having their leaves opposite, and yet there is no resemblance in their flowers. This consideration alone, would seem sufficient for establishing the superior importance of the characters of reproduction over those of vegetation. The seed unites in itself the characters both of reproduction and vegetation. The embryo is the commencement of the new plant, and it offers us the first characters of vegetation; but its situation in the fruit, the number, form, and consistence of its envelope, are characters which belong to fructification.

In separating or bringing together plants, we should, as far as possible, make use of prominent characters which the eye can see without the help of the microscope; but if experience teaches us that the characters most constant and proper for the explanation of physiological phenomena can only be discovered by such aid, it is

---

* That is, two long and two short stamens.

necessary to resort to this instrument, in order to establish the natural relations of plants.*

Having considered the meaning of *individual, species, genus,* and *family,* and of the characteristics by which these are grouped together, let us take a general view of the subject. It is evident, by the formation of species, genera, and families, that every *species* should offer the essential characters of the family and genus to which it belongs; while the marks which distinguish this species from another species of its genus, will be such as do not belong to the whole genus or family. The *different genera* in families are also distinguished by characters which do not belong to the whole family; every individual, then, will possess its *specific* character, its *generic* character, and its *family* character.

The specific character is less important than the generic, as it is mostly founded on the characters of the organs of vegetation, which we have seen are isolated, and less important than the coexistent characters. We often find, in the analysis of plants, a great difficulty in determining their species, from the want of definite marks of distinction.

Generic characters are mostly of the coexistent kind, and are more valuable than the specific characters. The distinctions of genera are usually much more apparent than those of species; as a rose can be more easily distinguished from a pink, than one species of rose from another species.

Families are grouped together by marks of resemblance found in genera. These family characters are, of all others, the most important. In the artificial classes and orders we depend on what we have before termed *factitious* characters. In species, genera, and families, the essential characters are also natural characters.

LECTURE XXIII.

USE OF BOTANICAL NAMES—ARTIFICIAL CLASSES AND ORDERS CONSIDERED IN GROUPS—CLASSES MONANDRIA AND DIANDRIA.

You have been taught the principles on which the Linnæan system is founded; we shall now examine each class separately, with the orders it contains, and the most remarkable plants and natural families which we shall meet with in our progress through this system.

We have observed, that this appears to be the best method yet discovered of classing new plants, and of ascertaining the botanical names of those which are already known by common names. If, in all countries, the common names were alike, there would be no need of any other; but the names of plants vary in different languages as much as other terms. Even in the same country, and often in the same neighbourhood, the common names of plants are different; but botanical names are the same, in all ages and coun-

---

* The foregoing rules and observations respecting characters for classification, are chiefly translated from Mirbel's " *Elemens de Botanique.*"

General view of the subject of classification—Which is the more important, the specific or generic character?—Why are generic characters most valuable?—How are families grouped together?—On what do artificial orders depend?—What are the essential characters in species, genera, and families?—Why are not the common names of plants sufficient for all purposes?

tries; without this uniformity no permanent improve\
made in the science.

Botanical names are chiefly taken from the Greek and
being the common languages of the learned world. A
botany were, for a long time, written in Latin;—the ori
of Linnæus are in that language. Although it is impoi ___ to the
interests of science that there should be such a medium, by which the
learned may communicate, it is also highly important to the general
improvement and happiness of mankind, that their discoveries should
be made accessible to all;—it would be useless to attempt to divest
botany of all its technical terms, and names borrowed from the dead
languages; in doing this we should destroy the science, and introduce confusion in the place of order. But such facilities are now
offered, that every young person can easily become acquainted with
the grand outlines of the vegetable world;—and, oh, how much are
the beauties of nature enhanced, when viewed with the eye of a
philosopher, and the emotions of a Christian!

Groups of Classes and Orders in the Linnæan System.

1st. *The first ten classes* are founded upon the *number* of stamens.
2d. *Eleventh* and *Twelfth*, upon the *number* and *insertion* of stamens.
3d. *Thirteenth* and *Fourteenth*, upon *number* and relative *length* of stamens.
4th. *Fifteenth, Sixteenth, Seventeenth,* and *Eighteenth*, upon *connexion* of stamens
by filaments or anthers.
5th. *Nineteenth* and *Twentieth*, upon *position* of stamens, relative to the pistil.
The *Twenty-first* class includes all plants which either have not stamens and pistils, or in which these organs are too minute to be seen without the help of a micro
scope.
The Orders are founded,
1st. Upon the number of pistils.
2d. Upon the seeds being covered or uncovered in the calyx.
3d. The relative length of the pods.
4th. The comparison between the disk and ray-florets of compound flowers.
5th. Number of stamens.
6th. The orders of the class Cryptogamia are distinguished by natural family characters.

Names of the Artificial Classes.

1. MONANDRIA, one stamen.
2. DIANDRIA, two stamens.
3. TRIANDRIA, three stamens.
4. TETRANDRIA, four stamens.
5. PENTANDRIA, five stamens.
6. HEXANDRIA, six stamens.
7. HEPTANDRIA, seven stamens.
8. OCTANDRIA, eight stamens.
9. ENNEANDRIA, nine stamens.
10. DECANDRIA, ten stamens.
11. ICOSANDRIA, over ten stamens, situated on the calyx.
12. POLYANDRIA, over ten stamens, situated on the receptacle.
13. DIDYNAMIA, four stamens, two long and two short, flowers *labiate.*
14. TETRADYNAMIA, six stamens, four long and two short, flowers *cruciform.*
15. MONADELPHIA, stamens united by their filaments into one set.
16. DIADELPHIA, stamens united by their filaments into two sets, flowers *papilionaceous.*
17. SYNGENESIA, five stamens united by their anthers, flowers *compound.*
18. GYNANDRIA, stamens growing on the pistil.
19. MONŒCIA, stamens and pistils on different flowers of the same plant.
20. DIŒCIA, stamens and pistils on different flowers of different plants.
21. CRYPTOGAMIA, stamens and pistils invisible.

Why are botanical names taken from the Greek and Latin?—Why cannot all the
terms in botany be translated into common language?—Repeat the distinctions in
the groups of the Linnæan classes?—On what are the orders founded?—Repeat the
names and characters of the artificial classes.

CLASS I.—MONANDRIA.

Order Monogynia.

Fig. 121.

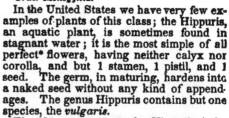

In the United States we have very few examples of plants of this class; the Hippuris, an aquatic plant, is sometimes found in stagnant water; it is the most simple of all perfect* flowers, having neither calyx nor corolla, and but 1 stamen, 1 pistil, and 1 seed. The germ, in maturing, hardens into a naked seed without any kind of appendages. The genus Hippuris contains but one species, the *vulgaris*.

Fig. 121, *a*, represents the Hippuris;† the stem is *erect* and *simple;* the leaves are *linear, acute,* and arranged in *whorls.* At *b*, is the flower of the Hippuris, showing an egg-shaped germ; a short filament crowned with a large anther composed of two lobes; the style is long and awl-shaped; the stigma is acute and inconspicuous; the germ is crowned by a border which resembles the upper part of a calyx.

The Marsh-samphire, (*Salicornia herbacea*,) with a bushy stem about a foot high and flowers in a short spike, grows in salt marshes near the sea-coast. It has a saltish taste, and is used for pickling It has been supposed that this was the plant alluded to by Shakspeare in his description of the cliffs of Dover :

> " How dreadful,
> And dizzy 'tis to cast one's eyes so low !
> Half way down,
> Hangs one that gathers Samphire : dreadful trade !"

It is probable, however, that the poet here refers to the Sea-Samphire, (*Crithmum maritimum*,) whose habit it is to grow on rocks near the sea; this, according to English botanists, is still found upon the Dover cliffs, from which those who gather it are let down in baskets. The Salicornia is found in great quantities on the coasts of the Mediterranean, where it is burned, and its ashes used in the manufacture of soda. It is also found at Onondaga Salt Springs, and on the sea-coast in North America.

Although the plants of this class are so very limited in the northern countries, some of the most valuable vegetable productions of the tropical regions are found here. The Arrow-root,‡ (*Maranta arundinacea*,) received its name from having been used by the Indians of South America, to extract the venom from wounds made by their poisoned arrows; from its roots, a substance is obtained, resembling starch, which is valued as nutritious for the sick. The *Curcuma*, sometimes called the Indian Crocus, furnishes from its root the *turmeric* imported from the East Indies; it is remarkable for the peculiar yellow colour of its bark, and is valuable as a chemical test of the presence of alkalies. It is an ingredient in the *curry*-powder.

The ginger, whose root is so extensively used in cooking and in medicine, was first known to the Arabians, and called by them Zinziber, which is now generally received as its generic name, though

---

* Although so destitute of other organs, it is called perfect, because it has stamens and pistils.

† See also Appendix, plate vi. fig. 7.   ‡ See Appendix, plate iii. fig. 4.

How many orders in the class Monandria?—Describe the Hippuris—Fig. 121—Marsh-Samphire—Arrow-root—Ginger.

Linnæus called it Ammomum. It belongs to the Natural Order *Cannæ*, which embraces several genera of aromatic plants. The distinguishing marks of this natural family are an herbaceous stem, very broad leaves, a germ with three corners, and a liliaceous flower which is beautiful and fragrant.

The red valerian (VALERIANA *rubra*) having but one stamen would belong to this class, but as other species of this plant have three stamens, this species is carried with the majority into the class Triandria.

Order Digynia,

Contains an American plant, BLITUM. At Fig. 121, *c*, is a flower of this genus; its calyx is deeply three-parted; it has no corolla; the germ resembles a berry, and is crowned by two styles, which give the plant its place in the order Digynia.

CLASS II.—DIANDRIA.

Order Monogynia.

Fig. 122.

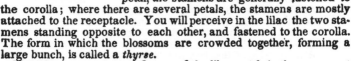

This, though more extensive than the preceding class, is somewhat limited. We can however, without difficulty, find examples for its illustration.

The lilac (Syringa) is cultivated in all parts of our country, and is exceeded in beauty and fragrance by few ornamental shrubs. The corolla is *salver* form, or with a tube which spreads out into a flat, four-parted border. You might, at first view, suppose the corolla to consist of several petals, but if you attempt to pull them out, they will all come off together, and you will plainly perceive there is but one piece, or that it is monopetalous. In flowers of one petal, the stamens are generally fastened to the corolla; where there are several petals, the stamens are mostly attached to the receptacle. You will perceive in the lilac the two stamens standing opposite to each other, and fastened to the corolla. The form in which the blossoms are crowded together, forming a large bunch, is called a *thyrse*.

Fig. 122, *a*, represents a flower of the lilac; at *b*, is the same, cut lengthwise to show the two stamens.

The lilac, although so common with us, is an exotic; the species most cultivated are the *vulgaris* or common, which has heart-shaped leaves, and the *persica*, or Persian, with narrower leaves.

The Jasmine, of which twenty-eight species are said to have been discovered, is an exotic of this class. The prim or privet (*Ligustrum*) is found growing wild in some parts of New England; though, in general, it is seen but little in the United States, except when cultivated. In England it is planted for fences; as it grows rapidly, it soon becomes useful for this purpose, and with its green leaves and white flowers, gives to the farms an air of neatness and taste.

The Sage, (*Salvia*,) on account of the form of the corolla, belongs to the natural family of the *labiate* flowers; these are, mostly, placed in the class Didynamia, having four stamens, two long and two short; but in some cases, the labiate flowers have but two stamens; this circumstance, according to the rules of classification, separates them from their natural family, and brings them under the class we are now considering. You may understand this better, if we compare it to taking a person from his relations, to place him among strangers. But this evil must sometimes be borne for the sake of some attendant good; we are also obliged to submit to the necessity of occasionally separating the flowers from their natural relations, because we cannot turn aside from our rules of classification to accommodate a few plants which deviate from the ordinary laws of nature. The sage seems to have made an effort to escape this misfortune, for it seems almost to have attained four stamens, by doubling its filaments, but two of these having no anthers cannot be considered as stamens; therefore the plant falls back into the second class, and is placed by the side of the lilac, to which it has no kind of resemblance, except in its two stamens. This plant, however, is not the only one of the labiate flowers which is removed from its natural family in the 13th class; for the rosemary and the mountain-mint accompany it into the second class; but these have not the two imperfect filaments which were remarked in the sage. The genus Salvia contains one hundred and fourteen species; the one most commonly cultivated with us is the *officinalis*, a shrub-like, perennial plant; to this we give more particularly the name of sage. Another species of the same genus is the *sclara*, called Clarry; this has larger and broader leaves than the common sage; it is cultivated for its medicinal properties.

A very small plant called Enchanter's night-shade, (*Circæa*,) may be found growing wild in shady places; it is a harmless, modest-looking plant, notwithstanding its name. It has a small white blossom, in the parts of which great uniformity as to number may be observed; it has two *stamens*, a corolla with two *petals*, a calyx with two *sepals*, capsule with two *cells*, each of which contains two *seeds*.

The symmetry of structure observable in the plant just described, is seen in many flowers; as those of two stamens often have this number in the other parts of the flower; the number is frequently doubled; as in the lilac, which has two stamens, and a four-parted corolla. In a plant with three stamens, the number three or six usually prevails in the divisions of the calyx, corolla, capsule, &c. A knowledge of this fact will assist you in determining the class of a plant; for example, if you have a flower whose calyx has five or ten divisions, and the corolla the same number, you may expect, if the flower is a perfect one, to find either five or ten stamens; or if the divisions of the flower be two, there will generally be two or four stamens; if three, either three or six stamens; if four, either four or eight stamens. The number five, as divisions of the calyx, corolla, and capsule, is generally united to five or ten stamens, and found in the fifth or tenth class.

Another native plant of the second class, is the *Veronica*. Of the seventy species which this genus is said to contain, no more than six o eight are common to North America. The Veronica and the Circæa both turn black when dried; although they do not add to the beauty of an herbarium, they are desirable in a collection of plants, as our country contains few specimens to illustrate the second class. At Fig. 122, *c*, is a representation of a flower of the Veronica; at *d*, is the *Circæa*.

Why is the sage removed from its place with the labiate flowers—Are there any marks of four stamens in the sage?—How many species of the genus Salvia?—What two are mentioned in particular?—Enchanter's night-shade—What is observed respecting the symmetry of structure in many flowers?—Veronica.

·Among the exotics of this order we find a singular plant, peculiar to the East Indies, the NYCTANTHES *arbor tristis*, or sorrowful tree; its boughs droop during the day, but through the night they are erect, and appear fresh and flourishing.

The Olive, (*Olea*,) is common on the rocks of Palestine; it may now, according to the accounts of travellers, be found upon the same spot which was called, eleven centuries before the Christian era, the mount of Olives, or mount Olivet.

Order Digynia.

In the second order of this class is the sweet scented spring-grass (ANTHOXANTHUM *odoratum*,) which is found in blossom in May; to this grass the pleasant smell of new made hay is chiefly owing; its odour is like that of clover. This plant is separated by the artificial system from the other grasses, on account of its having but two stamens. This is the kind of grass used in this country as a substitute for the Leghorn grass, in the manufacture of hats. The first hat of the kind was made a few years since by an ingenious female in the town of Wethersfield, Connecticut; since which time, many hats, not inferior to the best Leghorn, have been made from the same material.

The Catalpa, an elegant tree, with flat, *cordate*, or heart-shaped leaves, is indigenous to the Southern United States; its white flowers, striped with purple, grow in panicles similar to the Horse-chestnut. Only one species is found in North America.

Order Trigynia.

This order contains the genus PIPER, one species of which, the *nigrum*, is the common black pepper. The cayenne pepper belongs to the genus CAPSICUM, which is found in the eighth class. The flowers of the Piper genus have neither calyx nor corolla, but the fruit is borne on a spadix.

We have in this lecture remarked upon the use of botanical terms; we have considered the few groups into which the classes of Linnæus may be arranged, with the names of the classes, and the characters of each;—and have given a sketch of the two first classes, with some examples under each of their orders. In doing this, we have been obliged to pass by many plants which had an equal claim to notice, but as knowledge must be gained by the observation of particular cases, we have thus selected a few examples, in order that you may be prepared to examine the others with pleasure and advantage.

LECTURE XXV.

CLASS III.—TRIANDRIA.

Order Monogynia.

IN the first order of this class we find among our common exotics the Crocus, which is particularly interesting as being one of the earliest flowers of our gardens, not unfrequently blossoming in the neighbourhood of a snow-bank. It has a bulbous root, long and narrow leaves, a spatha, and six petals. Besides the CROCUS *vernus*, or spring crocus, which often appears even in our own climate as early as March, there is of this genus a very distinct species, the CROCUS *officinalis*, or the true saffron, which appears among the late flowers of autumn. The following beautiful lines, respecting these flowers, are from the pen of one* whose early and fervent piety, marked him as a fit inhabitant of a purer sphere;—a Christian philosopher, he could see an invisible hand directing the operations of nature.

> " Say, what *impels*, amid surrounding snow
> Congealed, the *Crocus' flamy bud* to grow ?
> Say, what *retards*, amid the summer's blaze,
> The *autumnal bulb*, till pale declining days ?
> The GOD OF SEASONS, whose pervading power
> Controls the Sun, or sheds the fleecy shower ;
> He bids each flower his quickening word obey :
> Or to each lingering bloom, enjoins delay."

The Iris, or Fleur-de-lis,† (pronounced by a corruption of the French language, *flower-de-luce*,) is very curious in its structure. It has no proper calyx, but a spatha; its corolla consists of six parts, alternately *reflexed*, or bent back, the pistil has three stigmas, which appear at first view like petals. The Iris is so named from Iris, the rainbow, on account of the various colours which it reflects, varying from different shades of purple, into blue, orange, yellow, and white. We have several native species of Iris, one of which, the common blue flag, is found in wet places. The flowers are purple, streaked with yellow; this is sometimes called Poison flag. The Crocus and Iris are found in the natural family of Jussieu called *Irideæ;* this family belongs to the division of monocotyledons, having sta-

Fig. 123.

mens around the germ, or *perigynous.* Linnæus calls the same plants *Ensatæ*, from the Latin word *ensis*, a sword, on account of the shape of their leaves, which are long, narrow, and pointed.

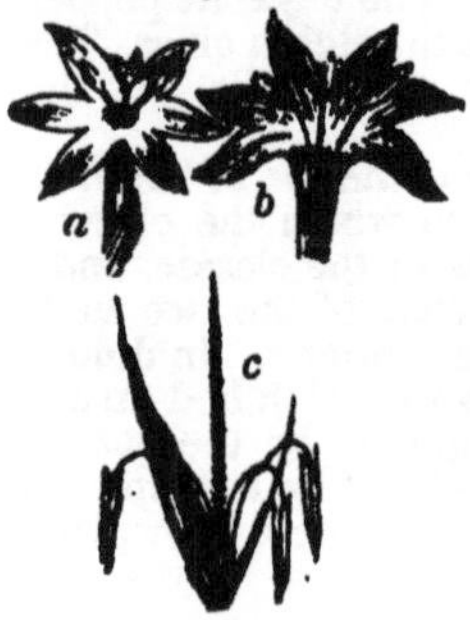

Fig. 123 represents the Ixia, (blackberry-lily;) *a*, is an entire flower; *b*, is the corolla cut lengthwise, to show the three stamens. The Ixia belongs to the same natural family as the Iris and Crocus. At *c*, is the flower of the matgrass, (*Nardus*,) having but one pistil; this is separated from the grass family, the greater part of which we shall meet with in the next order of this class.

Order Digynia.—The Grasses.

The 2d *Order* of the third class contains the family of the grasses, (*Gramina;*) they are distinguished by a straight, hollow, and jointed stem, or *culm;* the long and linear leaves are placed at each joint of the stalk, in alternate order, enclosing it like a sheath. The flower is found in what is called an ear or head; it consists of a corolla of two green husks, enclosed by a glume calyx of two husks or valves. These husks constitute the *chaff*, which is separated from the seed by an operation called thrashing.

These little flowers are also furnished with a nectary; they are green, like the rest of the plant, and you will need a microscope to

---

* Henry Kirke White.
† See Appendix, Plate vi. Fig. 6. At Plate vi. Fig. 5, is another plant of this class and order.

What is said of the Iris ?—In what natural families did Jussieu and Linnæus place the Crocus and Iris—Explain Fig. 123—Describe the grass family—The culm—glume.

view them accurately; they are best observed in a mature stage of the plant, when their husks being expanded, discover *three filaments*, containing each a large double anther; the *two pistils* have a kind of reflexed, feathered stigma.   They have no seed vessel; each seed is contained within the husks, which gradually open; and unless the seed is gathered in season, it falls to the ground.   This facility for the distribution of the seed is one cause of the very general diffusion of grasses.

The *roots* of grasses are fibrous, and increase in proportion as the leaves are trodden down, or consumed; and the stalks which support the flower are seldom eaten by cattle, so that the seeds are suffered to ripen.   Some grasses which grow on very high mountains, where the heat is not sufficient to ripen the seed, are propagated by suckers or shoots, which rise from the root, spread along the ground, and then take root; grasses of this kind are called *stoloniferous*, which means bearing shoots.   Some others are propagated in a manner not less wonderful; for the seeds begin to grow while in the flower itself, and new plants are there formed, with little leaves and roots; they then fall to the ground, where they take root.   Such grasses are called *viviparous*, which signifies producing their offspring alive, either by bulbs instead of seeds, or by seeds germinating on the plant.   The seeds of the grasses have but one lobe, or are not naturally divided into parts, like the apple seed and the bean; therefore these are said to be *monocotyledonous*.

The stems of gramineous plants, like those of all the monocotyledons, are of that kind which grow internally, or from the centre outward, and are therefore called *endogenous.*

With regard to the duration of the grass-like plants, some are *annual ;* as wheat, rye, and oats, whose roots die after the grain or seed is matured.   The meadow grasses are *perennial ;* their herbage dying in autumn, and the roots sending out new leaves in the spring.

The family of grasses is one of the most natural of all the vegetable tribes: the plants which compose it, seem, at the first glance, to be so similar, that it would appear impossible to separate them into *species*, much less into *genera ;* but scientific research and close observation present us with differences sufficient to form a basis for the establishment of a great number of genera.   The *essential character* of the oat (*Avena*) consists in the jointed, twisted awn or beard, which grows from the back of the blossom; the oat is also remarkable for its graceful *panicle.*   The rye (*Secale*) has two flowers within the same husk.   The wheat (*Triticum*) has three flowers within the same husk; the interior valve of the corolla of the wheat is usually bearded.   The filaments in the rye and wheat are *exsert*, that is, they hang out beyond the corolla; from which circumstance these grains are more exposed to injury from heavy rains than those whose filaments are shorter.

Perhaps, in the whole of the vegetable kingdom, although there are many plants of much greater brilliancy of appearance, there are none which are so important to man as the grass family.

Linnæus, who was distinguished for the liveliness of his fancy, no less than the clearness of his reasoning powers, seemed to delight in tracing analogies between plants and men : establishing among the

Filaments—pistils—Roots of grasses—Manner in which grasses are propagated—Seeds—How do the stems of the grasses grow?—What is said of the duration of grass-like plants?—What is remarked of the separation of the grasses into genera and species ?—Describe the oat, the rye, and wheat—What is said of the importance of the grass family?

former a kind of aristocracy, he called grasses, the plebeians of the
vegetable kingdom. To them, indeed, belong neither brilliancy or
appearance, nor delicacy of constitution; numerous, humble, and
rustic, and at the same time giving to man and beast the sustenance
necessary to preserve life, the grasses may well be compared to the
unassuming farmer, and mechanic, to whom society is indebted for
its existence and prosperity, far more than to the idle fop or bluster-
ing politician.

The grasses are supposed to include nearly one sixth part of the
whole vegetable world; they cover the earth as with a green carpet,
and furnish food for man and beast. Some of these, most valuable
as furnishing food for cattle, are herds-grass, (*Phleum pratense,*) .
meadow-grass, (*Poa,*) orchard-grass, (*Dactylis,*) and oats. Those
which are used in various ways as food for man, are wheat, rye,
barley, and Indian-corn; the latter botanically called ZEA *mays,* although of the natural family of the grasses, having a culm-like
stalk, and other distinguishing characteristics of grass-like plants,
is placed in the class Monœcia, because the stamens and pistils are
separated in different flowers, growing from the same root. The
styles, long, slender, and *exserted,* form what is called the *silk;* they
are thus favourably situated for receiving the fertilizing pollen which
is showered down from the staminate flowers.

The fruit of corn, wheat, rye, &c., is called *grain.* Grain, then,
consists of the seed with its pericarp; these are not easily distinguished from each other till the grain is ground into flour; the pericarp separating from the seed, then forms what is called the *bran;*
and the seed, the *flour* or *meal.*

The Sugar-cane (SACCHARUM *officinarum*)* is of the grass family;
it is supposed to have been brought from the south of Europe to the
West Indies. The stem or culm, which sometimes grows to the
height of twenty feet, affords the juice from which the sugar is made.

The Bamboo, (ARUNDO *bambos,*) of the East Indies, a species of
reed which is said to attain, in some situations, the height of sixty
feet, is also of this class.

The Sedge (*Carex*) is a gramineous plant, but it bears staminate
and pistillate flowers, and is therefore placed in the class Monœcia.
The carexes† constitute a very numerous family of plants.

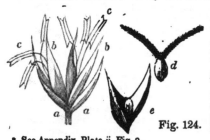

Fig. 124 represents two
magnified flowers of the
orchard grass, (*Dactylis
glomerata;*)‡ at *a,* is a
calyx§ composed of two
valves; these are *com-
pressed, keeled‖ acute;* one
valve is shorter than the
valves of the flowers, the
other longer; the calyx is
common to the two flowers;
b, shows the valves of the

Fig. 124.

* See Appendix, Plate ii. Fig. 2.
† The plural of *carex,* according to the Latin termination, is *carices.*
‡ Glomerata signifies a cluster, alluding to the crowded panicles of flowers.
§ The parts of the calyx, and also of the corolla, are sometimes called glumes; they
are all much alike in appearance, being merely a set of sheaths, for the purpose of
protecting the stamens: they are not distinguished by any difference in colour from
the leaves or stem. The anthers, which are usually yellow, are the only part of the
flower of the grasses which is coloured.   ‖ Resembling the keel of a boat.

What did Linnæus call the grasses?—Which are among the most valuable grasses
for cattle?—Which for the use of man?—What is said of Indian corn?—What is
grain?—Sugar-cane—Bamboo—Sedge—What does Fig. 124 represent?

corollas; they are *oblong* and *acute;* c, represents the stamens. which are three in each flower; the filaments are of the length of the corolla; the anthers are two-forked or *bifid; d,* is the pistil, having an egg-shaped germ, and two spreading and feathery styles; at *e,* is the seed, not having any proper pericarp, but enclosed by the two scales of the corolla; it is single and naked.

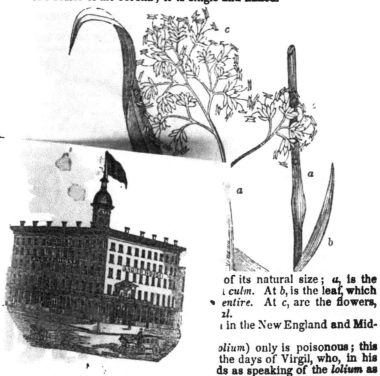

of its natural size; *a,* is the culm. At *b,* is the leaf which *entire.* At *c,* are the flowers, *nl.*

in the New England and Mid-

olium) only is poisonous; this the days of Virgil, who, in his ds as speaking of the *lolium* as

TRANDRIA.

number of stamens are found in the plants of this class, as in those of the 13th class, Didynamia. In the fourth class, the stamens are of *equal* length, but in the 13th, they grow in two pairs of *unequal* length. In this class we meet with no large natural family; the genera which compose it appearing little united by natural relations.

Order Monogynia.

As an example of this order, may be mentioned the HOUSTONIA *cærulea,* which is known by different common names; as *Innocence, Venus's Pride,* and *Blue Houstonia.* It is a very delicate little flower, appearing early in

Fig. 126.

* See Appendix, Plate iv. Fig. 6, for a representation of one of the grass tribe.

the spring, in grassy fields and meadows; the colour varies from
sky-blue (which gives its specific name *carulea*) to a pure white. It
has a small calyx, with four divisions, and a monopetalous corolla
of four divisions, which gives it the appearance of a cruciform plant.

The common Plantain, (*Plantago,*—see Fig. 126, *a,*) is found
here; it is a plant by no means useless, although it exhibits nothing
interesting to gratify the sight. The leaves are sometimes used in
external applications for medicinal purposes; they are also, when
young and tender, boiled and used for greens in some parts of the
United States. The flowers of the plantain grow on a spike; they
are very small, but each one has a calyx and corolla; these are four-
parted; the filaments are long, and the pericarp is ovate, with two
cells. Canary birds are very fond of the seeds of the plantain.

Aggregate flowers. We find in this class what Linnæus called the
aggregate flowers, such as have many flowers on the same receptacle; they have a general resemblance to the compound flowers i
the class Syngenesia, but differ from them in having but four sta
mens, with anthers separate, while the Syngenesious plants have
five united anthers. The aggregate flowers are not often yellow
like many of the compound flowers, but are usually either blue
white, red, or purple. The Button-bush, (*Cephalanthus,*) of about
five feet in height, affords a good example of this natural order. The
inflorescence is white, appearing in heads of a globular form, each
consisting of many perfect little florets; each head has its own 4-cleft
calyx, but there is no general calyx, or involucrum, for the whole
Only one species of this genus, the *occidentalis,**** is known, and this
is entirely confined to North America. The Teasel (*Dipsacus*)
belongs to the aggregate flowers; its inflorescence is in heads of the
form of a cone. The receptacle is furnished with narrow, stiff
leaves in the wild Teasel, (*sylvestris ;*) in the cultivated species, (*ful-
lonum,*) these bristly leaves are hooked, and are used by clothiers to
raise a nap or furze on woollen cloth. The *Cornus,* so called from
the Latin *cornu,* a horn, on account of the hardness of the wood, is
a genus composed mostly of shrub-like plants, with flowers growing
in flat clusters, or *cymes,* like the elder. The *florida,* a species of
Cornus, often called box-wood, sometimes dog-wood, is a beautiful
ornament of our woods. It may be considered either a large shrub
or a small tree; it grows from the height of fifteen to thirty feet. Its
real corollas are very small, and are clustered together in the manner which is called, in botany, an *aggregate.* This aggregate of
flowers is surrounded by that kind of calyx called an *involucrum,*
which, in this plant, consists of four very large leaves, usually white,
but sometimes of a pale rose-colour; to the latter circumstance is
owing its specific name *florida,* or florid. You would, no doubt, on
the first sight of this plant, mistake the large leaves of the involucrum for the petals. At Fig. 126, *b,* is a representation of a species
of the cornus; the style is about the same length as the petals; these
are four is number, oblong and equal.

At *c,* Fig. 126, is the *Cissus,*† or false grape; its calyx is very

* From *occidens,* the west, being found on the western continent.
† Mirbel thus names the plant whose flower is here described, and places it in the
class Tetrandria. Eaton describes it under the name of Ampelopsis, and places it in
the class Pentandria. Although it may occasionally be found with five stamens, its
four petals and four divisions of the calyx, seem to indicate that the fifth stamen is
but an accidental circumstance; this seems to have been the opinion of Mirbel and
some others.

small; petals spreading and reflexed; filaments shorter than the petals, and crowned with large cordate anthers.

Another very common genus in this class is the Bed-straw, (*Galium,*) an herbaceous plant, with very small white flowers; the leaves grow in whorls. In different species, the leaves thus clustered together stand around the stem in *fours, fives, sixes,* and *eights.* Some species exhibit a peculiar roughness upon the stems and leaves. This genus, with some others of a similar appearance, were arranged by Linnæus in a natural order, called *Stellatæ,** star-like plants; the leaves radiating from the stem, as rays of light from a star.

Among the exotics of this class are the SANTALUM, which produces the *sandal-wood,* and the Madder, (RUBIA *tinctoria,*) the root of which produces a beautiful scarlet colour. The latter plant is said to have the singular property of tinging, with its red colour, the bones of the animals that feed upon it. Jussieu has arranged this, and some of the plants whose leaves grow in whorls, under the order *Rubiaceæ.* The Silver-tree (PROTEA *argentea*) has soft leaves resembling satin, of a silver colour. Another species of *Protea,* the *aurea* or golden, has gold coloured leaves, which are edged with scarlet. Both these trees are natives of the Cape of Good Hope and have never been found in any other locality.

Order Digynia.

HAMAMELIS is a shrub from 6 to 12 feet high, and is found in woods throughout the United States. Its flowers are yellow, and grow in axillary clusters. You will often meet with this plant by the road-sides on the skirts of woods; and may know it from the fact o. its being in blossom after it has lost its leaves, in autumn, and even in winter. Its common name is Witch-hazel; it probably originated from the superstitious idea, which was long entertained, that a twig from this tree, called a divining rod, in the hands of particular individuals, had the property of being attracted towards gold or silver buried in the earth. Some botanists, however, ascribe the common name of this plant to its peculiarity, as to the season of blooming. By the subdividers of the Orders of Jussieu, viz. De Candolle and Lindley, this is taken from the order Berberides, and stands alone in an order, called from its generic name *Hamamelideæ.*

Order Tetragynia.

We find here the holly, (*Ilex ;*) this is an evergreen, with a smooth, grayish bark; shining, thorny leaves; whitish flowers; and scarlet berries; this plant is very common in England for fences; its verdure is not impaired by the most severe winter.

* From *stella,* a star.

Bed-straw—What plants are placed in Linnæus's natural order *Stellatæ,* and Jussieu's order *Rubiaceæ?*—Madder—Protea—Hamamelis—Ilex.

LECTURE XXV.

CLASS V.—PENTANDRIA.

Fig. 127.

THE class which we are about to examine is said to comprehend more than one tenth part of all known species of plants. It differs from the class Syngenesia in having its *five stamens separate*, while the Syngenesious plants have the same number of stamens united by means of their anthers. Plants with five stamens, including those which have anthers united, are said to constitute one fourth part of the vegetable kingdom.

Order Monogynia.

Asperifoliæ, or Boragineæ.

Here we find a group of plants called by Linnæus *Asperifoliæ*, a name derived from two Latin words, *asper*, rough, and *folium*, leaf, signifying rough-leaved plants. These have monopetalous corollas, with five stamens and five naked seeds. The seeds are *dicotyledons*. Jussieu forms these into the order *Boragineæ*, from a genus called Borago. "The change in the corolla of these plants, in general from a bright red to a vivid blue as the flower expands, apparently caused by the sudden loss of some acid principle, is a very curious phenomenon."[*]

The *Cynoglossum* is, perhaps, as common as any of the *asperifoliæ*, or rough-leaved plants. Its common name is hound's-tongue, so called from its soft oval leaves. Although the Cynoglossum is classed with the rough-leaved plants, its pubescence gives to its leaves a softness appearing to the touch like velvet; it is about two feet high, the flowers are of a reddish purple, growing in panicles.[†] The Lungwort, (*Pulmonaria*,) which also belongs to this natural family, has two species in North America with smooth leaves. The Mouse-ear (*Myosotis*) is valued for its medicinal properties; a species, the *arvensis*, or Forget-me-not, is an interesting little blue flower. The Gromwell (*Lithospermum*) is a rough plant with white flowers; the bark of the plant contains so much silex or flinty matter, as to injure the sickles of the reapers, when it grows in the field with the grain. The name, Lithospermum, is from the Greek, *lithos*, a stone, and *sperma*, a seed, in allusion to the hardness of the seeds. The *Borago* is an exotic very common in our gardens. The corolla is wheel-shaped, of a beautiful blue colour, having its throat closed with five small protuberances; the stamens are attached to the tube of the corolla. You must take off the corolla carefully, and you will see the little scales which choked up the throat of the corolla, and the manner in which the five stamens adhere to it.

Luridæ, or Solaneæ.

We next meet with a family of plants, named by Linnæus, *Luridæ*, from their pale or livid colour. Jussieu called them the *Solaneæ*,

[*] Smith.
[†] It is said that the leaves of this plant, if strewed about apartments infested with rats and mice, will expel these vermin.

---

from the name of the genus Solanum. The general characters of
these plants are a monopetalous corolla, of a lurid or pale appearance; five stamens attached to the base of the corolla, and alternating with its divisions; leaves alternate. The common potato (So-
LANUM *tuberosum*) is of this natural family; the flowers of this plant
are large, and the organs very plain for analysis. There is a peculiarity in the appearance of the anthers which it is well to notice;
these are of an oblong form, thick, and partly united at the top, and
open at the summit by two pores. The potato was not known in
Europe until after the discovery of America. In the year 1597, Sir
Walter Raleigh, on his return from this country, distributed a few
potatoes in Ireland, where they became numerous, and the cultivation
of them soon extended into England. It is said that the root of the
potato is white or red, according to the colour of the flower. The
little green balls, upon the stalks of this plant, are the pericarps, and
contain the seed; but this plant is usually produced from the root.
The little knobs called eyes, which you may notice upon the tubers
of the potato, are a kind of germ or bud; in planting, the whole root
is not always put into the ground, but cut into as many pieces as there
are eyes, each of which produces a plant.* In the same genus with
the potato is found the Tomato and the Egg-plant. In the natural
order Solaneæ is the DATURA *stramonium*, a large, ill-looking, nauseous scented weed; with a funnel-form, plaited corolla, either white
or purple; with broad, dark green leaves; when the corolla falls off,
and the germ matures, it then becomes a large, ovate, thorny pericarp, often called Thorn-apple; it continues to blossom during the
summer; is found by the sides of roads, around old buildings, and
in waste grounds. Yet even this disagreeable plant has its uses;
on account of its narcotic, and other active properties, it is highly
valuable in medicine.

In the group of plants we are now considering, is the tobacco,
(NICOTIANA *tabacum*.)   This is a native of America; it was imported
into Europe about the middle of the 16th century. It was presented
to Catherine de Medicis, Queen of France, as a plant from the New
World, possessing extraordinary virtues. The generic name, Nicotiana, is derived from *Nicot*, the name of the person who carried it
to France. King James I. of England, had such a dislike to the
fumes of this plant, that he wrote a pamphlet against its use, called
" A Counter-blast to Tobacco." It is highly narcotic, the excessive
use of it producing sleep, like opium. The oil of tobacco, when applied to a wound, is said to be equally fatal as the poison of a viper.

The Mandrake (ATROPA *mandragora*) was much used by the ancients as an opiate; they had many absurd notions respecting this
plant; they fancied in its roots, which are very large and of a peculiar appearance, a resemblance to the human form, and thought
that some judgment from heaven would follow those who took them
out of the ground. This superstition is not unlike that which is discovered, even in the present day, by those who are unwilling to sow
fennel, through fear of " sowing sorrow." Perhaps those very persons who would fear to perform an act so innocent as the taking a
root from the ground, or putting seeds into it, would have no dread
of the anger of God for the violation of his commands.

* This is more properly a *continuation* of the plant, than a *reproduction*;—it is
found that the vegetable thus continued, appears, in process of time. to degenerate,
and it is necessary to renew the race by reproducing it from seed.

Describe the potato—What other plants are in the genus Solanum?—Datura--To-
bacco—Mandrake.

The ATROPA *mandragora* must be distinguished from the American mandrake ;* the latter bears a fruit which is pleasant to the taste, and quite inoffensive ; its botanical name is Podophyllum ; and it is found in the class Polyandria. You can see in this instance the importance of botanical names. The common name, *mandrake*, has been given to two plants essentially different ; but by the use of scientific names, there is no danger of one being taken for the other, by those who know any thing of botany.

Before leaving this extensive natural order, we will notice the Mullein, (*Verbascum*,) which you must have seen too often to need any description of its general appearance ;† but though its *natural characters* may so far have attracted your attention, that you know a mullein from every other plant, you may not have examined its different parts with a view to scientific arrangement ;—it has, like all the plants of this natural order, a five-parted calyx, wheel-shaped corolla with five unequal divisions. The stamens are *declined*, or turned downward, and bearded. The capsule is two-celled and many-seeded. The leaves are *oblong, acuminate*, and *decurrent*, or with their bases extending downward around the stem ; they are downy on both sides. The flowers are arranged along the stem, in such a manner as to constitute what is called a *spike*. The technical name of the commôn mullein is VERBASCUM *thapsus ;* a species smaller and more delicate than the common mullein, is often found in woods ; this is the VERBASCUM *blattaria*. This genus is less active in its medicinal qualities than most others of the same family ; it is said to possess anodyne properties, and to be intoxicating to fish.‡

Lysimachia or *Primulaceæ.*§

The fifth class contains, in its first order, a family with wheel-form corollas. Its most important genus is the Lysimachia or Loose-strife, (see Fig. 127, *a ;*) several species of it may be found in blossom in June and July, along the banks of little brooks, and in low meadow grounds. The *racemosa*, or cluster-flowered loose-strife, is from one to two feet in height ; it bears a profusion of fine yellow blossoms, in a loose raceme. It sometimes bears bulbs in the axils of the leaves, and small branches. These bulbs, like those of the crocus and onion, contain the rudiments of a new plant.

The *Primula*, from which this natural family was named by Professor Lindley, is a beautiful genus ; most of its species blossom early, whence its name, *primula*, from *primus*, first. The primula is the proper *primrose ;* it received its name in England, where it is very common. The *Primula vulgaris*, is the common English primrose ;—then there is the cowslip, (*veris*,) and oxlip, (*elatior*) and Scottish primrose, (*scotica*,) all different species of the same genus. These are cultivated in our gardens, as also the *auricula*, (often improperly called polyanthos ;) we have but one native species of primula, which is much known ; this is the *farinosa*, commonly called bird's-eye primrose. When we read in the British poets about primroses and cowslips, we must remember that they are not the same flowers which we usually call by these names.

The English cowslip, (*Primula veris*,) has the segments of its

* Sometimes called may-apple.
† By *general appearance* we mean, what the French botanists call the *port* of the plant, or what is technically called its *habit*.
‡ Smith.
§ See Appendix, Plate vii. Fig. 9, for a plant of this family.

What other plant has the same common name ?—Describe the mull.in—Different species of Verbascum—Lysimachia—Primula.

corolla spotted with a rich, yellow colour, which Shakspeare seemed to suppose contained the fragrance of the flower. Thus in the "Midsummer Night's Dream," the Fairy says,

> "I serve the fairy queen,
> To dew her orbs upon the green:
> The *cowslips* tall, her pensioners be;
> In their gold coats spots you see;
> Those be rubies, fairy favours,
> In those freckles live their savours:
> I must go seek some dew-drops here,
> And hang a pearl in every cowslip's ear."

The American cowslip belongs to the genus *Caltha*, of the class Polyandria.

Miscellaneous Examples of Plants in this Class and Order.

The coffee-plant (Coffea *arabica*) is in this class and order. This is a native of Arabia; it is used to a great extent by the Turks and Arabs, to counteract the narcotic effects of opium, which they use in large quantities. It is remarked by a physician, that the question is often asked, which is the least detrimental to health, tea or coffee; he says, "The Turks, who drink great quantities of coffee, and the Chinese, who make equally as free use of tea, do not exhibit such peculiar effects as render it easy to decide, whether they are, in reality, deleterious to the human system."

The trumpet-honeysuckle (*Lonicera*) belongs to this part of the artificial system, (Fig. 127, *b*;) it has a very minute, five-cleft calyx, which is *superior*, or above the germ: the corolla is of one petal, and *tubular;* the tube is *oblong;* the *limb* of the corolla is deeply divided into five *revolute* segments, one of which seems separated from the others; the filaments are *exserted;* the anthers are *oblong*.

Before closing our remarks upon this order, we will remind you that the wine-grape is found here. The general characters of the grape (Vitis) are a calyx five-toothed; petals adhering at the top; a round five-seeded pericarp. The stamens and pistils are, in some species, diœcious, or on separate plants; this, according to our principles of classification, would carry the genus into the class Diœcia; but as some species have perfect flowers containing five stamens, and one pistil, and as it is never permitted to place in different classes the different species of a genus, we take the diœcious ones, which are less numerous than the pentandrous, into the fifth class.

The regions which produce the wine-grape have a mean annual temperature* of 50° on the northern border, and 59° on the southern. Lines of temperature have been fixed by Humboldt, by remarking the peculiar vegetables in different latitudes. He has traced the northern limit of the wine-grape, where the mean annual temperature is about 50°, across the United States to the Pacific Ocean; not, however, in a straight line, for climate, although chiefly dependant on latitude, is yet much modified by other circumstances; and on

---

* By mean annual temperature is meant a medium between the extremes of heat and cold. In a climate where the thermometer in summer would rise to 100 degrees, and in winter sink to zero or 0, the medium would be 50 degrees: this is probably not far from the mean annual temperature of our climate. The mean annual temperature at the equator is reckoned to be about 84 degrees.

Coffee—Trumpet-honeysuckle—What are the general characters of the grape genus?—Temperature of the regions which produce the wine-grape—What do you understand by mean annual temperature? (*see note*)—Within what degrees of mean annual temperature is the wine-grape produced?—What is the natural limit of the wine-grape?

13*

the western coast of America, we find in latitude 50° a similar climate to the 43d degree of latitude on the eastern coast. Thus, the wine-grape may grow in 50° of latitude near the lakes, the Mississippi, and Pacific Ocean; while, in the eastern part of New York and New England, it would not thrive beyond the 43d degree of latitude.

We find, on the eastern side of the Atlantic, the region of the wine-grape, including France, and the southern countries of Europe, extending as high as latitude 50°.

The southern limit of the wine-grape is traced from Raleigh, in the United States, in latitude 35°, to Europe, where it passes between Rome and Florence, in latitude 44°; this line is the boundary between the grape region and that of the olive and fig, which require a warmer climate.

The banks of the Rhine produce excellent grapes, which are brought down the river in great quantities to the seaports. The festival of the *Vintage*, or the gathering of the grapes, which, like our *Thanksgiving* season, is intended as a manifestation of gratitude for the fruits of the earth, was celebrated with much joy by the ancient Romans, and is still observed by the people of Italy; it occurs with them about the beginning of September; in France and the south of Germany, it is later.

The Falernian wine was the most celebrated among the Romans; some of the Latin poets spoke of it oftener than we should expect from those whose intellectual taste might seem to elevate them above any very great attention to the gratification of the external senses. The variety of wines in the days of Virgil was so great, that he said he might as well attempt to count the sand on the shore, or the billows of the ocean in a storm, as to make a catalogue of them.

The vines of Italy are often trained upon trees, particularly upon the lofty elm. In France, the vine is supported by short seedlings, about the length of bean-poles. The appearance exhibited by a luxuriant vineyard is truly rich and beautiful; of those of France and Italy, it may well be said,

> " The vine her curling tendrils shoots,
> Hangs out her clusters, glowing to the south.
> And scarcely wishes for a warmer sky."

It is said the Persian vine-dressers conduct the vines up the walls of their vineyards, and curl them over on the other side; this they do, by tying small stones to the extremity of the tendrils. This practice may illustrate a passage in Genesis: " *Joseph is a fruitful bough; even a fruitful bough by a well; whose branches run over the wall.*" " The vine, particularly in Turkey and Greece, is frequently made to intwine on trellises around a well, where, in the heat of the day, families collect and sit under their shade."

In this class and order is the violet, a genus which contains many native species. The garden-violet is the *Viola tri-colour.* It has a variety of common names, as pansy, heart's-ease, &c. Pansy is a corruption of the French *pensée*, a thought; thus Shakspeare, in the character of Ophelia, says:

> " There's rosemary—that's for *remembrance;*
> And these are pansies—
> That's for *thought.*"

How does the climate of the western coast of America correspond to that of the eastern coast?—Crossing the Atlantic, where do we find the northern and southern limits of the wine-grape?—Vintage—Wines—Vineyards—Illustration of a passage in Genesis—Violet.

Shakspeare also calls the same flower, "*Love in idleness*. will find the blue violet (*Viola cærulia*) among the first flov spring. Our meadows present a great variety of beautiful a grant violets.

The genus *Capsicum* affords the Cayenne pepper and the red pepper of our gardens. The pericarps, when ripe, are of a bright red; the seeds, which are attached to a central column, are heating and stimulating. A draught of hot cider and molasses, with a pod or two of red pepper steeped in it, was long held in high repute, in New England, as a remedy for colds. The green peppers are used for pickles. We might enumerate many other interesting plants which belong to this order, but our limits will not permit. The family of the *Convolvuli*, or the morning-glory tribe, and of the *Caprifoliæ*, or bush-honeysuckle tribe, are composed of genera of *pentandrous* plants.

LECTURE XXVI.

CLASS PENTANDRIA—*Continued.*

Order Digynia.

IN this order of the fifth class, is the family *Gentianæ*, which affords some delicate flowers, as well as medicinal articles. The fringed gentian a beautiful plant with a blue flower. This genus sometimes an irregularity in the number of stamens. In the natural family, called *Atriplices*, from the genus *Atriplex*, (sea-orache,) is the pig-weed, *Chenopodium;* this plant, notwithstanding its humble appearance, is dignified with a high-sounding name. It is grouped by natural characters with the beet and dock, flowers destitute of beauty. According to the late arrangement orders by De Candolle and Lindley, we find the order *diæ*, in which is the pig-weed, water-hemp, and several other plants, placed by Jussieu in his order *Atriplices*.

Umbelliferous Plants.

We meet, in this order of the class Pentandria, with a family of plants closely allied by natural characters; these are called *umbelliferous* from the Latin *umbella*, an umbrella, on account of the manner in which the peduncles grow out from the main stem.* Among the plants of this family, which are used for food, are the carrot, parsnip, celery, and parsley; the aromatics are dill, fennel, caraway, coriander, and sweet cicely. Poison hemlock, (*Conium*,) water parsnip, (*Sium*,) water cow-bane, are among the poisonous plants of this tribe. The water cow-bane (CICUTA *virosa*) grows in ponds and marshes. Cows are often killed in the spring by eating it, but as the summer advances, the smell becomes stronger, and they carefully avoid it. Linnæus relates, that in a tour made into Lapland, for scientific purposes, he was told of a disease among the cattle of Torneo, which killed a great many in the spring, when they first began to feed in pastures. The inhabitants were unable to account for this circum stance; but the Swedish botanist examining the pastures, discovered a marsh where the CICUTA *virosa* grew in abundance; he ac-

---

* See Plate ii. Fig. 3, for a plant of this family.

Capsicum—Gentianæ—Family Atriplices—Chenopodiæ—What is the origin of the word umbelliferous?—What are some of the plants of this family?—What is said of the water cow-bane?

quainted the people with the poisonous qualities of the plant, and thus enabled them to provide against the danger by fencing in the marsh. The poison hemlock (CONIUM maculatum) has a peculiarly unpleasant, nauseous smell; its stalk is large and spotted, from whence its specific name maculatum, which signifies spotted. This plant is supposed to be the poison so fatally administered by the Athenians to Socrates and Phocion.

The umbellate plants which grow on dry ground are aromatic; as dill, and fennel; those which grow in wet places, or the aquatic species, are among the most deadly poisons; as water parsnip, &c. Plants of this family are not in general so beautiful to the sight, nor so interesting, as objects of botanical analysis, as many others.*

In order to assist you in analyzing plants of this family, we will illustrate their botanical characters by a sketch of the coriander.

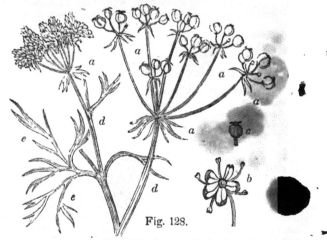

Fig. 128.

1. CALYX, a; this is of that kind called an involucrum; the leaves which you see at the foot of the universal umbel, form what is called the general involucrum; the leaves which are at the foot of the partial umbel, form a partial involucrum. Both of these involucrums are pinnatifid, or have the leaves divided.

2. COROLLA, b; this is represented as magnified; you can see that it has five petals, inflected or bent inwards.

3. STAMENS, five, anthers somewhat divided.

4. PISTILS, two, reflexed or bent back, as may be seen on the seed c, where the stigmas are permanent.

5. PERICARP, is wanting in all umbellate plants.

6. SEED, c, is round, with its two styles at the summit; it consists of two carpels.

* Botanists in general shrink from the study of the Umbelliferæ; nor have these plants much beauty in the eyes of amateurs; but they will repay the trouble of a careful observation. The late M. Cusson of Montpelier bestowed more pains upon them than any other botanist has ever done; but the world has, as yet, been favoured with only a part of his remarks. His labours met with a most ungrateful check, in the unkindness and mortifying stupidity of his wife, who, in his absence from home, is recorded to have destroyed his whole herbarium, scraping off the dried specimens for the sake of the paper on which they were pasted!—"Sir James Edward Smith's Introduction to Botany."

What is said of the poison hemlock ?—Describe Fig. 128.

7. STEM, *d*, is herbaceous, branched.
8. LEAVES, *e*, narrow, pinnatifid *
9. FLOWERS, terminal, umbelled.†

In distinguishing the genera of umbelliferous plants, the figure, margin, and angles of the seeds are much regarded. The seeds of the carrot are bristly, those of the poison hemlock marked with ridges, those of the parsnip flat.

Order Trigynia.

This order contains the elder, (*Sambucus*,) a shrub which ornaments the fields during the summer, with its clusters of delicate white flowers. From the appearance of the blossom you might suppose it to be umbelliferous; the stalks do at first radiate from one common centre, but afterward they are unequally sub-divided; this arrangement of flowers is called a *cyme*. The dark, rich purple berries of the elder, and the peculiarity of its pithy stem, are among its distinguishing, natural characters.

The snow-ball, *Viburnum*, has a natural affinity with the elder: the flowers in its *cymes* are more thickly clustered together. Both are distinguished by their flat corollas, which resemble a circular piece of paper, with five divisions notched on the border. The only general difference between the snow-ball and the elder is, that the former has a berry or pericarp, with one seed, the latter with three. The snow-ball which is cultivated in shrubberies is an exotic; but there is a native species of viburnum, the *oxycoccus*, which produces showy flowers in the spring, and is well worth a place in pleasure-grounds.

Order Tetragynia.

Here we find the grass of Parnassus, (*Parnassia*.) This is an interesting flower; the leaves are white, and beautifully veined with yellow; the stem produces but one flower; the nectaries are remarkable for their beauty and singular appearance; they are five in number, heart-form, and hollow, surrounded with thirteen little threads, each one terminating with a round, glandular substance. The plant is said to be a native of Mount Parnassus, in Greece, so celebrated in mythology, as the dwelling of the muses.

Order Pentagynia.

In the fifth order we find the flax, *Linum*, so called from a Celtic word, *lin*, a thread. It has a showy, blue flower, with an erect stem; a field of flax in blossom presents a very beautiful appearance. The cultivated species is said to be of Egyptian origin. It is from the liber or inner bark of the stem of this plant, that all linen goods, and the finest lawn and cambric, are manufactured. We owe to it, in one sense, our literature; as the paper of which our books are made, is mostly from linen rags. The fibres of the stem are not only thus important to the comfort of man, by contributing to his clothing, and to his intellectual improvement in furnishing a method of disseminating knowledge, but the seeds are highly valuable for their oil, called linseed oil. This is used in medicine. The delightful performances of the painter are executed by means of colours prepared with oil, from the seed of the flax, laid upon the canvass made from the fibres of its stems.

* The leaves of Umbelliferous plants are mostly compound, and sheathing at the base.
† The description of this plant is given on the authority of Nuttall, who calls it the American coriander, which he says is found in the neighbourhood of the Red River. The cultivated coriander has a one-leafed involucrum.

Order Polygynia.

The thirteenth order, containing plants with more than ten pistils, occurs next to the fifth; there being no plants in the class Pentandria with six, seven, eight, or nine pistils. The yellow root (*Zanthoriza*) is a native of the Southern States. It has 5 stamens, 13 pistils, no calyx, 5 petals, 5 nectaries, and 5 capsules; the flowers are purple, growing in panicles. It is a low shrub, with a yellow root, sometimes used by diers.

Our explanation of the class Pentandria has necessarily been somewhat tedious, on account of the number and importance of the plants which it contains, few of which, in comparison with the whole we have been able to notice. We do not, however, expect to make you practical botanists by introducing to your observation a few interesting plants;—this can only be done by gathering flowers, and examining them according to those rules of analysis which we have endeavoured to explain in the most simple manner. If you *study* flowers, you will read about them with pleasure and profit; if not, remarks upon them will convey little instruction. Sciences may be unfolded, every facility which books and teaching can give may be placed before the youthful mind; but that *mind* must itself be active, or the germs of knowledge will no more take root and expand, than the seeds of plants would vegetate if thrown upon the bare face of a granite rock.

LECTURE XXVII.

CLASS HEXANDRIA, CLASS HEPTANDRIA.

CLASS VI.—HEXANDRIA.

Of all the artificial classes, none presents us with so great a number of splendid genera as Hexandria; most of them are distinguished by *bulbous roots, monocotyledonous seeds,* and *endogenous stems;* the palms and some other plants of this class have fibrous roots in connexion with the last two characters; these are inseparable, the nature of the stem, or the manner of its growth, depending on the structure of the seed.

Order Monogynia.

Liliaceous plants, or the family of the Liliaceæ.

The most prominent group of plants in this class and order, is the lily tribe, comprehending not only the genus of the lily, but the tulip, crown-imperial, hyacinth, and many other of our most beautiful exotics, as well as many native plants. The liliaceous flowers have no calyx; the perianth is coloured, and petal-like; it is usually called the corolla. The number of stamens is generally 6, sometimes but 3; in the latter case the plant is in the class Triandria; the stamens are opposite the divisions of the corolla. The germ is triangular, 3-celled, superior. The root is bulbous. The leaves have parallel veins.

Zanthoriza—Remarks on closing the examination of the class Pentandria—Class Hexandria—Natural characters which distinguish plants of this class—General remarks upon the Liliaceæ.

You have already been made acquainted with the lily, as it was one of the first flowers you were taught to analyze. Pliny says the "lily is the next in nobility to the rose."[*] Linnæus called the liliaceous flowers "*Nobles* of the vegetable kingdom;" he also called the palm-trees "*Princes* of India," and the grasses *Plebeians*.

Fig. 129.

But in our republican country, where aristocratic distinctions among men are discarded, we will not attempt to introduce orders of nobility among the plants. In the lily, which has 6 stamens, there are 6 petals; 3 of these are exterior, 3 interior; the capsule is 3-sided, with 3 cells, and 3 valves; the seeds are arranged in 6 rows. This proportion of numbers seems to forbid the idea that this plant was produced without the agency of a designing mind. We are not always, however, to expect the same symmetry in plants, as has been here remarked. It is in the natural, as in the moral world, that, although we see around us such proofs of order and system, as manifest the superintending care of one Almighty Being, yet we meet with irregularities which we cannot comprehend; but, although we may admire the , we are not to say that even what seems *disorder*, is formed without a plan.

> Shall little haughty ignorance pronounce
> His works unwise, of which the smallest part
> the narrow visions of his mind?"

The tulip , but its three-parted stigma is attached to a three-cornered germ. The corolla of the tulip is more expanded at the base than that of the lily. The stem of the tulip is never more one-flowered, while that of the lily usually has a number of
In no plant is the variation made by culture, greater than lip; it is said, that of one single species, (Tulipa *gesneriana*,) hundred varieties are cultivated in Holland. About the middle of the seventeenth century, the rage for tulips was so great that one were sold for four thousand dollars, and one variety, called the Viceroi, for ten thousand dollars; but this extraordinary traffic was checked by a law, that no tulip or other flower should be sold for a sum exceeding one hundred and seventy-five dollars. The amateurs of this flower may truly be said to have had the *tulip-mania*, to have rendered such a law necessary. The Crown-imperial[†] is a majestic flower, and presents, in the regularity of its parts, the curious appearance of its nectaries, and the liquid secretion which takes place in them, facts of great interest both in the departments of botanical classification and physiology. But we find in the fetid odour of this splendid flower, a circumstance which leads us to prefer, as an ornament for our parlours, or as a gift to a friend, the humble mignionette, or the lowly violet.

[*] "*Lilium nobilitate proximum est.*" A French poet, in the following lines, gives the lily a rank above the rose.

> "Noble fils du soleil, le *lys majesteux*,
> Vers l'astre paternal dont il brave les feux
> Elève avec orgueil sa tête souveraine;
> Il est *roi* des fleurs, la *rose est la reine*."

[†] This plant is represented at Plate vii. Fig. 4, of the Appendix; the Yucca *aloifolia*, which belongs to the same natural family, is represented at Plate ii. Fig. 1. The Narcissus is represented at Plate vii. Fig. 7. The Agave, of the Narcissi family, is represented at Plate vii. Fig. 2. The Pine-apple, belonging to this class and order, is represented at Plate v. Fig. 3.

What is said of the lily?—Tulip—Tulip mania—Crown-imperial.

This simple fact might suggest to the young, that in order to be desirable to others, they must be agreeable; the mere circumstance of a fine person, cannot long render tolerable, the society of one who possesses neither useful nor amiable qualities.

The Family of Palms

The palms have mostly a liliaceous corolla with 6 stamens; but some are monœcious, and others, diœcious; while a part have their stamens and pistils within the same corolla and belong to the class Hexandria.

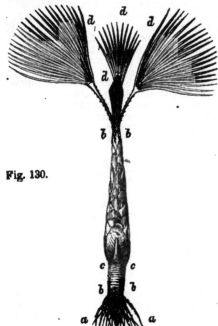

Fig. 130.

Fig. 130 represents a young palm tree, (*Chamærops humilis ;*)* at *a*, is the fibrous root; *b c*, represents the oldest part of the stipe, showing, by the lines and dots, the place of insertion of the first leaves; *c b*, represents the upper part of stipe, still covered with the sheathing bases of the petioles; *d*, represents the crowning, terminal leaves—these are petioled, fan-shaped, and plaited when young; the petioles are armed with prickles. Palms live to a great age; they are the product of tropical regions, and afford the date, cocoa-nut, and other valuable fruits.

Miscellaneous Examples of Plants in the 6th Class and 1st Order.

In this class and order is the Spiderwort, (*Tradescantia.*) It has 6 stamens, 3 petals, 3 sepals, and the capsule is 3-celled. The leaves are ensiform and very long. It remains in blossom nearly the whole summer, and is well worth cultivation, both for its cheerful appearance, and constant botanical characters. The Snow-drop is of the same natural, as well as artificial order, as the Spiderwort.

You may be surprised to find, in company with so many elegant flowers, the onion and bulrush; but you must recollect that the title to admission into this class and order is 6 stamens and 1 pistil; and no plant, however humble, with these characteristics, is excluded

---

* Although we have described this plant under the class Hexandria, in conformity with the classification of some writers, it is questionable whether it does not rather belong to Diœcia. In the Appendix, at Plate i. Fig. 1, is a representation of the *Areca*, which belongs to the Palm-tribe, and at Plate iii. Fig. 3, is a representation of the same palm-tree as seen at Fig. 130.

Palms—Describe Fig. 130—Spiderwort—Humble plants placed with those which are beautiful.

from a place beside the proud tulip and the noble lily. The onion belongs to the natural order of Jussieu, *Asphodeli.**

The Asphodel, which gives name to the family, was, among the ancients, a funereal plant; it was made to grow around the tombs, and a belief prevailed that the *manes* of the departed were nourished by its roots. An inscription upon a very ancient tomb commences thus, " *I am nourished by the Asphodel.*" This plant was supposed by the ancient poets, to grow in abundance upon the borders of the infernal regions. Fig. 129 represents a flower of the Asphodel family, (*Eucomis.*)

The genus *Scilla* is an exotic, containing the squill, a medicinal plant, and the hare-bell of English poets; the latter is SCILLA *nutans,* or nodding; it abounds in the woods and glens of Scotland, and has a very slender scape. Thus Scott, in the " Lady of the Lake," says of Ellen Douglas,

> " E'en the slight hare-bell raised its head
> from her airy tread."

The flower which call hare-bell, is the *Campanula rotundifolia*; this is very common near waterfalls, and upon rocks in other situations. The barberry (*Berberis*) is common in New-England; its stamens possess an unusual degree of irritability; they recline upon the petals, but when the bases of the filaments are touched by any substance, they instantly spring towards the pistil.

You may have observed, that although we have remarked upon the *beauty* of some flowers to be found in this class, nothing has been said of their *utility*; the truth is, that the former, as is too often the case with external beauty, constitutes their chief merit: when we compare the advantages which the world derives from the costly race of showy tulips, with the utility of the humble flax, we feel that though we may admire the one, reason would teach us to prefer the other. May you from this derive a moral lesson, which shall suggest to your minds some truths applicable to our own race as well as the plants.

The genus *Convallaria* contains the lily of the valley, and many other delicate and interesting species. Among these are Solomon's seal. This name is supposed to have been taken from certain marks on its roots, resembling the impressions made by a seal. It was formerly much celebrated for medicinal properties.†

Order Digynia.

We here find the Rice (*Oryza;*) this belongs to the family of grasses, which you have already met with in the class Triandria; but this plant having six stamens, is separated by the artificial system from the tribe to which it is allied by natural characters. No plant in the world appears of such general utility as an article of food. It is the prevailing grain of Asia, Africa, the southern parts of America, and is exported into every part of North America and Europe.

Order Trigynia.

We here find the genus *Rumex,* which contains the dock and sorrel;

* The *Dracæna draco,* belonging to this family, is represented in Plate I. Fig. 3, of the Appendix.
† Gerard, a very ancient botanist, has the following curious passage. "The root of Solomon's seal stamped, while it is fresh and greene, and applied, taketh away in one night, or two at the most, any bruse, black or blew spots gotten by fals, or woman's wilfulness, in stumbling upon their hasty husband's fists, or such like."

14

the flowers have no proper corolla, but the six stamens and three pistils are surrounded by a six-leaved calyx, or what, in this case, may be called a perianth.

The *Colchicum* or meadow-saffron of England is a medicinal plant, in some repute among physicians. The root is a large, egg-shaped bulb; in spring several narrow leaves arise, but the flower does not appear till September. The germ lies buried in the root all winter, and is raised in spring, to perfect its seeds before the next season. The flowers are pale purple.

CLASS VII.—HEPTANDRIA.

Order Monogynia.

The first order of this class contains the chick winter-green, (*Trientalis ;*) this plant has a calyx with 7 leaves, or sepals, and the corolla is 7-parted. One species is said to defend its stamens against injury from rain, by closing its petals and hanging down its head in wet weather.

Fig 131.

The cultivated Horse-chestnut, *Æsculus,* (Fig. 131,) is a native of the northern part of Asia, and was introduced into Europe about the year 1500; it was not probably brought to America until some time after the settlement of this country by Europeans. It is a small tree which produces white flowers, variegated with red, crowded together in the form of a panicle, the whole resembling a pyramid. In appearance it is very showy, and the more agreeable to us, as we have so few trees whose flowers are conspicuous. The blossom is very irregular in its parts, that is, its other divisions do not correspond with the usual number of stamens; the stamens, however, do not vary as to number. The seeds have a resemblance to chestnuts, but their taste is bitter. There are several native species of this plant in the southern and western states. The horse-chestnut exhibits in its buds, in a very conspicuous manner, the woolly envelope which surrounds the young flowers, the scales which cover this envelope, and the varnish which covers the whole. The stems and branches of this tree afford good subjects for studying the formation and growth of woody or exogenous stems.

Order Tetragynia.

There is but one plant with four pistils known in the class Heptandria; this alone constitutes the fourth order; its common name is lizard's-tail, (*Saururus.*) It has arrow-shaped leaves, flowers destitute of a corolla, and growing upon a spike; it is to be found in stagnant waters.

Order Heptagynia.

The Septas, a native of the Cape of Good Hope, is considered as the most perfect plant in this class; it has 7 stamens, 7 pistils, 7 petals, a calyx 7-parted, and 7 germs, (one to each pistil,) which germs become 7 capsules, or seed vessels.

Heptandria is the smallest of all the classes; we do not find here, as in most of the other classes, any natural families of plants; but the few genera which it contains differ not only in natural characters from other plants, but they seem to have no general points of resemblance among themselves.

Colchicum—What plant is in the 1st order of the 7th class ?—What is said of the Horse-chestnut ?—Saururus—What example is given of the order Heptagynia ?—Remarks upon the class Heptandria.

LECTURE XXVIII.

CLASSES OCTANDRIA AND ENNEANDRIA.

CLASS VIII.—OCTANDRIA.

Order Monogynia.

Fig. 132.

THE eighth class, although not large, contains some beautiful and useful plants. One of the first which we shall notice is the scabish, (*Œnothera*,) sometimes called evening primrose. Many species of this are common to our country; some grow to the height of five feet. The flowers are generally of a pale yellow, and in some species they remain closed during the greater part of the day, and open as the sun is near setting. This process of their opening is very curious, the calyx suddenly springs out and turns itself back quite to the stem, and the petals, being thus released from the confinement in which they had been held, immediately expand. There are flowers which thus hail the *setting* sun, though many salute it at its *rising*. The flowers of the Œno-thera are thickly clustered on a spike, and it is said that "each one, after expanding once, fades, and never again blos-soms."* This singular flower has been observed in dark nights to throw out a light resembling that of phosphorus. The regularity of its parts render it a good example of the eighth class; the different parts of its corolla preserve in their divisions the number four, or half the number of stamens. It has 4 large, yellow petals, the stig-ma is 4-cleft, capsule 4-celled, 4-valved, the seeds are affixed to a 4-sided receptacle.

The evening primrose belongs to an order of dicotyledonous plants called Onagræ;† the characters of which, are four petals above the calyx; stamens inserted in the same manner, and equal or double the number of petals; the fruit a capsule or berry. To this natural order belongs the willow herb, (*Epilobium,*) a very branching plant with red flowers and feathery seeds. The cranber-ry (*Oxycoccus*) also belongs to the same family, but having ten stamens, is placed in the class Decandria; a natural affinity being made to yield to the artificial system. The fruit of the cranberry consists of large scarlet berries, which contain tartaric acid. The flowers are white, they have a four-toothed calyx, and corolla four-parted. It is found in swamps in various parts of North America.

The ladies' ear-drop, *Fuschsia,* (see fig. 131,) is a beautiful exotic. It has a funnel-form calyx, of a brilliant red colour; the petals are almost concealed by the calyx, they are purple, and rolled round the stamens, which are long, extending themselves beyond the col-oured calyx. This plant is a native of Mexico and South America, except one species, from the Island of New Zealand. Ten species are said, by horticulturists, to be cultivated; but some of them are, probably, rather varieties, than distinct species.

The heath‡ (*Erica*) is not known to be indigenous to this coun-try; many species have been introduced. The common heath

* W. Barton.
† The common French name for the evening primrose, is *Onagré*.
‡ The term *heath* is said to have originated from an old Saxon word, alluding to the *heat* which the plant affords as fuel; it is used in England for heating ovens.

Evening Primrose—What are the characteristics of the natural order *Onagræ*, and what plants belong to it?—Ladies' ear-drop—Heath.

(*Erica cinerea*) has bell-form flowers, small and delicate, with the colour pink, or varying into other colours; the flowers intermixed with the delicate green leaves produce a fine effect. The kind of soil necessary to the growth of the heath, is the peat earth, so common in England and Scotland, in which countries this plant abounds; thus Scott says of his Lady of the Lake,

> "A foot more light, a step more true,
> Ne'er from the heath-flower brush'd the dew."

In the Highlands of Scotland, the poor make use of the heath to thatch the roofs of their cottages; their beds are also made of it The field in which this plant grows is termed a heath or heather.

> "The Erica here,
> That o'er the Caledonian hills sublime,
> Spreads its dark mantle, where the bees delight
> To seek their purest honey, flourishes;
> Sometimes with bells like amethysts, and then
> Paler, and shaded, like the maiden's cheek,
> With gradual blushes; other while, as white,
> As frost that hangs upon the wintry spray."

The Daphne is a rare plant; one species is called the Lace-bark tree, from the resemblance of its inner bark or *liber* to net-work or lace. This bark is very beautiful, consisting of layers which may be pulled out into a fine white web, three or four feet wide; this is sometimes used for ladies' dresses, and may even be washed without injury. Charles I. of England, was presented by the governor of Jamaica with a cravat made of this web. The plant is a native of the West Indies.

The Nasturtion (*Tropæolum*) is a very commonly cultivated exotic. It has not a regularity of parts; the divisions are not four or eight, which we might expect from its eight stamens, but the calyx is either four or five-parted, and the corolla is five-petalled. The fruit consists of three seeds; these are used for pickles. "The generic name (*Tropæolum*) signifies a *trophy-plant*; this alludes to its use for decorating triumphal arches, or to the resemblance of its peltate leaves to shields as well as its flowers to golden helmets pierced through and stained with blood."*

Order Trigynia.

This order contains the Buckwheat, (*Polygonum*,) which was classed by Linnæus in the same natural order as the dock, pigweed, &c., "having flowers destitute of beauty and gay colouring." The genus is extensive, containing many plants which are considered as common weeds. The *fagopyrum* is the true buckwheat; the meal obtained by grinding its seed, is much esteemed for cakes; these are called slap-jacks in New-England, in England, crumpits. The Polygonum is variable in its number of stamens; the seed is a triangular nut.

Order Tetragynia.

We here find the beautiful plant, Paris, which is said to have been named after a prince of ancient Troy, remarkable for his beauty. In every part of the flower there is the most perfect regularity; the numbers four and eight prevailing in the divisions. It has 8 stamens, 4 pistils, 4 petals, 4 sepals, a 4-sided and 4-celled pericarp, which contains 8 seeds, and 4 large spreading leaves, at a little distance below the flower. The colour of the whole is green. The plant is said to be narcotic. It is a native of England.

* Sir J. E. Smith.

CLASS IX.—ENNEANDRIA.

Order Monogynia.

Fig. 133.

This is also a very small class. In the *first* Order we find the genus *Laurus*. which includes the cinnamon, bay, sassafras, camphor, spice-bush, &c. The bay (*Laurus nobilis*) is a native of Italy; the Romans considered it a favourite of the Muses. The emperor Tiberius wore it not only as a triumphal crown, but as a protection against thunder; it being thought that Jupiter had a particular regard for the plant. The laurel, as well as the olive, was considered as an emblem of peace; it was sometimes called laurus *pacifera*, the peace-bearing laurel. Branches of laurel carried among contending armies, were considered as a signal for the cessation of arms. Poets crowned with laurel, were called *laureates*. Camphor is the produce of the LAURUS *camphora*, a large tree which grows in Japan. "The LAURUS *cinnamomum* is a tree which grows to the height of twenty feet; it sends out numerous branches crowned with a smooth bark. The leaves are of a bright green, standing in opposite pairs. The petals are six, of a greenish white colour. The fruit is a pulpy pericarp enclosing a nut. This tree is a native of Ceylon, where it grows very common in woods and hedges. The imported cinnamon is the inner bark (*liber*) of the tree; it is remarkable that the leaves, fruit, and root, all yield oil of very different qualities. That produced from the leaves is called the *oil of cloves;* that obtained from the fruit is of a thick consistence, very fragrant, and is made into candles for the use of the king; the bark of the roots affords an aromatic oil, called the *oil of camphor*. The Sassafras-tree (LAURUS *sassafras*) is a native American plant; when first introduced into Europe, it sold for a great price, the oil being highly valued for medicinal uses. It grows on the borders of streams and in woods; it is often no larger than a shrub; its flowers are yellow; its fruit, blue-berries. The LAURUS *benzoin*, called Spice-bush, has scarlet berries, and is an aromatic plant."[*]

Fig. 133, *a*,[†] represents a flower of the Butomas, (*flowering rush ;*) the petals are six; they are ovate. The *umbellatus* is the only species known; the flowers grow in rose-coloured umbels. It is found in wet grounds, and near the margin of lakes and ponds.

Order Trigynia.

The *third Order* presents us with but one genus; but this renders the order important; it is the Rhubarb, (*Rheum.*) In one species, the RHEUM *tartaricum*, the leaves are acid, and on this account, when young, they are used for making pies. This plant is a native of Tartary, but now common in our gardens. The RHEUM *palmatum* is the plant which produces the medicinal rhubarb; this is obtained from the roots, which are thick, fleshy, and yellow. This plant is cultivated in England, and is remarkable for the rapidity of its growth. An English writer,[‡] asserts that its stem has been known to grow more than eleven feet in three months; its leaves are five feet in circum-

---

[*] Woodville.　　　[†] See also Appendix, Plate viii. Fig. 4.　　　[‡] Woodville.

Class Enneandria—Different species of the genus Laurus—Describe the different species of Laurus--Butomas—What genus is found in the order Trigynia?

14*

ference; the root grows to a great size; some roots have been imported from Turkey which weighed more than seventy pounds. At Fig. 133, *b*, is a flower of the genus *Rheum*.

We have dwelt somewhat at length upon exotics, because they are seldom described in botanical works in common use. If you become interested in the study of plants, you will naturally wish to know something about those which you are in the habit of using for food, or medicine, or to which, as in the laurel of the ancients, allusions are often made in the books which you read. But you cannot become practical botanists without much observation of our native plants. You must seek them in their own homes, in the clefts of rocks, by the side of brooks, and in the shady woods; it is there you will find nature in her unvitiated simplicity. We do not go to the crowded city to find men exhibiting, undisguisedly, the feelings of the heart. The flower transplanted from its rural abodes, exhibits in the splendid green-house, a physical metamorphosis, not less remarkable than the moral change which luxury too often produces upon the character of man.

LECTURE XXIX.

CLASS X.—DECANDRIA.

PLANTS of this class have ten stamens, but this circumstance alone would not distinguish them from some of the other classes; the number of stamens must not only be ten, but these must be distinct from each other; that is, neither united by their filaments below, nor by their anthers above. Other classes, Monadelphia, Diadelphia, Gynandria, and the two classes with the stamens and pistils on separate flowers, may also have ten stamens; but circumstances respecting the situation of these organs distinguish these classes from each other.

Order Monogynia.

Fig. 134.

In the *first Order* of the tenth class, we find some plants with papilionaceous corollas; these, because their filaments are not united, are separated from the natural family to which they belong, and which are mostly in the class Diadelphia. Among those which are thus removed from the class where from their general appearance they might have been looked for, is the wild indigo, (*Baptisia*,) a handsome plant with yellow flowers, two or three feet in height, and very branching; the stem and leaves are of a bluish green. This is found in dry sandy woods; it was used as a substitute for indigo during the time of the American revolution.

The Cassia *fistula*, a native of the Indies contains in its legume a pulp which is much valued in medicine, and known by the name of Cassia. The CASSIA *senna* furnishes the senna used in medicine; this species grows in Egypt and Arabia. One species, the CASSIA *marylandica* is called American senna, on account of its medicinal

---

Concluding remarks—Are there any classes except the tenth, in which the flowers have ten stamens?—Order Monogynia—Wild Indigo—Cassia.

qualities. Another species, *nictitans*, has small yellow flowers, and beautiful pinnate leaves, which remain folded at night; it shrinks back from the touch, for which reason it is called the American sensitive plant.

A plant, called by the Indians, Red-bud, (Cercis *canadensis*,) belongs to this class. It is a large tree, appearing as early as April, loaded with clusters of fine crimson flowers; the leaves, which are large and heart-shaped, do not appear as early as the blossoms. The beautiful aspect of the tree attracts to it many insects, particularly humblebees. A botanist* says, "I have often observed hundreds of the common humblebees lying dead under these trees while in flower." This is not the only example of fatal consequences which result from trusting too much to external appearances! This tree is not improperly called Judas' tree.

The three genera of plants which we have now noticed, bear fruit in that kind of pod called a *legume;* this is the case in general with the papilionaceous flowers.

The rue (*Ruta*) is an exotic, which gives name to one of Jussieu's natural orders, called *Rutaceæ;* these plants have a monosepalous calyx; five , alternating with the lobes of the calyx; the germ is large and superior, (See Fig. 134, *a.*)

At *b*, Fig. 134, is a representation of a flower of the Saxifraga, a very extensive genus; one species of which, an exotic, sometimes called beefsteak geranium, is much cultivated as a green-house plant; it is very hardy; its leaves are roundish and hairy; it sends forth creeping shoots.

This class and order presents us with the *Wintergreen* tribe; plants which are more or less shrubby, with monopetalous, bell-form corollas and evergreen leaves. In shady woods, where the soil is loose and rich, we find, in June and July, the spicy wintergreen, (*Gaultheria*,) a perennial plant which grows to the height of eight or ten inches; the pleasant taste of the leaves and fruit of this plant, is well known to the children of this country; the drooping blossom is very delicate and beautiful, consisting of a bell-form corolla, (not unlike the lily of the valley,) the colour of which is tinged with pink. Though you may have often enjoyed eating the fruit and leaves of the wintergreen, you will experience a delight which this mere pleasure of sense could not have afforded, when in your botanical rambles in the woods you chance to meet with this plant in blossom, with its little flowers just peeping out from a bed of dry leaves; you may then have the pleasure of a beautiful object of sight, with the intellectual gratification of tracing those characters which give it a definite place in scientific arrangement. Among the wintergreen tribe are two genera, Pyrola and Chimaphila, which by some botanists have been included under one; but they appear to be sufficiently distinct from each other to constitute a separate genus. These plants were classed by Linnæus in the natural order Bicornes, or two horns, alluding to the two protuberances, like straight horns, which appear on their anthers.

A great proportion of the plants in the first order of the tenth class are to be found in shady woods in June and July. We can here enumerate but few of them. We will, however, mention the Monotropa, a most curious little plant;—several stems of a few inches in height, form a cluster; each stem supports a single flower,

* W. P. C. Barton.

Cercis—Natural order *Rutaceæ*—Saxifraga—Wintergreen tribe—Monotropa, or Indian pipe.

resembling a tobacco pipe. The stems are scaly, but without leaves ;
the whole plant is perfectly white, and looks as if made of wax; it
is sometimes called Indian-pipe. You must look for this in shady
woods near the roots of old trees, in June or July.

'Rhododendron, or, as it is sometimes called, mountain laurel or
rose-bay, an evergreen with large and beautiful oval leaves, is found
growing on the sides of mountains, or in wet swamps of cedar; it
flourishes beneath the shade of trees; the pink and white flowers
appear in large showy clusters, and continue in' bloom for a long
period; they have a 5-toothed calyx, a 5-cleft, funnel-form, somewhat irregular corolla, stamens 10, sometimes half the number, capsule 5-celled, 5-valved. At Fig. 134, c, is a flower of the genus *Ledum*,
which is found in the same family as the Rhododendron; it has a
very small calyx, and a flat, five-parted corolla.

Connected by natural relations to the two genera above mentioned, is the American laurel, (*Kalmia*,) a splendid shrub, sometimes
found ten or thirteen feet high. On the Catskill mountains, it is said
to have been seen twenty feet in height; the flowers grow in that
kind of cluster called a *corymb ;* they are either white or red ; but
this fair and beautiful shrub is of a poisonous nature, particula
fatal to sheep who are attracted towards it; one species o
Kalmia is on this account called sheep-laurel.

Among the plants which have a place in this part of th
system, is the DIONÆA *muscipula*[*], or Venus' fly-tra
native of North Carolina; the leaves spring fr m
leaf has, at its extremity, a kind of append
doubled; this is bordered on its edges by gl
and containing a liquid that attracts insects ;
unfortunate insect alight upon the leaf, than
it closes, and the little prisoner is crushed to d
the sweets it had imprudently attempted to sei
overcome by the closeness of the grasp, has ex
unfolds itself. Although we may account for t
attributing it to the irritability of the plant, we ha
the difficulty by adducing a cause which itself remain
plained. We shall in a future lecture make some remar
irritability, or, as it is sometimes called, sensibility of pl

Order Digynia.

This order contains the *Hydrangea*, an elegant East dia
tic ; a species of this plant, a shrub with white flowers, i said t
been found on the banks of the Schuylkill river.

The Pink tribe, of the natural order *Caryophylleæ*, s compo
plants belonging to this class, some of which have three styles, o
have five, but the greater part have two, and therefore belong to
2d order. The exotic genus *Dianthus*, containing the carnation, an
other garden-pinks, and sweet-william, is a great favourite with florists, who gravely tell us what varieties we ought most to admire ; as
if fashion, and not nature, were to regulate our emotions. The seed
of the carnation often produces a different kind of flower from its
parent. A writer on the culture of flowers, observes, that a florist
may consider himself fortunate, if, in the course of his life, he should
be able to raise *six* superior carnations ;—but the hope that such
success may crown his labours, he thinks a sufficient stimulus to continued exertions. Such contracted views of nature and of the pur-

* See Appendix, Plate iii. Fig. o.

Mountain-laurel—Kalmia, or sheep-laurel—Dionæa—Pink tribe.

suits most ennobling to man, are too contemptible to need a comment. To degrade the beautiful and innocent employment of cultivating plants, by rivalries to produce a flower that may claim to be *distingué*, shows that the serpent still lingers in Eden. Let the flower-garden be a retreat from low and grovelling competitions, the promoter of innocence, of benevolence to man, and devotion to God.

Order *Trigynia.*

We here find the genus SILENE, one species of which is called the catch-fly; another, the *nocturna*, or night-blooming, is,

> "That Silene who declines
> The garish noontide's blazing light;
> But when the evening crescent shines,
> Gives all her sweetness to the night."

Another genus, the sandwort, is the

> "*Arenaria*, who creeps
> Among the loose and liquid sands."

Order *Pentagynia.*

The corn-cockle (*Agrostemma*) is very common in corn-fields; although troublesome, and regarded as but a weed, it is a handsome pink-like plant, bearing a purple blossom. In its generic character it differs little from the genus which contains the pink, except in having five pistils instead of two, on which account it is placed in the fifth order.

Here is also found the Sorrel, (*Oxalis*,) which produces the oxalic acid, similar in its properties to the acid obtained from lemons; it is poisonous, and not known as a medicinal article, but is important in the arts.

Order *Decagynia.*

In this order is the Poke-weed, (*Phytolacca*,) a very common plant, found on the borders of fields and road-sides; the fruit consists of large, dark berries, often used by children for the purpose of colouring purple. The young shoots are tender, and are sometimes eaten as a substitute for asparagus. The flower of this plant presents us with 10 stamens, 10 styles, a calyx with 5 white sepals resembling a corolla, a berry superior, (above the germ,) with 10 cells, and 10 seeds.

We have completed our review of the first groups of classes, or those which depend upon the *number* of stamens; in our next lecture we shall consider the two classes which depend on the *number* and *insertion* of the stamens.

Plants in the order Trigynia—Order Pentagynia—Describe the Poke-weed.

LECTURE XXX.

CLASS XI.—ICOSANDRIA.

Fig. 135.

Had we followed the classification which has, until recently, been admitted by writers on botany, we should have met with the class Dodecandria, from *Dodeka*, 12, and *andria*, stamen; it was not, as you might infer from the name, confined to 12 stamens, but contained from 10 to 20, without any regard to their insertion. This class produced much confusion in the science; for it is found that plants having more than ten stamens, frequently vary as to their *number ;*—there being no difficulty in distributing all plants of this class in the two next, it has, by consent of most botanists, been left out of the system; and the plants which it contained, are arranged under Icosandria, if the stamens are on the *calyx*, and Polyandria, if the stamens are inserted upon the *receptacle*. The *manner of insertion is always the same* in the same genus, and therefore there can be no confusion with respect to determining the classes upon this principle.

You will observe, that this omission of one class, changes the numbers of the remaining classes; as Icosandria, which was formerly the twelfth, is now the eleventh, and so on with the other classes. It is on account of these changes, that we wish you to learn the classes by their appropriate names, as Monandria, Diandria, rather than to confine yourselves merely to the numbers, as 1st, 2d, &c. Besides, the name of each class is generally expressive of its character, and will, when you understand its derivation, convey to you the idea of this character, which, by the number alone, could not be done ; for example, the term *tenth class*, conveys no idea but that of mere number; but the classical name Decandria, from *Deka*, ten. and *andria*, stamens, reminds you of the circumstance on which the. class is founded.

The name Icosandria, from *eikosi*, 20, and *andria*, stamens, seems not, however, exactly well chosen to represent the eleventh class, which is not confined to twenty stamens, having sometimes as few as ten, and in some cases nearly a hundred stamens. An American botanist* has proposed to call the class *Calycandria*, from calyx and *andria*, as the insertion of the stamens on the calyx is the essential circumstance on which the class depends ; this change has been approved, but the old name is still used. Thus, with respect to the name given to the great American continent, all allow it should have been Columbia, after Columbus, its discoverer ; but when once custom has sanctioned a name, it becomes very difficult to overcome this authority.

Order Monogynia.

We meet here with the *Prickly-pear tribe*, (*Cacteæ*,) in which the Cactus is the most important genus. Jussieu included in this natural order, the currant and gooseberry ; but Lindley has formed them

* Darlington.

into a separate order, called *Grossulaceæ*, from *Grossularia*, the gooseberry. The species of Cactus are very numerous; among the most splendid is the night-blooming Cereus, (CACTUS *grandiflorus*,) having flowers nearly a foot in diameter, with the calyx yellow, and the petals white. The blossoms begin to expand soon after the setting of the sun, and close before its rising, never again to open. Another species, (*speciossissimus*,) with flowers like crimson velvet, is still more superb than the *grandiflorus*. The different species of this genus are distinguished by a diversity of common names; when they are of a *round* form, they are called *Melon thistles*; when more cylindrical and erect, *Torch thistles*; when creeping, with lateral flowers, *Cereuses*; and when composed of a stem resembling flattened leaves, *Prickly pears*.

Plants of the Cactus tribe are mostly destitute of leaves, but the stems often appear like a series of thick fleshy leaves, one growing from the top of another. The beautiful die, called cochineal, is obtained from an insect of this name, which feeds upon the *Cactus cochinillifer*. The Cactus opuntia, or true prickly pear, is found native in the United States.[*]

The family *Amygdalæ* of Lindley, comprehends the peach and almond of the genus Amygdalus, with the plum, cherry, and pomegranate. These, which were placed by Jussieu in his order Rosaceæ, or rose-like plants, seem very properly separated. The characteristics of this tribe are a calyx 5-toothed, petals 5; stamens about 20, situated on the calyx; ovary superior, one-celled. The fruit a drupe. Trees or shrubs. The leaves and kernel contain prussic acid.[†]

PRUNUS is the genus which contains the various kinds of the plum, cherry, and sloe; this genus, according to ancient writers, was brought from Syria into Greece, and from thence into Italy. The Roman poets often notice its fruit. We have several native species of it.

The pomegranate (PUNICA) is a shrubby tree, which is a native of Spain, Italy, and Barbary, and flowers from June till September. The Greek writers were acquainted with it, and we are told by Pliny, that its fruit was sold in the neighbourhood of Carthage. It is cultivated in England and in the United States; not for its fruit, which does not come to perfection so far north, but on account of its large and beautiful scarlet flowers, which render it an ornamental plant. At Fig. 135, *a*, is the flower of the pomegranate, (*Punica granatum*;) *b*, represents the stamens of the same, as adhering to the calyx.

The genus AMYGDALUS contains the peach and the almond. The latter is a native of warm countries, and seems to have been known in the remotest times of antiquity.

Order Di-pentagynia.

The four following orders in the class Icosandria, are included under one, called Di-pentagynia, signifying two and five pistils.

We find here an important natural order, the Pomaceæ[‡] or apple tribe. This is included in Jussieu's Rosaceæ, or rose-like plants; but although the flowers of the apple genus have a strong resemblance to that of the rose, the difference in the fruit seems to render

---

[*] For illustrations of this family, see Plate 9 Figures 2, 5, and 7.
[†] Now known in chemistry as *hydrocyanic* acid.
[‡] So called from *Pomum*, an apple.

this division proper. In this tribe, the most important genus is Pyrus, which contains the apple and pear. The *varieties* of these fruits are the effects of cultivation, not the produce of different species. By means of grafting, or inoculation, good fruit may be produced upon a tree which before produced a poorer kind.

Jussieu divided his natural order Rosaceæ into the following sections ; the *Pomaceæ*, with fruit fleshy, like the apple and pear ; the *Rosæ*, having urn-form calyxes ; *Amygdalæ*, having drupe-like fruits.

Order Polygynia.

The rose tribe (*Rosaceæ*) resemble the apple tribe, in the appearance of the blossom, but the fruit, instead of being a Pome, consists, either of nuts containing one-seeded acines, as the rose, or of berries, as the strawberry.   The leaves have two stipules at their base. The rose unchanged by cultivation has but five petals.   We have few indigenous species of this genus ; among these, are the small wild rose, the sweet brier, and swamp rose.   *Red* and *white* roses are remarkable in English history as emblems of the houses of *York* and *Lancaster ;* when those families contended for the crown, in the reign of Henry the Sixth, the white rose distinguished the partisans of the house of York, and the red those of Lancaster. Among the nations of the East, particularly in Persia, the rose flourishes in great beauty and is highly valued.   The Persians poetically imagine a peculiar sympathy between the rose and the nightingale

The Blackberry (*Rubus*) has a flower resembling the rose in general aspect ; there are several species of the Rubus, one which produces the common blackberry, another the red raspberry, another the black raspberry, and another the dewberry.   One species, the *odoratus*, produces large and beautiful red flowers, the fruit of which is dry and not eatable.

The Strawberry belongs to the same natural and artificial order as the Rose.   The gathering of strawberries in the fields, is among the rural enjoyments of children, which in after life are recollected with pleasure, not unfrequently mingled with melancholy reflections, upon the contrast of that happy season, with the sorrows with which maturer years are often shaded. The fruit of the strawberry, as was remarked in the classification of fruits, is not properly a berry, but a collection of seeds, imbedded in a fleshy receptacle.

Icosandria furnishes us with a great variety of fine fruits, more perhaps than any other of the artificial classes.   A great proportion of the genera to be found in this class, are natives of the United States.

LECTURE XXXI.

CLASS XII.—POLYANDRIA.

In this class we find the stamens separate from the calyx, and attached to the receptacle or top of the flower-stem.   The number of stamens varies from twenty to some hundreds. This class does not, like the one we have last examined, contain many delicious fruits, but abounds in poisonous and active vegetables.   The mode of insertion of the stamens is to be regarded in considering the wholesome

Order Pomaceæ—Pyrus, varieties by grafting—Order Rosaceæ divided into sections—Rose tribe—Blackberry—Strawberry—Class Polyandria.

Fig. 136.

qualities of plants; it is asserted that no plant with the stamens on the calyx is poisonous; we know that many with the stamens upon the receptacle are so.

Order Monogynia.

We find in the first order some flowers of a curious appearance, as the Mandrake, or May-apple, (*Podophyllum ;*) the distinction between this and the mandrake of the ancients, was remarked under the class Pentandria. This plant is very common in moist, shady places, where you may often see great numbers growing together; each stem supports a large white flower, and two large, peltate, palmate leaves; its yellow fruit is eaten by many as a delicacy; the root is medicinal.

The Side-saddle flower (*Sarracenia*) is a curious and elegant plant; it has large leaves proceeding directly from the root. These leaves form a kind of cup, capable of containing a gill or more of water, with which liquid they are usually filled. The stem is of that kind called a scape, growing to the height of one or two feet, bearing one large purple flower. This plant is found in swamps; its common name, Side-saddle flower, is given in reference to the form of its leaf. It is sometimes called Adam's cup, in reference also to the shape of the leaf. No foreign plant, as an object of curiosity, can exceed this native of our own swamps; it is well worth the trouble of cultivation by those who are fond of collecting rare plants.[*]

The white Pond lily (*Nymphæa*)[†] is a splendid American plant, very fragrant, and with a larger leaf than almost any other northern plant. This flower closes at evening and sinks under the water; at the return of day, its blossoms rise above the surface and expand. The yellow Pond lily, (*Nuphar*,) though less showy, is equally curious in its structure.

In this artificial class and order is the Tea-tree, (THEA ;) of this plant there are two species, the bohea tea, (*bohea*,) and the green tea, (*viridis*.) It is a small evergreen-tree or shrub, much branched, and covered with a rough, dark-coloured bark. The flowers are white; the leaves are lanceolate and veined; the capsule or seed vessel is three-celled, opening; the seeds are three, oblong and brown. This shrub is a native of China and Japan. Some suppose that all the teas are taken from the same species, and that the different flavour and appearance of them depend upon the nature of the soil and culture, and the method of preparing the leaves. On account of the secret and jealous policy of the Chinese, the natural history of the Tea plant is less known than might be expected from its very general use. The Chinese begin in February to gather the tea leaves, when they are young and yet unexpanded. The second collection is made in April, and the third in June. The first gathering, which consists only of the young and tender leaves, is the *Imperial Tea;* the other two kinds are less odorous: the last collected is the coarsest and cheapest kind. Tea was introduced into Europe by the Dutch East India Company, in the year 1666, when it sold for

[*] See Plate iii. Fig. 5.
[†] An extensive locality of this plant exists upon the Saratoga lake. I have seen its surface for a quarter of a mile whitened by these lilies, occasionally intermixed with the yellow lilies, and the rich blue of the Pontederia, another beautiful aquatic plant.

Order Monogynia—Podophyllum—Sarracenia—Pond lilies—Tea-tree.

sixty shillings a pound, and for many years its great price limited its use to the most wealthy.

The poppy (*Papaver*) is a fine example of this class and order. Its numerous stamens standing upon the receptacle around the base of the germ, and its large stigma, with the two sepals of a caducous calyx, are conspicuous characters. Single poppies have but four petals; but the change of stamens to petals is very common in this flower, and most of the cultivated poppies are double. From the *papaver somniferum* is obtained the opium of commerce. The juice which issues from incisions in the green capsules, is dried in the sun, and usually made into cakes. Six hundred thousand pounds of this drug are said to be annually exported from the banks of the Ganges. The narcotic property of opium renders it highly valuable as a medicine. Why it is that certain substances, acting upon the human system, have power to affect the mind, no physiologist has yet been able to explain. But in the power of fermented liquors to produce changes in the mind, or of opium to lull its faculties into temporary oblivion, there is nothing more wonderful, than that the presence of light should produce vision, or the vibrations of the air, sound. All are equally beyond our knowledge; we may trace a series of organic changes, but the last link of the chain, that which connects body and soul, is concealed from our observation. Though narcotics can for a time,

> " Rase out the written troubles of the brain,
> And, with a sweet oblivious antidote,
> Cleanse the full bosom of that perilous stuff
> Which weighs upon the heart,"

yet, they who attempt to drown sorrow by artificial means, whether of the intoxicating bowl or the stupifying opium, find their sensibilities return with aggravated terrors. When properly used to allay bodily anguish, the product of the poppy may be considered one of our greatest blessings; but like all our blessings, it may, by its abuse, be made a curse.

The genus *Citrus*, which contains the orange and lemon, is found here. Jussieu places this in his order *Aurantia*, or golden fruits. The fruit is a berry with a thick coat. It furnishes *citric acid*.

Few valuable fruits, with the exception of this genus, are found in the class Polyandria.

Order Di-pentagynia.

The four orders following Monogynia, are, as in the preceding class, united into one, called as before, Di-pentagynia, having from two to five styles.

We find here some plants of a poisonous nature, as the Larkspur, Monk's-hood, and the Columbine; these belong to the natural order *Ranunculaceæ*, which contains also the Ranunculus or crow-foot, the anemone and gold-thread, (*Coptis*.)

In the same natural and artificial order we find the Peony, (*Pæonia*,) a large and showy flower, which, in its native state, has a calyx with 5 sepals, a corolla with 5 petals; 2 or three germs, each crowned by a stigma; the capsules or carpels are the same in number as the germs; each contains several seeds; this flower is remarkable for becoming double by cultivation.

Order Polygynia.

This order is divided into two sections: 1st, flowers with no ca-

---

Poppy—Opium—Power of opium and fermented liquors to affect the mind—Genus Citrus—Order Di-pentagynia—Natural order Ranunculaceæ—Peony—Order Polygynia.

.yx or perianth; 2d, with a perianth. In the first section we find several interesting native plants. The Clematis or Virgin's bower is a beautiful climbing plant, which supports itself by winding its petiole or leaf-bearing stems around other plants; the flowers are white and clustered in corymbs; the seed has a long silk-like fringe, which gives it a fine appearance after the blossoms have faded. This plant contains many species, and is cultivated both in this country and in Europe. At fig. 136, *a*, is a flower of the Clematis; *b*, represents its receptacle with numerous styles proceeding from it, and the petal and stamens separated, showing them to be inserted upon the receptacle.

The HELLEBORE (*Helleborus*) is an exotic much spoken of by classical writers. Hippocrates, one of the most ancient physicians, remarked upon its qualities; it grew about Mount Olympus, and was early known as a very poisonous plant.

The Magnolia and Tulip-tree are among the most splendid trees North America; they are said also to be common to China. e region of the Magnolia grandiflora extends from South Caro to the isthmus of Darien. In some cases these trees rise to the t of 90 feet before sending off any considerable branches; the ding top is then clothed with deep green, oblong-oval leaves, - laurel; these are, at most seasons, enlivened by large and nt white flowers.

class Polyandria, though not important for its fruits, con ome valuable medicinal plants, besides those which we have

LECTURE XXXII.

CLASS DIDYNAMIA AND TETRADYNAMIA.

two classes which are to afford subjects for our present ob- ions, are founded upon the number and *relative length* of the ens. In distinguishing their orders, the number of styles is not arded, but new circumstances of distinction are introduced, viz. the seeds being enclosed in a pericarp, or destitute of this covering, and the comparative length of pods.

CLASS XIII.—DIDYNAMIA.

This class has flowers with 4 stamens, two of which are longer than the other two; the stamens stand in pairs; the outer pair being longer, the inner pair shorter and converging.

Fig. 137.

The class contains two orders, Gymnospermia, (seeds naked or without a pericarp,) and Angiospermia, (seeds enclosed in a per ricarp.)

The *labiate* flowers are found in this class; these are monopetalous, and irregular in their outline. The term labiate is derived from the Latin *labia*, signifying lips; the flowers being divided at the top into two parts, resembling the lips of an animal. This tribe

Clematis—Hellebore—Magnolia—What classes are now considered?—How are their orders distinguished?—Labiate flowers.

is divided into *ringent,* or gaping, and *personate;* or closed. These terms have been used in an indefinite manner. Linnæus called the whole tribe *ringent ;* these he subdivided into labiate and personate. This division is illogical, since the *specific* term *labiate,* having lips, has a more general signification than the *generic* term *ringent,* lips gaping.

A few of the labiate flowers having but two stamens, are placed in the class Diandria, as the sage and mountain-mint. Yet they have, besides their two perfect stamens, the rudiments of two others, as if nature had designed them for didynamous plants. Linnæus remarks, that the insects most fond of frequenting these plants have but *two perfect wings ;* while the rudiments of two other wings may be found concealed under a little membrane;—How wonderful are the sympathies of nature !

When you examine a labiate flower, as balm or catmint, you will observe that the arched upper lip of the petals covers the stamens, and that the lower lip hangs down, so that you can see the inside of the corolla. If you pull out the corolla, you will find the stamens attached to it, as they usually are to monopetalous corollas. The corolla shows an aperture at the base through which the pistil ascended.

The labiate plants inhabit hills and plains exposed to the sun. The *aroma* which escapes from their flowers, denotes their stimulating medicinal properties. Their action upon the animal economy differs according to the quantity of *essential oil* and of *bitter principle* which they contain ; when the former prevails, as in mint, they are aromatic and stimulating ; when the bitter principle is in excess, as in germander, they act as tonics, and strengthen the digestive organs.

The pericarp of the labiate flowers belongs to Mirbel's class of fruits, called *cenobion.*

Order Gymnospermia.

The plants in this order have *labiate* corollas of the ringent kind ; the seeds are *four,* lying uncovered in the calyx ; the flowers grow in *whorls ;* the stem is four-angled, and the leaves opposite. The calyx is either five-parted, or the upper part consists of two divisions, called lips.

At Fig. 137 is a flower of the genus Tencrium, (*germander ;*) the corolla is ringent, the upper lip two-cleft, the lower lip three-cleft ; the stamens and pistils are *incurved ;* the stamens are *exsert* through the cleavage on the upper side ;—*b,* shows the pistil with its four uncovered, or gymnospermous seeds.

The ringent flowers generally grow in whorls at the upper part of an angular stem, the leaves standing opposite. These plants are never poisonous. Among them we find many aromatic plants the peppermint, lavender, savory, marjorum, thyme, &c.; also many medicinal herbs, as pennyroyal, catmint, horehound, &c.; the scullcap, (*Scutellaria,*) which has been said to be a remedy for the hydrophobia, the modest *Isanthus,* (blue gentian,) and a little flower of a most beautiful blue colour, called blue curls, (*Trichostema.*)

Order Angiospermia.

The second *order* contains those plants which have many seeds, contained in a capsule. Plants of this order appear to have an affinity with some families of the class Pentandria. Many in addition

How divided ?—Are all labiate flowers in the class Didynamia ?—What is said of the properties of these plants ?—What kind of pericarps have the labiate flowers ?— What plants in the order Gymnospermia ?—Describe Fig. 137—What is said of the ringent flowers ?—How is the order Angiospermia distinguished ?

to the four stamens, have a fifth filament, which appears to be the rudiment of another stamen ; sometimes the irregular corolla varies into a regular form, with five divisions. Among those which exhibit the imperfect fifth stamen, are the trumpet-flower, fox-glove, and Penstemon.

In this order the *personate* corollas are to be found, or labiate flowers with closed lips. Fig. 137, *c*, represents a flower of this kind, at *d*, is the pistil showing the capsule, or that the seeds are *angiosperm*. It should be observed, that in this order some few flowers may be found with bell-form and funnel-form corollas. Plants of this order differ much in their natural characters, from those of the order *Gymnospermia*. None of them are used in preparations for food, as are the thyme and savory of the first order, but many of them possess powerful medicinal properties, as the fox-glove,[*] and the cancer-root, (*Epiphegus*.) They are in general a beautiful collection of plants; few flowers are more splendid than the Gerardia and the trumpet-flower. The Martynia is an exotic of easy cultivation, bearing a fine blossom, while its pericarp furnishes an excelent pickle.

As plants of this class are numerous in every part of the United States, you will have no difficulty in procuring them for analysis they are not usually found in blossom until the middle of summer.

CLASS XIV.—TETRADYNAMIA.

Fig. 138.

In this class we find the *cruciform* plants, or such as have four petals in the form of a cross; the stamens are six, four of which are longer than the remaining two. The cruciform tribe forms the natural order *Cruciferæ*, having flowers with a calyx of four sepals, and a corolla of four petals; each petal is fastened to the receptacle or bottom of the calyx by a narrow part called a claw; the whole exhibiting the form of a cross; hence the term cruciform, from *crux*, a cross. In the centre of the flower is a single pistil, long and cylindrical; the stigma is oblong and divided into two parts, which are reflexed or bent back on each side. Each petal is placed between two leaves of the calyx; this condition is always seen in flowers where the number of the number of leaves of the calyx. The cruciform flowers have six stamens, two of which standing opposite to each other are shorter than the other four, which always stand in pairs. This inequality in their length determines them to be in the class Tetradynamia. The germ soon becomes a long pod called a *silique*, or a short thick one, called *silicula:* this difference in the length of the pods constitutes the distinction of the two orders of the class in which they are placed. The cabbage, mustard, radish, and stockgilly-flower belong to this family. They are found, on a chemical analysis, to contain some sulphur.

[*] See Plate vii. Fig. 6.

Personate flowers—Class Tetradynamia—Describe the cruciform plants.
15*

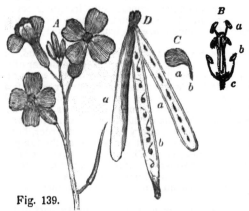

Fig. 139.

A flower of the cruciform tribe is represented at *A*, Fig. 139;—at *B* are seen the six stamens arranged in two sets, the four at *a* being longer than the two at *b ;* at *c* are two glands between the short stamens and the germ;—At *C* is a petal consisting of *a,* the border, and *b,* the claw; at *D* is the pod, which is a silique; *a* represents the valves; *b* the seeds, as alternately fastened to the edges of the partition, (dissepiment,) which divides this kind of pericarp into two cells. The cruciform plants have *dicotyledonous* seeds, polypetalous corollas, and the stamens are *hypogynous*. They are herbs, with leaves alternate. The flowers are usually yellow or white, seldom purple.

Plants of the class Tetradynamia are never poisonous ; they furnish many important vegetables for the table; their properties are antiscorbutic. The orders in this class are two, depending on the comparative length of the pods; this distinction is less definite than that which marks the orders of the class Didynamia.

Order Siliculosa.

The *first Order* contains plants which produce a short and round pod called a *silicula ;* a distinction in this order is made between such plants as have pods with a notch at the top, and such as have none, or are *entire*. The Pepper-grass, (*Lepidium,*) and the shepherd's purse, (*Thlaspi,*) afford examples of this order. At Fig. 138, *d,* is a representation of the silicula or pod of the Thlaspi. The plants found here, belong to the natural family *Siliquosæ,* the properties of which are nutritious and medicinal.

Order Siliquosæ.

The *second Order* contains cruciform plants with long and narrow pods; as the radish and mustard. The cabbage (*Brassica*) is an exotic; the turnip is a species of the same genus. At Fig. 138, *a,* is the wall-flower, (*Cheiranthus ;*) the calyx consists of four oblong sepals; the petals are obovate, spreading with claws as long as the calyx. At *b,* appear the six stamens divested of the petals; the germ is cylindrical, as long as the stamens ; *c,* shows the silique or pod; the valves are concave, and a thin membranous partition divides the silique into two parts.

In this lecture we have pointed out the most important characters of the two classes which depend upon considerations derived from the number and comparative length of the stamens. Both classes we found to have two orders, not as in the preceding classes, depending upon the styles; but in the one class, on the situation of the seed as lying in the calyx, or enclosed in a seed vessel; in the other class, from the comparative length of the pericarp or pod.

Describe Fig. 139—How many orders in the class Tetradynamia?—Order Siliculosa—Order Siliquosæ—Recapitulation

LECTURE XXXIII.

CLASS XV.—MONADELPHIA.

WE are now to considei the *brotherhoods*, as the names of the 15th and 16th classes signify; Monadelphia, meaning one, and Diadelphia two brotherhoods, in allusion to the manner in which the filaments are connected in *one* or *two sets*. The orders in these classes depend upon the number of stamens.

In the class Monadelphia, we include all such plants as have·their filaments united in one set, forming a tube at the bottom of the corolla ; in this respect, this class differs from the preceding ones, where the stamens are entirely separate; here you will observe that the *anthers* are separate, though the *filaments* are joined. We cannot in this class, as in the two preceding ones, point out any prevailing form of the corolla. The mark of distinction here is in some cases rather doubtful, the filaments being sometimes broad at their base, and yet not entirely connected.

Fig. 140.

You will recollect, that the orders depend upon the number of stamens. We have no first order here, for the character of the class is, *filaments united*, and one filament could not possess this requisite of union.

Order Triandria.

This is the first order in this class; the name, you will recollect, is the same as that of the third class, signifying three stamens ; but here they are united by their filaments, forming a tube. We find in this order a handsome plant, called blue-eyed grass, (*Sisyrinchium ;*) the three filaments have the appearance of being but one ; the corolla is tubular and 6-cleft, style 1, capsule 3-celled ; it belongs to the natural order *Iridæ*. The Mexican tiger-flower, genus *Tigridia*, is a splendid plant of this artificial order, and the natural order *Iridæ*. Its spotted flowers have given rise to the name which it bears.

Order Pentandria.

The *fifth Order* next occurs ; this presents us with the passion-flower, (*Passiflora*,) a climbing plant peculiar to the warm countries of America. "Its immensely long, and often woody branches, attain the summits of the loftiest trees, or trail upon the ground, adorned with perennially green or falling leaves, sometimes palmate or lobed like fingers, at others appearing like the laurel. They sustain themselves by means of undivided tendrils; and send out a succession of the most curious and splendid flowers, of which no other part of the world offers any counterpart."* Of this genus a number of species produce fruits of great excellence ; this fruit in South America is called Purchas. Sixty species of Passiflora are collected at the Linnæan garden near New York.† The generic characters of the passion-flower are a 5-parted, coloured calyx, 5 petals inserted upon the calyx, 5 stamens and three pistils, the nectary, a triple crown of filaments. The very singular appearance of this flower in the arrangement of its stamens in the form of a cross, and its triple crown,

---

* Nuttall.　　　　　　　　† See Prince's Horticulture.

The brotherhoods—Monadelphia—Orders—Order Triandria—What is said of the Passion-flower—Generic character and name.

has suggested the idea of its being emblematic of the passion or suffering of our Saviour; this is supposed to have given rise to its name. This plant has been placed in the class Gynandria, on the supposition that its stamens stood upon the pistil. An English botanist[*] thinks it belongs to the class Pentandria, and order Trigynia. Its situation in the class and order under which we have described it, is, however, that generally assigned it by American botanists.

In this order is the Stork's-bill geranium, (*Erodium ;*) it is an exotic, and belongs to the natural order *Geraniæ.*

Order Heptandria.

The *seventh Order* contains the genus *Pelargonium,* which includes the greater number of green-house Geraniums; it is taken from the tenth order, and placed here, because, though its flowers have 10 filaments, only 7 of them bear anthers, or are perfect. The flower of this genus is somewhat irregular. Among the varieties of the Pelargonium now cultivated in the United States, are,

The *Fairy-queen geranium,* with striped flowers, large and handsome leaves.

The *Fiery-flowered,* with cordate leaves, and black and scarlet flowers.

The *Balm-scented,* with leaves deeply five-lobed, the flowers dark red and black.

The *Grandiflorum* has an erect stem, little branched, with smooth leaves, from five to seven-lobed; as its name implies, the flowers are large.

The *Large-bracted* has an erect stem; leaves cordate, or heart-shaped, flowers large and white, with some streaks of purple.

Frequent-flowering, or *fish,* a shrubby, brown stem, with flat, cordate, five-lobed leaves, and red flowers, with spots of black and deep red.

Peppermint-scented, or *Velvet-leaved,* a shrubby stem, much branched; leaves cordate, five-lobed, soft to the touch like velvet, flowers small, white and purple.

Nutmeg-scented, or fragrant, an erect stem, much branched, leaves small, cordate and three-lobed, flowers small and pale, tinged with blue.

Royal purple, stem branched; flat cordate leaves, five-lobed; flowers large and of a bright purple.

Another genus of the Geranium family is called the Hoarea—this contains several varieties, differing chiefly from the Pelargonium in having a tuberous root, with *radical* leaves; most of the species are yellow. The plants of the natural family Geraniæ are mostly natives of the Cape of Good Hope, a region to which we are indebted for many of our finest exotics.

Order Decandria.

The *tenth Order* contains the genus *Geranium,* which differs from the Pelargonium, in having a regular calyx and corolla, and also in producing 10 perfect stamens, which vary in length, every alternate one being longer; 5 glands adhere to the base of the five long filaments. We have few native species of this plant; the common Crane's-bill, (GERANIUM *maculatum,*) with large, showy, purple flowers, is found in meadows during the first summer months. At

[*] Smith.

Stork's-bill geranium—Pelargonium?—Order Decandria.

Fig 140, *a*, is a flower of the genus Geranium. The three families, Erodium, Pelargonium, and Geranium, were, formerly, all united in one genus; but the difference in the number of stamens seems decidedly to separate them, not only into different genera, but different orders.

Order Polyandria.

The *thirteenth Order* (many stamens) is made up entirely of a group of genera which compose the natural order COLUMNIFERÆ of Linnæus; the stamens are united in the form of a *column*, (see Fig. 140, *b*;) by Jussieu they have been collected into an order, under the name of *Malvaceæ*, so called from the genus *Malva*. The peculiar characteristics of the whole group are, a calyx often double, 5 regular petals, stamens numerous, united by their filaments into a tube, and rising like a column in the centre of the flower; in the centre of this tube are the styles, forming an inner bundle, the number of these is various, though often found to be eight. The number of seed vessels, each of which contains one seed, equals the number of styles; these are arranged in a circle. Among the plants which compose this family, are the hollyhock, the mallows, and the cotton, (*Gossypium*.) The CAMELIA *japonica*, or Japan rose, a very splendid flower, equal in size to the largest rose, is found here. The rich colouring of its corolla contrasts beautifully with its dark green leaves.

Most of the native species of the class Monadelphia may easily be procured for analysis, in the season of flowers. The hollyhock is in almost every garden; the common mallows grows wild about dwellings; the lavatera, a hardy and cheerful-looking plant, though an exotic, spreads with great rapidity over our gardens and shrubberies.

The plants of this class vary in size, from the low mallows to some of the largest trees that have yet been discovered; "the Silk cotton tree (BOMBAX *pentandrum*) is so large, and spreads its branches so widely, that twenty thousand persons might stand under them. This tree is a native of Africa and America. The Adansonia, a native of Senegal in Africa, is said to grow to the size of 70 feet in circumference; this tree also attains great age. In 1749, the learned Adanson saw two of these trees in the neighbourhood of Gorrea, upon one of which was inscribed the date of the fourteenth, and upon the other that of the fifteenth century! yet there were good reasons to suppose that the trees were not young when the dates were cut. It may be conjectured that they have sometimes attained to the age of eight or nine hundred years! an immense period of time for the existence of any species of organized bodies."[*]

Having now considered the Class Monadelphia in its most important particulars, we will pass to the next class, which, in common with this, is founded upon the union of the filaments.

[*] B. S. Barton.

LECTURE XXXIII

CLASS XVI.—DIADELPHIA.

Fig. 14 î.

THIS is the class of *two* brotherhoods, the stamens being united by their filaments into two sets. The flowers of this class are *Papilionaceous*, or butterfly-shaped; this peculiar form of their corollas is an important mark of distinction.

Two circumstances should be noted here, in order to prevent you from falling into error with respect to this class.

1st. There are some plants with filaments united in *one* set, but with flowers papilionaceous; these are retained in Diadelphia, though there be no apparent division in the brotherhood or set.

2d. Though the flower be *papilionaceous*, if it have ten *separate* stamens, it is placed in the 10th class; this is the case with the cassia and wild indigo.

Linnæus, in reference to the form of the flowers, arranged this tribe under a natural order *Papilionaceæ;*—Jussieu, regarding the fruit, called the same *Leguminosæ.*

Papilionaceous Flowers.

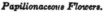

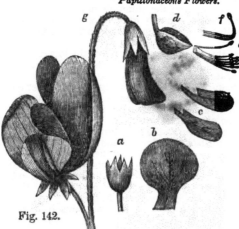

Fig. 142.

Fig. 142 represents the sweet pea (*Lathyrus odoratus;*) at *a*, is the five-toothed calyx; at *b*, is the upper petal, called the banner; at *c*, are the wings, or two side petals; at *d*, is the keel, formed of two petals united by their edges; at *c*, are the ten stamens, nine united and one separate; at *f*, is the pistil, the base of which, in process of time, becomes the pod or *legume.*

The flowers of the leguminous plants are so peculiar in appearance, that they are easily recognised. They are called by botanists, irregular. The rose, pink, and bell-flower, are regular in their form; that is, there is a symmetry and equality in their parts. There may be slight inequalities in regular corollas; as in the lily we sometimes see some petals a little longer than the others; this is an exception to the general rule. It is often owing to a want of discrimination between rules and exceptions, that young persons find difficulties in

Class Diadelphia—What two circumstances to be observed with respect to this class?—Natural order Papilionaceæ—Leguminosæ—Describe Fig. 142—Regular corollas.

understanding a science, thinking, very erroneously, that the knowledge of the one is as important as that of the other. If a clear conception of general rules be established in the mind, the exceptions will be easily learned. Irregular corollas differ so widely from the regular ones, that you will be in little danger of mistaking them for exceptions to the general rule; they constitute, indeed, a différent natural family, though, according to the artificial method of classification, they may often be placed near to regular corollas. Irregular corollas are various in their forms; the papilionaceous, which we are now considering, seem, as they stand upon their stem, to consist of an upper and under part. In examining a natural flower of this kind, a pea for example, you should first observe the calyx; this is monosepalous, that is, consisting of one sepal, ending in five distinct leafy points, (see Fig. 142, *a ;*) the two upper ones wider than the three under ones. The peduncle is slender and flexible, (see Fig. 142, *g ;*) thus the flower readily avoids a current of air by turning its back to the wind and rain.

In examining the corolla you will see that it is *polypetalous.* The first piece, or large petal, covering the others and occupying the upper part of the corolla, is called the *standard* or *banner.* This petal is evidently designed to protect the stamens and other parts of the flower from injuries by the weather. Upon taking off the banner, you will find that it is inserted by a little process or projecting part of the side-pieces, so that it cannot be easily separated by winds. The banner being taken off, the two side-pieces, or wings, are exposed to view; they are strongly inserted into the remaining part of the corolla, and their use appears to be that of protecting the sides of the flower. Upon taking off the wings, you will discover the last piece of the corolla, called, on account of its form, the keel, (*carina,*) or boat. This covers and protects the stamens and pistils. Upon drawing the keel downward, you will find the ten stamens, double in number to the petals; these stamens are joined together by the sides of their filaments, forming a cylinder which surrounds the pistil. One of the stamens, however, does not adhere to the rest; but as the flower fades and the fruit increases, it separates and leaves an opening at the upper side, through which the germ can extend itself by gradually opening the cylinder. In the early stage of the flower, this stamen will seem not to be separated; but by carefully moving it with a pin or needle, its filament will be found unconnected with the other nine.

The germ of the papilionaceous plant extends itself into that kind of pod called a *legume.* It is distinguished from the silique of the cruciform family, by having no partition in the legume. Besides, the seeds grow to one side only; but in the silique pod they are alternately attached to both edges of the partition. The legume opens lengthwise and rolls backwards; in the silique, the valves separate and diverge from the base upward. The seeds of this family have a marked scar, black spot or line, called the *hilum,* by which they adhere to the pod. Near this scar there is a minute opening into the body of the seed, through which moisture is imbibed at the period of its first growth or germination. The proper germ, or that part of the seed which is to be the future plant, continues to swell, and at length bursts through the coats of the seed, presenting between the divided halves, or cotyledons, the first true leaves, and the root.

Irregular corollas—In what manner should you proceed to examine a papilionaceous flower?—Distinction between the legume and silique—What is said of the seeds of the leguminous tribe?

The orders in the class Diadelphia, like those of the preceding class, are founded upon the number of stamens.

Order Pent-Octandria.

We could not expect from the character of the class, "stamens united into two sets," to find any plants with but one stamen. Those with five or eight stamens are all placed in one order called Pent-octandria, (five and eight stamens;) here we find the Corydalis, an elegant plant with bulbous roots; the corolla is rather ringent than papilionaceous. Fumaria is nearly allied to Corydalis by natural characters. In some cases the stamens have very broad bases, and scarcely seem united in this class. We find here POLYGALA, one species of which is called Seneca snake-root; this not only produces a beautiful flower, but is valuable in medicine. We have many species of this genus in our woods and meadows.

Order Decandria.—Leguminous Plants.

The *tenth Order* is wholly composed of plants with leguminous pods; the general character of these plants is, a calyx, often 5-parted, corolla 5-petalled, inserted on the calyx, and consisting of a banner, two wings and a keel; stamens generally 10, mostly united into two sets, 9 and 1; germ free; style 1; legume generally 2-valved, 1-celled, sometimes transversely divided into many cells; seeds affixed to the edge on one side.

At Fig. 141, *a*, is a flower of this kind; *b*, shows the stamens divested of their petals; *c*, shows the pistil, the germ already exhibiting the form and appearance of the legume.

In this large family of plants with leguminous pods are many genera of great importance in the vegetable kingdom; but when we are able to give striking natural characters, there seems to be less need of particularizing each genus. The form of the corolla and the nature of the fruit, with few exceptions, settle the character of this class.

The most savage nations usually pay some attention to Diadelphous plants. When Ferdinand de Soto marched his army into Florida, before the middle of the 16th century, he found the granaries of the natives " well stored with Indian corn and certain *leguminous seeds*;" which were probably the Lima bean, (*Dolichos*,) or some species of that genus, for the natives still continue to cultivate them.

The bean and pea tribes are found here. They consist of several different genera, as the vetch plants, Vicia, in which are many cultivated species, and the indigenous one, *Americana*. The Phaseolus, or kidney-bean, has its native as well as exotic species. The pea, so much valued as a table vegetable, belongs to the genus Pisum, a species of which, called Beach-pea, is found upon the shores of lakes and the sea-coast. The rattle-box (*Crotolaria*) with its inflated pericarp, is a favourite with children, who find it on sandy plains; it is a low pubescent plant with yellow blossoms. Of clover (*Trifolium*) there are many species, as the red, yellow, white, &c. The locust tribe contains many ornamental shrubs and trees.

The indigo (*Indigofera* tinctoria) of warmer climates, the red sandal-wood of the East Indies, the liquorice, and the sensitive plant, are all of this class. The gum-arabic is obtained from the acacia of the Nile, (*Mimosa nilotica*.) The liquorice of commerce is ob-

---

Order Pent-Octandria—Corydalis—Polygala—Order Decandria—General character of plants of this order—Savages cultivate these plants—Bean and pea tribe—Indigo, liquorice, &c.

tained by boiling the roots of the Glycirrhiza, a native of Italy and France. The tamarind is a native of tropical regions. The Arabians and Africans allay their thirst by the cooling freshness of the pulp contained in its legumes. Some plants of this class seem to possess active properties; the seeds of the Lupine are said to be poisonous. A traveller states, that the banks of the Nile are often visited in the night by the hippopotamus or river-horse, a large animal which does great damage to the gardens and fields; and that the inhabitants destroy the animal by placing a quantity of the Lupine seeds near where he is expected; these he devours greedily, they soon swell in his stomach, and distend it so much as to cause death.

The Furze (ULEX *Europæus*) is a very common plant in Europe, though not found so far north as Sweden. It is a flower of beautiful appearance; so much so, that Linnæus, as is said, when he first beheld it, fell upon his knees, in a transport of gratitude, and thanked the Author of nature for thus beautifying the earth.

A class called Polyadelphia, or many brotherhoods, having stamens united in *more than two sets*, was formerly admitted, but it was thought to be unnecessary, and the genera which it contained have been transferred to the class Polyandria; the St. John's wort (Hypericum) is among the plants which were in the rejected class Polyadelphia as has its numerous stamens in three clusters, not united by their ments; but all the species of the Hypericum are not thus divide o separate parcels of stamens. This distinction, as the of a class, is very properly laid aside; and the plants in the former 18th class, Polyadelphia, (*many brother*w placed in the 12th class, Polyandria, (*many stamens.*) two lectures, we have treated of two classes distin- ion of their filaments. In one class, Monadelphia, cter was that of filaments united in one set, forming class, no particular form of the corolla was found to l, unless we except the las der, in which the hollyhock av seve as an examp having a double calyx of an qual umber of divisions, a corolla of five heart-shaped petals, ut ed into one piece around the column formed by the united filaments.

In the class Diadelphia we found the marks of distinction to be, 1st. The union of the filaments into two sets;

2d. The papilionaceous corolla; and,

3d. The nature of the fruits, consisting of that kind of pod called a legume, and thus forming one great natural family of *Leguminous* plants, which furnish many of the most delicious table vegetables; such as peas, beans, &c.

LECTURE XXXIV.

CLASS XVII.—SYNGENESIA.

WE have now arrived at a class which contains a large portion of the vegetable tribes, particularly of those plants which blossom in the last summer months, and in autumn.

Furze—Class Polyadelphia, why rejected?—Recapitulation of the last two lectures —Class Syngenesia.

Fig. 143.

The term Syngenesia signifies a *union* of *anthers ;* this circumstance, you can readily conceive, forms a difference between this class and those which are distinguished by a union of filaments; in the one case, the tops of the stamens, or the anthers, are united, while the lower parts are separate; in the other case, the tops are separate, while the filaments, or lower parts of the stamens are united.

The number of stamens in plants of this class is mostly 5, distinguished from the fifth class not only by the compound character of the flowers, but by a union of anthers. In some cases, plants with five stamens have their anthers united, but having no other resemblance to those of the class Syngenesia, they are retained in the fifth class: the violet and impatiens are examples of this irregularity. This is an instance in which the artificial arrangement is made to bend to natural resemblances.

The term compound relates to the arrangement of the flowers, which are so closely connected as to have the appearance of one single flower. From the union of their stamens, these flowers are also called Syngenesious. The compound flowers have, by botanists, been distinguished under the three heads of *semi-flosculous,* (having *ligulate* florets ;) *flosculous,* (having *tubular* florets ;) and *radiated,* having *tubular* florets in the centre and ligulate at the circumference; the latter florets are called *rays.*

The *semi-flosculous* division contains a milky juice, which is bitter and of a narcotic quality; as the lettuce (*Lactuca*) and dandelion; their florets are all of one colour. The *flosculous* division usually exhibit in the leaves and roots a predominance of the bitter principle, as the burdock, (Arctium ; their florets are also of one colour. The *radiated* division is mostly composed of plants called Corymbiferous, (from *corymb* and *fero*, to bear,) because their flowers are corymbs, as the Chrysanthemum, Aster, &c. This division includes many beautiful flowers, with splendid colours ; and also affords many medicinal plants, as tansey and bone-set, (Eupatorium.) The colour of the florets in the disk and ray is often different in these flowers.

The compound flowers begin to blossom in the latter part of summer, and are found bordering upon the verge of winter. The dandelion is among the earliest flowers of spring, and one of the latest of autumn. The daisy is found in almost every spot which exhibits any marks of fertility; these are not single flowers, like the violet or rose, but crowded clusters of little florets.

The sun-flower is so large and conspicuous as doubtless to have frequently attracted your notice. If you examine one carefully, you will find it to be composed of more than a hundred little flowers, each as perfect in its kind as a lily, having a corolla, stamens, pistil, and seed. We distinguish the sun-flower into two parts,—the *disk*, which is the middle of the flower, and supposed to have resemblance to the middle or body of the sun ; the *ray* is the border of the flower, or those florets which spread out from the disk, as rays of light diverge from the sun. The florets in this, as in other compound flowers,

What does Syngenesia signify ?—What are the characteristics of this class ?—How are the compound flowers divided ?—Describe these divisions—Dandelion and daisy - Describe the sun-flower.

do not all begin to expand at the same time, they usually begin at the disk and proceed inwards towards the centre. If you examine with a microscope, one of the florets of the disk, you will perceive it to be tubular, containing one pistil surrounded by five stamens, which are separate; but the five anthers grow together, forming a tube around the pistil. It is this union of anthers which gives to it a place in the class Syngenesia. The florets of the ray are called neutral, having neither stamens nor pistils; the circumstance of neutral florets in the ray, places the sun-flower in the order *Frustraneu,* of the 17th class.

Although the term compound is confined to the flowers of the class Syngenesia, the real circumstance on which the class is founded is not the compound character of the flower, but the union of the anthers. A Clover blossom may in one sense be called compound, as it is a collection of many little flowers united; but each little floret of the clover has its own calyx; there is no general calyx enclosing the whole, as in most of the Syngenesious plants, but the florets are arranged in such a manner as to form a head; the anthers are separate, the filaments connected at their sides; and this latter circumstance, together with the papilionaceous form of the corolla, places the clover in the class Diadelphia.

Most of the Syngenesious flowers are composed of two sorts of florets, either *tubular,* with a toothed margin; or *strap*-shaped, (*ligulate,*) flat, but being also toothed at the edge; the latter are sometimes called Semi-florets, or half flowers.

Analysis of the Daisy.

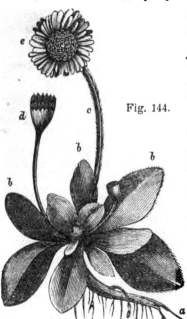

Fig. 144.

Fig. 144 represents the mountain daisy: we will consider its different parts.

1. The *Root,* (*a;*) this is *fibrous;* see the small thread-like parts issuing from the main root, or *radix;* from these fibres sometimes spring out little tubercles, it is then said to be *fibrous-tubercled.*

2. The *Leaves,* (*b;*) these spring from the root, and are hence called *radical;* being undivided, they are called *simple.* In form, they are somewhat oval, with the narrow end towards the stem; this form is called *obovate.* The leaves are said to be *ciliate,* on account of the hairs upon their margin.

3. The *Stem* (*c*) is called a *scape,* because it springs directly from the root, and bears no leaves; it is simple and pubescent.

4. The *Calyx* (*d*) is *hemispherical;* it is *common,* that is, enclosing many florets; the leafets of the calyx, sometimes called scales, are *equal.*

Is a clover blossom a compound flower?—Two sorts of florets in most of the compound flowers—Describe Fig. 144.

5. The *Corolla* (*e*) is *compound*, having many florets on one receptacle, *radiate*, having rays; the florets of the disk are *tubular*, (Fig. 145, *a*,) they have both stamens and pistils; they are funnel-shaped, and five-toothed; the florets of the ray (*b*) are flat, and have pistils without stamens.

6. The *Stamens* (*c*) are *five, united* at the summits by their anthers, forming a tube.

7. The *pistil* in the disk florets passes up through the tube formed by the anthers, (*d ;*) the stigma is parted into two divisions, which are *reflexed ;* the pistil in the ray florets passes up through the tube.

8. The plant has no *pericarp* or seed vessel; the seeds grow upon the receptacle, (*e ;*) they are single and shaped

Fig. 145.

somewhat like an egg; they are also naked, that is, destitute of the downy plume called egret, which is seen upon the dandelion, and many other of the syngenesious plants.

9. The *receptacle* is *conical*, or resembles in shape a sugar-loaf; it is dotted with little holes; these are the places in which the seeds were fixed; the appearance of the receptacle, whether naked or chaffy, is very important to be observed in the syngenesious plants, it sometimes constitutes a distinction between genera. The seed belongs to Mirbel's genus of fruits, *Cypsela*.

The botanical name of the daisy is BELLIS *perennis*. It belongs to the class 17th, Syngenesia, because the anthers are united; order 2d, Superflua, because the pistils in the ray are superfluous, having no stamens. The generic name, *Bellis*, is from an ancient Latin word, *belles*, handsome; from which comes also the French word *bel ;* the specific name, *perennis*, signifies that it is a perennial plant, or one whose roots live several years.

The common name, daisy, is derived from a property which many petals of the syngenesious plants possess of folding themselves at the setting of the sun, and expanding them with its rising. The poet Chaucer, who lived in the fourteenth century, is said to have first noticed this circumstance, and to have called the flower Day's-eye.

The orders of the class Syngenesia are founded on the situation of the several kinds of florets. We will, however, before explaining the orders, remind you of the distinction made in these florets.

1. *Perfect*, such as have both stamens and pistils.
2. *Barren*, or *staminate*, having only stamens.
3. *Fertile* or *pistillate*, having only pistils.
4. *Neutral*, destitute of either stamens or pistils.

They are also distinguished into *ligulate* and *tubular*.

The five orders in this class depend on the various situations of these different kinds of florets.

Order Æqualis.

The *first Order* contains those compound flowers which have all the florets perfect; this order is divided into three sections.

1st. Containing such as have *ligulate* florets; as *:e dandelion, let-
*uce, and vegetable-oyster.

2d. Florets *tubulous*, with flowers in a head; as the thistle, and
false saffron, (*Carthamus.*)

3. Florets *tubulous*, without rays; as, boneset, or thoroughwort,
(*Eupatorium.*)

You will find no difficulty in procuring for analysis, either dandelions or thistles; boneset is also abundant; therefore, for farther
investigation of this order we will refer you to the plants themselves,
aided by the generic and specific descriptions provided to assist
you in analyzing plants.

Order Superflua.

The *second Order* presents us with such compound flowers as
have the florets of the disk *perfect*, and those of the ray only *pistillate*, each pistil producing a perfect seed. The term *superflua* is
used, because the pistils in the ray, being unaccompanied with stamens, are said to be unnecessary, or superfluous.

This order is divided into two sections.

1st. Flowers *without rays*, or the ray florets indistinct; here we
find the tansey and the life-everlasting; of the latter there are many
species.

The ARTEMISIA, a genus which includes the wormwood and
southern-wood, both exotics, has but few native species. The name
Artemisia is often improperly given to an ornamental plant which
belongs to the genus Chrysanthemum. "The genus Artemisia was
named in honour of Artemis, the wife of *Mausolus*, whose monu-
ment was one of the wonders of the world, (hence our word Mauso-
leum.) Pliny observes that women have had, also, the glory of
giving names to plants."[*]

The 2d section of the order Superflua, includes such flowers as
have ligulate petals, arranged around the disk of the flower; these
are called rays. The receptacles in this section are naked, that is,
the top of the stem is found, on removing the different parts of the
blossom, to be smooth, without any hairs or down, this you may see
on the dandelion after the petals have fallen off. We here find the
star-flower, (ASTER,) a genus in which 120 species have already been
discovered; more than 60 of them are natives of the United States.
These are not seen in blossom until June and July; they appear in
flower until the approach of winter. Many of these flowers are
highly beautiful; the different species present a great variety of rich
and delicate colouring, from the dark blue, purple, and red, to a pale
blue, a light violet and pink, and in many cases, a pure white. In
some, the yellow prevails; sometimes they are variegated, and often
the disk and ray are of different colours. After having once become familiar with the Aster genus, you will seldom fail to distinguish it; but it is often difficult to determine the species. If you
meet with obstacles in this, you must not consider your time as lost;
comparison and research strengthen the mind, and the greater the
difficulties you overcome, the greater will be the advantage, in thus
accustoming yourselves to nice comparisons, and close investigations.

The golden rod (SOLIDAGO) is a numerous genus; the different
species are mostly yellow; in one section of these plants the flowers

* Thornton's British Flora.

are arranged in one-sided racemes, in another they form small and irregular clusters. The numerous species are in most cases so faintly distinguished, as to require some patience and application to trace out the specific differences.

The genus CHRYSANTHEMUM contains the common daisy, sometimes called ox-eye; it also includes many splendid foreign plants, mostly of Chinese origin. The Dahlia is at present a favourite with florists, who enumerate nearly a hundred splendid varieties.

Order Frustranea.

The *third Order* has the *disk florets perfect;* those of the *ray* are *neutral,* having neither stamens nor styles, though an imperfect seed is sometimes seen at the base of the florets; the name *Frustranea* alludes to this *imperfect seed.* We find here the Sun-flower, (HELIANTHUS;) this is a very good plant to examine, as the organs are large, and develop clearly the peculiar character of the class Syngenesia.

Fig. 143, *a,* represents the flower of the Coreopsis; *b,* a floret of the disk, with its bifid stigma above the tube formed by the united anthers; *c,* shows a ray floret, which is neutral.

In this order is the CENTAUREA *benedicta,* or blessed thistle, a native of Spain, which received its name on account of some extraordinary virtues which it was thought to possess; it was esteemed a remedy for the plague, with which warm countries are often afflicted. At present this plant is not much valued in medicine.

Order Necessaria.

The *fourth Order* includes plants in which the *rays* only are *fertile* or pistillate, and the *disk florets are barren* or staminate. We find here the marygold, (CALENDULA.)

Order Segregata.

The *fifth Order* contains a few genera, with each floret having a calyx proper to itself, besides a common calyx including the whole of the florets which make up the flower; this may be called a doubly-compound flower. The only plant of this order yet discovered in the United States is the elephant's-foot, (ELEPHANTOPUS,) a low, hairy-leaved plant, with purple, ligulate florets.

We have now completed a survey of the orders of the class Syngenesia, the plants which it contains are almost wholly referred to the natural order Compositæ or compound flowers: by Jussieu, they are subdivided into the three following orders.

Division of Compound Flowers by Jussieu.

1st, with florets all ligulate and perfect; leaves alternate, having milky juice; corollas mostly yellow. This includes the dandelion and lettuce.

2d order includes all compound flowers with tubular corollas; with receptacles fleshy and chaffy; egret stiff and bristly; leaves often with harsh prickles; flowers in a head. This includes the thistle, burdock, and false saffron.

3d order includes such compound flowers as have their inflorescence clustered in a corymb; as the life-everlasting, boneset, and aster.

The plants of the class Syngenesia are, in general, easily recognised at the first glance; there is something about them besides their

---

Chrysanthemum—Dahlia—Order Frustranea—Sun-flower—Coreopsis—Blessed thistle—Order Necessaria—Order Segregata—Elephant's foot—Order Compositæ—Jussieu's division of compound flowers.

compound character which distinguishes them from all other plants. One botanist observes, that they have a kind of " weed-like appear. ance, notwithstanding the beauty of their colouring; the stems and .eaves are often rough, and they seem to have been less completely reclaimed from their savage state, than most other plants, with the exception of the Cryptogamous class."*

Few plants of this class are poisonous; for though milky plants are generally so, those of this class are exceptions. The lettuce however contains a narcotic principle, and opium may be made from it. The dandelion, the thorough-wort, the chamomile, and wormwood, with many other plants of this class, are valued for medicinal properties.

The Syngenesious plants are particularly abundant in our own country, and you will never find difficulty in procuring specimens. If you commence botanical studies with the flowers of spring, nature gradually presents you with those that are more difficult to investigate. This class, it has been before remarked, are chiefly in blossom in the latter part of the season. Being previously prepared by a knowledge of the general principles of classification, and observations of plants, you will no doubt derive pleasure from the study of the class Syngenesia; though were you to commence a course of botany with these plants, you would feel as if thrown amidst a chaos of facts, without any clew to their classification.

LECTURE XXXV.

CLASS XVIII.—GYNANDRIA.

Fig. 146.

We shall now examine a class in which an entirely new circumstance from any yet considered, is regarded as forming its essential character. This circumstance is the *situation of the stamens upon the pistil;* the stamens appearing to grow out of that organ. In some cases the stamens proceed from the germ, in others, from the style. There is sometimes difficulty in deciding as to the number of stamens, for they are not here, as in other classes, distinct organs, but in some cases mere collections of glutinous pollen, called *pollinia.*

Order Monandria.

The orders in this class, as in Monadelphia and Diadelphia, depend on the number of stamens, or of those peculiar collections of pollen which are called stamens. The first order of the 18th class contains such plants as have but one stamen, or two masses of glutinous pollen, equal to one stamen; this order is divided into sections, with reference to the manner in which the anther is attached to the style; as, whether it is easily separated, whether the anther grows upon the top of the stigma, and also to the shape of the masses of pollen, which are called the anther.

* Barton.

Plants of this class valued for medicinal properties—Found in the latter part of the season—Class Gynandria—Orders.

Orchis tribe of Plants.

The natural order, Orchideæ, is composed of genera which be-
ong to the class Gynandria; the principal of these is the Orchis
genus, the different species of which are mostly perennial, and grow
in moist and shady places; some are parasites, adhering to the bark
ot trees by their fleshy, fibrous roots. The roots sometimes consist of two solid bulbs, in other cases, they are oblong, fleshy substances, tapering towards the ends like the fingers of the hand. The
name Orchis is derived from a Greek word, signifying an olive-ber-
ry, on account of the root being round, like that fruit. The distinguishing characters of this tribe, are a corolla, above the germ, 5
petals, 3 external and 2 internal. There is also in each corolla, a
petal-like organ called the lip, which varies in form and di
anthers always 1 or 2, and from 1 to 4-celled, sessile, or si
the side or apex of the style; the pollen is easily remove
cells in glutinous masses; the styles are simple, with visc
of various forms and positions. The capsules are 1-celle
3-keeled; the seeds are numerous and dust-like, the leav
stem like the leaves of grasses. The stems or scapes
and the flowers are arranged in spikes or racemes.

This natural order has *monocotyledonous* seeds, and s
gynous, or above the germ. The flowers are remarkab
irregular, and we might add, *grotesque* appearance; so
the figure of a fly, others of a spider, a bird, and ev
human figure. It would seem too that the freaks of these
ble beings are not designed for our observation, for they
peculiar in their choice of habitations as in their external
preferring wildness, barrenness, and desolation to the fostering
of man, or the most luxuriant soil. It is in forests of the equatorial
regions, that these plants appear in the greatest perfection. The
aromatic *vanilla* is obtained from the fruit of a climbing orchis of
those regions.

The Orchis genus has a nectary in the shape of a horn; its corolla is somewhat ringent, the upper petal vaulted, the lip is spreading, the 2 masses of pollen are concealed at the sides, by little sacs,
or hooded hollows of the stigma.

Fig. 146 represents a flower of this genus; *a*, shows the two
masses of pollen, brought out from the cells of the anther, which is
attached to the pistil.

Order Diandria

The 2d order contains the ladies' slipper, (CYPRIPEDIUM;) the nectary or lip is large, inflated, and resembles a slipper. We have
several species of this curious plant, some of which are yellow,
some white, and others purple.

Order Pentandria.

The 5th order contains the milk-weed, (ASCLEPIAS;) this by some
botanists is placed in the fifth class, on the supposition that the stamens do not proceed from the pistil.

Order Hexandria.

The 6th order contains the Virginia snake-root, (ARISTOLOCHIA
serpentaria,) a perennial plant, with brown fibrous roots; it is found
in shady woods, from New-England to Florida: the root is highly
valued in medicine; it possesses an aromatic smell, somewhat similar to spruce. It is said to have been found, by a chemical analysis,

to contain "pure camphor, a resin, a bitter extractive, and a strong essential oil." It was used by the Indians as a remedy for the bite of a snake; from this circumstance is derived its name. This plant in its medicinal properties differs essentially from the POLYGALA *senega*, or Seneca snake-root, and the mistaking one for the other, might, in critical stages of disease, be attended with fatal consequences.*

Order Decandria.

In the 10th order we find the wild ginger, (ASARUM;) this is a native plant, so low that its flowers are almost concealed in the ground; the roots are creeping and aromatic, having the taste and smell of the snake-root, (Aristolochia.)

We have now completed our view of the class Gynandria; although many species of it are indigenous to this country, you will not so readily procure specimens of this, as of most other native plants. The ladies'-slipper, milk-weed, and dogsbane, you can often find, but many of the plants of this family, particularly the Orchis tribe, opposing all attempts at cultivation, are to be found only in the depths of the forest, or places little frequented by man; like the aboriginal inhabitants of America, they seem to prefer their own native wilds to the refinements and luxuries of civilized life.

LECTURE XXXVI.

CLASSES MONŒCIA AND DIŒCIA.

IN all the classes hitherto examined, we have found *perfect flowers.* Our present inquiry is to be directed to two classes, in which the flowers are *imperfect,* or *both stamen and pistil are not found in the same individual flower.* The stamens are infertile, or disappear without any fruit; the pistils contain the germ, and being fertilized by the pollen of the infertile flowers, produce the fruit.

CLASS MONŒCIA.

Fig. 147.

The class Monœcia (one house) contains plants where, growing from the same root, we find some flowers containing only stamens, others only pistils. The orders in this class are determined by the number of stamens in each flower.

Order Monandria.

In the first order is the Bread-fruit tree, (ARTOCARPUS,) which grows to the height of forty feet, having fruit of the size of a large water-melon, hanging from its boughs like apples; it is a native of the East Indies; when roasted it resembles white bread, and is much valued for food.

This plant belongs to the natural order Urticæ, in which are the Fig and Mulberry.

* A physician prescribed for a sick child the Seneca snake-root, (POLYGALA *senega*;) the ignorant apothecary sent the Virginia snake-root, (ARISTOLOCHIA *serpentaria*.) The physician having fortunately remained to inspect the medicine which he had ordered, the mistake was seasonably discovered. This instance shows the importance of botanical knowledge, particularly in those who attempt to deal in medicine. Had the mother of the child understood botany, the mistake would have been discovered although the physician had not been present.

Wild ginger—Concluding remarks—In what respect do the two next classes differ from the preceding ones?—Class Monœcia—Order Monandria—Bread-fruit.

Order Triandria.

In the third order we find a very common plant, called cat-tail, (TYPHA;) this grows in swampy meadows, and stagnant waters, to the height of four or five feet. The long, brown spike, which grows at the summit of the stem (giving rise, from its peculiar appearance, to the name cat-tail) is the *catkin;* the upper part consists of sta minate flowers, having neither calyx nor corolla; the three stamens arising from a chaffy receptacle. The pistillate flowers form the lower part of the spike; each one produces a seed, supported in a kind of bristle. This plant is sometimes used by the poorer class of people for beds, but is considered by physicians as unhealthful on account of certain properties inherent in its substance. The leaves and stems of the Typha are employed for bottoming chairs and making mats; the young stalks are said to answer as a substitute for asparagus; the pollen of the flowers, which is very abundant and inflammable, is recommended by a French writer to be employed on the stage for fire.*

The sedge, or CAREX, is a genus consisting of nearly 140 known species. Though a grass-like plant, it is separated from the family of grasses, which are mostly in the 3d class, on account of the monœcious character of its flowers. A treatise upon this genus, called Caricography, has been lately published by an American botanist.* This extensive genus belongs to the natural order Cyperoideæ, so called from Cyperus, one of the most important genera of the order. This tribe of coarse grasses inhabit marshy grounds; though resembling the true grasses in their general aspect, they differ from them in having stems without joints, and often triangular. Unlike the grasses, they are of little utility; they spread rapidly, and often destroy the best pastures, by overrunning them. A species of Cyperus, the papyrus, which grows in abundance on the banks of the Nile, was used by the ancients in the manufacture of a kind of thick paper. A thin fibrous membrane beneath the bark was obtained, and several thicknesses being glued together, the whole was pressed into sheets. Fragile as it was, this parchment is still to be seen in ancient records, and offers to the observation o. the curious, the autographs of Egyptians, Greeks, and Romans. (See Plate 6, Fig. 5.)

The Indian corn (ZEA *mays*) is found in this order. The top or panicle consists of staminate flowers only, and of course never produces corn; the pistillate flowers grow in a spike enclosed in a husk; each pistil produces a seed, called corn; the pistils are very long, forming what is called silk. This genus belongs to the natural order Gramineæ.

Order Tetrandria.

We here meet with the mulberry, (Morus,) whose leaves furnish nourishment to the silk-worm. The white mulberry, MORUS *alba*, is the species which is chiefly used for this purpose. This plant belongs to the same natural order as the bread-fruit and fig.

Order Pentandria.

The 5th order contains the genus AMARANTHUS, in which is a very common weed, seeming to have some analogy to the pig-weed, not only in natural properties, but in being dignified with a name which

---

* See Plate i. Fig. 6.   † Professor Dewey.

forms a striking contrast with its mean appearance. This genus, however, contains some elegant, foreign species; one of which, AMARANTHUS *melancholicus,* has received the whimsical name of Love-lies-bleeding; probably from the circumstance of its long, red flower-stalks drooping and often reclining upon the ground. Another species, called Prince's feather, is always erect. The Cock's-comb is a well known plant of this genus. The Amaranth, whether from its being a good word to fall in with poetical measure, or from some fancied intrinsic beauty, has ever been a favourite with poets. Milton says of the angels,

> ————————" To the ground,
> With solemn admiration, down they cast
> Their crowns inwove with amaranth and gold;
> Immortal amaranth, a flower which once
> In Paradise, fast by the tree of life,
> Began to bloom, but soon for man's offence,
> To Heaven removed.
> With flowers that never fade, the spirits elect
> Bind their resplendent locks, inwreathed with beams."

In Portugal and other warm countries, the Globe Amaranth is used for adorning the churches in winter.

Order Polyandria.

This order contains many of the most useful and beautiful of our forest trees, forming the natural order, *Amentaceæ.* Fig. 147 represents a branch of the Corylus, (Hazle-nut;) at *a,* are the aments or catkins, formed wholly of staminate flowers; at *b,* is a bract or scale of the ament with adhering stamens; at *c,* are the pistillate flowers surrounded with scales; at *d,* is a pistillate flower, having two styles. The oak, beach, walnut, chestnut, birch, &c., bear their staminate flowers in nodding aments; their pistillate flowers are

Fig. 148

surrounded with scales for calyxes. The stems of these plants are woody and exogenous; you will recollect that such stems increase in diameter by new wood being formed around the old, and that this new wood is formed from the cambium which flows downward between the wood and bark. Fig. 148 shows a portion of the trunk of an oak, supporting the stem of a twining plant. As the oak is a dicotyledonous tree, its trunk is annually increased by new layers which are developed between the bark and wood;—hence it will be seen, that if any foreign substance encircles the trunk, it must, in time, produce a protuberance. The cambium from which the new layers are formed, is interrupted in descending, and accumulates just above the interposing body, forming the swellings that appear there, as are represented in the cut. Walking canes are often made of stems thus knotted. The Celastris scandens is one of the most common twining plants of our woods.

This order contains the genus CALLA, of which we have some native species, and which includes the elegant exotic, CALLA *ethiopica,* or Egyptian lily. In this genus, the flowers having neither calyx nor corolla, grow upon that kind of receptacle which is called a spadix; the staminate and pistillate flowers are intermixed, the

---

Different species of the Amaranthus—Order Polyandria—What is said of the natural order Amentaceæ?—Explain Fig. 148—Calla—Different species.

anthers have no filaments, but are sessile; the berries are one-celled, many-seeded, and crowned with a short style.   This spadix thus covered with the fructification, stands erect, surrounded by a

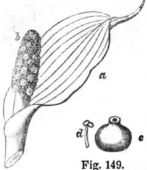

spreading, ovate spatha; this, in the Egyptian lily, is of pure white, presenting a very showy appearance. Without attention to the structure of the plant, you would probably suppose the spatha to be the corolla. The leaves are sagittate, or arrow-form.   The CALLA *palustris*,* a very common American plant, is represented at Fig. 149: at *a*, is the *spatha*, which is *ovate*, *cuspidate*, and *spreading;* at *b*, is the spadix covered with the fructification, the staminate and pistillate flowers being intermixed and uncovered; at *c*, is a pistil magnified, showing the style to be very short and the stigma obtuse; at *d*, is a stamen bearing two anthers.

Fig. 149.

The Wild-turnip is nearly allied to the Calla; they belong to the same family, *Aroides*, distinguished by peculiar characteristics; such as the mode of infloresence, fleshy and tuberous roots, and large, sword-shaped, or arrow-shaped leaves.

The arrow-head (*Sagittaria*) is unlike most of the Monœcious plants in general appearance; it has three sepals and three white petals; it is not unlike the spider-wort in the form of its flowers. Many species of this delicate-looking plant may be found in autumn, in ditches and stagnant waters.

Order Monadelphia.

The 15th order, or that in which the *filaments* are united in a column, presents us with the Cucumber tribe, (*Cucurbitaceæ;*) this includes not only the proper CUCUMIS, or cucumber, which is an exotic, but some native genera of similar plants; we find here the gourd, squash, watermelon, and pumpkin.   These plants have mostly a yellow, 5-cleft corolla; calyx 5-parted, 3 filaments united into a tube; a large berry-like fruit, called a Pepo; this, in the melon, is ribbed, and in the cucumber uneven and watery.   We find in the same artificial order a very different family of plants, called *Coniferous*, or cone-bearing plants; these have the staminate flowers in aments, each furnished with a scale or perianth supporting the stamens; the pistillate flowers are in strobilums, each furnished with a hard scale.   The stems are woody, the leaves evergreen, and the juice resinous.   To this natural family belong the pine and cypress.

The character of trees may be studied to advantage at four different seasons; in winter, when the forms of the ramification can be seen in the naked boughs, and the leaf and flower buds examined in their inert state; in spring, when in blossom; in summer, when the foliage is in perfection; and in autumn, when, during the first stages of decay, the mellowness and variety of teints afford beautiful subjects for the pencil of the painter, and for those who love the study of nature under all her forms.

* From *paluster*, signifying swampy, or growing in marshy places.

Describe Fig. 149—Family Aroides—Arrow-head—Order Monadelphia—Cucumber tribe—General character—Cone-bearing plants—Best periods of studying plants.

Fig. 150.

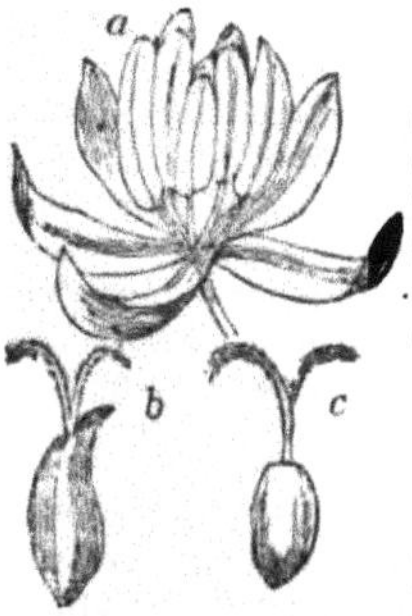

CLASS DIŒCIA.

The class Diœcia (two houses) has staminate and pistillate flowers on separate plants. The distinction with regard to the orders, as in the preceding class, is derived from the number of stamens.

There are no plants of the first order, or with one stamen.

Order Diandria.

The 2d *Order* contains the willow, (SALIX,) which has long and slender aments, both of staminate and pistillate flowers, the two kinds being on separate trees.

The order TRIANDRIA contains the fig, (FICUS,) remarkable for containing the flower within the fruit; this is botanically considered as a juicy receptacle, within which are concealed the minute flowers and seeds. The fig is peculiar to warm countries.

TETRANDRIA contains a parasitic plant, the *Mistletoe;* only one species is indigenous to this country. The Druids* considered this plant as sacred to the sylvan deities. Tradition relates, that where Druidism prevailed, the houses were decked with this plant, that the sylvan spirits might repair to them.

The order PENTANDRIA contains the hemp, hop, &c. Fig. 150 represents the pistillate and staminate flowers of the hemp, (*Cannabis sativa;*) at *a*, is the barren or staminate flower, containing five stamens, and having its calyx deeply five-parted; the corolla is wanting. At *b*, is a fertile or pistillate flower with its calyx opening laterally [illegible] shows the same flower divested of its calyx; the seed is a [illegible] crowned with two styles. The hemp belongs to the [illegible]*ea*, (from *Urtica*, a nettle;) the fibres of its stems [illegible] cloth, cordage, and thread. The hop produ[illegible]its fertile fl[illegible] large cones formed of membranous, imbri[illegible]ed scales; th[illegible]s have a peculiar odour, which is said to [illegible]oduce a narcotic [illegible]pon the brain. The use of the flowers of [illegible] hop to produc[illegible]entation in beer are well known. This [illegible]nt contains a sma[illegible]ion of the nitrate of potash, (saltpetre.)

[illegible]EXANDRIA contai[illegible]oney-locust and green-brier.

[illegible]CTANDRIA has th[illegible]r, (POPULUS,) similar in natural character [illegible] willow.

[illegible]PHIA, [illegible] order, contains the red-cedar and the yew, w[illegible]ne-bearing family, with the pine and cypress.

We [illegible]leted our remarks upon two classes which have imp[illegible]owers. Our review of these has been brief, when compared to the many interesting facts which presented themselves, in association with the various important plants which we have passed in rapid succession.

* The Druids, it is supposed, derived their name from *drus*, a Greek word, signifying oak, as it was in groves of this tree that the priests celebrated their mysterious rites, and sacrificed human victims to their sanguinary deities.

Class Diœcia—Willow—Fig—Mistletoe—Hemp—Hop—Order Hexandria—Octandria—Monadelphia.

LECTURE XXXVII.

CLASS XXI.—CRYPTOGAMIA.

Fig. 151.

THE twenty preceding classes include the Phenogamous plants; we are now to consider the Cryptogamous class;—we here find the stamens and pistils either wholly concealed from observation, or only manifest upon the strictest scrutiny. These plants constitute the first class of Jussieu's method called *acotyledonous;* their seed being destitute of any cotyledon.

As we proceed in this last of the Linnæan classes, we shall find all our former principles of arrangement fail us, and it might almost seem as if we had entered upon a new science. The class Cryptogamia includes all plants which do not find a place in some of the other classes.

Ferns, mosses, lichens, and mushrooms, constitute the principal part of this class. At Fig. 151, *a*, is a fern, of the genus Asplenium, which bears its fruit on the back of the fronds; at *b*, is a moss of the genus Hypnum, showing two of its flowers borne on slender pedicels; at *c*, is a genus of the Lichen family; at *d*, is the Agaricus, one of the most common of the mushrooms.

Some writer has said, that Linnæus, having arranged the plants which would admit of classification, took the remainder and cast them all into a heap together, which he called Cryptogamous;—he did not, however, rest satisfied in thus throwing them together, but subdivided this miscellaneous collection into orders; or we might more properly say, that he gave names to those divisions already marked out by nature.

Of these orders, which are natural families brought together on account of general resemblances and analogies, without reference to any one principle, there are six.

Order Filices, or Ferns.

The 1st *Order* contains the Ferns; their plume-like leaves are

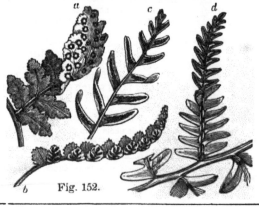

Fig. 152.

---

called *fronds*. The fruit mostly disposed in dots or lines, grows on the back, summit, or near the base of the *frond*. At Fig. 152, is a delineation of some of the various modes in which the fructification of ferns appears; *a*, is the genus POLYPODIUM or *polypody*, with capsules in roundish spots on the back of the frond; *b*, ASPLENIUM, capsules in lines nearly parallel, diverging from the centre of the frond; *c*, BLECHNUM, capsules in uninterrupted lines running parallel to the midrib of the frond on both sides; *d*, PTERIS, or brake, capsules forming lines on the edge of the leaf.

Some ferns bear their fruit in a peculiar appendage, as a spike or protuberance in the axils, or at the base of the leaves; no appearance of flowers in these plants is ever presented. When the brown or white dust-like spots are examined with a microscope, they are found to consist of clusters of very small capsules, at first entire, but afterward bursting elastically and irregularly. Besides attention to the situation and form of the capsules, it is necessary to observe the membrane which envelopes them; this is called their *involucrum*.* The seed is as minute as the finest powder, and so light as to be wafted by the air to any distance or height; we thus often see ferns growing high on the trunks of trees, or on the summits of old buildings. Some ferns grow to a great height in southern latitudes, almost like trees. At the southern extremity of Van Diemen's Land, a species has been found, whose trunks attained to the height of five or sixteen feet. One species in our country, ONOCLEA *sensi-*, called the sensitive fern, is said to wither on being touched by human hand, though the touch of other substances does not produce the same phenomenon.

The number of species of ferns which are already known, amounts to about seven hundred. They generally abound in moist and shady situations, but are sometimes found on rocks and dry places, and on the trunks and branches of old trees. The frond, or leaf of the fern, is often *pinnate*, or divided like a feather; sometimes it is undivided, and resembles a palm-leaf.

The EQUISETUM *hyemale* is known to housekeepers under the name of *scouring-rush*. The quantity of silex contained in the cuticle, renders it a good substitute for scouring-sand.

Order Musci, or Mosses.

The 2d *Order* contains the mosses, which are little herbs with distinct stems; their conical, membranous corolla is called a *calyptra*, or veil, its summit being the stigma; this veil clothes the capsules, which before the seeds, called *sporules*, ripen, is elevated on a footstalk. The capsule, called *theca*, is of one cell, and one valve, opening by a vertical lid; the seeds are very numerous and minute. In some genera the veil is wanting, this serves as a distinction in the order. The barren flower of mosses consists of a number of nearly cylindrical, almost sessile anthers; the fertile flowers have one perfect pistil, seldom more, accompanied by several barren pistils. Both stamens and pistils are intermixed with numerous, succulent threads. You may here observe (Fig. 153) the different parts of mosses; *a*, represents the *theca*; *b*, the *pedicel*, or stem; *c*, the *sheath*, which,

---

* Also called *indusium*. The capsules are the *thecæ*; a collection of them, *sori*; the seeds are *sporules*.

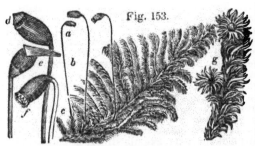

Fig. 153.

before the pedicel grew up, served as a kind of calyx to protect the embryo fruit; *d*, the *operculum*, or lid which, before the theca is ripe, is covered by the calyptra; *e*, the *calyptra*, or veil; *f*, the *fringe*, or teeth, which, when the theca is ripe, and has thrown off its other parts, often appear around 'ts edge; *g*, the barren or staminate flower of a moss.

The mosses are generally perennial and evergreen, and capable of growing in colder climates than most other vegetables. In Spitzbergen, the rocks which rise from the surrounding ice are thickly clothed with moss. A botanist who travelled in Greenland, counted more than twenty different species of moss without rising from a rock where he was seated.

All the parts of the mosses which have been described, are not seen without the assistance of a good microscope. It is not to be expected that young botanists will be fond of this department of the science, although those who become acquainted with it, discover much enthusiasm in its pursuit. The following interesting remarks on Cryptogamous plants are taken from an English writer.

"Mosses and Ferns, by the inconsiderate mind, are deemed a useless or insignificant part of the creation. That they are not, is evident from this, that He who made them has formed nothing in vain, but on the contrary has pronounced all his creation to be *good* Many of their uses we know; that they have many more which we know not, is unquestionable, since there is probably no one thing in the universe, of which we can dare to assert, that we know all its uses. Thus much we are certain of, with respect to mosses, that as they flourish most in winter, and at that time cover the ground with a beautiful green carpet, in many places which would otherwise be naked, and when little verdure is elsewhere to be seen; so at the same time, they shelter and preserve the seeds, roots, germs, and embryo plants of many vegetables, which would otherwise perish. They furnish materials for birds to build their nests with, they afford a warm winter's retreat for some quadrupeds, such as bears, dormice, and the like, and for numberless insects which are the food of birds and fishes, and these again the food or delight of men. Many of them grow on rocks and barren places, and by rotting away, afford the first principles of vegetation to other plants, which never else could have taken root there. Others grow in bogs and marshes, and by continual increase and decay, fill up and convert them into fertile pastures, or into peat-bogs, the source of inexhaustible fuel to the polar regions.

"They are applicable also to many domestic purposes. The *Lycopodiums* are some of them used in the dying of yarn, and in medicine; the *Sphagnum* (peat-moss) and *Polytrichum*, furnish convenient beds for the Laplanders, and the *Hypnums* are used in the ti-

hng of houses, stopping crevices in walls, packing brittle wares, and
the roots of plants, for distant conveyance.

" To which may be added, that all in general contribute entertain-
ment and agreeable instruction to the contemplative mind of the
naturalist, at a season when few other plants offer themselves to his
view.

" The Fungi have been suspected by some to be, like sponges and
corals, the habitations of some unknown living beings, and being alkaline, have been classed in the animal kingdom; but they are known
to produce seeds, from which perfect plants have been raised; and
the celebrated Hedwig, by great dexterity of dissection, and by using
microscopes of very highly magnifying powers, assures us that he
has discovered both stamens and pistils, not only in this order of
plants, but in the other orders of the Cryptogamous family."*

Order Hepaticæ, or Liverworts.

The 3d *Order* contains the Liverworts, which are more succulent
or juicy than the mosses; they have four-valved thecæ, which circumstance, and that of their not opening with a lid, distinguish them
from the mosses. Their name, Hepaticæ, signifies liver; but it is
not yet known whether they received that name on account of some
supposed virtue in curing diseases of the liver, or whether it was
because they were thought to resemble the lobes or divisions of that
organ. One of the most common genera of this order is the *Junger
mannia ;* you may here see (Fig. 154) a species of this, the *compla
nata*, with its parts, as represented under a magnifier.

Fig. 154.

a, is a plant of
natural size, in
fruit. *b*, the fruit
magnified, show-
ing the sheath, the
peduncle rising
from it, and the
theca at top, not
yet burst. *c*, the
open capsule
splitting and dis-
charging the seeds. *d*, the theca empty, showing its four valves.

Order Algæ, or Sea-Weeds.

The 4th *Order* includes the sea-weeds and frog-spittle; these have
leathery fronds, with fine dust-like seed, enclosed in inflated portions
of the frond. They are almost always aquatics; generally green or
reddish. One genus of this family is the *Fucus.* The Fucus *nutans*,
sometimes called the gulf-weed, is very abundant in the Gulf of Florida, and is found in various parts of the ocean, forming masses or
floating fields, many miles in extent. The plant seems to possess
no distinct root, though it perhaps originally vegetated on some sea-
beaten shore, from whence it was by accident thrown upon the
ocean's wave.

The Fucus *giganteus* is said to have a frond of immense length ·
from whence its specific name, signifying gigantic. You are here

* Notwithstanding the weight which Thornton, author of the above quotation,
gives to the opinion of Hedwig and others, it is, at present, much doubted by natura,
ists, whether the Fungi have organs analogous to stamens and pistils.   .

Describe Fig. 154—Liverworts—Derivation of the name—Sea-weeds—Fuci—Gulf-
weed.

presented (Fig. 155) with a delineation of three kinds of Fuci.* *a*, is FUCUS *nodosus*, (knobbed fucus;) this has forked fronds. The knobs which appear in the fronds are air-bladders, which Fig. 155. render it peculiarly buoyant upon the water. This is often more than six feet long. *b*, FUCUS *vesiculosus*, (bladder fucus;) here the air-bladders are mostly axillary, and at the sides of the midrib. It varies in length from one to four feet. On account of its mucilaginous property it forms a good manure; in some of the countries of Lapland it is boiled with meal, and given for food to cattle. *c*, FUCUS *serratus;* this has a beautiful *serrate* frond.

The Fuci,† on burning, afford an impure soda, called *kelp*.

Order Lichenes, or LICHENS.

The 5*th Order* contains the LICHENS; these are various in texture, form, and colour; they are leathery, woody, leaf-like, white, yellow, green, and black. When wet, they often appear like green herbage; some are seen on stones, or old fences and buildings; others with strong, green filaments, are suspended from branches of trees, and improperly called mosses. The fruit of the Lichen consists of saucer-like bodies, called *apothecia*, in which the seeds are contained; this may be seen in the following delineation. Fig. 156, *a*, represents

a lichen, of a leaf-like appearance; here the apothecia imbedded in the leaves are very apparent. *b*, is a lichen resembling a drinking-glass. *c*, is the rein-deer moss, furnishing almost Fig. 156. the sole food of that useful animal, so important to the existence of the Laplander. In the middle of Europe it grows only to the height of two or three inches; but in Lapland it sometimes attains to the height of one foot and a half.

Many of the lichens are useful on account of their colouring matter. *Litmus*, which is so common as a chemical test for acids and alkalies, is obtained from a species of white lichen, called *Orchal*, or *Archil*, this is also used for giving a crimson colour to wool and silk. The powder called *cudbear*, used for dying purple, is obtained from

---

* See also Appendix, Plate viii. Fig. 8, 9, 10, 11.
† Fuci is the plural of Fucus.

Three kinds of Fuci—Kelp—Lichens—Explain Fig. 156—Uses of the lichens.

a lichen. The order Lichen has sometimes been included under one genus called Lichen, and placed in the order *Algæ.*

Order Fungi, or MUSHROOMS.

The *6th Order* contains the *Mushrooms,* or fungus plants ; these never exhibit any appearance of green herbage ; they are generally corky, fleshy, or mould-like, varying much in form and colour. The fruit of some is external, of others internal. They are often of very quick growth and short duration. The genus *Agaricus,* which contains the common eatable mushroom, has a convex, scaly, white head, called a *pileus ;* this is supported on a stalk called a *stipe.* On the under surface of the pileus, or cup, are seen many flesh-coloured membranes called *gills.* These gills, in the young state of the mushroom, are concealed by a wrapper called a *volva,* which is considered as a kind of calyx. As the mushroom becomes older, the volva bursts and remains upon the stipe, while the pileus, released from its confinement, extends upwards and exhibits an uneven appearance upon its edge, caused by its separation from the volva.

Fig. 157.

Fig. 157 represents the most important parts of the mushroom ; *a,* the gills running from the stipe to the circumference, under the pileus ; *b,* a young mushroom, with the pileus of a globular form, and not separated from the volva ; *c,* the volva, or wrapper, bursting and separating from the pileus so as to exhibit the gills beneath ; *d,* part of the volva remaining upon the stipe in a circular form, and called an *annulus,* or ring.

"If the mushroom be left for a time on a plate of glass, a powder will be found deposited ; this is the seed,[*] or organic germ. That these are capable of germination, is evident to cultivators, who now form mushroom beds, by strewing the decayed plants on prepared beds of manure."[†]

A species of the genus Agaricus is common in Italy, and much valued for food ; it is of a fine red or orange-colour ; the ancient Romans esteemed it as a great luxury. The genus *Boletus* contains the *touchwood,* or *spunk,* which is sometimes used as tinder. The LYCOPERDON contains the puff-ball.

The Cryptogamous plants are probably the least understood of all the visible works of nature. Philosophers have asserted that some of this race do not belong to the vegetable, but to the animal kingdom ; having discovered insects in mushrooms, they say, like the sponge and the corals, these should be classed among animal productions. Few, however, at present, entertain this belief ; and the fact of their having been raised from seed sprinkled on the earth, proves them to be of vegetable growth. A curious field of inquiry presents itself in the consideration of the difference between animal and vegetable life. This we shall hereafter partially examine ; not, however, expecting to decide upon this subject, for in our researches

* Called *sporules.*   † Nuttall.

into the natural world we are continually led to exclaim, "the ways of the Almighty are unsearchable, and past finding out!"

After what has been remarked upon the difficulty of analyzing these plants, the young pupil will not be likely to expect too much from attempts to investigate them. It is well for mankind that there are philosophers, whom the enthusiasm of scientific pursuits will lead to spend years, even a whole life, in searching into the fructification of a moss, or mushroom, or in examining into the natural history of a gnat or spider;* as thus, discoveries are continually brought forward, which add to the general stock of knowledge. This is a kind of martyrdom in the cause of science, to which but few seem called by the powerful impulses of their own minds. Females, in particular, are not expected to enter into the recesses of the temple of science; it is but of late that they have been encouraged to approach even to its portals, and to venture a glance upon the mysteries within.

We have now completed our view of the vegetable world, according to the order in which the different tribes of plants have presented themselves. As we followed in the train of classification, we have endeavoured to notice the most conspicuous genera, and to trace their natural relations while considering their artificial arrangement.

In many cases, departing from the plan of general remarks, we have traced the natural history of some one genus, believing this method more likely to make a permanent impression, than merely general views. In reading the history of nations, we often feel less interested in the fate of a whole people, than in that of some prominent individual; the mind presented with general ideas only, has no opportunity of forming images, which are but an aggregate of particulars. It is in natural as in civil history,—general remarks upon the beauty and utility of the vegetable world, or the curious structure of plants, make but slight impressions. But by contemplating the peculiarities of some one tribe, genus, or species, the mind seizes upon something definite, and reason, imagination, and feeling are easily awakened; thus the impression made is permanent. When you now look back upon the view you have taken of the vegetable world, and consider what impressions are most lively in your minds, you will probably find them to be respecting some peculiarities of individual plants. Of this tendency of the mind we should avail ourselves by connecting these particular impressions with facts which lead to general principles. Narrow indeed would be our mental vision, were it confined to single unconnected observations, laid up indiscriminately in the storehouse of thought; but our minds, not by our own will, but by a faculty received directly from our Creator, instinctively generalize and arrange their mass of single observations; and we, with scarcely an effort, perform that operation in the world of thought within us, which the great Linnæus effected in the vegetable kingdom.

* I have been gravely assured by a naturalist of distinction, that the study of *spiders* is one of the most *elegant* and *delightful* of all pursuits.

Enthusiasm of some naturalists—View of classification completed—Tendency of the mind to generalize.

PART IV.

LECTURE XXXVIII

THE FLOWERING SEASON OF PLANTS.

Vernal and Summer Flowers.

On entering the fourth division of our course, we find before us an open field, freed in a great measure from the technicalities of science, and presenting a smooth and delightful path. Hitherto, we have been clearing our way through difficulties, and overcoming obstacles; first, we were obliged to learn to analyze plants according to the strict rules of botanical science; next to examine the organs of plants, anatomically and physiologically; we then investigated the principles of classification, as exhibited both in the natural and artificial methods, and followed the arrangements of plants as presented in these different methods.

The language of Botany is now familiar to the diligent student, who can enjoy the pleasant reflection, that by his own industry and application, he has elevated his mind to that state, in which it may, with little further effort, enjoy the pleasant views of the vegetable kingdom which now present themselves. Thus, the traveller, having toiled to gain some acclivity, looks complacently around, enjoying the beautiful view before him in proportion to the efforts made to attain it.

We will now suppose the dreary season of winter yielding to the gentle influences of spring, and organized nature awakening to new life and beauty ;—for animals, no less than plants, seem vivified and quickened by the returning warmth of this delightful season. How many wandering through life, " with brute, unconscious gaze," have never made the inquiry, " *what causes Spring?*" With the greater part of mankind the ordinary phenomena of nature excite no interest; it is only when something *unexpected* occurs, that they think, either of first or second *causes.* But it should be the main object of education to teach youth to reflect, to seek the connexion between cause and effect; and especially, to look through second causes to the Great Being who is the *First Cause* of all—" himself *uncaused.*"

But to return to the question, " what causes Spring?" or to state it in another form, by what means does the Almighty produce the changes which this season presents? To answer this, we must refer to *astronomical geography,* which, pointing out the course of the sun, shows us, that having journeyed to his utmost southern boundary, he returns, crosses the equator, and with rapid strides advances towards the northern hemisphere, beaming more directly upon us, and increasing the temperature of the atmosphere ; to *chemistry* we owe our knowledge of the effects of caloric on bodies; *physiological botany* shows us the sap or vegetable blood expanding by the influence of caloric, and every exhaling and inhaling organ of the plant commencing operations under the same powerful influence. The earth, released from the icy bonds of frost, turns kindly to the mute, but living children of her bosom, and imparts the maternal nourishment, which, rushing through every fibre of the vegetable being, invigorates it with health and strength.

Remarks introductory to the fourth part—What causes spring?

From the first appearance of vegetation in the spring, until the commencement of winter, nature presents an ever varying scene. The phenomenon of the *flowering* of plants,* is, in many respects similar to that of the *putting forth of leaves ;*† in both, the same causes either hasten or retard this period. The putting forth of leaves, and the blossoming of flowers, differ, however, in one circumstance; the leaves begin by the upper leaf-buds; the flowers by the lower flower-buds; *stipes, panicles,* and *thyrses,* begin to blossom gradually from the base to the summit, *cymes* and *umbels* blossom from the outside to the centre.

In plants of the north, transported to the south, the period of the putting forth of leaves, and blossoming, is hastened; in those of the south, carried to the north, it is retarded. Even in their native soil, this period varies in some degree in different seasons. With greater warmth of temperature, we have an earlier appearance of vegetation; yet in general this variation is so slight, that botanists are able, by observation, to fix with a sufficient degree of accuracy, the time of the flowering of plants in particular latitudes and climates.

The progress of vegetation varying little from latitude 40° to 43° north, the remarks we make on this subject may apply to that region of country extending south to the mouth of the Hudson, north to the mouth of the Mohawk, eastward to the Atlantic, and westward to the Pacific Ocean.

In Ohio, and the western part of New York, the climate, on account of the influence of the lakes, and the cold, eastern winds from the Atlantic being broken by ranges of mountains, is milder, and vegetation is somewhat earlier than in New England in the same latitude.

In some cases, a plant puts forth leaves and blossoms at the same time; but usually, the leaves appear before the flowers, probably having a greater force to draw up the sap than the flowers, in which it rises by slow degrees. We see little appearance of vegetable life as early as March; sometimes snow covers the ground nearly, or quite through the month; but if we examine the trees and shrubs, even then, we may perceive, by the swelling of their buds, that they have already felt the vivifying influence of heat, and that a little increase of temperature will cause the embryo flower, or leaf, to burst its prison and come forth.

Vernal Flowers.

In April, the leaves of trees and shrubs begin to put forth; a few flowers show themselves, amid the damp, chilly atmosphere with which they are surrounded. Among the most interesting of these harbingers of spring is the HEPATICA *triloba,* or liver-leaf; a lowly, modest flower of a pale blue colour, with beautifully formed, three-lobed leaves.

The low anemone, (ANEMONE *nemorosa,*)‡ with its pale blossoms, is found in shady woods and damp pastures. The bright yellow flowers of the colt's-foot (*Tusilago*) brave the cold winds of early spring, while the reluctant leaves wait for warmer breezes.

* This is called *florescentia.*   † *Foliation.*
‡ This little flower I have seen raising its head amid surrounding snows, on the banks of the Poesten-kiln, a streamlet which flows into the Hudson, near Troy.

Most species of the poplar are now in blossom; also the Salix, or willow, which is of the same class; this genus includes the weeping willow, or Salix *tristis*,* sometimes called Salix Babylonica, alluded to in a beautiful passage in the Psalms, which represents the children of Israel, when carried into captivity, as sitting down by the waters of Babylon to weep, and hanging their harps on

"Willow trees that wither'd there."

Among the forest trees now in blossom, are the maple and the elm. In the meadows and moist grounds is the American cowslip, (CAL-THA *palustris*,) a fine example of the class Polyandria; and the adder's tongue, (ERYTHRONIUM,) having a beautiful liliaceous flower; this affords a good example of the class Hexandria.

In woods, and by the sides of brooks, is to be seen the Sanguinaria, or blood-root, which bears a white blossom, more elegant and ornamental for a garden than many flowers which are brought from foreign countries, and affording from its root a highly valuable medicine.

The CLAYTONIA, or spring beauty, is also to be found at this season; the dandelion, too, is found among the earliest flowers of spring. The garden violet, which is an exotic, appears also at this time; the VIOLA *rotundifolia*, or yellow violet, with roundish leaves lying close to the ground, is found in the fields. Besides these, are found several species of Carex, a coarse kind of grass; the trailing arbutus, EPIGEA *repens*, and the TRILLIUM, which we remarked under the class Hexandria, as a flower exhibiting great uniformity in its divisions.

In May, many species of the *Viola* appear; there is sometimes a difficulty in determining between these species; the distinctive marks seem often to be blended; we are in such cases obliged to place our plant under that species to which it seems to have most resemblance.

One of the most *interesting* flowers of this season, found in woods and meadows, is the ANEMONE *Virginiana*, the Wind-flower, a name given, as some say, because the flower expands only in windy weather; its petals are large and usually white, the stem grows to the height of two or three feet, and contains one terminal flower. Several other species of the Anemone are in blossom about this time.

The *Xylosteum*, or fly-honeysuckle, may be found, by the side of brooks; this is a shrub with blossoms growing in pairs; the UVULARIA, a plant of the lily family, having a yellow blossom, grows in the woods; the strawberry is now found, with its numerous stamens growing on the calyx; it has also many styles, each one bearing a seed.

The ARONIA is an early flower; a species of this, the shad-blossom, is not unfrequently found in April; this is a large shrub, often growing upon the banks of brooks, with white petals, clustering together in the form of a raceme.

Many of the mosses are now in blossom; these, we trust, you have learned to consider as presenting much that is interesting to those who understand their structure; but you will not be called on to examine the mosses in the commencement of your botanical studies, neither will they be likely to force themselves upon your notice. You no doubt were surprised to learn that they have flowers, and are considered as deserving attention; but you must recollect that

----

* *Tristis* (Latin) signifies pensive, or sad.

they are the workmanship of the same hand that created the host of heaven.

The ARUM, or wild-turnip, is now in blossom; it is found in shady places. The root is valuable in medicine. The CALLA *palustris*, or water arum, abounds in wet grounds.

The AQUILEGIA, or wild columbine, with its horned nectaries, is found hanging in rich clusters from the clefts of rocks. The early garden flowers are the snow-drop, crocus, crown-imperial, violet, primula, polyanthus, daffodil, and others of the narcissus genus.

Flowers of Summer.

The plants which are now in blossom are very numerous; we will mention a few of the most common and interesting.

A well-known shrub, the elder, (SAMBUCUS,) is now found along the sides of hedges, or on the margin of brooks, and in the meadows; the RUBUS, or raspberry, the RANUNCULUS, or butter-cup, the CYNOGLOSSUM, or hound's-tongue, and the TRIFOLIUM, or clover. It is recorded in history that when St. Patrick went as a missionary to preach the Gospel to the pagan Irish, " he illustrated the doctrine of the Trinity, by showing them a trifolium or three-leaved grass with one stalk; this operating to their conviction, the Shamrock, which is a bundle of this grass, was ever afterward worn upon this Saint's anniversary, to commemorate this event."

In the meadows is seen at this time the GERANIUM *maculatum*, a showy flower, and almost the only American Geranium; in the woods, the splendid ladies'-slipper, (CYPRIPEDIUM,) and the wild mandrake, (PODOPHYLLUM,) a flower of curious appearance.

The genus CONVALLARIA, of which the Solomon's seal is an example, may now be found; it is usually white, of a funnel-form corolla. Some other species, as the lily-of-the-valley, have a bell-form corolla. The various species of VACCINIUM, of which the whortleberry is an example, are now in blossom; the woods are ornamented by the snowy white Cornus, or dog-wood flowers.

In the early part of June the foliage of the trees usually appears in perfection; among the earliest are the willow, poplar, and alder; next are the bass-wood, horse-chestnut, oak, beech, ash, walnut, and mulberry, which are not all usually in full leaf before the middle of June.

At the summer solstice a new race of blossoms appears; as the roses, pinks, and lilies, with many other exotics. The Iris is found in stagnant waters and in gardens. Among native plants we now find the ASCLEPIAS, or milk-weed, of which there are some very showy, and some delicate species. The little bell-flower (CAMPANULA) may be seen nodding over the brows of the rocks.

The brilliant laurel (KALMIA) is now in bloom. The climbing virgin's-bower (CLEMATIS) hangs in graceful clusters of white flowers from the boughs of shrubs and trees growing by the side of brooks. The curious side-saddle flower, (SARRACENIA,) which was described under the class Polyandria, is now to be found in swamps and wet grounds. The mullein, with its long yellow spike, is very conspicuous in old fields and by the road-side.

More flowers are in blossom about the time of the summer solstice than during any period of the year, until the blossoming of the autumnal plants. The hot breath of summer seems to wither the expanding flowers, the earlier ones fade away, and the late ones do

not immediately come forward;—it would seem as if the earth, hav
ing poured forth in rapid succession innumerable treasures, now required a suspension of her efforts; but with recovered energy, she
soon begins to spread forth new beauties, and to deck herself in her
most gorgeous attire.

LECTURE XXXIX.

AUTUMNAL FLOWERS—EVERGREENS—ANCIENT SUPERSTITION RESPECTING
PLANTS—VARIOUS PHENOMENA OF PLANTS.

THE autumnal flowers differ in appearance from those which we
find in the earliest part of the season. Few examples of the compound flowers occur until the latter part of July, and beginning of
August;—this is fortunate for students just commencing the analysis of plants; were they to find only the compound flowers at first,
they would be discouraged; but nature seems kindly to lead them on
step by step, reserving the more difficult plants until they have had
an opportunity of becoming familiar with the easier classes.

There is little difficulty in learning to distinguish the different families of compound flowers; as an *Aster* from a *Solidago* or a *Helianthus*. But some of these families contain many species; and the
chief difficulty consists, not in finding the *genus*, but in determining
the *species*. Indeed it is not to be concealed, that there is, in this
part of botanical science, some confusion among writers; and the
student must not be discouraged if he is not always able to find his
plant exactly to coincide with any other species described.

Among the fine flowers which autumn presents, are the scarlet
LOBELIA, or cardinal flower; the yellow GERARDIA, (false fox-glove,)
and the noble sun-flower, (*Helianthus*.) The LINNÆA *borealis* is
found in September; at this time the white pond-lily, (NYMPHÆA,)
one of the most splendid of American flowers, is seen whitening the
surface of the lakes and ponds, sometimes alternating with the yellow water-lily, (NUPHAR,) a flower of less striking elegance than the
former, but perhaps not less curious in its form.

Another aquatic plant, which, although it blossoms in summer,
continues in flower until late in the autumn, is the SAGITTARIA, or arrow-head, with a calyx of 3 sepals, and three white petals. The *Eupatorium*, or thorough-wort, which blossoms in autumn, has no external beauty to recommend it, but as a remedy in diseases, perhaps
no plant is more useful.

Among the exotics which grace the decline of the year, are the
splendid *dahlias*; the gay *chrysanthemums* blossom only on the verge
of winter, but they require protection from frosts. We see among
the last blossoms of the season, the aster, and some other compound
flowers; these seem for a time to endure the autumnal blasts, but
they gradually give way to the reign of winter; while the desolate
fields and meadows present but a gloomy contrast to their former
. verdant and glowing appearance.

Evergreens.

During the season of winter in our climate, no flowers appear, ex-

Autumnal flowers—Are they proper for first lessons in analysis?—Which is most
difficult to ascertain, the *genus* or *species*?—Various flowers of autumn—Last flowers
of autumn—What flowers appear in winter?

attended with an uncommon quantity of seed on these shrubs, whence their unusual fruitfulness is a sign of severe winter."

Besides the above, there are several plants, especially those with compound yellow flowers, which during the whole day turn their flowers towards the sun, viz. to the East in the morning, to the South at noon, and to the West towards evening. This is very observable in the sowthistle, *Sonchus arvensis ;* and it is a well known fact. that a great part of the plants in a serene sky expand their flowers and as it were with cheerful looks behold the light of the sun; but before rain they shut them up, as the tulip.

The flowers of the chick-wintergreen (*Trientalis*) droop in the night, lest rain or moisture should injure the fertilizing pollen.

One species of *woodsorrel* shuts up or doubles its leaves before storms and tempests, but in a serene sky expands or unfolds them, so that husbandmen can foretel tempests from it. It is also well known that the sensitive plants, and *cassia*, observe the same rule.

Besides affording prognostics of weather, many plants fold themselves up at particular hours, with such regularity as to have acquired names from this property. The following are among the more remarkable plants of this description.

Goatsbeard. The flowers of both species of *Tragopogon* open in the morning at the approach of the sun, and, without regard to the state of the weather, regularly shut about noon. Hence it is gener-ally known by the name of *go-to-bed-at-noon.*

The *four o'clock,* (*Mirabilis*,) sometimes called Princess' leaf, is an elegant shrub in its native clime, the Malay islands. It opens its flowers at four in the evening, and does not close them till the same hour in the morning. It is said people transplant them from the woods into their gardens, and use them as a dial or clock, especially in cloudy weather.

The *Evening Primrose* (*Œnothera*) is well known from its remarkable property of regularity, shutting with a loud popping noise about sunrise, and opening at sunset. After six o'clock, these flowers regularly report the approach of night.

The *Tamarind-tree,* the water-lily, (*Nymphæa,*) the mary-gold, the false sensitive-plant, and several others of the Diadelphia class, in serene weather expand their leaves in the daytime, and contract them during the night. According to some botanists, the tamarind-tree infolds within its leaves the flowers or fruit every night, in order to guard them from the cold or rain.

The flower of the garden lettuce opens at seven o'clock, and shuts at ten.

"A species of serpentine aloes, whose large and beautiful flower exhales a strong odour of the Vanilla during the time of its expansion, which is very short, is cultivated in the imperial garden of Paris. It does not blossom until towards the month of July, and about five o'clock in the evening, at which time it gradually opens its petals, expands them, droops and dies. By ten o'clock the same night it is totally withered, to the great astonishment of the spectators, who flock in crowds to see it.

"The *cereus,* a native of Jamaica and Vera Cruz, exhibits an exquisitely beautiful flower, and emits a highly fragrant odour for a few hours in the night, and then closes to expand no more. The flower is nearly a foot in diameter, the inside of the calyx of a splendid yellow, and the numerous petals are of a pure white.

Plants which turn towards the sun—Plants which hang their heads at night and in storms—The go-to-bed-at-noon—The four o'clock—Evening primrose—Tamarind-tree, &c.—Aloes—Night-blooming Cereus, &c.

The St. John's-wort blossoms near that saint's day. Lychnis, called the great candlestick, or candle, (CANDE gens,) was supposed to be lighted up for St. John the Ba was a burning and a shining light. The white lily expa. the time of the *annunciation*, affording another coinciden⌣⌣ ᴏr the blossoming of white flowers at the festivals consecrated to the mother of Christ. The roses of summer are said to fade about the period of St. Mary Magdalen's day.

The passion flower is said to blossom about Holy Rood day. Allusions to this day being frequently found among writers of forn er days, it may be well to inform you that according to the legends of the Romish church, the cross on which our Saviour was crucified was discovered in the year 326, by Helena, the mother of Constantine, who is said to have built a church on the spot where it lay The word *Rood* signifies the Cross; thus this day is the day of the Holy Cross.

It was during the middle ages, when the minds of men were influenced by the blindest superstition, that they thus imagined every operation of nature to be emblematical of something connected with their religious faith. Although these superstitions are trifling and absurd, they are interesting as connected with the annals of the human mind, and as showing us the origin of many names of plants. Had the superstitious monks and nuns, who were the authors of these conceits, and at that time the most learned part of the community, been possessed of as much knowledge as most children in our country, they would have known that plants bloom earlier or later, according to various circumstances of climate; and that a flower which in Italy blossoms as early as February, might not appear in England before April; while the day of the Saint which the flower was supposed to commemorate, would occur at the same time in both places.

Phenomena of Plants, arising from changes in the atmosphere.

Plants exhibit some phenomena which are supposed to arise from the state of the atmosphere; accurate observers of nature have made remarks upon these changes, as prognosticating certain changes of weather. Lord Bacon, who was remarkably attentive to all the appearances and changes of natural objects, is the author of the following observations.

"*Chickweed, (Anagallis.)* When the flower expands boldly and fully, no rain will happen for four hours or upwards: if it continues in that open state, no rain will disturb the summer's day; when it half conceals its miniature flower, the day is generally showery; but if it entirely shuts up or veils the white flower with its green mantle, let the traveller put on his great-coat, and the ploughman, with his beast of draught, expect rest from their labour.

"*Siberian Sowthistle, (Sonchus.)* If the flowers of this plant keep open all night, rain will certainly fall the next day.

"*Trefoil, (Hedysarum.)* The different species of trefoil always contract their leaves at the approach of a storm; hence these plants have been termed the husbandman's Barometer.

"*African Mary-gold.* If this plant opens not its flowers in the morning about seven o'clock, you may be sure it will rain that day, unless it thunders.

"*White thorns* and *dog-rose bushes.* Wet summers are generally

it requires, the degree of light which seems necessary, and the kind of exposure, as to winds, which appears most favourable.

Plants vary much in their susceptibility of naturalization. The horse-chestnut, which is now common in the middle and northern United States, was originally brought from the tropical regions. In these regions, however, it usually grows in grounds somewhat above the level of the sea, and therefore its habit, as to temperature, renders it in some degree fitted for more northern countries. Orange and lemon-trees cannot be brought to bear the roughness of our climate, without some protection.

In many cases, perennial plants by this change of climate are converted into annual ones ; as if fearing the inclemencies of a cold winter, they pass through their successive stages of existence with rapidity, and accomplish in one summer what they had been accustomed to require years to perform. The nasturtion was originally a perennial shrub, flourishing without cultivation on the banks of the Peruvian streams ; yet, transferred to this country, it is an annual herbaceous plant, which completes its term of existence in a few months.

The acclimating of some plants is with difficulty accomplished, and it is by slow removals that they can be made to grow in foreign situations. Rice by a slow progress has advanced from Carolina to Virginia, and it is now cultivated in New Jersey. The habits of Indian corn, aided by climate and culture, have suffered a still more remarkable change. After having been for several years raised in Canada, it arrives to perfection in a few weeks, and on that account is employed by us as an early corn ; but that which has been long cultivated in Virginia, will not ripen in a New England summer ; yet originally, the early corn of Canada and that of Virginia were the same, both in habit and other properties.

While merely ornamental or curious plants can with difficulty be made to vegetate freely in foreign situations, the vegetables most useful to man are disseminated and cultivated. The delicate exotic flowers often disappoint our expectations ; but the wheat, the potato, and corn, which are also exotics, seldom are withheld from the labour of the husbandman.

Thus should earthly parents, imitating their " Father in heaven," first provide their children with what is *useful* both for body and mind, leaving the *ornamental* to be bestowed or not, as circumstances may render proper.

Agents which affect the growth of Plants.

Of the various substances by which vegetables are nourished, *water* is thought the most important. Some plants grow and mature with their roots immersed in water, without any soil ; most of the marine plants are of this description.

Atmospheric air is necessary to the health and vigour of plants ; if a plant is placed under a glass into which no air can enter, it withers and dies.

Most plants are found by analysis to contain a certain portion of *salts*, such as nitre and muriate of soda,* or common salt. It appears that the root absorbs them from the soil by which it is nourished.

---

* According to modern chemistry, chloride of sodium.

No plants can grow without some degree of *heat*, though some require a greater portion of it than others.

Plants may be made to grow without *light*, but they will not exhibit the verdure, or any of the properties of health. The atmosphere, which is contaminated by the respiration of animals, is restored to purity by the vegetation of plants; but secluded from light, vegetables are no longer capable of converting a portion of the fixed air to their use, or of supplying the atmosphere with the oxygen, on which its importance in supporting animal life chiefly depends. By the action of light, the carbon of the fixed air* is interwoven with the texture of the plants. The aromatic plants, the clove, cinnamon, and the Peruvian bark, all owe their chief excellences to the intense light of the equatorial regions.

Gases of different kinds affect vegetation very differently. *Carbonic* acid gas, though prejudicial to the germination of the seed, has been found, when properly applied, to hasten the process of vegetation in the plant. Pure carbonic acid gas destroys vegetable life; thus, a growing plant placed over wort in a state of fermentation, dies in a few hours. Dr. Priestly, a celebrated chemist, proved that this gas is of great utility to the growth of plants vegetating in the sun, and that whatever promotes the increase of it in their atmosphere, at least within a certain degree, assists vegetation. In the shade, an excess of carbonic acid gas is found to be hurtful to plants.

Most kinds of manure afford large portions of carbonic acid gas. *Oxygen* gas is essential to the germination of the seed and to the growth of the plant. Flower-buds confined in an atmosphere deprived of oxygen, fade without expanding.

Neither *Nitrogen* nor *Hydrogen*, when unmixed with other substances, afford an atmosphere favourable to vegetation.

Habitation of Plants.

Vegetation is not scattered by chance over the surface of the globe but we perceive that the Creator has regulated its distribution according to certain fixed principles; we find not only a wonderful adaptation of plants to the physical necessities of animals in general, but that they are also varied to correspond to the peculiar wants of animals in different climates.

First, we would notice the *herbs* which cover the surface of the earth; had their stems been hard and woody, the greater part of the earth would have been inaccessible to the foot of man, until the vegetation was first destroyed by fire, or by some other means. Can we imagine that the grass and herbs which now afford a soft carpet for our feet, came *by chance* to grow thus, rather than hard and woody, like the trees? Can we imagine, that *by chance* the prevailing colour of vegetation is *green*, that colour upon which, above all, the eye rests with the most agreeable sensations? Suppose the grass and herbs to have been red or yellow, and with our present organs of sight, how painful would be the sensations excited by these bright colours! Instead of beholding the face of nature with delight, we should turn from it, and vainly seek some object on which the eye might repose.

Woody shrubs occasionally alternate with herbs, but they are so placed as not to offer obstructions to the foot of man; they often grow out of the clefts of rocks, affording a means of climbing almost

* Carbonic acid gas.

perpendicular precipices. *Large trees* are not usually placed so near each other as to prevent a passage between them; their lowest branches are mostly at a height sufficient to admit men and beasts under them, and thus few forests are impenetrable.

In cold countries, whether occasioned by distance from the equator, or elevation by means of mountains and table lands, we find the pine, fir, and cedar, and other resinous plants, which furnish man with light and fuel during the dreary season of winter. The leaves of these trees are mostly filiform, or long and narrow, thus fitted for reverberating the heat like the hair of animals, and for resisting the impetuosity of winds which often prevail in those regions.

In warm countries, trees present, in their foliage, a resource from the scorching rays of the sun; their leaves serving as fans and umbrellas. The leaf of the banana being broad and long, like an apron, it has acquired the name of Adam's fig-leaf. The leaves of the cocoa-tree are said to be from twelve to fifteen feet long and from seven to eight broad. A traveller remarks, that one leaf of the talipot-tree is capable of covering from fifteen to twenty persons. The soldiers, he says, use it for a covering to their tents. He observes, that it seems an inestimable blessing of Providence, in a country burnt up by the sun, and inundated by rains for six months of the year. In our climate, during the warm season, Providence bestows upon us a variety of juicy and acid fruits, cherries, peaches, plums, melons, and berries; nuts and many fruits are fitted for preservation during the winter, so that we are never destitute of some of these bounties.

A remarkable instance of the care of Providence in providing for the wants of man, appears in what is related of a plant* found amidst the burning deserts of Africa; the leaf of which is said to be in the form of a pitcher, and to possess the property of secreting moisture to such a degree as to form a quantity of water sufficient for a draught to a thirsty person; the end of the leaf is folded over the throat, as if to prevent the evaporation of the fluid. Various other plants, in hot regions, furnish refreshing draughts, or cooling fruits, for the thirsty traveller.

These remarks might be pursued to an extent as great as the vastness of the vegetable kingdom, and the wants of man; we have merely glanced at the subject of the adaptation of plants to the wants of animal life, hoping that these few suggestions may lead you to trace, from your own observation of the works of nature, the operations of that *great designing Mind*, which rules and governs all with infinite wisdom and benevolence.

The earth, then, we find to be covered with a multitude of species of plants, differing not more by their external forms than by their internal structure, and each endowed with peculiar habits and instincts.

. Some species seem adapted to the mountains, some to the valleys, and others to the plains; some require an *argillaceous* or *clayey* soil, others a *calcareous* soil or one impregnated with *lime*, others a *quartzose* or *sandy* soil, and some will only grow where the earth contains soda or marine salts. Many plants will grow only in water; we find here such as are peculiar to the marsh, the lake, the river, and the sea. Many plants require a very elevated tempera-

* Probably the *Nepenthes* distillatoria.

Trees—Trees of cold countries—Trees of warm countries—Fruits of our climate—A plant found in the deserts of Africa—Reflection—Plants adapted to various soils &c

ture, some will grow only in mild and temperate climates, and others only in the midst of frost and snows.

Thus every country where man is to be found, has its vegetation. Some species, with respect to localities, are confined to narrow limits.

A species of ORIGANUM (the *Tournefortii*) was discovered by Tournefort, in 1700, upon one single rock in the little island of Amorgos, in the Greek Archipelago; eighty years afterward, the plant was found in the same island, and upon the same rock, and has never been discovered in any other situation. Some plants confine themselves within certain longitudes, scarcely varying to the right or left. The MENZIESIA *pallifolia*, a species of heath, confined between ten and fifteen degrees of west longitude, is found in Portugal, Spain, and Ireland. Latitude and elevation, by reason of mountains and table lands, produce a greater variety in the appearance of vegetation than almost any other causes.

Few plants are found to endure extreme cold. Botanists formerly estimated, that at Spitsbergen, in north latitude about 80°, there are but about 30 species of plants,* in Lapland, in 70°, 539 species; Madagascar, at the tropic of Capricorn, 5000; and at the equator a much greater number. These estimates fall very far short of the number of species now known, but they may give some idea of the difference in the vegetation of cold and warm climates.

Geographical situation of Plants.

Every country exhibits a botanical character peculiar to itself. Linnæus, in his bold and graphic language, said,† "A practical botanist can usually at the first glance distinguish the plants of Africa, Asia, America, and the Alps; but it is not easy to tell how he is able to do this. There is a certain character of sullenness, gloom, and obscurity in the plants of Africa; something lofty and elevated in those of Asia, sweet and smiling in those of America; while those of the Alps seem rigid and stinted."

In investigating the geographical situation of the vegetable kingdom, we see the powerful effects of light and heat. Feeble in the polar regions, vegetation acquires strength as we approach towards the equator, where the light of the sun is vivid, and its heat permanent and intense.

The centre of the *frigid zone* is entirely destitute of vegetation. After passing the arctic circle, we find on the borders of the temperate zone a few species of plants, chiefly lichens, mosses, and ferns, also a few shrubs and berries. In the heat of a polar summer, the growth of plants is rapid; Lapland is the only country within this zone where any kind of grain can be raised.

The productions of the *temperate zone* gradually alter in character as we approach the tropics. Humboldt has divided the temperate zone, with respect to productions, into three regions; the *cold*, the *temperate*, and *warm regions*. In the cold region, *grain* may be raised to advantage, and *berries* grow in abundance. In the temperate region, the *wine, grape, grain*, and fruits of many kinds, are cultivated in their greatest perfection. The warm region produces *olives, figs, oranges*, and *lemons*.

* That is, exclusive of the Cryptogamous plants.
† "Primo intuitu distinguit sæpius exercitatus botanicus plantas Africæ, Asiæ, Americæ, Alpiumque, sed non facile dicerit ipse ex qua nota. Nescis quae facies torva, sic ea, obscuris Afris; quae superba, exaltata Asiaticis; quae laeta, glabra Americanis; quae coarctata, indura Alpinis."

Some plants have a confined locality—Few endure extreme cold—Every country has its own botanical character—Plants of the frigid zone—Temperate zone.

The variety of plants in the torrid zone is very great. Trees are more numerous, in proportion to other plants, than in the temperate zones; the same tribes which are there slender and humble plants, here spread into lofty trees, many of which are adorned with large and beautiful flowers. The richest fruits and spices, and the most valuable medicinal plants, are found here. In ascending the mountains of the torrid zone, as the temperature varies, each section has its own distinct plants; and we find in succession the production of every region from the equator to the poles.

As the mountains of the torrid zone afford every variety of climate between their base and their summit, so they are capable of producing all the vegetables of every climate; but, as latitude increases, temperature diminishes, so, generally speaking, the productions, as we proceed from the tropic northward or southward, correspond with the elevation at which the same plants will grow upon a mountain within the tropics. Every plant requires, other circumstances being the same, the same mean annual temperature;[*] for example: the plantain-tree and sugar-cane require a mean annual heat of from 82 to 83 degrees; but 70 degrees of mean annual heat is not found beyond the 27th degree of latitude; consequently, the plantain and sugar-cane will not ripen in the open air in a higher latitude; and this Baron Humboldt has found to correspond with the height of 3000 feet under the equator. Cotton will not flourish without 68 degrees of heat, which is not found beyond 34 degrees of latitude, which corresponds with about 3600 feet of elevation at the equator. The same reasoning applies to all other plants, with the exceptions arising from warm valleys, moisture of air, and richness of soil.

	Feet above the level of the sea.
The highest spot on which man ever trod	19,400.
The highest limit of the lichen plant	18,225.
The lowest limit of perpetual snow under the equator	15,730.
The highest limit of pines under the equator	12,801.
The highest limit of trees under the equator	11,125.
The highest limit of oaks under the equator	10,500.
The highest limit of the Peruvian bark-tree	9,500.
The lowest limit of pines under the equator	5,685.
The highest limit of palms and bananas	3,260

LECTURE XLI.

PLANTS AS AFFECTED BY CULTIVATION—CHANGE OF THE ORGANS—DISEASES—ECONOMICAL USES.

WE have before remarked upon the permanence of *species*, and that though they may in some respects be varied by cultivation, yet their distinctive characters will not be wholly lost. The differences which exist in species are expressed by the terms *races*, *varieties*, and *variations*.

Races are those differences in a species which are of a striking kind, and continued from the parent to its offspring, by being propagated by the seed. They are produced by strewing pollen of one species

---

* For explanation of mean annual temperature, see note, page 149.

Torrid zone—Production of every region found in ascending mountains of the torrid zone—Elevation produces similar effects on vegetation, as distance from the equator—Permanence of species—Races.

upon the pistils of another; the seed thus formed will produce a plant resembling both.

Varieties are a less important distinction than races; they are not continued by means of the seed, but produced by grafting or continuation of the plant under some new circumstances.

Variations denote the slightest kinds of difference; they are occasioned by peculiarities of climate, soil, moisture, dryness, &c.

Degeneration or change of the Organs of Plants.

The organs of plants, owing to peculiar causes, often experience a metamorphosis, and instead of their usual appearance exhibit anomalies, or vegetable deformities.

We here use the term deformity, as signifying any variation from the ordinary course of nature. The causes which produce these changes are,

1st. The adhesion of parts usually separate; thus we often see flowers, leaves, and fruits united, and appearing double.

Some writers, among whom is the celebrated French botanist De Candolle, assert, that a single petal which forms the corolla of many flowers, as the stramonium or the blue-bell, is in reality composed of several petals which become soldered, or cohere together before the flower expands. The same writers consider a monosepalous calyx to be composed of several little leaves thus united before their development.

2d. Changes are occasioned by a want of sufficient vigour in the plant to bring all parts to maturity. Some of the seeds thus often fail for the want of nourishment; many plants which in one flower produce several seeds, often ripen no more than one. The horse-chestnut has six seeds, but seldom matures more than two; in the blossom of the oak where six seeds are produced, but one acorn is perfected.

3d. In some cases organs appear from certain changes to be incapable of performing their original offices, and thus exhibit deformities; as where a bud, which, for want of sufficient nourishment, or some other cause, does not develop itself into a leaf, but forms a permanent protuberance or swelling upon the stem. The prickly pear exhibits a thick and expanded stem, which is formed of leaves imperfectly developed.

4th. The stamens and pistils, through excess of nourishment, swell and become petals; all double flowers are formed in this manner. The poppy in its natural state has many stamens, and but four petals; but you often see double poppies, with scarcely the vestige of a stamen left; the same change may be observed in the rose, which naturally has but five petals and many stamens and pistils, but in a very full double rose, scarcely any appearance of either stamen or pistil is to be seen. The stamens, more frequently than the pistils, meet with this metamorphosis, as they appear to be more intimately connected with the petals.

5th. The petioles or foot-stalks often change to leaves. This may be seen in an Arabian plant, Acacia *nilotica*, which furnishes the gum arabic. This tree at first exhibits upon one petiole six or eight pair of leaves; this number every year becomes less, until all the leaves disappear; the petiole then retaining all the nourishment which before was distributed to the leaves, flattens and expands, and appears in the form of a thick leaf.

6th. The peduncles and petioles sometimes change into tendrils, as in the vine; this plant at first throws out many large leaves and clusters of flowers; but the food not being sufficient to support such a profuse vegetation, the new leaves and clusters of flowers appear smaller; the nourishment becoming still more scanty, at length neither flower nor leaf is developed, and the peduncle and petiole become tendrils, which, by attaching themselves to some firm bodies, serve to sustain the rich fruit which is perfected on the lower parts of the branch.

7th. The last change we shall notice is the transformation of buds into thorns. When a plant forms more buds than it can nourish, some of them do not develop branches and leaves, but becoming hardened by the accumulation of sap, which is insufficient for their full perfection, they then exhibit the short, indurated process, called a thorn. It is said that wild plants, by rich cultivation, do, in time, become divested of their thorns, which change into what they seemed originally destined for, viz., leaves and branches.

Prickles, such as may be seen upon the rose, gooseberry, and other plants, do not change by cultivation, for these are a natural appendage, originating from the bark; while the thorn may be found connected with the wood, of which it seems to make a part.

Diseases of Plants.[*]

The diseases of plants (for these organized beings are, like animals, subject to disease and death) may, in many cases, arise from causes within the knowledge of the attentive naturalist.

1st. We notice constitutional diseases. Of this class are the varied colours of some leaves, such as the box and holly; this is supposed to be owing to certain juices which, by changing their elements, vary the colour of the leaf.

2d. Plants become diseased by being subjected to too great or too scanty a supply of food, as light, heat, water, air, and soil. Excess of *light* causes an escape of oxygen, and a too rapid deposite of carbon; the sap, incapable of sustaining so great a degree of action, becomes exhausted, the plant withers, and the leaves fall off. In this situation the food should be either increased by watering, or the vegetation retarded, by diminishing the light. Excess of *heat* absorbs the juices of the plant; deficiency of heat produces dropsy, and the plant losing its leaves, ultimately decomposes. More water is evaporated by a plant than is retained for its nourishment; therefore the absorption by the roots should be in proportion to the evaporation by the leaves.

3d. *External injuries* often affect the health of plants. *Rains* injure the wood by penetrating through apertures in the bark; the bark itself seems from its nature better fitted to bear the action of the weather. *Winds*, when violent, are mechanically destructive to vegetables; when moderate, the agitation which they produce is thought to be advantageous, by favouring the descent of the cambium, and promoting a more free circulation of the other juices.

Smoke is injurious to plants, it being composed of particles which, though invisible to our sight, are yet too gross to be absorbed by the minute pores of the leaves; it serves, therefore, to obstruct these

---

* This constitutes a department of Botany called *pathology;* a term derived from two Greek words, *pathos*, disease, and *logos*, account of.

6th, peduncles and petioles become vines—7th, buds, how transformed—Prickles—Diseases of plants—1st, Constitutional—2d, Light and heat—3d, External injuries—Rains—Wind—Smoke.

pores, and prevent their exhaling the oxygen gas which is necessary for the decomposition of the carbonic acid and the consequent deposition of carbon.

4th. Plants sustain injuries from *animals*, which produce diseases. Insects in particular make their way into the bark and external coats of the plant and deposite their eggs; these eggs when hatched produce larvæ, which, by their peculiar juices, often rot the wood. These insects are called *cynips*. One kind produces the hard protuberances on trees of different kinds, which are called gall-nuts, or nut-galls; others, which are softer and more spongy, are called apple-galls or berry galls. Another kind of insect, called *cochineal*, attaches itself to the bark of trees, and preys upon the juices. One species of the cochineal is of a brilliant scarlet colour and much valued for its use in dying; this species feeds on the *Cactus cochinillifer*, a Mexican plant.

5th. Diseases are produced by plants preying upon each other, either by fastening themselves upon their surfaces, or by so near a location as to deprive others of their necessary food. *Parasites* fasten themselves upon the surfaces of other plants; they are distinguished into two kinds, the *false* and *true parasites*; the former adhere to the plant without feeding on its juices, as mosses and lichens. These derive their nourishment from the atmosphere; but they injure the tree by harbouring insects, and attracting moisture which often rots the part of the stem on which they grow. The mistletoe is a true parasite, whose root, piercing the bark of trees, plants itself in the alburnum, and absorbs food from it, in the same manner as if it were fixed in the soil. The Puccospora is a very curious parasite which is sometimes found upon the leaves of shrubs, but more frequently upon the branches and leaves of trees. Mushrooms are of the class of false parasites. *Smut* is a black fungus, which fastens itself upon the ears of oats and other grain. The *rot* is a fungus excrescence which preys upon the seed; if seeds which have this disease fastened upon them are sown, the rot will be propagated also. *Ergot* is a disease mostly confined to rye. *Rust* is chiefly confined to the grasses; both are of the fungi family.

6th. Diseases resulting from age. Plants differ from animals in one important circumstance; the latter develop their organs at once; these organs in process of time become indurated and obstructed, until they at length decay from old age. Plants, on the contrary, renew themselves every year; that is, they form new vessels to convey the juices, new leaves to elaborate them, and new buds to produce flowers and fruits. Plants do not, then, like animals, seem destined to die with old age; or there does not seem to be in perennial plants any prescribed term of existence. The producing of fruit appears to exhaust the vital energy of the plant, in annuals in one year, in biennials in two, in perennials in a longer or shorter period, according to their natural constitution, and the quantity of fruit which they produce. Apple-trees, which bear heavy loads of fruit, are very short-lived in comparison with the oak, which perfects from each flower but one of six seeds, and this fruit is but a small acorn.

There are some trees now known to exist, which are supposed to be of great age; in the Island of Teneriffe is the DRACÆNA *draco*, which, according to many circumstances, appears to have some thousand years of age. In England, at Blenheim Park, it is said,

---

4th, Animals—5th, Parasites—6th, Diseases resulting from age—Aged trees

may be seen trunks of trees which shaded the bower of fair Rosa mond, and which it is supposed are not less than a thousand years old.

At Hartford, in Connecticut, is the Charter-oak, which was a hollow tree in the days of James II., nearly two hundred years ago. In the hollow of this tree was concealed the charter of the state, when the King of England, through his agents, attempted to deprive the colonists of that guarantee of their civil rights. This oak must, even at that period, have been an aged tree.

Economical uses of various Plants.

We perceive among the various species of vegetable beings, some which seem destined only to *beautify* and *enliven* the earth; others, with little or no beauty, are valuable only for their *utility*; and in some instances we find utility and beauty united; roses, lilies, tulips, carnations, and most of the green-house and garden plants, belong to the first-mentioned class. Trees are not only beautiful, but many of them are highly useful, affording fuel, shelter, and shade, nuts, berries, and other fruits; their bark is used in tanning, for medicine and spices; and their sap and secretions furnish sugar and various medicinal extracts.

Trees, with respect to their wood, may be divided, 1st, into such as have hard wood, as the oak, elm, apple, &c.; 2d, such as have soft wood, as the poplar and willow; 3d, such as have resinous wood, as the pine and fir; 4th, such as are evergreens, but not resinous, as the evergreen oak of the south of Europe.

Hard wood is considered best for fuel; as it contains the greatest quantity of carbon, it causes a more intense and permanent heat; resinous wood containing more hydrogen, burns with a more brilliant flame.

The fermented juice of the grape produces wine. Grain of different kinds produce gin, whiskey, &c. Apples, by their fermentation, produce cider; this liquor, concentrated by distillation, produces brandy and alcohol. The vineyards of Italy and France, and of some of the Atlantic islands, are the most celebrated for their wine. In America, the vine does not flourish in the same luxuriance as upon the eastern continent.

Grasses are the palms of cold climates; they are of the class of monocotyledons, and have endogenous stems. Some are perennial, some annual; the meadow grasses are of the former kind. The grains, Indian corn, and rice, are annual. There are certain grasses which are called artificial, because they do not spring up without cultivation; of this kind is clover. Gramineous plants, although very important, as furnishing from their leaves food for cattle, are yet more especially useful for their seeds, which furnish food for man.

Some plants furnish oils, which are of important uses in various ways. Of the fixed and volatile oils we have already spoken. The fixed oils are extracted from plants called oleaginous; they may be considered under three heads: 1st, olive-oil, produced from the olive in warm countries; 2d, nut-oil, of temperate climates, as obtained from walnuts, &c.; 3d, oil obtained from the seed of oleaginous, or oily plants, as the flax.

Tuberous roots, as the turnip, potato, carrot, beet, parsnip, &c., furnish important articles of food.

Asparagus, when young, is esteemed a luxury; the rhubarb plant is used in making pies; celery, onions, and even garlic, are esteem-

---

ed valuable for food and seasoning. Many of the labiate plants, as thyme, sage, &c., are used in cookery. The Cruciform family afford the cabbage, cauliflower, turnips, &c. ; the Leguminous family, beans, peas, &c.

The Cucurbitaceæ furnish us with melons, squashes, and cucumbers ; umbelliferous plants, with the aromatics, caraway, coriander, &c., which are useful in medicine and confectionary.

The plants chiefly used in domestic economy differ in different climates and countries; some, as many kinds of grain and grasses, are in common use in all countries; while others, as the bread-fruit and plantain, are only used in the few countries which produce them. The bread-corn of the temperate climates, is chiefly wheat, rice, and maize ; rice is a substitute for these in warm countries, and barley in cold countries.

The *esculent* roots of the old world, are chiefly the yam, carrot, and turnip; of the new, the potato.

The *pot-herbs*, such as the cabbage, sea-kale, and others of the cruciform family, are used in temperate climates ; in hot climates they are little used. Legumes furnish an important article of food in most parts of the old world, and in North America.

LECTURE LII.

HISTORY OF BOTANY, FROM THE CREATION OF THE WORLD, TO THE REVIVAL OF LETTERS IN THE REIGN OF **CHARLEMAGNE, A. D.** 770.

We now propose to give a brief sketch of the progress of botanical knowledge; and as this is closely connected with other branches of natural science, a history of the advancement of the one will necessarily be, in some degree, a record of the march of the others. Natural Philosophy, Chemistry, and Botany, were all nursed in the same cradle, and thus grew and gained strength side by side ; though Botany (at first rude and imperfect) may be considered the elder sister.

After becoming familiar with a science, the mind naturally seeks for information respecting its origin, and the progress by which it advanced from the first rude conceptions which might have been formed, to its gradual development and comparative perfection. The history of the progress of a science makes a part of the science itself; we are interested in the various efforts of philosophers, their experience and observations, and the trains of reasoning by which they arrived at those conclusions which constitute the basis of the science.

In Botany, as in the other sciences, physical wants were the first guides ; man at first sought to find in vegetables, food, then remedies for diseases, and lastly, amusement and instruction.

The first account of plants may be traced to the history of the creation by Moses. It was on the third day of this great work that God said, " Let the earth bring forth grass, the herb yielding seed, and the fruit-tree yielding fruit after his kind, *whose seed is in itself*, upon the earth : and it was so ; and the earth brought forth grass, and the herb yielding seed after his kind, and the tree yielding fruit, .

Melons—Umbelliferous plants—Bread-corn—Pot-herbs—Legumes—History of botanical science--Why do we wish to learn the progress of science ?--First account of plants traced to the history of the creation.

whose seed was in itself, after his kind; and *God saw that it was good.*" After this, it is recorded that God gave to Adam every herb and every *tree bearing fruit;* the latter was for him exclusively, but to the beasts of the earth, and the fowls of the air, and to every thing wherein there is life, he also gave the *green herb* for meat.

It is recorded that Adam gave names to all the beasts of the field, and the fowls of the air; and Milton imagines, that to Eve was assigned the pleasant task of giving names to flowers, and numbering the tribes of plants. When our first parents, as a punishment for their disobedience, are about to leave their delightful Eden, Eve, in the language of the poet, with bitter regret, exclaims:

> "Must I thus leave thee, Paradise?
> *   *   *   *   Oh flowers
> That never will in other climate grow,
> *   *   which I bred up with tender hand,
> From the first opening bud, and *gave ye names;*
> Who now shall rear ye to the sun, or rank
> *Your tribes?*"

The Bible, and the poems of Homer, afford us the only vestiges of the botanical knowledge of the earliest ages of the world. Great advantages were afforded to the Jews for obtaining a knowledge of plants, in their long wanderings over the face of the earth, before they settled in Judea. When in possession of this fertile country, they extended their intercourse with foreign nations; the vessels of Solomon frequented the shores of the Red Sea, the Persian Gulf, and the East Indian islands. In the Book of Kings it is said, " God gave Solomon wisdom and understanding above all the children of the East country, and all the wisdom of Egypt, for he was wiser than all men. He spake proverbs and songs; he also spake of *trees*, from the *cedar-tree* that is in Lebanon, even unto the *hyssop*, that springeth out of the wall; and people from all countries came to hear his wisdom."

The Magi, or " wise men of the East," cultivated the sciences to a great extent; but they kept their discoveries in mysterious concealment, in order the better to tyrannize over the minds of the people. Their researches were in a great measure lost to the world. Greece, however, received from Asia and Egypt the first elements of knowledge.

The philosophers of Greece, too eager to learn nature at one glance, were not satisfied with the slow process of observation and experiment, and to ascend from particular facts to general principles; but they believed themselves able, by the force of their own genius, to build up systems which would explain all phenomena; supposing that man had in his mind preconceived ideas of what nature ought to be. This error in the philosophy of the ancients for a long time obstructed the progress of all science; and it was not until laying aside this false notion, and admitting that the only sure method of learning nature is to study her works, that the labours of philosophers began to be followed by important discoveries.

The greater part of the ancient Greek philosophers asserted, that plants were organized like animals, that they possessed sensible and rational souls capable of desires and fears, pleasure and pain. Pythagoras of Samos, who travelled in Egypt, and was there instructed by the priests of the goddess Isis, is said by Pliny to have been

Milton imagines that Eve gave names to the plants, and numbered their tribes—What is known of the progress of botany during the earliest ages of the world—Solomon is said to have spoken of trees and other plants—The Magi—Philosophers of Greece—Pythagoras.

the first of the Greek writers who composed a treatise on the properties of plants. A disciple of his, Empedocles, seemed to have some correct ideas of vegetable physiology. He called the seeds the eggs of plants; the roots, their heads and mouths; and considered that the two sexes were combined in the same individual.

Several men of the name of Hippocrates wrote upon the medicinal properties of plants; but their descriptions, being destitute of system, are vague, and cannot be applied to plants with any degree of certainty.

Aristotle, perceiving that the course taken by preceding philosophers had not conducted them to the true knowledge of things, partially renounced their false ideas, and rested more upon observation and experience. In his researches, he was favoured by Alexander, of whom he had been the preceptor. That conqueror, in the midst of pride, and the fury of passion, still possessed the love of true glory, and a desire that his conquests might serve to promote the improvement of the human mind; he allowed to Aristotle, in the prosecution of his scientific inquiries, every facility that wealth and power could bestow.

Aristotle believed, that in nature there was a regular progress from inorganized matter upwards to man, and from man upwards to the Deity; that beings were connected together by certain affinities, composing an immense chain, of which the links were all connected. But,

> "Lives the man whose universal eye
> Has swept at once the unbounded scheme of things?
> Has any seen
> The mighty chain of beings, lessening down
> From infinite perfection, to the brink
> Of dreary nothing, desolate abyss?"

This idea of a regular chain of beings, presenting itself with such grandeur and simplicity, has had many admirers; but facts do not seem to correspond with this theory. In the vegetable kingdom we should find it impossible to trace a regular gradation from the oak to a moss (if we were to make these the extremes of the chain of vegetable substances,) and say exactly in what part of the scale each family of plants should be placed; it would rather seem, in many cases, as if the links of the chain had been broken or disunited.

Aristotle considered plants as intermediate between inorganized matter and animals. Plants, he said, are not distinguished from animals in being destitute of the seat of life, the heart; because of this the reptiles and inferior orders of animals are also destitute; but plants have no consciousness of themselves, or organs of sense to know what is out of themselves; animals possess these faculties; therefore, Aristotle says, they are different. We think it would have been difficult for him to have discovered any evidence of consciousness in the sponge, or any marks by which it might appear that this animal substance (for such it is thought to be) has the knowledge of any thing external to itself. However great may be the veneration entertained for the opinions of Aristotle, we believe his distinction between plants and animals will at this time find no supporters. This philosopher published his works on natural history about 384 years before Christ.

Theophrastus, the friend and pupil of Aristotle, published a great number of learned works; among others "A History of Plants," and "The Causes of Vegetation." He treated separately of *aquatic* plants, of *parasites*, of *culinary* herbs, and of *flowering plants;* he remarked upon the uses of each plant, the place where it grew, and whether it was woody or herbaceous. He had no idea of genera or species; his names were merely local, and his descriptions generally indefinite. His views upon the *physiology of plants*, were superior to his *descriptions* of them; he remarked upon their different external organs; distinguished the seed lobes (Cotyledons) from the leaves; gave just ideas upon their functions, and upon the offices of the root. He explained their anatomy as well as possible without the assistance of the microscope, which (as the science of optics was then unknown) had not been invented. Theophrastus seemed too much inclined to compare the structure of vegetables to that of animals; imagining that he found in plants, bones, veins, and arteries. A shrub which grows in the Antilles is named Theophrasta, in honour of this ancient botanist.

Dioscorides, a physician of Greek extraction, about the commencement of the Christian era, travelled over Greece, Asia Minor, and Italy, in order to observe the plants of those countries; his works were written in Greek; he divided plants into four classes, viz.: 1st, *aromatic*, 2d, *vinous*, 3d, *medicinal*, and 4th, *alimentary*, or *nutritious*. The labours of this botanist were of little value in after times, on account of want of method in his descriptions. He gave the names and properties of 600 plants; but having no idea of species or genera, his work was but a chaos of facts, which were so imperfectly expressed, as to render it impossible to apply them to use.

The elder Pliny, who lived in the reign of Nero, treated of the history of plants, but he neglected nature, and derived his science from the works of his predecessors.- False systems of philosophy seemed to fetter the noblest minds, and prevent their pursuing those methods of investigation which would have led to a true knowledge of nature. The genius of Pliny was vast and active; he consecrated to scientific researches and literary works, the leisure which public duties left him. His "History of the World," which was a compilation of all the knowledge of the ancients, upon the subject of natural history, the only one of his writings which has escaped the ravages of time and barbarians, is but a small portion of his labours. He is considered faulty in recording both truth and error, often transmitting them without observation or criticism, and sometimes favouring absurd traditions; but his work is justly admired for the greatness of its plan, which embraced the whole of nature, for the elegance of its style, and for the wonderful art with which the highest considerations of practical philosophy are associated with natural history. In the year 79 after Christ, Pliny fell a sacrifice to his desire of knowledge, in an eruption of Mount Vesuvius; wishing to contemplate as near as possible so sublime a spectacle, he perished, suffocated by the sulphureous exhalations.

Galen, in the second century, wrote upon the medicinal qualities of plants, but gave no descriptions. The love of the sciences seemed, in the prosperous days of Rome, to be extinguished; "Mistress of the world," corrupted by victories, and by tyrants, she had abandoned herself to luxury. The false philosophy of the vanquished Greeks reigned in the schools of victorious Rome, chasing away every trace of true knowledge. Religious fanaticism had also its

Dioscorides—Pliny—Galen—Condition of science in the most prosperous days of Rome.

influence; pretended Christians, as well as Pagans, destroyed libraries and the monuments of literature, sacred and profane.

At this time the barbarians of the north and west precipitated themselves upon a country weakened by effeminate habits. Italy, ravaged by the Huns and Vandals, became successively the prey of the Heruli, of the Goths and Lombards. These people, nursed in war, abhorred the sciences and arts, and believing they enervated courage, allowed not their children to cultivate them.

The Latin ceased to be the common language, and a corrupt mixture of barbarous languages took its place. The population was greatly diminished; the country, formerly fertile and cultivated, became overgrown with forests, and inhabited by wild beasts.

In this dark period, Botany shared the fate of other sciences. The monks, strangers to the first elements of literature, and yet passing for the lights of·their age, spoke in a barbarous language of the plants of Theophrastus and Pliny, commented upon writings they were incapable of comprehending, and mingled with their errors respecting facts, the most shameful superstitions

---

LECTURE XLIII.

HISTORY OF BOTANY, FROM THE EIGHTH CENTURY TO THE DISCOVERY OF AMERICA.

THE state of science was thus gloomy in the empire of the West, when Charlemagne, a monarch endowed with a genius for learning and civilization, vainly endeavoured to relight the torch of human knowledge in this barbarous age. The renown of Charlemagne extended to Asia; he entered into a correspondence with the famous Calif of the Saracens, Haroun Alraschid, a man who greatly contributed towards polishing and enlightening the Arabians; and who preferred the friendship of the king of France to that of all the princes of Europe, because none, like Charlemagne, possessed a desire for intellectual greatness. After the death of Charlemagne, which took place in the year 814, Europe became involved in still greater mental darkness than before.

When the Western empire, weakened by luxury and effeminacy, had fallen an easy prey into the hands of barbarians, the empire of the East, though feeble, yet preserved the precious deposites of ancient literature; but the greater part of the learned, occupied with the subtleties of scholastic theology, made no effort to enlarge the boundaries of natural science. Religious intolerance drove from the empire many enlightened men, who, banished by the emperor Theodosius, carried among the Arabs the taste for Greek and Latin literature, and founded schools upon the shores of the Euphrates, where they taught rhetoric, languages, and medicine.

The Arabs, fond of mysteries, and led by their genius and ardent imaginations to the cultivation of poetry and works of fiction, seemed to have little taste for sciences which required assiduous application and patient investigation. Urged on by fanaticism, under Mahomet they were the conquerors and scourges of the civilized world. Alexandria experienced their ruthless violence. This city,

---

Barbarians ravage Italy—Language corrupted—Botany shared the fate of other sciences—Charlemagne—Decline of learning in the Empire of the East—Literature carried among the Arabs.

by turns the asylum and tomb of letters, had witnessed under the
first of the Cesars the destruction of the library collected by the
Ptolemies; under Aurelian, that founded by Augustus; under Theodosius, that which Antony had given to Cleopatra; and for the
fourth time in possession of an immense collection of books, acquired through her love for philosophy, this city saw her magnificent library reduced to ashes by the victorious Saracens.

This barbarous but noble race at length became imbued with the
love of science; a succession of califs, (among whom was Haroun Alraschid, already spoken of as the friend of Charlemagne,) by
their devotion to learning, rendered Bagdad the most enlightened
city of the earth.   Their learned men began to construct maps of
conquered countries, and to describe objects of natural history;
distant voyages extended and multiplied their commercial relations;
and mathematics, medicine, and natural history, were cultivated
with ardour.

When the Arabs had conquered Spain, they carried thither letters
and arts, and their school became celebrated throughout the world.
In the 11th century the French, Italians, Germans, and English,
went to them to learn the elements of science. The Arabians preserved their superiority in the sciences at least, if not in literature,
until towards the close of the 15th century. But when this people,
divested gradually of their European conquests, were at last driven
from Spain into Africa, they seemed, as if by instinct, to replunge
into the savage ignorance from whence they had been drawn by the
efforts of a few great minds.

The Arabs had considered plants more as physicians and agricul-
turists, than as botanists; but although their descriptions of plants
were imperfect, their labours were not useless to botanical science.
They discovered many plants of Persia, India, and China, which
were unknown to the ancients. They, however, fell into the error
of dwelling more upon the works of Aristotle, Theophrastus, Dioscorides, and Pliny, than of observing nature; almost believing that
nature herself must be wrong, when she deviated from those celebrated philosophers.

The *Crusades*, commencing at the close of the 11th century, and
continuing until towards the middle of the 13th, prove the barbarity
of the times; yet we cannot doubt that these distant and romantic
expeditions were, in part, suggested by the desire of change and the
vague wish to see and to know new things, and hastened the awakening of the human mind from the sleep of ages.

The 12th and 13th centuries witnessed in Italy the revival of a
taste for letters and the fine arts. The commerce of that country
was flourishing, the people made long voyages by sea, and in the
accounts which they published, spoke of the vegetable productions
of the countries they had visited, in such a manner as excited the
curiosity of the nations of Europe.

About this period, it is supposed, herbariums, or collections of
dried plants, began to be preserved. This was an important era in
botanical science; for nature is ever true and incapable of leading
into error, while descriptions, or even drawings, may often give false
views of natural objects.

The science of Botany was not enriched by a single work of any
merit, from the fall of the Roman empire, a period which marked

_____

Destruction of the Alexandrian Library—Bagdad famous for learning—Schools of
Arabs in Spain—Remarks upon the Arabian botanists—The Crusades—Revival of literature—Herbariums made.

the decay of literature, until the 15th century. Those, in the dark ages, who pretended to any knowledge of plants, only quoted from the Greek and Roman writers, but they were ignorant even of the languages in which their works were written. In the 15th century, Italy was governed by wise princes, who were influenced by a desire to promote knowledge among their people. They invited to their country learned men from Greece, from whom they might learn the language of Homer and Aristotle.

At this time the Turks threatened Constantinople, and that capital of the empire of the East at length fell into their hands. The literature of Greece now took refuge in Italy; the ancient languages were revived, and at this time, translations of ancient writers, with learned commentaries, were given. But these labours, although exercising an important influence upon literature, were not equally fortunate with respect to the progress of natural history. The learned writings of antiquity were accurately studied, but, blinded by the brilliancy of great names, men of learning looked not upon nature; they had yet to learn, that without examining and comparing real objects, there can be no solid foundation in natural history.

At the period of which we are now speaking, a physician of Germany published some indifferent descriptions of plants, accompanied by a few engravings. This connexion of drawing and botany, although the whole was badly executed, was considered as an important improvement in the science.

While Italy was thus a second time enriched with the literary treasures of Greece, Spain and Portugal were becoming enlightened by intercourse with foreign nations. The Portuguese extended their voyages to the western coasts of Africa and the Cape de Verd islands; the Cape of Good Hope was at length discovered, and Vasco de Gama, sailing around it, reached the East Indies. It was at this period that Christopher Columbus discovered the NEW WORLD.

This event, so important to the old world, is to us who inhabit this pleasant and favoured country, one of deep interest. Ages passed on after the creation of the world, and America remained, with regard to the eastern continent, as though she existed not. The lofty Andes raised their snowy heads to the clouds, the majestic Amazon rolled onward to the Atlantic, our lakes spread out their vast expanse of waters, our Hudson and Connecticut received their tributary streams, and bore them to the ocean;—but to what people were these grandeurs presented, and what were the changes in the moral world, while nature thus moved on in her unchanging course ?—History is silent! But while in the old world empires had been rising, continuing for centuries stationary, and then decaying, succeeded, and succeeded by others pursuing the same track; were no moral changes going on in the American continent? Have no mighty nations ever existed here; have no arts or letters been cultivated; was the savage Indian for thousands of years sole lord of one half of the world?—And when, and how, did the first inhabitants of this continent come from Asia, where man was placed at his creation? These are inquiries which naturally arise, on tracing the historic page through so long a period of time, until suddenly this new world bursts upon our vision! But, though many speculations have from time to time appeared, respecting the probable history of America, before its discovery by Columbus, the subject is still shrouded in darkness and obscurity.

Constantinople taken by the Turks, and the literature of Greece transferred to Italy—New world discovered—What was the history of America before this period?

LECTURE XLIV.

HISTORY OF BOTANY FROM THE BEGINNING OF THE SIXTEENTH CENTURY TO
THE TIME OF LINNÆUS.

WE have now traced the progress of botanical knowledge, from
the earliest periods of the world, to the discovery of America.
About this time, botanic gardens began to be cultivated; these afforded new opportunities for investigation, by comprehending the
vegetables of all countries within such limits as enabled the botanist
to compare them, and to watch their growth and different stages of
development.

From the days of Theophrastus until the beginning of the 16th
century, Botany, instead of becoming more perfect, had been rendered more obscure. This was not owing to want of attention or
labour, but to the false rules of philosophy which had so long prevailed.

At length the cause of the evil seemed to be discovered. Many
writers protested against the erroneous opinions of their times; they
said, "our blind respect for the ancients is an insurmountable ob-
stacle to the progress of Botany. We expect to find everywhere the
plants of Theophrastus, Dioscorides, and Pliny; whereas they did
not know one hundredth part of the plants which cover the globe.
The first of them never went out of Greece; the second left only unconnected notes, treating without order upon the medicinal qualities
of plants; and Pliny copied these notes without comment or criticism. We cannot apply to the plants of Germany or France, the
names under which the ancients described those of Italy, Greece, and
Asia; before studying the plants of foreign countries, we ought to
know those of our own. Of what use are disputes about the nature
and qualities of species, when we are not able to distinguish one from
another? The true method of doing this, is to explore the plains,
valleys, and mountains, to examine and compare the plants of our
own and foreign countries. Libraries alone are insufficient to make
botanists."

These reflections led to a happy revolution, not only in this science, but in all others; it may be called the era of true philosophy.*
Yet the principles which were now discovered, were not much applied to science until the time of Bacon, Newton, Linnæus, and
Locke; and it remained for the late Dr. Thomas Brown, of Edin-
burgh, to show that the human mind itself is subject to the same
general laws of inquiry which now regulate investigations in the
physical sciences.

Up to the period of which we are now speaking, plants had only
been described in alphabetical order; about this time, some German
botanists attempted a *collection of individual plants into species;* this
improvement was received with much approbation.

* Lord Bacon is generally considered as having first taught the proper method of
studying the sciences, viz.: by ascending from facts to principles; this is called the
method of induction. It has recently been asserted by an able writer in one of our
first American periodicals, that Bacon was not the author of the inductive philosophy,
but that he borrowed his rules of philosophizing from Aristotle, whose real principles
had for ages been misunderstood. It is to be hoped that men of talents will not so
far depart from the true rules of philosophizing, as to devote that time in contending
about their author, which might be profitably applied in the application of these rules
to the investigation of truth and nature.

Botanic gardens first cultivated—Botanists began to discover the obstacles to the
progress of science—Era of true philosophy—Improvements of German botanists

These species were arranged according to certain general resemblances, or natural relations; thus we see that natural methods were prior to any attempts at an artificial system.

In the beginning of the 16th century, we find the names of many who were engaged in investigating the vegetable kingdom. Some are commemorated by the names of plants; Leonard Fuschs of Germany, by the plant Fuschsia; Lobelius, physician to James I., by the Lobelia; and Lonicer, by the Lonicera.

Lobelius distinguished the cotyledons of seeds, divided monocotyledonous from dicotyledonous plants, and attempted to form families by grouping species according to their natural relations. Zaluzian of Bohemia laboured to perfect the natural groups of former botanists; he is the first of the moderns who positively affirmed the existence of stamens and pistils in all species of plants, and suggested the necessity of these organs.

But, notwithstanding the labours of many learned men, little real improvement would have been made in the science of Botany, had there not, at that time, existed some minds of superior genius, who turned their attention to tracing some proper method of classification. These were Gesner, Clusius, Cæsalpinus, and Bauhin; of the latter name were two brothers, both of whom are deservedly celebrated.

Gesner, called the Pliny of Germany, born in 1516, was of an obscure and humble origin, but possessed of a powerful and penetrating mind. He attempted to make a general collection of the objects of natural history; he explored the Alps, and discovered many plants until then unknown. He is distinguished from those who had gone before him, in his suggestions that there existed in the vegetable kingdom, groups, or *genera*, each one composed of many species, united by similar characters of the flowers and fruit. Soon after the publication of this opinion, botanists began to understand that the different families of plants have among themselves natural relations, founded upon resemblances and affinities, and that the most obvious are not always the most important. These are fundamental truths; and the *distinction of species, the establishment of genera, and of natural families*, seemed to follow of course, after these principles were once established. The *Tulipa gesneriana*, and genus Gesneria, have been dedicated to this botanist.

Clusius was born in 1526; his parents had destined him for the profession of law, but his decided taste for Botany induced him to abandon this profession. He was learned in the ancient and modern languages, but his enthusiasm for natural history induced him to lay aside every other pursuit. He travelled over almost all the west of Europe, in order to make discoveries in the vegetable kingdom; and soon excelled all the botanists of the age in the knowledge of both native plants and exotics. He had the direction of the imperial garden at Vienna, and afterward was public professor of Botany at Leyden. His enthusiasm for this science terminated only with his life. Before his time, the art of describing plants with precision and accuracy was unknown; but, unlike the descriptions of his predecessors, his were neither faulty from superfluous terms, nor from the omission of important circumstances.

Cæsalpinus, a native of Florence, who was contemporary with Clusius, proposed to form *species into classes*. The characters which he employed for this purpose, were, the *duration* and *size* of *plants;* *presence,* or *absence of flowers ;* the number of cotyledons; the situation of the seed, as erect or pendent; the adherence of the pericarp to the seeds; the number of cells in the pericarp, and the *number of seeds* which they contained ; the *adherence* of the *calyx* to the *germ';* and the *nature of the root,* whether *bulbous,* or *fibrous.* This method was too imperfect to be followed, having neither the simplicity nor the unity to render its application useful.

John Bauhin, though younger than Gesner, was his friend and pupil; he composed a general history of plants; this was a work evincing great learning and accurate investigations. Gaspard Bauhin, the younger brother, no less active and learned, conceived the design of a work which should contain a history of all known plants, together with the different names which other writers had applied to the same plant. Clusius and the elder Bauhin had imagined something like a genus of plants, formed by the grouping of similar species, but Gaspard Bauhin expressed this more decidedly in remarks upon generic distinctions. His work, the result of forty years' labour, was of great assistance to Linnæus, in perfecting our present system of Botany.

We find, in looking back upon the labours of botanists during the 16th century, that more had been accomplished than during any former period; the character of novelty and originality exhibited in these researches, is highly creditable to those who thus led the way in the march of improvement.

The 17th century, in its commencement, was not favourable to the sciences. Europe was agitated by continual wars, and the arts of peace were neglected; but in the last part of that age, a taste for natural history revived ; men of highly gifted minds applied themselves to the study of Botany, and many undertook long voyages, with the sole design of examining foreign plants. Botanists were astonished at the great number of interesting plants discovered by travellers, in the region of South Africa, around the Cape of Good Hope, and in the East India Islands.

Two Dutch botanists of the name of Commelin, who wrote about this period, are commemorated in the beautiful genus Commelina, first discovered in America. Bonnet* of Geneva, a close observer of facts, wrote upon the "Nature and Offices of Leaves;" and a work entitled, "Contemplation of Nature, or the Regeneration of Beings." Two writers of the name of Camararius are distinguished in the annals of the science for learning and ingenuity. Gaertner of Germany wrote upon fruits, or, as he termed this department of the science, Carpology. He dissected the fruits of more than a thousand plants, the figures of which he designed and engraved. To Gleditsch, professor of Botany at Frankfort, is dedicated the genus Gleditscha. Rudbeck the younger, who preceded Linnæus as professor of Botany in Upsal, was, by the latter, commemorated in the genus Rudbeckia.

At this period, the plants of our own country began to excite the curiosity of scientific Europeans. Louis XIV. sent to America, Plumier, a man celebrated for his mathematical and botanical knowledge, and who was styled, botanist to the King. He made three voyages, and gave drawings and descriptions of more American species than any other traveller had done.

* Pronounced Bonnay.

About this time, the practice of naming newly-discovered plants after distinguished botanists became common.   History now presents us with many who were distinguished by their efforts in the cause of science, but a notice of each individual would carry us beyond our limits.

Botanists now began to study *the stamens and pistils of plants;* and it was suggested that the science would remain imperfect as long as species and genera were undefined.   Orders and classes also were recommended, and natural resemblances and affinities studied. A work was written upon the umbelliferous plants;* this was the first attempt at describing in one mass any single group of plants by characters peculiar to the whole.   This was followed by several attempts to form a natural method of classification; among the most approved of these methods was that of Ray, who published a work called " *A General History* of Plants;" in this he divided all Plants into 33 classes, 27 of which were composed of *herbs,* the rest of *trees.*

The first botanist who th⬛⬛⬛ of classing plants without any reference to their being eith⬛⬛rbs or trees, was a German, of the name of Rivannus, who proposed to consider, as the foundation of classification, the *absence or presence of flowers; the manner in which they were situated, or their inflorescence; the number of petals; the ⬛gular or irregular form of the corolla; the adherence or non-adhe⬛nce of the calyx to the germ; the nature of the pericarp; the number of seeds, and of cotyledons.*

A botanist of the name of Magnol, at this time, was honoured by having his name given to the splendid Magnolia, an American plant, which then began to be known in Europe.

Joseph Pitton de Tournefort was born in 1656.   While very young, he discovered an enthusiastic fondness for botanical pursuits: he had been destined by his friends for a profession; but his genius seemed so strongly bent upon the study of nature, that he was at length permitted to indulge without restraint in his favourite pursuits. He ranged over the Alps and Pyrenees, and many provinces of France, collecting the flowery treasures offered by those fertile regions; often in peril from banditti, and exposing his life to hazards in climbing terrific precipices, or amidst the glaciers of the mountains.

The method of Tournefort, which was founded upon the form of the corolla, although imperfect, greatly assisted the progress of *that botanist* who stands unrivalled in this department of Natural History. You do not need to be told that we here refer to Linnæus.

You will observe that the attempts of botanists, until this time, had been chiefly directed towards the attainment of some proper method for the arrangement of plants; the attention of some investigating minds was now turned towards their *Anatomy and Physiology.* Since the days of the first Greek naturalists, these departments of botanical science had lain neglected; but the confused opinions of the ancients now served to suggest experiments, which resulted in new observations and solid discoveries.

The invention of the microscope threw light upon the mysteries of

---

* The author of this was Robert Morrison, a Scotchman. These *monographs,* or descriptions of single families, are now of great value; no botanist can thoroughly investigate the whole vegetable kingdom; but by close attention to one department, important discoveries may be made.

Various improvements in Botany—Ray—Rivinus—Magnol—Tournefort—Attention of botanists turned towards anatomy and physiology—Microscope.

nature, which, without this instrument, must ever have remained in obscurity; by its assistance botanists studied the internal structure of vegetables; they described the *heart, wood,* and *pith;* they perceived the newly *formed bud,* yet invisible to the naked eye; the *future plant existing in the bulbous roots,* and *even in the seed; pores* were discovered, which were found to be the organs of the expiration and inspiration of gases, thrown out as noxious, or inhaled as nutritious.* The importance of the stamen and pistils as essential to the perfection of the seed of vegetables began to be suspected

As yet, however, the science of Botany lay in scattered fragments .of various imperfect and contending systems; much labour had been bestowed, and great improvements made, but there was no central point around which these improvements might be collected.

The learned world were sensible of the deficiency; but it required genius, great observation of nature, and courage to stem the tide of popular prejudices, in him who should come forward to attempt the work of reform.

Charles Von Linnæus, an inhabi___of Sweden, suddenly emerging from obscurity, offered to the world a system of Botany, so far superior to all others, as to leave no room for dispute as to its comparative merit. All preceding systems were immediately laid aside, and the classification of Linnæus was received with scarcely a dissenting voice. What this system was, you have not now to learn, since it was the alphabet of your botanical studies. Linnæus extended the principles of his classification to the animal and mineral kingdom; in the language of an eminent botanist,† "His magic pen turned the wilds of Lapland into fairy fields, and the animals of Sweden came to be classed by him as they went to Adam in the garden of Eden to receive each his particular name."

LECTURE XLV.

HISTORY OF BOTANY FROM THE TIME OF LINNÆUS TO THE PRESENT.

LINNÆUS was born in 1707; his father was a clergyman, and had designed his son for the same sacred office; but seeing him leave his studies to gather flowers, he inferred that he possessed a weak and trifling mind, unfit for close investigation; he was about to put him to a mechanical employment, when some discerning persons perceiving in his devotion to the works of nature the germ of a great and lofty mind, placed him in a situation favourable to the development of his peculiar talents, where he was allowed, without restraint, to study the book of nature,

" This elder Scripture, writ by God's own hand."

Linnæus formed anew the language of botanical science; every organ of the plant he defined with precision, and gave it an appropriate name; every important modification was designated by a particular term. Thus comparisons became easy, and confusion was avoided. The characters of plants appeared in a new light. Each species took, besides the name of the genus to which it belonged, a specific name which recalled some peculiarity distinctive of the

* Leuwenhoek, Grew, Malpighi, and Camerarius, are among the first of the moderns who investigated the internal structure of vegetables.
† Sir James E. Smith.

species. Before that time the species, instead of being thus designated, required in some cases a whole sentence to express the name.

But what most tended to render the works of Linnæus popular, was his artificial system, in which he had made the stamens and pistils subservient to a most simple and clear arrangement; he remarked the *different insertion of the stamens; their union by means of their filaments had been before observed, but he employed them in a manner entirely original.*

This "*Northern Light,*" as he has sometimes been termed, contributed to the progress of physiology both by his own discoveries, and by improving upon the suggestions of those who had gone before him. In the details of science, he was no less accurate than bold and comprehensive in his general views. The world knew not which to admire the most, the multiplicity, the novelty, or the profound views of this modern Aristotle. His school became the resort of men of science from all Europe; and he seemed to have acquired that influence over the human mind, which had been peculiar to the ancient philosophers of Greece. The defects of this great man, for human nature is never without its imperfections, were, that he sometimes carried too far a favourite idea; endowed with a brilliant imagination, he was at times somewhat blinded by the beauty of his conceptions, and strove to reconcile nature to the visions of his own fancy.

We have, in our investigations of the artificial system, occasionally pointed out some imperfections, particularly in the separation of natural families; but no means of remedying these have yet been found, and after the lapse of near a century, with the exception of a few alterations, we still receive this system as left by its author.

Linnæus died in 1778; he is honoured among the scientific by a title far more proud that any hereditary distinctions, that of "*Prince of Naturalists.*" The most important works of this great man are, "Philosophy of Botany," "Genera and Species of Plants," "System Nature," and "Flowers of Sweden, Lapland," &c. The Linnæa realis was dedicated to him by Gronovius. Ten years after his death, a society, distinguished by his name, was founded in London; is now in possession of his library, herbariums, collections of ects and shells, with numerous manuscripts. Sir James Edward ith was the founder of this society, and its first, and only president until his death, which has recently occurred. He translated the itings of Linnæus, and illustrated them by his own comments.

The study of plants, after the discoveries and classifications of Linnæus, became, in a degree, general. The knowledge of vegetable physiology began to be usefully applied to agriculture. Duhamel, France, very successfully laboured to exhibit the connexion between the science of Botany and the cultivation of plants. Bossuet, of Geneva, proved by experiments that the *vascular system of plants is tubular and transparent; and that leaves perform the office of respiration.*

Grew of England, had, before this period, ascertained the existence of the *cambium*, and Duhamel afterward proved that it was distinct from the sap and proper juices. The latter opposed the idea, till then entertained, that earth and water were the only food of plants; he proved that the various solids and fluids diffused in the soil and atmosphere, are all important to vegetation.

The observations of Priestley, Saussure, and others, aided by the discoveries made in pneumatic chemistry, of the existence of oxygen, hydrogen, and carbonic acid gases, formed a new era in the history of vegetable physiology. It was proved that vegetables do ultimately consist of *oxygen, hydrogen,* and *carbon,* sometimes of a small quantity of nitrogen, combined with mineral salts, and often some silex, sulphur, and iron. These elementary substances were found to be diffused through air and water, and the animal and vegetable substances which the latter holds in solution : the green parts of vegetables were observed to exhale oxygen in the light, and carbonic acid gas in the dark; and the carbon left by the decomposition of the carbonic acid, was shown to be incorporated into the vegetable substance, giving to the wood its strength and hardness.

The naturalist whose labours, in point of utility, will best bear a comparison with those of Linnæus, is Bernard de Jussieu. He was remarkable for the extent of his knowledge, the penetration of his genius, and the solidity of his judgment. He is said to have been unambitious. The love of truth and science were with him sufficient excitements to the most severe labour. "Many of our contemporaries," says Mirbel, "knew this sage; they say that never have they seen so much knowledge combined with so high a degree of candour and modesty." To this botanist we are indebted for a natural method of classification, superior to those of his predecessors.

Jussieu proposed a method of classing plants according to certain distinctions in the seed, which were found to be universal; this was perfected and published by his nephew, Antoine-Laurent de Jussieu, and is now generally received as the best mode of natural classification which has yet been discovered. This method is called natural, because it aims to bring into groups such genera of plants as resemble each other in medicinal and other properties; while the system of Linnæus is called artificial, because, by a certain rule, plants which have no such resemblance in their properties are brought together. We therefore find in one of the Linnæan classes the poisonous flag and the nutritious grass, the grain which supports life and the darnel which destroys it; in another, the healthful potato and the poison mandrake, the deadly hemlock and the grateful coriander. Throughout this system we meet with similar contrasts in the qualities of the plants which are collected into the same classes. Nor are their external appearances less unlike; for here the oleander and pigweed, the tulip and the dock, meet in the same classes. This system, it should always be remembered, is not the whole science of Botany, but is the key to the natural method, by which alone, we should find great difficulty in ascertaining the names of plants; it is, as it were, a stepping-stone by which we must ascend to the valuable knowledge which cannot well be reached in any other way. The more practical a botanist becomes, the less need he has for this assistance; the eye becomes quick to seize on natural characters without reference to the dictionary, as the artificial system is aptly termed. Thus a pupil, in studying a language, may, in time, be able to dispense with his dictionary; though he could not have proceeded, at first, without its assistance. For more particular explanations of Jussieu's method, you are referred to the comparison of that with the method of Linnæus and Tournefort in the remarks on classification, and to the Natural Orders contained in the appendix.

Priestley, &c.—Character of Jussieu—Natural method of classing plants.

Adanson, previous to the time of the younger Jussieu, had published a system of classification, in which he arranged plants according to the resemblance observed in all their organs. In one class, all which had similar *roots* were placed; in another, all which had similar *stems ;* a third was arranged by resemblance of *leaves*, in their forms and situations; but the most important distinctions he considered as founded upon the *organs of fructification*. The name of this ingenious botanist is preserved in the Adansonia, or calabash-tree, of Africa, which is considered as the colossus of the vegetable kingdom.

Among other botanists, we would notice Louis Richard, who wrote in French an interesting account of the Orchideæ of Europe, and assisted in compiling from ancient works a very useful botanical dictionary.

Des Fontaines first showed that the stems of monocotyledonous and of dicotyledonous plants differ from each other in their structure and modes of growth; he divided them into *endogenous*, growing inwardly, as the palms, and *exogenous*, growing outwardly, as the oak.

France is distinguished for the number and accuracy of its naturalists. Mirbel, a distinguished professor of Botany in Paris, has pursued his inquiries into the anatomical structure, and the physical operations of plants, to an extent not exceeded by any other naturalist; his "*Elemens de Botanique*" is a splendid work, which forms a very important and valuable addition to a botanical library.

The Baron Humboldt spent five years in investigating the vegetable productions of the equatorial regions in America, and his remarks on vegetables, as a criterion of climate, are original and interesting.

Josephine, the first wife of Napoleon, was distinguished for her fondness of this study; other ladies of distinction, stimulated by her example, cultivated plants with reference to scientific observations. In England, Mrs. Wakefield, and the industrious and enlightened Mrs. Marcet, (author of Conversations on Natural Philosophy, Chemistry, &c.) have distinguished themselves as the authors of useful treatises on Botany.*

De Candolle's "Elementary Theory of Botany," is highly valued as a scientific and able performance; but it is useful, rather for those who have already attained a knowledge of the elements of Botany, than for the beginner in the science. The natural method of Jussieu has been modified and improved by the labours of De Candolle, Mirbel, Lindley, and Robert Brown.

In turning from Europe to the United States, we find the state of literature flourishing, and a taste for the natural sciences becoming extensively diffused. The names of many of our scientific men stand high in Europe, as well as in their own country. Among these are Silliman, who established the first scientific journal,† and encouraged others to pursue the course of investigation which he himself has followed so successfully. Eaton has laboured to bring science within the reach of every inquirer; not only by rendering the

* Mrs. Somerville, from the extended views of science which she has exhibited, may, perhaps, be called *the* scientific woman of her age.
† Except the Mineralogical Journal of Bruce, which ceased after the appearance of a few numbers.

Adanson—Richard—Mirbel—Humboldt—Females who have interested themselves in the study of Botany—De Candolle—Silliman—Eaton.

labours of others of more general utility, but adding to the common stock the result of years of inquiry and observation.

To go back to the infancy of Botany in the United States, we find the name of Bartram stands recorded in history, as that of the first native of our country who was conspicuous for botanical researches. Houston investigated the region of Canada, and described many of its plants; in honour of him is named the little flower HOUSTONIA *cærulea*, which is abundant in New England. Clayton made a list of Virginian plants, and is commemorated in the beautiful CLAYTONIA *virginica*. Kalm, a pupil of Linnæus, whose name is given to the KALMIA, (American laurel,) spent three years in America, and returned to Europe laden with botanical treasures; the sight of the American plants brought by his pupil, many of which were entirely new to him, is said to have produced such an effect upon Linnæus, that although lying ill of the gout, and unable to move, his spirits were rekindled, and in the delight of his mind he forgot his bodily anguish, and recovered from his disease.*

Although American works on Botany are not wanting, the author of these Lectures found no one book, either foreign or American, which seemed designed to conduct the pupil through a full and connected course of study. To bring together in one volume the Elements of Vegetable Anatomy and Physiology, the principles on which the Natural and Artificial Classification depend, and to teach these systems by a full exposition of them, and by a Flora of Plants, for practice in analytical Botany—these have been the objects in view in the preparation of this work. Its publication, we hope, has removed the obstacles which formerly impeded the progress of botanical information, in schools, and among our own sex. We have seen that even children may become botanists, and lay aside their toys to divert themselves by distinguishing the organs of plants and tracing out their classification.

Of all sciences, perhaps no one is settled on a firmer foundation

---

* Among the earliest botanists of North America, were Colden, Michaux, and Muhlenberg; Pursh was the first who finished a system of North American plants, so arranged as to be useful to the student. Some of the first teachers of the science were Barton, Hosack, and Mitchill. The first public lecturer on Botany, was Professor Amos Eaton. Dr. Bigelow gave a course of lectures in Boston, in the year 1813, and soon after published his Boston Flora. Professor Ives and Dr. Tully did much in New England towards awakening a zeal for the science, in the years 1815 and 1816; and at a later period, Dr. Sumner has pursued and illustrated the study with much ardour and success.

Want of books was a great impediment to the progress of the science when Eaton published his Manual of Botany; this book gave a new impulse to the progress of the science; its familiar method and simple style induced many to commence the study. This was followed by many other works describing plants, and several elementary works; of the former class were Nuttall's Genera, Elliott's Southern Plants, Barton's Flora of Philadelphia, Darlington's, Torrey's, and Bigelow's Floras; these furnished descriptions of most American plants, not included in the works of Pursh. Among Elementary books are "Barton's Elements," a large work containing much that is interesting in the physiology of Plants; "Lock's Botany," a small book, but exhibiting a plan of arrangement simple and methodical; "Sumner's Compendium of Botany," written in a beautiful and pure style; and more recently, "Nuttall's Elementary Work," which gives in popular language more facts with regard to plants, than almost any other work of the kind; a small work entitled "Catechism of Botany," by Miss Jane Welsh, was the first attempt by an American lady to illustrate the science. Professor Lindley's late work, entitled "Introduction to the Natural System of Botany," though it may be highly useful to the advanced student, cannot be studied with advantage except by the practical botanist. Beck's Botany is a neat and beautiful introduction to the natural system, and his descriptions of Genera and Species are valuable.

---

than that of Botany; the improvements of future years, we are not able to anticipate; but it is probable that as discoveries and improvements are made, they will cluster around the principles already established; each taking its proper place in the various departments now arranged for the reception of scientific truths.

The spirit of our government is highly favourable to the promotion and dissemination of knowledge; and although Europe may boast of many stars, which irradiate her firmament of letters, shining with brilliant lustre amidst the surrounding darkness of ignorance may we not justly feel a national pride in that more *general diffu sion of intellectual light, which is radiating from every part, and to every part of the American republic!*

LECTURE XLVI.

GENERAL VIEW OF NATURE—ORGANIZED AND INORGANIZED BODIES—CLASSIFICATION OF ANIMALS.

HAVING considered the vegetable kingdom under its various aspects, it may be proper, before closing our course of botanical study, to take a general view of that external world of matter, of which the part we have examined, extended and diversified as it is, constitutes but a very small portion. The science you have been investigating, with some others, constitutes a general branch of knowledge termed *Natural science.* The study of nature presents, in a lively and forcible manner, the power and wisdom of the Creator; and offers to the enlightened mind, a never-failing source of the most pure and refined enjoyment. Those who know nothing of this source of happiness, cannot appreciate its value; they may inquire the use of studying into the nature of objects, without any reference to the enjoyment of the senses, to personal gain or honour. A celebrated naturalist[*] observes: "The rich and the great imagine, that every one is miserable, and out of the world, who does not live as they do; but they are the persons who, living far from nature and from God, live out of the world. Misled by the prejudices of a faulty education, I have pursued a vain felicity amid the false glories of arms, the favour of the great, and sometimes in frivolous and dangerous pleasures. I have never been happy but when I trusted in God: opposed to THEE, the AUTHOR of all things, power is weakness! supported by THEE, weakness becomes strength! When the rude northern blasts have ravaged the earth, THOU callest forth the feeblest of winds; at the sound of THY voice, the zephyr breathes, the verdure revives, the gentle cowslip and the humble violet cover the bosom of the bleak earth with a mantle of gold and purple."

To the pious reflections of this French writer, we will add the following quotation from an English author,[†] the energies of whose rich and cultivated intellect were devoted to the cause of religion, who viewed nature as a philosopher, but what is far better, as a Christian. Happy indeed, are those in whom philosophy and Christianity are blended, and delightful is the intercourse, even in this world, between minds thus enlightened and purified!

[*] St. Pierre.
[†] Rev. Legh Richmond.

" There is a peculiar sweetness in the recollection of those hours which we have spent with friends of a kindred spirit, amidst the beauties of created nature. The Christian can alone find that congeniality in associates, who not only possess a lively and cultivated sense of the high beauty which landscape scenery presents to the eye, but who can also see creation's God in every feature of the prospect. The painter can imitate, the poet describe, and the tourist talk with ecstacy of the sublime and beautiful objects which constitute the scene before him; but he can only be said to enjoy them aright, whose talents, taste, and affections are consecrated to the glory of Him by whom 'all things were made, and without whom was not any thing made that was made.' When the pencil that traces the rich and animated landscape of mountains, lakes, and trees, is guided by a grateful heart as well as by a skilful hand, then the picture becomes no less an acceptable offering to God, than a source of well-directed pleasure to the mind of man. And when the poet, in harmonious numbers, makes hill and dale responsive to his song, happy is it if his soul be in unison with the harp of David, and if he can call on all created nature to join in one universal chorus of gratitude and praise. The Christian traveller best enjoys scenes like these. In every wonder he sees the hand that made it—in every landscape, the beauty that adorns it—in rivers, fields, and forests, the Providence that ministers to the wants of man—in every surrounding object he sees an emblem of his own spiritual condition, himself a stranger, and a pilgrim, journeying on through a country of wonders and beauties; alternately investigating, admiring, and praising the works of his Maker, and anticipating a holy and happy eternity to be spent in the Paradise of God, where the prospects are ever new, and the landscapes never fade from the sight!"

> " Oh! for the expanded mind that soars on high,
> Ranging afar with Meditation's eye!
> That climbs the heights of yonder starry road,
> Rising through nature up to nature's God.

> " Oh! for a soul to trace a Saviour's power,
> In each sweet form that decks the blooming flower:
> And as we wander such fair scenes among,
> To make the Rose of Sharon all our song."

Naturalists, to the great discredit of science, have formerly shown an unhappy tendency to skepticism; enabled to comprehend some of the great operations of nature, they presumed to set up their own reason against the revelation of God, and impiously refused to believe any thing which could not be explained according to the principles of human science. Searching into the elements which compose the human body, and observing the dispersion of the same, and their incorporation into other substances, they affirmed that it was "a thing impossible for God to raise the dead." Well might we, in addressing such a philosopher, say, with the Apostle, "Thou fool!" Cannot he who formed all things of nothing, reanimate the sleeping dust, and recall the spirit to its own body? Happily, this melancholy perversion of human learning seems to have passed away, and we now see many of the most enlightened investigators of the principles of science among the most humble disciples of Jesus.*

* In the character of Dr. Mason Good, as exhibited in his biography, written by Olinthus Gregory, we find this union of science with deep and fervent piety most happily exemplified.

By the word *Nature*, derived from a term signifying *born* or *produced*, in a general sense we mean all the works of God. Using a figure of speech called *Metonomy*, we often put the effect for the cause; as when we speak of the "works of nature," meaning what the Almighty has brought forth: or we often mean by nature the Deity himself; as when we say that "nature produces plants and animals."

With respect to the *heavenly bodies*, which manifest themselves to us with so much magnificence, we know them to be *matter*, because we observe them to be subject to the laws which govern matter; as we have been able, by the discoveries of astronomers, to understand their various revolutions; we have, in general, clearer ideas of their motions than even of our own planet; it is more easy for us to imagine them as moving, than that our firm earth is whirling with inconceivable velocity. Were it possible for us to conceive the quantity of matter which even one world as large as our sun contains, the thought would be overwhelming; and of all the worlds which we behold at one view in a serene night, what finite being could imagine their united extent? They are suspended over our heads, each one pursuing its destined course; why do we not fear that some one may be precipitated upon our little world, and crush it to atoms? It is because we know that they are all upheld by that Power which "created the heavens and the earth," and who governs the universe by regular laws. This universe is as infinite as the God who formed it; our sun, with all its systems, is but a point lost in immensity. Astronomers have proved that the fixed stars are at such an immense distance from us, that moving at the rate of 500 miles an hour, we should not reach the nearest of them in 700,000 years, a distance more than 200,000 times greater than that of the sun from the earth. The same space probably separates all the fixed stars. Around those stars revolve millions of opaque globes, as our earth revolves around the sun, which is also one of the fixed stars. The satellites describe around the primary planets almost circular orbits; they are carried with their primaries around the sun in their annual motion; the sun himself, with all his numerous train of primary planets, each with its satellites, revolves around the common centre of gravity of the fixed stars, of which himself constitutes a part; and these are supposed to revolve around the centre of the universe. Here *may* be the throne of the Almighty Creator and Director of all these stupendous objects.

Yet we need not fear that we shall be forgotten in the immensity of creation; the same Being who created and rules the host of heaven, made the little moss and the lilies of the field, which are so beautifully arrayed. If God condescends to care for them, he will not neglect us, who are made in his own image, and destined to an immortal existence.

Turning our thoughts from the heavenly host to our own little globe, and considering the matter which exists upon it, we find two great classes of substances; 1st, *inorganized*, and 2d, *organized*.

The 1st *class of substances*, viz., such as are *inorganized*, comprehends all matter destitute of a living principle; such as *fluids, gases, and minerals*. The particles which compose them are entirely subject to chemical and mechanical laws.

The 2d *class*, viz., *organized* substances, includes animals and vegetables; the particles constituting them are in a perpetual state of motion. They are supported by air and food, endowed with life, and subject to death; the active power or life which operates in them we call the *vital principle*. This vital principle eludes the researches of man; all that we know of it is in its effects, enabling the organized body to resist putrefaction, and, to a certain degree, to maintain a temperature different from surrounding bodies. Deprived of this vital principle, both animals and vegetables become subject to chemical decomposition; their solid parts are dissolved, and they return to the earth from whence they were taken.

If you dig up a stone, and remove it from one place to another, it will suffer no alteration; if you dig up a plant, it will wither and die If you break a mineral to pieces, every fragment will be a perfect specimen of its kind; it will only be altered in shape and size; but if you tear off a branch from a plant, or if a limb be taken from an animal, they will both immediately begin to decay; the vital principle being extinguished, putrefaction and dissolution follow.

We should never have been able to predict, from the appearance of the stone, the plant, and animal, that they were thus differently constituted; by observations, we find that the productions and mode of growth have been attended with different circumstances. We find that the stone has grown by a gradual accumulation of particles, independent of each other, and can only be destroyed by chemical or mechanical force; the plant and animal have, on the contrary, grown by nourishment, been possessed of parts mutually dependant, and contributing to the existence of each other.

So far, our observation teaches us the distinction between organized and inorganized beings; though it does not teach us in what the internal power of life consists. God permits us to know much, in order to lead us to industry in the attainment of knowledge; but he places boundaries beyond which we may not pass, that we may be humble.

COMPARISON OF ORGANIC AND INORGANIC BODIES.

INORGANIC BODIES.	ORGANIC BODIES.
Structure.	
Their parts always analogous to, and not depending on each other: thus a fragment of stone is as much a stone as the block or rock to which it belonged.	Their parts are mutually dependant; thus *stem, leaf, flower,* &c. do not constitute a vegetable being, except as they are united; it is the same with the different parts of an animal.
Origin.	
Molecular attraction, modified by time and space, or by the art of man, (as in chemistry;) they are *made.*	Owe their existence to beings similar to themselves, produced either from eggs, or brought into existence in a living state; they are *hatched* or *born.*
Development.	
They grow by the addition of new particles; they are hence said to increase by *juxtaposition* or *accretion.*	They develop by assimilating to their nature, or converting to their sustenance, foreign substances which they absorb, or receive internally; they increase by *nourishment.*
Termination.	
They are limited to no particular form, (except in the case of crystals;) they have no life, and are not subject to death; they *decompose.*	They have a determinate form and duration; their existence terminates either by old age, or disease; they *die.*

Vital principle—Difference between a stone and a plant—*Structure* of inorganic bodies—Of organic bodies—*Origin* of inorganic bodies—Of organic bodies—*Development* of inorganic bodies—Of organic bodies—*Termination* of inorganic bodies—Of organic bodies.

Having considered the distinction between inorganic and organic substances, we will proceed to a division which may be more familiar to you; that by which the matter upon our globe is ranged under three kingdoms—the ANIMAL, VEGETABLE, and MINERAL.

We find it somewhat difficult to explain the difference between the different kinds of organized beings, viz. animals and vegetables; the lines of distinction often seem to fade so gradually, that we can not well decide where the animal ends, and the vegetable begins.

This difficulty may seem at first somewhat strange, as you may perhaps never have been at a loss to tell an animal from a vegetable: you would certainly know how to distinguish between a nightingale and a rose, or between an ox and an oak; but these are animals and vegetables in a comparatively perfect state.

The perfect animal you see has the power to move about, to seek the nourishment most agreeable; you perceive it uttering audible sounds, possessing sensation and apparent consciousness. The plant, on the contrary, is confined to a particular spot, having no other nourishment than substances which themselves come in contact with it; exhibiting no consciousness, nor, to common observation, any sensation. It is only when we examine with close attention the various phenomena in the vegetable and animal kingdoms, that we learn to doubt as to the exact boundaries by which they are separated.

The division of nature into three kingdoms, animal, vegetable, and mineral, is very ancient, and appears at first to be clear and precise.

Minerals destitute of life increase by the accumulation of new particles.

Vegetables grow, produce seeds which contain the elements of future plants like themselves, and then die.

Animals unite to the properties of vegetables, the feeling of their own existence; or as Linnæus has said, "*Stones grow, vegetables grow* and *live, animals grow, live,* and *feel.*" Although this simple view of the works of creation is pleasing, it is not satisfactory; because we are not able to decide where, in the vast series of organized beings, sensation ceases.

That you may the better understand what is meant by the gradations of animal life, we will present you with a sketch of the classification of animals. The study of this department of nature you have already been told is termed Zoology.

A very general and simple classification of animals is as follows:—

"VERTEBRAL ANIMALS, having backbones.

AVERTEBRAL ANIMALS, destitute of backbones.

VERTEBRAL animals are divided into,

1. *Quadrupeds.* The science of which has no popular name. It includes four-footed animals; as ox, dog, mouse.

2. *Birds.* The science of which is called *ornithology.* It includes the feathered tribe; as pigeon, goose, wren.

3. *Amphibious Animals.* The science of which is called *amphibiology.* It includes those cold-blooded animals which are capable of living on dry land, or in the water; as tortoise, lizard, serpent, frog.

4. *Fishes.* The science of which is called *ichthyology.* It includes all aquatic animals which have gills and fins; as shad, trout, sturgeon, eel.

Three kingdoms of nature—Distinction between the different kinds of organized beings—The perfect animal—The plant—Minerals—Vegetables—Animals—Zoology—Division of animals into two classes—How many classes of Vertebral animals?

AVERTEBRAL animals are divided into,

5. *Insects.* The science of which is called *entomology.* It includes all animals with jointed bodies, which have jointed limbs: as flies, spiders, lobsters.

6. *Vermes.* The science of which is called *serminthology.* It includes all soft animals of the avertebral division, which have no jointed limbs, with or without hard coverings; as angle-worms, snails, oysters, polypi, and infusory animals."

The system of Zoology most approved, is the one taught by Linnæus, with some improvements made by the great French naturalist, Cuvier; according to this mode of classification, the animal kingdom is divided into four grand divisions, viz.:—

VERTEBRAL, MOLLUSCOUS, ARTICULATED, and RADIATED. These are subdivided into classes and orders.

Vertebral Animals.

CLASS I. *Mammalia,* or such as are at first nourished by milk. This class have lungs, and peculiar organs for imbibing their food during their first stage of existence.

The *First Order* is called *Bi-mani,* (from *bis,* two, *mani,* hands ;) this order includes *man* only; we find here no generic or specific differences, but the following *varieties.*

1st. *Caucasian* race, anciently inhabiting the country about the Caspian and Black seas, from whom we are descended.

2d. The *Mongolian,* the ancient inhabitants about the Pacific Ocean, from whom the Chinese are descended.

3d. The *Ethiopian,* or negro race.

The *Second Order* contains the *Quadru-mani,* (from *quatuor,* four, and *mani,* hands.) These have thumbs or toes, separate on each of the four feet. We here find the ourang-outang, (sometimes called the wild-man,) and the monkey.

The *Third Order* contains *Carnivorous* animals, or *flesh-feeders* having no separate thumbs, or great toes without nails; as the dog and cat.

The *Fourth Order* contains the *Gnawers,* having no canine teeth, (those which are called eye-teeth,) feeding almost wholly on vegetable substances; as the rat and squirrel.

The *Fifth Order* is *Edentata,* or animals *wanting teeth ;* as the sloth and armadillo.

The *Sixth Order, Pachyderma,* contains thick-skinned animals with hoofs; as the elephant, horse, and hog.

The *Seventh Order* contains the *Ruminating* animals, such as chew the cud, having front teeth (incisors) below only, and feet with hoofs cloven, or divided; as the ox, sheep, and camel.

The *Eighth Order, Cete,* contains *Aquatic* animals, (such as live in water,) having no kind of feet, or whose feet are fin-like limbs; as the whale and dolphin.

We have enumerated all the orders of the class Mammalia, as it is the one in which man is placed; we shall now notice the remaining classes of animals, without going into so minute a detail of their orders.

CLASS II, contains *Birds,* (*Aves,*) which are distinguished by having the body covered with feathers and down, long naked jaws, two wings formed for flight, and bi-ped, (from *bis,* two, and *pedes,* feet.)

How are Avertebral animals divided?—Cuvier's four grand divisions—1st class of Vertebral animals—Order bi-mani—Varieties in this order—Order quadru-mani—Third order—Fourth order—Fifth order—Sixth order—Seventh order—Eighth order—Class 2d.

The orders in this class are chiefly distinguished from each other by the peculiar make of the bill and feet.

CLASS III, *Amphibia,* contains *Amphibious* animals, including what are commonly called *reptiles.* It is divided into four orders:

1st. With shells over their back, and four feet; as the tortoise and turtle.

2d. Covered with scales, and having four feet; as the crocodile and lizard.

3d. Body naked, destitute of feet; as serpents.

4th. The body naked, and having two or four feet; as the frog, and toad.

CLASS IV, contains *Fishes,* (*Pisces,*) natives of the water, unable to exist for any length of time out of it; swift in their motions, and voracious in their appetites; breathing by means of gills, which are generally united in a long arch; swimming by means of radiate fins, and mostly covered with scales.

Molluscous Animals.

CLASS V. *Molluscous* animals have soft bodies without bones; their muscles are attached to a calcareous covering called a shell, which is supposed to be formed by the secretions of the animal. This class are destitute of most of the organs of sense; the nauti lus and cuttle-fish are of the highest order of molluscous animals The oyster and clam are destitute of heads; they have a shell o two pieces, which are therefore termed *bi-valved.*

Articulated Animals.

We proceed next to those animals called *Articulated;* these have jointed trunks, and mostly jointed limbs. They possess the faculty of *locomotion,* or changing place; some have feet, and others are destitute of them; the latter move by trailing along their bodies.

CLASS VI, *Annelida,* contains such animals as have red blood, without a bony skeleton; bodies soft and long, the covering divided into transverse rings; they live mostly in water; some of them secrete calcareous matter, which forms a hard covering, or shell; as the earth or angle-worm, and leech.

CLASS VII, *Crustacea,* contains animals without blood, with jointed limbs fastened to a calcareous *crust;* they breathe by a kind of gills.

CLASS VIII, *Arachnida,* contains spider-like animals, without blood, or horns with jointed limbs. They breathe by little openings, which lead to organs resembling lungs, or by small pipes distributed over the whole body; these do not pass through any important change of state, as insects do; they have mostly six or eight eyes, and eight feet, and feed chiefly on living animals; examples of this class are the spider and scorpion.

CLASS IX, *Insecta,* or insects, without blood, having jointed limbs and horns; they breathe by two pipes, running parallel to each other through the whole body; they have two horns; they are mostly winged, having one or two pairs; a few are without wings; mostly with six feet. They possess all the senses which belong to any class of animals, except that of hearing.

The winged insects pass through several changes or metamor phoses. The butterfly is first an *egg;* this, when hatched, is long and cylindrical, and divided into numerous rings, having many short legs, jaws, and several small eyes; this is the *larva,* or caterpillar.

At length it casts off its skin, and appears in another form without limbs. It neither takes nourishment, moves, nor gives any signs of life; this is the *chrysalis.* In process of time, by examining it closely, the imperfect form of the butterfly may be seen through the envelope; this it soon bursts, and a *perfect butterfly* appears. When about to pass into the chrysalis state, of which they appear to have warning, the insect selects some place where it may repose safely during its temporary death.* The silk-worm spins a silken web for a shroud to wrap itself in, and from this all our silks are made.

Radiated Animals.

Fig. 158.

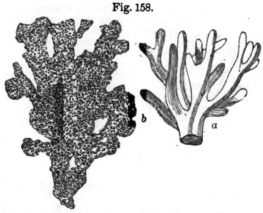

CLASS X, *Zoophites,* or animal plants. Here we find the lowest beings in the animal kingdom. Some of the orders of this class contain animals which have neither heart, brains, nerves, nor any apparent means of breathing. These are sometimes called *animal plants ;* many of them, as the *corals,* are fixed to rocks, and change place. The term coral includes under it many species; the red coral used for ornaments, is the most beautiful. The substance of coral, when subjected to chemical analysis, is found to consist chiefly of *carbonate* of lime; the hard crust which envelops the animal substance, is an excretion formed by it in the same way as the shells of the oyster and lobster are produced, or as nails grow upon the fingers and toes of the human body. The quantity of this carbonate of lime, elaborated by the little coral animal, is truly wonderful; islands are formed, and harbours blocked up by it. Fig. 158, *a,* represents a branching coral; the dots show the apertures by which the animal receives its nourishment. Some of the zoophites are fixed by a kind of root, to the bottom of the sea; some, as the *sea-nettle,* which appears like the segment of a circle, are carried about by the motion of the waters, without any voluntary motion, as are also the *sea-daisy, sea-marygold,* and the *sea-carnation,* so named from an apparent resemblance to those plants. We find here the *sea-fan,* the *sea-pen,* and the *madrepore,* the latter of which are often thrown together in vast quantities.

* May not this be considered as a lesson to man to anticipate and provide for the change in his existence, which, his bodily infirmities and daily observation teach him, is to be his own lot?

The sponge also belongs to this class of strange animal substances; it consists of a fibrous mass, containing a jelly-like substance, which when touched, discovers a slight sensation, the only sign of life manifested by it. There are many species of sponge; those most valued in the arts are found in the Mediterranean sea and Indian ocean. Some grow upon rocks, and are found covering the interior of submarine caves. The *Spongia parasitica* is seen growing upon the back and legs of a species of crab; sometimes as many as forty individual sponges extend themselves over the crab, impeding the motion of its joints, spreading like a cloak over its back, or forming for its head grotesque and towering ornaments, from which the poor crab vainly attempts to disencumber itself.

Some species of the sponge grow to a very large size; one has been found in the East Indies in the form of a cup, capable of containing ten gallons of water. The fibrous part of the sponge is the skeleton of the animal; the large apertures (see fig. 158, *b*,) serve to carry out fluids from within; while the water by which the animal is nourished, is imbibed by minute pores: this continual circulation of water is one of the most important functions of the living sponge.

These animals resemble plants in their manner of producing others; they form a species of germ, like the bud growing upon the stalk; this falls off from the stem, and becomes a perfect animal. If a part of one of these animals is separated from the rest, it will itself be as perfect a living animal as was the whole before. A polypus can be divided into as many animals as it contains atoms; some of this order are very properly called hydras, (many-headed.) Besides these, there is another order of animal substances, *infusoria*, which appear like a homogeneous mass, having no appearance of limbs whatever; these are either angular, oval or globular.

LECTURE XLVII.

COMPARISON BETWEEN ANIMALS AND PLANTS.

In our last lecture, after a glance upwards to the heavenly bodies, returned to our globe, and considered its various substances; here we found two classes of bodies, inorganized and organized substances; the former including minerals, the latter embracing the animal and vegetable kingdoms. We then took a brief view of the animal creation.

At the head of the animal kingdom, we found man, sufficiently resembling brute animals in his material frame to constitute part of an extensive class, embracing the ape, elephant, and dog; yet between the lowest degree of intelligence in the human race, and the highest faculties of brutes, there is a line of distinction marked by the hand of the Almighty, in characters too obvious for doubt. God said, "Let us make man in our own image, and he breathed into him the breath of life, and man became a living *soul.*"

Some writers have attempted to show that man differs only from the inferior order of animals in possessing a greater variety of instincts. But however wonderful may appear the instinctive perception of brutes, they are destitute of reason, and incapable of being

---

Sponge—Manner in which these animals are reproduced—Recapitulation—Man at the head of the animal kingdom—How resembling inferior animals.

the subjects of moral government; we must, therefore, both from our own observation and the declarations of scripture, infer, that the faculties of man differ not in *degree* only, but distinctly in their *nature*, from those of all other beings upon our globe.

" Man, (says Buffon,) by his form and the perfection of his organs, and as the only being on earth endowed with reason, seems properly placed at the head of the kingdom of nature. All, in him, announces the lord of the earth; his form marks his superiority, over all living beings; he stands erect, in the attitude of command; he can gaze upon the heavens; on his face is imprinted the character of dignity; the image of his soul is painted upon his features, and the excellence of his nature penetrates through his material organs, and animates the expression of his countenance."

In the orders of animals nearest to man, we find the senses of sight, touch, taste and smell, equally perfect as those possessed by him, and in some cases they are even more acute; but as we proceed downwards through the gradations of animal existence, we perceive the number and acuteness of the senses to diminish—we find some beings with but four senses, some with three, others with two, and lastly, in Zoophytes, we find only the sense of touch, and that so faintly exhibited as almost to lead us to doubt its existence.

Let us now return to the distinction between animals and vegetables. You now perceive that although you would find no difficulty with regard to a nightingale and a rose, to discover to which of the king doms of nature they belong; yet with respect to a sponge or coral, a mushroom or lichen, it would be somewhat difficult, without a previous knowledge of their classification, to say which is called animal, and which vegetable, or to give the distinctions between them. We have seen among the zoophites, that the polypus, like a vegetable, may be increased by cutting shoots and ingrafting them upon other animals.

With respect to *sensation*, some plants seem to possess this, apparently even in a greater degree than some of the last orders of animals;—the *sensitive plant* shrinks from the touch; the *Dionea* suddenly closes its leaves upon the insect which touches them; the leaves of plants follow the direction of light, in order to present their upper surfaces to its influence; as you may observe in flower pots placed by a window. The seed of a plant, in whatever situation it may be placed in the earth, always sends its root downward, and its stem upwards; in these cases, does there not seem as much appearance of sensation and instinct, and even more, than in the lower orders of animals?

We find, then, that the possession, or want of instinct, does not constitute a mark of distinction between animals and plants.

Some have attempted to draw a line of distinction, by considering that locomotion, or the power of changing place, belongs to animals only; but this criterion seems to fail, since we find animals fixed to the bottom of the sea, or growing upon rocks, and plants moving upon the surface of the water.

Another mark of distinction has been given, in the supposed presence of *nitrogen* in animals, detected by a peculiar odour when animal substances are burning, similar to what we perceive in the combustion of bones; but nitrogen having been discovered in some vegetables, this proof is no longer considered infallible.

It appears then, from a comparison between animals and vegetables, that these beings are closely connected by the essential charac-

How differing from them.

ters of organization; that it seems impossible to distinguish them by any trait that belongs exclusively to either; that the connexion between them appears the most striking in the least perfect species of both kingdoms; and that as we recede from this point, the differences become more numerous and more marked.

We may illustrate this view, by imagining two ascending chains, rising from one common point, each side of the chain becoming more and more unlike in proportion to the intervening distance from the centre. From this same central point, also proceeds the chain of inorganized substances; some imperfect animals resembling plants in their outward form, some, both of animals and plants, resembling minerals in their hard and calcareous coverings and shapeless forms.

Having thus learned the almost imperceptible gradations by which the animal and vegetable kingdoms are blended, we must, in stating the important differences which exist between animals and plants, consider the imperfect species of both kinds as exceptions to any general rule, and confine ourselves to perfect animals and plants.

1st. *Plants differ from animals with respect to the elements which compose them; carbon, hydrogen, and oxygen, form the base of vegetable substances; animals exhibit the same elements, with this important distinction, that carbon prevails in plants, and nitrogen in animal food.*

2d. *They differ in their food; plants are nourished with inorganized matter, absorbed with water, which holds in solution various substances; animals are mostly nourished either by vegetable or other animals.*

3d. *Plants throw off oxygen gas, and inhale carbonic acid; animals, in respiration, inhale oxygen gas and throw off carbonic acid.*

4th. *Although plants and animals both possess a principle of life, it is in the one case much more limited than in the other; exhibiting itself in plants by a feeble power of contraction or irritability; in animals appearing in sensation, muscular movement, and voluntary motion.*

We see, then, many important *differences* between perfect animals and perfect plants. We have, in numerous instances, pointed out striking *analogies* between the two great divisions of organized bodies: this subject might be greatly enlarged; but we have already, amid the multitude of interesting facts and reflections presented by the vegetable creation, far exceeded the bounds originally prescribed. A few remarks on the inorganic matter upon and around the earth, and our course of Lectures is closed.

Inorganic bodies form the solid base of the globe. *Minerals* are spread upon the face of the earth, or lie buried beneath its surface. They form vast masses of rocks, chains of mountains, and the ground upon which we tread. The *Water* occupies a still greater surface of the earth than the land; it is filled with life and animation; the treasures and wonders of the deep seem almost unbounded. The *Air*, lighter than earth and water, extending on all sides about forty miles in height, surrounds the whole globe, separating us from the unknown elements which exist beyond it. *Heat*, or *Caloric*, is a subtle fluid which pervades all matter, in an increasing proportion from solids to fluids, and fluids to gases. *Light*, reflecting its hues from terrestrial objects, produces, by the decomposition of its rays all the beautiful variety of colouring.

Result of the comparison between animals and vegetables—Chains of beings proceeding from one point—Differences between animals and plants—Different kinds of inorganic matter—The Deity manifested in his works.

Wherever we turn our eyes, we behold wonders; "if we go up to heaven, God is there;" "the firmament showeth forth his handy-work;" if we contemplate the earth on which we are placed, with its varied tribes of beings, and the provision made for their comfort and subsistence, we realize, that it is indeed God, "who maketh the grass to grow on the mountains, and herbs for the use of man."

The universe, how vast! exceeding far
The bounds of human thought; millions of suns,
With their attendant worlds moving around
Some common centre, gravitation strange!
Beyond the power of finite minds to scan!
Can He, who in the highest heav'n sublime,
Enthron'd in glory, guides these mighty orbs—
Can He behold this little spot of earth,
Lost midst the grandeur of the heav'nly host:
Can God bestow one thought on fall'n man?
  Turn, child of ignorance and narrow views,
Thy wilder'd sight from off these dazzling scenes;
Turn to thy earth, and trace the wonders there.
Who pencils, with variegated hues,
The lowly flower that decks the rippling stream,
Or gorgeously attires the lily race?
Who with attentive care, each year provides
A germ to renovate the fading plant
And gives soft show'rs and vivifying warmth.
Kindling within the embryo inert
The little spark of life, unseen by all,
Save him who gave it, and whose care preserved?
Who teaches, when this principle of life,
Thus animated, swells the germ within,
And bursts its tomb, rising to light and air—
Who teaches root and stem to find their place,
Each one to seek its proper element?
  Who gilds the insect's wings, and leads it forth
To feast on sweets and bask in sunny ray?
None can the life of plant or insect give,
Save God alone;—He rules and watches all;
Scorns not the least of all His works; much less
Man, made in his image, destin'd to exist
When e'en yon brilliant worlds shall cease to be.
Then how should man, rejoicing in his God
Delight in his perfections, shadow'd forth
In every little flow'r and blade of grass!
Each opening bud, and care-perfected seed,
Is as a page, where we may read of God.

PART V.

APPENDIX

TO THE

LECTURES ON BOTANY,

CONTAINING

I. ILLUSTRATIONS OF THE HABITS OF PLANTS.
(*With Eight Engravings.*)
II. NATURAL ORDERS.
III. DESCRIPTIONS OF GENERA.
IV. DESCRIPTIONS OF SPECIES.
V. VOCABULARY.
VI. SYMBOLICAL LANGUAGE OF FLOWERS.
VII. ALPHABETICAL INDEX.
VIII. COMMON NAMES OF PLANTS

SECTION I.

ILLUSTRATIONS OF THE HABITS OF PLANTS.

WITH EIGHT ENGRAVINGS.

THE following Wood Engravings, copied from the elegant work of C. F. Brisseau Mirbel, entitled "*Elemens de Botanique*," are added to this volume, in order to exercise the pupil in the study of the *habits* of plants. The author above alluded to, thus remarks, [we give a translation of his words:]

"In order to learn any part of Natural History, the student must *see much*, and exercise himself that he may *see clearly*; this demands zeal and perseverance. A thousand characters offer themselves to the eye of the naturalist, which are unseen by others; this is, because these characters become striking only by comparison, and the art of comparison supposes knowledge already acquired. In placing before the eye of the pupil figures representing the most striking characters of objects, we take the surest method of helping him forward. We cannot vary too much the forms we offer him.

"The following designs present examples of the plants of all climates, and such as are found in all classes. The minute and extended analyses which will be found in the explanations of some of these plants, are made for the benefit of those pupils who love to push their investigations beyond the mere elements of science; such will soon learn to make observations for themselves, and to test those of others by a comparison with nature.

"The relative size of the different plants represented, is preserved as far as possible, but it was in many cases impossible to give an accurate idea of this, in grouping the figures."

PLATE I.

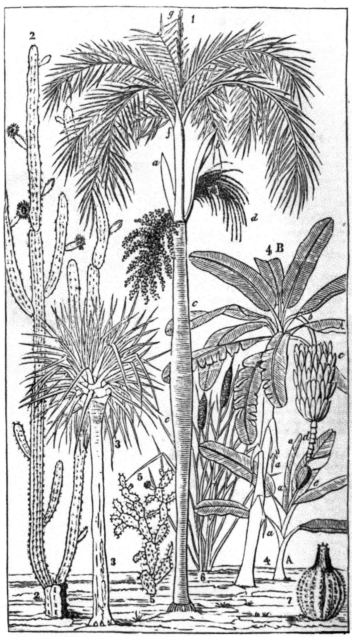

1 Areca oleracea. 2 Cactus peruvianus. 3 Dracœna draco. 4 Musa paradisiaca. 5 Cactus opuntia. 6 Typha latifolia. 7 Cactus melocactus.

Fig. 1. ARECA *oleracea*. Cabbage-tree. [Family of the *Palms*.] This tree is monœcious. It grows to the height of 120 feet. This is a young plant, little more than 20 feet in height. The stipe is slender, simple, and vertical. Leaves terminal, very long, pinnate; petioles sheathing; leafets elongated, lanceolate; spathas monophyllous, growing from the axils of the lower leaves, which fall off; flowers in panicles, the staminate and pistillate flowers enclosed by different spathas. *a*, Spatha shut; *b*, spatha opened laterally; *c*, stipe, which is fusiform ;* *d*, panicle of staminate flowers, which were contained in the spatha before it opened ; *e*, panicle of pistillate flowers, entirely separated from its spatha; *f*, part of the stipe, formed at its superfices by the base of the developed leaves, and in the interior by the young, tender, and succulent leaves, which form a white compact head. These are eaten by the people of the West Indies as a salad, cooked as we prepare cabbage; the name Areca is given in the East Indies, where this tree flourishes. *g*, is a young leaf folded like a fan. The areca-nut is chewed by the people of India. It is said to resemble the nutmeg. This plant belongs to Monœcia Monodelphia.

Fig. 2. CACTUS *peruvianus*. (Family of the *Cacti*.) The name Cacti was given by the Greek botanist, Theophrastus, who first discovered the plant. A succulent plant, becoming woody by age; it rises to the height of thirty feet. It grows among the rocks in Peru, near the sea. The stem is vertical, articulated, branching, spinose, with seven or eight prominent angles. Branches erect; spines acicular, fasciculated, divergent, placed at intervals upon the ridges of the stem and branches. Flowers lateral, cauline, solitary, sub-sessile, it belongs to Icosandria Monogynia.

Fig. 3. DRACÆNA *draco*. Dragon-tree. (Family *Asphodel*.) A tree of Africa and the Indies, the diameter of whose trunk is very great in comparison to its height. Stipe cylindrical, vertical, marked with transverse cicatrices left by the leaf in falling. Leaves terminal, alternate, crowded, semi-amplexicaulis, ensiform, cuspidate; the upper ones erect, the lower ones pendent, the intermediate ones spreading or reflexed; a red, resinous extract, obtained from this plant, and called Dragon's blood, is sold in the shops. The ancient Greeks introduced it into medicine. This plant is classed in Hexandria Monogynia.

Fig. 4. MUSA *paradisiaca*, or the Banana tribe. (Family *Musæ*.) The name Musa is said to have been given by Linnæus in honour of Antonius Musa, the physician of Augustus, who wrote on botany. This is an herbaceous plant, with a perennial bulbous root; it grows to the height of 15 or 20 feet. It is a native of the East Indies, but has been long cultivated in South America. The leaves are radical, petioled, at first convolute; petioles long, large, sheathing, forming by their brim a thick and smooth stem resembling a stipe. The lamina of the leaf is sometimes 9 feet in length and two in breadth, oblong, entire; the sides thick and strong, with the veins at right angles to them, and to the midrib. Scape cylindrical, naked, sheathed. Spike terminal, pendent. Flowers semi-verticillate, bracted; the fertile flowers at the base of the spike, the infertile at the summit. *A*, is a young Banana; *a a*, central leaves, convolute. *B*, a Banana bearing fruit; *b*, remains of old leaves; *c*, the scape; *c, d, e*, pendent spike; *c*, the fruit, (classed by Mirbel in the genus berry;) *d*, portion of the axis from which the flowers have fallen; *e*, steril flowers, crowded into a compact head, terminal, enveloped by their bracts. This plant is by some placed in the class Hexandria, by others in the now obsolete class Polygamia; but Mirbel, very properly, I think, considers it as belonging to the class Monœcia. The spikes of fruit sometimes weigh from thirty to forty pounds each. The fruit when ripe is yellow. Each berry is about eight inches in length, and one in diameter.

Fig. 5. CACTUS *opuntia*. Prickly-pear. (Family of the *Cacti*.) A succulent plant with a woody stem, first described and named by Theophrastus, as a spiny, edible plant. It is a native of southern latitudes, where it grows to the height of eight or ten feet. Stem thick, compressed, ramose, articulated, spinose; the joints are ovate. Leaves very small, cylindrical, subulate, caducous. Spines fasciculated, divergent, growing at the base of the leaves.

Fig. 6. TYPHA *latifolia*. Cat-tail. (Family *Typhæ*.) The name from the Greek *tiphos*, a lake, because it grows in marshy places. An herbaceous plant, monœcious, with a perennial root, growing to the height of eight or ten feet in marshy grounds, in Europe and North America. Stem vertical, simple, aphyllous at its summit, surrounded at the lower part with sheathing petioles. Leaves very long, riband-like. Flowers in a terminal, crowded, cylindrical spike. Barren flowers superior, and separated from the fertile flowers by a short interruption. This plant belongs to Monœcia Triandria.

Fig. 7. CACTUS *melocactus*. (Family of the *Cacti*.) Succulent plant from the Antilles, perennial, melon-form, with fifteen or twenty sides, garnished with fascicles of divergent spines.

---

* Mirbel, whose description I follow, defines fusiform as tapering at both ends and swelled towards the middle ; thus he considers the Radish root as fusiform, while the carrot he calls conical.

EXPLANATION OF PLATE II.

Fig. 1. YUCCA *aloifolia.* (Family of the *Liliaceæ.*) Adam's Needle. A tree of ten or twelve feet in height, indigenous in the West Indies. Stype cylindric, erect, sometimes two or three-forked. Leaves terminal, alternate, crowded, semi-amplexicaulis, ensiform; the upper ones erect, the lower ones pendent, the intermediate, spreading or reflexed. Panicle simple, terminal, pyramidal. Flowers pendent. Perianth simple, six-sepalled, campanulate. This plant belongs to Hexandria Monogynia. It is the majestic lily of the tropics. The name Yucca is from Jucca, the Indian appellation.

Fig. 2. SACCHARUM *officinale.* (Family of the *Grasses.*) Sugar-cane. An herbaceous, perennial plant, which grows to the height of ten or twelve feet. Culm is vertical, cylindrical, solid. Leaves sheathing, elongated, ensiform. Panicle large, silky. The name Saccharum is from the Arabic, *soukar,* sugar. This plant is thought to be a native of India, but it is now cultivated in most warm countries. With most of the grass-like plants, it belongs to Triandria Digynia.

Fig. 3. FERULA *tingitana.* (Family of the *Umbelliferæ.*) Giant-fennel. Herbaceous plant, biennial, 8 or 9 feet in height. Stem cylindrical, vertical. Leaves alternate, large, decompound, with very small leafets. Petioles with a large base, amplexicaulis. Panicle terminal, composed of umbels. This plant grows in Spain and Barbary; it belongs to Pentandria Digynia, where the umbelliferous tribe is mostly classed. A species of this genus, FERULA *assafœtida,* produces from its root the medicinal gum, assafœtida; from another species, the galbanum is obtained.

Fig. 4. CYMBIDIUM *echinocarpon.* (Family of the *Orchideæ.*) A parasitic plant of South America which grows to the height of two or three feet. Stems compressed. Leaves opposite, oval, acute. Capsule bristly. This plant belongs to Gynandria Monandria. A species *C.* pulchellum* (grass-pink) is very common in our region.

---

* It may be proper to inform the student, that where several species of a genus are mentioned, it is very common to designate the name of the genus by the initial letter; thus *C.* stands for Cymbidium.

PLATE II.

1 Yucca aloifolia.   2 Saccharum officinale.   3 Ferula tingitana.   4 Cymbidium echinocarpon.

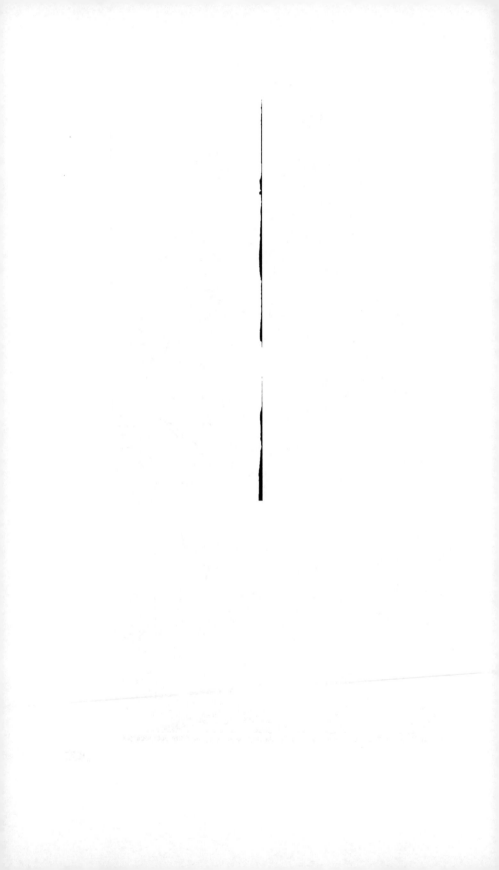

PLATE III.

1 Populus fastigiata. 2 Salix babylonica. 3 Chamærops humilis. 4 Maranta arundinaceæ Sarracenia purpurea. 5 Dionæa muscipula. 7 Phallus impudicus. 8 Agaricus cretaceus 9 Boletus

PLATE IV.

1 Carica papaya. 2 Crescentia cujete. 3 Vanilla aromatica. 4 Nepenthes distillatoria.
5 Sempervivum tectorum. 6 Panicum italicum 7 Clathrus cancellatus.

EXPLANATION OF PLATE III.

Fig. 1. POPULUS *fastigiata*.* (Family *Amentaceæ*.) Diœcious-tree. It was orginally carried from the Levant into France, and is known in the United States as the Lombardy poplar. Trunk vertical. Branches erect, fastigiate. The staminate flowers only are known in this country.

Fig. 2. SALIX *babylonica*. Weeping-willow. (Family *Amentaceæ*.) A Diœcious tree growing to the height of 35 feet; it was originally from the Levant. The fertile tree only exists in this country. Stem branching; the branches are supple, pendent. Leaves alternate, lanceolate.

Fig. 3. CHAMÆROPS *humilis*. (Family of the *Palms*.) Diœcious tree, whose height varies from 4 to 30 feet. It grows in Barbary, Spain, and Italy. Its fruit is called wild dates.

Fig. 4. MARANTA *arundinacea*. Arrow-root. (Family *Canneæ*.) Perennial plant, four feet high; native of South America. Stem herbaceous, slender, branching. Leaves entire, oval-lanceolate, petioled. Petioles short, sheathing. Flowers terminal. The root of this plant affords a substance resembling starch in many of its properties; this is much valued for its nutritious qualities. The plant belongs to Monandria Monogynia.

Fig. 5. SARRACENIA *purpurea*.† (Family undetermined.) Side-saddle flower; an herbaceous plant peculiar to marshes of North America. Leaves radical, ascidiate. Calyx five-sepalled. Corolla five-petalled.

Fig. 6. DIONÆA *muscipula*. Venus' fly-trap. (Family uncertain.)‡ Perennial, herbaceous. Scape vertical, about eight inches high. Leaves radical, radiating from the centre, petioled. Petiole cruciform. Leaf round, folds itself up suddenly on being touched. Flowers corymbed. Decandria Monogynia.

Fig. 7. PHALLUS *impudicus*. (Family of the *Fungi*.) Mushroom called morel. *A*, young plant still enclosed in its volva. *B*, a plant perfectly developed; *a*, volva which has burst to make room for the pedicel, *b*; *c*, pileus; *d*, umbo, a central part of the hat, which is pierced in its turn.

Fig. 8. AGARICUS *cretaceus*. (Family of the *Fungi*.) Mushroom without a volva. *a*, pedicel; *b*, neck; *c*, pileus; *d*, interior surface, forming a layer for the seeds to rest in; *e*, umbo.

Fig. 9. BOLETUS *salicinus*. Parasite. (Family *Fungi*.) Pileus dimidiate, sessile.

---

* The *dilitata* of most authors.
† Lindley establishes a family, *Sarracenia*, in which this is the only genus; he considers it to be allied to Papaveraceæ, on account of its dilated stigma, its indefinite number of stamens, and small embryo lying at the base of copious albumen. He also thinks it nearly related to Droseraceæ, or to whatever family the Dionæa may be placed in. The pitcher-form leaf of the Sarracenia is analogous to the dilated foot-stalk of the Dionæa, and the lid of the pitcher in the former leaf is represented by the irritable lamina in the latter. In the structure of its leaves, the Sarracenia is related to the family Nepenthæ, containing the pitcher-plant.
‡ Referred by Lindley to Droseraceæ.

EXPLANATION OF PLATE IV.

Fig. 1. CARICA *papaya*. Papaw-tree. (Family unknown.)* The name carica is from Caria, where the tree was first cultivated. Diœcious. 20 feet high. It is a native of the East and West Indies and Guinea—Fig. 1. A fertile plant. Trunk very simple, vertical, cylindric, marked with cicatrices produced by the fall of leaves. Leaves terminal, large, seven-lobed, petioled. Petioles two or three feet long. Flowers grow at the base of the petioles. Berries large, furrowed, depressed in the centre. The green fruit is eaten by the Indians in the same manner as we use the turnip. The buds are used for sweetmeats. The ripe fruit is eaten for a dessert, like melons.

Fig. 2. CRESCENTIA *cujete*. Calabash-tree. A tree 16 feet in height; native of South America and the West Indies. Trunk thick. Branches horizontal or reflexed. Leaves fasciculate, obovate, cruciform, fascicles alternate. Flowers rameus, sometimes cauline, solitary. Calyx campanulate, bi-lobed. Corolla large, sub-campanulate. Berries large, resembling the pumpkin in figure; the epicarp cortical, like that of the gourd.

Fig. 3. VANILLA *aromatica*. (Family of the *Orchideæ*.) This plant is sometimes called Epidendron vanilla, the generic name being derived from *epi*, upon, and *dendron*, a tree, because the plant grows parasitically on the trunks and branches of trees. It is perennial, climbing, parasitic; a native of South America. Stems cylindric; flowers ramose, producing roots at every joint, which fasten themselves to the bark of trees. Leaves alternate, oval, oblong, acute, thick. Flowers in terminal spikes, which are lax and pendent. Perianth simple, six-lobed. Capsule fusiform, containing small black seeds which have an aromatic taste and fragrant smell; they are used as perfumes. This plant belongs to Gynandria Monandria.

Fig. 4. NEPENTHES *distillatoria*. (Family unknown.)† A perennial plant of the Indies. Stem simple, with leaves towards the base. Leaves alternate, large, oval, lanceolate, contracting at the base into petioles which are semi-amplexicaulis, and terminated at the summit by a tendril which supports an ascidium; this is cylindric, and furnished with an operculum which opens and shuts according to the state of the atmosphere. Flowers terminal, panicled.

Fig. 5. SEMPERVIVUM *tectorum*. House-leek tribe.‡ The generic name is derived from the Latin, *semper*, always, *vivire*, to live, and the specific name from *tectum*, house. This is a perennial, herbaceous plant, which grows to the height of sixteen inches. The stem is simple, vertical, foliated. Leaves succulent, oblong, alternate; radical leaves cordate. Flowers in close panicles. Polyandria Polygynia.

Fig. 6. PANICUM *italicum*. (Family of the *Grasses*.) An herbaceous, annual plant, two feet in height, a native of India. Culm erect. Leaves elongate, lanceolate, sheathing. Spike elongated, compounded of numerous spikelets.

Fig. 7. CLATHRUS *cancellatus*. Mushroom. (Family of the *Fungi*.) *A*, young plant enclosed in its volva. *B*, another more advanced; *a*, volva ruptured; *b*, peridium beginning to appear. *C*, a plant entirely developed. The peridium is globular and cancellated.

---

* Lindley forms of this a distinct family, called *Papayaceæ*. He considers it as allied to the Passion-flower tribe, in its fruit; and to the Fig tribe, in the separation of stamens and pistils, and in its milky juice, which resembles that found in some species of Ficus.
† Formed by Lindley into a new family, *Nepenthea*.
‡ Belonging to the Crassulaceæ of Lindley; allied to the Cacti and Euphorbiæ.

EXPLANATION OF PLATE V.

Fig. 1. PANDANUS.* Screw-pine. Diœcious tree of South America, 24 feet in height. Fertile plant. Stype cylindric, rectilinear, vertical, branches at the summit. Leaves terminal, crowded, spiral, elongated, amplexicaulis, acuminate, bordered with spinose teeth. Fruit sorose, peduncled, axillary, large, round, woody, composed of a great number of small pericarps of an hexagonal figure. The name Pandanus is from the Malay word, *pandang.* The common name is given from the direction of the grain of the bark, which runs spirally.

Fig. 2. RHIZOPHORA *mangle.*† A low tree of South America, which grows in salt marshes, and at the mouths of rivers near the sea. It puts forth two kinds of branches, the one bearing leaves, and forming the head of the tree; the other aphyllous, stoloniferous, and inclining downwards, at length taking root and producing new shoots which become perfect plants. Branches opposite. Leaves opposite. Seeds germinating in the fruit still suspended from the branches, and producing clavate radicles twelve or fourteen inches in length; these, detaching themselves from the cotyledon which remains enclosed in the pericarp, fall, and planting themselves in the earth, develop a new trunk and branches. *a,* shows a shoot germinating.

Fig. 3. BROMELIA *ananas.*‡ Pineapple. An herbaceous, perennial plant, 4 feet high; it is a native of South America and the West Indies. Leaves radical, coriaceous, channelled, ensiform, long, denticulate. Teeth spinose. Scape short. Sorose, ovate, succulent, surmounted with a crown of leaves. This plant belongs to Hexandria Monogynia.

Fig. 4. THEOPHRASTA *americana.* (Family of the *Apocineæ.*)§ Shrub of South America, four feet high. Trunk very simple, spinose. Leaves crowning, verticillate, elongated, obcrenulate, denticulate. Fruit spherical.

---

* Belonging to the family Pandaneæ of Brown and De Candolle; somewhat allied to Typhæ in its fructification, and to the Palms in its arborescent stem.
† The Mangrove tribe, or Rhizophoreæ of Brown and De Candolle; described as "natives of the shores of the tropics, where they root in the mud, and form a dense thicket to the verge of the ocean."
‡ Of the family Bromeliaceæ, or Pineapple tribe; Lindley says, "the habit of the Bromeliaceæ is peculiar; they are hard, dry-leaved plants, having a calyx, the rigidity of which is strongly contrasted with the delicate texture of the petals."
§ Lindley follows Brown in placing this in the order Myrsineæ. He considers it as nearly related to Primulaceæ through some of the genera of that order, and to Sapoteæ through the genus Jacquinia.

EXPLANATION OF PLATE VI.

Fig. 1. CASUARINA. (Family *Coniferæ*.)* A large tree of New Holland. Trunk thick, head branched; branches flexible, pendent, verticillate, articulated. Monœcia Monandria.

Fig. 2. AGAVE *americana*.† (Family *Narcissi*.) A succulent plant which grows in South America. Leaves radical, crowded, more than four feet long, tapering gradually to a point, channelled, bordered with spinose teeth. Scape more than 20 feet high, cylindric, rectilinear, vertical, with scattering, scale-like, appressed leaves. Panicle simple, pyramidal. Flowers erect, numerous, grouped at the extremity of the long peduncle. This magnificent plant belongs to Hexandria Monogynia.

Fig. 3. STIZOLOBIUM *altissimum*. (Family *Leguminosæ*.) A climbing plant which ascends the loftiest trees of the equatorial region. Stem flexible. Leaves alternate, pinnate, trifoliate. Peduncle axillary, filiform, very long, pendent, terminated by an umbel of large and beautiful flowers. Legume acinaciform, wrinkled. Diadelphia Decandria.

Fig. 4. PASSIFLORA *quadrangularis*.‡ Climbing plant of warm regions of America. Stem quadrangular, slender, cirrose. Leaves alternate, petioled, oblong-oval. Tendrils axillary. Flowers large, axillary. Berries large, ellipsoid.

Fig. 5. CYPERUS *papyrus*. Herbaceous plant, perennial, aquatic; fifteen feet high; a native of Egypt. Stem erect, three-sided, aphyllous, sheathing at the base; umbels large, terminal, compound, with an involucrum and an involucel. Triandria Monogynia.

Fig. 6. IRIS *germanica*.§ (Family *Irideæ*.) Herbaceous plant of Europe, three or four feet high, with a perennial root. Leaves radical, equitant, compressed, ensiform. Stem leafy, branching at its summit. Flowers terminal. Perianth simple, six-lobed; three lobes exterior, reflexed; three lobes interior, erect. Triandria Monogynia.

Fig. 7. HIPPURIS *vulgaris*. Perennial plant growing in wet grounds. Stem cylindrical, very simple. Leaves linear, verticillate. Flowers very small, verticillate. Monandria Monogynia.

---

* Mirbel establishes a natural order, Casuarineæ, in which he places this genus; Lindley considers it as belonging to Myriceæ, or the Gale tribe; he says, "the nearest approach made by these plants is to the Elm tribe, (Ulmaceæ,) and to the Birch tribe, (Betulineæ,) from the former of which they are readily known by their amentaceous flowers, and want of a perianth; from the latter they are distinguished by their erect ovules, aromatic leaves, and one-celled ovary. Casuarina has the habit of a gigantic Equisetam, (fern,) and can scarcely be compared with any other dicotyledonous tree." Brown considers the genus Casuarina as approximating to Coniferæ, where it was placed by Jussieu, whose arrangement we have followed.

† By Lindley, this is placed in his natural order Bromeliaceæ, called Bromeliæ by Jussieu. The habit of Agave is similar to that of Aloe in the order Asphodeleæ.

‡ Botanists are much divided with respect to that place in the natural method which the Passion-flower tribe should occupy. Jussieu and De Candolle, in view of the organization of the fruit, consider it as nearly allied to Cucurbitaceæ. A separate order, Passifloreæ, is now established among botanists, for this interesting tribe of plants. Jussieu considered that the parts taken for petals, are nothing but inner divisions of the calyx, usually in a coloured state, and wanting in some species. Lindley considers the outer species of the floral envelopes as the calyx, and the inner as the corolla, for two principal reasons; first, they have the ordinary position and appearance of calyx and corolla, the outer being green, the inner coloured; second, there is no essential difference between the calyx and corolla, except one being the outer, the other the inner of the floral envelopes. "The nature of the filamentous appendages, or rays as they are called," says Lindley, "which proceed from the orifice of the tube, and of the processes which lie between the petals and stamens, is ambiguous. I am disposed to refer them to a peculiar form of petals rather than to stamens. There can be no doubt, at least, of their being of an intermediate nature between petals and stamens." The zealous Catholics who discovered them in the woods of South America, attached to the form of their corolla ideas connected with their religious faith.

§ The Irideæ differ from the Narcissi and Amaryllideæ in being triandrous, with the anthers turned outwards; from Orchideæ, to which they are in some respects nearly allied, in not being gynandrous, and in all their anthers being distinct.

PLATE V.

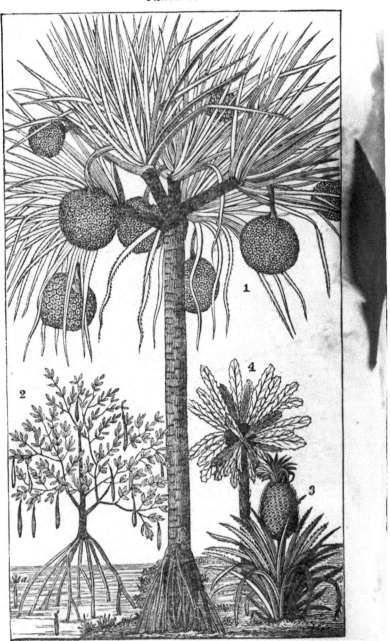

1 Pandanus.   2 Rhizophora mangle   3 Bromelia ananas.   4 Theophrasta americana.

PLATE VII.

1 Pinus pinea. 2 Abies picea. 3 Cycas circinalis. 4 Fritillaria imperialis. 5 Lycopodium cernuum 6 Digitalis purpurea. 7 Narcissus poeticus. 8 Lycopodium alopecuroides. 9 Dodecatheon meadia·

PLATE VI.

1 Casuarina. 2 Agave americana. 3 Stizolobium altissimum. 4 Passiflora quadrangularis. 5 Cyperus papyrus. 6 Iris germanica. 7 Hippuris vulgaris

Fig. 1. Pinus pinea. Stone-pine. The fir tribe. (Family *Coniferæ*.) A native of the south of Europe. The head low and branching. Leaves of a sea-green colour, acicular, forming an egret upon the summits of the branches. Strobilums large, ovate, thick; served up in desserts in Italy and France. This tree, according to Loudon, forms a distinguished ornament of the villas of Rome and Florence.

Fig. 2. Abies picea. Fir-tree. (*Coniferæ*.) Trunk rectilinear, vertical. Branches forming a pyramid; sub-verticillate, very open. Boughs pendent. Leaves small, linear, acute. Strobilums cylindrical, pendent. A tree common to mountainous regions in the north of Europe, and in the United States.

Fig. 3. Cycas circinalis.* A small diœcious tree of India, resembling the palms in its aspect. Stipe vertical, cylindric. Leaves pinnate; leafets lanceolate-linear. Petioles spinose. Spines leafy. Staminate flowers in a catkin. Pistillate flowers in spikes. A fertile plant showing the fructification at *a*. The pith of this plant affords an article called Sago, superior to that brought from the West Indies under that name. This was placed by Linnæus in the family of the Palms, and afterward classed among Ferns. According to Mirbel's drawing and description, the first arrangement was most natural.

Fig. 4. Fritillaria imperialis. Crown-imperial. (*Liliaceæ*.) Bulbous plant, two or three feet in height; a native of Persia. Leaves radical, elongated, ensiform. Scape naked, vertical. Flowers large, terminal, peduncled, umbelled, pendent. Perianth six-sepalled, campanulate. Bracts numerous, elongated, leafy, erect, crowning.

Fig. 5. Lycopodium cernuum.† Stem erect, branching. Leaves scattered, setaceous, inflated. Spikes small, ovate, drooping. Cryptogamous.

Fig. 6. Digitalis purpurea.‡ Fox-glove. (*Scrophulariæ*.) Biennial, native of mountainous and sandy regions of Europe. Stem generally simple, leafy below. Leaves alternate, oval-lanceolate; the radical leaves larger. Flowers in a spike, unilateral, peduncled, pendent. Corolla tubular, campanulate.

Fig. 7. Narcissus poeticus.§ (*Narcissi* or *Amaryllideæ*.) Bulbous plant, ten or twelve inches in height. Native in the meadows of Italy and the south of France. Leaves radical, erect, riband-like. Scape naked, uni-flowered. Flower drooping, spathaceous.

Fig. 8. Lycopodium alopecuroides. Native of South America. Branches fall and take root at their extremities. Leaves linear, subulate.

Fig. 9. Dodecatheon meadia. (*Primulaceæ*.) Herbaceous plant, eight inches high; originally a native of Virginia. Leaves radical, spreading, oblong. Scape naked, erect. Flowers pedicelled, umbelled, pendent. Corolla five-parted, the divisions reflexed.

---

* This plant is the principal genus of an order not recognised by Jussieu, the Cycadeæ, first proposed by Ventenat and established by M. Richard. In the cylindrical stem and pinnate leaves, this order resembles the Palms; in many other characteristics, particularly in the organization of the fruit, it approximates to the Coniferæ; in the mode of developing leaves, it bears a relation to the Ferns.

† This genus belongs to the natural order Lycopodiaceæ, being, according to Lindley, " intermediate between Ferns and Coniferæ on the one hand, and Ferns and Mosses on the other; related to the first of those tribes in the want of stamens and pistils; to the second, in the aspect of the stems of some of the larger kinds; and to the last, in their whole appearance." M Brogniart supposes that in the primitive ages of the world, these plants attained a gigantic size, equal to the largest forest trees of the present day; this opinion arises from discoveries made in coal mines, where, along with Ferns, are found what appears like remains of species of this tribe. At present their habit resembles that of the Mosses; they are usually low, prostrate plants.

‡ Lindley says, Digitalis forms a connecting link between Scrophulariæ and Solaneæ in its relation to Verbascum, both genera having alternate leaves.

§ This order is allied to Asphodeleæ and Liliaceæ, in the appearance of various organs, but distinguished from them by its inferior germ. The corona or nectariferous cup of the Narcissus is considered by Lindley, to be nothing more than an organ formed of an extra number of stamens, developed in a petaloid state. The same author remarks, that " there is in this whole order a strong tendency to form another set of staminiferous organs between the perianth, and those stamens that actually develop."

EXPLANATION OF PLATE VIII.

Fig. 1. VALLISNERIA *spiralis*. A diœcious aquatic plant of Europe, America, and New Holland. Leaves radical, riband-like. A, staminate flower. Peduncles short, terminated by a spike; ovate, spathaceous, remaining under water until the period for fertilizing the pistillate flowers. B, fertile plant, peduncles very long, spiral, uni-flowered. Flower spathaceous, floating. This singular plant, in which the two kinds of flowers are entirely separate, is fertilized by a curious provision of nature. When arrived at a mature state, the spiral peduncles of the pistillate flowers untwist themselves, and the flowers rise to the surface of the water; the short spike of stam-inate flowers breaks off from its peduncle; the flowers light upon the other plant, and shower their pollen over it. After this period, the pistillate flowers disappear be-low the surface of the water, where their fruit is produced.

Fig. 2. PISTIA *stratiotes*. The Duckweed tribe. A floating, stoloniferous plant. Leaves radical, spreading, flabelliform.

Fig. 3. TRAPA *natans*.* (*Onagræ*.) An aquatic plant. Stem sub-merged, pro-ducing radical filaments of two sorts; the one simple, filiform; the other ramified and pinnate; they appear to be transformed leaves. The leaves are terminal, diverg-ing; petioles broad, dentate. A, a plant soon after germination; a, the fruit; b, peti-ole from one of the two cotyledons which remain enclosed in the fruit; c, the other cotyledon; d, root; e, stem. B, a plant more developed.

Fig. 4. BUTOMUS *umbellatus*.† Flowering-rush tribe. A plant which grows on the border of lakes and rivers. Leaves radical, erect, riband-like, pointed at the sum-mit. Scape rectilinear. Umbel simple, terminal, involucred.

Fig. 5. POTAMOGETON *compressum*.‡ An annual, aquatic plant, common in brooks and ditches. Stem compressed, slender, leafy. Leaves alternate, linear. Spikes terminal, interrupted. Flower whorled.

Fig. 6. NELUMBO *nucifera*.§ An aquatic, perennial plant found in Egypt, India, and America. Leaves radical, peduncled, peltate, round, concave. Peduncle one-flowered. Calyx caducous. Corolla of many spreading petals. Stamens numerous; style, very short; stigma, like a cup; a, young leaves; b, flower; c, fruit.

Fig. 7. JUNCUS *conglomeratus*.‖ The Rush tribe. (*Juncæ*.) Stem very simple, aphyllous, rectilinear, vertical, terminating in a point. Panicle crowded, unilateral.

Fig. 8. FUCUS *articulatus*.¶ The Sea-weed tribe. (*Algæ*.) A marine plant of the Atlantic Ocean. Frond cartilaginous, dichotomous, moniliform, articulated, each joint containing fruit.

Fig. 9. FUCUS *digitatus*. Stem simple, cylindric. Frond compressed, digitate, flabelliform.

Fig. 10. FUCUS *natans*. A marine plant which, detaching itself from the rocks where it originates, floats in vast quantities upon the surface of the sea, forming islands which retard navigation. Stem filiform. Frond branching, lanceolate, den-tate.

Fig. 11. FUCUS *obtusatus*. A marine plant of Cape Van-Diemen. Frond com-pressed, coriaceous, branching, linear.

PLATE VIII.

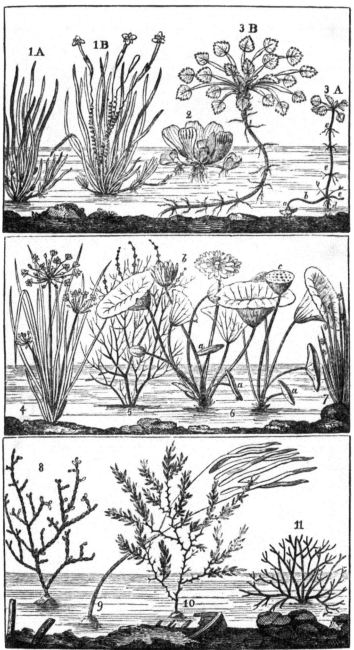

1 Vallisneria spiralis. 2 Pista stratiotes. 3 Trapa natans. 4 Butomus umbellatus. 5 Potamogeton compressum. 6 Nelumbo nucifera. 7 Juncus conglomeratus. 8 Fucus articulatus. 9 Fucus digitatus. 10 Fucus natans. 11 Fucus obtusatus.

SECTION II.
NATURAL ORDERS.

THE following arrangement of *Natural Orders*, is that of Jussieu, as approved by Mirbel, and adopted at the *Jardin Des Plantes* at Paris. Many of the subdivisions of Brown, De Candolle, and Lindley, are noticed under their proper heads. These orders are introduced that the student, by reference to them in the analysis of plants, may gain general ideas of the agreements which exist among the different vegetable tribes. The author would recommend to teachers, to give the advanced pupil these orders as an exercise for occasional recitations, dwelling chiefly on the most important divisions.

CLASS I. *Acotyledons.*

Embryo destitute of cotyledons, and a separate albumen.

1. FUNGI, or *Mushroom-like plants.* These are either parasitical, or spring from the ground naked or enclosed in a *volva.* The substance of mushrooms is fleshy, fungous, or mucilaginous. They are round or flat; some have a *pileus,* (signifying hat) They have neither leaves nor flowers. Instead of *anthers,* they have a scattered, external or internal powder. Instead of pistils they have organs, which resemble thin plates, wrinkles, pores, tubes, &c. In these organs exists a substance analogous to seeds, called *sporules,* which germinates and reproduces the species. The different species of fungi are known by the common names of toad-stool, puff-ball, &c. The medicinal qualities of this order are, *tonic* when dry, narcotic when juicy. Some are eatable, others poisonous.

2. MUSCI. *Moss-like plants* These are little herbaceous plants, often resembling trees in miniature. They grow in humid situations, and are found in the most northern latitudes which are known to produce vegetation. They resemble the Hepaticæ in their general appearance, but the latter are destitute of the *operculum* or lid which covers the seed vessel of the mosses.

3. ALGÆ. *Sea-weed-like plants.* Aquatic; differently coloured, herbaceous, cartilaginous or membranous; seeds contained in conceptacles, or in the substance of the plant. These plants are found both in salt waters, and in ponds, ditches and rivers. They are often mere tufts of fine filaments. Examples: Sea-rock weed, *Fucus,* and *Conferva.* (Plate 8. Figs. 8, 9, 10.)

4. LICHENS.* Seldom vegetating on the earth, sometimes upon living plants, as leaves and bark, often upon stone and dead wood; sometimes pulverulent, dry, or coriaceous; sometimes thick, woody, or fungous. Colour various. In dry places. Some used in dying; some, food for the arctic rein-deer.

5. FILI'CES.† *Fern-like plants.* Roots fibrous, leaves radical, *circinate* when young. Capsules collected in clusters (*sori*) upon the frond or leaf. Examples: Common fern, scouring rush, &c.

6. HEPAT'ICÆ. *Liverwort plants.* Succulent; some grow in earth, some in water, and others are parasites. Resemble the mosses in their general appearance.

7. NAIA'DES ‡ *Duck-meat Tribe.* Floating plants with very cellular stems, and leaves scarcely to be distinguished. Astringent.

CLASS II. *Monocotyledons.*

Stamens hypogynous (below the germ.) *Embryo with one cotyledon. The characters of this class are:—stamen inferior; calyx inferior, when present; stamen seldom indefinite; leaves mostly alternate and sheathing.*

8. AROI'DEÆ. *The Arum tribe.* Inflorescence a spadix, surrounded by a spatha. Leaves petioled, sheathing at the base with parallel or branching veins. Roots often tuberous. Properties: acrid and heating. Examples: Wild-turnip and *Calla.*

9. TY'PHÆ. *Cat-tail tribe.* Growing in marshes or ditches. Leaves rigid, ensiform, with parallel veins. (See Plate 1. Fig. 6.)

10. CYPEROI'DEÆ. *Sedge-grass tribe.* Stem herbaceous, simple. Leaves grass-like. Petiole sheathing. Flowers glume-like, in spikes. Roots fibrous.

* Mirbel makes of this order a division called *Hypoxylea.*
† Mirbel makes of this order a division called *Lycopodiaceæ.*
‡ a *Pistiaceæ* of Lindley

Ovary one-seeded, often surrounded by bristles. Examples: *Carex, Cyperus, Scirpus, &c.*

11. GRAMINE'Æ. *The Grasses.* This is a very important family. The flowers have generally three stamens and one germ. The embryo is small and attached to a farinaceous albumen. In germinating, the cotyledon remains attached to the albumen and nourishes the plume. The roots are fibrous and capillar The culms are cylindrical, hollow, or pithy. The flower and calyx consist of scales, called glumes. The *chaffy flower, single seed, mealy albumen, situation of the embryo, and method of germination,* distinguish, in a peculiar manner, this family. Properties: farinaceous, valuable as food for men and animals. Examples: wheat, meadow-grass, sugar-cane. (See Plate 2. Fig. 2, and Plate 4. Fig. 6.)

CLASS III. *Monocotyledons.*

Stamens perigynous (around the germ.) *Fruits with three cells. Embryo small, with a lar ge albumen.*

12. PAL''MÆ. *The Palm tribe.* This family is a native of warm climates. The flowers are often diœcious. (See ARECA *oleracea*, Plate 1. Fig. 1.) The number of stamens is usually six; the filaments are often united at the base. The germ is superior; corolla deeply parted into six segments, the three outer ones being smallest. The germ is superior. The fruit is a berry or a fibrous drupe, the albumen of which is, at first, tender and eatable, and at last becomes hard. The stems of palms are usually undivided, lofty, and round; they are not composed of concentric circles, being endogenous or growing internally; they are scaly from the remains of the indurated foot-stalks of leaves. The leaves of palms appear in a terminal tuft, alternate and sheathing.

13. LILIA'CEÆ. *Lily-like plants.* Six petals spreading gradually from the base, and exhibiting a bell-form appearance, but differing from the campanulate flowers in being polypetalous. The number of stamens is generally six, sometimes but three, usually alternate with the petals. The germ is always of a triangular form, and contains three cells; the roots are mostly bulbous. The calyx is usually wanting; the stems are simple, without branches; the leaves entire, and nerved   To this family belong the tulip, lily, crown-imperial, dog-tooth violet, &c. Plants of this natural family usually belong to the artificial class, Hexandria; the Crocus and Ixia, having 3 stamens, belong to the class Triandria. (Plate 7. Fig. 4.)

14. ASPAR''AGI. *Asparagus-like plants.* Corolla, monopetalous, 6-parted. Stamens six. Fruit a berry, superior, 3-celled. Roots fasciculated. Examples: Asparagus and Convallaria. (See Plate 1. Fig. 3, for a plant of this family.)

15. NARCIS'SI. *Roots mostly bulbous.* Leaves sessile, elongated, alternate; radical leaves sheathing. Flowers with spathas; panicled, corymbed or solitary. Perianth, which is usually called a corolla, 6-parted. Stamens 6, inserted into the tube of the perianth. Style 1. Stigma simple or 3-parted, Capsule 3-celled 3-valved, or 3-parted. Seed with a perisperm. Examples: Narcissus and Galanthus.

16. IRIDE'Æ. *Iris-like plants.* Root tuberous. Leaves sessile, alternate, equitant, compressed, ensiform. Flowers with spathas. Perianth petal-like, 6-parted, 3 internal, 3 external. Stamens 3. Style 1. Stigmas 3, often petaloid. Capsule 3-celled, 3-valved, many-seeded. Examples: Iris, Gladiolus. Roots useful in dropsical complaints, antiscorbutic. (Plate 6. Fig. 6.)

17. JUN:E'Æ. *The Rush tribe.* Flowers imperfect, glumaceous. Leaves fistular, or flat and channelled, with parallel veins. Examples: Juncus, Luzula. The leaves are used for bottoming chairs. Medicinal properties doubtful. (Plate 8. Fig. 7.)

18. BROME'LIÆ. *Pine-apple tribe.* Leaves radical, ensiform, caniculate. Scape short. Fruit a sorose, ovate succulent, surmounted with a crown of leaves. Examples: Bromelia, Agave. (See Plate 5. Fig. 3.)

19. ASPHODE'LI. *Asphodel tribe.* Stamens 6; corolla 6-parted; germ 3-celled. Roots bulbous, or fasciculated. Examples: Onion, Hyacinth. Properties: acrid and stimulating.

20. COMMELI'NEÆ. *The Spider-wort tribe.* Examples: Tradescantia, Commelina. Herbaceous plants. Leaves usually sheathing at the base. This family is taken from Junceæ.

21. ALISMA'CEÆ. *The Arrow-head tribe.* Examples: Sagittaria, Alisma Taken from Junceæ. (Plate 3. Figs. 4, 5.)

22. COLCHICE'Æ. *Colchicum tribe.* Emetic and cathartic. Examples: Colchicum, Melanthium. This order is by some called *Melanthaceæ.*

CLASS IV. *Monocotyledons.*
Stamens epigynous, (above the germ.)

23. ORCHIDE'Æ. *Orchis-like plants.* Roots fibrous or tuberous. Stem simple. Leaves mostly radical, sheathing; cauline ones sessile. Flowers bracted, commonly in a spike, seldom solitary. Perianth irregular, 6-parted, 3 divisions external, 3 internal, and 9-petaloid; a lower one in the form of a lip, often spurred. Stamens 3, adnate to the style in part or wholly; two are usually abortive. Style thick. Stigma oblique, viscid. Examples: Orchis, Cypripedium, Neottia. Properties: farinaceous and emollient.

24. MU'SÆ. *Banana tribe.* Examples: Plantain-tree, (*Musa,*) Bread-fruit, (*Artocarpus.*) (See Plate 2. Fig. 4.)

25. CAN''NÆ. *The Indian reed-tribe.* This is subdivided into *Marantaceæ,* the arrow-root tribe, and *Amomæ,* or *Scitaminaceæ,* the ginger tribe. Properties: aromatic, and carminative. (Plate 3. Fig. 4.)

26. HYDROCHAR''IDES. *Tape-grass tribe.* Floating plants. Examples: *Hydrocharis, Vallisneria.* (Plate 8. Fig. 1.)

CLASS V. *Dicotyledons.*
Apetalous—Stamens epigynous. Calyx superior. Monosepalous, (above the germ.)

27. ARISTOLO'CHIÆ. *Wild ginger tribe.* Perennial. Flowers Gynandrous. Examples: Virginia snake-root, (*Aristolochia,*) Wild ginger, (*Asarum.*)

CLASS VI. *Dicotyledons.*
Stamens perigynous, (around the germ.) *Perianth single,* in some cases resembling a calyx, in others a corolla.

28. ELEAG''NÆ. Flowers diœcious. Fruit a drupe or nut. Leaves alternate. Trees or shrubs. Examples: Pepperage-tree and Eleagnus.

29. HYMELE'Æ. Under-shrubs. Stamens 8. Style 1. Fruit, a drupeole. Cotyledons large, fleshy. Perisperm, thin. Examples: Leather-wood and *Daphne.* Bark caustic when chewed.

30. PROTE'Æ. *Silver-tree tribe.* Deciduous shrubs from the Cape of Good Hope. Example: *Protea.*

31. LAU'RI, (or *Laurineæ.*) *The Laurel tribe.* Trees. Flowers Enneandrous; 4 to 6 cleft. Fruit a berry or drupe. The American plants of this family are the spice-bush, (Laurus benzoin,) and Sassafras. Medicinal properties various and important.

32. POLYGO'NEÆ. *The Dock tribe.* Herbaceous. Leaves alternate, at first revolute, petioled. Flowers panicled, or in a spike. Fruit a nut, usually triangular, as in the buckwheat. Seed with farinaceous albumen. Examples: Dock, rhubarb, buckwheat.

33. ATRIP''LICES. *Pig-weed tribe.* Flowers with little beauty. Herbs or small shrubs. The beet, poke-weed, and pig-weed, are examples of this family. The pig-weed is by some arranged in a new order, Chenopodeæ; and the poke-weed in another, Phytolacceæ.

CLASS VII. *Dicotyledons.*
Stamens, (beneath the germ.)

34 AMARAN''THI. *Coxcomb-like plants.* Stem herbaceous. Leaves entire. Flowers small, numerous, often bracted, sometimes imperfect, in a head, raceme or spike. Perianth often coloured, monosepalous. Pericarp either a pyxide or utricle. Example: Amaranthus.

35. PLANTAGIN''EÆ. *Plantain tribe.* Herbaceous. Leaves many-nerved. Flowers sessile, bracted in a spike. Stamens 4. Pyxide 4-celled, many-seeded. Example: Plantain. Useful as a pot herb. *Emollient.*

36. NYCTA'GINES. *Mirabilis tribe.* The principal family in this order is the Four o'clock, (Mirabilis.) Properties: cathartic and emetic.

37. PLUMBA'GINES. *Marsh rosemary tribe.* Herbs or under-shrubs. Leaves alternate or clustered. Corolla regular. Stamens 5, ovary 1-celled; ovule, pendulous. Fruit, a utricle. Properties: astringent, tonic. Example: Statice.

CLASS VIII. *Dicotyledons.*
Corollas monopetalous. hypogynous, (below the germ,) *regular or irregular, bearing the stamens, which generally alternate with its segments when of equal number; germ superior.*

38. LYSIMACH''IÆ, (or *Primulaceæ.*) *The Loose strife, or primrose tribe.* A fami-

ly comprising many showy flowers, but belonging to genera which differ much in the appearance of their inflorescence. Examples: Trientalis, Primula, Lysimachia.

39. PEDICULA'RES, (or *Rhinanthea.*) This family contains genera of plants which appear to have little natural resemblance, as Rhinanthus, Pedicularis, Bartsia, &c.

40. ACAN''THI, (or *Acanthaceæ.*) Contains no important genera. Examples. Malabar-nut, (*Justicia,*) and *Ruellia.*

41. JASMI'NEÆ. *Lilac tribe.* Trees or shrubs. Leaves generally opposite. Flowers in a thyrse or corymb. Stamens 2. Pericarp 2-celled, 2-seeded, a berry or drupe, or capsular. Example: Lilac, (*Syringa.*)

42. VI'TICES, (or *Verbenaceæ.*) *The Verbena tribe.* Properties: secernent stimulant.

43. LABIA'TÆ. *Mint-like plants.* A very extensive family; of importance in seasoning food, as Sage, Summer-savory, and thyme; medicinal, as Catnip, Mint, Horehound, &c.

44. SCROPHULA'RLÆ, (or *Personeæ.*) Flowers with personate corollas, as snapdragon, (Anterhinum.) Scrophularia, and Digitalis. Properties: narcotic.

45. SOLA'NEÆ. *Potato-like plants.* Stamens 5. Pericarp sometimes a berry, sometimes a pyxide or a capsule. Examples: Potato, Tomato, Red pepper, (*Capsicum.*) Narcotic, stimulating.

46. BORAGI'NEÆ. *Borage-like plants.* Leaves often rough, or pubescent. Examples: *Borago, Myosotis.* Properties: emollient.

47. CONVOL''VULI. *Convolvulus tribe.* Stem often twining. Peduncles axillary or terminal. Calyx 5-parted. Corolla 5-lobed. Stamens 5. Some (as the sweet potato, Convolvulus batatus) are edible, some (as Convolvulus panduratus) are medicinal.

48. POLEMO'NLÆ. *Phlox-like plants.* Herbs. Calyx 5-parted. Corolla 5-lobed, regular, stamens 5. Examples: Phlox Polemonium.

49. BIGNO'NIÆ. *Trumpet-flower tribe.* Mostly trees or shrubs, often climbing or twining. Examples: Bignonia, Catalpa.

50. GENTIA'NÆ. *The Gentian tribe.* Calyx monosepalous, 5 to 10-divided. Corolla with usually as many lobes as the divisions of the calyx. Herbs, seldom shrubs. Leaves opposite without stipules. A division of this family, *Spigeliaceæ,* contains the Carolina Pink, (Spigelia,) used in medicine as a vermifuge. The Frasera, or American Columbo root, which is very bitter, is valued as a cathartic.

51. SAPO'TÆ. *West India plum.* A family of little importance.

52. APOCY'NEÆ. *Dog-bane tribe.* Herbs or small shrubs. Leaves opposite. Calyx 5-parted. Corolla 5-parted. Stamens 5, inserted on the corolla. Pericarp a double follicle, Follicle many-seeded. A division of this fruit, *Asclepiadæ,* or milk-weed plants, have a milky juice.

CLASS IX. *Dicotyledons.*

Corolla monopetalous, perigynous, (around the germ.)

53. EBENA'CEÆ. *The Ebony tribe.* Example: *Diospyros.*

54. KLENA'CEÆ. *The Persimon tribe.* Example: Sarcolæna, a foreign plant.

55. RHODODEN''DRÆ. *The rose-bay tribe.* Herbs and shrubs. Flowers often bracted Inflorescence various. Included by De Candolle in the next order.

56. ERI'CEÆ. *Heath-like plants.* Shrubs, or under-shrubs. Leaves evergreen, rigid, entire, whorled, or opposite, without stipules. Examples: *Arbutus Gaultheria, Kalmia.*

57. GUAIACA'NÆ. Example: *Lignum vitæ.* The gum-guaiacum of medicine is from a plant of this family.

58. CAMPANULA'CEÆ. *The Bell-flower tribe.* Calyx usually 5-parted, Corolla 5-lobed, inserted into the top of the calyx, withering on the fruit. Stamens 5. Leaves simple, or deeply divided. Examples: Campanula, Lobelia. Lindley makes a subdivision, *Lobelia'ceæ,* in which is the genus Lobelia, a species of which, called the Indian tobacco, is powerfully medicinal, and often improperly used by ignorant practitioners.

CLASS X. *Dicotyledons.*

Corollas monopetalous, epigynous, (above the germ,) *anthers united.*

59. CICHORA'CEÆ. Flowers Syngenesious. Calyx divided into hairs or pappus.

Corolla either ligulate, or tubular. Stamens 5, alternating with the teeth of the corolla; filaments distinct, anthers forming a cylinder by their coherence. Ovary inferior 1-celled, with a single erect ovule. Style single; stigmas 2. Fruit an achenium. Seed solitary, erect. Examples: Dandelion and Lettuce. Antiscorbutic, and mild anodyne.

60. CINAROCEPH"ALÆ. Examples: Thistle and Burdock, differs little from the preceding.

61. CORYMBIF"ERÆ. *Thorough-wort plants.* Examples: Eupatorium and Rudbeckia. Very valuable for their medicinal qualities. The compound flowers are by some writers classed under the general head *Compositæ*, and subdivided into numerous sections, viz: *Carduaceæ*, or the Thistle tribe, *Astereæ*, or the Aster tribe, *Eupatorineæ*, or the Thorough-wort tribe, *Jacobea*, or the Colt's-foot tribe, and *Heliantheæ*, or the Sunflower tribe.

CLASS XI. *Dicotyledons.*
Corolla monopetalous, epigynous, (above the germ,) *anthers distinct.*

62. DIPSA'CEÆ. *Teasel plants.* Flowers densely capitate. Leaves opposite or whorled. Herbs or under-shrubs. Examples: Teasel, Button-bush.

63. RUBIA'CEÆ *Bed-straw tribe.* Leaves whorled, very entire. Flowers axillary or terminal. Stamens 4, ovary simple, fruit a dieresil, 2-seeded. Examples: Galium, Rubia. Some of this family are of use in dying.

64. CAPRIFO'LIÆ. *Elder, Snow-ball, and Honey-suckle-like plants.* Shrubs. Ovary cohering with the calyx; fruit crowned by its limb. Leaves opposite. Flowers terminal, corymbose or axillary. Examples: Viburnum, Lonicera, Symphoria.

CLASS XII. *Dicotyledons.*
Corolla polypetalous ; stamens epigynous, (above the germ.)

65. ARA'LEÆ. *Ginseng tribe.* Calyx superior. Stamens 5 or 6, or 10 or 12, arising from within the border of the calyx; ovary with many cells; ovules solitary, pendulous; styles equal in number to the cells. Trees, shrubs, or herbs, resembling umbelliferous plants in their habit. Examples: Spikenard, (Aralia,) Ginseng, (Panax.)

66. UMBELLIF"ERÆ. *Parsley-like plants.* Stem herbaceous. Leaves mostly pinnate or pinnatifid. Flowers in umbels. Calyx adhering to the germ. Corolla 5-petalled. Stamens 5. Style and stigma 2. Fruit a cremocarp. Seeds closed, remaining after maturity, suspended to a central axis. Examples: Dill, Fennel, Parsley, Caraway. Uses and medicinal qualities various. The following subdivision has been made of this tribe: 1st, *Hydrocotoliæ; umbels* simple or imperfect. Examples: Water-hemlock, (Cicuta,) Water-parsnip, (Sium,) Fool's-parsley, and Angelica; 2d, *Campylospermæ;* Sweet cicely, and Hemlock, (Conium.)

CLASS XIII. *Dicotyledons.*
Corolla polypetalous ; stamens hypogynous, (under the germ.)

67. RANUNCULA'CEÆ. A very large order, containing the Virgin's-bower, Ranunculus, Anemone, Hepatica, &c. Calyx with many definite sepals, or many-parted. Stamens and pistils numerous. Fruit often consists of dry nuts or carpels. Herbs or under-shrubs. Leaves simple, often variously lobed and subdivided, petioled. Some of this family, as the gold thread,(Coptis,) are highly astringent, some are valuable as dies, and some are beautiful as ornamental flowers.

68. PAPAVERA'CEÆ. *Poppy-like plants.* Lactescent. Stem herbaceous. Leaves alternate. Flowers solitary, in a spike or umbel. Calyx 2-3 sepalled, caducous. Stamens numerous. Examples: Poppy, Blood-root. Properties: narcotic, anodyne.

69. CRUCIF"ERÆ. *Plants with cruciform corollas,* as cabbage, turnip, radish. Stem herbaceous. Leaves alternate. Flowers corymbed, panicled or in a spike. Calyx 4-sepalled. Corolla 4-petalled. Stamens 6, solitary, 4 disposed in two pairs. Glands nectariferous. Fruit a silique. Chiefly useful as garden vegetables. This order is subdivided into *Siliculosæ,* pods short, and *Siliquosæ,* pods long.

70. CAPPAR'IDES. A small order. Cruciform plants. Examples: Cleome, Gynandropsis.

71. SAPIN'DI. Example: Soap-berry, (Sapindus.)

72. ACE'RA, (or *Acerineæ,*) *Maple tribe.* Trees, with opposite, simple, rarely

pinnate leaves. Flowers often polygamous, sometimes apetalous. Examples: Acer, Negundo.

73. MALPI'GHIÆ. Example: Barbadoes cherry, (Malpighi.)

74. HYPER'ICÆ. *St. John's-wort tribe.* Herbs or shrubs, with a resinous juice. Leaves opposite, entire, dotted, occasionally alternate and crenate. Flowers generally yellow. Examples: Hypericum, Ascyrum. Some species are said to be healing for wounds.

75. GUTIFE'RÆ. Example: Cambogia.

76. AURAN''TIA. *Orange tribe.* Examples: Orange, Lemon. Properties: refrigerant, tonic.

77. ME'LIÆ. *Tea.* Astringent, anodyne.

78. VITES. *The Vine family.* Stem woody, sarmentose, cirrifferous. Leaves alternate, stipuled. Tendrils and peduncles opposite. Flowers in a thyrse. Calyx 5-toothed. Corolla 5-petalled. Stamens 5. Pericarp a berry. Example: Vitis, the grape. Another order has been substituted by De Candolle, called *Ampelidea*, which contains Ampelopsis and Vitis.

79. GERA'NIÆ. *Geranium tribe.* Stamens 10, monadelphous. Stigmas 5. Fruit a dieresil; 5 carpels, each 2-seeded.

80. MALVA'CEÆ. *Holly-hock tribe.* Leaves alternate, stipuled. Calyx 5-parted. Corolla 5-petalled. Stamens indefinite, monadelphous. Dieresil with many carpels. Carpels many-seeded. Examples: Holly-hock, Lavatera, Mallows.

81. MAGNO'LIÆ. *Tulip-tree tribe.* Trees or shrubs. Leaves alternate, coriaceous. Flowers large, solitary, often odoriferous. Examples: Magnolia, Liriodendron.

82. ANNO'NÆ. Example: Anona, custard-apple.

83. MENISPER''MÆ. Example: Menisperm, moon-seed.

84. BERBER'IDES. Example: Hamamelis, witch-hazel; Berberis or barberry Flowers panicled. Pericarp a capsule or berry.

85. TILI'ACEÆ. *Bass-wood plants.* Trees. Leaves alternate, stipuled. Flowers corymbed. Example: Tilia, bass-wood, or lime-tree.

86. CIS''TI. *Rock-rose plants.* Small shrubs. Example: Cistus.

87. RUTA'CEÆ. *Rue plants.* Leaves compound. Stamens 6. Fruit a dieresil or regmate. Example: Rue, (Ruta.)

88. CARYOPHY'LLEÆ. *Pink-like plants.* Herbaceous. Leaves opposite. Flowers often terminal, sometimes axillary. Fruit a capsule. Example: Dianthus.

CLASS XIV. *Dicotyledons.*

Corolla polypetalous; stamens epigynous, (around the germ.)

89. SEMPERVI'VÆ. *House-leek plants.* Emollient.

90. SAXIFRA'GÆ. *Saxifrage plants.*

91. CAC''TI. *Prickly-pear tribe.*

92. PORTULAC'CEÆ. *Purslane tribe.* Example: Portulacca. Properties: emollient.

93. FICOI'DEÆ. Example: Mesembryanthemum, ice-plant.

94. ONA'GRÆ. *Willow-herb plants.* Example: Epilobium.

95. MYR''TI. Example: Myrtus.

96. MELAS''TOMÆ. Example: Rhexia, deer-grass.

97. SALICA'RIÆ. Examples: Lythrum, Cuphea.

98. ROSA'CEÆ. *Rose and apple tribe.* Stamens numerous. Pericarp a pyridion. Examples: Rosa, Pyrus.

99. LEGUMINO'SÆ. *Pea tribe.* Stamens 10, diadelphous. Flower papilionaceous. Fruit a legume. Examples: Lupinus, Trifolium.

100. TEREBINTA'CEÆ. Example: Rhus, the sumach.

101. RHAM''NI. Examples: Buckthorn, (Rhamnus,) and Ceonothus.

CLASS XV. *Dicotyledons.*

Stamens and pistils diclinous, or on different flowers.

102. EUPHOR''BIÆ. Example: Euphorbia, or spurge.

103. CUCUR''BITACEÆ *Melon-like plants.* Stem herbaceous. Stamens 5. Fruit a pepo. Examples: Watermelon, Cucumber.

104. URTICE'Æ. Example: Hop, (Humulus.)

105. AMENTA'CEÆ. *Trees with inflorescence in an ament or catkin.* Examples: Oak, Willow.

106. CONIF''ERÆ *Cone-bearing trees.* Examples: Pine, Cedar.

THE FLORA,

OR

PRACTICAL BOTANIST'S COMPANION:

CONTAINING

GENERIC AND SPECIFIC

DESCRIPTIONS

OF THE

PLANTS OF THE UNITED STATES,

CULTIVATED AND EXOTIC.

GENERA OF PLANTS.

CLASS I. MONANDRIA.

ORDER I. MONOGYNIA.

SALICOR"NIA. Calyx inflated, entire, 3 or 4-sided, obconic; corolla 0; style 2-cleft; seed 1, enclosed in the calyx. (samphire.)

HIPPU'RIS. Calyx superior, obsolete, with a 2-lobed margin; corolla 0; seed 1; stigma simple; style in the groove of the anther. (mare's-tail.)

CAN"NA. Anthers adhering to the petal-like filaments; styles thick, club-shaped; stigma obtuse, linear. (Indian-reed.)

THA'LIA. Anther attached to the filaments; style depressed; stigma gaping.

ORDER II. DIGYNIA.

BLI'TUM. Calyx 3-cleft, or 3-parted, berry-like; corolla 0; seed 1, immersed in the calyx. (blite.)

CAL"ITRI'CHE. Calyx 0; petals 2, resembling a calyx; seeds 4, naked.

CORISPER"MUM. Calyx 2-leaved; corolla 0; seed 1.

CLASS II. DIANDRIA.

ORDER I. MONOGYNIA.

A. *Corolla 1-petalled, inferior, regular; seeds in a drupe or nut.*

OR"NUS. Calyx 4-parted; corolla 4-parted; petals long and ligulate. Two barren filaments; nut winged; fruit a capsule. (flowering ash.)

ELYTRA'RIA. Calyx 4 or 5-parted; corolla 5-cleft; capsule 5-valved, 2-seeded; seeds adhering to a dissepiment, contrary to the valves.

LIGUS"TRUM. Calyx 4-toothed; corolla with 4 ovate divisions; berry 1 or 2-celled, 2 or 4-seeded. (prim.)

CHIONAN"THUS. Calyx 4-parted; corolla 4-parted, sometimes more, tube short, with very long divisions; nucleus of the drupe striate-fibrous. (fringe tree.)

OLE'A. Corolla 4-cleft, with obovate divisions; drupe 1-seeded.

JASMI'NUM. Corolla salver-form, 5 to 8-cleft; berry 2-seeded, each seed solitary, arilled. (jasmine.) Ex.

SYRIN"GA. Corolla salver-form; capsule 2-celled. (lilac.) Ex.

B. *Corolla 1-petalled, inferior, irregular; seeds in capsules.*

VERON"ICA. Calyx 4-parted; corolla cleft into 4 lobes, lower division smaller; capsule obcordate, few-seeded, 2-celled. (speedwell.)

LEPTAN"DRIA. Calyx 5-parted, segments acuminate; corolla tubular-campanulate, border 4-lobed, a little ringent, lower segment narrow; stamens and at length the pistils much exserted; capsule ovate, acuminate, opening at the top. (culver's physic.)

GRATIO'LA. Calyx 5-parted, often with 2 bracts at the base; corolla irregular, resupinate, 2-lipped, upper lip 2-lobed, lower one equally 3-cleft; stigma 2-lipped; capsule 2-celled, 2-valved. (hedge-hyssop.)

LINDER"NIA. Calyx 4-parted; corolla resupinate, tubular, 2-lipped, upper lip short, reflexed, emarginate, lower one trifid, unequal, filaments 4, the 2 longer ones forked and barren; capsule 2-celled, 2-valved, the dissepiment parallel to the valves.

CATAL"PA. Corolla 4 or 5-cleft, somewhat inflated, bell-form; calyx 2-parted, or 2-leaved; stigma 2-lipped; capsule cylindric, 2-celled. (catalpa-tree.)

HEMIAN"THUS. Calyx tubular, cleft on the under side, border 4-toothed; upper lip of the corolla obsolete, lower 3-parted, intermediate segments ligulate, longer, incurved; stamens with 2-cleft filaments, the side branches bearing anthers; capsule 1-celled, 2-valved, many-seeded.

JUSTI'CIA. Calyx 5-parted, often with 2 bracts at the base; corolla irregular, labiate, upper lip emarginate, lower lip 3-cleft.

UTRICULA'RIA. Calyx 2-leaved, equal; corolla ringent, lower lip spurred at the base; filaments incurved; stigma divided.

PINGUIC"ULA. Corolla labiate, spurred; calyx 5-cleft; capsule 5-celled.

MICRAN"THEMUM. Calyx 4-parted; corolla 4-parted, the upper lip smaller; filament incurved, shorter than the corolla; capsule 1-celled, 2-valved; seeds striate.

C. *Corolla 1-petalled, inferior, irregular; seeds naked.*

MONAR"DA. Calyx cylindric, striated, 5-toothed; corolla ringent, tubular, upper lip lance-linear, involving the filaments, lower lip reflexed, 3-lobed. (Oswego tea, mountain mint.)

LYCO'PUS. Calyx tubular, 5-cleft or 5-toothed; corolla tubular, 4-cleft, nearly equal, upper division broader and emarginate; stamens distant; seeds 4, retuse. (water horehound.)

SAL"VIA. Calyx tubular, striated, 2-lipped, under lip 2 to 3-toothed, lower lip 2-cleft; corolla ringent, upper lip concave, lower lip broad, 3-lobed, the middle lobe the largest, notched; stamens with two spreading branches, one of which bears a 1-celled anther; germ 4-cleft; style thread-shaped, curved; seeds 4, in the bottom of the calyx. (sage.)

COLLINSO'NIA. Calyx tubular, 3-lipped; upper lip 3-toothed; corolla funnel-form, unequal, under lip many-cleft, capillary; one perfect seed. (horse-balm.)

ROSMARI'NUS. Corolla ringent, upper lip 2-parted; filaments long, curved, simple, with a tooth. (rosemary.) Ex.

D. *Corolla superior.*

CIRCÆ'A. Calyx 2-leaved or 2-parted; corolla 2-petalled; capsule hispid, 2-celled, not gaping; cells 1 or 2-seeded, seeds oblong. (enchanter's nightshade.)

ORDER II. DYGYNIA.

ANTHOXAN"THUM. Calyx of two, egg-shaped, pointed, concave, chaffy scales, 1-flowered; corolla of two equal husks, shorter than the calyx, awned on the back; an internal corolla or nectary, consisting of two, egg-shaped, minute scales; stamens longer than the corolla; anther oblong, forked at both ends; germ superior; seed 1. (sweet vernal grass.)

CRYP"TA. PEPLIS. Calyx 2-sepalled, inferior; corolla 2 or 3-petalled; closed; styles none; stigmas 2 or 3, very minute; capsule 2 or 3-celled, 2 or 3-valved; cells 4 or 5-seeded; seeds sub-cylindric, striate, incurved. (mud-purslane.) *The corolla appears like a capsule with an aperture at the top, when examined without a lens.*

ERIAN"THUS. Glumes 2, valves nearly equal, villose at the base; paleas 2, unequal; inner glume bearing a long awn near the summit. S.

ORDER III. TRIGYNIA.

PI'PER. Spadix simple and slender; calyx and corolla wanting. Fruit a berry, globose. (pepper.) Ex.

CLASS III. TRIANDRIA.

ORDER I. MONOGYNIA.

A. *Flowers superior.*

I'RIS. Calyx spatha, 2 or 3-valved; corolla 6-parted, divisions alternately reflexed; stigmas 3, petal-like; style short; capsule 3-celled. (flower-de-luce, iris or flag.) Ex.

FE'DIA. Calyx 3 to 6-toothed, permanent; corolla tubular, 5-cleft or 5-parted; nut or capsule 2 or 3-celled; seed naked, or crowned with a tooth. (lamb-lettuce.)

DILA'TRIS. Perianth superior, 6-parted, segments unequal; stamens 3; style declining; stigma minutely 3-lobed; capsule 3-celled, truncated, many-seeded. (red-root.)

PLECTRI'TIS. Calyx with the margin straight, entire; corolla with the tube short-spurred at the base, gibbose before, limb 5-cleft, 2-lipped; capsule 1-celled, 2-winged.

VALERIANEL"LA. Calyx with the limb toothed, persistent; corolla not spurred, regular, 5-lobed; stigma sub-trifid; capsule 3-celled, sub-membranous, crowned with the limb of the calyx. (wild lamb-lettuce.)

GLADIO'LUS. Spatha 2-valved; corolla 6-parted, two divisions much smaller, upper divisions broadest; style long and

slender; stigmas 3; capsule triangular, 3-celled, many-seeded.

VALERIA'NA. Corolla monopetalous, 5-cleft, horned at the base; seed 1, crowned with a feathery pappus. Ex.

CRO'CUS. Spatha radical; corolla funnel-form, with a long slender tube; stigma deep-gashed, crested.

IX"IA. Spatha 2 or 3-valved, ovate, short; corolla 6-parted or 6-petalled, sometimes tubular; stamens straight or incurved; stigmas sub-filiform. (blackberry-lily.)

PHYLLAC"TIS. Involucrum 1-leaved, sheathing; calyx marginal, minute; corolla 3-cleft; seed 1; styles and stamens exsert.

APTE'RIA. Calyx 0; corolla monopetalous, wingless, bell-tubular, with the margin 6-toothed, alternately smaller; capsule 1-celled, valveless; seeds numerous and minute.

BOERHAA'VIA. Calyx tubular, margin entire; corolla 1-petalled, campanulate, plaited; nut 1, invested by the permanent calyx, naked or tubercled, 5-grooved, obconic.

CALYME'NIA. See ALLIO'NIA.

B. *Flowers inferior.*

COMMELI'NA. Sheath cordate; calyx 3-leaved; corolla 3-petalled, sub-equal; 3 barren filaments—sometimes the whole 6 filaments bear anthers; stigma simple; nectaries 3, cross-form, inserted on peculiar filaments; capsule sub-globose, 3-celled; cells 2-seeded or empty. (day-flower.)

SCHOL"LERA. Spatha 1-flowered; corolla long-tubular, border 6-parted, with the stamens on the divisions; filaments equal.

HETERANTHE'RA. Calyx a 2 or 3-flowered spatha; corolla dull-colored, membranaceous, with a long slender tube, border 4 to 6-parted; anthers 3, 2 of them attached to the divisions of the corolla, the other attached to the top of the style, and much larger; stigma sub-capitate, 1-sided; capsule 3-celled, many-seeded, dehiscent at the angles. *This genus is intermediate between the classes Triandria and Gynandria.*

SYE'NA. Calyx 3-leaved; corolla 3-petalled; anthers oblong; capsule 1-valved, 3-celled. S.

STI'PULICI'DA. Calyx 5-parted; corolla 5-petalled; stigmas 3; capsule 1-celled, 3-valved. S.

POLYCNE'MUM. Calyx 3-sepalled; petals calyx-like; capsule 1-seeded, membranaceous, not opening, covered by the calyx. Stamens vary from 1 to 5. S.

C. *Flowers with glumes. Sedges.*

DULI'CHIUM. Spikes sub-racemed, axillary; spikelets linear-lanceolate, sub-compressed; glumes distichus, sheathing; style very long, bifid, base persistent; nut bristled at the base.

CYPE'RUS. Scales imbricated 2-ways; ovary without bristles; spikelets compressed, distinct.

LIMNET'IS. Glume 2-valved, compressed, one keeled and longer than the other;

24*

paleas 2-valved, awnless, unequal; flowers in unilateral spikes, somewhat imbricated in 2-rows. (salt-grass.)

KYLLIN"GA. Flowers distinct, disposed in a roundish, sessile, sub-imbricated spike; glumes 2-valved, 1-flowered; paleas 2, longer than the glumes. (bog-rush.)

ORYZOP"SIS. Glume 1-flowered, 2-valved; valves membranaceous, nearly equal, lax, obovate, awnless; paleas 2, coriaceous, cylindric ovate, hairy at the base, the lower one awned at the tip; scales linear-elongated; flowers panicled. (mountain rice.)

RHYNCHOS"PORA. Glumes fascicled into a spike, the lower ones empty; seed 1, crowned with a persistent style; bristles surrounding the base. (false bog-rush.)

SCHŒ'NUS. Glumes fascicled into a spike, the lower ones empty; style deciduous; seeds naked.

SCIR"PUS. Glume 1-valved, 1-flowered, imbricated on all sides; seed or nut naked or bristled at the base. (club-rush.)

MIE'GIA. Flowers polygamous, panicled; glume 2-valved, many-flowered, short and unequal; paleas 2, unequal, the larger one acuminate; style very short, 3-cleft, plumose; nectary 3-parted, divisions lanceolate, acute, as long as the germ, large, acuminate. (cane.) S.

NAR"DUS. Glume 1-flowered, 1-valved; palea 1, included in the calyx; stigma 1. Flowers spiked, alternating and sessile. (mat-grass.)

XY'RIS. Calyx a cartilaginous glume, 2 or 3-valved, in a head; corolla 4-petalled, equal, crenate; capsule 3-valved, many-seeded. (yellow-eyed grass.)

ORDER II. DIGYNIA.

A. *Spikelets 1-flowered; corollas without abortive rudiments of flowers at the base.* [*Calyx and corolla different in texture.*]

LEER"SIA. Calyx 0; corolla 2-valved, closed; valves compressed, boat-shaped; nectary obovate, entire, collateral; stamens varying in number. (cut-grass.)

MIL"IUM. Glume 2-valved, naked, beardless, paleas 2, oblong, concave, shorter than the glume, awnless; seed 2-horned; flowers panicled. (cane.)

PAS"PALUM. Glume in 2 membranaceous, equal, roundish valves; paleas awnless, cartilaginous, of the form of the glume; stigma plumose, colored; nectary lateral; flowers in unilateral spikes.

PIPTATHE'RUM. Glume membranaceous, longer than the cartilaginous, oval paleas; lower valve awned at the tip; nectary ovate, entire; seed coated.

ARIS"TIDA. Glume 2-valved, membranaceous, unequal; paleas 2-valved, pedicelled, sub-cylindric, lower valve coriaceous, involute, 3-awned at the tip, upper valve very minute or obsolete; nectary lateral. (beard-grass.)

STI'PA. Glume 2-valved, membranaceous; paleas 2-valved, coriaceous, valves involute, truncate; awn terminal, long, caducous; seed coated. (feather-grass.)

SAC"CHARUM. Calyx involucred, with long wool at the base, 2-valved; corolla 1 or 2-valved; stamens 1 to 3. (sugar-cane.)

[*Calyx and corolla of similar texture— flowers in spreading panicles.*]

AGROS"TIS. Calyx herbaceous, 2-valved, 1-flowered, valves acute, a little less than the corolla; corolla 2-valved, membranaceous, often hairy at the base; stigmas longitudinally hispid or plumose, florets spreading; nectary lateral; seed coated. (redtop.)

MUHLENBERG"IA. Glumes 2, very minute, unequal, one scarcely perceptible; paleas 2, many times longer than the glumes, linear-lanceolate, nerved, hairy at base, the lower one terminating in a long awn.

ARUN"DO. Calyx 2-valved, unequal, membranaceous, surrounded with hair at the base, lower valve mucronate or slightly awned. Sometimes there is a pencil-form rudiment at the base of the upper valve. (reed.)

[*Calyx and corolla of similar texture; flowers in compact panicles, often spike-form.*]

PHLE'UM. Calyx hard, 2-valved, equal, sessile, linear, truncate, bicuspidate; corolla enclosed in the calyx, 2-valved, awnless, truncate. (timothy-grass.)

ALOPECU'RUS. Glumes 2, equal, generally connate at the base; paleas 1-valved, utricle-like, cleft on one side, awned below the middle. (foxtail-grass.)

B. *Spikelets 1-flowered; corolla with 1 or 2 abortive rudiments of flowers at the base.* [*Calyx and corolla of similar texture.*]

PHALA'RIS. Calyx membranaceous, 2-valved, valves keeled, nerved, equal in length, including the 2-valved, pilose corolla; the corolla is shorter than the calyx and coriaceous; rudiments opposite, sessile, resembling valves; nectary lateral. (riband-grass, canary-grass.)

C. *Spikelets many-flowered. [Flowers in panicles; corolla unarmed.]*

PHRAGMI'TES. Calyx 5 to 7-flowered; the florets on villose pedicels, lower valve elongated, acuminate, involute, upper one somewhat conduplicate.

BRI'ZA. Spikelets heart-ovate, many-flowered; calyx chaffy, shorter than the 2-ranked florets; corolla ventricose, lower valve cordate, upper one orbicular, short. (quake-grass.)

PO'A. Spikelets oblong or linear, compressed, many-flowered; calyx shorter than the florets; corolla herbaceous, awnless, often scarious at the base; lower valve scarious at the margin. (spear-grass.)

SOR"GHUM. Florets in pairs, one perfect, with a 3-valved corolla and sessile, the other staminate or neutral, and pedicelled. (broom-corn.) Ex.

[*Flowers in panicles; corolla armed or mucronate.*]

DAC"TYLIS. Spikelets aggregated in unilateral heads, many-flowered, calyx shorter than the florets, with one large glume, keeled, pointed; corolla with the lower valve keeled, emarginate, mucronate, upper valve sub-conduplicate. (orchard-grass.)

KŒLE'RIA. Glume 2 to 4-flowered,shorter than the florets; lower palea mucronate, or with a short bristle a little below the tip.

FESTU'CA. Spikelets oblong, more or less terete, at length compressed, acute at each end; florets sub-terete; glumes unequal, shorter than the lower florets, keeled, acute; paleas lanceolate, the lower one acuminate or rounded at the extremity. (fescue-grass.)

BRO'MUS. Glumes 2, shorter than the florets; spikelets terete, 2-ranked, many-flowered; outer paleas bifid, awned below the tip, inner one pectinate-ciliate, sub-conduplicate. (chess, broom-grass.)

DANTHO'NIA. Glumes 2 to 5-flowered, longer than the florets, cuspidate; palea, bearded at the base, lower one 2-toothed, with a twisted awn between the teeth, upper one obtuse, entire; panicle spiked. (wild-oats.)

AI'RA. Glumes rarely 3-flowered, beardless, 2-valved, equal to the florets, or shorter; one of the florets peduncled; paleas 2, equal, enclosing the seed when ripe, outer one usually awned; panicle compound.

AVE'NA. Calyx 2-valved; 2, 3, or many-flowered; corolla with valves mostly bearded at the base, lower one torn, with a twisted awn on the back; glumes membranaceous, and somewhat follicle-like; seed coated. (oats.) Ex.

[*Flowers in spikes.*]

LO'LIUM. Calyx 1-leafed, permanent, many-flowered; florets in many-flowered, 2-rowed, simple, sessile spikelets on a rachis, lower valve of the corolla herbaceous-membranaceous, mucronate, or bristled at the tip. (darnel-grass.)

TRIT"ICUM. Calyx 2-valved, about 3-flowered; florets sessile on the teeth of the rachis, obtusish and pointed; glumes beardless, or interruptedly bearded. (wheat.) Ex.

SESLE'RIA. Glumes 2 to 5-flowered; paleas 2; stigmas sub-glandular; base of the spike bracted or involucred. (moor-grass.) S.

MELI'CA. Glumes unequal, 2 to 5-flowered, membranaceous, nearly as long as the florets, of which the upper are incomplete and abortive; scales truncate, fimbriate; seed loose, not furrowed; panicle simple or compound. (melic-grass.)

SECA'LE. Calyx 2-valved, 2 or 3-flowered; spikelets sessile on the teeth of the rachis, with the terminal floret abortive; calyx 2-valved; glumes subulate, opposite, shorter than the florets; corolla with the lower valve long awned. (rye.) Ex.

D. *Flowers polygamous.* [*In panicles.*]

PAN"ICUM. Calyx 2-valved, 2-flowered; the lower glume generally very small; the

lower floret abortive, 1 or 2-valved, the lower valve resembling the calyx, the upper one membranaceous; perfect floret with cartilaginous valves, unarmed. (cockfoot grass, panic-grass.)

HOL"CUS. Glumes 2-valved, 2-flowered, nearly equal; paleas 2; florets dissimilar, polygamous, one awned, the other awnless, without any imperfect one between them; panicles contracted.

[*In spikes.*]

HOR"DEUM. Spikelets 3 at each joint of the rachis, 1 or 2-flowered, all perfect, or the lateral ones abortive; glume lateral, subulate; perfect flower with a 2-valved corolla, lower valve ending in a bristle; seed coated. (barley.) Ex.

ANDROP"OGON. Spikelets in pairs, involucred, the lower one staminate or neutral, on a bearded pedicel; glumes and paleas often very minute, or wanting; upper spikelets sessile, 1-flowered, perfect; glumes sub-coriaceous; paleas 2, shorter than the glumes, membranaceous, lower valve generally awned. (beard-grass.)

PENNISE'TUM. Inflorescence, a compound cylindrical spike; spikelets 2-flowered, invested with an involucrum of 2 or more bristles; glumes 2, unequal, herbaceous; superior florets perfect; paleas cartilaginous. (bristled panic.)

ROTBOL'LIA. Rachis jointed, somewhat terete, often filiform; calyx ovate-lanceolate, flat, 1 or 2-valved, 1 or 2-flowered; florets alternate on a flexuous rachis. S.

MONOCK'RA. Flowers in 2 rows on one side of the rachis; glumes 3, many-flowered, valves awned below the tip. Perfect flowers have 2 paleas, unequal, outer valve awned below the tip. Neutral flowers have 2 awnless paleas. S.

ORDER III. TRIGYNIA.

MOLLU'GO. Calyx 5-leaved, colored within; corolla 0; capsule 3-celled, 3-valved. (carpet-weed.)

LECHE'A. Calyx 3-sepalled; petals 3, linear; stigmas 3, plumose; capsule 3-celled, 3-valved, with 3 inner valvelets; seed 1.

MON"TIA. Calyx 2 to 3-sepalled; petals 5, sub-connate at the base, 3 rather small; stamens on the claws of the petals, mostly before the 3 smallest; styles short, reflexed; capsules 1-celled, 3-valved, 3-seeded.

POLYCAR"PON. Calyx 5-sepalled; petals 5, very short, emarginate, permanent; capsule ovate, 1-celled, 3-valved. S.

PROSERPINA'CA. Calyx 3-parted, superior; corolla 0; fruit a hard nut, 3-sided, 3-seeded, crowned by the calyx. (mermaid-weed.)

CLASS IV. TETRANDRIA.
ORDER I. MONOGYNIA.

A. *Flowers superior.* [1-*petalled.*]

ALLIO'NIA. Common calyx oblong, simple, 3-flowered; proper calyx obsolete; corolla irregular; receptacle naked.

CEPHALAN"THUS. Inflorescence in a head; general calyx none, proper calyx superior, minute, angular, 4-cleft; corolla funnel-form; receptacle globular. hairy; capsule 2 to 4-celled; seed solitary, oblong. (button-bush.)

DIP''SACUS. Flowers in an ovate or roundish head; common calyx polyphillous, foliaceous, proper calyx monophyllous, superior; corolla tubular, 4-cleft; seed solitary; receptacle conic, chaffy. (teasel.)

GA'LIUM. Calyx 4-toothed; corolla flat, 4-cleft; fruit dry; seeds 2, roundish; leaves stellate. (bedstraw.)

RU'BIA. Calyx small, 4-toothed, superior; corolla bell-form; filaments shorter than the corolla; fruit pulpy. (madder.)

SCABIO'SA. Involucre many-leaved; calyx double, superior; corolla tubular; filaments longer than the limb of the corolla; seed naked, crowned by the calyx.

DIO'DIA. Calyx with the tube ovate, 2 or 4-toothed; corolla funnel-form, 4-lobed; capsule ovate, 2-celled.

HEDYO'TIS. Calyx 4-toothed; corolla tubular, bearded at the throat, 4-parted; capsule ovate, 2-celled, many-seeded.

HOUSTO'NIA. Calyx 4-toothed; corolla salver-form, 4-cleft; capsule 2-celled, many-seeded, opening transversely. (innocence.)

SPERMACO'CE. Corolla funnel-form, 4-cleft; capsule 2-celled; seeds 2, 2-toothed. S.

MITCHEL''LA. Calyx 4-toothed; corollas 2 on each germ, funnel-form; tube cylindric, limb 4-parted, spreading, villose on the inside; stamens scarcely exsert, stigma 4-cleft; berry double, 4-seeded. (partridge-berry.)

LINNÆ'A. Calyx double, that of the fruit 2-leaved, inferior, that of the stigma globose; berry 3-celled, dry. (twin-flower.)

COR''NUS. Calyx 4-toothed; drupe with a 2-celled nut. Some species have a 4-leaved involucrum. (dogwood, false box.)

LUDWIG'IA. Calyx 4-parted, persistent; corolla sometimes 3; capsule quadrangular, 4-celled, inferior, many-seeded.

ELEAG''NUS. Calyx 4-cleft. bell-form, colored within; anthers sub-sessile; style short; drupe 1-seeded, marked with 8 furrows. S.

POLYPRE'MUM. Calyx 4-parted; corolla 4 cleft, wheel-form, bearded at the orifice, stamens included; capsule compressed, 2-celled, many-seeded. S.

B. *Flowers inferior.*

PLANTA'GO. Calyx 4-cleft; corolla 4-cleft, reflex; capsule 2-celled, opening transversely; stamens exsert, very long. (plantain, ribwort.)

CENTAUREL''LA. Calyx 4-parted; corolla sub-campanulate, segments somewhat erect; stigma thick; capsule 1-celled, 2-valved, many seeded, surrounded by the persistent calyx and corolla.

EX''ACUM. Calyx deeply 4-parted; corolla 4-cleft, tube globose; capsule bisulcate, 2-celled, many-seeded.

SWER''TIA. Calyx flat, 4 or 5-parted; corolla 4 or 5-parted, tube short, border spreading, with 2 pores at the base of each; style short; stigma 2; capsule 1-celled, 2-valved.

FRASE'RA. Calyx deeply 4-parted; corolla 4-parted, spreading; segments oval, with a bearded orbicular gland in the mid-

dle of each; capsule compressed, partly emarginate, 1-celled; seeds few, large, imbricate, with a membranous margin.

AMMANN''IA. Calyx bell-tubular, plaited, 8-toothed; petals 4 or none, on the calyx; capsule 4-celled, many-seeded (sometimes 2-celled and 8-stamened). (tooth-cup.)

PTE'LEA. Calyx 4-parted; petals 4, spreading; stigmas 2; samara compressed, orbicular, 2-celled, 2-seeded.

LY'CIUM. Corolla tubular, having the throat closed by the beards of the filaments; stamens often 5; berry 2-celled, many seeded. (matrimony.)-

KRAME'RIA. Calyx none; corolla 4 or 5-petalled; nectary 4-leaved; anthers perforated; drupe prickly. S.

CAL''LICAR''PA. Calyx 4-cleft; corolla 4-cleft; berry 4-seeded. S.

RIVI'NA. Calyx 0; perianth 4-parted, persistent; stamens 4, 8, and 12; style 1; berry 1-seeded; seeds lentiform, scabrous.

ICTO'DES. General calyx a spatha; spadix simple, covered with flowers; perianth corolla-like, deeply 4-parted, permanent. becoming thick and spongy; style pyramidform, 4-sided; stigma simple. minute; berries globose, 2-seeded, enclosed in the spongy spadix receptacle. (skunk cabbage.)

SANGUISOR''BA. Calyx colored, 4-lobed, with 2 scales at the base; capsule 4-sided, 1 or 2 celled. (burnet saxifrage.)

ALCHEMIL''LA. Calyx 8-cleft, segments spreading, alternately smaller; style lateral from the base of the germ; fruit surrounded by the calyx. (ladies' mantle.)

ORDER II. DIGYNIA.

HAMAME'LIS. Involucrum 3-leaved; perianth 4-leaved or 4-cleft; petals 4, very long, linear; nut 2-celled, 2-horned. (witch hazel.) Flowers in autumn, and perfects its seed the following spring.

APHA'NES. Calyx 2-cleft, alternate, segments minute; petals none; stamens minute; styles 2; seeds 2, covered by the converging calyx, one of them often abortive; stamens 1, 2, and 4. S.

ORDER IV. TETRAGYNIA.

I'LEX. Calyx minute, 4, or 5-toothed; corolla 4-parted; style 0; stigmas 4; berry 4-celled, cells 1-seeded. (holly.)

NEMOPAN''THES. Flowers abortively dicious or polygamous; calyx small, scarcely conspicuous; petals 5, distinct, oblong-linear, deciduous; ovary hemispheric, style none; stigmas 3-4 sessile; berry sub-globose, 3-4 celled.

RUP''PIA. Calyx 0; corolla 0; seeds 4, pedicelled.

SAGI'NA. Calyx 4 or 5-parted; petals 4—5, or none; stamens 4 or 5; capsule 4 or 5-valved, 1-celled, many-seeded.

POTAMOGE'TON. Flowers on a spadix; calyx and corolla 0; nuts 4, 1-seeded, sessile.

CLASS V. PENTANDRIA.
ORDER I. MONOGYNIA.

A. *Flowers 1-petalled, inferior; seeds naked in the bottom of the calyx.*
ROUGH-LEAVED PLANTS.

MYOSO'TIS. Calyx half 5-cleft, or 5-cleft; corolla salver-form, curved, 5-cleft, vaulted, the lobes slightly emarginate, throat closed with 5 convex converging scales; seeds smooth or echinate. (scorpion-grass.)

CYNOGLOS"SUM. Calyx 5-parted; corolla short, funnel-form, vaulted, throat closed by 5 converging, convex processes; seeds depressed, affixed laterally to the style. (hound-tongue.)

LYCOP"SIS. Calyx 5-cleft; corolla funnel-form, throat closed with ovate, converging scales; seeds perforated at the base.

BORA'GO. Corolla wheel-form, the throat closed with rays. (borage.) Ex.

ANCHU'SA. Calyx 5-parted; corolla funnel-form, vaulted, throat closed; seeds marked at the base, and their surface generally veined. (bugloss.) Ex.

SYM"PHYTUM. Limb, or upper part of the corolla tubular, swelling, the throat closed with subulate rays. (comfrey.) Ex.

HELIOTRO'PIUM. Calyx tubular, 5-toothed; corolla salver-form, 5-cleft, with teeth or folds between the divisions, throat open; spikes recurved, involute. (turnsole.)

LITHOSPER"MUM. Calyx 5-parted, segments acute; corolla funnel-form, border 5-lobed, orifice naked; stamens within the tube of the corolla, stigma obtuse; seeds hard and shining. (gromwell.)

ONOSMO'DIUM. Calyx deeply 5-parted, segments linear; corolla somewhat tubular, campanulate, border ventricose, half 5-cleft, segments connivent, acute; anthers sessile, included; styles much exserted; nuts imperforate, shining.

ROCHEL"IA. Calyx 5-parted; corolla salver-form, throat closed with converging scales; nuts prickly, compressed, affixed laterally to the style.

BATSCH"IA. Calyx deep 5-parted; corolla salver-form, with a bearded ring within the straight tube. (false bugloss.)

PULMONA'RIA. Calyx prismatic, 5-cornered, 5-toothed; corolla funnel-form, border 5-lobed; tube cylindrical. (lung-wort.)

ECH"IUM. Calyx 5-parted; segments subulate, erect; corolla bell-form, with an unequal 5-lobed border, the lower segment acute and reflexed. (viper's bugloss.)

B. *Flowers* 1-*petalled, inferior; seeds covered.* [*Capsule* 1-*celled.*]

ANAGAL"LIS. Calyx 5-parted; corolla wheel-form, deeply 5-lobed; capsule opening transversely, globose, many seeded; stamens hairy. (scarlet pimpernel.)

LYSIMA'CHIA. Calyx 5-cleft; corolla wheel-form, 5-cleft; capsule 1-celled, globular, 5 or 10-valved, mucronate; stigma obtuse. (In some species the filaments are united at the base.) (loose strife.)

PRIMU'LA. Umbellets involucred; calyx tubular, 5-toothed; corolla salver-form, 5-lobed, tube cylindric, throat open, divisions of corolla emarginate; capsule 1-celled, with a 10-cleft mouth; stigma globular. (primrose, cowslip.)

DODECATH"EON. Calyx 5-cleft; corolla wheel-form, reflexed, 5-cleft; stamens in the tube; stigma obtuse; capsule oblong, opening at the apex. (false cowslip.)

SAMO'LUS. Calyx 5-cleft, semi-superior, persistent; corolla salver-form, 5-lobed; stamens 5, antheriferous, opposite the segments of the corolla, and, 5 scales alternate with them, sterile; capsule 1-celled, 5-toothed, many-seeded.

MENYAN"THES. Calyx 5-parted; corolla funnel-form, limb spreading, 5-lobed, hairy within; stigma capitate. (buck-bean.)

VILLAR"SIA. Calyx 5-parted, 5-lobed, limb spreading, ciliate on the margin; stigma 2-lobed; glands 5, alternating with the stamens; capsule 2-valved, many-seeded. (water-shield.)

HOTTO'NIA. Calyx 5-parted; corolla salver-form, 5-lobed; stamens in the tube of the corolla; capsule globose. (feather-leaf.)

SABBA'TIA. Calyx from 5 to 12-parted; corolla wheel-form, from 5 to 12-parted; stigmas 2, spiral; capsule 2-valved, many-seeded. (centaury.)

HYDROPHYL"LUM. Calyx 5-parted; corolla funnel-form, 5-cleft, with 5 longitudinal grooves inside. (water-leaf.)

ELLIS"IA. Calyx deeply 5-parted; corolla smaller, funnel-form, 5-cleft, naked within; stamens not exsert, filaments smooth, anthers roundish; stigma 2-cleft; capsule 2-celled, 2-valved, 2-seeded; seeds punctate; capsule on the spreading calyx. *S.*

NEMOPHI'LA. Calyx 10-parted, alternate lobes reflexed; corolla sub-campanulate, 5-lobed, the lobes emarginate, with nectariferous cavities at the base; stamens shorter than the corolla; capsule fleshy, 2-valved, 4-seeded.

ANDROSA'CE. Flowers in an involucred umbel; calyx 5-cleft, permanent; corolla salver-form, 5-lobed, tube ovate, orifices glandular; capsule 1-celled, globe-ovate, 5-valved, many-seeded. *S.*

[*Capsule* 2 *to* 5-*celled.*]

VERBAS"CUM. Calyx 9-parted; corolla wheel-form, 5-lobed, somewhat irregular; stamens declined, hairy; capsules 2-celled, 2-valved, valves inflexed when ripened, many-seeded. (mullein.)

NICOTIA'NA. Calyx urceolate, sub-tubular, 5-cleft; corolla funnel-form, 5-cleft, limb plaited; stigma notched, capitate; stamens inclined; capsules 2-celled, 2 to 4-valved. (tobacco.)

OPHIORHI'ZA. Calyx 5-cleft; corolla 5-cleft, funnel-form; germ 2-cleft; stigmas 2; capsule mitre-form, 2-lobed, 2-celled, many-seeded. *S.*

CONVOL"VULUS. Calyx 5-parted, with or without 2 bracts; corolla funnel-form, plaited; stigma 2-cleft or double; cells of the capsule, 2 or 3; each 1 or 2-seeded. (bind-weed.)

IPO'MÆA. Calyx 5-cleft, naked; corolla funnel or bell-form, with 5 folds; stigma globe-headed, papillose; capsule 2 or 3-celled, many-seeded. (cypress-vine, morning-glory.)

LISIAN"THUS. Corolla tubular, ventricose, segments recurved; calyx bell-form, keeled; stigma in a 2-lobed head; capsule 2-celled, 2-furrowed, 2-valved; seeds numerous, sub-imbricate, surrounded with a very small membranaceous margin. *S.*

PHACE′LIA. Calyx 5-parted; corolla somewhat bell-form, 5-cleft; stamens exserted; capsule 2-celled, 2-valved, 4-seeded.

PHLOX. Calyx prismatic, 5-cleft, segments converging; corolla salver-form, 5-lobed, with a tube somewhat curved; filaments unequal in length, attached to the inside of the tube of the corolla; stigma 3-cleft; cells 1-seeded; seeds oblong, concave. (lichnidia.)

DATU′RA. Calyx tubular, angled, caducous, with a permanent orbicular base; corolla funnel-form, plaited; capsule 4-valved, 2-celled, and each cell half-divided; generally thorny. (thorn-apple.)

AZA′LEA. Calyx 5-parted; corolla tubular, 5-cleft, somewhat oblique; stamens on the receptacle, declined; stigma declined, obtuse, usually ending with 5 short capillæ; capsule 5-celled, 5-valved, opening at the top. (wild honeysuckle.)

DIAPEN″SIA. Capsule 3-celled, 3-valved, many-seeded; corolla salver-form; calyx 5-parted, bracted at base; stigmas 3.

VIN″CA. Corolla salver-form, twisted, border 5-cleft, with oblique divisions; throat 5-angled; seed naked, oblong; follicle-like capsules 2, erect, terete, narrow. Ex.

PHYSA′LIS. Calyx 5-cleft; corolla wheel-form, 5-cleft; stamens converging; berry globose. (winter-cherry.)

SOLA′NUM. Calyx 5 to 10-parted, permanent; corolla bell or wheel-form, 5-lobed, plaited; anthers thickened, partly united, with two pores at the top; berry containing many seeds, 2 to 6-celled. (potato, nightshade, bitter-sweet.)

ATRO′PA. Corolla bell-form; stamens distant; berry globose, 2-celled, sitting on the calyx. (deadly nightshade.) Ex.

CAP″SICUM. Corolla wheel-form; berry juiceless, inflated; anthers converging; calyx angular. (red pepper.)

ANDROCE′RA. Calyx inflated, 5-cleft, caducous; corolla sub-ringent, wheel-form, 5-cleft; stamens unconnected, unequal, declined, one larger, and horn-formed; anthers with 2 terminal pores; style simple, declined; stigma not distinct; berry dry; seed without margins, rugose. *S.*

SPIGE′LIA. Calyx 5-parted; corolla funnel-form, border 5-cleft, equal; anthers convergent; capsule 2-celled, 4-valved, many-seeded. Ex.

HYOCYA′MUS. Calyx tubular, 5-cleft; corolla funnel-form, irregular, lobes obtuse; stigma capitate; capsule ovate, covered with a lid.

POLEMO′NIUM. Capsule bell-form, 6-cleft; corolla wheel-form, 5-parted; stamens inserted upon the 5 valves which close the orifice of the corolla.

C. *Flowers* 1-*petalled, superior.* [*Seeds in a capsule.*]

CAMPAN″ULA. Calyx mostly 5-cleft; corolla bell-form, closed at the bottom by valves bearing the flattened stamens; stigma 3 to 5-cleft; capsules 3 to 5-celled, opening by lateral pores. (bell-flower.)

LOBE′LIA. Calyx 5-cleft; corolla irregular, often irregularly slitted; anthers cohering and somewhat curved; stigma 2-lobed; capsule 2 or 3-celled. (cardinal-flower, wild tobacco.)

PINCKNE′YA. Capsule 2-celled, bearing the partition in the middle of the valves; calyx with 1 or two segments resembling bracts; filaments inserted at the base of the tube; seeds winged. *S.*

PSYCHO′TRIA. Tube of the calyx ovate, crowned; limb short, 5-lobed, 5-toothed or sub-entire; corolla funnel-form, short, 5-cleft; throat bearded; stigma 2-cleft; berry drupe-like. *S.*

DIERVIL″LA. Calyx oblong, 5-cleft, with 2 bracts; corolla 5-cleft, twice as long as the calyx, funnel-form; border 5-cleft, spreading; stigma capitate; capsule oblong, 4-celled, naked, many-seeded. (bush honeysuckle.)

CHIOCOC″CA. Calyx 5-toothed; corolla funnel-form, equal; berries compressed, twinned, 2-seeded; seeds oblong, compressed.

SYMPHO′RIA. Calyx minute, 4-toothed; corolla tubular, short, 4 or 5-lobed; stigma globose; berry crowned by the calyx; 4-celled, 4-seeded. (snow-berry.)

LONICE′RA. Calyx 5-toothed; corolla tubular, long, 5-cleft, unequal; stamens exsert; stigmas globose; berry 2 or 3-celled, distinct; seeds many. (trumpet honeysuckle.)

XYLOS″TEUM. Calyx 5-toothed, with 2 conate bracts; corolla tubular, border 5-parted, nearly equal; berries in pairs, united at their bases, or combined in one; 2-celled. (fly honeysuckle, twin-berry.)

TRIOS″TEUM. Calyx 5-cleft, with linear divisions; corolla tubular, 5-lobed, gibbous at the base; berry 3-celled, 3-seeded. (fever root.)

MIRAB″ILIS. Corolla funnel-form, coarctate below; calyx inferior; germ between the calyx and corolla; stigma globular. (four o'clock.) Ex.

D. *Flowers* 4 *to* 6-*petalled, inferior.* [*Seed in a capsule.*]

I′TEA. Calyx 5-cleft, bell-form; petals linear, reflexed, spreading, inserted into the calyx; stigma capitate, 2-lobed; capsule 2-celled, 2-valved, many-seeded.

IMPA′TIENS. Calyx 2-leaved, deciduous; corolla irregular, spurred; anthers cohering at the top; capsule 5-valved, bursting elastically when ripe. (ladies' slipper, jewel-weed.)

VI′OLA. Calyx 5-leaved, or deeply 5-cleft; corolla irregular, with a horn behind (sometimes the horn is wanting, or a mere prominence); anthers attached by a membranous tip, or slightly cohering; capsule 1-celled, 3-valved. (violet.)

IONI′DIUM. IONIA. (From Viola.) Calyx 5-sepalled, with bases extended; corolla somewhat 2-lipped, not spurred; stigma simple; capsule 1-celled, 3-valved, seeds on the middle valve.

SOLE′A. Sepals 5, not auricled at the base, decurrent into a pedicel; petals nearly equal, the lower a little larger, and somewhat gibbous at the base; filaments with short, broad claws at the base.

CLAYTO'NIA. Calyx 2-leaved, or 2-parted, the leaves valve-like; corolla 3-petalled, emarginate; stigma 3-cleft; capsule 1-celled, 3-valved, 3 to 5-seeded. (spring beauty.)

CEANO'THUS. Petals scale-like, vaulted; claws long, standing in the 5-cleft, cupform calyx; stigmas 3; berry or capsule dry, 3-grained, 3-celled, 3-seeded, 3-parted, opening on the inner side. (New-Jersey tea.)

EUON"YMUS. Calyx 4 to 6-lobed, flat, covered at the base by a peltate disk; petals 4 or 6, spreading, inserted into the disk; capsule 3 to 5-celled.

CELAS"TRUS. Calyx 5-lobed, flat; corolla spreading; capsule obtusely 3-angled, 3-celled, berry-like; valves bearing their partitions on their centres; cells 1 or 2-seeded; stamens standing around a glandular 5-toothed disk; style thick; stigma 3-cleft; seeds calyptred or arilled. (staff-tree, false bitter-sweet.)

PICKERIN"GA. Calyx small, 5-parted, inferior; petals 5; anthers sagittate; style simple; capsule 1-celled, sub-globose, many-seeded. S.

ZI'ZIPHUS. Calyx 5-cleft; petals 5, resembling scales, inserted into the glandular disk of the calyx; styles 2; drupe 2-celled, one cell often empty. S.

GOMPHRE'NA. Calyx 5-leaved, colored, exterior one 3-leaved; 2 leafets converging, keeled; petals 5, villose (or rather no corolla); nectary cylindric, 5-toothed; capsule opening transversely, 1-seeded; style semi-bifid. (bachelor's button.) Ex.

CYRIL"LA. Calyx minute, sub-terminate, 5-parted; petals 5, stellate; stigmas 2 or 3; capsule 2-celled, 2-seeded, not opening; seed ovate, attached to the summit of a columella by a filament.

GA'LAX. Calyx 5-parted, permanent; corolla twice as long as the calyx, 5-petalled, affixed to the base of the stamens; anther-bearing tube, 10-cleft, the 5 shorter segments bearing the anthers; stigma 3-lobed; capsule 3-celled, 3-valved; seeds many, affixed to the columella.

[*Seed in a berry.*]

VI'TIS. Calyx 5-toothed, minute; petals cohering at the tip, hood-like, withering; style 0; stigma obtuse, capitate; berry 5-seeded, globular, often dioecious; seeds sub-cordate. (grape-vine.)

AMPELOP"SIS. (Cissus.) Calyx minute, 4 or 5-toothed; petals 4 or 5, unconnected above, caducous; germ surrounded with a glandular disk; berry 4 or 5-seeded.

RHAM"NUS. Calyx urceolate, 4 or 5-cleft; petals alternating with the lobes of the calyx, or wanting; stigma 2 or 4-cleft; berry 2 or 4-celled.

E. *Flowers 5-petalled, superior.*

RI'BES. Calyx bell-form, 5-cleft (sometimes flat); corolla and stamens inserted on the calyx; style 2-cleft; berry many-seeded. (currant, gooseberry.)

HEDE'RA. Petals oblong: berry 5-seeded, surrounded by the calyx; style simple. (European ivy.) Ex.

F. *Flowers incomplete.*

HAMILTO'NIA. Polygamous; perianth turbinate, campanulate, 5-cleft; corolla 0; nectary with the disk 5-toothed; style 1; stigmas 2 or 3, germ immersed in the nectary; drupe1-seeded, enclosed in the adhering base of the calyx. (oil nut.)

SIPHONY'CHIA. Sepals 5, linear; nectaries 5; style filiform; utricle 1-seeded, enclosed in the calyx.

THE'SIUM. Perianth 4 or 5-cleft; stamens 4 or 5, villous externally; nut 1-seeded, covered by the persistent perianth. (false-toad flax.)

ANY'CHIA. Sepals 5, united at the base, slightly concave, sub-saccate at the apex; sub-mucronate on the back; petals none; stamens 2 to 5; inserted on the base of the sepals; styles very short, distinct, or united at the base; utricle enclosed in the calyx; 1-seeded.

GLAUX. Calyx campanulate, 5-lobed, colored; capsule globose, 5-valved, 5-seeded surrounded by the calyx.

CELO'SIA. Sepals 3, like a 5-petalled corolla; stamens united at the base by a plaited nectary; capsules opening horizontally; style 2 or 3-cleft. Ex.

ORDER II. DIGYNIA.

A. *Corolla 1-petalled, inferior.*

GENTIA'NA. Calyx 4 or 5-cleft; corolla with a tubular base, bell-form, without pores, 4 or 5-cleft; stigmas 2, sub-sessile; capsule 1-celled, oblong; columellas 2, longitudinal; stamens but 4, when the divisions of corolla are 4. (gentian.)

CUSCU"TA. Calyx 4 or 5-cleft; corolla 4 or 5-cleft, sub-campanulate, withering; capsule 2-celled, dividing transversely at the base; seeds binate. (dodder.)

GELSEMI'NUM. Calyx small, 5-leaved; corolla funnel-form; border spreading, 5-lobed, nearly equal; capsule 2-celled; seeds flat.

HYDRO'LEA. Calyx 5-petalled; corolla wheel-form, or bell-form; anthers cordate; style long, diverging; stigma peltate; capsule 2-celled. S.

DICHON"DRA. Calyx 5-parted, with spatulate segments; corolla short, bell-form, 5-parted; stigma peltate. capitate; capsule compressed, 2-celled, 4-seeded. S.

EVOL"VULUS. Calyx 5-parted; corolla bell-form; styles 2, 2-cleft; stigma simple.

SWER"TIA. Corolla rotate, with 2 pores at the base of each segment.

B. *Corolla 5-petalled.*

PA'NAX. Polygamous, umbelled; involucrum, many-leaved; calyx 5-toothed, in the perfect flower, superior; berry cordate, 2 or 3-seeded; calyx in the staminate flower entire. (ginseng.)

HEUCHE'RA. Calyx inferior, 5-cleft; corolla on the calyx; petals small; capsule 2-beaked, 2-celled, many-seeded. (alum-root.)

C. *Corolla wanting.*

SALSO'LA. Perianth inferior, 5-cleft, persistent, enveloping the fruit with its base, and crowning it with its broad, scarious limb. (salt-wort.)

ATRI'PLEX. Polygamous; calyx 5-leaved, 5-parted, inferior; style 2-parted; seed 1; in the pistillate flowers the calyx is 2-parted. (orach.)

PLANE'RA. Calyx membranous, bell-form, 4 to 5-cleft; corolla 0; stigmas 2; capsule globose, membranous, 1-celled, 1-seeded; stamens 4 to 6; polygamous. *S.*

KO'CHIA. Calyx inferior, bell-form, 5-cleft, forming a permanent band around the fruit, somewhat resembling 5 petals; corolla 0; style short; stigmas 2 or 3, long, simple. *S.*

CEL"TIS. Perfect or polygamous; perianth inferior, 5-lobed; drupe globose, 1-seeded. (nettle-tree.)

CHENOPO'DIUM. Calyx 5-parted, obtusely 5-angled; inferior, style deeply cleft; seed 1, lens-like, horizontal, invested by the calyx. (pigweed, oak of Jerusalem.)

UL"MUS. Calyx bell-form, withering; border 4 or 5-cleft; seed 1, enclosed in a flat, membranaceous samara; stamens vary from 4 to 8. (elm.)

BE'TA. Calyx 5-leaved; seed kidney-form within the fleshy substance of the base of the calyx. (beet.) Ex.

D. *Plants umbelliferous ; flowers 5-petalled, superior ; seeds 2.*

ERYN"GIUM. Fruit ovate, with bristles; petals oblong, equal, inflected; flowers aggregate, forming a head.

ŒNAN"THE. Carpels 3-ribbed (rarely 5-ribbed); styles permanent; germ oblong-ovate, corticate, solid; apex denticulate; perianth slenderly 5-toothed; petals of the disk and florets of the umbel, cordate-inflexed, sub-equal; those of the ray florets large and deformed; general involucre mostly wanting. (waterdrop wort.)

SANIC"ULA. Seeds with hooked prickles, oblong, solid; umbels nearly simple, capitate; flowers polygamous; involucre few-flowered; calyx 3-parted, permanent.

DAU'CUS. Seeds striate on their joining sides, outer sides convex, having hispid ribs; involucrum pinnatifid; flowers sub-radiate, abortive in the disk. (carrot.)

URASPER"MUM. [Osmorhiza.]* Seeds sublinear, solid, acute-angled, not striate; ribs 5-acute; angles a little furrowed, hispid; the joining-sides furrowed, and attached to a 2-cleft columella-like receptacle; style subulate, permanent, rendering the seed caudate; involucrum none, or few-leaved; fruit, stiped, oblanceolate, polished, part of it hispid. (sweet cicely.)

[*Seeds with wing-like ribs.*]

HERAC"LEUM. Seeds with winged margins, and 3 ribs on the back, obtuse, 3 grooves on their outer sides; germ oval, emarginate at the apex; petals emarginate, inflexed; general involucre 0; partial involucre 3 to 7-leaved; flowers somewhat radiated. (cow parsley.)

---

* Where two or more generic names are given, the pupil will understand that those within the brackets are synonymes of the other name

CNI'DIUM. Involucre 1-leaved or 0; fruit ovate, solid; ribs 5, acute, somewhat winged; intervals sulcate, striate.

FER"ULA. Calyx minute entire; petals oblong, sub-equal; fruit sub-oval, compressed, flat, wing-margined; carpels with 3 dorsal lines; intervals and joining sides striate; universal involucre caducous; partial ones many-leaved.

ANGEL"ICA. Seeds with 3 ribs on their backs, and winged margins; intervals between the ribs grooved; germ oval, corticate; general involucrum none. (angelica.)

PASTINA'CA. Seeds emarginate at the apex, somewhat winged; ribs 3, besides the wings; intervals striate; joining-sides 2-striate; germ oval, compressed; perianth calyx entire; petals entire, incurved, sub-equal; involucrum none. (parsnip.)

TRE"POCAR"PUS. Calyx 5-toothed; te subulate; petals obcordate, with inflexe margins; fruit pyramid-angled; commissure thick furrowed in the middle, filleted within.

PEUCEDA'NUM. Calyx minute, 5-toothed; petals oblong, incurved, equal; fruit oval, compressed, surrounded with a winged margin, having 5 striæ on each carpel, and elevated intervals, joining-sides flat. (sulphur-wort.)

ARCHEMO'RA. Calyx with the margin 5-toothed; petals obcordate, with inflexed divisions; fruit compressed at the back; sides dilate into a membranaceous margin, rather broader than the seeds; leaves pinnatifid; no general involucre, but many-leaved partial ones; flowers white.

THAS"PIUM. Calyx 5-toothed at the margin; petals oval, tapering into long inflexed apexes; fruit not contracted at the side, sub-oval; wings sub-equal, filleted at the joining edges; carpel terete; no general involucre, partial ones 3-leaved.

SES"ELI. Umbel globose; margin of the calyx 5-toothed; petals obovate; fruit oblong or oval, crowned with a reflexed style.

SELI'NUM. Fruit oval-oblong, compressed, flat, striate in the middle; involucrum reflected; petals cordate, equal; calyx entire.

ANE'THUM. Seeds flat or convex, 5-ribbed; germ lenticular, compressed; calyx and petals entire; involucrums none. (fennel, dill.) Ex.

[*Seeds with 3 ribs nearly equal.*]

HYDROCOT"YLE. Umbel simple; fruit compressed, sub-rotund. (marsh pennywort.)

CRITH"MUM. Fruit elliptical, ribbed, crowned; petals elliptical, acute, incurved, equal; styles short or thick, with swelled bases. Ex.

CA'RUM. Seeds oblong-ovate, striate; petals carinate, emarginate, inflexed; involucrum about 1-leaved. (caraway.) Ex.

[*Seeds with 5 ribs nearly equal.*]

CO'NIUM. Seeds 5-ribbed; ribs at first crenate, with flat intervals between them; germ ovate, gibbous; perianth entire; petals unequal, cordate, inflexed; general involucrum about 3 to 5-leaved; partial ones mostly 3-leaved, unilateral. (poison hemlock.)

CICU'TA. Seeds gibbous-convex, ribs 5,

obtuse, converging, with intervening tuber-culate-grooves and prominences; joining-sides flat; germ sub-globose, compressed laterally; calyx obsolete, 5-toothed; petals cordate. inflexed; partial involucrums 5 or 6-leaved, or wanting. (water hemlock,)

SI'UM. Fruit somewhat prismatic, with 5 obtuse ribs; perianth minute: petals cordate, inflexed; involucres many-leaved, entire. (water-parsnip.)

CRYPTOTÆ'NIA. Calyx with the margin obsolete; petals obovate, sub-entire, nar-rowing into an inflexed point; fruit con-tracted at the side, linear-oblong, with a short slender foot-stem, and crowned with a straight style; and having many edging fillets, concealed, or. nearly so; fruit-cover-ing bifid at the apex; plant glabrous; leaves 3-cleft, segments curve-toothed.

ZI'ZIA. Margin of the calyx obsolete or very short, 5-toothed; petals oval, tapering into a long point: fruit contracted from the side, roundish, or oval; carpels terete-con-vex, a little flattish before; flowers yellow, rarely white or dark-purple.

BUPLEU'RUM. Calyx none; flowers reg-ular; petals 5; styles very short, spreading; stigmas minute, simple; fruit egg-shaped, obtuse.

SI'SON. Fruit ovate, striate; involucrams generally 4-leaved.

A'PIUM. Seeds convex externally; ribs 5, small, a little prominent; germ sub-globose; perianth entire; petals equal, roundish, inflexed at the apex; involucram 1 to 3-leaved or wanting. (celery, parsley.) Ex.

CORIAN'DRUM. Seeds sub-spherical; germ spherical; perianth 5-toothed; petals cordate, inflexed, outer ones largest; invo-lucrum 1-leaved or wanting. (coriander.) Ex.

LIGUS'TICUM. Germ oblong, with 5 acute ribs; intervals sulcate; universal and par-tial involucres. (lovage.)

ÆTHU'SA. Fruit ovate, sub-solid, having bark; ribs acute and turgid; intervals acute-angled; joining-sides flat, striate; in-volucrum 1-side, or none. (fools' parsley.)

ORDER III. TRIGYNIA.

A. *Flowers superior.*

VIBUR'NUM. Calyx 5-parted or 5-toothed, small; corolla bell form, 5-cleft, with spread-ing or reflexed lobes; stigmas almost ses-sile; berry or drupe 1-seeded. (snow-ball, sheep-berry, high cranberry.)

SAMBU'CUS. Calyx 5-parted or 5-cleft, small; corolla sub-urceolated, 5-cleft; stig-ma minute, sessile; berry globose, 1-celled, 3-seeded.

B. *Flowers inferior.*

RHUS. Calyx 5-parted; petals 5; berry 1-seeded, small, sub-globular. (sumach, poi-son-ivy.)

STAPHYLE'A. Calyx 5-parted, colored; petals 5 on the margin of a glandular 5-an-gled disk; capsules inflated, connate; nuts globular and cicatrized, 1 or 2 remaining in each capsule, though several appear as ru-diments while in bloom. (bladder-nut.)

LEPU'ROPE'TALON. Calyx 5-parted; pet-

als 5, resembling scales, inserted into the calyx; capsule free near the summit, 1-celled, 1-valved. S.

ORDER IV. TETRAGYNIA.

PARNAS'SIA. Calyx inferior, permanent, 5-parted; corolla 5-petalled; nectaries 5-fringed, with stamen-like divisions; stig-mas sessile; capsule 4-valved, 1 or 2-celled; seed membranaceous-margined. (parnas-sus grass, flowering plantain.)

ORDER V. PENTAGYNIA.

ARA'LIA. Umbellets involucred; peri-anth 5-toothed, superior; petals 5; stigmas sessile, sub-globose; berry crowned, 5-celled; cells 1-seeded. (spikenard, wild-sarsaparilla.)

LI'NUM. Calyx 5-leaved or 5-parted, per-manent; corolla 5-petalled, inferior, with claws; capsule 5 or 10-valved, 10-celled; seeds solitary, ovate, compressed; filaments spreading or united at the base. (flax.)

SIBBAL'DIA. Calyx 10-cleft, with the al-ternating segments narrower; petals 5, in-serted in the calyx; styles attached to the germ laterally; nuts 5, in the bottom of the calyx.

STAT'ICE. Calyx funnel-form, plaited, scarious; petals 5; stamens inserted on the petals; styles 5; flowers in spikes or heads; capsule 1-seeded, without valves.

ORDER VI. HEXAGYNIA.

DROS'ERA. Calyx inferior, deeply 5 cleft, permanent; petals 5, marescent; anthers adnate; styles 6, or 1 deeply divided; cap-sule round, 1 or 3-celled, many-seeded; valves equalling the number of stigmas. (sundew.) The leaves of all the species are beset with glandular hairs resembling dew.

ORDER XII. POLYGYNIA.

XANTHORHI'ZA. Calyx 0; petals 5; nec-taries 5, pedicelled; capsule half 2-valved, 1-seeded, about 5 in number. (yellow-root.)

MYOSU'RUS. Calyx inferior, of few, lan-ceolate, colored sepals; petals 5, with tu-bular, honey-bearing claws; filaments as long as the calyx; calyx spurred at the base.

CLASS VI. HEXANDRIA.

ORDER I. MONOGYNIA.

A. *Flowers complete, having a calyx and corolla.*

TRADESCAN'TIA. Calyx inferior, 3-leav-ed; corolla 3-petalled; filaments with joint-ed beards; capsules 3-celled, many-seeded. (spider-wort.)

LEON'TICE. CAULOPHYLLUM. Calyx of 3-6 sepals, naked externally; petals 6, un-guiculate, with a scale on each claw; ova-ry superior, ventricose obovoid, obliquely beaked; seeds 2-4, globose, inserted in the bottom of the capsule, which is ruptured at an early period.

BER'BERIS. Calyx inferior, 6-leaved; petals 6, with 2 glands at the claw of each; style 0; berry 1-celled, 2 or 4-seeded; stigma umbilicate; stamens spring up on being irritated. (barberry.)

CLEO'ME. Calyx 4-leaved, inferior; petals 4, ascending to one side; glands 3, one at each sinuate division of the calyx, except the lowest; stamens from 6 to 20, or more; capsule stipid or sessile, silique-like, often 1-celled, 2-valved. Does not belong to the class Tetradynamia by its natural or artificial characters. It has *no silique*, though the capsule appears, like a silique, until it is opened. (false mustard.)

GYNANDROP"SIS. Sepals 4, distinct, spreading; petals 4; receptacle linear, elongated; stamens with the lower part of the filaments, adnate to the receptacle its whole length; pod linear-oblong, raised on a long stipe, which rises from the top of the receptacle.

ISOME'RIS. Sepals 4, united below, somewhat spreading, marescent; petals 4, oblong, sessile, regular; receptacle fleshy, sub-hemispherical, produced into a small dilated appendage on the upper side; stamens equal, much exserted; capsule large, obovate, elliptical, coriaceous, indehiscent, stipitate, crowned with the very short subulate style; seeds several, very large, smooth.

LEON"TICE. Calyx of 6 sepals, caducous; petals 6, having a scale at the base; nectaries 5, inserted upon the claws of petals; anthers adnate to the filaments, 2-celled. (pappoose root.)

PRI'NOS. Calyx minute, 6-cleft; corolla sub-rotate, monopetalous, 6-parted; berry 6-seeded. (winter-berry.)

FLŒR'KIA. Calyx 3-leaved; petals 3, shorter than the sepals; seeds 2 or 3, superior.

TILLAND"SIA. Calyx 3-cleft, sub-convolute, permanent; corolla 3-cleft, bell-form, somewhat tubular; capsule 1 to 3-celled; seed comose.

DIPHYL"LIA. Sepals 3, caducous; petals 6, opposite the divisions of the calyx; anthers adhering to the filaments; berry 1-celled; seeds 2 or 3, roundish.

B. *Flowers issuing from a spatha.*

AMARYL"LIS. Corolla superior, 6-petalled, unequal; filaments unequal, declined, inserted in the throat of the tube. (atamask lily.) S.

AL"LIUM. Spatha many-flowered; corolla inferior, 6-parted, very deeply divided; divisions ovate, spreading; capsule 3-celled, 3-valved, many-seeded; flowers in close umbels or heads. (leek, garlic, onion, cives.)

HYPOX"IS. Glume-like spatha 2-valved; corolla superior, 6-parted, permanent; capsule elongated, narrow at the base, 3-celled, many-seeded; seed roundish. (star-grass.)

PONTEDE'RIA. Corolla inferior, 6-cleft, 2-lipped, with 3 longitudinal perforations below; capsule with utricles, fleshy, 3-celled, many-seeded; 3 stamens, commonly inserted on the tip, and 3 on the tube of the corolla. (pickerel weed.)

PANCRA'TIUM. Flower funnel-shaped, with a long tube; nectary 12-cleft, bearing the stamens. S.

BRODLÆ'A. Corolla inferior, bell-form, 6-parted; filaments inserted in the throat of the corolla; germ pedicelled; capsule 3-celled, many-seeded. S.

CRI'NUM. Corolla superior, funnel-form, half 6-cleft, tube filiform; border spreading, recurved; segments subulate, channelled; filaments inserted on the throat of the corolla, separate. S.

GALAN"THUS. Petals 3, concave, superior; nectaries (or inner petals) 3, small, emarginate; stigma simple. (snowdrop.) Ex.

NARCIS"SUS. Corolla bell-form, 1-leafed spreading, 6-parted, or 6-petalled, equal, superior; nectary bell-form, 1-leafed, enclosing the stamens. (jonquil, daffodil.) Ex.

C. *Flowers with a single, corolla-like, perianth.*

ALE'TRIS. Corolla tubular-ovate, 6-cleft, wrinkled; stamens inserted upon the orifice; style 3-sided, 3-parted; calyx half superior, 3-celled, many-seeded. (false aloe.)

LOPHIO'LA. Corolla 6-cleft, persistent, woolly, bearded inside; anthers erect; filaments naked; stigma simple; capsule opening at the summit.

AGA'VE. Corolla superior, tubular, funnel-form, 6-parted; stamens longer than the corolla, erect; capsule triangular, many-seeded.

PHALAN"GIUM. Corolla inferior, 6-petalled, spreading; filaments smooth; capsule ovate; seeds angular.

NARTHE'CIUM. Corolla 6-parted, colored; filaments hairy; capsule prismatic, 3-celled; seed appendaged at each end. (false asphodel.)

STREPTO'PUS. Corolla 6-cleft, cylindrical, segments with a nectariferous pore at the base; anthers longer than the filaments; stigma very short; berry sub-globose, smooth, 3-celled; seeds few.

HEMEROCAL"LIS. Corolla 6-parted, tubular, funnel-form; stamens declined; stigma small, simple, somewhat villose. (day-lily.) Ex.

ORNITHOG"ALUM. Corolla 6-petalled, inferior, erect, permanent; spreading above the middle; filaments dilated, or subulate at the base; capsule roundish, angled, 3-celled; seed roundish, naked. (star of Bethlehem.)

LIL'IUM. Corolla liliaceous, inferior, 6-petalled; petals with a longitudinal line from the middle to the base; stamens shorter than the style; stigma undivided; capsule sub-triangular, with the valves connected by hairs crossing as in a sieve. (lily.)

CLINTO'NIA. Perianth 6-parted, campanulate; stamens 6, inserted at the base; style compressed; stigma 2-lobed, compressed; berry 2-celled, cells many-seeded.

ERYTHRO'NIUM. Corolla liliaceous, inferior, 6-petalled; petals reflexed, having 2 pores and 2 tubercle-form nectaries at the base of the 3 inner, alternate petals; cap-ule somewhat stiped; seeds ovate. (dog-tooth violet, or adder-tongue.)

UVULA'RIA. Corolla inferior, 6-petalled. with a nectariferous hollow at the base of each petal; filaments very short, growing to the anthers; stigmas reflex; capsule 3-cornered, 3-celled, 3-valved, with transverse partitions; seeds many, sub-globose, arilled at the hilum. (bell-wort.)

CONVALLA'RIA. [SMILACI'NA, POLYGON-A'TUM, DRACÆ'NA.] Corolla inferior, 6-cleft; berry globose, 3-celled, spotted before ripening. (Solomon's seal.)

ASPAR''AGUS. Corolla inferior, 6-parted, erect, the three inner divisions reflexed at the apex; style very short; stigmas 3; berry 3-celled, cells 2-seeded. (asparagus.)

POLYAN''THES. Corolla funnel-form, incurved; filaments inserted in the throat; stigma 3-cleft; germ within the bottom of the corolla. (tuberose.) Ex.

HYACIN''THUS. Corolla roundish or bell-form, equal, 6-cleft; 3 nectariferous pores at the top of the germ; stamens inserted in the middle of the corolla; cells somewhat 2-seeded. (hyacinth.) Ex.

TU'LIPA. Corolla 6-petalled, liliaceous; style 0; stigma thick; capsule oblong, 3-sided. (tulip.) Ex.

ASPHODE'LUS. Corolla 6-parted, spreading; nectary covering the germ with 5-valves. (king's-spear, or asphodel.) Ex.

YUC''CA. Corolla inferior, bell-form; style 0; capsule oblong, 3-celled, opening at the summit; seeds flat. (Adam's needle.) S.

FRITILLA'RIA.Corolla inferior, 6-petalled, bell-form, with a nectariferous cavity above the claw of each; stamens of the length of the corolla; seeds flat. (crown imperial.) S.

SCIL''LA. Corolla 6-petalled, spreading, caducous; filament thread-form, attached to the base of the petals. (squilla.) S.

C. *Flowers with a single, calyx-like peri-
    anth, without a spatha.*

A'CORUS. Receptacle spadix-like, cylindric, covered with florets; calyx 6-parted, naked; corolla 0 (or calyx 0, corolla 6-parted or 6-petalled); style 0; stigma small; capsule 3-celled, 3-seeded. (sweet-flag.)

JUN''CUS. Glume or outer calyx 2-valved; perianth inferior, 6-leaved, glume-like, permanent; stigmas 3; capsule 1 or 2-celled, 3-valved, many-seeded; seeds attached to a partition in the middle of each valve. (rush-grass, bulrush.)

ORON''TIUM. Spadix cylindrical, crowded with flowers; perianth 6-petalled, naked; stigma 0; capsule bladder-like, 1-seeded. (flowering arum.)

LUZU'LA. Perianth 6-parted, glumaceous; capsule superior, 3-celled, 3-valved; cells 1-seeded. (false rush-grass.)

ORDER II. DIGYNIA.

ORY'ZA. Calyx-glume 2-valved, 1-flowered; corolla 2-valved, adhering to the seed. (rice.) Ex.

OXY'RIA. Perianth simple, 4-sepalled, 2 inner ones largest; corolla none; nut 3-sided, with a broad membranaceous margin; stamens 2 to 6; stigma large, plumose.

NEC''TRIS. Calyx inferior, 6-sepalled; corolla none; carpels 3, not opening. Calyx considered as 6-parted, 3 inner divisions petal-like, obtuse, and smaller; capsule bladder-like, 1 or 2-celled, 1 or 2-seeded.

ORDER III. TRIGYNIA.

VERA'TRUM. Polygamous; calyx 0; corolla 6-parted, expanding; segments sessile, without glands; stamens inserted upon the receptacle; capsules 3, united, many-seeded.

TRIL''LIUM. Calyx 3-leaved, inferior, spreading; corolla 3-petalled; styles 0; stigmas 3; berry 3-celled, many-seeded. (false wake-robin.)

RU'MEX. Calyx 3-leaved; petals 3, valve-like, converging (or calyx 6-sepalled, and corolla 0); stigmas many-cleft; seed 1, naked, 3-sided. (dock, field-sorrel.)

MELAN''THIUM. Polygamous; perianth rotate, 6-parted; segments with 2 glands at the base of each; claws staminiferous; capsule sub-ovate, 3-celled; apex 3-cleft; seeds many, membranaceous, winged. (black-flower.)

ZIGADE'NUS. Perianth 6-leaved, colored, spreading, with 2 glands above the narrow base of each leaf; stamens inserted in contact with the germ; capsule 3-celled, many-seeded.

HELO'NIAS. Perianth 6-parted, spreading, without glands; styles 3, distinct; capsule 3-celled, 3-horned; cells few-seeded.

XEROPHYL''LUM. Perianth sub-rotate, deeply 6-parted; stigmas 3, revolute; capsule sub-globose, 3-celled; cells 2-seeded, opening at the top.

TOFIEL''DIA. Perianth 6-parted, with a small 3-parted involucre; capsule 3 to 6-celled; cells many-seeded.

SCHEUCHZE'RIA. Perianth 6-parted; anthers linear; stigmas sessile, lateral; capsule inflated, 2-valved, 1 to 2-seeded.

TRIGLO'CHIN. Perianth of 6 deciduous leaves, 3 inserted above the rest; stamens very short; capsules 3 to 6, united by a longitudinal receptacle. (arrow-grass.)

MEDEO'LA. GYRO'MIA. Perianth 6-parted, revolute; stigmas 3-divaricate, united at the base; berry 3-celled; cells 3 to 6-seeded. (Indian cucumber.)

SA'BAL. Flowers perfect; spathas partial; filaments free, thickened at the base; 1 to 3-seeded, seeds bony. (false fan-palm.) S.

CHAMÆ'ROPS. Flowers polygamous; spatha compressed; spadix branched; perianth 3-parted; corolla 3-petalled; filaments partly united; drupe 3-celled, 2 of them often empty. The staminate flowers grow on distinct plants. (fan-palm.) S.

CAL''OCHOR''TUS. Corolla 6-parted, spreading, 3 inner segments larger, with the upper side woolly; filaments short, inserted on the base of the petals; anthers arrow-form; stigmas reflexed; capsule 3-celled. S.

NOLI'NA. Corolla 6-parted, spreading, segments nearly equal; styles short; stigmas recurved; capsule 3-sided, 3-celled; seed 1, convex. S.

ORDER XIII. POLYGYNIA.

ALIS″MA. Calyx 3-leaved; petals 3; capsules numerous, 1-seeded, not opening.

CLASS VII. HEPTANDRIA.

ORDER I. MONOGYNIA.

TRIENTA′LIS. Calyx 7-leaved; corolla 7-parted, equal, flat; berry juiceless, 1-celled, many-seeded; number of stamens variable. (chick-wintergreen.)

Æ′SCULUS. Calyx inflated, 4 or 5-toothed; corolla 4 or 5-petalled, inserted on the calyx, unequal, pubescent; capsule 3-celled; seeds large, solitary, chestnut-form. (horse-chestnut.) *S.*

ORDER III. TRIGYNIA.

FRANKE′NIA. Sepals 5, united in a furrowed tube, persistent, equal; petals 5, unguiculate, with appendages at the base of the limb; capsule 1-celled, many-seeded. *S.*

ORDER IV. TETRAGYNIA.

SAURU′RUS. Calyx in an ament or spike, with 1-flowered scales; corolla 0; anthers adnate to the filaments; germs 4; berries or capsules 4, 1-seeded; stamens 6, 7, 8, or more. (lizard-tail.)

CLASS VIII. OCTANDRIA.

ORDER I. MONOGYNIA.

A. *Flowers superior.*

RHEX″IA. Calyx ventricose-ovate at the base, limb 4-cleft; petals 4, ovate; capsule included in the calyx, 4-celled; seeds numerous, cochleate. (deer-grass.)

GAU′RA. Calyx 4-cleft, tubular; corolla 4-petalled, ascending toward the upper side; nut 4-cornered, seeded. (Virginian loose-strife.)

ŒNO′THERA. Calyx 4-cleft, tubular, caducous, divisions deflected; petals 4, inserted on the calyx; stigma 4-cleft; capsule 4-celled, 4-valved; seeds not feathered, affixed to a central 4-sided columella. (scabish, or evening-primrose.)

EPILO′BIUM. Calyx 4-cleft, tubular; corolla 1-petalled; capsule oblong and of great length; seeds feathered. (willow-herb.)

OXYCOC″CUS. Calyx superior, 4-toothed; corolla 4-parted, the divisions sub-linear, revolute; filaments converging; anthers tubular, 2-parted, berry many-seeded. (cranberry.)

FU′SCHSIA. Calyx funnel-form, colored, superior, caducous; petals (or nectaries) 4, sitting in the throat of the calyx, alternating with its divisions; stigma 4-sided, capitate; berry oblong, 4-celled; seeds numerous. (ear-drop.)

CLARK″IA. Calyx 4-cleft, tubular; corolla 4-petalled, 3-lobed, cruciform; petals with claws; stamens 4; stigma petal-like, 4-lobed; capsule 4-celled. (beautiful clarkia, false tree-primrose.) *S.*

B. *Flowers inferior.*

MENZIE′SIA. Calyx deeply 5-cleft; corolla 1-petalled, ovate, 4 to 5-cleft; stamens inserted into the receptacle; capsule 4-celled; seeds numerous, oblong.

DIR″CA. Perianth colored, campanulate, border obsolete; stamens unequal, exserted; berry 1-seeded. (leather-wood.)

JEFFERSO′NIA. Calyx 4-sepalled; petals 8; capsule obovate, opening below the top. (twin-leaf.)

DODO′NÆA. Sepals 4, deciduous; petals 0; style 1, filiform.

A′CER. Polygamous (sometimes hexandrous); calyx 5-cleft; corolla 4 or 5-petalled, or wanting; samaras 2, united at the base, 1-seeded, often 1 rudiment of a seed. (maple.)

ERI′CA. Calyx 4 leaved, permanent; corolla 4-cleft, permanent; filaments inserted on the receptacle; anthers bifid; capsules membranaceous, 4 to 8-celled, the partitions form the margins of the valves; seeds many in each cell. (heath.) Ex.

DAPH″NE. Calyx 0; corolla 4-cleft, withering, including the stamens; drupe 1-seeded. (mezereon.) Ex. *Wit.*

TROPÆO′LUM. Calyx 4 or 5-cleft, colored, spurred; petals 4 or 5, unequal; nuts leathery, sulcate. (nasturtion.) Ex.

ELLIOT″TIA. Calyx 4-toothed, inferior; corolla deeply 4-parted; stigma capitate. (false-spiked alder.) *S.*

AMY′RIS. Flowers perfect; calyx 4-toothed; petals wedge-form, longer than the stamens; germ 1-celled; stigma sessile.

ORDER II. DIGYNIA.

CHRYSOSPLE′NIUM. Calyx superior, 4 or 5-cleft, colored; corolla 0; capsule 2-beaked, 1-celled, many-seeded. (golden saxifrage, water-carpet.)

ORDER III. TRIGYNIA.

POL″YGO′NUM. Calyx inferior, 5-parted, colored; corolla 0; seed 1, angular, covered with the calyx; stamens and pistils vary in number. The calyx in some species might be taken for a corolla. (knot-grass, water-pepper, buck-wheat, heart's-ease.)

BRUNICH″IA. Calyx tubular, inflated, 5-cleft, angular at the base; corolla 0; styles short; stigma 2-cleft; seed 1; stamens 8 to 10.

SAPIN″DUS. Calyx of 4 sepals; corolla of 4 petals; capsule fleshy, ventricose. (soap-berry.)

CARDIOSPER″MUM. Calyx 4-sepalled; petals 4; nectary 4-leaved, unequal; capsule membranaceous, inflated, 3-lobed, 3-celled; seeds round, marked at the hilum with a heart-formed spot. (heart-seed.) *S.*

ORDER IV. TETRAGYNIA.

ADOX″A. Calyx inferior, 2 or 3-cleft; corolla 4 or 5-cleft; berry 1-celled, 4 or 5-seeded, attached to the calyx; flowers lateral; stamens 8 to 10. *S.*

BRYOPHYL″LUM. Sepals 4; petals 4, connate into a cylinder; seeds many.

CLASS IX. ENNEANDRIA.

ORDER I. MONOGYNIA.

LAU′RUS. Calyx 4 to 6-parted; corolla 0; nectaries 3, each a 2-bristled or 2-lobed gland, surrounding the germ; drupe 1-seeded; stamens vary from 3 to 14, but they are generally in two series of 6 each, with 3 of the inner series barren, often dioecious. The calyx may be taken for a corolla. (sassafras, spice-bush.)

ORDER II. DIGYNIA.

ERIGO′NUM. Perianth bell-form, 5-cleft; seed triangular, covered by the calyx; flowers involucred. *S.*

PLEE′A. Calyx none; corolla 6-parted, spreading; segments linear, acute; capsule roundish, 3-angled, 3-celled, partitions obsolete; seeds numerous, minute, sub-terete and caudate, attached to the margin of the valves.

ORDER III. TRIGYNIA.

RHE′UM. Perianth 6-cleft, permanent; seed 1 to 3-sided. (rhubarb.)

CLASS X. DECANDRIA.

ORDER I. MONOGYNIA.

A. *Flowers polypetalous, irregular (mostly papilionaceous).*

CAS″SIA. Calyx 5-leaved; corolla 5-petalled; anthers 3, lower ones beaked, and on longer and filaments; legume membranaceous.

Calyx 4 or 5-cleft, half-way (sometimes 4-toothed), somewhat 2-lipped; corolla papilionaceous; wings of the length of the reflexed banner; stamens caducous; legume inflated, smooth, many-seeded. (wild indigo.)

CER″CIS. Calyx 5-toothed, gibbous below; corolla papilionaceous, wings longer than the banner; keel 2-petalled; legume compressed; seed-bearing suture margined; seeds obovate. (Judas-tree.)

SOPHO′RA. Calyx 5-toothed; pod many-seeded, not winged. *S.*

THER″MIA. Calyx oblong, 2-lipped, convex behind; banner reflexed; keel obtuse; pod linear, many-seeded. (false lupine.) *S.*

VIRGIL″IA. Calyx 5-cleft; petals equal; stigma beardless; pod compressed, oblong, many-seeded. *S.*

POMA′RIA. Calyx turbinate, 5-parted, caducous; petals 5, with short claws; filaments hirsute below; legume 1-celled, 2-seeded. *S.*

RHODO′RA. Calyx 5-toothed; corolla 3-petalled, or 2-petalled, with the upper one deeply parted; stamens declined; capsule 5-celled, 5-valved, opening at the top.

B. *Flowers polypetalous, regular.*

PYRO′LA. Calyx 5-parted; styles 5; styles longer than the stamens; anthers with 2 pores at the base before, and at the top after the opening of the flower; capsule 5-celled, dehiscent at the angles near the base. (shin-leaf.)

CHIMAPH″ILA. Calyx 5-parted; petals 5; anthers beaked, with 2 pores at the base before, and at the top after the opening of the flower; style immersed; stigma thick, orbiculate; capsule 5-celled, dehiscent at the angles near the summit. (prince's pine, pipsissiwa.)

LEIOPHYL″LUM. Calyx 5-parted; corolla flat, 5-parted or 5-petalled; stamens longer

than the corolla, with lateral anthers opening longitudinally on their insides; capsule 5-celled, dehiscent at the top, 5-valved; valves ovate with margins inflexed, remote, straight; columella sub-ovate, terete, rugose; seeds small, not winged; leaves always glabrous. (sleek-leaf.)

CLE′THRA. Calyx 5-parted, permanent; corolla 5-petalled; style permanent; stigma short, 3-cleft; capsule 3-celled, 3-valved, enclosed by the calyx. Spiked. (sweet pepper-bush.)

RU′TA. Calyx 5-parted; petals concave; receptacle surrounded by 10 nectariferous dots; capsule lobed; petals sometimes 4, and stamens 8. (rue.) *Ex.*

LE′DUM. Calyx minute, 5-toothed; corolla 5-petalled, spreading; stamens exserted; anthers opening by 2 terminal pores; capsule sub ovate, 5-celled, 5-valved, opening at the base. (Labrador tea.)

MYLOCA′RIUM. Calyx 5-toothed; petals 5; stigma sessile; capsule superior, winged, 3-celled, 1-seeded, seed subulate. (buck-wheat-tree.) *S.*

ME′LIA. Calyx minute, 5-parted; petals 5; nectary 10-toothed, cylindric; drupe 5-celled, 5-seeded. (pride of China.) *Ś.*

JUSSI′ÆU. Calyx 4 or 5-parted, superior, persistent; petals 4 or 5, ovate; capsule many-seeded, seeds minute. *S.*

SWIETE′NIA. Calyx 5-cleft; petals 5; capsule 5-celled, opening at the base, woody; seeds winged. (mahogany-tree.) *S.*

TRIB″ULUS. Calyx 5-parted; petals 5, spreading; styles none; stigma partly 5-cleft; capsules generally 5, gibbous, sub-spinose, 2 or 3-seeded. (caltrops.) *S.*

PROSO′PIS. Calyx hemispherical, 4-toothed; petals 5, lance-linear, recurved at the apex; filaments capillary, adnate at the base; stigma simple; legume long, many-seeded. *S.*

LIMNAN″THES. Sepals 5, united at the base; petals 5, cuneiform, retuse, longer than the sepals; ovaries 5; styles united into one, nearly to the top. *S.*

LIMO′NIA. Calyx 4 or 5-cleft, urceolate, marescent; petals 4 or 5; stamens 8 to 10; filaments distinct, subulate; anthers cordate, oblong; receptacle elevated, forming a short stipe to the ovary; style 1; stigma somewhat lobed; fruit orange-form, 4 or 5-celled, or, by abortive growth, fewer; seeds solitary in each cell.

DIONÆ′A. Calyx 5-parted or 5-leaved; petals 5; stigma fringed; capsule roundish, gibbous, 1-celled, many-seeded; petals sometimes 6. (Venus' fly-trap.) *S.*

C. *Flowers monopetalous.*

ARBU′TUS. Calyx inferior, 5-parted, minute; corolla ovate, pellucid at the base; border small, 5-cleft, revolute; filaments hairy; capsule 5-celled. (bear-berry.)

EPIGÆ′A. Calyx double, outer 3-leaved, inner 5-parted (or calyx 5-parted, with 3 bracts); corolla salver-form; border 5-parted, spreading; tube villose within; capsule 5-celled, many-seeded; receptacle 5-parted. (trailing arbutus.)

DALIBAR''DA. Calyx 5 or 8-cleft, inferior; corolla 5-petalled; styles long, caducous, 5 to 8; berry composed of dry grains. (dry strawberry.)

GE''UM. Calyx inferior, 10-cleft, 5 alternate divisions smaller; corolla 5-petalled; seeds with a bent awn; receptacle columnar, villous. (avens, or herb-bennet.)

STYL''IPUS. Calyx inferior, 5-cleft, divisions equal; petals 5, oval, distant; stamens permanent, on a glandular ring; seeds compressed, ovate, glabrous, with scattering pubescence, sub-margined; receptacle columnar, villose, becoming elongated; awns geniculate.

POTENTIL''LA. Calyx flat, inferior, 10-cleft, 5 alternate divisions smaller; corolla 5-petalled; petals roundish or obovate; seeds awnless, roundish, rugose, fixed to a dry, small receptacle. (five-finger, cinquefoil.)

SIEVER''SIA. Calyx with a concave tube, and 5-cleft limb, and 5 bracts outside; petals 5; carpels numerous, caudate; style persistent; seeds ascending.

FRAGA''RIA. Calyx inferior, 10-cleft, 5 alternate divisions smaller; corolla 5-petalled; receptacle ovate, berry-like; acines naked, immersed in the receptacle, caducous. (strawberry.)

DRY''AS. Calyx 8 to 9-parted, tube concave; petals 8 to 9; carpels many, crowned by a terminal style.

CALYCAN''THUS. Lobes of the calyx in many rows, imbricate, lanceolate, colored; corolla 0; stamens unequal; acines many. S.

CLASS XII. POLYANDRIA.
ORDER I. MONOGYNIA.

TIL''IA. Calyx 5 or 6-parted, inferior, caducous; corolla 5 or 6-petalled; capsule 5 or 6-celled, globular, coriaceous, dehiscent at the base, 1-seeded; 4 of the cells sometimes empty. (bass-wood.)

CORCHO''RUS. Sepals 4 or 5; petals 4 or 5, rather shorter than the sepals inferior; style very short, deciduous; stigmas 2 to 5; capsule pod-like or roundish; seeds commonly numerous in each cell.

PORTULAC''CA. Calyx 2-cleft, inferior; corolla 5-petalled; capsule 1-celled, opening transversely; columella 5, filiform. (purslane.)

CHELIDO''NIUM. Calyx 2-leaved, caducous; corolla 4-petalled; silique-like, capsule 1-celled, 2-valved, linear; seeds crested, many. (celandine.)

POLANIS''IA. See Cleo'me.

CIS''TUS. [Helianthemum.] Sepals 5, 2 smaller; petals 5; capsule 1-celled, 3-valved; valves septiferous in the middle. (rock-rose, frost-weed.)

HUDSO''NIA. Calyx tubular, 5-parted, unequal, inferior, petals 5; capsule 1-celled, 3-valved, 1 to 2-seeded.

TALI''NUM. Calyx of 2 ovate sepals; petals 5; capsule 1-celled, 3-valved, many-seeded.

CALANDRIN''IA. Sepals 2, inferior, persistent, united at the base; petals 3 to 5 without claws; stamens 4 to 15; style short,

stigmas 3, thickish, short; capsule 3-valved, many-seeded; seeds turgid, smooth and shining.

MECONOP''SIS. Petals 4; stigma 4 to 6-rayed; capsule prickly, 4 to 6-valved.

ARGEMO''NE. Petals 4 to 6; stigma 4 to 7-lobed; capsule obovate, 1-celled, opening at the summit by valves. (prickly poppy.)

SANGUINA''RIA. Calyx caducous, 2-leaved; corolla about 8-petalled; stigma sessile, twinned, 2-grooved; capsule pod-like, ovate 1-celled, 2-valved, acute at each end; valves caducous; columella 2, permanent. (bloodroot.)

GLAU''CIUM. Calyx 2-sepalled, caducous; corolla 4-petalled; capsule 1-celled, linear; seeds many, punctate.

PODOPHYL''LUM. Calyx 3-leaved, minute; corolla about 9-petalled; stigma large, crenate, sessile; berry 1-celled, crowned with the stigma, large, many-seeded; columella 1-sided. (wild mandrake.)

ACTÆ''A. [Cimcifuga.] Calyx 4-leaved, deciduous; petals 4, often wanting; stigma sessile, capitate; berry superior, 1-celled, many-seeded; seeds hemispherical. (necklace-weed, baneberry.)

MACRO''TIS. Calyx about 4-leaved, becoming colored before expanding, caducous; corolla many minute petals, very caducous, or wanting; stigma simple, sessile, curving towards the gibbous side of the germ; capsule 2-valved, dehiscent at its straight suture. (cohosh, blacksnake-root, bug-bane.)

SARRACE''NIA. Calyx double, permanent, 3 or 5-leaved; corolla 5-petalled, caducous; stigma peltate, permanent, very large, covering the stamens; capsule 5-celled, 5-valved, many-seeded. (side-saddle flower.)

NU''PHAR. Calyx 5 or 6-leaved; petals many, minute, inserted on the receptacle with the stamens, nectariferous; stigma with a broad disk, and radiate furrows, sessile; pericarp berry-like, many-celled, many-seeded. (water-lily, yellow pond-lily.)

NYMPHÆ''A. Calyx 4 to 7-leaved; corolla many-petalled, petals about equalling the length of the calyx leaves attached to the germs beneath the stamens; stigma with a broad disk, marked with radiated lines; pericarp berry-like, many-celled, many-seeded. (pond-lily.)

PAPA''VER. Calyx 2-leaved, caducous; corolla 4-petalled; stigma a broad disk, with radiating lines; capsule 1-celled, dehiscent by pores under the permanent stigma. (poppy.) Ex.

THE''A. Calyx 4 or 6-leaved; corolla 6 or 9-petalled; capsule 3-seeded. (tea.) Ex.

CIT''RUS. Calyx 5-cleft; petals 5, oblong; filaments dilated at the base, in several parcels; berry 9 or 18-celled; polyadelphous (orange, lemon.) Ex.

† CHRY''SEIS. Receptacle dilated, salver-formed; limb expanding, entire; calyx mitre-form, deciduous; corolla 4-petalled, inserted by the claws in the throat of the receptacle, and bearing the stamens; capsule silique-form, 2-valved; seeds affixed to the margins of the valves.

BEJA'RIA. BAFA'RIA. Calyx 7-cleft; petals 7; stamens 14; capsule 7-celled, many-seeded. *S.*

LEWIS'IA. Calyx from 7 to 9-sepalled; petals 14 to 18; stamens 14 to 18; style about 3-cleft; stigmas 2-cleft; capsules 3-celled, many-seeded; seeds shining. *S.*

ORDER II. DIGYNIA, INCLUDING ORDER V. PENTAGYNIA.

DELPHIN"IUM. Calyx 0; corolla 5-petalled, unequal; nectary 2-cleft, horned behind; capsules 1 or 3, pod-like. (larkspur.)

ACONI'TUM. Calyx 0; petals 5, upper one valved; nectaries 2, hooded, peduncled, recurved; capsule 3 or 5, pod-like. (monk's-hood.)

AQUILE'GIA. Calyx 0; petals 5, caducous; nectaries 5, alternating with the petals, and terminating downward in a spur-like nectary; capsules 5, erect; acuminated with the permanent styles, many-seeded. By some, the nectaries are considered as petals, and the corolla as a colored calyx. (columbine.)

ASCY'RUM. Sepals 4, the 2 inner larger and cordate; petals 4; stamens scarcely united at the base. (St. Peter's wort.)

CALLIGO'NUM. Calyx 5-parted; corolla 0; filaments numerous, united at the base; germ superior, 4-sided, nut winged. *S.*

RESE'DA. Perfect flower apetalous, surrounded by several fringed, petal-like, barren flowers; involucre spreading, many-leaved. (mignonette.) Ex.

RHIZOPHO'RA. Calyx 4-parted; corolla 4-parted; stigmas 2; seed 1, very long, base fleshy.

HYPER"ICUM. Calyx 5-parted; divisions equal, sub-ovate; corolla 5-petalled; filaments often united at the base in 3 or 5 sets; styles 2 to 5; capsules membranaceous, roundish, with a number of cells equal to the number of styles. The bases of the filaments are often in groups, when they are not united. (St. John's wort.)

PÆO'NIA. Calyx 5-leaved; petals 5; styles 0; stigmas 2 or 3; capsules pod-like, many-seeded. Remarkable for the multiplication of petals by rich culture. (peony.) Ex.

ELO'DEA. Sepals 5, equal, somewhat united at the base; petals 5, deciduous, equal; stamens 9 to 15, polyadelphous-parcels alternating with glands; styles 3, distinct; capsule oblong, membranaceous, 3-celled.

NIGEL"LA. Calyx 0; petals 5; nectaries 5, 3-cleft, within the corolla; capsules 5, convex. (lady-in-the-green, fennel flower.)

DENDROM"ECON. Sepals 2; petals 4; stamens numerous; stigmas 2, sessile; capsule pod-shaped, furrowed; valves thick and coriaceous, almost woody, opening from the base to the apex; seeds rather large and numerous; pyriform, smooth. *S.*

ORDER XII. POLYGYNIA.

ASIMI'NA. PORCEL"IA. Calyx 3-parted; petals 6, spreading, ovate, oblong, the inner smaller; anthers sub-sessile; berries several, ovate. (custard apple.)

TROL"LIUS. Sepals colored, 5 to 15, deciduous, petaloid; petals 5 to 20, small; capsules many, cylindrical, sessile, many-seeded. (globe-flower.)

HYDROPEL"TIS. Sepals 3 to 4; petals 3 to 4; ovaries 6 to 18; seeds pendulous, ovate, globose. (water-shield.)

HYDRAS"TIS. Calyx 3-leaved, petaloid; leafets ovate; petals 0; berry composed of many 1-seeded grains. (orange-root.)

NELUM"BIUM. Calyx petaloid, of 4 or 6 sepals; petals many, deeply immersed in the upper surface of a turbinate receptacle.

ILLI'CIUM. Sepals 6; petals numerous, in 3 series; capsules many, disposed in a circle, 2-valved, 1-seeded. (anise-tree.) *S.*

CLEMA'TIS. Petals 3, 4, 5, or 6; seeds compressed; styles permanent, becoming long, plumose tails. Some species are dioicious. (virgin's bower.)

THALIC"TRUM. Petals 4 or 5; filaments very long; seeds without tails, striate, terete. Some species are ditecious. (meadow-rue.)

ANEM"ONE. Petals 5 to 9; seeds numerous, naked. (wind-flower, rue, anemone.)

COP"TIS. Petals 5 or 6, caducous; nectaries small, 5 or 6, cowled; capsules oblong, 5 to 8, stiped, stellate, beaked, many-seeded. (gold-thread.) By some the nectaries are mistaken for corollas, and the corollas for calyxes.

CAL"THA. Petals 5 to 9, orbicular; capsules numerous (5 to 10), many-seeded, compressed; 1-celled, spreading; nectaries 0; pistils variable in number. (American cowslip.)

HELLEBO'RUS. Petals 5 or more; nectary 2-lipped, tubular; capsules 5 or 6, many-seeded, erectish, compressed. (bellebore.) Ex.

MAGNO'LIA. Calyx 3-leaved; corolla 6 to 9-petalled; capsules numerous, imbricate on a strobile-like spike, 2-valved; seeds arilled, pendulous on long cords; berry-like. (magnolia, or beaver-tree.)

LIRIODEN"DRON. Calyx 3-leaved; corolla 6 or 9-petalled, liliaceous; seeds in a sub-lanceolate samara, imbricate on a strobile-like spike. (tulip-tree, or white-wood.)

HEPAT"ICA. Calyx 3-leaved, a little distance below the corolla, entire; petals 6 to 9; seeds without tails. (liverleaf.)

RANUN"CULUS. Calyx 5-leaved; petals 5, with claws, and a nectariferous pore or scale on the inside of each; seeds without tails, naked, numerous. (crow-foot.) Some mistake an extra tegument for a capsule.

SEMPERVI'VUM. Calyx 9 to 12-parted; petals 8 to 12; capsules 12, many-seeded; stamens 16 or 20. (house-leek.) Ex.

PLATYS"TEMON. Sepals 3, pilose; petals 6; stamens numerous; ovaries 10-14 distinct; stigmas sessile; carpels 10-14, linear, indehiscent articulated or transversely strangulated between each seed. *S.*

ADO'NIS. Calyx 4 to 5-leaved; petals 5 or more, without nectariferous pores; seeds awnless. (pheasant's eye.) Ex.

Anno'na. Calyx 3-sepalled, thickened together at the base, concave, sub-cordate, acutish; petals 6, thickish, inner smaller or none; anthers sub-sessile, at the apex, angled, dilated, covering its receptacles; germs united into a sessile berry with the back muricate, scaly or reticulate, pulpy within, having 2 one-seeded cells. *S.*

CLASS XIII. DIDYNAMIA.

ORDER I. GYMNOSPERMIA.

A. *Calyx 5-cleft, with the divisions or teeth nearly equal.*

Teu'crium. Corolla deep-cleft on the upper side and without an upper lip, lower lip 3-cleft, the middle division rounded; stamens and pistils incurved; stamens exsert through the cleavage on the upper side of the corolla. (wood-sage, wild germander.)

Men''tha. Corolla nearly equal, 4-lobed; broadest division emarginate; stamens erect, distant. (spearmint, peppermint.)

Isan''thus. Calyx somewhat bell-form; corolla 5-parted; tube straight, narrow; divisions ovate, equal; stamens nearly equal; stigma linear. recurved. (blue gentian.)

Hedeo'ma. Calyx 2-lipped, gibbous at the base; upper lip with 3 lanceolate teeth; lower lip with 2 subulate ones; corolla ringent; 2 short stamens barren. (pennyroyal.)

Cuni'la. Calyx cylindrical, 10-striate, 5-toothed; corolla ringent, with the upper lip erect, flat, and emarginate; 2 barren stamens, the 2 fertile ones with the style exserted; stigmas divided. (dittany.) On account of their barren stamens, this and the preceding genus have been classed under Diandria.

Nepe'ta. Calyx dry, striate; corolla with a longish tube; under lip with the middle division crenate; throat with a reflexed margin; stamens approximate. (catmint.)

La'mium. Upper lip of the corolla vaulted, entire; lower lip 2-lobed, toothed on each side.

Sta'chys. Calyx with its divisions awned; corolla with the upper lip vaulted, the lower lip 3-lobed; the middle division largest, emarginate; the lateral divisions reflexed; stamens reflexed towards the sides after discharging the pollen. (woundwort, hedge-nettle.)

Leonu'rus. Calyx 5-angled, 5-toothed; corolla with the upper lip erect, villose, flat, entire; lower lip 3-parted; middle division undivided; lobes of the anthers parallel, having shining dots. (mother-wort.)

Verbe'na. Calyx with one of the teeth truncate; corolla funnel-form, with a curved tube; border 5-cleft, nearly equal; seeds 2 or 4, with an extra vanishing tegument; sometimes 2 stamens are barren. (vervain.)

Marru'bium. Calyx salver-form, rigid, marked with 10 lines; corolla with the upper lip cleft, linear, straight. (horehound.)

Glecho'ma. Calyx 5-cleft; corolla double the length of the calyx; upper lip 2-cleft;

lower lip 3-cleft, with middle segment emarginate; each pair of anthers approaching so as to exhibit the form of a cross. (ground-ivy, gill-overground)

Pycnan''themum. Involucrum bract-like, many-leaved, under small heads of flowers; calyx tubular, striate; corolla with the upper lip sub-entire; lower lip 3-cleft; middle segment longer; stamens distant, nearly equal; cells of the anthers parallel. (mountain mint.)

Aju'ga. Upper lip of corolla very small, 2-toothed; stamens longer than the upper lip; anthers reniform. *S.*

Ballo'ta. Calyx 5-toothed, salver-form, 10-striate; upper lip of the corolla crenate, concave; seed ovate, 3-sided. (false mother-wort.) Ex.

Hysso'pus. Lower lip of the corolla 3-petalled; middle lobe sub-crenate; stamens straight and distant. (hyssop.)

Galeop''sis. Calyx 5-cleft, awned; upper lip of the corolla vaulted, sub-crenate; lower lip with 3 unequal lobes, having 2 teeth on its upper side. (flowering nettle.)

Hyp''tis. Calyx 5-toothed; corolla 2-lipped, the upper one 2-lobed, lower one 3-lobed, with the middle lobe calyx-like; stamens inserted in the large part of the tube and declined. *S.*

Leu'cas. Calyx tubular, striate, 6 to 10-toothed; upper lip entire, lower lip long, 3-lobed; middle segment largest; anthers beardless, spreading; stigma 2-cleft, shorter than the upper lip. *S.*

Synan''dra. Calyx 4-cleft; segments unequal, subulate, inclined; upper lip of the corolla entire, vaulted, lower lip with 3 unequal lobes; throat inflated, naked; filaments downy. *S.*

Lavandu'la. Calyx ovate, sub-dentate; bracted; corolla resupinate; stamens in the tube. (lavender.) Ex.

Satur'ja. Calyx tubular, striate; corolla with divisions nearly equal; stamens distant. (savory.) Ex.

Moluccel''la. Calyx bell-form, much larger than the corolla, spinose. (shell-flower.) Ex.

B. *Calyx 2-lipped.*

Origa'num. Calyxes collected into a 4-sided, strobile-like cone, with broad intervening bracts; corolla with the upper lip erect, flat, straight, emarginate, under lip 3-parted, divisions nearly equal. (marjoram.)

Prunel''la. Calyx with the upper lip dilated; filaments 2-forked, with an anther on one of the points; stigma 2-cleft. (self-heal or heal-all.)

Scutella'ria. Calyx with an entire mouth, which is closed with a helmet-form lid after the corolla falls out; tube of the corolla bent. (scull-cap.)

Tri'choste．．Calyx resupinate; corolla with the upper lip falcate, the under lip 3-parted, with the middle division small, oblong; filaments very long, exsert, incurved or coiled. (blue-curls.)

Clinipo'dium. Involucre of many, linear,

acuminate bracts; leafets placed under the whorls of flowers; upper lip of the corolla erect..emarginate, lower one the longest, emarginate. (field thyme.)

DRACOCEPH″ALUM. Calyx sub-equal, 5-cleft; orifice of the corolla inflated; upper lip concave, notched; stamens unconnected. (dragon-head.)

OCY′MUM, Calyx with the upper lip orbiculate, lower lip 4-cleft; corolla resupinate; one lip 4-cleft, the other undivided. A process at the base of the outer filaments. (sweet basil.) Ex.

THY′MUS. Calyx sub-campanulate, the throat closed with hairs; corolla with the upper lip flat, emarginate, lower lip longer. (thyme.) Ex.

MELIS″SA. Calyx dry, flattish above, with the upper lip sub-fastigiate; corolla with the upper lip somewhat vaulted, 3-cleft, lower lip with the middle lobe cordate. (balm.) Ex.

MACBRI′DEA. Calyx top form, 3-cleft, 2 segments large; corolla 2-lipped, the upper entire, the under 3-parted; anthers 2-lobed; the lobes spreading, fringed with small spines. S.

CALAMIN″THA. Calyx closed with hairs after flowering; throat of the corolla somewhat inflated, upper lip emarginate; lower one 3-parted. S.

CERAN″THERA. Calyx 2-lipped, the upper lip emarginate, the lower one 2-cleft; upper lip of the corolla 2-lobed, the lower one 3-parted, stamens exsert; anthers horizontal, awned at each end. S.

TUL″LIA. Calyx with the upper lip 3-toothed, lower one 2-toothed: teeth appendaged; corolla 2-lipped, with the upper lip very entire, lower one 3-parted, middle division largest. S.

ORDER II. ANGIOSPERMIA.

A. Calyx 2 or 3-cleft.

OBOLA′RIA. Calyx bract-like; corolla 4-cleft, bell-form; capsule 1-celled, 2-valved, many-seeded; stamens proceeding from the divisions of the corolla; stigma 2-cleft or emarginate. (penny-wort.)

CASTILLE′JA. Calyx spathe-form, upper lip 2-cleft, lower one wanting; corolla 2-lipped, lower one very short, 3-cleft, with 2-glands between the divisions; capsule 2-celled. S.

PHRY′MA. Calyx cylindric, upper lip longer, 3-cleft, lower lip 2-toothed; upper lip of the corolla emarginate, smaller; seed solitary. (lop-seed.)

B. Calyx 4 or 5-cleft.

EUCHRO′MA. Calyx inflated, 2 or 4-cleft; corolla 2-lipped, upper lip long, linear, embracing the style and stamens; anthers linear, with unequal lobes, cohering so as to form an oblong disk; capsule ovate, compressed, 2-celled; seeds numerous, surrounded with an inflated membrane.

BART″SIA. Calyx lobed, emarginate, colored; corolla less than calyx, upper lip longest, concave, entire, lower lip 3-cleft and reflexed; anthers with equal lobes, not

cohering; capsule 2-celled; seed angled. (painted cup.)

MELAMPY′RUM. Corolla with the upper lip compressed, the margin folded back, lower lip grooved, 3-cleft, sub-equal; capsule 2-celled, oblique, dehiscent on one side; seeds 2, cylindric, gibbous, cartilaginous, and smooth. (cow-wheat.)

SCHWAL″BEA. Calyx ventricose, tubular, upper segment shortest, lower large and emarginate; corolla ringent, upper lip entire, arched; capsule 2-celled, 2-valved; seeds imbricate, winged. (chaff-seed.)

RHINAN″THUS. Calyx inflated, 4-toothed; corolla ringent, upper lip compressed, lower lip flat, 3-lobed; capsule 2-celled, obtuse, compressed. (yellow-rattle.)

LANTA′NA. Flowers capitate; calyx 4-toothed; corolla unequally 4-parted; throat open; stamens within the tube; stigma hooked; drupes aggregated. S.

ORTHOCAR″PUS. Calyx tubular, 4-cleft; corolla 2-lipped, closed, upper lip smaller, compressed, margin inflexed, lower lip concave, 3-toothed; capsule 2-celled, 2-valved. S.

EUPHRA′SIA. Calyx cylindric; corolla 2-lipped, the upper lip 2-cleft, lower lip 3-lobed, with the divisions 2-cleft; lower anthers lobed, spinose. (eye-bright.) S.

C. Calyx 4 or 5-cleft, or 5-toothed; plant without green herbage.

OROBAN″CHE. Corolla ringent; capsule ovate, acute, 1-celled; seeds numerous; a gland beneath the base of the germ.

EPIPH″EGUS. Polygamous; calyx abbreviated, 5-toothed; corolla of the barren flowers ringent, compressed, 4-cleft, lower lip flat, of the fertile flowers minute, 4-toothed, caducous; capsule truncate, oblique, 1-celled, imperfectly 2-valved, opening on one side. (beech-drops, cancer-root.)

D. Calyx 5-leaved, or 5-cleft; plant with green herbage.

SCROPHULA′RIA. Corolla sub-globose, resupinate, short bi-labiate, with an internal, intermediate scale; capsule 2-celled.

BIGNO′NIA. Calyx 5-toothed, cup-form, sub-coriaceous; corolla bell-form, 5-lobed, ventricose beneath; capsule silique-like, 2-celled; seed membrane winged, (trumpet-flower.)

BUCHNE′RA. Calyx 5-toothed; corolla with a slender tube, and the limb in 5 equal divisions, the lobes cordate; capsule 2-celled. (blue hearts.)

ANTIRRHI′NUM. Calyx 5-leaved or deeply 5-parted, the two lower divisions remote; corolla personate or ringent, spurred or with a prominent base; the throat closed with a prominent palate; capsule ovate, 2-valved, dehiscent at the apex, with reflexed teeth. (snap-dragon, toad flax.)

GERAR″DIA. Calyx 5-cleft or 5-toothed; corolla sub-campanulate, unequally 5-lobed, segments mostly rounded; capsule 2-celled, dehiscent at the top. (false foxglove.)

PEDICULA′RIS. Calyx ventricose, 5-cleft,

or obliquely truncate; corolla ringent, upper lip arched, emarginate and compressed; capsule 2-celled, mucronate, oblique; seeds numerous, angular, coated; leaves many-cleft. (louse-wort, high heal-all.)

MIMU'LUS. Calyx prismatic, 5-toothed; corolla ringent, upper lip folded back upon its side, lower lip with a prominent palate; stigma thick, 2-cleft; capsule 2-celled, many-seeded; seeds minute. (monkey flower.)

CHELO'NE. Calyx 5-cleft or 5-leaved, 3-bracted; corolla ringent, inflated; the upper lip emarginate-obtuse, under lip slightly 3-cleft; the rudiment of a smooth filament between, and shorter than the two tallest stamens; anthers woolly; capsule 2-celled, 2-valved; seeds with membranous margins. (snake head.)

PENTSTE'MON. Calyx 5-cleft or 5-leaved; corolla ringent, inflated; the rudiment of a bearded filament between, and longer than the two tallest stamens; anthers smooth; capsule 2-celled, 2-valved, ovate; seeds numerous, angular. Taken from the last genus. (beard tongue.)

ZAPA'NIA. Flowers capitate; calyx 5-toothed; corolla 5-lobed; stigma peltately capitate, oblique; seeds 2, at first enclosed in an evanescent pericarp. (fog-fruit.)

AVICEN''NIA. Calyx 5-parted; corolla 2-lipped, the upper lip square; capsule coriaceous, rhomboid, 1-seeded, seeds germinating within the capsule.

HERPES''TIS. Calyx unequal, bi-bracted at the base; corolla tubular, somewhat 2-lipped; stamens included; capsule 2-valved, 2-celled; dissepiment parallel with the valves.

LIMOSEL''LA. Calyx 5-cleft; corolla 4-5-lobed, equal; stamens approaching by pairs; capsule 2-valved, partly 2-celled, many-seeded. (mad wort.)

RUEL''LIA. Calyx often 2-bracted; corolla somewhat bell-form, border 5-lobed; stamens approaching by pairs; capsule smaller at the ends, toothed, dehiscent. (ruel.)

COLLIN''SIA. Calyx 5-cleft; corolla 2-lipped, throat closed, upper lip 2-cleft, lower lip 3-cleft; the bag-like, keeled segment closed over the declined stamens and style; capsule globose, seeds 2-3-umbilicate.

CONRAD''IA. Calyx 5-cleft, foliaceous; corolla monopetalous, cylindrical, sub-equal, 5-toothed at the apex, teeth reflexed; stamens 4, scarcely declined, sub-equal, long-exsert; style very long; stigma minute; capsule short ovate, 2-celled, many-seeded.

MARTYN''IA. Calyx 5-cleft; corolla ringent, with a ventricose tube; capsule 4-celled, 2 valved; each of the valves terminating in a long, hooked beak. (unicorn plant.) S.

CAPRA'RIA. Calyx 5-parted; corolla bell-form, 5-parted, acute; capsule 2-valved, 2-celled, many-seeded. S.

SEYME'RIA. Calyx deeply 5-parted; corolla sub-campanulate, 5-lobed; stamens near the throat; style declined; capsule inflated, ovate, acute. S.

SESA'MUM. Calyx 5-parted; corolla bell form, 5-cleft; the lower lobe largest. The rudiment of a fifth stamen; stigma lanceolate; capsule 4-angled, 4-celled. (oily grain.)

DIGITA'LIS. Calyx 5-parted; corolla bell-form, ventricose, 5-cleft; stigma simple or bilamellate; capsule ovate, 2-celled; flowers racemed. (fox-glove.) Ex.

CLASS XIV. TETRADYNAMIA.

ORDER I. SILICULOSA.

THLAS''PI. Calyx spreading, equal at the base; filaments distinct, without teeth; silicle compressed, emarginate, obcordate, many-seeded; valves resemble two boats with the keels outward. (shepherd's purse.)

LEPID''IUM. Calyx spreading; corolla regular; silicle emarginate, cordate or oval; cells 1-seeded; valves carinate, dehiscent; partition contrary; cotyledons incumbent. (pepper-grass.)

COCHLEA'RIA. Silicle thick, rugose, many-seeded, 2-valved; valves gibbous, obtuse; partition nearly parallel to the valves. (horse-radish, water-radish.)

CAK''ILE. [Bunias.] Panicle compressed, of 2 single-seeded joints; the upper joint with an erect single seed, inferior with a pendulous seed. (sea-rocket.)

DRA'BA. Silicle entire, oval or oblong; valves flat or convex; cells many-seeded; seeds not margined; filaments without teeth; style 0; cotyledons accumbent. (whitloe-grass.)

ALYS''SUM. Calyx equal at the base; petals entire; stamens mostly toothed; silicle orbicular, or illiptical; valves flat, or convex in the centre; seeds 2 to 4 in each cell, compressed, sometimes membranously winged; cotyledons accumbent. (gold-of-pleasure.)

CAMELI'NA. Silicle subovate, many-seeded; valves thick; cotyledons incumbent.

PLATYSPET''ALUM. Silicle oval, many-seeded; valves convex; styles very short; calyx a little spread; laminas of the petals dilated.

SUBULA'RIA. Silicle entire, ovate, concave (convex without); stigma sub-sessile; seed linear, 2-plaited; cotyledons incumbent.

PLATYSPER''NUM. Silicle oval, compressed at the back, flat, stigma sessile; seeds few with broad margins; scapes numerous, 1-flowered.

CRAM''BE. Silicle globose, stalked, coriaceous, 1-celled, without valves, deciduous; seed solitary. (sea-kale.) Ex.

LUNA'RIA. Silicle entire, oval, flat, compressed, pedicelled; valves equalling the partition, parallel, flat; calyx consists of colored, sack-like leafets. (honesty, or satin-flower.) Ex.

VESICA'RIA. Silicle globose, inflated, with hemispheric valves; seeds more than 8, sometimes margined; petals entire.

ISA'TIS. Silicle compressed, oblong, ligulate, without valves, 1-seeded; partition like lattice-work. (woad.) Ex.

IBE'RIS. Corolla irregular, the two outer

petals longest; silicle many-seeded, emarginate. (candy-tuft.) Ex.

THY'SANOCAR"PUS. Silicle obovate, plano-convex, broad-winged at both margins, emarginate at the apex, 1-celled, one seed ed; seed broad-obovate, pendulous. Flowers small, and white.

CORONO'PUS. Silicle reniform, compressed, wrinkled; cells valveless, 1-seeded. S.

ORDER II. SILIQUOSA.

DENTA'RIA. Silique lanceolate; valves flat, nerveless, often opening elastically; receptacles not winged; funicle dilated; seeds in a single series, ovate, not margined; cotyledons accumbent. (tooth-root.)

NASTUR"TIUM. Silique teretish, abbreviated or declined; stigma somewhat 2-lobed; calyx equal at the base, spreading; seeds small, irregularly in two series, without margins.

TURRI'TIS. Calyx converging, erect; silique very long, striate, 2-edged; valves keeled or nerved; seeds arranged in a double series; cotyledons accumbent. (tower mustard.)

CARDAM"INE. Calyx leaves spreading but little; stigma entire; a single gland between each of the short stamens and the calyx; silique with truncate margins, linear, long, bursting elastically with revolute valves, narrower, but equalling the length of the partitions; seed with a slender funicle, not margined. (American water-cress.)

STREPTAN"THUS. Calyx erect, colored; petals dilated; having twisted, channeled claws; glands none; stamens with filaments subulate, and thickened at the base; silique very long, angled, compressed; seeds in one series, flat, margined; cotyledons accumbent.

AR"ABIS. Glands 4, one within each leafet of the erect calyx, of the size of the reflexed scale; silique compressed, torulose, sub-divaricate; valves flat, 1-nerved; seeds arranged in a single series; cotyledons accumbent. (wall-cress.)

CHEIRAN"THUS. Calyx closed, two of the leafets gibbous at the base; petals dilated; silique, when young, with a glandular tooth each side; stigma 2-lobed; seed flat, sometimes margined. (stock-july flower, wall-flower.)

PHŒNICAU'LIS. Calyx colored, nearly equal at the base; much shorter than the entire unguiculate petals; silique ensiform, acuminate, flat, not opening elastically; cells about 3-seeded; valves with a prominent central nerve; seeds large, in a single series, not margined.

SINA'PIS. Calyx spreading; corolla with straight claws; glands between the short stamens and the pistil, and between the long stamens and the calyx; partition extending beyond the valves of the silique, ensiform; seeds in a single series. (mustard.)

RAPHA'NUS. Calyx closed, setose; silique torose, terete, not opening by valves, 1 or 2-celled; glands between the short stamens and pistil, and between the long stamens and the calyx. (radish.)

WA'REA. Silique. 2-celled, stiped, flat, with a seed-bearing margin on both sides; seed flattish, striate; petals with long claws, spreading; calyx deflected, caducous, colored.

BRAS"SICA. Calyx erect, converging; partition extending beyond the valves of the silique; seed globose; glands between the short stamens and pistil, and between the long stamens and calyx. (cabbage, turnip.) Ex.

BARBARE'A. Silique 4-edged; cotyledons accumbent; seeds in a single row; calyx equal at the base, erect; shorter filaments with intermediate glands. (water-radish.)

SISYM"BRIUM. Calyx mostly spreading, equal at the base; silique sub-terete; cotyledons incumbent, sometimes oblique, flat. (hedge-mustard.)

ERY'SI"MUM. NASTUR"TIUM. Silique subterete, often short; valves concave, nerveless, not keeled; calyx equal, spreading; cotyledons accumbent. (English water-cress.)

HES"PERIS. Calyx closed, furrowed at the base, shorter than the claws of the petals; petals bent obliquely, linear or obovate; silique 4-sided, 2-edged; stigma sub-sessile of 2 lobes; cotyledons incumbent. (rocket.)

CLASS XV. MONADELPHIA.

ORDER III. TETANDRIA.

SISIRYN"CHIUM. Spatha 2-leaved; perianth 0; corolla superior, 6-cleft or 6-petalled, tubular; style 1; stigma 3-cleft; capsule 3-celled. (blue-eyed grass.)

TAMARIN"DUS. Petals 3, ascending; 3 filaments longer; legume 1 to 3-celled, pulpy inside. (tamarind.) Ex.

TIGRI'DIA. Calyx 0; petals 6; tube made by the union of the filaments, long. (tiger-flower.) Ex.

ORDER V. PENTANDRIA.

PASSIFLO'RA. Calyx 5-parted, colored; corolla 5-petalled, on the calyx; nectary a triple, filamentous crown within the petals; gourd-like berry, pedicelled. (passion-flower.) S.

ERO'DIUM. Calyx 5-leaved; corolla 5-petalled; nectariferous scales 5, alternating with the filaments; arils 5, 1-seeded, awned; beaked at the base of the receptacle; awn spiral, bearded within. Taken from geranium. (stork's bill.) Ex.

OPLOTHE'CA. Calyx double, outer 2-leaved, convolute, truncate, scarious; inner calyx 1-leafed, muricate, somewhat 5-cleft, downy, longer than the outer calyx; nectary cylindric, 5-toothed, stamens in the nectary; stigma single, hairy; capsule bladder-like, enclosed in the calyx, 1 seeded. S.

ACHYRAN"THES. Calyx double, permanent, membranaceous; outer calyx 3-leaved, inner 5-leaved, unequal; seed 1, covered by the converging calyx. *S.*

PHILOX"ERUS. Calyx 5-parted; corolla none; stamens united at the base into a small entire cup, shorter than the germ; anthers 1-celled; stigmas 2; bladder-like capsule membranaceous, 1-seeded, valveless. *S.*

MALACHODEN"DRON. Calyx bracted; petals 5-6; limb crenulate; germ 5-striate; stigmas capitate; capsules 5, united, seed 1. *S.*

ORDER VII. HEPTANDRIA.

. PELARGO'NIUM. Calyx 5-parted, upper division broader, ending in a capillary nectariferous tube; corolla 5-petalled, irregular; the 2 upper petals usually broader, with colored veins; filaments 10, 3 of them usually without anthers; arils 5, each 1-seeded, awned; some of the awns spiral. (stork geranium.) Ex.

ORDER VIII. OCTANDRIA.

PIS"TIA. Spatha ligulate, hooded; corolla 0; filament lateral; anthers 3 to 8; style 1; capsule 1-celled, many-seeded. *S.*

ORDER X. DECANDRIA.

GERA'NIUM. Calyx 5-leaved; corolla 5-petalled, regular; nectariferous glands 5, adhering to the base of the 5 alternating long filaments; arils 5, 1-seeded, awned, beaked at the elongated top of the receptacle; awn naked or smooth within, straight. (cranebill, false crowfoot, herb robert.)

ACA'CIA. Polygamous; calyx tubular, 5-toothed; petals 5; stamens 5 to 10, exsert; pod 1-celled, 2-valved.

DARLINGTO'NIA. Calyx bell-form, 5-7-toothed; petals 5, distinct; stamens 5 to 10, sub-exsert; legume bivalve, juiceless, small-seeded, lanceolate-falcate.

SCHRANK"IA. MIMO'SA. Polygamous; calyx 5-toothed, tubular; petals 5; stamens 8 to 10, exsert; pod 4-valved.

ORDER XII. POLYANDRIA.

SI'DA. ABU'TILLON. Calyx simple, angular, 5-cleft; style many-parted; capsules many, arranged circularly, 1 celled, 1 or 3-seeded; pedicel articulate under the apex. (Indian mallows.)

AL"THÆA. Calyx double, outer one 6 or 9 cleft; capsules many, arranged circularly, 1 seeded. (hollyhock.)

MAL"VA. Calyx double, outer one 3-leaved, inner one 5-cleft; capsules many, arranged circularly, 1-celled, 1-seeded. (mallows.)

MALVAVIS"CUS. Calyx surrounded by a many-leaved involucre; petals erect, convolute; stigmas 10; carpels 5, 1-seeded, sometimes sub-distinct, and often united in 5-celled fruit.

HIBIS"CUS. Calyx double, outer one many-leaved; inner one about 5-cleft; stigmas 5; capsule 5 or 10-celled, many-seeded. (marsh mallows.)

LAVATE'RA. Calyx double, outer one 3-cleft; capsules many, seeds numerous. Ex.

STUART"IA. Calyx 5-parted; petals 5;

stigma 5 lobed; capsule 5-celled, 5-valved; cells 1 or 2-seeded; seeds long, ovate. *S.*

HO'PEA. Calyx superior, 5-cleft; petals 5; stamens united in 5 groups; style 1; drupe 3-celled. (yellow-leaf.) *S.*

NUTTAL"LIA. Calyx 5-cleft, simple; capsules many, 1-seeded, annular. *S.*

HALE'SIA. Calyx superior, 4-toothed; corolla 4-sided, winged, covered with bark; 2 to 4-celled, 2 to 4-seeded. (snow-drop-tree) *S.*

MAL"OPE. Calyx double, the exterior one 3-leaved; capsules clustered without order, 1-seeded. *S.*

STY'RAX. Calyx inferior, bell-form, 5-toothed; corolla 5 to 7-parted; stamens 6 to 16, united at the base, standing in the throat of the corolla; anthers oblong, linear. *S.*

GORDO'NIA. Calyx connate at the base, simple, 5-leaved; style 5-sided; stigmas 5; capsule 5-celled, 5-valved; receptacle columnar; cells 2-seeded; seeds winged. *S.*

GOSSYP"IUM. Calyx double, outer one 3-cleft; capsule 4-celled; seeds involved in a tomentose mass. (cotton.) Ex.

CLASS XVI. DIADELPHIA.

ORDER V. PENTANDRIA, TO ORDER VIII. OCTANDRIA, OR PENTOCTANDRIA.

CORYDA'LIS. Calyx 2-leaved; corolla ringent, 1 or 2-spurred at the base; filaments 2, membranaceous, each with 3 anthers; capsules silique-like, 2-valved, compressed, many-seeded. In some species the stamens are separate, with broad membranaceous bases. (colic-weed.)

DICLY'TRA. Petals 4, 2 outer ones equally spurred at the base; pod 2-valved, many-seeded. (Dutchman's breeches.)

FUMA'RIA. Calyx 2-leaved, caducous; corolla irregular, spurred, or gibbous at the base of one petal; filaments 2, each with 3 anthers; capsules or silicle drupe-like, 1-celled, 1-seeded, not opening by valves; seeds affixed to the side of the cell. (fumitory.)

ADLU'MIA. Sepals 2; petals united in a spongy persistent; monopetalous corol, bigibbous at the base, 4-lobed at the apex; capsule pod-shaped, linear-oblong, many-seeded.

PETALOS"TEMON. Petals 5, nearly equal, 4 petals alternating with the stamens, and forming with them a cleft tube; legume included in the calyx, 1-seeded. *S.*

POLYG"ALA. Calyx 5-leaved, permanent, unequal, 2 of the leafets wing-like, larger, colored; corolla irregular (or rather calyx 3-leaved, corolla imperfectly papilionaceous); capsule obcordate, 2-celled, 2-valved; keel of the corolla sometimes appendaged; seeds hairy (snake-root, milkwort, low centaury, mountain-flax).

ORDER X. DECANDRIA.

A. *Legume without transverse divisions or partitions; seeds numerous.*

PI'SUM. Calyx with the divisions leaf-like, about equal; banner protruding 2-folds;

style compressed, carinate, villose above; legume without down at the suture. (pea.)

LATHY'RUS. Calyx with the 2 upper divisions shorter; style flat, villose above, broader toward the top; stems mostly winged, leafets 2 or more, terminated by a divided tendril. (sweet pea.) *S.*

VI'CIA. Calyx emarginate above, 2-toothed, 3 straight long teeth below; banner emarginate; style bearded transversely on the lower side beneath the stigma. (vetch.)

ER''VUM. Calyx 5-cleft, segments linear, acute, nearly equalling the corolla; stigma glabrous; legume oblong. 2-4 seeded. (creeping-vetch.)

ASTRO'PHIA. Calyx campanulate, 5-cleft, the 2 upper segments a little shorter; style flat, linear, pubescent along the inside; legume broadly-oblong, compressed, few-seeded.

ORO'BUS. Style linear; corolla long; calyx obtuse at the base, upper segments deeper, often shorter. (bitter vetch.) *S.*

PHA'CA. Keel obtuse; style not pubescent; stigma capitate; legume 1-celled, inflated. *S.*

PHASEO'LUS. Keel, stamens, and style, spirally twisted together; legume compressed, falcate; seeds sub-compressed, reniform. (bean.)

STROPHOS''TYLES. GLY'CINE. Keel, stamens, and style, spirally twisted together; legume terete, with a longitudinal half-breadth partition attached to one edge; seed reniform, sub cylindric. (wild bean.)

A'PIOS. GLYCI'NE. Calyx somewhat 2-lipped, truncate, 1-toothed; keel of the corolla falcate, bending back the apex of the banner; germ sheathed at the base; legume coriaceous, many-seeded. (ground-nut.)

AMPHICAR''PA. Calyx bell-form, 4-toothed, obtuse, and naked at the base; petals oblong, banner broader, close pressed upon other petals, sub-sessile; stigma capitate; legume flat, stiped; seeds 2 to 4. (wild bean-vine.)

ROBIN''IA. Calyx small, bell-form, 4-cleft, upper division 2-parted; banner large, reflexed, roundish; legume compressed, elongated, many-seeded; seeds compressed, small. (locust tree.) *S.*

GALAC''TIA. Calyx 4-toothed, with 4 bracts at the base; petals oblong, standard incumbent; anthers oblong; stigma obtuse; germ on a naked stipe; legume terete, many-seeded.

VEXILLA'RIA. Calyx surrounded at the base by 2 longer bracts, 5-cleft; corolla resupinate; standard large, covering the wings; style dilated at the apex; legume linear, compressed, straight, 2-valved, many-seeded. (butterfly-weed.)

ASTRAG''ALUS. Keel obtuse; legume more or less completely 2-celled; lower suture inflexed. (milk-vetch.)

GALE'GA. TEPHRO'SIA. Calyx with subulate teeth, nearly equal; standard large, roundish, pubescent without, reflexed, spreading; legume compressed, linear, many-seeded. (goat's rue.)

MEDICA'GO. Keel of the corolla deflected from the standard; legume compressed, spiral. (lucerne clover.)

COLU'TEA. Calyx 5-cleft, with the keel obtuse; style bearded on its back through its whole length; legume inflated, opening on the upper suture at the base. (bladder senna, bush locust.) *Ex.*

GLYCYRRHI'ZA. Calyx tubular, equal, 5-parted; standard at the base; the sides reflexed; wings spreading; legume ovate; flowers in a raceme. (liquorice.) *S.*

LUPINAS''TER. Calyx bell-form, 5-toothed; teeth setaceous, one under the keel; stigma hooked; legume terete, without joints. *S.*

OXYTRO'PIS. Keel mucronate; legume with the upper suture inflexed. *S.*

INDIGOFE'RA. Calyx spreading; keel with a subulate spur both sides; legume linear, small, terete or quandrangular. (indigo.) *S.*

TRI'GONEL''LA. Banner and wings sub-equal, spreading, resembling a 3-petalled corolla; legume often curved, compressed (fenu-greek.) *S.*

B. *Legume without transverse divisions or partitions; seeds few, or single.*

MELILO'TUS. Flowers racemed; calyx tubular, 5-toothed; keel simple, shorter than the wings and banner; legume rugose, longer than the calyx, or about as long. (melilot clover.)

TRIFO'LIUM. Flowers sub-capitate; legume included in the calyx, not opening by valves, 1 to 4-seeded; leaves always ternate. (clover.)

DO'LICHOS. Banner with two oblong, parallel, callous processes at the base, compressing the wings beneath them. (cowhage.) *S.*

DA'LEA. Calyx half 5-cleft; corolla partly papilionaceous; wings and keel adnate to the undivided column of stamens; legume 1-seeded, included in the calyx. *S.*

PSORA'LEA. Calyx 4-cleft, lower segments elongated; legume the length of the calyx, 1-seeded, beaked. *S.*

C. *Fruit or loment in several joints, or in a single-seeded piece.*

HEDYSA'RUM. Calyx 4-cleft; keel of corolla transversely obtuse; loment many-jointed; joints 1-seeded, truncate, compressed, generally hispid; plants mostly with ternate leaves. (bush clover.)

LESPEDE'ZA. Calyx 5-parted, 2-bracted, divisions nearly equal; keel obtuse; legume 1-seeded; leaves always ternate. (bush clover.)

ÆSCHYNOM''ENE. Calyx 5-cleft, upper lip 2-cleft, lower lip 3-cleft; stamens in 2 equal sets; loment compressed, one suture straight, the other lobed.

STYLOSAN''THES. Calyx tubular, very long, bearing the corolla; loment 1-2 jointed, hooked.

DESMO'DIUM. Calyx with 2 bracts at the base, obscurely bi-labiate toward the middle, upper lip bifid, lower one 3-parted; corolla papilionaceous; standard roundish; keel obtuse, not truncate; wings longer than the keel; stamens diadelphous (3 and 1); filaments sub-persistent; legume with many joints; joints compressed, 1-seeded, membranaceous or coriaceous, scarcely dehiscent.

CI'CER. Calyx 5-parted, of the length of the corolla, 4 upper divisions resting on the banner; legume turgid, 2-seeded. (chick-pea.) *Ex.*

ZOR"NIA. Calyx inferior, bell-form, 2-lipped; banner cordate, revolute; anthers half oblong, half globose; loment jointed, hispid. *S.*

CORONIL"LA. Calyx 2-lipped; petals with claws; loment teretish, jointed flowers in umbels; seeds generally cylindric. (coronilla.) *Ex.*

SESBA'NIA. Calyx 5-toothed; legume terete jointed. *Ex.*

D. *Stamens united in one set.*

AMOR"PHA. Calyx somewhat bell-form, 4 or 5-cleft; banner ovate, concave; wings and keel 0; legume 1 or 2-seeded, falcate. (false indigo.)

LUPI'NUS. Calyx 2-lipped; anthers, 5 oblong and 5 roundish; legume coriaceous, torulose. (lupine.)

CROTALA'RIA. Corolla with the banner cordate; large keel acuminate, the membrane formed by the united filament, has a fissure on the back; style curved; legume pedicelled, turgid. (rattle-box.)

GENIS"TA. Calyx 3-lipped; upper lip with 2, lower lip with 3 teeth. (dyer's broom.)

SPAR"TIUM. Stigma longitudinal, pubescent above; filaments adhering to the ovary; calyx lengthened at the base. (Spanish broom.) *Ex.*

U'LEX. Calyx 2-leaved, 2-bracted; stamens all united; legume about the length of the calyx, spinose. (furze.) *S.*

ARA'CHIS. Calyx 2-lipped; corolla inverted; legume gibbous, torulose, veiny, coriaceous. (pea-nut.) *Ex.*

PITCH"ERIA. Calyx tubular, somewhat 2-lipped, 5-cleft. divisions subulate, upper lip bifid, equal to the lower one; wings narrow, subulate, 1-toothed; style filiform, ascending; legume oblong, 2-seeded, scarcely exceeding the calyx. *S.*

ERYTHRI'NA. Calyx 2-lipped; banner long, lanceolate; legume torulose, many-seeded. (coral-tree.) *S.*

CLASS XVII. SYNGENESIA.

ORDER I. POLYGAMIA ÆQUALIS.

A. *Florets ligulate.*

CICHO'RIUM. Calyx calycled; egret plumose. sessile, unequal; receptacle somewhat chaffy. (succory or endive.)

LEON"TODON. Calyx double, imbricate, with flexible leafets; receptacle naked; egret stiped. (dandelion.)

PRENAN"THES. Florets from 5 to 20, in a simple series (or in one circular row); calyx calycled; receptacle naked; egret simple, sub-sessile. (white lettuce.)

LACTU'CA. Calyx imbricate, cylindric, with the margin of the scales membranaceous; receptacle naked; egret simple, stiped; seed smooth. (lettuce.)

HIERA'CIUM. Calyx imbricate, ovate, egret simple, sessile; receptacle naked, punctate, or sub-pilose. [From white becoming yellowish.] (hawk-weed.)

APAR'GIA. Calyx imbricate; receptacle naked, punctate; egret plumose, sessile, unequal. (false hawk-weed.)

SON"CHUS. Calyx imbricate, swelling at the base; receptacle naked; egret simple, sessile. (swine thistle.)

KRI'GIA. Calyx many-leaved, simple, receptacle naked; egret double, exterior 5 to 8-leaved, interior of 5, 8, or 24 scabrous bristles. (dwarf dandelion.)

TROXI'IMON. Calyx oblong, cone-like, many-sepalled, sepals unequal, imbricate; receptacle naked; egret sessile, pilose. *S.*

APO'GON. Calyx 8-sepalled, in a double series; receptacle naked; egret 0. *S.*

CHONDRIL"LA. Receptacle naked; egret pilose, stiped; calyx calycled; florets in many series. *S.*

TRAGOPO'GON. Calyx simple, many-leaved; receptacle naked; egret plumose and stiped. (goat's-beard, vegetable oyster.) *Ex.*

B. *Florets tubulous; flowers capitate.*

ARC"TIUM. Calyx globose, with scales hooked at the apex; egret chaff-bristly; receptacle chaffy. (burdock.)

CNI'CUS. Calyx swelling, imbricate, with prickly scales; receptacle villose; egret caducous, plumose. (thistle.)

CAR'DU'US. Calyx ovate, imbricate, with prickly scales; receptacle villose; egret pilose. (comb-tooth thistle.)

CARTHA'MUS. Calyx ovate, imbricate, with scales, ovatish, leafy at the apex; egret chaff-hairy, or none; receptacle chaff-bristly. (false saffron.) *Ex.*

SAUSSU'REA. Involucre sub-cylindric; scales imbricate, beardless; receptacle setose or chaffy; egret in 2 series, outer series short, filiform, inner one long and plumose; anthers cordate, sub-entire; bony akenes, glabrous.

CYNA'RA. Receptacle bristly; calyx dilated, imbricate, scales with fleshy bases, emarginate and pointed; egret plumose, sessile. (garden artichoke.) *Ex.*

AMMO'BIUM. Anthers with 2 bristles at the base; chaffs of the receptacle distinct; egret with toothed edge; sepals imbricated, colored, radiated.

ONOPOR"DON. Calyx ventricose, imbricate, with spreading, spinous scales; receptacle alveolate; egret capillary, deciduous, scabrous. (cotton thistle.)

LIA'TRIS. Calyx oblong, imbricate; receptacle naked; egret plumose, persistent (mostly colored); seed pubescent, striate.

VERNO'NIA. Calyx imbricate, ovate; egret double, exterior short, chaffy, interior capillary; receptacle naked; stigma 2 cleft.

STOKE'SIA. Involucre foliaceous, sub imbricate; florets of the ray funnel-form and regular; receptacle naked; egret 4-bristled. *S.*

STE'VIA. Receptacle naked; egret chaff-bristled; involucre cylindric, from a simple series of leafets. *S.*

BRICKEL"LIA. Receptacle naked, dotted; egret hairy or scabrous; akenes nearly glabrous, 10 streaked; involucre many-leaved, imbricate. *S.*

C. *Florets tubulous; flower discoid.*

EUPATO'RIUM. Calyx imbricated (rarely simple), oblong; style long, cloven half way down; egret pilose, scabrous, or rough papillose; receptacle naked; seed smooth and glandular, 5-striate. (boneset, thoroughwort, joepye.)

MIKA'NIA. Calyx 4-6-leaved, 4-6-flowered; receptacle naked; egret pilose.

KUH'NIA. Calyx imbricate, cylindric; receptacle naked; egret plumose, sessile; seed pubescent, striate. (false boneset.)

POLYP"TERIS. Involucrum many-leaved, leaves oval; egret chaffy, many-leaved, the chaff broad-subulate, cuspidate, rigid, as long as the seed.

CHRYSOCO'MA. Calyx imbricate, oblong; receptacle naked; egret hairy, scabrous; seed pubescent. (golden-locks.)

CACA'LIA. Calyx cylindric, scaly at the base; receptacle naked; egret hairy. (wild-caraway.)

SPARGANOPH"ORUS. Calyx sub-globose, imbricate; scales secured at the point; receptacle naked; seed crowned with a cartilaginous, shining cup. (water-crown-cup.) *S.*

MARSHAL"LIA. Involucrum imbricate; scales sub-lanceolate, incumbent; receptacle chaffy; egret 5, membranaceous, acuminate; nerveless scales.

MELANANTHE'RA. Involucrum imbricate; leafets ovate, close-pressed, sub-equal; receptacle chaffy; scales keeled, the lower part embracing the florets; egret consisting of 4 or 5 unequal, unarmed awns.

SANTOLI'NA. Calyx imbricate, hemispherical; scales keeled, with scarious points. *S.*

AGERA'TUM. Egret with 5 somewhat awned scales; leaves of the calyx oblong, in a double row; corolla 4 or 5-cleft; receptacle naked. Ex.

ORDER II. POLYGAMIA SUPERFLUA.

A. *Flowers discoid; the ray-florets being obsolete.*

TANACE'TUM. Calyx imbricate, hemispheric; scales acuminate; rays obsolete, 3-cleft; egret somewhat marginal; receptacle naked; flowers corymbed. (tansey.)

ARTEMI'SIA. Calyx imbricate, ovate, with scales rounded, converging; ray-florets sub-ulate; egret 0; receptacle somewhat villose, or nakedish; flowers mostly rounded. (wormwood, southern-wood.) *S.*

GNAPHA'LIUM. Calyx imbricate, with the marginal scales rounded, scarious, shortish, glossy, colored; receptacle naked; egret pilose or plumose, scabrous; florets of the ray subulate, of the disk entire. Sometimes all the florets are perfect. (life everlasting.)

CONY'ZA. Involucre imbricate, the scales appressed; receptacle naked; marginal florets fertile, 3-cleft; egret simple, capillary; acines hairy.

BAC"CHARIS. Calyx imbricate, cylindric; scales ovate, sub-coriaceous; fertile florets mixed with the perfect; receptacle naked; egret hairy. (groundsel-tree.)

PTEROCAU'LON. Involucre imbricate, with close-pressed, downy, sub-scarious scales; receptacle naked; perfect and pistillate florets intermixed, the pistillate ones slender, border 3-toothed, perfect ones with a 5-cleft border; egret hairy, scabrous; akenes angled, hairy. *S.*

B. *Flowers radiate; the ligulate ray-florets very manifest.* [*Receptacle naked.*]

ERI'GERON. Calyx imbricate, sub-hemispherical; florets of the ray very numerous and narrow; egret double, outer minute, inner hairy, of few rays.

INU'LA. Calyx imbricate, generally squarrose; egret simple, scabrous, sometimes a minute, exterior, chaffy one; anthers ending in 2 bristles at the base; ray-florets numerous, always yellow. (elecampane.)

AS"TER. Calyx imbricate, the inferior scales generally spreading; egret simple, pilose; receptacle often deep-pitted; florets of the ray more than 10, except in a few species; color purple or white, never yellow. (star-flower.)

SOLIDA'GO. Calyx oblong or sub-cylindric, with oblong, narrow, pointed, straight scales, imbricate, closed upon the flower; ray-florets about 5, and fewer than 10, lanceolate, 2-toothed, equal to, or shorter than the calyx; filaments capillary, very short; style thread-form, equalling the length of the stamens; stigma-cleft, spreading; egret simple, pilose, scabrous; receptacle furrowed with dots or punctures; seeds oblong, ovate; yellow. (golden-rod.)

TUSSILA'GO. Calyx simple, swelling, scales equal, and equalling the disk, sub-membranous; pistillate florets ligulate or without teeth; egret simple, sessile; sometimes polygamous. (colt's-foot)

CHRYSAN"THEMUM. Calyx hemispherical, imbricate, with the scales membranous at the margin; egret none, or a narrow margin. (ox-eyed daisy, fever-few.) *h*

BEL"LIS. Calyx hemispherical; scales equal; egret 0; receptacle conical; seed ovate. (garden daisy.) Ex.

TAGE'TES. Calyx simple, 1-leafed, 5-toothed, tubular; florets of the ray about 5, permanent; egret 5 erect awns. (marygold.) Ex.

MATRICA'RIA. Involucre flat, imbricate, with scales having scarious margins; receptacle naked, terete; egret none. Ex.

TRI'CO'PHYL"LUM. Involucre oblong-cylindric, many-leaved, equal; ray-florets oblong; receptacle naked; egret chaffy, minute, 5 to 8-leaved; leafets obtuse, awnless. *S.*

PEC"TIS. Involucrum 5-leaved; ray-florets 5; receptacle naked; egret 3-5 awns.

ARNI'CA. Calyx hemispherical, leafets equal, longer than the disk; receptacle na ked; egret simple, hairy; florets of the ray yellow, often destitute of anthers.

SENE'CIO. Calyx sub-cylindric, equal, scaly at the base; scales withered at the points; receptacle naked; egret simple; rays sometimes wanting. (fire-weed.)

CINERA'RIA. Involucre simple, many-leaved, equal; egret simple.

BOLTO'NIA. Calyx imbricate; rays numerous; receptacle conic, punctate; seeds flat; egret consisting of minute bristles, with 2 elongated and opposite bristles. (false chamomile.)

CHRYSOP"SIS. Calyx imbricated; ray-florets mostly yellow; receptacle naked; egret double, outer one chaffy, minute, inner one scabrous, many-rayed. *S.*

DAH"LIA. Receptacle chaffy; egret none; calyx double, outer one many-leaved, inner one 1-leaved, 8-parted. Ex.

[*Receptacle chaffy or hairy.*]

AN"THEMIS. Calyx hemispherical; scales with scarious margins, nearly equal; egret 0, or a membranous margin; florets of the ray more than 5; receptacle chaffs flat, with a rigid, acuminate apex; seed crowned with a membranous border or egret. (may-weed, chamomile.)

ACHILLE'A. Calyx imbricate, ovate, unequal; egret 0; florets of the ray 5 to 10, roundish, dilated; flowers corymbed. (yarrow.)

HELIOP"SIS. Calyx imbricate, with ovate-linear lined scales; ray-florets linear, large; receptacle chaffy, conic; the chaffs lanceolate; seeds 4-sided; egret 0. (sun-ray.)

HELE'NIUM. Calyx 1-leafed, many-parted; egret 5-awned, chaffy leaves; receptacle globose, naked in the disk, and chaffy in the ray only; florets of the ray half 3-cleft; seed villose; leaves decurrent. (false sunflower.)

VERBESI'NA. Calyx many-leaved; leafets disposed in a double series; rays about 5; receptacle chaffy; egret awned. (crown-beard.)

ECLIP"TA. Involucrum many-leaved, the leaves nearly equal; florets of the disk 4-cleft; egret none; receptacle bristly. *S.*

SIEGESBEC"KIA. Outer involucrum 5-leaved, spreading, inner one many-leaved, 5-angled, nearly equal; rays only on one side of the flower; receptacle chaffy; egret 0; akenes somewhat 4-sided. *S.*

ZIN"NIA. Calyx ovate, cylindric; rays 5, entire, permanent; receptacle chaffy; egret 2, erect awns. (blood marygold.) *S.*

ORDER III. POLYGAMIA FRUSTRANIA.

HE'LIAN"THUS. Calyx imbricate, sub-squarrose, leafy; receptacle flat, chaffy; egret 2-leaved, chaff-like, caducous. (sunflower, Jerusalem artichoke.)

RUDBECK"IA. Calyx consisting of a double series of leafets or scales; receptacle chaffy, conic; egret a 4-toothed margin, or 0. (cone-flower.)

BI'DENS. Calyx sub-equal, leafy or scaly at the base; rays often wanting; receptacle chaffy, flat; egret of 2 or 4 awns; seed quadrangular. (burr-marygold.)

COREOP"SIS. Calyx double, each series many-leaved, the interior equal and colored; receptacle chaffy; scales flat; seed compressed, emarginate. *S.*

CENTAU"REA. Calyx various, mostly imbricate, roundish; egret simple, various; receptacle bristly; corollas of the ray funnel-shape, longer, irregular. (blue-bottle, blessed thistle.) Ex.

LEP"TOPO'DA. Involucrum simple, many-parted; rays 20 or more, 3-cleft, widening at the top; receptacle naked, hemispherical; egret consists of 8 to 10 awnless, chaff-like valves. *S.*

GALAR"DIA. Involucre many-leaved, flat, sub-equal; rays 3-cleft, widening toward the top; receptacle bristly, hemispherical; egret chaffy; leafets 8 to 10, awned. *S.*

ACTINOME'RIS. Calyx simple, many-leaved, foliaceous, sub-equal, remote, elongated, 4 to 8; receptacle small, chaffy; seed compressed; margin crowned with 2 persistent awns.

ORDER IV. POLYGAMIA NECESSARIA.

CALEN"DULA. Calyx many-leaved, equal; receptacle naked; egret none; seeds of the disk membranaceous. (pot marygold.) Ex.

SILPH"IUM. Calyx squarrose, scales broad and leafy; receptacle chaffy; seed flat, obcordate, emarginate, bidentate.

POLYM"NIA. Calyx double, exterior 4 to 5-leaved, interior 10-leaved; leafets concave; receptacle chaffy; egret 0.

GYMNOSTY'LES. Involucre many-leaved, the leafets in a single series; pistillate florets, apetalous; akenes compressed, somewhat toothed at the summit, awned with the permanent styles. *S.*

PARTHE'NIUM. Involucrum 5-leaved; rays very small; receptacle chaffy, minute; outer scales dilated; akenes obovate, minutely 3-awned.

I'VA. Calyx about 5-parted; florets of the ray 5; receptacle having seeds obovate, naked. (high-water shrub.)

CHRY"SOGO'NUM. Involucre 5-leaved; receptacle chaffy; egret 1-leaved, 3-toothed; akenes surrounded by a 4-leaved calycle. *S.*

ORDER V. POLYGAMIA SEGREGATA.

ELEPHANTO'PUS. Partial calyx 4-flowered; florets 5-cleft, ligulate, perfect; receptacle naked; down setaceous. (elephant-foot.)

ECHI'NOPS. Proper calyx 1-flowered; corolla perfect, tubular; receptacle setose (globe-thistle.)

CLASS XVIII. GYNANDRIA.

ORDER I. MONANDRIA.

A. *Anthers adnate, sub-terminal, not caducous; masses of pollen affixed by the base, and made up of angular particles.*

OR"CHIS. Corolla ringent-like, upper petal vaulted; lip dilated, spurred beneath; masses of pollen 2, adnate, terminal. (orchis.)

PLATAN"THERA. Corolla vaulted; lips narrow, entire, spurred at the base; cells of the anther widely divided at the base by the broad interposed stigma; pollinia pedicelled; glands of the pedicels naked.

HABENA'RIA. Corolla ringent; lip spurred at the base beneath; stripes of the pollinia with naked and distinct glands; cells of the stalks adnate, or separated.

B. *Anther parallel with the stigma, not caducous; masses of pollen affixed to the summit of the stigma, and made up of farinaceous or angular particles.*

GOODYE'RA. Corolla ringent-like, the lower petals placed under the gibbous lip, which is divided above; style free; constituent particles of the masses of pollen angular. (rattlesnake-leaf, scrophula weed.)

NEOT"TIA. Corolla ringent, the 2 lower petals placed beneath the lip, which is beardless, interior petals converging; style wingless; pollen farinaceous.

LISTE'RA. Corolla irregular; lip 2-lobed, sessile, with no calli; column apterous; anther fixed by the base. (tway-blade.)

CRANI'CHIS. Corolla 5-petalled, resupinate, sub-ringent lip, vaulted behind.

C. *Anther inserted, terminal, not caducous; masses of pollen farinaceous or angular.*

POGO'NIA. Petals 5, distinct, without glands; lip sessile, cowled, crested internally; pollen farinaceous. (snake-mouth.)

CYMBID"IUM. Petals 5, distinct; lip behind, or inverted, unguiculate; the lamina bearded; style free; pollen angular. (grass pink.)

ARETHU'SA. Petals 5, connate at the base; lip below growing to the style, cowled above, crested within; pollen angular. (arethusa.)

TRIPHO'RA. Petals 5, distant, equal and connivent, without glands; lip unguiculate, cucullate; column spatulate, flattened, apterous. (three-bird-orchis.)

TIPULA'RIA. Segments of the perianth spreading; lip entire, sessile, with a conspicuous spur at the base beneath; column or style without wings, lengthened, free; anthers resembling a lid, permanent; pollinia (or masses of pollen) 4, parallel. (limodore.)

LIPA'RIS. MALAX"IS. Corolla spreading; petals 5; lip flat, expanded, entire, turned various ways; 'column or style winged; pollinia 4, parallel, affixed to the summit of the stigma.

MICROS"TYLIS. Lip flat, sagittate or deeply cordate; column very small, round; pollinia 4, loose.

CORALLORHI'ZA. Lip produced behind, adnate with the spur, or free; pollinia 4, oblique, not parallel. (coral-root.)

APLEC"TRUM. Lip unguiculate, not produced at the base; anther below the summit of the column; pollinia 4, oblique, lenslike.

CALYP"SO. Segments of the perianth ascending; petals 1-sided, lip ventricose,

spurred beneath, near the end, column petaloid, dilated; pollinia 4.

EPIDEN"DRUM. Pollinia 4, parallel, each mass with an elastic filament at the base; style united with the claw of the lip into a tube. (vanilla plant.)

ORDER II. DIANDRIA.

CYPRIPE'DIUM. Calyx colored, 4-leaved, spreading; corolla 0 (by some the calyx is called a corolla); nectary large, hollow, inflated; style with a terminal lobe, and petallike appendage on the upper side. (ladies' slipper.)*

ORDER V. PENTANDRIA.

Plants bearing seeds in follicles, and pollen in masses called pollinia.

ASCLE'PIAS. Petals 5, reflexed; nectaries 5, concave, erect, containing little horns; each stamen with a pair of pendulous masses of pollen suspended from the top of the stigma; follicle smooth. (milk-weed, silkweed.)†

APOC"YNUM. Corolla bell-form; stamens with converging anthers, proceeding from the middle of the stigma, and alternating with 5 nectaries; stigma thick, almost sessile; follicles in pairs, long, linear. (dogbane, Indian hemp.)

ACERA'TEA. Corolla reflexed; 5 concave, short nectaries; each stamen with a pair of pendulous masses of pollen; follicle smooth; corolla with purple tips, much longer than the calyx.

ECHI'TES. Follicles 2, distinct, terete; seed crowned with a pappus; corolla funnel-shaped, with the border 5-parted; anthers adhering in the middle to the stigma; scales 5, fleshy, surrounding the base of the germ. S.

GONOLO'BUS. Corolla wheel-form, 5-parted; nectary cylindric, fleshy, 5-lobed; anthers opening transversely, terminated by a membrane; pollinia 5 pairs, not separating into grains; stigma flat; follicle 2, ventricose; seeds comose. (false choke-dog.)

PODOSTIG"MA. Stigma on a stipe; masses of pollen 10, smooth, pendulous; nectary 5-leaved; leaves compressed; corolla bell-form; follicles smooth. S.

PERIPLO'CA. Calyx 5-cleft; corolla rotate, 5-parted; orifice surrounded with an urceolate crown, terminating in 6 filiform awns; style 1; stigma 5-cornered; pollinia solitary, composed of 4 grains; follicles 2, divaricate. (milk-vine.)

ANSO'NIA. Follicles 2, erect; corolla funnel-shaped, with the throat closed; seeds terete, naked, with the summit obliquely truncate.

HOY'A. Corolla 5-cleft; pollen masses fixed by the base, conniving, compressed; stigma depressed with an obtuse wart; follicles smooth; seeds concave. Ex.

---

* The ladies' slipper of the garden belongs to the genus Impatiens, of the class Pentandria.

† The genera in this order are, by many botanists, placed in the class Pentandria.

ENSLE'NIA. Calyx small, 5-parted, permanent; corolla 5-parted; segments converging, erect; nectary 5-parted, petal-like, divided almost to the base; segments truncate, flat, each terminated by 2 central filaments; each stamen with a pair of pendulous, cylindric masses of pollen, suspended from the top of the conic stigma; follicles in pairs, small.

ORDER VI. HEXANDRIA.

ARISTOLO'CHIA. Calyx 0; corolla superior, 1-petalled, ligulate, inflated at the base; capsule 6-celled, many seeded. (birthwort.)

ORDER XII. POLYANDRIA.

AS"ARUM. Calyx sub-campanulate, 3 to 4-cleft; corolla 0; anthers adnate to the middle of the filaments; capsule inferior, 5-celled, crowned with the calyx. (wild ginger.)

CLASS XIX. MONŒCIA.

ORDER I. MONANDRIA.

ZOS"TERA. Stamens and pistils inserted in 2 rows upon one side of a spadix; spatha foliaceous. Staminate flowers with the anthers ovate, sessile, alternating with the germs. Pistillate flowers with the germ ovate; style 2-cleft; drupe with 1 seed. (grass-wrack.)

CAULIN"IA. Staminate flowers: calyx 0; corolla 0; anthers sessile. Pistillate flowers: calyx and corolla wanting; style filiform; stigma 2-cleft; capsule 1-seeded; flowers axillary. (river-nymph.)

ZANNICHEL"LIA. Staminate flowers: calyx and corolla wanting. Pistillate flowers: perianth single, of 1 leaf; ovaries 4 or more; style 1; stigma peltate; capsule sessile.

CHA'RA. Staminate flowers: calyx 0; corolla 0; anthers globose, sessile. Pistillate flowers: calyx 0; corolla 0; style 0; stigmas 5; berry 1-celled, many-seeded. (chara.) N.

NA'JAS. Staminate flowers: calyx cylindric, 2-cleft; stamen filamentous, long; anther 4-valved, valves spreading. Pistillate flowers: calyx 0; style with 2 stigmas; nut 4-seeded.

EUPHOR"BIA. Rarely a perianth; involucre monophyllous, campanulate, 8 to 10-toothed, the inner segments membranaceous. Staminate flowers 12 or more; calyx and corolla generally wanting. Pistillate flowers: solitary, central, stipitate; calyx and corolla 0; capsule 3-lobed. (spurge.)

ORDER II. DIANDRIA.

LEM"NA. Staminate flowers: perianth of 1 leaf; stamens on the base of the germ. Pistillate flowers: perianth of 1 leaf; stigma funnel-form; capsule 1-celled, from 1 to 5-seeded. (duck's meat.)

PODOS"TEMUM. Staminate flowers: calyx 0; corolla 0; stamens affixed to a pedicel. Pistillate flowers: calyx 0; corolla 0; germ ovate; stigma 1, sessile; capsule 2-celled, 2-valved, many-seeded; seeds minute. (thread-foot.)

ORDER III. TRIANDRIA.

TY'PHA. Ament cylindric, dense-flowered. Staminate flowers: calyx obsolete, 3-leaved; corolla 0; stamens 3 together, on a chaffy or hairy receptacle, united below into 1. Pistillate flowers: below the staminate; calyx 0; corolla 0; seed 1, pedicelled; the pedicels surrounded at the base with long hairs resembling egret. (cat-tail, or reed mace.)

SPARGA'NIUM. Ament globose. Staminate flowers: calyx 3-sepalled; corolla 0. Pistillate flowers: calyx 3-sepalled; corolla 0; stigma 2-cleft; drupe juiceless, 1 or 2-seeded. (burr-reed.)

SCLE'RIA. Staminate flowers: glume 2 or 6-valved, many-flowered; paleas awnless. Pistillate flowers: calyx 2 or 6-valved, 1-flowered; paleas none; stigmas 1 to 3; not colored, sub-globose. (whip-grass.)

CA'REX. Aments imbricate, usually in cylindric spikes. Staminate flowers: calyx-scales single; corolla 0. Pistillate flowers: calyx-scales single; corolla inflated, monopetalous, 2-toothed at the apex; stigmas 2 or 3; nut 3-sided, enclosed in the inflated, permanent corolla, which becomes an utriculus-like permanent aril; sometimes diœcious. (sedge.)

TRIP"SACUM. Staminate flowers: glume 2-flowered, outer one staminate, inner one neutral; corolla a membranaceous glume. Fertile flowers: glume 1 or 2-flowered, surrounded by a 1-leafed involucrum perforated near the base; paleas with numerous thin membranaceous valves; styles 2; seed 1. (sesame grass.)

COMPTO'NIA. Staminate flowers: ament cylindric, with calyx-scales 1-flowered; corolla 2-petalled or none; filaments 2-forked. Pistillate flowers: spike or ament oval; corolla 6-petalled (the corolla may be called a calyx); styles 2; nut oval, 1-celled. (sweet-fern)

CO'IX. Staminate flowers: in remote spikes; calyx-glume 2-flowered, awnless; corolla-glume awnless. Pistillate flowers: calyx-glume 2-flowered; corolla-glume awnless; style 2-parted; seed covered with the bone-like calyx. (Job's tear.) Ex.

ZE'A. Staminate flowers: calyx-glume 2-flowered, awnless; corolla-glume awnless. Pistillate flowers: calyx-glume 2-valved (number of valves increased by cultivation); style 1, very long, filiform, pendulous; seed solitary, immersed in an oblong receptacle. (Indian corn.) S.

TRA'GIA. Staminate flowers: calyx 3-parted; corolla none. Pistillate flowers: calyx 5-parted; corolla none; style 3-cleft; capsule tricoccus, 3-celled; seed solitary. S. [Grasses found in class 3d vary into this order.]

ORDER IV. TETRANDRIA, TO ORDER VI. HEXANDRIA.

AL"NUS. Staminate flowers: ament composed of wedge-form, truncate, 3-flowered receptacles; calyx a scale, 3-lobed; corolla 4-parted. Pistillate flowers: calyx 2-flowered scales, somewhat 3-cleft; corolla 0; seed compressed, ovate, wingless. (alder.)

ERIOCAU'LON. Involucre many-leaved;

florets many, in an imbricate head; partial perianth superior, 2 or 3-sepalled. Staminate flowers central, with monopetalous-cleft corollas. Pistillate flowers marginal, with 2-petalled corollas; stigmas 2 or 3; capsule 2 or 3-celled, 2 or 3-lobed; pericarp 1-seeded, crowned with the corolla. (pipe wort.)

XAN"THIUM. Monœcious. Staminate flowers: involucrum imbricate; anthers approximate, but not united; receptacle chaffy. Pistillate flowers: involucrum 2-leaved, 1-flowered; corolla none; drupe muricate, 2-cleft; nut 2-celled. (sea-burdock.)

UR"TICA. Staminate flowers: calyx 4-leaved; corolla 0; nectary central, cyathiform. Pistillate flowers: calyx 2-leaved, 2-valved; corolla 0; seed 1, glossy. (nettle.)

MO'RUS. Staminate flowers: calyx 4-parted; corolla 0. Pistillate flowers: calyx 4-leaved; corolla 0; styles 2, calyx becoming berry-like; seed 1. (mulberry.) S.

BUX"US. Staminate flowers: calyx 3-leaved; petals 2. Pistillate flowers: calyx 4-leaved; petals 3; styles 3. (box.) Ex.

AMARAN"THUS. Staminate flowers: calyx 3 or 5-leaved; corolla 0; stamens 3 or 5. ... ate flowers: calyx and corolla as ... ...ute; styles 3; capsule 1-celled, ...u anstransversely; seed 1. (amaranth, ...ocksetmb.)

... ...'SIA. Staminate flowers: common ... ...ved; anthers in contact, but not ... ...rolla 1-petalled, 5-cleft, funnel-...ptacle naked. Pistillate flowers: ...d, entire, or 5-toothed, 1-flow-...ola 0; nut covered with the indu-...lyx, 1-seeded. (hog-weed.)

...A N"DRA. Calyx about 4-sepalled; ...e; filaments sub-clavate; styles ... 3-horned, 3-celled; cells 2-seed-

...S"DRA. Calyx 9-sepalled; sepals ...ew; corolla 0; anthers sub-sessile, ...sew at the tips; berries 1-seeded, in-...un elongated receptacle. S.

...FFA'RIA. Polygamous. Perfect flow-...lyx 4-cleft, inferior; corolla none; ...lastic; style 1; seed 1. Pistillate ... calyx 2-sepalled; nut covered ...in the dry elongated calyx. (pellitory.)

ORDER VI. HEXANDRIA.

ZIZA'NIA. Staminate and pistillate flowers mixed. Staminate flowers: calyx 0; corolla glume 2-valved, awned. Pistillate flowers: glume 2-valved, hooded, awned; style 2 parted; seed inverted in the plaited, glume-like corolla. (wild-rice.)

HYDROCHLO'A. Pistillate flowers: glumes none; paleas awnless. Staminate flowers: glumes none; paleas 2, awnless; stigmas 2, very long; seed 1, reniform.

Co'cos. See specific description.

ORDER XII. POLYANDRIA.

A. Stems not woody.

SAGITTA'RIA. Staminate flowers: calyx 3-leaved; corolla 3-petalled; filaments mostly 24. Pistillate flowers: calyx and corolla as in the staminate; germs many; capsules

aggregate, 1-seeded, not opening. (arrow-head.)

CERATOPHYL"LUM. Staminate flowers: calyx many-parted; corolla none; stamens 16-20; anthers tricuspidate. Pistillate flowers: corolla none; style 1, filiform; nut 1-seeded. (horn-wort.)

MY'RIOPHYL"LUM. Flowers monœcious, or rarely perfect. Staminate florets: calyx 4-parted; petals 4-lobed, alternating with the calyx, ovate, caducous; stamens 4-6, or 8. Pistillate flowers: calyx adhering to the ovary; limb 4-lobed; petals none; nuts 4, compressed or sub-globose, 1-seeded. (water milfoil.)

A'RUM. Spatha cucullate, 1-leafed; spadix not entirely covered with the fructification, being more or less naked above, with pistillate flowers beneath, and staminate in the middle (sometimes a few are staminate beneath; berry mostly 1-seeded, generally cirrose-glandular beneath). (Indian, or wild turnip, wakerobin.)

RENSSELAE'RIA. Spatha convolute; spadix covered with flowers, fertile at base, sterile above; perianth 0; berry 1-seeded. (spear arum.)

CAL"LA. Spatha ovate, becoming expanded; spadix covered with the fructification; stamens intermixed. Staminate flowers: calyx and corolla 0; anthers sessile. Pistillate flowers: calyx and corolla 0; berries 1-celled, many-seeded, crowned with the short style. (water-arum.)

POTE'RIUM. Staminate flowers: calyx 4-leaved; corolla 4-parted; stamens 30 to 50. Pistillate flowers: calyx and corolla like the staminate; pistils 2; berry from the indurated tube of the corolla. (burnet.)

B. Stems woody.

QUER"CUS. Staminate flowers: ament loose; calyx sub 5-cleft; corolla 0; stamens 5 to 10. Pistillate flowers: calyx 1-leafed, entire, scabrous, being a woody cup; style 1; stigmas 2 to 5; nut or acorn 1-celled, 1-seeded, coriaceous, surrounded at the base by the permanent calyx. (oak.)

CORY'LUS. Staminate flowers: ament cylindric, imbricate; calyx a 3-cleft scale; pericarp none; stamens about 8. Pistillate flowers: calyx 2-parted, laciniate; stigmas 2; nut ovate, surrounded by and included in the permanent leaf-like calyx. (hazle-nut.)

FA'GUS. Staminate flowers: ament roundish; calyx 5 or 6-cleft, bell-form; stamens 5 to 12. Pistillate flowers: calyx 5-toothed, setose; germs 2; nuts 2, enclosed in the calyx, becoming coriaceous, echinate. (beech.)

CASTA'NEA. Polygamous. Staminate flowers: ament naked, linear; corolla or calyx 1-leafed, 5 or 6-parted; stamens 10 to 20. Pistillate flowers: calyx 5 or 6-leaved, or 5 or 6-lobed, muricate; germs 3; nuts 3, with coriaceous putamen enclosed in the calyx, becoming echinate. (chestnut.)

BE'TULA. Staminate flowers: ament cylindric, imbricate; scales peltate, 3-flowered; stamens 10 to 12. Pistillate flowers: calyx a 2 or 3-flowered scale; seed 1, winged. (birch.)

PLATA'NUS. Ament globose. Staminate flowers: corolla none, or scarcely apparent; anthers growing around the filaments. Pistillate flowers: calyx many-leaved; style with a recurved stigma; seed roundish, crowned with the mucronate style, with egret-like hairs at the base. (button-wood, false sycamore.)

JU'GLANS. Staminate flowers: ament imbricate; calyx a scale, generally 5-parted; corolla 4 or 5-parted; stamens 18 to 36. Pistillate flowers: calyx 4-cleft; corolla 4-cleft or 4-parted; styles 1 or 2; drupe partly spongy; nut rugose and irregularly furrowed. (butternut, black-walnut.)

CAR'YA. Staminate flowers: ament imbricate; calyx of 3 parted scales; corolla 0; stamens 4 to 6. Pistillate flowers: calyx 4-cleft, superior; corolla 0; styles 0; stigma disk-like, 4-lobed; pericarp 4-valved; nut sub quadrangular, even. (hickory, walnut.)

LIQUIDAN"BER. Staminate flowers: ament conical, surrounded by a 4-leaved involucre; perianth none. Pistillate flowers: ament globose; perianth 1-leafed, urceolate, 2-flowered; styles 2; capsules 2, 1-celled, many-seeded. (sweet gum-tree.) S.

OS"TRYA. Staminate flowers: ament cylindrical; scales 1 flowered; perianth 0; filaments branched. Pistillate flowers: ament naked; capsule inflated, imbricate. (iron wood, hop, horn-beam.)

CARPI'NUS. Staminate flowers: ament long-cylindric; scales ciliate at the base; stamens 8 to 14, somewhat bearded at the top. Pistillate flowers: strobilum loose; scales leafy, 2-flowered; stigmas 2; nut long, sulcate, 1-seeded. (horn-beam.)

ORDER XV. MONADELPHIA.

JATRO'PHA. Staminate flowers: calyx 5-leaved or wanting; corolla funnel-form; stamens 10, alternately shorter. Pistillate flowers: calyx 0; corolla 5 petalled, spreading; style 3-cleft; capsule 3-celled, seed 1. (physic-nut.) S.

CRO'TON. Staminate flowers: calyx 5-toothed; petals 5 or wanting; stamens 10 to 15. Pistillate flowers: calyx 5-leaved or none; corolla none; styles 2-cleft; capsule 3-grained, 3-celled, 3-seeded. S.

STILLIN"GIA. Staminate flowers: involucre hemispherical, many-flowered or wanting; perianth tubular, eroded; stamens 2 or 3, exsert. Pistillate flowers: calyx 1-flowered, inferior; style 3-cleft; capsule 3-grained. (tallow-tree.) S.

MELO'THRIA. Staminate flowers: calyx 1-flowered, 3 to 5-toothed; corolla bell-form; filaments 3. Pistillate flowers: calyx and corolla superior; style 1; stigmas 3; berry 3-celled, many-seeded. (creeping cucumber.)

MOMOR"DICA. Staminate flowers: calyx 5 or 6-cleft; corolla 5 or 6-parted; filaments 3. Pistillate flowers: style 3-cleft; berry gourd-like, and bursting elastically; seeds compressed. (balsam apple, wild cucumber.)

CU'CUMIS. Staminate flowers: calyx 5-toothed; corolla 5-parted; filaments 3. Pistillate flowers: calyx and corolla like the staminate; stigmas 3, thick, 2-parted; berry with pointed seeds. (cucumber, musk-melon.) Ex.

CUCUR"BITA. Staminate flowers: calyx 5-toothed; corolla 5-cleft; filaments 3. Pistillate flowers: calyx and corolla like the staminate; pistil 3-cleft; berry large, 3 to 5 celled; seeds thickened at the margin. (gourd, squash, pumpkin, water-melon.)

RICI'NUS. Staminate flowers: calyx 5-parted; stamens numerous. Pistillate flowers: calyx 3-parted; styles 3, 2-cleft; capsules echinate, 3-celled, 3-seeded. (palma christi, or castor oil plant) Ex.

SIC"YOS. Staminate flowers: ament imbricate; calyx 5-toothed, teeth subulate; corolla 5-parted; filaments 3, or perhaps 5 in 3 sets. Pistillate flowers: style 3-parted; stigmas thick, 3-parted; fruit 1-seeded, often spinose. (single-seed cucumber.)

PI'NUS. Staminate flowers: calyx 4-leaved, peltate; corolla 0; stamens many; anthers naked, 2, sessile, 1-celled. Pistillate flowers: calyx in strobilums or cones; scales close-imbricate, 2-flowered; pistil 1; nut with a membranaceous wing, or a samara. (pine.)

CUPRES"SUS. Staminate flowers: ament ovate, imbricate; calyx a peltate scale; corolla 0; anthers 4, sessile. Pistillate flowers: ament strobilaceous; calyx a 1-flowered, peltate scale; corolla 0; germs 4 to 8, under each scale of the calyx; nuts angular, compressed. (white cedar.)

THU'JA. Staminate flowers: ament imbricate; calyx and corolla 0; anthers 4, sessile. Pistillate flowers: strobilum with scales 2-flowered; corolla 0; nut 1, wing. (arbor vitæ.)

PHYLLAN"THUS. Staminate flowers: calyx 5 or 6-parted; filaments often columnar; anthers 3. Pistillate flower resembling the staminate; nectary a 12-angled margin; styles 3; capsules mostly 3-grained. (leaf-flower.)

ACALY'PHA. Staminate flower: calyx 3 to 4-parted; corolla 0; stamens 8-16. Pistillate flower: calyx 3-leaved; corolla 0; styles 3; capsule 3-celled. (three-seed mercury.)

CLASS XX. DIŒCIA.

ORDER II. DIANDRIA.

SA'LIX. Staminate flowers: ament cylindric; calyx a 1-flowered scale, with a nectariferous gland at the base; stamens 1 to 6. Pistillate flowers: ament and calyx like the staminate; stigmas 2, generally 2-cleft; capsule 1-celled, 2-valved; seeds many, with egret-like down. (willow.)

FRAX"INUS. Polygamous. Perfect flowers: calyx 0, or 3 or 4-parted; corolla 0, or 4-petalled; pistil 1; samara 1-seeded, with a lanceolate wing. Pistillate flowers: calyx, corolla, and pistils, same as perfect. (ash.)

VALLISNE'RIA. Staminate flowers: spatha ovate, 2-parted; spadix covered with minute flowers; calyx 3-parted. Pistillate flowers: spatha 2-cleft, 1-flowered; calyx 3-parted, superior; corolla 3-petalled; stigmas ligulate, 2-cleft; capsules without valves, 1-celled; seeds numerous, attached to the sides. (tape-grass.)

CERATIO'LA. Calyx bud-like, imbricated with 6 to 8 scales; corolla 0; stamens 2, exsert; stigmas 4 to 6, 2 of them longer; berry with 2 long seeds.

ORDER III. TRIANDRIA.

EMPE'TRUM. Calyx 3-parted, persistent. Staminate flowers: petals 3, marescent; stamens 3; filaments long; anthers 2-parted. Pistillate flowers: germ superior, depressed; style 0, or very short; stigmas 9, reflexed, spreading; berry round, 1-celled. 2 to 6-seeded; seeds bony.

FI'CUS. Common receptacle fleshy (becoming the fruit), enclosing the apetalous florets; both staminate and pistillate, either in the same, or in distinct individuals. Staminate flowers: calyx 3-parted. Pistillate flowers: calyx 5-parted; pistil 1, lateral; seed 1, covered with the closed, permanent, somewhat fleshy calyx. (fig-tree.)

ORDER IV. TETRANDRIA.

VIS"CUM. Staminate flowers: calyx 4-parted; corolla 0; anthers sessile, adhering to the calyx. Pistillate flowers: calyx 4-leaved, superior; corolla 0; style 0; berry 1-seeded, globose; seed cordate; parasitic, adhering to trees. (mistletoe.)

MYRI'CA. Ament ovate, oblong; scales lunulate. Staminate flowers: stamens 4 to 6; anthers 4-valved. Pistillate flowers: germ 1; stigmas 2; drupe 1-celled, 1-seeded. (bay-berry.)

BROUSSONE'TIA. Staminate flowers: ament cylindrical; calyx 4-parted. Pistillate flowers: ament globose; calyx tubular, 3 or 4-toothed; germ club-shaped; seed covered with the calyx. (paper mulberry.) S.

ORDER V. PENTANDRIA.

XANTHOX"YLUM. Staminate flowers: calyx 5-parted; corolla 0; stamens 3 to 5. Pistillate flowers: pistils 3 to 5; capsules equal to the number of pistils, 1-seeded. (prickly-ash, or toothache-tree.)

HU'MULUS. Staminate flowers: calyx 5-leaved; corolla 0; anthers with 2 pores at the extremity. Pistillate flowers: calyx 1-leafed, entire, oblique, spreading; styles 6; seed 1, within the leaf-like calyx; inflorescence strobile-form. (hop.)

NYS"SA. Perfect flower: calyx 5-parted; corolla 0; pistil 1; drupe inferior; nut 1-seeded. Staminate flowers 5, 8, 10, or 12, inserted around a peltate gland. (pepperidge-tree.)

HAMILTO'NIA. Perfect flowers: calyx sub-campanulate, superior, 5-cleft; corolla 0; nectary with a 5-toothed disk; stamens 5; pistil 1; fruit a drupe. Staminate flowers vary only in having no pistil. (American oil-nut.)

ACNI'DA. Staminate flowers: calyx 5-parted; corolla 0. Pistillate flowers: calyx 3-parted; corolla 0; styles 0; stigmas 6, sessile; capsule 1-seeded. (water hemp.)

CAN"NABIS. Staminate flowers: calyx 5-parted. Pistillate flowers: calyx 5-leaved, entire, gaping laterally; styles 2; nut 2-valved, within the closed calyx. (hemp.) Ex.

SPINA'CIA. Staminate flowers: calyx 5-parted; corolla 0. Pistillate flowers: styles 4; seed 1, within the indurated calyx. (spinach.) Ex.

NEGUN"DO. Calyx minute, unequally 4-5-toothed; petals none; anthers 4-5, linear, sessile; pedicels of the staminate flowers capillary, fascicled from lateral aggregate buds; fertile flowers in racemes; samaras in pairs, diverging, 1-seeded; leaves compound, pinnately 3-5 foliate.

ORDER VI. HEXANDRIA.

SMI'LAX. Staminate flowers: calyx 6-leaved; corolla 0; anthers adnate to the filaments. Pistillate flowers: style minute; stigmas 3; berry 3-celled, superior, 1-3-seeded. (green-brier.)

DIOSCORE'A. Staminate flowers: calyx 6-parted; corolla 0; styles 3; capsule 3-celled, triangular, compressed; cells 2-seeded; seeds with membranaceous margins. (yam-root.) S.

GLEDITSCH"IA. Perfect flowers: calyx 6 or 8-parted, deciduous, 3 or 4 of the exterior segments smaller; corolla 0; stamens 5 or 6, seldom 8; legume flatly compressed, 1 or many-seeded. Staminate flowers: calyx sub-turbinate, 5-8-parted; 3 to 5 of the segments interior; stamens 6 to 8. (honey-locust.) S.

ORDER VIII. OCTANDRIA.

POP"ULUS. Staminate flowers: ament cylindric; calyx a torn scale; corolla turbinate, oblique, entire, supporting 8 to 30 stamens. Pistillate flowers: ament, calyx, and corolla, like the staminate; stigma 4 or 6-lobed; capsule 2-celled, 2-valved, many-seeded; seed with egret-like hairs; leaves having a tremulous motion. (poplar, balm of Gilead.)

DIOSPY'ROS. Calyx 4 to 6-cleft, dilated; corolla urceolate, 4-6-cleft. Staminate flowers: stamens 8 to 16; filaments often with 2 anthers. Pistillate flowers: stigmas 4-5; berry 8-12-seeded. (date plum.) S.

HIPPO'PHÆ. Staminate flowers: perianth 4-cleft; stamens 8, alternating with 8 glands. Pistillate flowers: perianth superior, campanulate; style 1; stigma oblique; berry 1-seeded. (sea-buck-thorn.)

ORDER IX. ENNEANDRIA.

UDO'RA. SERPIC"ULA. ELO'DEA. Spatha 2-parted; perianth 6-parted, 3 inner segments petaloid. Staminate flowers: stamens 9, 3 of them interior. Pistillate flowers: tube of the perianth very long; barren filaments 3; utricle about 3-seeded; seeds cylindric. (ditch-moss.)

HY'DROCHA'RIS. Staminate flowers: spatha 2-leaved; calyx 3-leaved; corolla 3-petalled; stamens 8-12, united at the base. Pistillate flowers: spatha 1-leaved, 1-flowered; calyx 3-leaved; corolla 3 petalled, with 6 glands between the petals; germ inferior; styles 6, 2-cleft; capsule 6-celled, many-seeded. *S.*

ORDER X. DECANDRIA

GYMNOCLA'DUS. Staminate flowers: calyx tubular, 5-cleft; petals 5. Pistillate flowers: style 1; legume 1-celled, pulpy within; seed roundish, large, and hard. (coffee-bean.)

CAR"ICA. Staminate flowers: calyx minute; corolla funnel-form, 5-cleft; stamens alternately shorter, enclosed in the tube of the corolla. Pistillate flowers: calyx 5-toothed; petals 5; stigmas 5; berry cucumber-form, grooved, 1-celled, many-seeded. (false papaw-tree.) *S.*

ORDER XII. POLYANDRIA.

MENISPER"MUM. Staminate flowers: calyx 2-bracted, about 6-leaved, glandular; petals 6-9, glandular, minute, retuse; stamens 16-24; anthers adnate to the filaments, 4-lobed, 2-celled. Pistillate flowers: germs and styles 3-6; drupes mostly solitary, 1-seeded; nut lunate, compressed.

CY'CAS. Staminate flowers: ament imbricated; scales spatulate, single; anthers globose, sessile, on a scale. Pistillate flowers: spadix compressed, 2-sided; perianth a scale.

ZA'MIA. Ament a strobile. Staminate flowers: scales obovate; anthers globose, sessile upon scales, opening by a fissure. Pistillate flowers: scales peltate; corolla none; germs 2; styles none; berries 2, 1-seeded.

DATIS"CA. Staminate flowers: calyx 5-leaved; corolla 0; anthers sessile, about 15. Pistillate flowers: calyx superior, 2-toothed; styles 3; capsules 3-angled, 3-horned, 1-celled, many-seeded. (false hemp.)

THALIC"TRUM. See class 12 : 12.

ORDER XV. MONODELPHIA.

JUNIPE'RUS. Staminate flowers: ament ovate, whorled; calyx a peltate scale; anthers 4 to 8. Pistillate flowers: calyx 3-parted; petals 3; styles 3; berry 1 or 2-seeded; nut long, 1-celled, with balsamy glands at the base. (red cedar.)

TAX"US. Staminate flowers: calyx consists of 4 to 6 imbricate scales; corolla 0; stamens 8 to 10; anthers peltate, 6 to 8-cleft. Pistillate flowers: style 0; receptacle succulent; nut or drupe fleshy, 1-seeded. (yew.)

CLASS XXI. CRYPTOGAMIA.

ORDER I. FILICES.

A. *Capsule having an elastic ring at right angles with its opening.*

POLYPO'DIUM. Capsules disposed in round, scattered fruit-dots (or clusters of capsules), on various parts of the lower surface of the frond; involucrum 1. (polypod.)

ACROS"TICHUM. Capsules numerous, covering the lower surface of the froud; involucrum none. The fertile leaves differ in shape from the barren. When old, the fruit often cover the whole frond. (fork-fern.)

HYPOPEL"TIS. Fruit-dots roundish, having a cup-form involucre beneath, divided into 5 or 6 irregular segments; capsules all sessile.

ASPID"IUM. Capsules in scattered, roundish fruit-dots, on various parts of the whole lower surface of the frond; involucrum a kidney-form, or round membrane, fastened to the frond in or near the centre of the fruit-dot, and opening on all sides, or to one side of the fruit-dot, and opening on the other. The involucrum, when a little opened, is often peltate. (shield-fern.)

ASPLE'NIUM. Sori in lines parallel to each other, situated exactly *upon* the secondary veins of the frond; involucres opening inward. (spleen-wort, walking-leaf.)

SCOLOPEN"DRIUM. Sori linear, transverse, scattered; involucrum double, occupying both sides of the sorus, superficial, at length opening longitudinally. (caterpillar-fern.)

PTE'RIS. Capsules arranged in a continued line along the very margin of the frond; involucres opening inward, being formed of the inflexed margins of the fronds. When the leaves are extremely small, the rows of capsules on opposite sides meet and cover the lower surface. (brake.)

A'DIAN"TUM. Capsules disposed in oblong fruit-dots, arranged along the margin of the frond; involucrum is formed by turning back the margin of the frond over the capsules, and it opens inward. The lines of oblong spots are generally along that margin, which may be considered the end of the leaf, or of the segments of the leaf. (maidenhair.)

ONO'CLEA. Fruit-dots indeterminate, presenting a berry-like appearance; capsules covering the whole lower surface of the frond; involucrum formed by turning in or rolling back the margin of the leaf, which opens inward, in maturity, toward the midrib, or remains closed. The fertile leaves are contracted, and narrower than the barren ones. (sensitive polypod.)

B. *Capsule without a ring—being cellular-reticulate, pellucid, sub-striate, radiate at the tip.*

LYGO'DIUM. Capsules sessile, ovate; 2 ranks of small spikes issuing from the margin of the frond, radiate-striate or wrinkled, opening on the inner side from the base to the summit; involucrum scale-like, covering each capsule.

SCHIZ Æ'A. Capsules with radiating furrows at the top, somewhat turbinate, bursting laterally, sessile; involucrum continuous, formed of the inflexed margin of the unilateral spikes.

OSMUN"DA. Capsules globose, pedicelled, radiate-striate or wrinkled, having a hinge at the joining of the 2 valves, which resembles part of the jointed ring of annulated ferns; the capsules either occupy the whole frond, to a limited extent, or a panicled raceme. The parts of the frond occupied by the fruit are always more contracted than the barren parts. (flowering fern.)

C. *Capsule without a ring—being adnate at the base, sub-globose, coriaceous, not cellular, somewhat 2-valved.*

OPHIOGLOS″SUM. Capsules round, 1-celled, opening transversely; they are placed on a somewhat jointed spike in two close rows. (adder-tongue fern.)

BOTRYCH″IUM. Capsules coriaceous, globose, 1-celled, smooth, adnate to each rachis of a compound raceme, separate; valves 2, connected behind, opening transversely. (grape fern.)

D. *Sub-order, APTERES—without pinnate, pinnatifid, or other winged leaves.*

LYCOPO′DIUM. Capsules mostly kidney-form, or roundish, 2 or 4-valved, opening elastically; they are placed under separate scales in a spike, or sometimes in the axils of the leaves; leafy, their stems being generally covered with 2, 3, or 4 rows of narrow, simple, entire leaves. (ground pine.)

EQUISE′TUM. Fruit placed under peltate polygons, being pileus-like bodies, which are arranged in whorls, forming a spike-form raceme; 4 to 7 spiral filaments surround the seed, which resemble green globules. Fertile plants mostly leafless; the stems of all are jointed with toothed sheaths at every joint, and usually longitudinally striated and hollow. (scouring-rush, horse-tail.)

SALVIN″IA. Involucres 4-9, imbricate, connate, resembling a 1-celled capsule; sporules inserted upon a central receptacle.

ORDER II. MUSCI.

FUNA′RIA. Teeth of the outer peristome[*] 16, cohering together at the apex and twisted obliquely; the inner peristome consists of 16 membranaceous hairs, opposite to the teeth, lying flatly.

POLYTRI′CHUM. Peristome very short; teeth 16, 32, or 64; mouth of the germ covered by a dry membrane, which is connected to it by the teeth of the peristome; calyptra very small, with a large villose or hairy covering.

ORDER III. HEPATICÆ.

MARCHAN″TIA. Receptacles pedicelled, radiate-lobed, disk-like, or bell-form, with the inside downward, to which the globose 4-valved capsules are attached with their apexes downward. The umbrella-like receptacle is elevated one or two inches by a stipe attached to the centre of its lower side, among the capsules and many pilose appendages. The frond is leafy, reticulate, furnished with a midrib, and beset with villose roots on the under side, which attach themselves to the stones in brooks, to damp earth, &c.

JUNGERMAN″NIA. Capsules 4-valved, globose, elevated by peduncles or stipes from within a bell-form calyx. The fronds are

made up of finer leaves than those of the Marchantia, and are often mistaken for mosses, among which they generally grow.

ORDER IV. ALGÆ.

A. *The section FUCOIDEÆ comprises those sea-weeds of the old genus Fucus, whose fronds are cartilaginous or leathery, and of an olive or copper color, becoming brown or black. They are composed of interwoven, longitudinal fibres. The floating vesicles appear like portions of the frond blown up in bubbles.*

FU′CUS. Receptacles tubercled; tubercles perforated, nourishing aggregated capsules within, intermixed with articulated fibres.

B. *The section FLORIDEÆ comprises those sea-weeds of the old genus Fucus, whose fronds are leathery, membranous, or gelatinous, and of a purple or rose color.*

HALYME′NIA. Frond membranaceous, leathery, nerveless, punctate; seed immersed throughout the whole frond, disposed in spots.

C. *The section ULVOIDEÆ comprises the plants of the old genus Ulva. Fronds membranaceous (broad, or in narrow slips), thin, of a grass-green color. Their substance consists of cells, with the fruit immersed in the frond. They grow on rocks, stones, shells, &c., in the sea; also in ditches, stagnant waters, damp woods, &c.*

UL′VA. Seeds in fours, immersed in every part of the membranaceous frond.

D. *The section CONFERVOIDEÆ comprises the plants of the old genus Conferva. Fruit capsular or naked granulations. Fronds filiform and geniculate, containing the fruit immersed in them, generally strung on threads; mostly of a grass-green or greenish color, sometimes purple. They grow in fresh-water streams, springs, ditches, and stagnant waters; sometimes in damp woods, and some in the sea.*

CONFER″VA. Filaments articulated, uniform, simple or branched, containing the seed within them. No external fruit.

E. *The section TREMELLINÆ comprises the old genus Tremella. Plants of this section are all gelatinous, hyaline, and covered with a membrane. They are globose, palmate, or filiform, and contain conferva-like filaments within. Color green or purplish. They resemble Confervoideæ in habit and place of growth.*

NOS″TOC. Filaments moniliform, constituted from coadunate globules. Fronds bullate, vesicular (at length becoming flattened), crowded with simple moniliform, curve-crisped filaments.

ORDER V. LICHENES.

GYROPHO′RA. Frond foliaceous, coriaceous-cartilaginous, peltate, monophyllous (when luxuriant, polyphyllous), free be-

[*] The *peristome* is the membrane which appears round the mouth of the capsule of mosses, under the lid.

neath; apothecia somewhat shield-form, sessile adnate, clothed with a dark membranaceous cartilage, including a somewhat solid parenchymous substance; disk warty or circinal, plicate and margined.

PARME'LIA. Frond coriaceous, sub-membranaceous, flat, expanded, close-pressed, orbicular, stellate and lobed, or multifid-laciniate; having fibres beneath; apothecia shield-form, sub-membranaceous, formed under side from the frond, free, with a central puncture by which it is affixed; disk concave, colored, covering the whole receptacle above, within similar, sub-cellular, and striate, cut round, inflexed with a frond-like margin.

CETRA'RIA. Frond cartilaginous or membranaceous, ascending or expanded; lobe laciniate, smooth and naked both sides; apothecia shield-like, obliquely attached to the margin of the frond, the lower free, being separated from it, the upper one sessile; seed bearing lamina forming the disk, colored, plano-concave, surrounded with a frond-like inflexed margin.

OENOMY'CE. Frond crusty or cartilaginous, foliaceous, laciniate, sub-imbricate free (rarely adnate); bearing sub-fistulous peduncles (*podetia*) both barren and fertile; receptacles (knobs) orbicular, without margins, at length convex and capitate, inflated or empty beneath, terminal attached to the peduncles by their peripheries; seed-bearing lamina forming the receptacle above, thickish, colored, similar within, convex, reflexed, and attached at the periphery, invested beneath with the woolly integument of the frond.

BÆOMY'CES. Frond crustaceous, flat, expanded, adnate; bearing soft, solid, fertile podetia; apothecia capitate, without margins, solid, terminal, sessile on the peduncles; seed-bearing lamina covering the whole receptacle and adnate to it, convex, reflexed, thickish, colored, similar within.

US"NEA. Frond sub-crustaceous, teretish, branched, mostly pendulous; central part hyaline, elastic, composed of fascicles of tubes; receptacles orbicular, terminal, peltate, formed wholly from the frond, covered all over with its cortical substance, similar, nearly of a uniform color; its periphery destitute of margin, but often surrounded by a ciliate edging.

ORDER VI. FUNGI.

LYCOPER"DON. Receptacles somewhat caulescent, at length bursting at the top, with scaly warts or prickles scattered over its surface, especially when young. Seminal dust green.

MU'COR. Receptacle membranaceous, globose, stiped, at first watery and pellucid, then opake; seeds naked, sub-cohering; very minute and fugaceous.

URE'DO. Receptacle 0; seminal dust under the cuticle of leaves and stems, when ruptured it is easily brushed off; the little masses of seeds uniform, mostly globose.

AGAR"ICUS. Destitute of a volva at the base of the stipe, with or without the ring; lamellæ either entire or with shorter ones intermixed, rarely simple, ramose; never veiny.

BOLE'TUS. Pileus various; tubes and pores terete, entire. A large genus.

SECTION IV.

SPECIES OF PLANTS.

ABI'ES. See *Pi'nus.*

ABU'TILON. See *Si'da.*

ACA'CIA. 15—10. (*Leguminosæ.*) [From the Greek *aka'so*, to sharpen.]

glandulo'sa, (w. Ju. ♃.) leaves bipinnate, leafets 12-paired, glands between each pair; spikes globose, solitary, peduncled, axillary; legume falcate; unarmed. *S.*

farnesia'na, (black thorn, y. ♄.) leaves bipinnate, leafets 8-paired; spikes globose, sessile. Flowers fragrant, legumes fusiform. *S.*

ACAL"YPHA. 19—15. (*Euphorbiæ.*) [From the Greek *a*, not, *kalos*, agreeable, *aphe*, to the touch.]

virgin"ica, (three-seeded mercury, g. Au. ☉.) pubescent; leaves on short petioles, lanceolate-oblong, remotely and obtusely serrate; involucre cordate, ovate, acuminate. toothed; fertile flowers at the base of the sterile spike. *Road-sides.* 12 to 18 i. Var. *car'olinia'na*, with longer petioles and broader leaves.

A'CER. 8—1. (*Acera.*) [Latin *acer*, acrid, referring to the juice of some of the species.]

da'sy car'pum, (white maple, silver maple, g-y. ♄). leaves palmate, 5-lobed, truncate at the base, unequally gash-toothed, glabrous and glaucous beneath, obtusely sinuate; flowers glomerate; pedicels short; germs downy. 50 f. Fruit a *samara.*

barba'tum, (hairy maple w-g. Ap. ♄.) leaves heart-ovate, short, 3-lobed, unequally serrate, glaucous beneath, and hairy at the nerves; peduncles hairy, staminate ones branching. pistillate ones simple; calyx bearded within; wings of the capsules erect; small. 15 f.

ni'grum, (sweet tree, black maple, y. Ap. ♄.) leaves palmate, 5-lobed, cordate, with the sinus at the base closed, lobes spreading, sinuate-toothed, downy beneath; flowers corymbed; capsules turgid, subglobose; wings diverging. Large tree, affording almost as much sugar as the sugar maple. 50 f.

spica'tum, (mountain maple bush, y-g. M. ♄.) leaves sub-5-lobed, acute, toothed, pubescent beneath; racemes compound, erect. 15 f. *Mountains.*

ru'brum, (red maple, soft maple, r. Ap. ♄.) leaves palmate, 5-lobed, cordate at the base, unequally gash-toothed, glaucous beneath, sinuses acute; flowers in about fives, in sessile umbels, with long pedicels; germs glabrous; stamens variable. Precocious. 50 f.

sacchari'num, (sugar maple, rock maple, hard maple, r. y. M. ♄.) leaves palmate, 5-lobed, at the base sub-cordate, acuminate, sinuate-toothed, glaucous beneath; pedun-

cles in a nodding corymb. Large tree. 50 f. *Fruit ovoid, smooth, the wings about an inch long.*

stria'tum, (striped maple, false dogwood, moosewood, g. M. ♄.) lower leaves roundish, upper ones 3-cuspidate-acuminate, sharply serrate, glabrous; racemes simple, pendant. Small tree, with a greenish, striped bark. 15 f.

ACER"ATES.* 18—5. (*Asclepiada.*) [From the Greek *a*, without, *keras*, horn.]

virid"i flo'ra, (green milkweed, g. Ju. ♃.) stem erect, simple, hairy; leaves oblong, on short. petioles, tomentose, obtuse; umbels lateral, solitary, sub-sessile, nodding, dense; flowered; umbels about 3; horns of the nectary wanting. Sandy fields. Stem 2 f.

ACHILLE'A. 17—2. (*Corymbifera.*) [From the Greek warrior *Achilles.*]

millefo'lium, (yarrow, milfoil, w. J. ♃.) leaves 2-pinnatifid, downy, the divisions linear, toothed, mucronate; calyx and stem furrowed; flowers in large, dense, terminal corymbs; rays about 5; disk-florets few; receptacle flat, chaffy, the chaff lanceolate-oblong. *Naturalized.* 15 i. *S.*

ACHYRAN"THES. 15—5. (*Amaranthi.*) [From the Greek *achu'ron*, chaff, and *anthos*, flower.]

re'pens, (forty knot, March. ♃.) stem procumbent, pubescent; leaves opposite, petioled, lanceolate. Flowers in heads.

ACNI'DA. 20—5. (*Chenopodea.*) [From the Greek *a*, wanting, *knide*, a sting.]

cannabi'na, (water hemp, w. g. Ju. ☉.) leaves ovate-lanceolate; capsules smooth, acutely angled. Marshes. Can. to Flor. Flowers small, green, in large panicles.

ACONI'TUM. 12—5. (*Ranunculaceæ.*) [From the Greek *akone*, rugged, in allusion to its habit.]

uncina'tum, (monk's hood, b. J. ♃.) stem flexuose; leaves palmate, 3 to 5-parted, divisions rhomb-lanceolate, gash-toothed; upper lip of the corolla lengthened, convex, beaked; stem twining, branching. *Grows on mountains and rough places. Cultivated.* 2 f.

napel"lus, (wolf's bane, b. J. ♃.) leaves shining, 5-parted, the divisions 3-parted, subdivisions linear; upper lip of the corolla lanceolate, ascending, 2-cleft, spur straight. obtuse. 2 f. Ex.

ACO'RUS. 6—1. (*Aroideæ*) [From *a*, without, and *kore*, the pupil, because it was esteemed good for disorders of the eyes.]

cal'amus, (sweet flag, g-y. J. ♃.) spike

---

* This genus is scarcely distinct from Asclepias.

27*

protruding from the side of an ensiform leaf; scape leafy above the spadix; leaves 3–4 feet long; water or wet grounds; root creeping, strongly aromatic. 2 f.

ACROS"IICHUM. 21—1. (*Filices.*) [From *akros*, highest, and *stikos*, order, from the row of leafets at the top.]
aure'um, (fork fern, Ju. ♃.) frond pinnate; leafets stiped, lance oblong, entire, acuminate, the upper ones bearing fruit. Very large, 4 or 5 feet high.

ACl'Æ'A. 12—1. (*Ranunculaceæ*) [From *Acteon*, the hunter]
america'na, (bane berry, w. ♃.) leaves twice and thrice ternate; racemes ovate; petals shorter than the stamens; berries ovate-oblong. Var. *alba* (red cohosh), petals truncate; pedicels of the fruit thicker than the peduncle; berries white. Var. *ru'bra*, petals acute, pedicels of the fruit slender, berries red. 2 f.
racemo'sa, leaves ternately decompound, leafets ovate-oblong, 2 to 4 inches long, nearly smooth, often 3-lobed; racemes compound, terminal, 6 to 12 inches long; many-flowered, petals minute; carpels dry, opening with 2 valves.

ACTINOME'RIS 17—3. (*Corymbiferæ.*) [From *aktin*, a ray, *meris*, part.]
helianthoi'des, (y.) leaves lanceolate, acute, serrate, white-villose beneath; corymb simple, compact; stem winged. *S.*
squarro'sa, (w. y. Au. ♃.) stem-winged, corymbose-paniculate; leaves lanceolate, acuminate at each end, serrate, roughish-pubescent; disk sub-globose, in fruit squarrose. 3 f.

ADIAN"TUM. 21—1. (*Filices.*) [From *a*, not, and *diaino*, to grow wet, because its leaves are not easily wet [
peda'tum, (maiden hair, J. ♃.) frond pedate, with pinnate branches; leafets halved, upper margin gashed, barren segments toothed, fertile ones entire; stipe capillary, very glabrous. Woods, 1 f.

ADLU'MIA. 16—5. (*Fumariaceæ.*) [In honor of John Adlum, a distinguished cultivator of this vine]
cirrho'sa, (climbing colic-weed, y. w. J. ♂.) stem climbing; leaves cirrous; racemes axillary, corymbed, nodding; corolla monopetalous, gibbous both sides of the base. Cultivated.

ADO'NIS. 12—12 (*Ranunculoceæ.*) [Said to have been consecrated by Venus to the memory of the beautiful Adonis.]
autumna'lis, (pheasant's eye, Au. ☉.) flowers 5 to 8-petalled; fruit cylindric; petals emarginate. Ex.

ADOX"A. 8—4. (*Saxifraga.*) [From the Greek *a*, without, and *doxa*, glory.]
moschatelli'na, (g.) peduncles 4-flowered; filaments united at the base in pairs; anthers round. *S.*

ÆSCIIYNOM'ENE. 16—10. (*Leguminosæ.*)
his"pida, (false sensitive plant, y-r. Ju. ☉.) stem herbaceous, erect; petioles and peduncles hispid; leaves in many pairs, leafets linear, obtuse; racemes simple, 3 to 5-flowered; legumes with 6 to 9 hispid joints. Marshes. Penn. to Car.

viscid"ula, (y. ♃.) stem procumbent, viscid, slender; leafets 7 to 9-obovate; peduncles about 2-flowered; legume hairy; joints deeply notched. Sandy grounds. 3 f. *S.*

Æ'SCULUS. 7—1. (*Hyppocastaneæ.*) [From the Latin *esca*, food]
hippocas"tanum, (horse chestnut, w. J. ♄.) leaves digitate, with about 7 divisions; corolla 5-petalled, spreading; flowers in a panicled pyramid. 15 f. Ex.
glabra, (buck-eye, y-w. May. ♄.) leaves quinate, smooth, leafets ovate-acuminate; corolla 4-petalled, spreading, with the claws as long as the calyx; stamens longer than the corolla; capsules echinate. Woods; a small tree with flowers in panicled racemes. Penn. to Miss.
macrostá'chya, (Ap. ♄.) leaves in 5 divisions, downy beneath; raceme very long; corolla 4-petalled, expanding; stamens long. Beautiful shrub. 6 f. By some called *pa'via*, and considered a separate genus.

ÆTHU'SA. 5—2. (*Umbelliferæ.*) [A Greek word signifying beggarly.]
divarica'ta, (w. ☉.) stem erect, slender; leaves biternate, segments narrow-linear; umbels terminal, without involucres, partial umbels, 3 to 5-flowered; fruit hispid.
cyna'pium, leaves bi- and tripinnate, dark green; segments ovate, lanceolate; umbels terminal. Road sides. Flowers white, in many-rayed umbels; very poisonous. 1 f. Fool's parsley.

AGARI'CUS. 21—6. (*Fungi.*) [The name is said to have been given in consequence of the resemblance of the plant to a mineral called Agaricus, which is soft and spongy in its texture.]
campes"tris, pileus fleshy, flattish, having dark yellow scales; lamella becoming yellowish red; stipe short; the ring-volva rather incomplete. This is the common eatable mushroom.

AGA'VE. 6—1. (*Bromeliæ.*) [From a Greek word, signifying beautiful]
virgin'ica, (y. g. ♃ S.) stemless, herbaceous; leaves with cartilaginous serratures; scape simple; flowers sessile. Scape 6. f. Flowers fragrant. Rocky banks. Penn. to Car. False aloe.]

AGER"ATUM. 17—1. (*Helianthea*) [From the Greek *a*, without, *geras*, old age, because it never changes color.]
mexica'num, (b. J. ☉) hispid; leaves cordate, ovate, crenate, rugose; corymb compound; chaffs of the egret lanceolate, awned. *Cultivated.* Mexico.

AGRIMO'NIA. 11—2. (*Rosaceæ.*) [From *agros*, a field, *monos*, alone.]
eupato'ria, (agrimony, y. Ju. ♃.) cauline leaves interruptedly pinnate, the terminal leafet petioled, leafets obovate, gash-toothed, almost glabrous; flowers sub-sessile; petals nearly twice as long as the calyx; fruit hispid. 2 f.
suaveo'lens, (y. Ju. ♃.) stem very hispid; leaves interruptedly pinnate, leafets numerous, lanceolate, acutely toothed, scabrous above, and pubescent beneath; fruit turbinate, smooth at the base. 5 f.
parviflo'ra, (dotted agrimony, y. Ju. ♃.)

hairy, leaves interruptedly pinnate, with the terminal ones sessile, leafets numerous, mostly linear-lanceolate, incisely serrate; spike virgate; flowers on very short pedicels; petals longer than the calyx; fruit roundish, divaricately hispid.

A'GROSTEM''MA. 10—5. (*Caryophyllea*.) [From the Greek *agros*, field, *stemma*, garland.]

githa'go, (cockle, r. J. ⚇.) hirsute; calyx longer than the corolla; petals entire.

corona'ria, (Au. ♂.) tomentose; leaves lance-ovate; petals emarginate. Rose campion. Ex.

AGROS''TIS. 3—2. (*Graminea*.) [From *agros*, field.]

stric''ta, (bentgrass, J. 2/.) panicle elongated, straight; glumes equal; paleas smaller than the glumes, unequal, with an awn at the base of the outer one longer than the flower.

lateriflo'ra, (Au. 2/.) culm erect, branched above, sending off shoots at the base; panicle lateral and terminal, dense; glumes acuminate; paleas longer than the glumes, equal, hairy at the base, awnless; root creeping; leaves broad, with scabrous margins and compressed sheaths. 2 f. Var. *filifor''mis*, a very slender panicle, and the paleas nearly equalling the glume. Sometimes the culm is sub-decumbent. Swamps.

vulga'ris, (red-top, J. 2/.) panicle with smoothish branches, spreading in maturity; outer valve of the corolla 3-nerved; stipule short, truncate. 18 i.

al''ba, (white-top, bonnet grass, J. 2/.) panicle with hispid, spreading, lax branches; outer valve of the corolla 5-nerved; stipule oblong. 18 i. Var. *decum''bens*, stem decumbent. This variety is considered as a distinct species by some, and called *stolonif'era*.

AI'RA. 3—2. (*Graminea*.) [From the Greek, a deadly instrument.]

flexuo'sa, (hair-grass, J. 2/.) panicle spreading, trichotomous; branches flexuous; glumes a little shorter than the florets, and about the length of the awn; leaves setaceous; culm nearly naked. 20 i.

a'quatica, (water hair-grass, M. 2/.) panicles spreading, half whorled; flowers beardless, obtuse, smooth, longer than the glumes; leaves flat; culm creeping. Water.

AJU''GA. 12—1. (*Labiatæ*.) [From *a*, without, *zugos*, yoke, not paired.]

chamæpi'thys, (y. J. ⚇.) leaves 3-cleft; flowers axillary, solitary, shorter than the leaves; stem diffused.

ALCHEMIL''LA. 4—1. (*Rosacæ*.) [A plant formerly in repute among the alchymists.]

alpi'na, (A. w. 2/.) leaves digitate, serrate, white, soft beneath. Ladies' mantle. High mountains. Ver. N. Hamp.

ALE'TRIS. 6—1. (*Asphodeli*.) [From a Greek word signifying meal, from a substance contained in the corolla.]

farino'sa, (Ju. y. 2/.) leaves radical, broad-lanceolate, smooth; flowers pedi-

celled, oblong-tubular panicled, nakedish; glo-decaying nearly smooth corolla sub-globular, N. Eng. to Car. White bag, beardless. A gitudinal spike. Root very varieties. Flowgrass, colic-root.

aure'a, (Aug. 2/.) flowers yellow lanceolate, sessile, sub-campanulate. N. J. to Carienate,

ALIS''MA. 6—13. (*Junci*.) [From the Greek *als*, the sea.]

planta'go, (water plantain, w. Ju. 2/.) leaves ovate-cordate, acute or obtuse, 5 to 9-nerved; flowers in a compound, verticillate panicle, fruit obtusely triangular. Var. *parviflo'ra*, flowers very small; leaves oval, 5 to 7-nerved, acuminate.

ALLIO'NIA. 4—1. (*Jasmineæ*.) [Named in honor of an Italian botanist.]

al''bida, (Ap.) leaves opposite, somewhat scabrous, lance-oblong; involucrum 5-cleft; corolla longer than the involucrum.

nyctagyn''ia (Ju. 2/.) stem erect; leaves broad-cordate, glabrous, acute; peduncles solitary.

AL''LIUM. 6—1. (*Asphodeli*.) [From *aleo*, to smell.]

ce'pa, (garden onion, Ju. 2/.) scape naked, swelling toward the base, longer than the terrete leaves.

schænopra'sum, (cives, Ju. 2/.) scape naked, equalling the leaves, which are terete-filiform. Ex.

vinea'le, (field garlic, p. J. 2/.) stem slender, a little leafy; cauline leaves rounded, fistulous; umbelliferous; umbels bearing bulbs; stamens alternately tri-cuspidate. Rose-colored. Introduced from N. Scotia.

canaden''se, scape naked, terete; leaves linear; head bulbiferous. Meadows. Flowers numerous, rose-colored. Can. to Vir.

sati'vum, (garlic, Ju. 2/.) stem flat-leaved, bulb-bearing; bulb compound; stamens tri-cuspidate.

AL''NUS. 19—4. (*Amentaceæ*.) [From *alno*, Italian for alder.]

serrula'ta, (alder. r-g. Ap. ♄.) leaves obovate, acuminate; veins and their axils hairy beneath; stipules oval, obtuse. 9 f.

glutino'sa, (Ap. ♄.) leaves round-wedge form, obtuse, glutinous, axils of the veins downy.

glau'ca, (Mar. ♄.) leaves oblong, acute, doubly serrate, glaucous beneath, axils of the veins naked; stipules naked. Black alder.

ALOPECU'RUS. 3—2. (*Graminea*.) [From *alopex*, a fox, and *aura*, a tail.]

praten''sis, (meadow grass, foxtail, J. 2/.) culm erect, smooth; spike cylindric; paleas as long as the glumes.

AL''THÆA. 15—12. (*Malvaceæ*.) [From *altheo*, to heal]

officina'lis, (marsh mallows, r. y. Ju. 2/.) leaves downy, oblong-ovate; obsoletely 3-lobed, toothed. Flowers large, purple, near salt marshes. 2 f.

rose'a, (hollyhock, ♂.) stem erect; leaves rough, heart-form, 5 to 7-angled, crenate. Ex.

ficifo'lia, (fig-hollyhock, ♂.) leaves 7-lobed, sub-palmate, obtuse. Ex.

ALYS''SUM. 14—1. (*Cruciferæ*.) [From the

protruding from the ... erly
leaf; scape leaf ... mad
leaves 3–4 f ... eaves
grounds; ... lowers
ic. 2 f. ... retuse.
ACRO ... i.) [Sig-
... ntandrous,
... ther, erect;
... aranth, Au.
... npound, com-
pact, ... nceolate, mu-
cronate, red. ... feather. Cultivated.*

the perianth when
Sandy woods
wers in a lon-
tier. Star.
sub-
71

spino'sus, (Au. ⊙.) racemes pentandrous, terminal, compound; axils spinose.

albus, (white coxcomb, g-w. Ju. ⊙.) glomerules axillary, triandrous; leaves obovate, retuse; stem 4-cornered, simple. Common garden weed.

melanchol''icus, (love lies bleeding, r. ⊙.) glomerules axillary, peduncled, roundish; leaves lance-ovate, colored. Ex.

tri-color, glomerules sessile; leaves lance-oblong, colored. Ex.

livid''us, glomerules triandrous, sub-spiked, roundish; leaves oval, retuse; stem erect. Ex.

AMARYL''LIS. 6—1. (*Narcissi.*)

atamas''co, (atamasco lily, w. and r. J. ♃.) spatha 2-cleft, acute; flower pedicelled; corolla bell-form, sub-equal, erect; stamens declined. *S*,

formosis''sima, (jacobea. ♃.) spatha 1-flowered; corolla ringent-like; petals declined. Ex.

undula'ta, (waved lily, Sept.) The flowers numerous on each stalk; petals pink, undulate. Ex.

AMBRO'SIA. 19—4. (*Urtices*) [The name *ambrosia*, food for the gods, seems strangely misapplied to a genus of plants possessing neither beauty nor valuable properties.]

ela'tior, (hog-weed, S. ⊙.) leaves doubly pinnatifid, smoothish; petioles long, ciliated; racemes terminal, panicled; stem wand-like.

tri'fida, (g. y. ⊙. S.) hirsute, rough; leaves very large, 3-lobed, serrate, the lobes oval-lanceolate, acuminate; fruit 6-spined below the summit; flowers in terminal panicles composed of long axillary spikes. 5 to 8 f.

artemisifo'lia, (⊙.) leaves doubly-pinnatifid, hoary beneath, at the summit pinnatifid; racemes terminal in threes, branches level-topped.

panicula'ta, (⊙.) leaves glabrous, doubly pinnatifid, pinnatifid at the summit; racemes terminal, solitary; branches level-topped.

his''pida, (w. ⊙.) white hispid on all parts; leaves 2-pinnatifid, divisions gashed; racemes sub-panicled, terminal. 1 f. *S*.

AMMANN''IA. 4—1. (*Salicariæ.*) [From Ammann, a Russian botanist.]

ramo''sior, (w-p. Au.) stem erect, thick, sub-terete; leaves lanceolate, dilated at the base; lower flowers completely whorled. Salt meadows. 6 i.

hu'milis, (w-r. Au. ⊙. ♂.) stem procumbent at the base, branched, slender, quadrangular; leaves lanceolate, slender at the base; flowers solitary. Stem red. 6 i.

AMMO'BIUM. 17—1. (*Cichoraceæ.*) [From the Greek *ammos*, sand, *bio*, to live, found growing in sand.]

ala'tum, (w. ♃.) stem-winged, leaves oblong, undulate, decurrent. Cultivated; brought from N. Holland.

AMOR''PHA. 16—10. (*Leguminosæ.*) [From the Greek *a*, wanting, *morphe*, shape.]

frutico'sa, (Ju. ♄.) smooth, sub-arborescent; leaves petioled, emarginate; spikes aggregated, long: calyx hoary, pedicelled, one of the teeth acuminate, the rest obtuse; legume few-seeded. N. J. to Car. and W. to Rocky Mountains. A shrub with spikes of purple flowers.

pubes'cens, (w. J. ♄.) small, shrubby; leaves on very short petioles, obtuse at each end, hairy; spikes long, panicled, hairy; calyx sub-sessile, with acuminate teeth. 3 f.

carolinia'na, (broom, ♄. Ju. b.) nearly glabrous; leafets elliptical or oblong, petiolate, dotted, the lowest pair approximate to the stem; flowers on very short pedicels; calyx villose on the margin, teeth short, the two upper obtuse, the three lower longer, and commonly equal, acuminate, or subulate, aristate. 4 to 5 f.

AMPELOP''SIS. 5—1. (*Vites.*) [From the Greek *ampelosa*, vine, and *ops*, resembling.]

quinquefo'lia,* (g. Ju. ♃.) stem climbing and rooting; leaves quinate, digitate, smooth, leafets petiolate, oblong, acuminate, toothed; racemes dichotomous. Var. *hirsuta*, leaves pubescent on both sides; leafets ovate, coarsely toothed.

corda'ta, (Ju. ♄.) stem climbing, with slender branches; leaves cordate, acuminate, toothed, and angular; nerves beneath pubescent; racemes dichotomous, few-flowered; panicles opposite the leaves. Banks of streams.

bipinna'ta, leaves doubly pinnate; lance-ovate, deeply toothed and lobed. Flowers in corymbs. Southern.

AMPHICAR''PA. 16—10. (*Leguminosæ.*) [From the Greek *amphi*, about, *karpos*, the fruit.]

mono'ica, (wild bean-vine, b. and w. Ju. ⊙.) stem slender, twining, hairy backward; leaves ternate, ovate, nearly smooth; stipules ovate, striate. Var. *comosa* has hirsute leaves. Twining. 4 f.

AMSO'NIA. 18—5. (*Apocyneæ.*)

latifo'lia, (b. Ap. ♃.) stem glabrous; leaves lance-oval, upper ones acuminate, pubescent on the under surface of the nerves. 2 f. *S*.

salicifo'lia, (b. M.) stem smooth; leaves lance-linear, acute at each end, very glabrous. *S*.

angustifo'lia, (b. Ap.) stem hairy; leaves narrow-linear, numerous, erect, hairy. *S* 2 f.

* Cissus hederacea of Mirbel.

Flowering ... ...

AMYG"DALUS 11—1. (*Rosaceæ.*) [Derived from a Greek word, which signifies to lacerate, alluding to the furrows upon the pericarp of the almond.)

per"sica, (peach, r. M. ♄.) serratures of the leaves all acute; flowers sessile, solitary. 15 f. Ex.

na'na, (flowering almond, ♄.) leaves ovate, tapering to the base, sharply serrate. 3 f. Ex.

commu'nis, (almond) leaves serrate, the lower ones glandular; flowers sessile, binate. Ex.

AMY'RIS. 8—1. (*Terebintaceæ.*) [From the Greek, signifying balm or ointment, so called from its use, or smell. Ex.]

florida'na, (w. ♄.) leaves ovate, sessile, entire, obtuse; flower sub panicled.

gileaden"sis, (balm of Gilead,) leaves ternate, entire; peduncles 1-flowered. Grows near the Red sea.

ANAGAL"LIS. 5—1. (*Jasmineæ.*) [From a Greek word, signifying to laugh, because by curing diseases it was thought to promote cheerfulness]

arven"sis (red chick-weed, scarlet pimpernel, r. J. ☉.), stem spreading, naked, procumbent; petals entire, flat, with hairs at the margin. S.

ANCHU'SA 5—1. (*Boragineæ.*) [Greek, *to strangle.*]

officina'lis, (bugloss, y. ♃.) leaves lanceolate; spikes imbricate, one-sided; bracts ovate. Ex.

ANDROCE'RA. · 5—1. (*Solaneæ.*) [From *andros* stamen. *keros* a horn, from the conform appearance of one of the authers.]

loba'ta, (J. ☉.) prickly, hirsute; leaves in pairs, lobe-pinnatifid, segments obtuse, obsoletely crenate, undulated; racemes lateral, many-flowered. S.

ANDROM"EDA. 10—1. (*Ericæ.*)

calycula'ta, (leather-leaf, w. M. ♄.) leaves lanceolate-oblong, obsoletely serulate, subrevolute, with scaly dots, rust-colored beneath; racemes terminal, leafy, turned one way; pedicels short, solitary, axillary; calyx acute, 2-bracted at the base, bracts broad-ovate, acuminate; corolla oblong-cylindric. Wet. 2 f.

arbo'rea, (w. Ju. ♄.) leaves oblong-oval, acuminate-serrate, smooth; panicles terminal, many-spiked; corolla ovate-oblong, pubescent. Mountains. A beautiful tree. 50 f. Sorrel-tree.

marian"a, (J. ♄.) leaves oval, entire, subacute at both ends, glabrous, leathery, paler beneath; flower-bearing branches almost leafless; peduncles fascicled; corolla ovate-cylindric; calyx leafy; anthers beardless; capsule ovate, resembling the form of a pine-apple. One variety has narrow-lanceolate leaves.

polifo'lia, (wild rosemary, r-w. M. ♄.) leaves linear-lanceolate, convex revolute, white, glaucous beneath and hoary-glaucous above; flowers aggregate, terminal; corolla sub-globose; anthers bearded toward the top. 1 f. Wet.

panicula'ta, (white bush, pepper bush, w. J. ♄.) pubescent; leaves obovate-lanceolate, acute, sub-entire; flower-bearing branches terminal, panicled, nakedish; glomerules peduncled; corolla sub-globular, pubescent; anthers obtuse, beardless. A shrub running into several varieties. Flowers small.

augustifo'lia, (A.) leaves linear-lanceolate, acute, with scaly dots; bracts 2 and minute, acute; racemes terminal, leafy, secund; peduncles solitary, axillary. S.

rhomboid"alis, flower-bearing branches 3-angled; leaves rhomboid and lanceolate, entire, glabrous, terminated by a gland; peduncles clusteréd, axillary. S.

axilla'ris, (w. Mar. ♄.) leaves oval-lanceolate, acuminate, coriaceous, lucid, with spiny serratures; racemes axillary, closely flowered; corolla oblong-ovate; anthers at the summits 2-horned. 3 f. S.

ANDROP"OGON. 3—2. (*Gramineæ*) [From *andros,* a man, and *pogon,* a beard, from the resemblance of little tufts of hair on the flower to a man's beard.]

scopa'rius, (broom grass, Au. ♃.) spikes simple, lateral, and terminal, pedunculate, in pairs; rachis hairy; abortive floret neuter; valves awned.

virgin"icus, (bent grass, ♃.) culm compressed; superior leaves and sheaths smooth; spikes short, 2 or 3 from each sheath, partly concealed at the base; rachis sub-terete; abortive flower, a mere pedicel without valves; perfect flowers monandrous.

fusca'tus, (fork spike, Au. ♃.) spikes digitate, generally by fours; abortive flower staminiferous, awnless, resembling the perfect one, the awn of which is sub-contorted.

nu'tans, (beard-grass, Au. ♃.) panicle oblong, branched; nodding spikelets by pairs; glumes hairy; awn contorted.

ANDROSA'CE. 5—1. (*Primulaceæ.*) [From *anex,* a man, and *sakos,* a shield, so called from its large, round, hollow leaf.]

occidenta"is, (☉) very slenderly pubescent; leaves (or involucres) oblong-spatulate, entire; perianths angled; capsule shorter than the calyx. S.

carina'ta, (w. J. ♃.) leaves crowded, lanceovate, acute, entire, keeled, margin ciliate; umbels few-flowered; leaflets of the involucre linear-oblong; corolla exceeding the ovate calyx; divisions obovate, entire. James' Peak. S.

septentrional"is, (☉.) leaves lanceolate, toothed, glabrous, shining; perianth angled, shorter than the corolla. S.

ANEM"ONE. 12—12. (*Ranunculaceæ.*) [From *anemos,* the wind, so called because the petals expand through the influence of the wind blowing upon the flower.]

virginia'na. (wind-flower, g-w. Ju. ♃.) stem dichotomous; leaves in threes, 3-cleft, upper ones opposite, leafets gash-lobate and serrate, acute; peduncles solitary 1-flowered, elongated; seed oblong, woolly, mucronate, in heads. 18 i.

nemoro'sa, (low anemone r-w. M. ♃.) stem 1-flowered; cauline leaves in threes, 5-parted, leafets wedge-form, gash-lobed, toothed, acute; corolla 5—6-petalled; seeds ovate, with a short style, hooked. A variety, *quinquefo'lia*, has lateral leafets, deeply 2-cleft. 6 i. *S.*

thalictroï'des, (rue anemone, w. M. 2f.) umbels involucred; radical leaves twice ternate, leafets sub-cordate, 3 toothed; involucrum 6 leaved; leafets petioled, uniform, umbel few-flowered; seed naked, striate; root tuberous. A variety, *uniflo'ra*, has a 1-flowered involucrum. 5 i. *S.*

pennsylva'nica, (w. Ju. 2f.) leaves 3-parted, segments 3-cleft, lobes oblong, toothed, acuminate; involucrum sessile, bearing several pedicels, one naked and 1-flowered, the others involucellate; petals 5; fruit pubescent, crowned with a long style. Meadows. Flowers large. Considered the same as A. dichotoma.

horten"sia, (garden anemone,) radical leaves digitate, divisions 3-cleft, cauline ones ternate, lanceolate, connate, sub-divided; seed woolly. Ex.

ANE'THUM. 5—2. (*Umbelliferæ*) [From the Greek *aneu*, to run, *theo*, afar, alluding to the spreading roots. Ex.]

grave'lens, (dill,) fruit compressed; plant annual.

fœnic"ulum, (fennel,) fruit ovate; plant perennial.

ANGEL"ICA. 5—2. (*Umbellifera.*) [*Angelic*, on account of its supposed virtues]

atropurpu'rea, (angelica, g-w. J. 2f.) stem smooth, colored; leaves ternate, partitions sub-quinate, leafets ovate, acute, gash-serrate, sub-lobed, 3 terminal ones confluent; petioles very large, inflated. Wet meadows. Root purplish. Aromatic angelica. 4 f.

triquina'ta, (w. Au. 2f.) stem terete, pubescent above; leaves ternate, very smooth, partitions quinate, leafets oblong; ovate, equally serrate, lower ones 2-lobed at the base. 4 f.

archangel"ica, (archangel, ♂.) leaves unequally lobed. A native of Lapland. Medicinal.

ANNO'NA. 12—12. (*Annonæ.*)

gla'bra, (Ju. r-y. ♄.) calyx large, bell form; peduncles 2-flowered, opposite the leaves; leaves lance-ovate, glabrous; fruit subconic, obtuse, smooth. 16 f. Evergreen tree. Carolina.

AN"THEMIS. 17—2. (*Corymbiferæ.*) [From the Greek *anthos*, a flower.]

cot"ula, (may-weed, w. J. ☉.) receptacle conic, chaff bristly, seed naked; leaves 2-pinnate, leafets subulate, 3-parted. 10 i.

no'bilis, (chamomile, w. Au. 2f.) leaves 2-pinnate; leafets 3-parted, linear, subulate, sub-villous; stem branching at the base. Fragrant. 4 i. Ex.

arven"sis, (wild chamomile, w-y. J. ♂.) leaves bipinnate, segments lanceolate, linear; receptacle conic; chaff lanceolate, akenes crowned with a margin.

AN"THOXAN"THUM. 2—2 (*Gramineæ*) [From the Greek *anthos*, a flower, *xanthos*, yellow]

odora'tum, (sweet vernal grass, M. 2f.) spike oblong-ovate; florets sub-peduncled, shorter than the An American variety, *altis"simum*, is larger and of a dark

green. An elegant substitute for the Leghorn grass. 10-18 i.

ANTIRRHI'NUM. 13—2. (*Bignonia.*) [From *anti*, against, *ris*, nose, said to be so named from an unpleasant odor in some of its species.]

canaden"se, (flax snap-dragon, w-b. Ju. ☉.) rising in a curve, glabrous, simple; leaves scattered irregularly, erect, narrow, linear, obtuse, remote; flowers racemed; scions procumbent. Flowers small.

lina'ria, (snap-dragon, y. Ju. 2f.) erect, glabrous; leaves scattered, lanceolate-linear, crowded together; spikes terminal, dense-flowered; calyx glabrous, shorter than the spur. Flowers large. Toad-flax. Naturalized. 12-18 i.

elat"ine, (y. Ju. ☉.) procumbent, hairy; leaves alternate, hastate, entire; peduncles solitary, axillary, very long. Flowers small, bluish white. Introduced.

trianthop"orum, leaves whorled, lanceolate, 3-parted; stem decumbent; racemes terminal, few-flowered. Flowers large. Ex.

ANY'CHIA. 5—1. (*Amaran'ti.*)

dichot"oma, (fork chickweed, w. Ju. 2f.) stem dichotomous, very branching, spread; leaves oval, lanceolate, glabrous, erect; 6 or 8 inches high, very slender; branches axillary; leaves obtusish; flowers mostly longer than the stipules.

APAR"GIA. 17—1. (*Cichoraceæ.*) [A Greek word, signifying succory.]

autumna'lis, (false hawk-weed, y. J. 2f.) scape branching; peduncles scaly; leaves lanceolate, toothed, or pinnatifid, smoothish. Flowers bright yellow; resembling the dandelion. Fields and road-sides. Introduced.

oron"tium, (Ju. ☉.) erect, branching, hairy; leaves alternate, lanceolate; flowers sub-spiked; involucre digitate, longer than the corolla. S.

tenel"lum, (b. Ju. ☉.) small, simple, glabrous; leaves opposite, linear, acute; flowers axillary, short-peduncled; involucre bell-form. S.

APHA'NES. 4—2. (*Rosaceæ.*) [From a Greek word, signifying low in stature]

arven"sis, (parsley-piert, ☉.) leaves 3-parted; divisions 3-cleft, hairy; flowers axillary, glomerate, monandrous. S.

A'PIOS. 16—10. (*Leguminosæ.*) [From the Greek *apios*, mild, in allusion to the root]

tubero'sa, (ground-nut, dark p. Ju. ♂.) stem twining; leaves pinnate, with 7 lance-ovate leafets; racemes shorter than the leaves; root tuberous, farinaceous, in taste resembling the cocoa-nut, and highly nutricious. Ex.

A'PIUM. 5—2. (*Umbelliferæ.*) [Supposed to be derived from the Greek *apes*, bees, because they are fond of the plant]

petroseli'num, (parsley, Ju. ♂.) cauline leaves linear; involucrum minute. Ex.

grave'lens, (celery, Ju. ♂.) stem channelled; cauline leaves wedge-form. Ex.

APLEC"TRUM. 18—1. (*Orchideæ.*) [From *a*, without, *plectron*, spur,]

hiema'lis, (g-p. M. 2f.) leaf solitary, ovate, striate; lip trifid, obtuse, with the palate ridged, central lobe rounded, crenulate. Shady woods. Flowers pendulous. 1 f.

APO'GON. 17—1. (*Cichoraceæ*.) [From *a*, without, *pogon*, beard.]

humil"is, (y. Ap. ⊙.) stem glabrous; radical leaves sessile, cauline leaves, ligulate, acute, leaves entire, glabrous.

APOC'YNUM. 18—5. (*Apocyneæ*.) [From *apo*, against, and *kunos*, a dog.]

androsæmifo'lium, (dog-bane, r-w. J. ♃.) stem erect and branching; leaves ovate; cymes lateral and terminal; tube of the corolla longer than the calyx, with a spreading limb. 3 f.

cannab"inum, (g-y. J. ♃.) leaves lanceolate, acute at each end, smooth on both sides; cymes paniculate; calyx as long as the tube of the corolla.

APTE'RIA. 3—1. (*Junci*.) [From the Greek *a*, without, *pteris*, wings.]

seta'cea, (w. and p. ⊙.) stem with minute, ovate, remote scales; spikes with bifid branches. 6 i. *S*.

AQUILE'GIA. 12—5. (*Ranunculaceæ*.) [From the Latin *aqua*, water, and *ago*, to gather, so called from the shape of its leaves, which retain water.]

canaden"sis, (wild columbine, r. y. Ap. ♃.) horns straight; stamens exsert; leaves decompound. Growing frequently in crevices of rocks. 15 i.

cæru'lia, (b. J. ♃.) horns twice as long as the petals; nectaries acute; segments of the leaves deeply lobed. 18-1. Southern.

vulga'ris, (garden columbine, J. ♃.) horns incurved; leafy; stem and leaves glabrous; leaves decompound. The nectariferous horns become numerous by culture; one hollow horn within another. 15 i. Ex.

brevisty'la, sub-pubescent; spur incurved, shorter than the limb; stipe short, inclined; stamens shorter than the corolla. Upper Canada.

formo'sa, (♃. r.) spur straight, much longer than the limb; sepals lanceolate, acute, 3 times the length of the petals; style as long as the sepals. Oregon.

AR"ABIS. 14—2. (*Cruciferæ*.) [Probably named in Arabia.]

lyra'ta, (w. A. ♂.) stem and upper leaves smooth and glaucous; radical leaves lyratepinnatifid, often pilose; stem branched at the base; pedicels much longer than the calyx. 10 i.

canaden'sis, (w. J. ♃.) stem leaves sessile, oblong-lanceolate, narrow at the base, pubescent; pedicels pubescent, reflexed in the fruit; siliques pendulous, sub-falcate, nerved. 2 f.

rhombo'idea, (spring cress, w. M. ♃.) leaves glabrous, rhomboidal, repand-toothed, the lower ones nearly round, on long petioles; root tuberous. 15 i. Wet.

denta'ta, (⊙. w. Ap.) stellately pubescent; radical leaves obovate, tapering at the base into a petiole, as long as the limb, irregularly sharp toothed; cauline ones, oblong, clasping; flowers minute; petals spatulate, scarcely longer than the calyx; silique short, spreading on very narrow pedicels, pointed with a nearly sessile stig-

ma; stem branched from the base. Arkansas, Mississippi.

sagitta'ta, (wall-cress, w. J. ⊙. ♂.) leaves sub-dentate, rough. with the pubescens often branched; radical ones ovate or oblong, attenuated into a petiole; stem leaves lanceolate, sagittate, cordate; pedicels as long as the calyx; siliques straight and erect. 18 i.

ARA'CHIS. 16—10 (*Leguminosæ*) [A Greek word, signifying a rooting plant.]

hypogæ'a, (pea-nut, false ground-nut, ⊙.) stem procumbent, pilose; leaves pinnate; flowers axillary; peduncles become long, and the fruit is ripened under ground.

ARA'LIA. 5—5. (*Araliæ*) [From *ara*, a bank in the sea, in allusion to the habit of the plant.]

racemo'sa, (spikenard, w. J. ♃.) spreading branches; petioles 3-parted, the partitions 3-5-leaved; leafets often heartform; branchlets axillary, leafy; umbels many, sub-panicled, leafless above. Damp. 4 f.

nudicau'lis, (g-w. J. ♃.) stem hardly a caulis; leaf solitary, terquinate; scape shorter than the leaf; umbels few. Wild sarsaparilla. 15 i. *S*.

spino'sa (shot-bush, angelica tree, w. y-w. Au. ♄.) stem and leaves thorny; leaves doubly pinnate; leafets slightly serrate; panicles branching; umbels numerous.

ARBU'TUS. 10—1. (*Ericæ*)

uvu-ur"si, (bear-berry, kinnikinnick, w-r. M. ♄.) stem procumbent; leaves wedgeobovate, entire; berry 5-seeded. Dry, barren sand-plains, &c. Very abundant about the great lakes.

alpi'na, (strawberry-tree, w. M. ♄.) stem procumbent; leaves obovate, acute, rugose, serrate; racemes terminal. Canada.

ARCHEMO'RA. 5—2. (*Umbelliferæ*.) [From *arche*, the conqueror, *moros*, a fool, from poisoning those who eat it.]

ambig"ua or *rigid"a*, (water drop-wort, w. Au. ♃.) leaves gash-pinnate, 3-5 pairs, acute, leafets lance-linear, often falcate and mostly entire; fruit ovate; stem smooth. 3-5 f.

ARC"TIUM. 17—1. (*Cinarocephalæ*) [From *arktos*, a bear, so called on acc.unt of its roughness.]

lap"pa, (burdock, r. Au. ♃.) cauline leaves heart-form, petioled, toothed; flowers panicled, globose; calyx smooth.

ARENA'RIA. 10—3. (*Caryophylleæ*) [From *arena*, sand.]

lateriflor"a, (sand-wort, w. J. ♃.) stem filiform, simple; leaves ovate, obtuse, subtriple-nerved; peduncles lateral, solitary, elongated, 2-cleft; pedicels alternately bracted; corolla longer than the calyx. 6 10 i.

gla'bra, (♃.) very smooth; stems numerous, erect, filiform; leaves subulate, linear, flat, spreading; pedicels 1-flowered, elongated, divaricate; sepals ovate, obtuse, shorter than the petals. Mountains. Flowers large, white. Stem 4-6 i. erect, slender.

serpyllifo'lia, (thyme-leaved sand-wort,

w. Ju. ☉.) stem dichotomous, spreading; leaves ovate, acute, subciliate; calyx acute, sub-striate; petals shorter than the calyx. 5 i.

stric"ta, (w. M. ♃.) glabrous, erect, many stems; leaves subulate-linear, erect; panicles few-flowered; petals much longer than the calyx, which is oval-lanceolate, striate. Dry. 6-12 i.

peploi'des, (sea chickweed, Ju. ♃.) glabrous; leaves ovate or oblong, acute, fleshy; flowers sub-solitary, short-peduncled; divisions of the calyx obtuse, exceeding the corolla. 8-12 i. Lower Canada.

pitche'ri (☉.) erect, slender, glabrous, fastigiately branched, few-flowered; leaves linear-filiform, obtuse, not fascicled; peduncles slightly glandular-pubescent; petals oblong, somewhat exceeding the lanceolate, nerved sepals. Texas. Arkansas.

ARETHU'SA. 18—1. (*Orchideæ*.)
bulbo'sa, (arethusa, r. J. ♃.) leafless; root globose; scape sheathed, 1-flowered; calyx with the superior divisions incurved, lips sub-crenulate; flowers large, sweet-scented. Damp.

ARGEMO'NE. 12—1. (*Papaveraceæ*.)
mexicana, (y. Ju. ☉.) leaves pinnatifid, spinose, gashed; flowers axillary. Var. *albiflora*. S.

ARIS"TIDA. 3—2. (*Gramineæ*.)
dichot"oma, (beard grass, poverty grass, ♂. ♃.) cespitose; culm dichotomous; flowers racemose-spiked; lateral awns very short, intermediate ones contorted. 8-12 i.

spicifor"mis (♃.) flowers crowded together, somewhat spiked; the middle awn villous at the base. 3 f. S.

gra'cilis, (♃.) stem very slender; flowers in spikes; spikelets few-flowered, somewhat remote, appressed; lateral awns short, erect, the intermediate ones longer, expanding. 1 f. S.

tuberculo'sa, culm erect, dichotomous, joints tumid with small tubercles in the axils; panicles rigid; glumes keeled, with long subulate points; paleas stiped; awns smooth, convolute. 3 f. S.

ARISTOLO'CHIA. 18—6. (*Aristolochia*.)
serpentu'ria, (p. J. ♃.) leaves heart-form, oblong, acuminate; stem zigzag, ascending; peduncles radical; lips of the corolla lanceolate. Virginia snake-root. A variety has very long, narrow leaves.

si'pho, (Dutchman's pipe, J. ♄) leaves heart-form, acute; stem twining; peduncles 1-flowered, furnished with an ovate bract; corolla ascending, the border 3-cleft, equal. A vine climbing over large trees. Flowers solitary, brown.

tomento'sa, (g-y.) stem twining; leaves nearly round, cordate, tomentose underneath; corolla villous; border 3-cleft, nearly equal. S.

hasta'ta, stem fluxuose, simple, erect; leaves somewhat cordate, hastate, acute; flow. rs on scapes; lip of the corolla ovate. S.

ARMENI'ACA. 11—1. (*Rosaceæ*.) [From Armenia.]

vulga'ris, (apricot, ♄.) leaves sub-cordate; stipules palmate. Var. *præ'cox*, early apricot. Fruit small, yellow. Var. *persicoi'des*, peach apricot. Fruit sub-compressed.

ARNI'CA. 17—2. (*Corymbiferæ*.)
nudicau'lis, (y. J. Ju. ♃.) hirsute; radical leaves opposite, decussate, broad-lanceolate, nerved, and toothed; stem nearly leafless, divided near the summit into a few 1-flowered branches. Flowers large. 2-3 f. Pine barrens. Leopard's bane.

plantagin"ea, (y. Ju. ♃.) glabrous; leaves entire, glabrous both sides, acute, 3-nerved; radical ones lance-spatulate, terminating in a narrow petiole at the base; cauline ones opposite, lanceolate, sessile; stem 1-flowered. 7 i.

ful'gens, (y. Ju. ♃.) hairy; radical leaves lanceolate, obtusish, tapering to the base; petioles 3-nerved; cauline leaves opposite, remote, linear; stem 1-flowered. 1 f. S.

clay'toni, (y. Ju. ♄.) hirsute; radical leaves decussately opposite, oblong ovate, sub-dentate; stem somewhat leafless; top divided into 1-flowered peduncles. 2 f. S.

ARO'NIA. 11—5. (*Rosaceæ*.) [A Greek word, signifying the medlar-tree.]
botrya'pium, (shad-bush, june-berry, w. Ap. ♃.) leaves oblong-oval, cuspidate, glabrous when mature, (when first expanded lanceolate and downy); flowers racemed; petals linear; germs pubescent; segments of the calyx glabrous.

arbutifo'lia, (M. ♄.) unarmed; leaves ovate-oblong, acute, serrulate, tomentose beneath; flowers in corymbs; calyx tomentose. Low thickets. 2-4 f. Redchokeberry.

ova'lis, leaves roundish-elliptical, ovate, smooth; flowers in racemes; petals obovate; germs and segments of the calyx pubescent. Swamps. A small shrub; berries black and eatable. Medlar-bush.

sanguin"ea, (bloody choke-berry. w. M ♄.) leaves oval, obtuse at both ends, mucronate, serratures very slender; racemes few-flowered; calyx glabrous; petals linear, obtuse. 3-6 f.

alnifo'lia, (♄.) smooth; leaves roundish, upper part toothed, pinnately-nerved, subglaucous beneath; raceme simple, elongated. Fruit black and sweet. S.

ARTEMI'SIA.* 17—2. (*Corymbiferæ*.) [From an ancient queen.]
pon"tica, (Roman artemisia,) leaves downy beneath. cauline ones bipinnate; leafets linear; branches simple; flowers roundish, peduncled, nodding. Ex.

absinth"ium, (wormwood, ♃.) stem branching, panioled; leaves hoary, radical ones triply pinnatifid, divisions lanceolate, toothed, obtuse; cauline ones 2-pinnatifid or pinnatifid, divisions lanceolate, acutish; floral ones undivided, lanceolate. Naturalized in most mountain districts of New England.

* The cultivated plant often called Artemisia, belongs to the genus Chrysanthemum

abrota'num, (southern-wood, ♃. and ♄.) stem straight; lower leaves bipinnate; upper ones hair-form, pinnate; calyx pubescent, hemispheric. Ex.

canaden"sis, (wild wormwood, w. y. Au. ♃.) sub-decumbent, scarcely pubescent; leaves flat, linear-pinnatifid; branchlets spike-flowered; flowers sub-hemispheric; involucre scarious. Receptacles smooth. 3-4 f.

cauda'ta, (♃.) stem simple, herbaceous, much branched. pyramidal; radical and cauline leaves bipinnate, pubescent; upper ones pinnate, with sub-setaceous, alternate, divaricate, somewhat convex segments; flowers pedicelled, erect, globe-ovate. 2 f.

A'RUM. 19—12. (*Aroideæ.*) [From *jaron,* a Hebrew word, signifying a dart, in allusion to the shape of the leaves.]

triphyl'lum, (Indian turnip, wild turnip, wakerobin, p. g. and w. M. ♃.) sub-caulescent; leaves ternate; leafets ovate, acuminate; spadix club-form; spatha ovate, acuminate, peduncled, with the lamina as long as the spadix. One variety,*vi'rens,* has a green spatha: another, *atropurpu'reum,* a dark purple spatha: another,*al'bum,* a white spatha. 1.3 f.

dracon"tium, (Ju. ♃.) stemless; leaves pedate; leafets lanceolate-oblong, entire; spadix subulate, longer than the oblong, convolute spatha. Banks of streams. Green-dragon.

atroru'bens, (brown dragon, M. ♃.) stemless; leaves ternate; leafets ovate, acuminate; spadix cylindrical; spatha sessile, ovate, acuminate, spreading horizontally above. Spatha dark-brown; disagreeable smell.

quina'tum, (♃.) stemless; leaves quinate, lanceolate, acuminate. S.

walte'ri, (Ap.) stemless; leaves sagittate, triangular, angles divaricate, acute. S.

ARUN"DO. 3—2. (*Gramineæ*) [Latin, signifying reed.]

canaden"sis, (Au. ♃.) panicle oblong, loose; glumes scabrous, pubescent, as long as the corolla; corolla awned on the back; hairs at the base equalling the valves; culm and leaves smooth. 3-4 f.

phragmi'tes, (reed-grass, Au. ♃.) spikelets 3 to 5-flowered; glumes shorter than the florets; paleas awnless, the lower linear lanceolate, with a long slender acumination, which is involute and resembles an awn.

aroi'des, (♃.) panicle sub-coarctate, incurved; glumes 2-flowered, glabrous, unequal; paleas membranaceous, of the length of the glumes; hairs equalling the paleas; leaves flat, scabrous. S.

AS"ARUM. 18—12. (*Aristolochia.*) [From *a,* not, *sairo,* to adorn, this flower not being admitted into the ancient coronal wreaths.]

canaden"se, (white snake-root, wild-ginger, g-p. M. ♃.) leaves broad-reniform, in pairs; calyx woolly, deeply 3-parted; the segments sub-lanceolate, reflexed.

arifo'lium, (Mar. ♃.) leaves sub-hastate,
28

cordate; calyx urceolate, border 3-cleft, converging. pubescent within. S.

ASCLE'PIAS. 18—5. (*Apocyneæ.*) [Supposed to have been named in honor of the founder of medical science, Æsculapius, or, as he is sometimes called in mythology, Asclepois]

A. *Leaves opposite.*

syri'aca, (common milkweed, w-p. Ju. ♃.) stem very simple; leaves lanceolate-oblong, gradually acute, downy beneath; umbels sub-nodding, downy, 3 to 5 feet high; flowers in large close clusters, sweet-scented. 3-5 f.

incarna'ta, (r. Ju. ♃.) stem erect, branching above, downy; leaves lanceolate, sub-downy both sides; umbels mostly double at their origin; the little horn of the nectary exsert. A variety, *pul'chra,* is more hairy. Var. *gla'bra,* almost glabrous. Var. *al'ba,* has white flowers. Damp. 3 f.

obtusifo'lia, (J. ♃.) stem single, erect; leaves clasping, oblong-obtuse, undulate on the margin, very smooth glands beneath; umbel terminal, long peduncled; horns of the nectary exsert. Stem 3 f. Leaves much waved on the margin. Flowers large, pale purple.

phytolacoï'des, (Ju. ♃.) stem erect, simple; leaves broad-lanceolate, acuminate, smooth, pale beneath; umbels many-flowered, lateral and terminal, solitary, on long peduncles, nodding; nectary 2-toothed. Wet; rocky grounds. Flowers large, greenish purple, 3 f.

quadrifo'lia, (w. p-w. M. ♃.) stem erect, simple, glabrous; leaves ovate, acuminate, petioled; those in the middle of the stem are largest, and in fours; umbels 2, terminal, lax-flowered; pedicels filiform. 18 i. Flowers small and sweetscented.

ama'na, (p. J. ♃.) stem simple, a little hairy on two sides; leaves sub-sessile, oblong-oval, pubescent beneath; terminal umbels and nectaries erect, appendages exsert. Damp.

purpuras"cens, (p. Ju. ♃.) stem simple; leaves ovate, villose beneath; umbels erect; horn of the nectaries resupinate. Shades. 2 f.

pul'chra, (r. Ju. ♃.) leaves lanceolate, hairy beneath; stem divided near the top; umbels erect, in pairs; flowers small; bark very showy.

variega'ta, (w. Ju. ♃.) stem simple, erect; leaves ovate, petioled, rugose, naked; umbels sub-sessile, pedicelled, tomentose. The umbels dense.

parviflo'ra, (w. Ju. ♃.) smoothish; stem weak, erect, simple; leaves petioled, oval-lanceolate, acute at both ends, membranaceous; umbels terminal, lax-flowered; pedicels capillary. The bark a good substitute for flax. 1-2 f.

B. *Leaves not opposite.*

verticilla'ta, (dwarf milkweed, g-y. w. Ju. ♃.) stem erect, very simple, marked with lines, and small pubescence; leaves very narrow-linear, straight, glabrous, whorled, scattered; horn in the nectary exsert. 2 f.

tubero'sa, (Ju. ♃.) stem erect, hairy, with spreading branches; nectary without horns; leaves oblong-lanceolate, sessile, alternate, somewhat crowded; umbels numerous, forming terminal corymbs. Sandy fields. Flowers large, bright-orange, in numerous, erect umbels. Medicinal. Pleurisy-root, butterfly-weed.

pauper''cula, (r-g. M. ♃.) leaves linear, lanceolate, very long, remote, glabrous with the margin pubescent; umbels few-flowered. 3-4 f.

conni'vens (J. ♃.) leaves oblong, oval, mucronate, slightly hairy, sessile; leaves of the nectary usually long, incurved, connivent at the summit. 1-2 f. *S.*

tomento'sa, (♃.) leaves oval, lanceolate, acute, tomentose; umbels sessile with the horns exsert. 1-2 f. *S.*

amplexicau'lis, (p. w. Ap. ♃.) very glabrous; stems decumbent; leaves sessile, cordate, strongly veined, glaucous, appressed; umbels terminal and axillary. 1-2 f. *S.*

nive'a, (Ju. ♃.) leaves ovate, lanceolate, nearly glabrous; umbels erect, lateral, solitary; stem simple. *S.*

angustifo'lia, (g. w. M. ♃.) leaves scattered, strap-shaped, slightly pubescent; umbels solitary, terminal; horns included. 8-18 i. *S.*

cine'rea, (J. ♃.) leaves long, linear, opposite; umbels few, terminal, naked; horns short. 3 f. *S.*

ASCY'RUM. 12—2. (*Hyperica.*) [From *a*, without, *skuros*, roughness.]

crux-andre'æ, (y. Ju. ♄.) stems numerous, subfruticose, terete, with erect branches; leaves ovate-linear, obtuse; inner petals sub-orbicular; pedicels with 2 bracts; flowers sessile; styles 1-2. Sandy fields. N. J. to Car. Flowers solitary, axillary, nearly sessile, pale yellow. This plant varies so much in the size and number of its leaves, and in the number of its styles, that it seems doubtful whether more than one species are not here included. Sand. St. Peter's wort.

amplexicau'le, (Ap. y. ♃.) erect, sparingly branched, with the branches compressed; leaves ovate, oblong, clasping; outer sepals cordate; styles 3 to 4. 1-2 f. *S.*

ASIMI'NA. 12—12. (*Annonæ.*) [From the Greek *asamenos*, sad.]

trilo'ba, (Ap. ♄.) leaves oblong, crenate, acuminate, and with the branches smoothish; flowers on short peduncles; outer petals roundish ovate, 4 times as long as the calyx. Banks of streams. N. Y. to Flor. Flowers solitary, dark brown; fruit large, fleshy, eatable, sweetish. 15-20 f. American papaw tree.

ASPAR''AGUS. 6—1. (*Asparagi.*) [A Greek word, signifying a young shoot.]

officina'lis, (asparagus, Ju. ♃.) stem herbaceous, unarmed, sub-erect, terete; leaves

bristle-form, soft; stipules sub-solitary. Naturalized. 4 f.

ASPHODE'LUS. 6—1. (*Asphodeli.*) [From the Greek *spodelos*, ashes, because it was formerly planted upon the graves of the dead.]

lu'teus, (asphodel, king's spear, ♃.) stem leafy; leaves 3-sided, striate. Ex.

ramo'sus, stem naked; leaves ensiform, carinate, smooth. Ex.

ASPID''IUM. 21—1. (*Filices.*) [From *aspides*, round like a shield; shield-form.]

margina'le, (Ju. ♃.) frond doubly-pinnate; lesser leafets oblong, obtuse, decurrent, crenate, more deeply crenate at the base; fruit-dots marginal; stipe chaffy. 2-3 f.

ASPLE'NIUM. 21—1. (*Filices.*) [*a*, without, *spleen*, the spleen, being used in the cure of this disease.]

rhizophyl'lum, (walking leaf, Ju. ♃.) frond lanceolate, stiped, sub-crenate, heart-form ears at the base; apex very long, linear-filiform, rooting. Var. *pin''natifi'dum*, leaves with the crenatures so deep as to become sub-pinnatifid.

ebe'num, (ebony spleen-wort, Ju. ♃.) frond pinnate; leafets sessile, lanceolate, serrulate, cordato at the base, auricled above. 6-10 i. Rocks and dry places.

AS''TER. 17—2. (*Corymbiferæ.*) [A Greek word, signifying star.]

A. Leaves entire.

ri'gidus, (p. y. Au. ♃.) leaves linear, mucronate, sub-carinate, rigid; margin roughciliate; the cauline leaves reflexed, the branch ones spreading, subulate; stem erect, somewhat branched above; branchlets 1-flowered, corymbed; calyx imbricate, twice as short as the disk; scales obtusish, carinate; rays about 10-flowered, reflexed. Hardly a foot high.

linariifo'lius, (p. y. Au. ♃.) leaves thick set, nerveless, linear, mucronate, dotted, carinate, rough, stiff, those on the branches recurved; stem sub-decumbent; branches level-topped, 1-flowered; calyx imbricate, of the length of the disk; stem rough, purplish.

multiflo'rus, (w-y. Au. to Nov. ♃.) leaves linear, smoothish; stem very branching, diffuse, pubescent; branchlets one way; calyx imbricate; scales oblong, scurvy, acute.

flexuo'sus, (y. w-p. Au. ♃.) very glabrous; leaves subulate, linear, somewhat fleshy, sub-reflexed; stem slender, very branching; branches and branchlets spreading, bristle-form, 1-flowered; scales of the peduncles divaricate, subulate; calyx imbricate, scales close-pressed, acute. Salt marshes.

cornifo'lius, (w. Au. ♃.) glabrous; leaves oblong-ovate, acuminate, short-petioled; margin rough; stem glabrous; panicle few-flowered; branches 2-flowered, calyx sub-imbricate.

amygdali'nus, (w. S. ♃.) leaves lanceolate, tapering to the base, acuminate; margin rough; stem simple, level-topped corymbed at the top; calyx lax-imbricate; scales lanceolate, obtuse; rays large.

nova-ang'liæ, (b-p. Au. ♃.) leaves linear-lanceolate, pilose, clasping, auricled at the base; stem sub-simple, pilose, straight, and stiff; flowers sub-sessile, terminal, crowded; scales of the calyx lax, colored, lanceolate, longer than the disk. In rich soil it grows 10 feet high. Flowers large.

cya'neus, (b. p. Au. ♃.) leaves linear-lanceolate, clasping, smooth; stem wand-like, panicled, very glabrous; branches racemed; scales of the calyx lax, lanceolate, equalling the disk, inner ones colored at the apex. 3-4 f. Flowers many and large. This is the handsomest of the asters.

tennifo'lius, (w. Au. ♃.) leaves linear-lanceolate, tapering to both ends; margins hispid; stem glabrous, branching, erect; branchlets 1-flowered; involucre imbricate; scales lanceolate, acute, lax.

hyssopifo'lius, (star-flower, w. y. p. Au. Oc. ♃.) leaves linear-lanceolate, 3-nerved, dotted, acute, margin scabrous; branches fastigiate, clustered; rays about 5-flowered; involucrum imbricate, twice as short as the disk. 1-2 f. Sandy fields and woods.

humil"is, (w. Au. ♃.) leaves sub-rhomboid, oval-lanceolate, acuminate at both ends, sub-petioled, glabrous; margin hispid; corymb divergingly dichotomous, nakedish; few-flowered; involucre lax-imbricate; rays 8-flowered; a foot high. Flowers large.

amygdali'nus, (w. ♃.) leaves lanceolate, tapering to the base, acuminate; margin rough; stem simple, level-top-corymbed; involucre lax-imbricate; scales lanceolate, obtuse; rays large.

ericoi'des, (w-y. Au. ♃.) leaves linear, very glabrous; those of the branchlets subulate, approximate; cauline ones elongated; involucre scurfy; leaves acute; stem glabrous; flowers small.

B. *Leaves more or less cordate and ovate, serrate, or toothed.*

diversifo'lius, (E. y. p. S. ♃) leaves nearly entire, undulate, pubescent, sub-scabrous; lower ones cordate, ovate, with winged petioles; upper ones lance-oblong; panicle loose, the branches slender, racemose. 3 f.

panicula'tus, (b-p. Au. to Nov. ♃.) leaves ovate-lanceolate, sub-serrate, petioled, glabrous; radical ones ovate, heart-form, serrate, rough, petioled; petioles naked; stem very branching, glabrous; branchlets pilose; calyx lax, sub-imbricate. 2-4 f. Flowers smallish, numerous.

cordifo'lius, (w. S. ♃.) leaves heart-form, pilose beneath, sharp-serrate, petioled; petioles winged; stem panicled, smoothish; panicles divaricate; calyx lax, sub-imbricate; flowers small.

corymbo'sus, (w. Au. ♃.) leaves ovate, sharp-serrate, acuminate, smoothish; lower one heart-form, petioled; petioles naked; stem glabrous, level-top-corymbed above;

branches pilose; calyx oblong, imbricate; scales obtuse, very close-pressed. 12-14 i. Flowers rather large.

undula'tus, (♃.) stem-leaves heart-oblong, clasping, undulate, scabrous, toothed near the summit; branches of the panicle expanding, few-flowered; involucre sub-squarrose; flowers large.

macro'phyl"lus, (w. b. Au. ♃.) leaves ovate, petioled, serrate, rough; upper ones ovate-heart-form, sessile; lower ones heart-form, petioled; petioles sub-margined; stem branching, diffused; involucre cylindric, closely imbricate; scales oblong, acute. 1-2 f. Flowers largish.

C. *Leaves lanceolate and ovate, lower ones serrate.*

amplexicau'lis, (b. S. ♃.) leaves ovate-oblong, acute, clasping, heart-form, serrate, glabrous; stem panicled, glabrous; branchlets 1-2-flowered; scales of the calyx lanceolate, closely imbricate; flowers middle sized.

versic'olor, (y-w. Au. ♃.) leaves sub-clasping, broad-lanceolate, sub-serrate, glabrous; radical ones serrate in the middle; stem very branching, glabrous; scales of the calyx lanceolate, lax, shorter than the disk; flowers many and large, elegant.

tardiflo'rus, (b. Oct. ♃.) leaves sessile, serrate, glabrous, spatulate-lanceolate, tapering to the base, deflected at the margin and both sides; branches divaricate; calyx lax, the leafets lanceolate-linear, sub-equal, glabrous; flowers not middle size.

conyzo'ides, (w. Ju. ♃.) leaves oblong, 3-nerved, narrow and acute at the base; upper ones sessile, sub-entire; lower ones petioled, serrate; stem simple, corymbed at the top; calyx cylindric, scurfy; rays 5, very short. About 12 inches high. Flowers small.

carolinia'nus, (p. Oct. ♃.) stem shrubby, flexous, much-branched, pubescent; leaves sessile, oblong-lanceolate, tapering at each end; scales of the calyx lance-linear, very pubescent, sub-squarrose. 10-12 f. S.

chinen"sis, (china aster, ☉.) leaves ovate, thickly toothed, petioled; cauline ones sessile, at the base wedge-form; floral ones lanceolate, entire; stem hispid; branches 1-flowered; calyx foliaceous. A variety has very full flowers; various colored, and very short rays. Ex.

prenan"thoi'des, (b. ♃.) leaves clasping, spatulate-lanceolate, acuminate, serrate in the middle, heart-form at the base; branchlets pilose; scales of the involucre lanceolate, scurfy.

em"inens, (y-r. S. ♃.) leaves lance-linear, acuminate, scabrous at the margin; lower ones sub-serrate; stem panicled; branchlets 1-flowered; involucre lax-imbricate, with lanceolate leaflets.

grandiflo'rus, (p. y. Oct. ♃.) leaves sub-clasping, linear, subulate, rigid reflex, with the margin ciliate and hispid; stem hairy; branches 1-flowered; involucre squarrose; the scales linear-lanceolate. 2-3 f. S.

squarro'sus, (b-y. S. ⚁.) leaves very numerous, sessile, ovate, acute, reflexed, rigid, margin hispid; stem branching, hairy; branches 1-flowered; scales of the involucre lanceolate, hairy, loose. 2 f. *S.*

sca'ber, (p. y. S. ⚁.) lower leaves petioled, oblong, cordate, acute, entire; upper ones sessile, clasping, lance-ovate, tapering to an acute point; all the leaves scabrous, undulate; panicle loose, long; the branches racemose. 3 f. *S.*

obova'tus, (w. y. M. ⚁.) leaves sessile, oval or obovate, obtuse, sub-rugose, very pubescent; corymb paniculate; scales of the involucre closely imbricate. 3 f. *S.*

ASTRAG''ALUS. 16—10. (*Leguminosæ.*) [A Greek word, signifying a leguminous plant.]

canaden''sis, (J. y. ⚁.) caulescent, diffuse; leaflets 10-12 pairs, with an odd one, smooth on both sides; legume sub-cylindrical, mucronate. Barren fields. 2 f.

glaux, (milk vetch, ♃.) caulescent, the little heads peduncled, imbricate, ovate; flowers erect; legume ovate, callous, inflated. Ex.

depres''sus, (trailing vetch, ♄.) sub-caulescent, procumbent; leafets obovate; raceme shorter than the petiole; legume terete, lanceolate, reflexed. Ex.

ASTRO'PHIA. 16—10. (*Leguminosæ.*)

littora'lis, (⚁.) silky, pubescent; racemes about 5-flowered; leaves pinnate, small; leaflets in 2 or 3 pairs, linear-spatulate; seeds globose, brown.

ATRI'PLEX. 5—2. (*Atriplices.*) [Latin, signifying dark.]

horten''sis, (garden orache, Ju. ♀.) stem erect, herbaceous; leaves triangular, dentate, green on both sides; calyx of the fruit ovate, reticulate, entire; flowers in racemes or spikes. Waste places. Flowers green. 3-4 f.

lacinia'ta, (♀.) stem erect, herbaceous; leaves triangular, deep-toothed, white beneath; calyx of the fruit rhomboid, 3-nerved, denticulate.

ATRO'PA. 5—1. (*Solaneæ.*) [From *Atropos*, the goddess of destiny, in allusion to its fatal effects.]

physalo'ides, (w. b. Ju. ♀.) stem very branching; calyx 5-angled, reticulate; berry fleshy, covered with the calyx; leaves sinuate-angled.

belladon''na, (deadly·night-shade, w. y. ⚁.) stem herbaceous; leaves ovate, entire.

AVE'NA. 3—2. (*Gramineæ.*) [From the Latin *aveo*, to covet, a favorite of cattle]

præ'cox, (dwarf oats, J ♀.) panicle oblong, in a dense raceme; florets as long as the glumes; awn exserted; leaves setaceous. Sandy fields.

steril''is, (animated oats, Ju. ♀.) panicled; calyx about 5-flowered; florets hairy, the middle ones awnless. The heads are set in motion, when moistened, by the untwisting of the awns. Ex.

sati'va, (oats, J. ♀.) panicled; 2-seeded; seeds smooth, one of them awned. First discovered in the island of Juan Fernandes. A variety is awnless, and has black seeds. Ex.

elat'ior, (J. ⚁.) panicle sub-contracted, nodding; glume 2-flowered; florets perfect, sub-awnless, staminate awned; culm geniculate, glabrous; root creeping. Introduced.

AVICEN''NIA. 13—2. (*Polemonia.*) [After an Arabic physician of repute.]

tomento'sa, (mangle, ♄.) flowers in sub-sessile clusters; leaves oblong, obtuse, tomentose beneath. 20 f. *S.*

AZA'LEA. 5—1. (*Rhododendra.*) [From *azaleos*, dry, growing in dry soil.]

nudiflo'ra, (early honeysuckle, r. M. ♄.) sub-naked-flowered; leaves lanceolate-oblong, or oval, smooth or pubescent, uniform-colored; nerves on the upper side downy, and beneath bristly; margin ciliate; flowers abundant, not viscous, their tubes longer than their divisions; teeth of the calyx short, oval, sub-rounded; stamens very much exsert. A variety, *coccin''ea*, has scarlet flowers and minute calyx; another, *car'nea*, has pale red flowers, with red bases and leafy calyx; another, *al''ba*, has white flowers, with a middling calyx; another, *papilionae'a*, has red flowers, with the lower divisions white, calyx leafy; another, *parti'ta*, has flesh-colored flowers, 5-parted to the base; another, *polyan''dria*, has rose-colored flowers, with from 10 to 20 stamens. Woods. 2-6 f.

visco'sa, (white honeysuckle, w. J. ♄.) leafy; branches hispid; leaves oblong-obovate, acute, glabrous, and one-colored; flowers viscous, tube twice as long as the divisions; teeth of the calyx very short, rounded; flowers very sweet-scented.

procum''bens, (Ju. ♄. r.) stems diffusely procumbent; leaves opposite, elliptical, glabrous, revolute on the margins; corolla bell-form, glabrous; filaments enclosed, equal. High mountains. Northern. Flowers small, in small terminal umbels or corymbs. 3-4 i.

canes''cens, (r. J. ♄.) sub-naked-flowered; leaves obovate-oblong, pubescent on the upper side, and downy beneath; nerves not bristle-bearing; flowers not viscous; tube of the corolla scarcely shorter than its divisions; teeth of the calyx very short, round-obtuse; stamens scarcely exsert. Catskill mountains.

arbores''cens, (r. ♄.) flowers leafy; leaves obovate, sub-obtuse, smooth both sides, glaucous beneath, ciliate on the margin; nerve almost smooth; flowers not viscous; tubes longer than the segments; calyx leafy, with oblong-acute segments; filaments exsert. 15 f.

nit''ida, (swamp honeysuckle, w. J. ♄.) leafy-flowered; branches smoothish; leaves few, oblanceolate, sub-mucronate, leathery, glabrous both sides, and the upper side shining; nerve bristle-bearing beneath; margin revolute-ciliate; flowers viscous; tube somewhat longer than the divisions; calyx very short; filaments exsert; leaves dark green. Swamps.

glau'ca, (fragrant honeysuckle, w. J. ♄.) leafy-flowered; branches hispid; leaves oblanceolate, acute, both sides glabrous, and glaucous beneath; nerve bristle-bearing, margin ciliate; flowers very viscous; tube of the corolla twice as long as its divisions; calyx very short; filaments about equal to the divisions of the corolla; rather lower than the other species; flowers abundant. Perhaps this is a variety of the *viscosa.*

bico'lor, (r. w. M. ♄.) naked-flowered; leaves obovate, covered on both sides with fine, whitish hairs; the nerve not bristled; flowers small, not viscid; the tube scarcely longer than the segments of the corolla; calyx very short; one segment narrow, and 4 times longer than the rest; filaments longer than the tube; smaller branches hairy and hispid. 2-8 f. *S.*

BAC''CHARIS. 17—2. (*Corymbiferæ.*) [Dedicated to Bacchus.]
halimifo'lia, (w. S. ♄.) leaves obovate and oval, incisely toothed near the summit; panicle compound, leafy; heads of flowers peduncled; egret of the fertile florets hairy; twice as long as the corolla. The whole plant is covered with a whitish dust. 16.12 f. Groundsel-tree.

BÆOMY'CES. 21—5. (*Algæ.*)
rose'us, crust uniform, warty, white; peduncle (podetia) short, cylindric; receptacle sub-globose, pale red. On the earth.

BALLO'TA. 13—1. (*Labiatæ.*) [From *ballo,* to put forth, *otos,* the ear.]
ni'gra, (black horehound, ♄.) leaves undivided, ovate, serrate; calyx dilated above, sub-truncate, with spreading teeth; flowers purple or white, in axillary whorls. 2-3 f.

BAPTI'SIA. 10—1. (*Leguminosæ.*) [From *bapto,* to dye]
tincto'ria, (wild indigo, y. Ju. ♃.) very glabrous and branching; leaves ternate, sub-sessile, leafets wedge-obovate, round-obtuse, becoming black in drying; stipules obsolete, oblong-acute, much shorter than the petioles; racemes terminal; legumes ovate, long-stiped. 2-3 f.
al''ba, (w. J. ♃.) branches spreading; leaves ternate, petioled; leafets lanceolate, wedge-form at the base, obtuse, mucronate, glabrous; stipules subulate, shorter than the petioles; racemes terminal. 2 f. *S.*
cæru'lea, (spiked indigo weed, b. Ju. ♃.) glabrous; leaves ternate, short-petioled; leafets oblong, wedge-form, obtuse; stipules lanceolate, acute, twice as long as the petioles; racemes spiked, elongated; legumes acuminate.

BARBARE'A. 14—2. (*Cruciferæ.*)
vulga'ris, (J. ♃. y.) lower leaves lyrate, the terminal lobes roundish; upper ones sessile, obovate, toothed; pod 4-sided, tapering into a slender style; flowers in corymbs, small. Bitter winter cress; found in old fields. 12-18 i.

BARTO'NIA. 11—1. (*Onagræ.*) [In honor of Dr. Barton, of Phil.]
lævicau'lis, (w. J.) petals 5, stamens 5; petalloid; bracts 0; stem very smooth; seeds winged.
parviflo'ra, (w. J.) petals 5; stamens 5-7,

petalloid; bracts 0; stem scabrous; seeds winged.

BART''SIA. 13—2. (*Scrophularia.*)
pal''lida, (white painted cup, w-y. Au. ♃.) leaves alternate, linear, undivided, upper ones lanceolate, floral ones sub-oval, sub-toothed at the summit, all are 3-nerved; teeth of the calyx acute.
acumina'ta, (♃.) leaves alternate, long-linear; floral leaves ovate, long-acuminate, 3-nerved, all undivided; flowers shorter than the bracts; teeth of the calyx acute.
tenuifo'lia, (y. Ju. ♃.) very hirsute; leaves alternate, linear, gash-pinnatifid, divisions filiform; bracts (yellow) membranaceous, oblong, obtuse, tooth-hastate at the base on each side, longer than the flowers; calyx short, hairy, with subulate teeth. 1 f.

BATSCH''IA. 5—1. (*Boragineæ.*) [In honor of Batsch, a German.]
canes''cens, (puccoon, Ju. ♃.) whitish-villose; leaves all oblong; calyx short; divisions of the corolla entire. Hills. Flowers axillary, crowded near the top of the stem, bright orange. The root is used by the Indians as a red dye.
gmeli'na, (r-y. Ap. ♃.) hirsute, floral leaves ovate; segments of the calyx long, sub-lanceolate. Dry woods. 10-16 i.
longiflo'ra, (y. Ju. ♃.) hirsute; erect; leaves approximating, long-linear, margin reflexed, fascicles fastigiate; tube of the corolla sub-pentangular; border flat, with fringed crenatures. *S.*
decum''bens, hirsute; stem decumbent; segments of the calyx and leaves linear; flowers scattered; lobes of the corolla fringed-crenate, shorter than the tube. *S.*

BEJA'RIA. 12—1. (*Rhodendra.*) [In honor of a Spanish botanist.]
racemo'sa, (w-r. J. ♄.) leaves lance-ovate, glabrous; flowers in a panicled raceme, terminal; stem hispid. 3 f. Sandy plains. *S.*

BEL''LIS. 17—2. (*Corymbiferæ.*) [From *bellus,* handsome.]
peren''nis, (daisy, w. and p. Ap. ♃.) leaves obovate, crenate; scape naked, 1-flowered. Ex.
integrifo'lia, caulescent; leaves entire; lower ones obovate, upper ones lanceolate; leafets of the calyx very acute, and acuminated with a hair. *S.*

BER''BERIS. 6—1. (*Berberides.*) [From *berberi,* Arabic, signifying wild.]
vulga'ris, (barberry, y. M. ♄.) branches punctate; prickles mostly in threes; leaves obovate, remotely serrate; flowers racemed.
canaden''sis, (Ju. ♄.) branches verrucose-dotted, with short tripple spines; leaves spatulate-oblong. remotely serrate, with somewhat bristly teeth; racemes sub-corymbose, few-flowered; petals emarginate; berries sub-globose, or oval. 2-3 f. Virginia, Georgia.

BE'TA. 5—2. (*Atriplices.*) [So called from the river Bœtis, in Spain, where it grows wild.]
vulga'ris, (beet, g. Au. ♂.) flowers heaped together; lower leaves ovate. Ex.
ci'cla, (white beet, ♂.) flowers in threes; radical leaves petioled, cauline ones sessile; lateral spikes very long. Ex.

BE'TULA.   19—12.   (*Amentaceæ.*)   [Latin, birch.]

populifo'lia, (white birch, poplar birch, Ju. ♄.) leaves deltoid, long-acuminate, unequally serrate, very glabrous; scales of the strobile with rounded, lateral lobes; petioles glabrous.   30-40 f.

papyra'cea, (paper birch, ♄.) leaves ovate, acuminate, doubly serrate; veins hirsute beneath; petiole glabrous; pistillate ament peduncled, nodding; scales with lateral, short, sub-orbicular lobes.   The bark used by the Indians for canoes.

len''ta, (black birch, M. ♄.) leaves heart-ovate, sharp-serrate, acuminate; nerves and petioles pilose beneath; scales of the strobile glabrous, with obtuse, equal lobes, having elevated veins.   Wood resembles mahogany.   Very sweet-scented.   80 f.

glandulo'sa, (scrub birch, M. ♄.) branches glandular-dotted, glabrous; leaves obovate, serrate, at the base entire, glabrous, sub-sessile; pistillate ament oblong, scales half 3-cleft; fruit orbicular, with a narrow margin.   2-8 f.

pu'mila, (dwarf birch, J. ♄.) branches pubescent, dotted; leaves orbicular-obovate, petioled, dense-pubescent beneath, pistillate ament cylindric.   2-3 f.

BI'DENS.   17—3.   (*Corymbiferæ.*)   [From *bis*, two, and *dens*, tooth.]

cer''nua, (y. Au. ☉. water beggar-ticks,) flowers sub-radiate, cernuous; outer involucre as long as the flower; leaves lanceolate, sub-connate, dentate.   Ponds and ditches.   1-2 f.

chrysanthemoï'des, (daisy beggar-ticks, Au. ☉.) flowers rayed, drooping rays erect, longer than the sub-equal involucre; leaves oblong, tapering both ends, toothed, connate; flowers large.

bipin''na'ta, (hemlock beggar-ticks, y. Ju. ☉.) flowers sub-rayed; outer involucre of the length of the inner; leaves doubly pinnate, leafets lanceolate, pinnatifid.

BIGNO'NIA.   13—2.   (*Polemoniæ.*)   [In honor of the Abbe Bignon.]

radi'cans, (trumpet flower, r. and y. Ju. ♄.) leaves pinnate, leafets ovate, toothed, acuminate; corymb terminal; tube of the corolla thrice as long as the calyx; stem rooting.   Most beautiful climbing shrub.   One variety, *flam''mca*, has yellowish scarlet flowers; another variety, *coccin''ea*, has bright scarlet flowers.   Cultivated.

crucif''era, (y. r. J. ♄.) leaves conjugate, cirrose; lower ones ternate; leafets heart-ovate, acuminate; racemes axillary; stem muricate.

BLI'TUM.   1—2.   (*Atriplices.*)   [From the Greek *bliton*, an insipid pot-herb.]

capita'tum, (strawberry blite, r. J. ☉.) heads in a terminal spike, not intermixed with leaves; leaves triangular, toothed.   15 l.

mariti'mum, (Aug. ☉.) stem erect; perianth membranaceous; clusters axillary, spiked, naked; leaves lanceolate, tapering to each end, gash-toothed.   Salt marshes.   1-2 f.

virga'tum, (slender blite, r. J. ☉.) late-

ral heads scattered, top ones leafy; leaves triangular-toothed.

BOERHAA'VIA.   2—1.   (*Nyctagines.*)   [In honor of Boerhaave, the celebrated physician.]

erec''ta, (w. p. J. ♃.) stem columnar, trichotomous, rough below, smooth above; flowers in corymb-panicles.   S.

BOLE'TUS.   21—6.   (*Fungi.*)   [From *boles*, a mass.]

igniarius, dilated, smooth, cuticle in ridges; pileus hard, becoming dark at the base, at the margin cinnamon color, beneath yellowish white.   Grows on trunks of trees.   General form like a horse's hoof.   It is called *touchwood*.

BOLTO'NIA.   17—2.   (*Corymbiferæ.*)

asteroides, (false aster, w. r. Au. ♃.) leaves very entire; flowers long-peduncled, seed oval, sub-awnless, glabrous.

glastifo'lia, (false camomile, w. Ju. ♃.) lower leaves serrate; flowers short-peduncled; akenes obcordate, apparently winged, pubescent; awns of the egret two, of equal length with each other.   Resembles an''themis cot''ula.

BORA'GO.   5—1.   (*Boraginæ.*)   [Formerly called *corago*, from *cor*, the heart, and *ago*, to affect, because it was thought to cheer the spirits.]

officina'lis, (borage, b. Ju. ☉.) leaves alternate; calyx spreading.   Ex.

africa'na, (☉.) leaves opposite, petioled, ovate; peduncle many-flowered.   Ex.

BOTRYCH''IUM.   21—1.   (*Filices.*)   [*Botrus*, a bunch of grapes, from the fructification resembling one.]

fumarioï'des, (grape fern, J. ♃.) stipe naked; frond smooth, radical, 3-parted, bipinnate; leafets lunate, crenate; spikes bipinnate.

virgin''icum, (rattlesnake fern, Ju.) somewhat hairy; scape bearing the frond in the middle; frond 3-parted, bipinnatifid; divisions incisely pinnatifid; segments obtuse, about 3-toothed; spikes decompound.   2 f.

BRAS''SICA.   14—2.   (*Cruciferæ.*)

ra'pa, (turnip, ♂.) root caulescent, orbicular, depressed, fleshy; radical leaves rough, cauline ones very entire, smooth.   Var. *ruta-baga*, has a turbinate, sub-fusiform root.   Ex.

olera'cea, (common cabbage, including all the varieties caused by culture, ♂.) root caulescent, terete, fleshy; leaves smooth, glaucous, repand-lobate.   Ex.

na'pus, (kale or cole, ♂.) root caulescent, fusiform; leaves smooth, upper ones heart-lanceolate, clasping, lower ones lyrate-toothed.

BRICKEL''LIA.   17—1.   (*Cinerocephalæ.*)

cordifo'lia, (p. Au.) involucre many-flowered; corolla tubular, 5-cleft; stamens attached to the corolla; akenes long, hairy; lower leaves cordate, acuminate, dentate, pubescent; upper ones obtuse; receptacle naked and dotted.

BRI'ZA.   3—2.   (*Gramineæ.*)   [From the Greek *britho*, to nod.]

me'dia, (quaking grass, rattlesnake grass, J. ♃.) panicle erect; spikelets heart-ovate, about 7-flowered; calyx smaller than the flowers.   1 f.   Probably introduced.

BRODIÆ'A. 6—1. (*Narcissa*.) [In honor of James Brodie.]
grandiflo'ra, (Ap. ♃.) umbels many-flowered; flowers pedicelled; stamens alternate, with membranaceous margins. Missouri hyacinth. *S.*

BRO'MUS. 3—2. (*Graminea*.) [*Bromos*, a species of wild oats.]
secali'nus, (chess, J. ☉.) panicle nodding, spikelets ovate, compressed; glumes naked, distinct; awns shorter, subulate, straightish-zigzag. Florets about 10 in each spikelet; leaves somewhat hairy. Common in rye and wheat fields.
pubes'cens, (broom grass, J. ♃.) culm hairy below; joints brown; stipules very short; panicle at length nodding, pubescent; glumes less than paleas, 8 to 12-flowered; paleas pubescent, one valve awned beneath the apex. Var. *cilia'tus* has a ciliate valve in each palea; 8 to 10-flowered. Var. *canaden"sis*, has one very hairy 7-nerved valve in each palea, and a short bristle.
mol"lis, (Ju.) panicle erect, compact; peduncles ramose; spikelets ovate; florets imbricate, depressed, nerved, pubescent; bristle straight, nearly as long as the paleas; leaves with short hairs.

BROUSSONE'TIA. 20—4. (*Urticea*.) [In honor of Broussonnet.]
papyrif'era, (M. ♄.) leaves sub-cordate, lobed or undivided; roots sending off suckers. 20 f. Paper mulberry. *Ex.*

BRUNNICH"IA. 8—3. (*Polygonea*.)
cirrho'sa, (♃.) climbing; leaves cordate, acute, glabrous, entire; panicles terminal; bracts ovate, mucronate. *S.*

BRYOPHYL"LUM. 8—4. (*Sempervivea*.)
calyci'num, (leaf plant, sprout leaf, r-g. Ju. ♃.) leaves remarkably thick and succulent, crenate-serrate, oval; petioled, channeled above; leafets in pairs; flowers long-cylindric, pendulous. *Ex.*

BUCHNE'RA. 13—2. (*Jasminea*.)
america'na, (blue-hearts, b. Au. ♃.) stem simple; leaves lanceolate, sub-dentate, rough, 3-nerved; flowers remote, spiked. 1 f.

BUPLEU'RUM. 5—2. (*Umbellifera*.) [Named from the stiff striated leaves of some of the species.]
rotundifo'lium, (hare's-ear, or thorough-wax, y. Ju. ☉.) leaves perfoliate, broadly egg-shape, alternate. 1 f. *Ex.*

BUX"US. 19—4. (*Euphorbia*.) [From the Greek, signifying hard.]
semperui'rens, (box, ♄.) leaves ovate, petioled, somewhat hairy at the margin; anthers ovate, arrow-form. Var. *angustifo'lia*, lanceolate leaves. *Suffrutico'sa*, leaves obovate, stem hardly woody. *Ex.*

CACA'LIA. 17—1. (*Corymbifera*.) [From *kakon*, bad, and *nos*, exceedingly, because it is bad for the soil.]
atriplicifo'lia, (wild caraway, w. Au. ♃.) stem herbaceous; leaves petioled, smooth, glaucous beneath; radical ones cordate, toothed; cauline ones rhomboidal; flowers corymbed, erect; involucrum 5-flowered. Low ground. 3-6 f.
coccin"ea, tassel-flower; from the East Indies. 18 i. The flowers of a scarlet color.

ova'ta, (w. Oc.) stem herbaceous; leaves ovate, obtuse; obtusely toothed, nerved, slightly glaucous beneath; the lower ones petioled; involucre 5-leaved; 5-flowered. 3-4 f. *S.*
lanceola'ta, (y. w.) stem herbaceous; leaves narrow-lanceolate, acute at each end, remotely-toothed, nerved, slightly glaucous beneath; involucre 5-leaved, 5-flowered. 4-6 f. *S.*

CAC"TUS. 11—1. (*Cacti*.) [A Greek word, signifying prickly.]
opun"tia, (prickly-pear, S. y. J. ♃.) proliferous; articulations compressed, ovate; bristle fascicular. The plant appears like a series of thick succulent leaves, one growing from the top of another. *Ex.*
phyl"lanthoi'des, (leaf flowered, prickly pear, r. Oc.) branches leaf-life, ensiform, compressed, obovate with spreading rounded teeth; spines few, setaceous, longer than the woody covering. 2 f. *Ex.*
vivipa'rus, (r. Ju. ♃.) roundish, manifold or cespitose; tubercles cylindric, bearded, grooved and proliferous above the furrows. *S.*
mammilla'ris, tubercles ovate, terete, bearded; flowers scarcely exserted; berries scarlet, about equal with the tubercles. *S.*
fe'rox, (y. & r. Ju.) proliferous; articulations large, nearly circular, spiny; spines double; larger ones radiate, persistent; fruit dry, spiny. *S.*
fragil"is, proliferous; articulations short, oblong, somewhat terete, fragile; spines double; flowers solitary, small; fruit dry, spiny. *S.*
cylin"dricus, (p. ♄.) very branching; terminal branches consisting of long cylindrical articulations; surface reticulated with decussate furrows. *S.*

CALAMIN"THA. 13—1. (*Labiata*.) [From *kalos*, beautiful, *mentha*, mint.]
grandiflo'ra, (mountain calamint, r. Ju.) suffruticose; leaves ovate, obtuse, crenate, smooth; whorls many-flowered, on short peduncles, shorter than the leaves. 12-18 i. *S.*
nepe'ta, (r.-w. J. ♃.) pubescent, very branching; whorls peduncled, dichotomous-corymbed, longer than the leaves; leaves ovate, obtuse, sub-serrate; down or wool of the calyx prominent. *S.*

CAK"ILE. 14—1. (*Crucifera*.) [From a Latin word, signifying noise, alluding to the rattling of the seeds.]
america'na, (p. Oct. ☉, American sea-rocket,) leaves fleshy, oblong, obtuse, margins toothed, joints of the pouch one-seeded; the upper ones ovate, acute. Sea-coast, shores of the great lakes. Plant fleshy, branched, decumbent. Flowers corymbed.

CALANDRIN"IA. 12—1. (*Portulacea*.) [From *kalos*, beautiful, *andrion*, stamen.]
specio'sa, (p.) glabrous, diffuse; leaves spatulate, acute, attenuate into a petiole; flowers racemed; peduncles shorter than the bracts; petals longer than the calyx. 4-5 i. California. Var. *grandiflo'ra*, the flowers, notwithstanding its name, are

smaller than those of the *disco'lor*, this last being one of the most splendid flowers growing in the open air.

CALEN"DULA. 17—4. (*Corymbiferæ*.) [So called because it flowers every month, from *calends*, month.]

officina'lis, (pot marygold, y. ☉.) seed keeled, muricate, incurved. Ex.

stella'ta, starry marygold. Barbary orange. 2 f.

pluvia'lis, (rainy marygold,) florets of the ray pure white inside, dark purple outside. *Aybrida*, dingy orange on the outside.

CAL"LA. 19—12. (*Aroideæ*.) [From *kalos*, beautiful.]

palus'tris, (water arum, w. J. 2f.) leaves sub-roundish, heart-form, acute; spatha ovate, cuspidate, spreading when mature. Grows in wet places.

ethio'pica, Egyptian lily. Ex.

CAL"LICAR"PA. 4—1. (*Vitices*.) [From *kalos*, beauty, *karpos*, fruit.]

america'na, (r. J.) leaves serrate, tomentose beneath. 3-4 f. S.

CAL"LITRI'CHE. 1—2. (*Onagræ*.) [From *kalos*, beauty, and *trichos*, hair, appearing like hair]

ver"na, (water chickweed, w. M. ☉.) upper leaves spatulate, obovate, lower ones linear, obtuse, and emarginate; flowers polygamous. In shallow streams. Stem floating. 2-3 f. Upper leaves in a tuft. Flower solitary, axillary.

CAL"OCHOR"TUS. 6—3. (*Narcissi*.)

ele'gans, (w. and p. 2f.) scape nearly 3-flowered, shorter than the single leaf; petals woolly within. S.

lute'us, (y. 2f.) stem forked, about 2-flowered; leaves setaceous, short; flowers large, inner petals the largest, glabrous at the apex, spotted, ciliate at the base; a mark in the claws of the downy petals; root bulbous.

CAL"THA. 12—12. (*Ranunculaceæ*.) [A Greek word, signifying yellow.]

palu'stris, (y. Ap. 2f. American cowslip,) stem erect; leaves cordate, sub-orbicular, acute-crenate. 12-18 i.

integer"rima, (M. 2f.) stem erect, corymbose; leaves orbicular-cordate, very entire, with the sinus closed; floral ones sessile, reniform, obsoletely crenate at the base; sepals oval, obtuse.

parnassifo'lia, (y. Ju. 2f.) stem erect, 1-flowered, 1-leaved; radical leaves petiolate, lanceolate-cordate, obtuse, many-nerved; sepals elliptical.

sagitta'ta, (w. J.) scape 1-flowered; leaves ovate, obtuse, entire, heart-sagittate at the base with inflexed auricles above; divisions of the nectary 9. Pistils 13 to 15. S.

CALYCAN"THUS. 11—12. (*Rosaceæ*.) [From *calyx*, and *anthos*, the flower being inserted into the calyx.]

lævig a'tus, (b-p. Ju. ♄.) lobes of the calyx lanceolate, calyx brownish purple; leaves oblong, or oval, gradually acuminate, somewhat rugose, smooth and green on both sides; branches erect, straight; flowers large, solitary, terminal. 4-6 f.

flor'i dus, (Carolina allspice, p. M. ♄.) divisions of the calyx lanceolate; leaves broad-oval, acute, tomentose beneath; branches spreading. 3-7 f. S.

CALLIGO'NUM. 12—4. (*Polygoneæ*.)

canes'cens, (Ju. ♄.) diœcious, leaves lanceolate; flowers axillary, crowded, spiked toward the ends of the branches. S.

CALYP"SO. 18—1. (*Orchideæ*.) [From the fabled nymph, Calypso.]

america'na, lip narrowed, sub-unguiculate at the base; spur semi-bifid, longer than the lip, with acute teeth; peduncle longer than the ovary. Scape 6-8 inches high, sheathed, 1-flowered; radical leaves roundish-ovate, nerved. Flowers large, purplish, resembling a Cypripedium.

CAMELI'NA. 14—1. (*Cruciferæ*.)

sa'tiva, (wild flax, gold-of-pleasure, y. J. ☉.) silicle obovate-pyriform. margined, tipped with the pointed style; leaves roughish, sub-entire. lanceolate, sagittate; flowers small, numerous, in corymbs. 2 f. Cultivated grounds. Introduced.

CAME'LLIA. 12—13 (*Meliæ.*) [From *Camellas*, a learned Jesuit.]

japon"ica, (Japan rose,) leaves ovate, acuminate, acutely serrate; flowers terminal, sub-solitary. By some, the Tea (Thea) is classed in the genus Camellia.

CAMPAN"ULA 5—1. (*Campanulaceæ*.) [Latin, *campanula*, a little bell.]

rotundifo'lia, (flax bell-flower, hair-bell, b. J. 2f.) glabrous; radical leaves heart-reniform, crenate; cauline ones linear, entire; panicle lax, few-flowered; flowers nodding.

america'na, (b. Au. 2f.) leaves ovate-lanceolate, long-acuminate; lower ones sub-cordate, with the petioles ciliate; flowers axillary, nearly sessile, in a terminal leafy raceme; corolla sub-rotate; style exsert. Cultivated. 2 f. *Antinella*

spec"ulum (b. Au. ☉.) stem branched; leaves oblong, sub-crenate; flowers solitary, scales at the base. Purple. South of Europe. 1 f. Venus' looking-glass.

amplexicau'lis, (clasping-bell, b. M. ☉.) stem simple, erect; leaves heart-form, crenate, clasping; flowers axillary sessile, glomerate. 12-18 i.

erinoï des, (prickly bell flower, w-b. J. ☉.) slender; stem simple, angular; angles, and the margin and nerves of the leaves, with reverse prickles; leaves linear-lanceolate, glabrous on the upper side; peduncles few; those on the top of the stem flexuose; axillary ones 1-flowered, filiform. 12 i.

uniflo'ra, pubescent; radical leaves round-obovate; cauline ones lance-linear, somewhat toothed; stem about 1-flowered.

me'dium, (canterbury bells, w. b. Au. ♂.) capsule 5-celled, covered; stem undivided, erect, leafy; flowers nerved. Ex.

CAN"NA. 1—1. (*Canna.*) [From the Hebrew, signifying a reed.]

flac"cida, (y. J. 2f.) inner limb of the corolla 3-cleft; segments flaccid. 2-3 f. S.

indica, Indian shot plant. 4 f. Scarlet. A native of the East Indies.

CAN"NABIS. 20—5. (*Urticæ.*) [From the Arabic *kannab*, to mow.]

sati'va, (hemp, g. Au. ☉.) stem pilose;

leaves petioled, digitate; leafets lanceolate, serrate, pilose; staminate flowers solitary, axillary; pistillate ones spiked. 4-10 f. Ex.

CAPRA'RIA. 12—2 (*Vitices.*) [From *capra*, a goat.]

pusil"la, (w. ⊕.) hairy, leaves opposite, cordate, repand-toothed, petioled; peduncles axillary, longer than the petioles. *S.*

CAP"SICUM. 5—1. (*Solanea.*) [From *kapto*, to bite, on account of its effect upon the tongue.]

an"nuum, (guinea pepper, red pepper, cayenne pepper, y-g. w, Au. ⊕.) stem herbaceous; peduncles solitary. From South America. 10-18 i.

bacca'tum, (bird pepper, ♄.) stem smooth; peduncles in pairs. Florida.

CARDAM"INE. 14—2. (*Crucifera.*) [From *kardia*, the heart, because it acts as a cordial.]

pennsylvan"ica, (American water-cress, w. M. 2/.) glabrous, branching; leaves pinnate, hairy; leafets roundish-oblong, obtuse, tooth angled; silique narrow, erect.

praten"sis, (field water-cress, r-p. M. 2/.) simple, glabrous, erect; leaves pinnate; radical leafets roundish, toothed; cauline ones lanceolate, sub-entire; racemes subcorymbed.

CARDIOSPER"MUM. 8—3. (*Sapindi.*)

halica'bum, (Au. ⊕.) glabrous; leafets incised and lobed; the terminal one rhomboidal. Balloon vine. East Indies. 5 f. Flowers white and green.

CAR"DUUS 17—1. (*Cinarocephala.*) [From *keiro*, to tear.]

pectina'tus, (p. ♂.) unarmed; leaves decurrent, lanceolate, pectinately pinnatifid; peduncles almost leafless, terminal, very long, about 1-flowered; flowers nodding, discharging the pollen; scales of the ...linear, spreading.

... (*Cyperoidea.*) [From Latin ... the upper spikes of these ...nstantly without seeds, com...taminate flowers.]

...escence diœcious.

...(barren sedge. M. 2/.) spikes ...terile 3-5; fertile about 6; ...ndrogynous;) fruit ovate, com...quetrous; margin ciliate-serrate; ...urved and bicuspidate. 8-12 i.

B. *Inflorescence monecious.*
† *Spikes androgynous.*
* *Spike solitary.*
a. *Stamens at the summit of the spikelets.*

frase'ri, (Ap. 2/.) spike simple, ovate; fruit ovate-sub-globose, entire at the point, longer than the oblong glume; leaves lanceolate, undulate, crenulate; scape-sheathed at the base. 1 f. This species has broader leaves than the common sedges, and produces fine flowers resembling small lilies.

polytrichoi'des, (M. 2/.) spike simple, fruit oblong-lanceolate, compressed, triquetrous, obtuse, emarginate; glumes oblong-obtuse, mucronate. 10 i. Wot.

** Spikes distinct (*not aggregated into a head*).
a. Stamens at the summit.
1. With 2 stigmas.

retroflex"a, (M. 2/.) spikes about 4, sub-approximate, ovate, the lowest one with a short bract; fruit ovate-lanceolate, bidentate, scabrous on the margin, spreading and reflexed, as long as the ovate-acute glume. 1 f. Woods, meadows.

ro'sea, (M. 2/.) spikes 4-6, remote, about 9-flowered, the lowest one with a setaceous bract overtopping the culm; fruit ovate, acuminate, diverging and radiate, scabrous on the distinct margin, twice as long as the ovate-obtuse glume. 12 i. Moist.

stipa'ta, (M. 2/.) spike compound, oblong; spikes numerous (10-15), oblong, aggregated, bracteate; bracts a little longer than the spikelets; fruit lanceolate, subterete, and smooth below, spreading, bidentate at the point, which is scabrous, twice as long as the glume. 1-3 f. Wet meadows.

2. *With 3 stigmas.*

peduncula'ta, (Ap. 2/.) spikes about 4, on long peduncles, very remote; fruit obovate, triquetrous, obtuse, smooth, entire at the orifice; glumes ovate, mucronate (purple and green). 6 i. Rocky Hills.

b. *Pistillate at the summit.*
1. *With 2 stigmas.*

scopa'ria, (M. 2/.) spikelets mostly 5, ovate, sessile, approximate, aggregate, lowest one bracteate; fruit ovate-lanceolate, margined, nerved, smooth, bi-cuspidate, longer than the lanceolate acuminate glume. 1-2 f. Swamps.

scirpoi'des, (M. 2/.) spikes 4, ovate, obtuse, approximate, uppermost one clavate; fruit ovate, bidentate, plano-convex, erect and a little spreading, but not reflexed, sub-cordate, serrulate, longer than the ovate-obtuse glume.

2. *With 3 stigmas.*

atra'ta, (J. 2/.) androgynous spikes 3, pedunculate, crowded, sub-pendulous in fruit, (black); fruit roundish-ovate, with a short beak, bidentate. 6 i.

c. *Summits of the highest and lowest spikelets staminate, the middle spikes wholly staminate.*
1. *With 2 stigmas.*

sicca'ta, (J. 2/.) terminal spikes obtuse; lower ones mostly in fours, ovate, somewhat acute; fruit ovate-lanceolate, acuminate, compressed, scabrous on the margin, bifid, nerved, nearly equal to the ovate-lanceolate scale. 12-18 i. Sandy plains.

†† *Terminal spikes androgynous; the rest pistillate; stigmas 3.*

vires"cens, (green sedge, M. 2/.) spikes 3, oblong, erect; upper one pedunculate, sterile below, the rest fertile, sub-sessile, and bracteate; fruit ovate, obtuse, costate, pubescent. 18-24 i. Dry woods. Var. *costa'ta*, has its fruit strongly ribbed, and its outer sheaths purplish-brown; leaves more numerous and larger.

formo'sa, (M. ♃.) spikes 4, oblong, thick, distant, on exsert peduncles, nodding, uppermost one sterile at the base; fruit oblong, triquetrous, somewhat inflated, rather acute at each end, orifice nearly entire, or 2-lobed, obscurely nerved, twice as long as the ovate-acute glume. 12-18 i. Wet.

††† *Staminate and pistillate spikes distinct.*
  * *Staminate spike solitary.*
  1. *With 2 stigmas.*

au'rea, (J, ♃.) fertile spikes mostly 3, oblong, loose-flowered, sub-pendulous, rather approximate, lower ones pedunculate; fruit obovate or pyriform, obtuse, nerved, entire at the orifice, longer than the ovate-acute glume. 4-10 i. Wet rocks.

2. *With 3 stigmas.*
§ *Pistillate spikes sessile, or with the peduncles enclosed.*

vesti'ta, (J. ♃.) sterile spike mostly solitary, (rarely germinate, with the upper one elongated), pedunculate, cylindrical-oblong; fertile spikes 2, ovate-oblong, sessile, sub-approximate, sometimes sterile at the summit; fruit ovate, sub-triquetrous, nerved, with a short rostrum pubescent, rather longer than the ovate-mucronate glume. 2 f. Wet.

tentacula'ta, (M. ♃.) fertile spikes 2-3, (rarely 4), sessile, ovate or ovate-cylindrical, approximate, horizontal; bracts very long; fruit ovate, ventricose, nerved, with a very long rostrum, orifice bidentate, longer than the lanceolate glume. 12-18 i. Wet.

§§ *Pistillate spikes on exsert peduncles, partly sheathed at the base.*

conoi'dea, (M. ♃.) fertile spikes 2-3, oblong, remote, rather loose, uppermost sub-sessile, lower ones on long peduncles; fruit oblong-conical, obtuse, nerved at the apex, as long as the awned glume. 6-12 i. Woods.

plantagin''ea,(Apr. May, ♃.) fertile spikes mostly 4, on peduncles scarcely exserted, loosely flowered; fruit oblong-cuneiform, triquetrous, recurved at the apex; culm sheathed at the apex; sheaths of the culm all leafless (colored); leaves broad. 8-12 i.

washingto'niana, (J. ♃.) sterile spike solitary, erect; fertile spikes oblong, cylindric, sub-sessile, sub-remote, erect; flowers somewhat scattered; fruit oval, acute at each end, compressed, shortly beaked, with a smooth and entire orifice, about equalling the ovate-oblong, acutish scale. 1 f. White Mountains, N. H.

§§§ *Pistillate spikes on long peduncles, nearly destitute of sheaths.*

umbella'ta, (M.♃.) cespitose; fertile spikes mostly 4, ovate, few-flowered, one sessile at the summit of the culm, the rest on radical peduncles and appearing sub umbellate; fruit ovate, acuminate-rostrate, sub pubescent, as long as the ovate-acuminate glume. 1-6 i. In small tufts on dry hills.

  ** *Staminate spikes, 2 or more.*
  1. *With 2 stigmas.*

cespito'sa (M. ♃.) sterile spike sub-solitary, (or germinate); fertile spikes mostly 3, cylindrical, obtuse, distant, the lower on a short exsert peduncle; bracts striate; fruit ovate, somewhat acute, densely fruited in about 8 rows; orifice minute, longer than the ovate (black and margined) glume; leaves spreading. 12-18 i. Mountain bogs.

2. *With 3 stigmas.*

(retror''sa M. ♃.) sterile spikes about 3, lower one often fertile at the base; fertile spikes about 5, approximate, (and clustered in a sub-corymbose manner), oblong-cylindrical, inclusely pedunculate, lowest one often remote; fruit ovate, inflated, reflexed, rostrate, half as long as the lanceolate glume. 2 f. Near ponds in clusters.

CAR''ICA. 20—10. (*Amentaceae*.) [First cultivated in Caria.]
papa'ya, leaves palmate, 7-lobed, middle lobe sinuate; divisions oblong, acute; staminate flowers corymbed. Papaw tree. Native of Guinea.

CARPI'NUS. 19—12. (*Amentaceae*.)
america'na, (May ♄.) leaves oblong-ovate, acuminate, unequally serrate; scales of the strobile 3-parted; the middle segment oblique, ovate-lanceolate, toothed on one side. Woods. Hornbeam.

CARTHA'MUS. 17—1. (*Cinarocephala*.) [From *katheiro*, cathartic.]
tincto'rious, (false saffron, safflower. y. J. ☉.) leaves oval, entire, serrate, aculeate. Ex.
caru'leus, (blue saffron. b. ♃.) stem about 1-flowered; leaves lance-ovate, spine-toothed.

CA'RUM 5—2. (*Umbellifera*.) [From *Caria*, a province in Asia.]
ca'rui. (caraway, w. ♂.) stem branching; leaves with ventricose sheaths; partial involucrum none. Ex.

CAR''YA. 19—12. (*Juglandeae*.) [From *carua*, a nut.]
al''ba, (shag walnut, shag-bark hickory, M. ♄.) leafets about 7; long-petioled, lance-oblong, acuminate, sharply serrate, villose beneath; the terminal leafet sessile; ament filiform, glabrous; fruit globose, a little depressed; nut compressed, oblique.
sulca'ta, (shell-bark hickory, Ap. ♄.) leafets about 9, oblanceolate, acuminate, serrate, pubescent beneath; the terminal leafets sub-sessile, tapering to the base; fruit roundish, 4-keeled; nut sub-globose, a little compressed, smooth, long-mucronate.
ama'ra, (bitter nut. Ap. ♄.) leafets about 9, ovate-oblong, acuminate, sharply serrate, glabrous both sides, the terminal leafets short-petioled; fruit sub-globose, with the sutures prominent above; nut smooth, subglobose, mucronate; putamen easily broken; nucleus bitter.
porci'na, (pig nut, broom hickory. M. ♄.) leafets about 7 lanceolate, acuminate, serrate, glabrous both sides; terminal leafets sub-sessile; fruit pear-form or globose; nut smooth; putamen very thick and hard; nucleus small. Var. *obcorda'ta* has an obcordate nut. Var. *ficifor''mis* has the fruit turbinate and nut oblong. 70-80 f.
aquat''ica, (Ap. ♄.) leafets about 11, narrow, obliquely lanceolate, acuminate, sub-serrate, glabrous, sessile; fruit peduncled,

ovate; sutures 4, prominent; nut roundish, compressed; putamen thin; nucleus bitter. 40-50 f. *S.*

olivæfor"mis, (pecan nut, Ap. ♃.) leafets numerous (13-15), lanceolate, sub-falcate, serrate; petioles not in pairs; fruit oblong, 4-sided; nut olive-shaped, smooth. *S.*

CAS"SIA. 10—1. (*Leguminosea.*) [From the Arabic *katsia*, to tear off, alluding to the peeling of the bark.]

marilan"dica, (wild senna, ♂. Au. ♃.) somewhat glabrous; leaves in 8 pairs, lance-oblong, mucronate; flowers in axillary racemes, and in terminal panicles; legumes linear, curved. River alluvion. 2-4 f.

chamæchris"ta, (cassia, partridge pea, E. y. Au. ☉.) somewhat glabrous; leaves linear, in many pairs, the glands on the petioles sub-pedicelled; two of the petals spotted; legumes pubescent. 8-16 i. Dry sand, &c.

nic"titans, (E. y. Ju. ☉.) spreading, pubescent; leaves in many pairs, linear; glands of the petioles pedicelled; peduncles short, supra-axillary, 2-3 flowered; flowers pentandrous; the leaves of this species, and of the chamæchrista possess a considerable degree of irritability. 12 i.

senn"a, (Egyptian senna, ☉) leaves in 6 pairs; petioles glandless; legume reniform. Ex.

to'ra, (y, Ju. ☉.) glabrous; leaves in 3 pairs, obovate-ciliate; terminal ones largest; a subulate gland between the lower pair; peduncles few-flowered, axillary; legumes curved. 3 f. *S.*

occidenta'lis, (y. M. ♃. ♃.) glabrous leaves in 5 pairs, ovate-lanceolate, acuminate, scabrous along the edges; peduncles clustered, few-flowered, axillary; legumes compressed, falcate. 12-18 i. *S.*

linea'ris, (y. J. ♃.) glabrous; leaves in 5 or 6 pairs, ovate, acute; peduncles axillary, few-flowered; legume terete. *S.*

as"pera, (8. ☉.) strigose, rough; leaves in many pairs, linear, lanceolate, ciliate; peduncles few-flowered, above the axils; stamens 7-9; three longer than the rest. 1-3 f. *S.*

CASTA'NEA. 19—12. (*Amentaceæ.*) [From *Castana*, a city of Thessaly.]

ves"ca, (chestnut. g. J. ♃.) leaves lance-oblong, sinuate, serrate, with the serratures mucronate; glabrous both sides. Large tree.

pu'mila, (chinquapin, g. J. ♃.) leaves oblong, acute, mucronate, serrate, with white down beneath. Small tree. Florida.

CASTILLE'JA. 13—2. (*Pediculares.*) [Named from a Spanish botanist.]

occidenta'lis, (♃.) stem simple, pilose; leaves linear-lanceolate, narrowing toward the apex, minutely pubescent, entire, acutish; flowers spiked, sessile; corolla scarcely exceeding the calyx; bracts 3-cleft; divisions erect.

CATAL"PA. 2—1. (*Bignonia.*) [An Indian name.]

cordifo'lia, (M. w. and y. ♃.) leaves simple, cordate, entire, in threes; flowers in panicles. 40-50 f.

CAULIN"IA. 19—1. (*Aroideæ*, or more properly *Fluviales*.)

flex'ilis, (water knot-grass, Au. ☉.) leaves in sixes, toothed at the apex, spreading. Immersed in ditches. Stem long; flowers small.

CEANO'THUS. 5—1. (*Rhamni.*) [From the Greek *keanothos.*]

america'nus, (New Jersey tea, w. J. ♃.) leaves ovate, acuminate, serrate, 3-nerved, pubescent beneath; panicles axillary, long-peduncled, sub-corymbed.

ova'lis, (w. ·♃.) leaves oval, with glandular serratures, 3-nerved; nerves pubescent beneath; panicle corymbose, abbreviated. Canada.

mi'crophyl"la, (w. Ju. ♃.) leaves very small, obovate, nearly entire, clustered, glabrous; racemes corymbose, terminal. 1-2 f. *S.*

CELAS"TRUS. 5—1. (*Rhamni.*) [From *kela*, a dart.]

scan"dens, (false bittersweet, staff-tree, y. w. J. ♃.) stem twining; leaves oblong, acuminate, serrate; racemes terminal. Retains its scarlet berries through the winter.

bulla'ta, (w. Ju. ♃.) unarmed; leaves ovate, acute, entire; panicle terminal. 20 f. *S.*

CELO'SIA. 5—1. (*Amaranti.*) [Greek *kelos*, singed, from the appearance of the flowers.]

crista'ta, (cockscomb, r. J.) leaves ovate, acuminate, stipules falcate; common peduncle striated; spike oblong, compressed.

CEL"TIS. 5—2. (*Amentaceæ.*)

occidenta'lis, (M. ♃. g-w.) leaves ovate, acuminate, equally serrate, unequal at the base, scabrous above, hairy beneath; flowers small, sub-solitary. Woods. Drupe purple. Nettle tree. Beaver wood.

austra'lis, 20 f.; flowers small; berries black. Lote-tree. Ex.

crassifo'lia, (hag-berry, w. M. ♃.) leaves ovate, acuminate, unequally serrate, unequally cordate at the base, sub-coriaceous; peduncles about 2-flowered. 20 f.

CENOMY'CE. 21—5. (*Algæ*)

pyxada'ta, frond foliaceous; divisions crenulate, ascending; peduncles all turbinate, cup-form, glabrous, at length warty-granulate, scabrous, greenish-gray; cups regular, afterward the margin is extended and proliferous; receptacles tawny.

coccif'era, frond foliaceous, minute; divisions round, crenate, naked beneath; peduncles long turbinate, naked, warty-scabrous, pale yellowish, cinereous and green, all bearing cups, which are wine-glass form; margin extended, fertile; receptacles rather large, at length roundish, scarlet.

CENTAU"REA. 17—3. (*Cinarocephalæ.*) [From *Chiron*, the *centaur*, who is said to have cured a wound in his foot with the plant.]

cya'nus, (great blue-bottle, b. w. r. J. ☉.) scales of the calyx serrate; leaves linear, entire, lower ones toothed. Naturalized.

america'na, (great American centaury, ☉.) stem branching; leaves sessile, lower ones oblong-ovate, upper ones lanceolate, acute; peduncles thick at the apex. 2 f.

benedic'ta, (blessed thistle, y. J. ☉.) scales of the involucre doubly armed with spikes, woolly, bracted; leaves decurrent, toothed, spiny. Ex.

ni'gra, (black knapweed, p. Au. 2{.) lower leaves angular-lyrate, upper ones ovate; scales of the involucre ovate, ciliated with capillary teeth. 2 f. Ex.

scabio'sa, (scabrous centaury, 2{.) leaves pinnatifid, roughish; divisions lanceolate, spreading, acute-pinnatifid at the base; involucre ciliate. Ex.

suave'lens, (yellow sultana, ☉.) leaves lyrate-pinnatifid; involucre smooth. Ex.

moscha'ta, (sweet sultana, ☉.) leaves slightly pinnatifid; lower divisions mostly entire; involucre smooth. Ex.

CENTAUREL''LA. 4—1. (*Gentianea.*)

panicula'ta, (Sept. ☉.) stem branched, smooth; peduncles opposite; leaves minute, subulate; flowers in panicles. Damp grounds. Flowers small, greenish-white. 4-8 i.

ver''na, (w. M·r. ☉.) stem simple, few-flowered; corolla thrice as long as the calyx; style as long as the germ. 4-8. S.

CEPHALAN''THUS. 4—1. (*Rubecæ.*) [From *kephale*, head, *anthos*, flower.]

occidenta'lis, (button bush, w. Ju. ♄.) leaves opposite, and in threes, oval, acuminate; inflorescence a round head. Swamps. Var. *pubes''cens*, has the leaves and the branchlets pubescent. 4-5 f.

CERATIO'LA. 20—2. (*Euphorbeæ*)

erico'des, (Au. ♄.) branchlets sub-tomentose; leaves whorled, narrow, linear, smooth. An evergreen shrub. 4.6 f.

CERAN''THERA. 13—1. (*Labiateæ.*) [From *keras*, horn, and *anthos*, flower; anthers bearing horns.]

linearifo'lia, leaves opposite, linear, sometimes clustered; stem glabrous, branching; racemes terminal; peduncles opposite. 12 i.

CERAS''TIUM. 10—5. (*Caryophyllea.*) [From *keras*, horn, alluding to the form of its capsule.]

vulga'tum, (mouse-ear, chickweed, w. Ap. ☉.) hirsute, viscid, cespitose; leaves ovate; petals oblong, about equal to the calyx; flowers longer than the peduncle. 6-10 i.

visco'sum, (sticky chickweed, w. J. ☉.) hairy and viscid, spreading; leaves oblong-lanceolate; flowers somewhat panicled, shorter than the pedicels. 4-6 i.

nu'tans, (w. J. ☉.) viscid and elongated; stems erect, deeply striate; leaves elongated, distant, linear-oblong, acute; petals oblong, bifid at the tip, longer than the calyx; peduncles much longer than the flowers. 6-12 i.

oblongifo'lium, (w. J. 2{.) cespitose; pubescent; stem erect, terete, even; leaves lanceolate-oblong, rather acute, shorter than the joints; flowers terminal, shorter than the pedicels; petals obovate, bifid at the tip, twice the length of the calyx. 6-12 i.

CERAS''US. (*See Pru'nus.*)

CERATOPHYL'LUM. 19—12. (*Onagra.*) [Named from the horned divisions of the leaves.]

demer'sum, (hornwort, Ju. 2{.) fruit armed with 3 spines; stem long, slender; leaves

verticillate in 8; flowers axillary, solitary very minute.

submer''sum, (2{.) leaves dichotomous in three pairs; fruit without spines. In water.

CER''CIS. 10—1. (*Leguminosa.*)

canaden''sis, (red-bud, judas-tree, r. M. ♄.) leaves round heart-form, acuminate, villose at the axils of the nerves; stipules minute; legumes short-stiped. Var. *pubes''-cens*, has roundish, acute leaves, pubescent beneath. 15.30 f.

CETRA'RIA. 21—5. (*Algæ.*) [From *cetra*, a buckler.]

island''ica, (the Iceland lichen, Iceland moss,) frond olive-chestnut-brown, at the base reddish-white, white beneath; divisions erectish, sub-linear, many-cleft, channelled, tooth-ciliate, the fertile ones dilated; receptacles close-pressed, flat. 1-colored; margin frond-like, elevated, entire. On sandy plains, as on the barren plains near Beaver pond, in New Haven, where it covers the earth very densely in many places.

CHAMÆ'ROPS. 6—2. (*Palmæ.*) [From *chamai*, on the ground, *ops*, appearing.]

serrula'ta, (E. Ju. ♄.) caudex creeping; stipes sharply serrate; fronds plaited, palmate. Fronds 2 f. S.

palmet''to, (Ju. ♄.) caudex arborescent; stipes unarmed; spathes doubled; fronds plaited-palmate; fronds 5-6 f. Florida.

CHA'RA. 19—1. (*Naiades.*) [From *chairo*, to rejoice, because it delights in water.]

vulga'ris, (feather-beds, Ju. ☉.) stem and branches naked at the base; branches terete, the joints leafy; leaves oblong, subulate; bracts shorter than the berry. Grows in ponds and ditches.

flex'ilis, (Au. ☉.) stem translucent, naked; branchlets jointless, leafless, compressed; berries lateral, naked.

CHEIRAN''THUS. 14—2. (*Cruciferæ.*) [From *cheir*, hand, *anthos*, flower, the blossoms resembling the fingers.]

chei'ri, (wall-flower, J. 2{.) leaves lanceolate, acute, glabrous; branches angled; stem somewhat of a woody texture. Ex.

an''nuus, (stock july-flower, Ju. ☉.) leaves lanceolate, sub-dentate, obtuse, hoary; silique cylindric, with an acute apex. Ex.

pallas''ii, (r. Ju. ♂.) stem simple, terete, somewhat glabrous; leaves glabrous, lanceolate-linear, tapering, repand-toothed.

inca'nus, (brompton stock, ♂.) leaves lanceolate, entire, obtuse, hoary; silique truncate, compressed at apex. Stem somewhat of a woody texture. Ex.

CHELIDO'NIUM. 12—1. (*Papaveraceæ.*) [From *chelidon*, a swallow, because it blossoms about the time this bird appears.]

ma'jus, (celandine, y. M. 2{.) umbels axillary, peduncled; leaves alternate, pinnate, lobed. Naturalized.

CHELO'NE. 13—2. (*Bignonia.*) [From *chelone*, a tortoise.]

gla'bra, (snake-head, w. and r. Ju. 2{.) leaves opposite, lance-oblong, acuminate, serrate; spikes terminal, dense-flowered. Var. *al''ba*, leaves sub-sessile; flowers white. Var. *purpu'rea*, leaves short-petioled; flowers purple. Var. *lanceola'ta*, leaves lanceolate, acuminate, serrate, sessile, pubescent beneath; segments of the calyx oblong. Damp.

lyo'ni, (p. Au. ♃.) glabrous, branching; leaves petioled, cordate-ovate, serrate; spikes terminal; flowers clustered. *S.*

CHENOPO'DIUM. 5—2. (*Atriplices.* [From *chen*, a goose, and *podos*, foot, so called from its supposed resemblance to a goose's foot.]

al"bum, (green pigweed, g. Ju. ⊙.) leaves rhomboid-ovate, erose, entire behind, the upper ones oblong. entire; seed smooth. Var.*vir'ide*,leaves lance-rhomboid, sinuate-toothed; racemes ramose, sub-foliaceous; stem very green. 2-4 f.

bo'trys, (oak-of-Jerusalem, g. J. ⊙.) leaves oblong, sinuate; racemes naked, many-cleft. Sweet-scented. 12 i.

?'brum, (red pigweed, r-g. Ju. ⊙.) ?es rhomboid-triangular, deeply toothed ? sinuate; racemes erect, compound, ?fy. 2-3 f.

ambrosioi'des, (sweet pigweed, g. Ju. ⊙.) leaves lanceolate, remotely toothed; flowers in interrupted sessile clusters; on slender, axillary, leafy branches. 1-2 f.

anthelmin"ticum, (wormseed, g. Au. ♃.) leaves oblong-lanceolate, toothed; spikes long, interrupted, leafless; odor strong. 12-24 i.

scopa'rium, (summer cypress,) leaves flat, lance-linear, margin ciliate; flowers glomerate, axillary. Ex.

CHIMAPH"ILA. 10—1. (*Ericæ.*) [From *cheima*, winter, and *philos*, a lover.]

macula'ta, (spotted wintergreen, w. Ju. ♃.) leaves lanceolate, rounded at the base, remotely serrate, marked with long spots; scape 2 3-flowered; filaments woolly.

umbella'ta, (prince's pine, bitter wintergreen, r. w. Ju. ♃.) leaves serrate, uniformly green, wedge-lanceolate, with an acute base; scape corymbed; filaments glabrous.

CHIOCOC"CA. 5—1 (*Rubiaceæ.*) [From *Chion*, snow, *kokkos*, berry.]

racemo'sa, (y. w. Ju. ♄.) leaves ovate, oblong, acute, flat; racemes axillary, peduncled, simple. *S.*

CHIONAN"THUS. 2—1. (*Jasmineæ.*) [From *chion*, snow, *anthos*, flower.]

virgin"ica, (fringe-tree, w. M. ♄.) panicle terminal, trifid; peduncles 3-flowered; leaves acute. Var. *monta'nus*,leaves oval-lanceolate, coriaceous, glabrous; panicle dense; drupe oval. Var. *mariti'mus*,leaves obovate-lanceolate, membranaceous, pubescent; panicle very lax; drupe elliptic; berries purplish-blue.

CHONDRIL"LA. 17—1. (*Chicoraceæ.*)

carolinia'na, (y. March. ♃.) leaves lance-oblong, glabrous; stem erect, few-flowered, peduncles long. 2 f.

CHRYSAN"THEMUM. 17—2. (*Corymbiferæ.*) [From *chrusos*, golden, *anthos*, flower.]

parthe'nium, leaves petioled, compound, flat; leafets ovate, gashed; peduncles branching, corymb?d; stem erect. Feverfew. Ex.

carina'tum, (r. w. Au. ⊙.) leaves bipinnate, fleshy, glabrous; scale of the calyx carinate. Three-colored daisy. Ex.

corona'rium, (Au. ⊙.) leaves bipinnatifid, acute. broader outward; stem branching.

Garden chrysanthemum, improperly called artemisia. Ex.

leucan"themum, (ox-eyed daisy, J. ♃.) leaves clasping, lanceolate, serrate, cut-toothed at the base; stem erect, branching. 12-20 i.

CHRYSEIS. 12—1. (*Papaveraceæ.*)

califor"nica, (y.) stem branching, leafy; torus obconic; calyx ovoid, with a very short abrupt acumination; petals bright yellow, with an orange spot at the base. *S.* Oregon.

CHRYSOBALA'NAS. 11—1. (*Rosaceæ.*) [From *chrusos*, gold. *balanus*, a nut, so called on account of the yellow color of the nut before it is dried.]

oblongifo'lius, (w. J. ♄.) leaves oblong, lanceolate, entire, glabrous, shining; flowers panicled; fruit oblong. 1-2 f. *S.*

CHRYSOCO'MA. 17—1. (*Corymbiferæ.*) [From *chrusos*, gold, *kome*, hair.]

virga'ta, herbaceous, smooth; leaves narrow, linear; stem branching; branches corymbed, fastigiate, virgate; scales of the calyx glutinous, appressed. 18 i. Golden locks. Flowers yellow.

nuda'ta, (y. S. ♃.) radical leaves spatulate, lanceolate; cauline ones linear, scattered; corymb compound, fastigiate; involucre oblong, 3-4-flowered. 2 f. *S.*

CHRY"SOGO'NUM. 17—4. (*Corymbiferæ.*)

virginia'num, (y. J. ♃.) low, woolly, villose; leaves oval-dentate, narrowing into the petiole. 6-12 i. *S.*

CHRYSOP"SIS. 17—2. (*Corymbiferæ.*) [From *chrusos*, golden, *ops*, appearance.]

maria'na, (y. Au. ♃.) hairy; leaves oblong-lanceolate, serrate; the upper ones sessile, acute; the lower ones spatulate, and generally obtuse; corymb simple; involucre viscid-pubescent. Florets of the ray 16-20. Sandy woods.

graminifo'lia, (y. S. ♃.) silky, leaves lanceolate-linear, acute, entire, nerved; corymb compound; stem leafy toward the summit. Var. *tenuifo'lia.* Silky or woolly leaves, narrow-linear, shining; stem few-flowered; scales of the involucre glabrous. 2 f.

pinifo'lia, (y. Oc. ♃.) very glabrous; stem rigid; leaves linear, crowded, rigid; corymb large; scales of the involucre woolly at the summit. 18.24 i. *S.*

tri'chophyl"la, (y. Au. ♃.) hairy; leaves oblong, obtuse, very entire, somewhat clasping, scabrous on the margin; corymb simple; scales of the involucre very narrow, glandular. 12-18 i. *S.*

gossyp"ina, (y. S. ♃.) woolly, hoary; leaves sessile, oblong, spatulate, obtuse, very entire; corymb fastigiate. 1-2 f. *S*

denta'ta, (y. S. ♃.) lanuginous; leaves cuneate, obtuse, deeply toothed; upper ones oblong, oval, entire; corymb simple. 2 f. *S.*

CHRYSOSPLE'NIUM. 8—2. (*Saxifrageæ.*) [From *chrusos*, gold, *asplenion*, spleenwort.]

oppositifo'lium, (golden saxifrage, y-r. M. ♃.) leaves opposite, roundish, slightly crenate, tapering for a little distance to the petiole. In rivulets, springs, &c.

27*

CI'CER. 16—10. *Leguminosæ.*) [From *cicer*, vetch.]

arieti'num, (☉.) peduncle 1-flowered; seeds globose; leaves serrate. Chickpea.

CICHO'RIUM. 17—1. (*Chicoraceæ.*) [An Egyptian name, signifying creeping.]

in'tybus, (succory or endive, b. Ju. 2{.) flowers axillary, in pairs, sessile; leaves runcinate. *July V. 1836 A. J. J.*

endiv'ia, (garden endive, b. Ju. ♂.) peduncles axillary in pairs; one long, 1-flowered, the other short, about 4-flowered; leaves oblong, denticulate. Var.*cris'pum*, has fringed leaves and solitary flowers. Ex.

CICU'TA. 5—2. (*Umbelliferæ.*) [From *cæcuta*, blind, because it destroys the sight of those who use it.]

macula'ta, (w. Ju. 2{.) serratures of the leaves mucronate; petioles membranaceous, 2-lobed at the apex. Damp. 3-6 f. Cow-bane.

bulbife'ra, (w. Au. 2{.) leaves ternate and biternate; bulbiferous; leaflets linear and linear-lanceolate, remotely toothed. Wet. 2-3 f.

viro'sa, (water hemlock, 2{.) umbels opposite to the leaves; petioles margined; obtuse; leaflets ternate, acutely serrate. Root containing a yellow juice. Ex.

CIMCIFU'GA. See ACTÆ'A, MACRO'TRYS.

CINERA'RIA. 17—2. (*Corymbiferæ.*) [From a Latin word, signifying ashes, from the appearance of the leaves.]

heterophyl'la, (ash-wort, y. M. 2{.) downy; radical leaves long-petioled, obovate-spatulate, also ovate, acutish and pinnatifid; cauline ones 2-3, linear, pinnatifid; flowers corymbed. 8 i.

CIRCÆ'A. 2—1. (*Onagræ.*) [From *Circe*, the enchantress.]

lutetia'na, (Aug. r-w. 2{.) stem erect; leaves ovate, remotely toothed, opaque, nearly smooth. 1-2 f. Enchanter's nightshade.

alpi'na, (r-w. Au. 2{.) stem branched, glabrous, often procumbent; leaves broad-cordate, membranaceous, acutely toothed, shining. 6-8 i.

CIS'TUS. 12—1. (*Cisti.*)

canaden'sis, (rock-rose, y. J. 2{.) without stipules, erect; leaves alternate, erect, linear-lanceolate, flat, tomentose beneath; racemes terminal, few-flowered; divisions of the calyx ovate-acuminate; capsules shorter than the calyx. 6-14 i.

cre'ticus, leaves spatulate-ovate; scales of the calyx lanceolate. Candia, where the juice of the plant is collected and sold under the name of *lada'num*.

corymbo'sum, (J. 2{.) without stipules, erect, ramose, minutely pubescent; leaves alternate, lanceolate, whitish downy beneath; corymb fastigiate, with numerous crowded flowers; divisions of the calyx ovate, acute; capsule longer than the calyx. 12 i.

carolin'ia'num, (J. 2{.) without stipules, hirsute,erect; leaves alternate, oblong-oval, sub-denticulate; bottom ones obovate, hirsute on both sides; peduncles few, terminal, with the calyx very villose; divisions of the calyx oblong, acute, shorter than the petals. S.

polifo'lium, (Ju. Au.) primary or petaliferous flowers terminating the stem, and the numerous short branches on filiform peduncles, many times longer than the flower; the broadly-cuneiform petals a little exceeding the calyx; secondary flowers very small, apetalous, 3 to 6 androus, clustered in lateral cymules on the foliferous branches, at first glomerate and nearly sessile, at length on pedicels as long as the calyx; leaves linear, or linear-oblong, with revolute margins; beneath tomentose-caulescent. S.

CIT'RUS. 12—1. (*Aurantiæ.*) [The Latin name for lemon.]

me'dica, (lemon-tree, w. J. ♄.) leaves ovate, acuminate, with linear, wingless petioles. Var. *li'mon* (lime-tree), bears smaller fruit, which is almost round. 4-10 f. Ex.

auran'tium, (orange-tree, w. ♄.) leaves oval, acuminate, with the petioles winged or margined. Ex.

limel'la, yields bergamot. Ex.

limo'num, yields citric acid. Ex.

CLARK'IA. 8—1. (*Onagræ.*)

pulchel'la, (r-p. J. ♂.) stem erect, terete; leaves alternate-linear, entire, glabrous; flower sub-sessile, large. 12-18 i. Cultivated. Beautiful Clarkia.

CLAYTO'NIA. 5—1. (*Portulaceæ.*) [In honor of Dr. John Clayton.]

virgin'ica, (w. r. A. 2{.) leaves linear-lanceolate; petals obovate, retuse; leaves of the calyx somewhat acute; root tuberous. Var. *latifo'lia*,leaves ovate-lanceolate; leaves of the calyx obtuse. 6-12 i. Spring-beauty.

carolin'ia'na, (Ap.) leaves ovate-lanceolate or oval, sub-spatulate at the base or abruptly decurrent into a petiole; radical leaves very few, spatulate; pedicels slender, nodding; sepals and petals very obtuse. Canada to Carolina.

CLEMA'TIS. 12—12. (*Ranunculaceæ.*) [From *klema*, a tendril.]

virgin'ica, (virgin's bower, w. Ju. ♄.) climbing; leaves ternate; leaflets ovate, sub-cordate, gash-toothed and lobate; flowers panicled, diœcious. 12-20 f.

ochroleu'ca, (w. y. J. 2{.) erect, simple, pubescent; leaves simple, ovate, entire, young leaves and calyx silky; flower terminal, peduncled, solitary, nodding. 12 i.

vital'ba, (traveller's-joy, w. Au.) leaves pinnate; flowers in clusters; seeds plumose. Ex.

vior'na, (blue Virginian climber, J. ♄.) climbing leaves pinnately divided; leaflets lance-ovate, entire, acute at both ends, 3-lobed; peduncles 1 flowered; petals thick acuminate, reflexed at the apex. Ex.

flam'mula, (sweet virgin's bower, ♄.) lower leaves laciniate; upper ones simple, entire, lanceolate. Ex.

viticel'la, (purple virgin's bower, p. Ju. 2{.) climbing leaves compound and decompound; leaflets oval, sub-lobate, entire; petals obovate, spreading. Ex.

holoseri'cea, (w. ♄.) climbing; leaves divided, ternate; segments oblong-lanceolate, entire, pubescent on both sides; corymbs trichotomous, few-flowered, diœcious; petals linear, longer than the stamens. Whole plant silky. *S.*

cylin''drica, (p-b. Ju. ♄.) climbing; leaves pinnate, decompound; leafets ovate, acute at each end, glabrous, simple, petioled; peduncles terminal, solitary; corolla nodding, cylindrical; petals coriaceous; awns of the carpels plumose. *S.*

reticula'ta, (p-r. Ju. ♄.) climbing; leaves pinnate in 4 pairs, leafets ovate, obtuse at each end, all entire, petioled, membranaceous, reticulately nerved on both sides; flowers solitary, petals coriaceous; awns of the carpels plumose. *S.*

linearilo'ba, peduncles 1-flowered; petals very acute; leaves divided, pinnate glabrous; leafets entire or 3-parted. *S.*

CLEO'ME. 6—1. (*Capparides.*)

dodecan''dra, (r. w. Ju. ⊕.) viscid-pubescent; leaves ternate; leafets elliptical oblong; flowers generally dodecandrous. 1 f.

serula'ta, (p-w. Au. ⊕.) glabrous; leaves ternate; leafets lanceolate, obsoletely serrulate; raceme elongated, bracts linear; stamens 6. 3-4 f. *S.*

CLE'THRA. 10—1. (*Erica.*)

alnifo'lia, (w. Au. ♄.) leaves wedge-obovate, acute, coarse-serrate, glabrous, both sides one color; racemes spiked, simple, bracted, hoary-tomentose. 4-8 f. Sweet pepper-bush.

tomento'sa, (w. Au. ♄.) leaves cuneate-obovate, acute, sub-serrate, white-tomentose beneath; racemes spiked, simple, bracted; villose tomentose. 2-4. f. *S.*

sca'bra, (w. Ju. ♄.) leaves broad-wedge-obovate, acute, coarse-serrate, scabrous on both sides, serratures uncinate; racemes spiked, sub-panicled, bracted, sub-tomentose. *S.*

panicula'ta, (w. Ju. ♄.) leaves narrow-wedge-lanceolate, glabrous on both sides, acute, serrate; serratures acuminate; panicle terminal, racemose, white-tomentose. *S.*

acumina'ta, (w. Au. ♄.) leaves oval, acuminate, serrate, glabrous on both sides, glaucous beneath; racemes spiked, bracted, white-tomentose. *S.*

CLINIPO'DIUM. 12—1. (*Labiatæ.*)

vulga're, (field thyme, r. p. Ju. ♃.) flowers in head-form whorls; bracts setaceous, hispid; stem simple. Rocky woods.

CLINTO'NIA. 6—1. (*Campanulacea.*) [Named in honor of Gov De Witt Clinton]

multiflo'ra. (M. ♄.) leaves radical, oblong-oval, with the margin and keel ciliate; scape pubescent; umbel terminal; pedicels with minute bracts at base. Referred by some to CONVALLA'RIA.

CNI'CUS. 17—1. (*Cinerocephala.*) [From *knao*, to scratch]

lanceola'tus, (common thistle, p. J. ♂.) leaves decurrent, hispid, pinnatifid; divisions 2-lobed, divaricate, spinose; calyx ovate, with spider-web-like pubescence; scales lanceolate, spinose, spreading. 2-4 f.

arven''sis, (Canada thistle, p. J. ♃.)

leaves sessile, pinnatifid, ciliate, spinose stem pancled; calyx ovate, mucronate scales broad-lanceolate, close-pressed; margin woolly. 2-3 f.

altis''simus, (tall thistle, w. p. Au. ♃.) leaves sessile, lance-oblong, scabrous, downy beneath, toothed, ciliate, radical ones pinnatifid; involucre bracted, ovate; scales lance-ovate, spinose, close-pressed. 3-8 f.

horrid''ulus, (w-y. Ju. ♂.) tall; leaves sessile, pinnatifid, acutely gashed, very spinose; bracts terminal. 1-flowered, many-leaved; leafets very spinose, spines in pairs; involucre unarmed. 2-3 f.

virginia'nus, (p. J. ♃.) stem simple; leaves sessile, lanceolate, hoary-tomentose beneath, remotely toothed, teeth spinous; flowers solitary; involucre globose, scales mucronate, appressed, carinate. 3-5 f.

odora'tus, (r. Ju. ♂.) woolly; stem 1 to 3-flowered; leaves clasping, lance-oblong, pinnatifid; segments irregularly lobed, ciliate, tipped with spines, color similar on both sides; involucre large, sub-globose, naked; scales close-pressed, lanceolate, acuminate, spinose. 1-2 f.

glutino'sus, (p. Au. ♂.) leaves pinnatifid, segments divaricate; involucre ovate; scales unarmed, glutinous. 4-6 f.

CNID'IUM. 5—2. (*Umbellifera.*)

canaden''se, (w. Ju. ♃.) stem angular, flexuous; leaves bipinnate, shining, leafets many-parted; segments lanceolate; involucrum many-leaved. Banks of streams.

atropurpu'reum, (p. J. ♃.) radical leaves sub-cordate, simple, serrate, cauline ones ternate; leafets ovate, acute, sub-cordate, middle one petioled; partial involucre dimidiate, 3-leaved. 2-3 f.

COCHLEA'RIA. 14—1. (*Crucifera.*) [From *cochleare*, a spoon.]

armora'cia, (horse-radish, w. J. ♃.) radical leaves lanceolate, crenate, cauline ones gashed. Naturalized. Ex.

officina'lis, (scurvy-grass,) radical leaves roundish, cauline ones oblong, sub-pinnate; silicles globose.

CO'COS. 19—6. (*Palma.*) [From the Portuguese *coqun*, monkey, the three holes at the end of the cocoa-nut shell giving it the appearance of a monkey's head.]

nucif'era, stem erect, vertical, crowned with long, pinnate leaves. Cocoa-nut. E. and W. Indies. The species *butyra'cea* affords the palm-oil.

CO'IX. 19—3. (*Graminea.*) [From *koix*, a palm-leaved tree.]

lach''ryma, (Job's tear, Ju. ⊕.) culm semi-terete above; flowers naked; fruit ovate.

COLLIN''SIA. 13—2. (*Scrophularia.*) [In honor of Zaccheus Collins, of Philadelphia.]

ver''na, (b. M. ⊕.) leaves opposite, ovate-oblong, sessile, obtuse, the lower ones with a long petiole; peduncles long, axillary, 1-flowered. Banks of streams.

COLLINSO'NIA. 2—1. (*Labiata.*)

canaden''sis, (y. Au. ♃.) leaves broad-cordate, ovate, glabrous; teeth of the calyx short, subulate; panicle terminal, compound. Woods. 2-3 f.

sca'bra, (r-y. Au. ♃.) leaves small, ovate, sub-cordate. somewhat hairy; teeth of the calyx short, subulate; panicle terminal, simple; stem hairy, rough. 2-3 f. *S.*

ova'lis, (y. Au. ♃.) leaves oblong-oval, acute at each end, glabrous; petioles long; teeth of the calyx short; panicle terminal, simple, naked; *stem glabrous. S.*

tubero'sa, (y. S. ♃.) leaves somewhat rhomboid-oval, acute at each end, glabrous; teeth of the calyx setaceous, longer than the tube; panicle compound, leafy; stem branching, somewhat hairy. 3-4 f. *S.*

anisa'ta, (y. Au. ♃.) leaves ovate, cordate, rugose, glabrous; nerves pubescent beneath; teeth of the calyx linear, nearly as long as the tube; panicle leafy, compound, pubescent; flowers tetrandrous; stem branching, pubescent. *S.*

puncta'ta, (y. S. ♃.) leaves ovate-lanceolate, acuminate, acute at the base, pubescent and dotted beneath; panicle compound. 2-6 f.

verticilla'ta, (M.) leaves verticillate, oval, and acuminate. Var. *purpuras'ceus*, flowers purplish; panicle short. 1 f. *S.*

COLU'TEA. 16—10. (*Leguminosa.*)

vesica'ria, (scena-herb, y. Ju.) leaves pinnate, leaflets ovate; stem herbaceous, decumbent, villose; legumes orbicular, inflated.

COMMELI'NA. 3—1. (*Junci.*) [In honor of Commelins, a family of Amsterdam, who advanced the science of botany in the seventeenth century.]

angustifo'lia, (day-flower, b. Ju. ♃.) assurgent, weak, somewhat glabrous; leaves lance-linear, very acute, flat, glabrous; sheaths sub-ciliate; bracts (or involucres) peduncled, solitary, short-cordate. 12 i.

virgin'ica, (b. Ju. ♃.) stiffly erect, all over pubescent; leaves long, lanceolate, sheaths red-bearded at the throat; bracts (or involucres) sub-sessile, lateral, and terminal; calyx petal-like, 3-leaved, nearly equal. 2 f.

cœles'tis, resembles, in most particulars the preceding species; the leaves are sheathing, broad at the base, rough on the edges. The flower is of a beautiful light blue, concealed by the foliaceous sheath before blossoming. Mexico. Blue commelina of the florists.

commu'nis, (b. Au. O.) corolla unequal; leaves ovate, lanceolate, acute; stem creeping, glabrous. *S.*

COMPTO'NIA. 19—3. (*Amentaceæ.*) [Lord Compton.]

asplenifo'lia, (sweet-fern, g. Ap. ♄.) leaves long-linear, alternately crenate-pinnatifid. 18-48 i.

CONFER''VA. 21—4. (*Algæ.*) [From *conferveo*, to knit together, so named from its supposed use in healing broken bones.]

ru'fa, threads ramose, capillary, straight, obsoletely geniculate; branches and branchlets opposite, remotish; length of the joints equalling the diameter. In the sea. Reddish yellow, shining, in fascicles; threads of the thickness of human hair, 2 inches and longer, flaccid, soft.

CO'NIUM. 5—3. (*Umbelliferæ.*) [From *koneo*, poisonous.]

macula'tum, (poison hemlock, w. Ju. ♃.)

stem very branching, spotted; leaves very compound; seed striate. Var. *crispat''ulum*, leaves crisped; ultimate divisions acuminate, or terminated in a bristle. 2-4 f.

CONRAD'IA. 13—2. (*Nyctagines.*) [Named after S. W. Conrad, Prof. Bot. Un. Phil.]

fuschsioi'des, (♃.) glabrous; calyx foliaceous, divisions exsert, denticulate; leaves petioled, lanceolate, lyre-pinnatifid, lobes denticulate outside. 4 f. Resembles gerardia quercifolia.

CONVALLA'RIA. 6—1. (*Asparagi.*) [From the Latin *convallis*, a valley, its usual place of growth.]

1. *Corolla deeply 4-parted, spreading; stamens 4; berry 2-celled.*

  (Flowers in a terminal raceme.)

bifo'lia, (dwarf solomon seal, w. J. ♃.) stem with two heart-oblong, sub-sessile, glabrous leaves; raceme simple, terminal; flowers tetrandrous. Var. *trifo'lia*, stem 3-leaved. 4-6 i.

2. *Corolla 6-parted, spreading; filaments divergent, attached to the base of the segments.*

  (Flowers in a terminal raceme.)

stella'ta, (w. M. ♃.) stem with alternate, clasping, oval-lanceolate leaves; raceme simple, terminal. 8-18 i.

trifo'lia. (w. J. ♃.) stem about 3-leaved; leaves alternate, ovate lanceolate, contracted at the base; raceme simple, terminal, few-flowered. 6-10 i.

cilia'ta, (w. ♃.) stem arched; leaves alternate, sessile, ovate. ciliate; panicle terminal, crowded.

racemo'sa, (spiked solomon's seal, y-w. M. ♃.) stem with alternate leaves; leaves sessile, oblong-oval, acuminate, nerved, pubescent; flowers in a terminal raceme-panicle. 18-24 i.

3. *Corolla sub-campanulate, deeply 6-parted; style elongated; berry 2-celled, many-seeded.*

borea'lis, (wild lily of the valley, dragoness plant, g. y. J. ♃.) sub-caulescent; leaves oval-obovate; margin ciliate; scape pubescent; umbel few-flowered, sub-corymbed, sometimes proliferous; pedicels naked, nodding. 6-10 i.

umbella'ta, (w. Ju. ♃.) leaves radical, oblong-ovate, with the margin and keel ciliate; scape pubescent; umbel terminal; pedicels bracteate. 8-12.

4. *Corolla 6-cleft, cylindric; filaments inserted on the upper part of the tube; berry 3-celled; cells 2-seeded.*

  (Flowers axillary.)

multiflo'ra, (giant solomon's seal, w. Ju. ♃.) stem terete; leaves alternate, clasping, oblong-ovate; peduncles axillary, some of them many-flowered. 2-3 f.

biflo'ra, (g-y. J. ♃.) stem terete, smooth; leaves alternate, sessile, elliptic-lanceolate, 3-nerved; peduncles axillary, solitary, few-flowered. 12-18 i.

pubes'cens, (w. M. ♃.) stem teretish, furrowed; leaves alternate, clasping-ovate, pubescent beneath; peduncles axillary, about 2-flowered. 18 i *canalic"ula'ta*, (clasping solomon's seal, w. Ju. ⚄.) stem channeled; leaves alternate, clasping, oblong, margin pubescent; peduncles axillary, about 2-flowered.

latifo'lia, (Ju. ⚄.) stem angled; leaves sessile, ovate, acuminate; peduncles one. or many-flowered. 4 f.

hir"ta, (⚄.) stem angular, hispid; leaves alternate, somewhat clasping, ovate, abruptly acuminate; peduncles axillary, 3-flowered.

maja'lis, (lily of the valley, w. J. ⚄.) scape naked, smooth; leaves oval-ovate. S. Cultivated.

CONVOL"VULUS. 5—1. (*Convolvuli*.) [From *convolvo*, to intwine.]

re'pens, (field bind-weed, w. and r. J. ⚄.) twining; leaves sagittate, with the apex acute and the lobes truncate, entire, (some obtuse); bracts acute, longer than the calyx, and shorter than the middle of the corolla; peduncle angled, exceeding the petiole.

pandura'tus, (mochoacan, w. and. r. Ju. ⚄.) twining, pubescent; leaves broad-cordate, entire or lobed, guitar-form; peduncles long; flowers fascicled; calyx glabrous, awnless; corolla tubular bell-form. Resembles rhubarb in its effects.

stans, (w. J. ⚄.) erect; leaves oval or oblong, sub-cordate, pubescent; peduncles 1-flowered, generally longer than the leaves. 6-12 i.

arven"sis, (bind-weed, w. J. ⚄.) stem climbing or prostrate; leaves sagitate; lobes acute, spreading; peduncles about 1-flowered; bracts minute, acute.

spitha'meus, (dwarf morning glory, w. J. ⚄.) erect; leaves oval. or oblong, sub-cordate, pubescent; peduncles 1-flowered, generally longer than the leaves. 9-12 i.

tri'color, (3-colored bind-weed, Ju. ⊙.) leaves lance-ovate, glabrous; stem declined; flowers solitary. Ex.

jala'pa, leaves ovate, sub-cordate, obtuse, villose. South America. The root affords the jalap of commerce.

bata'tus, (sweet potato, Carolina potato, w-r. Ju. ⚄.) creeping, tuberous; leaves cordate, hastate, angular-lobed, 5-nerved, smoothish; peduncles long; flowers fascicled; corolla sub-campanulate. Cultivated.

purpu'reus, (common morning-glory, b. p. J. ⊙.) pubescent; leaves cordate, entire; peduncles 2 to 5-flowered; pedicels nodding, thickened; divisions of the calyx lanceolate; capsules glabrous. Cultivated.

CONY'ZA. 17—2 (*Corymbiferæ*.) [From *Konis*, dust, or *konops*, a gnat; the powder destroys fleas.]

camphora'ta, or *marylan"dica*, (plowman's wort, p. Au. ⚄.) herbaceous, slightly pubescent; leaves on petioles, ovate-lanceolate, very acute, denticulate; corymbs terminal, shorter than the leaves; scales of the involucre acute, as long as the florets.

COP"TIS. 12—12. (*Ranunculaceæ*.)

trifo'lia, (gold-thread. w. M. ⚄.) scape 1-flowered; leaves ternate; roots long, filiform, golden yellow; very bitter. 2-4 i.

asplenifo'lia, (⚄.) leaves biternate, leafets sub-pinnatifid; scape 2-flowered.

occiden"talis, (false gold-thread, y. ⚄.) evergreen; leaves gash 3-lobed, or obsoletely 3-leaved, sub-coriaceous; scape very short, about 3-flowered.

CORALLORHI'ZA. 18—1. (*Orchideæ*.) [From *korallion*, coral, and *riza*, root.]

odontorhi'za, (coral-teeth, p. w. Ju. ⚄.) lip entire, oval, obtuse, margin crenate; spur obsolete, adnate to the germ; capsule sub-globose. 12 i.

ver"na, (corol-root, w. y. M. ⚄.)́ petals linear-lanceolate, spreading: lip oblong, without spots, bidentate at the base, apex recurved, ovate; spur obsolete, adnate. 5-6 i.

multiflo'ra, (p. Ju.) scape many-flowered, (15-30), lip cuneate-oval, 3-parted. recurved, spotted; spur conspicuous, adnate. 12-20 i.

CORCHO'RUS. 12—1. (*Tiliaceæ*.)

siliquo'sus, branching; leaves ovate or lanceolate, acute, equally serrate; capsules pod-shaped, linear, 2-valved, nearly glabrous. Alabama.

COREOP"SIS. 17—3. (*Corymbiferæ*.) [From *koris* insect, *opsis*, resembling.]

tripte'ris, (tickseed sunflower. y. ⚄.) glabrous; leaves petioled, lanceolate, entire, radical ones pinnate, cauline ones ternate; rays entire; seeds obovate.

tincto'ria, (elegant coreopsis, y-p.) radical leaves sub-bipinnate, leafets sub-oval, entire, glabrous; cauline ones sub-pinnate, leafets linear; rays 2-colored, seeds naked. 1-4 f. Missouri.

ro'sea, (tickweed, y. r. M.) small, smooth; stem simple; leaves linear, entire, opposite, and undivided; axils leafy; flowers few, long-peduncled, dichotomous, terminal; rays unequally 3-toothed. 1 f.

trichosper"ma, (y. Ju. ♂.) glabrous, dichotomous; leaves opposite, divided, quinate-pinnate, lanceolate-serrate; outer leafets of the involucre ciliate-serrate; rays entire; akenes wedge-form, about 4-toothed. 2 f.

dichot"oma, (y. S. ⊙.) stem glabrous, na, kedish, and dichotomous above; leaves mostly alternate, undivided, entire, narrowing into the petioles; akenes obovate, 2-bristled, scabrous, with a torn margin. 2 f.

as"pera, leaves lanceolate-linear, rough, upper ones alternate, lower ones opposite; stem 1-flowered.

palma'ta, (y.) stem simple, 1-3 flowered; leaves alternate, sessile, sub-coriaceous, palmate, 3-lobed; margin scabrous; double involucre 8-parted; akenes oblong-elliptic, naked. 12 i.

lanceola'ta, (y. S. ⚄.) leaves opposite, undivided, sessile, lanceolate-linear, entire, ciliate; peduncles long, naked; akenes orbicular, scabrous, winged, 2-toothed at the summit, emarginate. S.

arista'ta, (y. Au. ⚄.) pubescent; leaves opposite, divided, quinate, pinnate, leafets serrate; rays entire, broad, oval; akenes cuneate-obovate, 2-awned; awns long, divaricate. S.

CORIAN"DRUM. 5—2. (*Umbelliferæ*.) [From *koris*, a bug, probably from its peculiar smell.]

sati'vum, (coriander, w. J. ⊙.) fruit globose; calyx and style permanent. Ex.

CORISPER"MUM. 1—2. (*Atriplices.*) [From *korus*, bug, *sperma*, seed.]
hyssopifo'lium, (Au. ♃.) spikes terminal, leaves unarmed, nerveless, linear. A variety, *america'num*, has spikes axillary; leaves nerved, mucronate.

COR"NUS. 4—1. (*Caprifolia.*)
canaden"sis, (dogweed, low cornel, w. M. ♃.) herbaceous; leaves at the top whorled, veiny; involucre ovate, acuminate; fruit globose. 4-8 i.

flori'da, (false-box, w-y. M. ♄.) leaves ovate, acuminate; involucre 4, very large, somewhat obcordate; fruit ovate. 15-30 f.

circina'ta, (w. J. ♄.) branches warty; leaves broad-oval, acuminate, white-downy beneath; cymes depressed. 6-8 f.

seri'cea, (red osier, red rod, w. J. ♄.) branches spreading; branchlets woolly; leaves ovate, acuminate, rounded at the base, rusty-pubescent beneath; cymes depressed, woolly. Var. *nerva'ta*, leaves tapering to the base, unequal, veins beneath very prominent; berries bright blue. 8-12 f.

sanguin"ea, (common dog-wood, w. M. ♄.) branches straight; leaves ovate, pubescent, both sides colored alike; cymes spreading; berries dark-brown; anthers yellow. 8-12 f.

al"ba, (white dog-wood, J. ♄.) branches recurved; branchlets glabrous; leaves ovate, acute, pubescent, hoary beneath; cymes depressed; berries bluish-white. 10 f.

panicula'ta, (bush dog-wood, w. J. ♄.) branches erect; leaves ovate-acuminate, oblong, tapering to the base, pubescence close-pressed, hoary beneath; flowers in a thyrsed cyme; berries white, globular, flattened. 8-12 f.

stric"ta, (w. J. ♄) branches straight, fastigiate; leaves ovate, color green both sides, glabrous when mature, a little downy beneath when young; panicled cyme convex. 8-12 f.

alternifo'lia, (w. M. ♄.) branches warty; leaves alternate, ovate, acute, hoary beneath; cymes depressed, spreading; berries purple. 18 f.

mas"cula, (cornelian cherry, M. ♄.) umbels equalling the involucre. Ex.

asperifo'lia, (w. J. ♄.) branches erect, pubescent; leaves oval-lanceolate, acuminate, scabrous above, tomentose beneath. 4-10 f. S.

CORONIL"LA. 16—10. (*Leguminosa.*)
va'ria, (r-p. Ju. ♃.) herbaceous, diffuse, glabrous; stipules small, acute; leafets 9 to 13, oblong, mucronate, lower ones of the stem near each other; umbels 16 to 20-flowered; legumes erect. 4 f.

glau'ca, (y. M. ♄.) leafets 7, very blunt; stipules lanceolate; umbels 10 or 12-flowered; peduncles longer than the leaves. Remarkably fragrant during the night, and almost scentless during the day. 3 f.

o'merus, (coronilla, y. ♄.) stem angled; woody; peduncles about 3-flowered; claws of the petals about thrice as long as the calyx. Ex.

CORONO'PUS. 14—1. (*Crucifera*.). [From *korone*, a crow, and *pous*, foot; the leaves resemble a bird's foot]
ruel"lii, (w. Ju. ⊕.) silicle entire; margin

muricate; style prominent; corymb few-flowered. S.

didy"ma, (swine's cress, Ju. ⊕.) silicles emarginate, in pairs, reticulate, rugose; style obsolete; corymb many-flowered. 1-2 f. Charleston.

CORYDA'LIS. 16—5. (*Corydales.*) [From *korus*, a helmet, alluding to the form of its flowers.]
cuculla'ria, (colic-weed, y. & w. M. ♂.) corolla 2-spurred; scape naked; raceme simple, 1-sided; nectaries divaricate, of the length of the corolla; style enclosed. 8-12 f. This plant is referred by some to DIELYTRA, by others to FUMARIA.

glau'ca, (r-y-g. J. ⊕.) stem erect, branched; leaves glaucous, decompound; segments cuneate, trifid; bracts oblong-acute, shorter than the pedicels; pod linear, flat, scarcely torulose. 1-4 f. S. Mch.

au'rea (y. M. ⊕.) stem branched, diffuse; leaves glaucous, doubly pinnate, lobes oblong-linear; bracts linear-lanceolate, acuminate, toothed, longer than the pedicels; pod terete, torulose. 8-12 i.

CORY"LUS. 19—12. (*Amentacea.*) [From *korus*, a nut.]
america'na, (hazel-nut, Ap. ♄.) leaves roundish, cordate, acuminate; calyx roundish-campanulate, larger than the sub-globose nut; border dilated; coarsely serrate. 3-5 f.

rostra'ta, (beaked hazel, Ap. ♄.) leaves oblong-ovate, acuminate; stipules lance-linear; involucre of the fruit bell-tabular, 2-parted; divisions gash-toothed, elongated beyond the nut into a beak. 2-3 f.

avella'na, (filbert, Ap. ♄.) stipules oblong, obtuse; involucre of the fruit campanulate, spreading at the apex, torn-toothed; leaves round-cordate, acuminate. Var. *max"ima*, has a gash-toothed involucre; nut depressed ovate. Ex.

CRAM"BE. 14—1. (*Crucifera.*) [A name given by Dioscorides to cabbage.]
mara'tima, (sea-kale, w. ♄.) stem foliaceous, smooth; leaf sinuate, glaucous; flowers corymbed, panicled.

CRANI'CHIS. 18—1. (*Orchidea.*)
multiflo'ra, root fascicled, villose; leaves oval-lanceolate, sub-sessile; scape many-flowered, pubescent toward the summit; inner petals connivent; lip vaulted, acuminate.

CRATÆ'GUS. 11—5. (*Rosacea*) [From *kratus*, strength, from the toughness of its wood.]
coccin"ea, (thorn-bush, w. M. ♄.) thorny; leaves long-petioled, ovate, acutely lobed, serrate, glabrous; petioles and pubescent calyx glandular; flowers pentagynous. Var. *vir"idis*, has lance-ovate leaves, subtrilobate; stem unarmed.

puncta'ta, (common thorn-tree, w. M. ♄.) thorny or unarmed; leaves wedge-obovate, sub-plicate, glabrous, serrate; calyx villose; divisions subulate, entire.

oxycan"tha, (quickset, w. M. ♄.) leaves obtuse, somewhat 3-cleft, serrate, glabrous peduncles and calyx somewhat glabrous segments of the calyx lanceolate, acute styles 2. Naturalized.

pyrifo'lia, (pear-leaf thorn, w. J. ♄.) thorny or unarmed; leaves oval ovate, gash-serrate, somewhat plaited and rather rough-haired; calyx a little villose; leafets lance-linear, serrate; styles 3.

crus-gal''li, (thorn-tree, w. M. ♄.) thorny; leaves wedge-obovate, sub-sessile, shining, leathery, serrate; corymbs compound; leafets of the calyx lanceolate, sub-serrate styles 2.

fla'va, (yellow-berried thorn, M. ♄.) thorny; leaves wedge-obovate, angled, glabrous, shining; petioles, calyx, and stipules, glandular; flowers sub-solitary; berries turbinate, 4-celled. 8·10 f.

lu'cida, (A. ♄.) thorny; leaves wedge-obovate, crenate, coriaceous, lucid; corymbs simple, few-flowered; styles 5. 10-12 f. *S.*

CRI'NUM. 6—1. (*Narcissi.*) [From *krinon*, a lily.]

america'num, leaves oblong-lanceolate, glabrous at the margin; flowers pedicelled, tube shorter than the limb. *S.*

CRITH''MUM. 5—2. (*Umbellifera.*)

mariti'mum, (sea samphire, w. ♃. Au.) leafets lanceolate; leaves twice ternate, glaucous, smooth, with a salt aromatic flavor. This is the true samphire of English botanists.

CRO'CUS. 3—1. (*Irida.*) [The ancients fabled that a youth, Crocus, was changed into this flower. Crocus also signifies saffron color.]

officina'lis, (saffron crocus, y. ♃.) leaves linear, with revolute margins; stigma exsert, with long linear segments. Var. *sati'vus*, having violet corollas. The stigma is of a deep orange color, and affords the saffron of commerce. Blossoms in September. Ex.

ver''nus, (spring crocus, stigma not exsert, with three short, wedge-shaped segments; tube hairy at the mouth. Color of the flower various, purple, yellow, &c. Blossoms in March. Ex. Var. *versico'lor*, feathered with purple. *bifto'rus*, the Scotch crocus, striped white and purple, the earliest in spring. Var. *susia'nus*, striped orange and dark purple; *sulphure'us*, very pale yellow; *lute'us*, the common yellow.

CROTALA'RIA. 16—10. (*Leguminosa.*) [From *krotalon*, a rattle.]

sagitta'lis, (rattle-box. y. Ju. ☉.) hairy, erect, branching; leaves simple. ovate-lanceolate; stipules lanceolate. acuminate, decurrent; racemes opposite the leaves, about 3-flowered; corolla smaller than the calyx. 12 i.

parviflo'ra, (y. J. ☉.) hirsute, erect, branching; leaves simple, lance-linear; stipules above decurrent, with two short teeth; racemes opposite to the leaves, corolla smaller than the calyx.

ova'lis, (y. Ju. ☉.) hirsute, diffuse, branching; leaves simple, oval, petioled; upper stipules scarcely decurrent, short; racemes opposite to the leaves, long; corolla as long as the calyx. 12 i. *S.*

lævig'a'ta, (Ju. ☉.) glabrous, erect, simple; leaves lance-oblong; stipules lance-

olate, acuminate, decurrent; racemes opposite to the leaves, 3-flowered. *S.*

CRO'TON. 19—15. (*Euphorbia.*) [From *kroteo*, a tick, from the form of its seed.]

marati'mum, leaves oval, sub-cordate, obtuse, pale above, hoary beneath; branches tomentose; pistillate spikes few-flowered.

laccif''erum, is the species from which the gum-lac is obtained; it is a southern plant.

tigli'um, leaves oval, acuminate, serrate; stem aborescent; this species affords a celebrated medicinal substance, called *croton* oil, an extract from the seeds. Ex.

tincto'rium, leaves rhomboid, stem herbaceous; from this plant is obtained the *litmus*, considered as one of the most delicate tests of the chemist. Ex.

CRYP''TA. 2—2. (*Portulacca.*) [From a Greek word, to conceal, the stamens being concealed in the capsular calyx.]

min''ima. (mud-purslane, w-g. 8.) stem dichotomous, decumbent, striate; leaves wedge-oval or obovate, opposite, sessile, entire, papillose above, with very minute stipules; flowers axillary, sessile, solitary. Very abundant on the shores of the Hudson, between low and high-water mark, about a mile below Albany.

CRYPTOTŒ'NIA. 5—2. (*Umbellifera.*) [From a Greek word, to conceal, in allusion to the concealed edgings of the fruit.]

canaden''sis, (w. J. ♃.) the lower umbels originate from the axils of the upper leaves; fruit oblong; stem glabrous; leaves ternate, smooth; leafets rhomb-ovate, acute, gash-toothed. 1·2 f.

CUCU'BALUS. 10—3. (*Caryophyllea.*)

be'hen, (campion, w. Ju. ♃.) glabrous, decumbent; leaves oblong-oval, acute, nerveless; calyx inflated, veiny.

CUCU'MIS. 19—15. (*Cucurbitacea.*) [From the Celtic *cuce*, a hollow vessel.]

angu'ria, (prickly cucumber,) leaves palmate-sinuate; fruit globose, echinate.

me'lo, (muskmelon, y. Ju. ♂.) angles of the leaves rounded; pome oblong, torulose. Sweet scented. Ex.

sati'vus, (cucumber, y. Ju. ♂.) angles of the leaves straight; pomaceous berry oblong, scabrous. Brought from Asia.

colocyn''this, (bitter apple, ♂.) leaves many cleft; fruit globose, glabrous, very bitter. Ex. Poisonous.

an''guinis, (snake cucumber, ♂.) leaves lobed, fruit cylindric, very smooth, long, contorted, plaited. Ex.

CUCUR''BITA. 19—15. (*Cucurbitacea.*) [The name signifies crooked.]

ovif''era, (egg-squash, ♂.) leaves cordate, angled, 5-lobed, denticulate, pubescent; pomaceous berry with fillet-like stripes lengthwise. Ex.

pep''o, (pumpkin, y. Ju. ♂.) leaves cordate, obtuse, sub-5-lobed, denticulate; pomaceous berry roundish or oblong, smooth. Var. *poti'ro*, has the fruit more or less flattened. From Asia.

citrul''lus, (watermelon, y. Au. ♂.) leaves 5-lobed; the lobes sinuate-pinnatifid, obtuse; pomaceous berry oval, smooth. Fruit watery, often striped. From Africa and the south of Asia.

lagena'ria, (gourd, calabash, w. Au. ☉.) leaves cornate, round-obtuse, pubescent, denticulate, with 2 glands at the base on the under side; pomaceous berry clavate, somewhat woody. Ex.

verruco'sa, (club squash, y. J. ☉.) leaves cordate, deeply 5-lobed; middle narrowed at the base, denticulate; pepo clavate, a little warty. Ex.

fœtidis"sima, (Ju. ♃.) stems procumbent, sulcate; leaves alternate, long-petioled, somewhat erect, triangular-cordate, scabrous, glaucous, thick; margin sinuate, undulate; fruit globose, smooth, sub-sessile; tendrils trichotomous. *S.*

CUNI'LA. 13—1. (*Labiata.*)
glabel"la, smooth; radical leaves nearly oval, cauline leaves oblong-linear, entire; flowers axillary, mostly solitary, on long peduncles. Limestone rocks. Niagara Falls. Stems 8 to 10 inches high, branched below. Corolla violet, longer than the calyx.

CUPHE'A. 11—1. (*Salicaria.*)
viscosis"sima, (wax-bush, p. J. ♄.) viscous; leaves opposite, petioled, ovate-oblong; flowers with 12 stamens, lateral, solitary; peduncles very short.

CUPRES"SUS. 19—15. (*Coniferæ.*)
thyoi'des, (white cedar, M. ♄.) branchlets compressed; leaves imbricate four ways, ovate, tuberoled at the base; strobile globular.

dis"ticha, (Feb.) leaves distichous, flat, deciduous; sterile florets paniculate, leafless; strobile spherical.

CUSCU'TA. 5—2. (*Convolvuli.*)
america'na, (dodder. w. Au. ☉.) flowers peduncled, umbelled, 5-cleft; stigma capitate. A bright yellow leafless vine, twining round other weeds, in damp places.

europe'a, (w. Au. ☉.) flowers sub-sessile; stigma acute; stamens 4 or 5. Ex.

CY'CAS. 20—12. (*Cycadeæ.*) [This plant is intermediate between the Pines and Ferns.]
circina'lis, (sago-plant,) frond pinnate; leafets lance-linear, acute, 1-nerved, flat. East Indies.

CYMBID"IUM. 18—1. (*Orchideæ*) [From *cymba,* a boat.]
pulchel"lum, (grass pink, r. Ju. ♃.) radical leaves ensiform, nerved; scape few-flowered; lip erect, slender at the base; lamina spread; disk concave, bearded. Var. *graminifo'lia,* leaves 1 2 lines broad; bracted ones acuminate. 12-18 i.

CYDO'NA. See PY'RUS.

CYNA'RA. 17—1. (*Cinarocephala.*)
scol"ymus, (garden artichoke, ♃.) leaves sub-spinose, pinnate; scales of the calyx ovate. Naturalized. Ex.

CYNOGLOS"SUM. 5—1. (*Boragineæ.*) [From *kuon,* a dog, and *glosse,* tongue.]
amplexicau'le, (wild comfrey, w. & b. J. ♃.) very hirsute; leaves oval-oblong, upper one clasping; corymbs terminal, leafless, long-peduncled.

sylvat"icum, (b. Ju. ♂.) nakedish; leaves spatulate-lanceolate, shining, scabrous beneath; racemes scattered. *S.*

officina'le, (hound-tongue, p. Ju. ♂.) very soft-pubescent; leaves broad-lanceolate, sessile; panicled racemes.

CYPE'RUS. 3—1. (*Cyperoideæ.*) [From *ku paros,* a round vessel, which the root resembles.]
inflex"us, (Au. ♄.) umbel 2 to 3-rayed, or conglomerated and simple; involucre 3-leaved, very long; spikelets collected into ovate heads, oblong, 8-flowered; glumes squarrose at the tip. 2-3 i.

flaves'cens, (yellow grass, Au. ♃.) spikelets linear-lanceolate, in fascicles of 3 to 4; glumes obtuse; style 2-cleft and lenticular; involucre 3-leaved, longer than the spikes. 6-8 i.

phymato'des, (Au.) umbel simple or decompound; involucre 3 to 9-leaved; three of the leaves very long; peduncles compressed; spikelets distichous, linear; lower ones branched, about 15-flowered; sides rather convex; glumes oblong, obtuse; radicles tuberous at the extremities. 1 f.

mariscoi'des, (Au.) umbel simple or 1 to 2-rayed; spikelets capitate, linear, 7 to 8-flowered; glumes loose, obtuse. 8-12 i.

strigo'sus, spikes oblong, loose; spikelets subulate, expanding, a little remote; small involucres generally wanting; partial umbels with alternate rays. 2-3 f.

CYPRIPE'DIUM. 18—2. (*Orchideæ.*) [From *kupris,* Venus, *podion,* slipper.]
pubes'cens, (yellow ladies' slipper, y. M. ♃.) stem leafy; lobe of the style triangular-oblong, obtuse; outer petals oblong-ovate, acuminate; inner ones very long, linear, contorted; lip compressed, shorter than the petals.

specta'bile, (gay ladies' slipper, w. and p. J. ♃.) stem leafy; lobe of the style oval-cordate, obtuse; outer petals broad oval, obtuse; lip longer than the petals, split.

acau'le, (low ladies' slipper, w. and p. M. ♃.) scape leafless, 1-flowered; radical leaves 2, oblong, obtuse; lobe of the style roundish-rhomboidal, acuminate, deflected; petals lanceolate; lip shorter than the petals, cleft before. 1 f.

can"didum, (white ladies' slipper, w. M. ♃.) stem leafy; leaves oblong-lanceolate; lobe of the style lanceolate, rather obtuse; lip compressed, shorter than the lanceolate segments of the perianth.

parviflo'rum, (common ladies' slipper, y-g. M. ♃.) stem leafy; lobe of the style triangular, acute; outer segments of the perianth ovate-oblong, acuminate; inner ones linear, contorted; lip compressed, shorter than the perianth. 12 i.

CYRIL"LA. 5—1. (*Erica.*) [After Dr. Cyrilli, a botanist of Naples.]
racemiflo'ra, (w. J. ♄.) leaves lanceolate, cuneate at the base, coriaceous, very smooth; petals thrice as long as the calyx. 15 f. Sandy woods. Carolina. Charleston. La.

DAC"TYLIS. 3—2. (*Gramineæ.*) [From *dac tulos,* a finger, from the appearance of its pericarp.]
glomera'ta, (J. ♃.) panicle glomerate, leaves carinate. 2-3 f.

DAH"LIA. 17—2. (*Corymbiferæ.*) [From *Dahl*, a Swedish botanist, and pupil of Linnæus.]
super"flua, root tuberous, leaves broad-lanceolate, serrate; 4-6 feet high. Varieties are numerous, exhibiting splendid and brilliant colors. Blossoms in autumn. A native of Mexico.
frustra'nea, (r. Oc. ♃.) rays barren: petiole wingless; leafets roughish beneath. 6 f. Mexico. Var. *coccin"ea*, (scarlet daily,) rachis of leaves winged; leaflets, ovate, acuminate, serrate, shining, and smooth beneath; outer calyx reflexed. Var. *auran"tia*, (orange daily,) rachis of leaves naked; leafets ovate-acuminate, serrate, roughish beneath; outer calyx spreading. Var. *lu'tea*, (yellow daily,) leaves pinnate, leafets linear, pinnatifid toothed. *Excel"sa*, the most remarkable of the new species. It is a tree Dahlia, and is said to grow in Mexico thirty feet high, with a trunk thick in proportion.

DA'LEA. 16—10. (*Leguminosæ.*) [In honor of Dr. Dale, who wrote on medicine about the year 1700.]
aure'a, (y. ♃.) erect; spikes dense, cylindric; bracts as large as the calyx; calyx villose; leafets obovate, pilose beneath.
laxiflo'ra, has white flowers upon panicled spikes.
alopecuroi'des, has blue flowers upon crowded spikes.
formo'sa, is a woody, branching plant, with purple flowers. This species furnishes green-house shrubs with pinnate leaves and papilionaceous flowers.

DALIBAR"DA. 11—1 . (*Rosaceæ.*) [In honor of M. Dalibard.]
fragaroi'des, (dry strawberry, y. M. ♃.) leaves ternate; leafets wedge-form, gash-serrate, ciliate; peduncles many-flowered; tube of the calyx obconic. 5-8 i.
re'pens, stem creeping; leaves simple, cordate, crenate; stipules linear, setaceous; peduncles 1-flowered; calyx reflexed, smooth without. Mountains. Flowers white, on long peduncles.

DANTHO'NIA. 3—2. (*Gramineæ.*) [Named in honor of M. Danthoin, a French botanist.]
spica'ta, (Ju. ♃.) panicle simple, appressed; spikelets 7-9, about 7-flowered; lower palea hairy; leaves subulate ; lower sheaths hairy at the throat.

DAPH"NE. 8—1. (*Thymelæa.*) [From the nymph Daphne.]
meze'reum, (mezereon, M. ♄.) flowers sessile, cauline, in threes; leaves lanceolate.
odo'ra, (sweet mezereon, w. Ap. ♄.) flowers small, in terminal heads; leaves scattered, lance-oblong, glabrous.

DARLINGTO'NIA. 15—10. (*Leguminosæ.*) [Named after Dr. W. Darlington of Penn.]
interme'dia, (♃.) glabrous, herbaceous, unarmed; leaves 8 or 9 pairs; leafets 20 to 24 pairs, oblong-linear, with glands between the lower leaves; little heads solita⬤ ped. duncled, axillary; legumes falcate.

DATIS"CA. 20—12. (*Urticeæ.*)
hir"ta, (false hemp, y. ♃.) stem hirsute;

leaves pinnate; leafets running together at the base. Flowers small, panicled.

DATU'RA. 5—1. (*Solaneæ.*)
stramo'nium, (thorn apple, w-p. Au. ⊙.) pericarps spinose, erect, ovate; leaves ovate, glabrous, angular-dentate.
arbo'rea, (great Peruvian datura, w. Oct.) flowers pentangular, about one foot in length, fragrant. Ex.
tat"ula, (purple thorn apple, b. Ju. ⊙.) pericarps spinose, erect, ovate; leaves cordate, glabrous, toothed. Stem reddish.
me'tel, (w. J. ⊙.) leaves cordate, nearly entire, pubescent; pericarps prickly, globose, nodding.

DAU'CUS. 5—2. (*Umbelliferæ.*)
caro'ta, (carrot, w. J. ♂.) seeds hispid; petioles nerved underside; divisions of the leafets narrow-linear, acute. 2-3 f.

DECO'DON. 11—1 (*Salicaria.*)
verticilla'tum, (swamp willow-herb, p. Aug. ♃.) leaves opposite, alternate, sometimes in threes, lanceolate, petiolated; flowers axillary, whorled; petals undulate; stem erect, pubescent. 2-3 f. Swamps.

DECUMA'RIA. 11—1. (*Myrti.*)
barba'ra, (w. Ju.) leaves ovate-oblong, acute at each end, slightly serrate.

DELPHIN"IUM. 12—2. (*Ranunculaceæ.*) [From *delphinos*, the dolphin, from the resemblance of the flower to a dolphin's head.]
azu'reum, (M. ♃.) petioles a little dilated at the base; leaves 3-5 parted, many cleft, lobes linear; raceme erect; petals densely bearded at the apex; flowers on short pedicels.
exalta'tum, (b. Ju. ♃.) petioles not dilated at the base; leaves flat, 3-7 cleft beyond the middle; lobes cuneate, 3-cleft at the apex, acuminate; lateral ones often 2-lobed; raceme erect; spur straight, about as long as the calyx; capsules 3.
tricor"ne, (b-w. M. ♃.) petiole scarcely dilated at the base, glabrous; leaves 5-parted; divisions 3-5 cleft, segments linear; nectary shorter than the corol; carpels arched, expanding from the base 8-12 l. S.
consol"idum, (larkspur, p. Ju. ⊙.) nectaries 1-leafed; stem sub-divided. Ex.
ela'tum, (bee-larkspur, ♃.) 6 f. A native of Siberia.
aja'cis, (rocket larkspur, b. Au. ⊙.) nectary 1-leafed, stem simple. 1 f. Ex.

DENDROM"ECON. 12—2. (*Papaveraceæ.*)
ri'gidum, (y. ♄.) glabrous, branching; leaves rigid and coriaceous, articulated with the stem, lanceolate or oblong, cuspidate, acuminate, strongly reticulate, denticulate on the margin; peduncles axillary, 1-flowered; flowers large; a shrub. California. Poppy-tree.

DENTA'RIA. 14—2. (*Cruciferæ.*) [Either from *dens*, a tooth, because its root is dentate; or from its supposed virtue in curing the toothache.]
diphyl"la, (tooth-root, w. M. ♃.) stem 2-leaved; leafets ternate, sub-ovate, unequally and incisely dentate; root toothed. 6-8 i.
lacinia'ta, (w. M. ♃.) leaves in threes, ternate; leafets 3-parted, segments oblong, gash-toothed; root tuberous, moniliform. 8 i.

hetorophyl"la, (p. J. ♃.) stem 2-leaved; leaves ternate, petiolate; leafets linear, sub-lanceolate, acute, entire, margin rough-ciliate; radical leafets ovate-oblong, incisely and coarsely toothed. Very small. Corymb about 9-flowered.

max"ima, (p. J. ♃.) leaves many, alternate, on long petioles, ternate; leafets sub-oval, incisely and acutely toothed, lateral ones lobed; axils naked; racemes lateral and terminal. 12-18 i.

multif'da, (p.) stem 2-leaved; leafets many-parted; segments linear, somewhat acute. 10 i. *S.*

DESMO'DIUM. 16—10. (*Leguminosæ.*)

marylan"dicum, (Ju. Aug. p. ⊕.) stem erect, pilose, branching; leaves ternate, leaflets oblong, villose beneath; stipules subulate; racemes paniculate; legumes 3-jointed, joints rhomboidal, reticulate, somewhat hairy.

obtu'sum, (Ju. Aug. ♃.) stem erect or ascending, pubescent; leaves ternate; leaflets ovate, obtuse, sub-cordate at base; stipules lanceolate-subulate; panicle terminal; joints of the legume semi-orbiculate, reticulate, hispid. 2-3 f.

akinia'num, (Ju. Aug. ♃.) stem erect, branching, pubescent; leaves ternate; leaflets ovate-oblong and sub-deltoid, acute, mucronate, scabrous beneath; stipules lanceolate-cuspidate, racemes paniculate, bracted; legumes with scabrous oval joints. 3 f.

cilia're, (Aug. ♃.) stem erect, branching, pubescent; leaves ternate on short petioles; leaflets small, oval-obtuse, pubescent underneath, fringed along the margin; racemes axillary and terminal; joints of the legume (2-3) oval, hispid.

lævига'tum, (Aug. ♃.) stem simple, erect, smooth, somewhat glaucous; leaves ternate, on long petioles; leaflets ovate, acute; panicle terminal; flowers in pairs on long pedicels; bracts ovate, acute, shorter than the flower-buds; lower segment of the calyx elongated; joints of the legume triangular. 3-4 f.

bracteo'sum,(Aug. ♃.) stem erect, smooth; leaves ternate; leaflets oblong-oval, acuminate, smooth; stipules subulate; racemes terminal, few-flowered; bracts ovate-acuminate, striate, glabrous; legume with sub-oval joints. 3-5 f.

DIAN"THUS. 10—2. (*Caryophylleæ.*) [From *dios*, Jove, and *anthos*, flower, from its superior elegance and fragrance.]

arme'ria, (pink, r. Ju. ⊕.) flowers aggregate, fascicled; scales of the calyx lanceolate, villose, equalling the tube. 1 f.

barba'tus, (sweet-william, r. and w. Ju. ♄.) flowers fascicled; scales of the calyx ovate-subulate, equalling the tube; leaves lanceolate. Ex. *s*

caryophyl"lus, (carnation or pink, and w. ♄.) flowers solitary; scales of the calyx sub-rhomboid, very short; petals crenate, beardless; leaves linear-subulate, channeled. By rich culture the stamens mostly change to petals. Ex.

arbor'eus, (tree pink,) a variety of the carnation.

chinen"sis, (china pink, Ju. ⊕.) flowers solitary; scales of the calyx subulate, spreading, leafy, equalling the tube; petals crenate; leaves lanceolate. Ex.

pluma'rius, (pheasant-eyed pink, r. and w. ♄.) flowers solitary; scales of the calyx sub-ovate, very short and obtuse, awnless; corolla many-cleft, with the throat hairy. Ex.

carolin"ia'nus, flowers aggregate; peduncles long; scales smaller than the tube *S.*

deltoi'des, (London-pride,) flowers small, panicled. 9 i.

DIAPEN"SIA. 5—1. (*Convolvuli.*)

lappon"ica, (w. Ju. ♃.) cespitose; leaves spatulate, glabrous; flowers peduncled; anthers simple; stem short; leaves crowded, fleshy, evergreen, entire. Mountains.

cuneifo'lia, (J. ♄.) creeping; leaves lance wedge-form, pubescent below; flowers sessile; anthers horizontal, beaked at the base.

DICHON"DRA. 5—2. (*Convolvuli.*) [From *dis*, two, *chondros*, seed.]

carolin"ien"sis, (p. J. ♄.) pubescent; leaves reniform-emarginate; calyx villose, ciliate, creeping. *S.*

DIELY'TRA. 16—6. (*Papaveraceæ.*)

formo'sa, (M. ♄.) scape naked; raceme many-flowered, nodding; segments of the leaves oblong, pinnatifid; spurs slightly curved, obtuse; stigmas 2-angled; root bulbous; flowers rose-colored. Hills.

exim"ia, (p-r. M. ♄.) scape naked, simple, few-flowered; leaves bipinnate; segments linear, glaucous beneath; spurs short, obtuse, stigma 4-angled, which di guishes it from the preceding species. Scape 6-8 i. Root tuberous rather than bulbous. See CORYDA'LIS.

canaden"sis, (g-w. p. Ap.) spurs short, rounded; wing of the inner petals projecting beyond the summit; raceme simple; 4-6-flowered.

DIERVIL"LA. 5—1. (*Caprifoliæ.*) [From M. Dierville, who first brought it from Arcadia.]

hu'milis, (bush honeysuckle, y. Ju. ♃.) peduncles axillary and terminal, dichotomous, 3-flowered; leaves ovate, serrate, acuminate. 2-3 f.

DIGITA'LIS. 13—2. (*Scrophulariæ.*) [From *digitus*, a finger.]

purpu'rea, (foxglove, p. Ju. ♂.) leafets of the calyx ovate, acute; corolla obtuse; upper lip entire; leaves lance-ovate, rugose. Ex.

interme'dia, (p. Ju.) sepals lanceolate, equal; corolla slightly pubescent, upper lip emarginate, 2-cleft; leaves pubescent at the margin and base.

DILA'TRIS. 3—1. (*Irideæ.*) [From *dis*, double, and *latris*, servant or attendant, because Bergius found two long, and one short stamen.]

tincto'ria, (red root, y. Ju. ♃.) leaves ensiform, shorter than the stem. Flowers in a corymbose panicle, woolly, yellow within. 2 f.

DIO'DIA. 4—1. (*Rubiaceæ.*) [From *diodos*, the way-side.]

virgin"ica, smooth; stem procumbent; leaves lanceolate, opposite, acute, scabrous on the margin; fruit crowned by the 2-lobed calyx; stem smooth, slender, and purple; flowers white, solitary. (2J. Sept.)

DIONÆ'A. 10—1. (Hypericea.) [From Dione, one of the names of Venus.]

muscip"ula, (Venus' fly-trap, w. 2J.) radical leaves, with terminal, ciliate appendages, somewhat resembling a rat-trap; this is suddenly closed, on being irritated. S.

DIOSCORE'A. 20—6. (Asparagi.) [From Dioscorides.]

villo'sa, (May, 2J.) leaves alternate, opposite, verticillate, cordate, acuminate, pubescent beneath, 3-nerved. Woods. Stem climbing; 12 feet high. Flowers small, in panicles. The yam-root of the Indies is obtained from a species of this plant.

quaterni'lia, (J. 2J.) leaves verticillate by fours, and alternate, cordate, acuminate, glabrous, 7-9-nerved; lateral nerves divided. Stem climbing.

DIOSPY'ROS. 20—8. (Rhododendra.)

virgin"ia'na, (persimmon, g-y. May, ♄.) leaves ovate, alternate, oblong, acuminate, reticulately veined, nearly smooth; petioles pubescent; flowers solitary, axillary; fruit as large as a common plum, golden yellow. Var. pubes"cens, leaves oblong, acute, pubescent beneath; petioles long; fruit bearing few seeds. S.

[DIPHY]L"LIA. 6—1. (Berberides.) [From dis, double, phullon, leaf.]

cymo'sa, (w. J. 2J.) very glabrous; leaves sub-palmate, angularly lobed, serrate; cyme many-flowered. S.

DIP"SACUS. 4—1. (Dipsaceæ.)

sylves"tris, (wild teasel, w-b. Ju. ♂.) leaves rarely connate, opposite; scales of the receptacle straight; involucrum curved upward. 3-4 f. S.

fullo'num, (teasel, w. Ju. ♂.) leaves sessile, serrate; chaff hooked. 3-6 f.

DIR"CA. 8—1. (Thymeleæ.) [From dirke, a fountain.]

palus"tris, (leather-wood, y. Ap. ♄.) leaves oval, alternate, petioled, entire, obtuse. Shrub. 2-4 f.

DODECATH"EON. 5—1. (Lycimachia.) [From dodeka, twelve, and theos, a divinity, signifying the twelve Roman divinities.]

me'dia, (false cowslip, p. M. 2J.) leaves oblong-oval, repandly-toothed; scape erect, simple, smooth; umbel many-flowered; flowers nodding; bracts numerous, oval. Flowers large. 1-12 i.

integrifo'lium, (b. J. 2J.) leaves sub-spatulate, entire; umbels few-flowered, straight; bracts linear.

DODO'NÆA. 8—1. (Sapindi.)

visco'sa, ( ) leaves viscous, ovate-oblong, cuneiform at the base. Florida.

DO'LICHOS. 16—10. (Leguminosa.)

multiflo'rus, (p-w. 2J.) stem twining, pubescent; leaves orbicular, short, acuminate, nearly glabrous when mature; racemes axillary, densely spiked, many-flowered, about as long as the petioles. 5-10 f. * Ark. Geo.

pui pu'reus, (wild cowhage, p. ☉.) twi-

ning; stem glabrous; corolla with spreading wings; petioles pubescent. S.

pru'riens, (cowhage, or cowitch, p. ☉.) twining; leaves hairy beneath; legumes in racemes; valves slightly keeled, hairy; peduncles in threes; legumes covered with stinging hairs. Ex.

luteo'lus, (w-y. Ju. ☉.) climbing-pubescent; leafets ovate, acuminate; peduncles longer then the leaves; spikes short, somewhat capitate; banner broad, reflexed; wings rhomboidal. 4 f. S.

DRA'BA. 14—1. (Cruciferæ.) [From drasso, to sneeze, from its effects upon the noses of those who eat it.]

carolin"ia'na, stem leafy at the base, hispid, naked and smooth at the top; leaves ovate, roundish, entire, hispid; pouch linear, smooth, longer than the pedicel. (Ap. ☉. 2-4 i. w.)

ara'bizans, (M. ♂.) stem leafy, somewhat branched, sub-pubescent; leaves lanceolate, acute, toothed; silicles acuminate, with the permanent style.

ver"na, (w. M. A. ☉.) scapes naked; leaves lanceolate, somewhat toothed; petals 2-parted; silicles elliptical.

DRACOCEPH"ALUM. 13—1. (Labiatæ.) [From drakon, dragon, kephale, head.]

virgin"ia'num, (dragon-head, p. Au. 2J.) spikes long, with the flowers crowded; bracts small, subulate; teeth of the calyx short, nearly equal; leaves sessile, opposite, linear-lanceolate, acutely serrate. 12 f.

canarien"se, (balm of Gilead,) flowers whorled; bracts lanceolate; leaves ternate-oblong. Ex.

corda'tum, (b. J. 2J.) stem and petioles pubescent; leaves cordate, obtusely crenate, somewhat hirsute above; spikes second; pedicels 2-bracted. S.

parviflo'rum, (w. Ju. ♂.) flowers verticillate, sub-capitate; leaves ovate-lanceolate, deeply serrate, petioled; bracts foliaceous, ovate, ciliate, serrate; serratures mucronate; teeth of the calyx unequal, scarcely shorter than the corol. S. The canes"cens, grandiflo'rum, and austria'cum, are exotics, and have large and splendid blue flowers.

DROSE'RA. 5—6. (Hypericea.) [From drosera, dewy.]

rotundifo'lia. (sundew, y-w. Au. 2J.) scape simple; leaves nearly orbicular, narrowed at the base; petioles long, downy. Wet or damp. 4-8 i.

longifo'lia, (y-w. Ju. 2J.) scape simple; leaves spatulate-obovate; petioles long, naked. 3-6 i. Swamps.

filifor'mis, (p. J. 2J.) scape sub-ramose, terete, glabrous; leaves very long, filiform; styles 6 to 9.

brevifo'lia, (w. r. J. 2J.) very small; scape rooting, simple; leaves short, wedgeform, scarcely petioled; petals oval. S.

DRY'AS. 11—12. (Rosaceæ.) [From the Dryads, fabled wood-nymphs.]

integrifo'lia, (w. Ju. 2J.) leaves very entire, acute at the base; peduncles 1-flowered.

octopet"ala, (mountain avens, w. Ju. 2J.) leaves ovate-oblong, coarsely toothed, rugose, white-tomentose beneath; peduncles one-flowered.

ECHI'TES. 18—5. (*Apocynæ.*) [From *echis*, a serpent, on account of the twisting form of its shoots.]

diffor"mis, (w-y. M. Au. ♃.) climbing; lower leaves nearly linear, upper ones oval-lanceolate, acuminate; raceme corymbed; stamens included. Beautiful climbers. *S.*

ECHI'NOPS. 17—5. (*Cinerocephalæ.*) [From *echinos*, beset with prickles like a hedge-hog.] *sphæroceph"alus*, (globe thistle, b.) leaves pinnatifid; stem branching. Austria.

ECH"IUM. 5—1. (*Boraginæ*) [From *echis*, a viper, because it was supposed to heal the stings of that reptile.]

vulga're, (blue thistle, b. M. ♂.) stem tuberculate, hispid; leaves lance-linear, hispid; spikes lateral; stamens longer than the corolla. 2-3 f.

ECLIP"TA. 17—2. (*Corymbiferæ.*) [From *ekleipo*, to be deficient, its wingless seed distinguishing it from Verbesina.]

erec"ta, (w. Ju. ☉.) erect, dichotomose, strigose; leaves lanceolate, attenuate at base, rarely serrate; peduncles by pairs, long; leaves of the involucrum ovate, acuminate. *S.*

procum"bens, (w. J. ☉.) procumbent or assurgent; leaves long-lanceolate, narrowed at the base, sparingly serrate; leaves of the involucrum acutely lanceolate; disk florets 4-cleft. *S.*

ELEAG"NUS. 4—1. (*Eleagni.*) [From *eleia*, the olive.]

argen"tea, (oleaster, J. ♄.) unarmed; leaves undulate, oval-oblong, covered with silvery scales; flowers aggregate, sub-solitary, nodding. Southern. The fruit resembles small olives.

angustifo'lius, narrow-leaved oleaster.

latifo'lius, broad leaves, green on the upper surface, silvery beneath.

ELEPHANTO'PUS. 17—5. (*Corymbiferæ.*) [From *elephos*, elephant, *pous*, foot.]

carolinia'nus, (elephant-foot, r. Au. ♃.) radical and cauline leaves oblong, narrowed at the base, pilose on both sides; stem erect, pilose, leafy. 2 f.

nudicau'lis, (r. Au. ♃.) radical leaves oval-lanceolate, crenate, serrate, sub scabrous, hairy beneath; stem hairy, rough, nearly naked. 1-2 f. *S.*

ELLIOT"TIA. 8—1. (*Ericæ.*) [In honor of Elliott, author of the Southern Flora.]

racemo'sa, (w. J. ♄.) leaves alternate, lanceolate, mucronate, entire, short-petioled, pubescent; racemes terminal. *S.*

ELLIS"IA. 5—1. [In honor of John Ellis.]

nycte'lea, (w. and b. J. ☉.) stem decumbent, branchy, leafy, brittle; leaves alternate, petioled, pinnatifid, roughish; flowers solitary. 6-8 i.

ambig"ua, (w. b. M. ☉.) stem decumbent, branching, glabrous, somewhat glaucous; leaves hirsute, lyrate, pinnatifid, sub-sessile; divisions sub-lanceolate, angularly toothed or lobed; racemes lateral and terminal. 4-6 i.

ELO'DEA. 12—5. (*Hyperica.*)

virginica, (Ju. Au. p. ♃.) leaves sessile, clasping; stamens united below the middle.

petiola'ta, (P. Au. ♃.) leaves attenuated into a petiole filaments united above the middle.

ELYTRA'RIA. 2—1. (*Acanthi.*)

virga'ta, (J. ♄.) leaves entire near the summit; scales under the flower ovate, villose along the margin. 12-18 i.

car"damon, furnishes the cardamon seeds of commerce. Highly aromatic. Ex.

EMPE'TRUM. 20—2. (*Ericæ.*) [From the Greek *en*, in, and *petron*, a stone.]

ni'grum, (M. ♄.) procumbent; branchlets glabrous; leaves imbricate, oblong-retuse, glabrous, with a revolute margin. A low shrub, found on the White Hills, with small and dense evergreen foliage, like that of the heaths. Flowers small, red; berries black.

ENSLE'NIA. 18—5. (*Apocynea.*) [In honor of A. Enslen, a botanist.]

al"bida, (Ju. y-w. ♃.) training; stem marked with an alternating pubescent line; leaves opposite, smooth, cordate-ovate, somewhat acuminate, sinuate at the base; corymbs axillary, many-flowered, long-peduncled; pedicels and calyx pubescent.

EPIDEN"DRUM. 18—1. (*Orchideæ.*) [From *epi*, upon, and *dendron*, tree.]

conop"sium, (air-plant. y. Au.) stem simple; leaves lanceolate, rigid, perennial; spikes erect; lamina of the lip 3-lobed, middle one retuse; inner petals narrow. Parasite.

vanil"la, climbing; leaves ovate, oblong, sessile, cauline. The vanilla plant. The pericarp, which is a pod, contains aromatic seeds. Ex.

EPIGÆ'A. 10—1. (*Ericæ.*) [From *epi*, upon, *ge*, the earth.]

re'pens, (trailing arbutus, r. and w. Ap. ♄.) stem creeping; branches and petioles very hirsute; leaves cordate-ovate, entire; corolla cylindric.

EPILO'BIUM. 8—1. (*Onagriæ.*) [From *epi*, upon, *lobos*, a pod.]

spica'tum, (willow herb, p. Ju. ♃.) leaves scattered, lance-linear, veiny. glabrous; flowers unequal; stamens declined. 4-6 f.

tetrago'num, (r. Ju. ♃.) leaves sessile, lanceolate oblong, denticulate, lower ones opposite; stigma undivided; stem 4-sided, nearly smooth; flowers in terminal racemes. Low grounds. 2 f.

colora'tum, (r. p. Ju. ♃.) stem terete, pubescent; leaves mostly opposite, lanceolate, acute, serrulate, sub-petiolate, smoothish, with colored veins. 3-4 f.

linea're, (w. r. Ju. ♃.) stem terete, pubescent, wand-like, branched above; cauline leaves opposite, branch leaves alternate, linear, very entire; flowers few, terminal, long-peduncled. 1-2 f.

palus"tre, (marsh willow-herb, p. Ju. ♃.) stem terete, branched, somewhat hirsute; leaves sessile, lanceolate, somewhat toothed, opposite and alternate, smooth; stigma undivided; fruit pubescent.

leptophyl"lum, stem branching, sub-scabrous; leaves alternate, sub-sessile, linear, narrow, entire, glabrous, 1-nerved, acute, narrowed at the base; flowers axillary, solitary, peduncled.

EPIPH"EGUS. 13—2. (*Pediculares*.) [From *epi*, upon, *phegas*, the beech.]

virginia'nus, (beech-drops, cancer-root, y. p. Ju. ♃.) stem very branching; flowers alternate, distant; calyx short, cup-form, shorter than the capsule. The whole plant is yellowish-white, and of a naked appearance. 8-12 i. Astringent.

EQUISE'TUM. 21—1. (*Filices*.) [From *equus*, a horse, *seta*, bristly.]

hyema'le, (scouring rush, Ju. ♃.) stems erect, very scabrous, bearing spikes at the apex; sheaths 2-colored, withering at the base and apex; teeth with caducous awns. 2-3 f.

arven"se, (horse-tail, Ap. ♃.) sterile stems somewhat decumbent, with simple, square, and scabrous branches; fertile ones erect, simple; sheaths incisely toothed, cylindrical; teeth acute.

scirpoi'des, (Ju. ♃.) stem simple, ascending, glabrous, filiform, bearing a spike at the top; sheaths 3-toothed; teeth withering, with caducous awns at the apex. 3-6 i.

uligino'sum, (♃.) stem erect, round, furrowed, nearly smooth, somewhat branched; branches from the middle joints unequal; sheaths serrate above; teeth even, acute, black.

ERIAN"THUS. 2—2. (*Gramineæ*.)

alopecuroi'des, (p. S. ♃.) hair-like involucre much longer than the glumes; awns straight. 6-10 f. *S.*

contor"tus, (Oc. ♃.) hairy involucre as long as the glume; inner valve of the paleas eared; awns spirally twisted. *S.*

ERI'CA. 8—1. (*Ericæ*.) [From *ereiko*, easy to break.]

pubes"cens, (downy heath, r. M.) corolla linear, pubescent, with the limb erect; capsule glabrous; leaves fringed. Ex.

cine'rea, (common heath, p. Au. ♄.) leaves narrow-linear, in threes; stem branched; flowers in dense clusters, drooping. Abundant on the heaths of England and Scotland.

cilia'ris, leaves in fours, ciliate; corolla egg-shaped, inflated. In boggy grounds. The heaths, though very common in Europe, are all exotics in America.

ERIGO'NUM. 9—1. (*Polygonæ*.) [From *erion*, wool, *gone*, joint.]

tomento'sum, (Ju. ♃.) leaves oval, wedge-form at the base, glabrous above, white-downy beneath; cauline leaves in threes and fours; fascicles of flowers axillary, solitary, sessile. 2 f. *S.*

ERI'GERON. 17—2. (*Corymbiferæ*.) [From *er*, the spring, *geron*, an old man, because in the spring it has a white, hoary blossom, resembling gray hair.]

bellidifo'lium, (w-p. M. ♃.) hairy, gray; radical leaves obovate, sub-serrate; stem leaves remote, oblong-ovate, amplexicaul, entire; stem 3-5 flowered; rays nearly twice as long as the hemispherical calyx. 12-18 i.

philadel"phicum, (w-p. J. ♃.) pubescent; leaves wedge-oblong, sub-serrate, cauline ones half-clasping; ray florets capillary, as long as the disk; stem branched above, many-flowered. 2-3 f.

purpu'reum, (O. p. Ju. ♃.) pubescent; leaves oblong, toothed, clasping, upper ones entire; peduncles thickened, corymbed, lower ones elongated; scales of the calyx hairy on the keel; rays twice as long as the calyx. 2 f.

strigo'sum, (O. w. Ju. ♂.) strigose-pilose, leaves lanceolate, tapering to both ends; in the middle are a few coarse teeth, or it is entire; flowers corymb-panicled. 2-3 f.

heterophyl"lum, (W. w. J. ♂.) radical leaves round-ovate, deeply toothed, petioled, cauline ones lanceolate, acute, serrate in the middle; corymb terminal. 2-3 f.

canaden"se, (flea-bane, pride-weed, O. w. Ju. ☉.) stem hispid, panicled; leaves lance-linear; ciliate calyx cylindric; rays crowded, short. Var.

nudicau'le, (E. w. y. J. ♃.) glabrous; radical leaves lance-spatulate, acute, slightly toothed; stem simple, nearly leafless. long; terminal corymb few-flowered; rays as long as the involucre. 2 f.

* *as"perum*, (W. w. Au.) hirsute-scabrous; stem slender, about 2-flowered; leaves lanceolate, acute, entire; calyx hemispherical. 12 i.

ERIOCAU'LON. 19—4. (*Junci*.) [From *erion*, wool, *kaulos*, a stem, because some of the species have a velvety stem.]

pellu'cidum, (pipe-wort, g. Au. ♃.) scape very slender, about 7-striped; leaves linear-subulate, channeled, glabrous, pellucid, 5-nerved, reticulate; head small, globose; scales of the involucre oval-obtuse. Grows in water. 6-12 i.

villo'sum, (♃.) scapes numerous, compressed, about 4-furrowed, villous; leaves short, subulate, linear, hairy; head small, spherical; corolla nearly black. 12 i. Charleston, S. C.

ERO'DIUM. 15—5. (*Gerania*.) [From *erodias*, a stork.]

cico'nium, (stork-bill geranium, ☉.) peduncled, many-flowered; leaves pinnate; leafets pinnatifid, toothed; petals oblong, obtuse; stem ascending. Ex.

cicuta'rium, (hemlock-geranium, p. Ap. ☉.) peduncles many-flowered; leaves pinnate; leafets sessile, pinnatifid, gashed; corolla larger than the calyx; stem prostrate, hirsute. Ex.

moscha'tum, (musk geranium, ☉.) peduncles many-flowered; leaves pinnate; leafets sub-petioled, oblong, gash-toothed; petals equalling the calyx; stem procumbent. Ex.

ER"VUM. 16—10. (*Leguminosæ*.) [From *ervum*, a field. Growing wild.]

hirsu'tum, (hairy tare, b-w. J. ☉.) leafets linear, obtuse, mucronate; peduncles 3-6 flowered, shorter than the leaves; legume oblong, hairy. 2-3 f. Stem diffuse; leaves cirrose.

ERYN"GIUM. 5—2. (*Umbelliferæ*.)

aquat"icum, (button snake-root, w-b. Au. ♃.) leaves ensiform, ciliate-spinose; 12-18 inches long; flowers in ovate heads at the end of the branches.

mariti'mum, radical leaves sub-rotund, plicate, spinose; heads of flowers peduncled. Sea-holly. Root medicinal. Ex.

ERY'SIMUM. 14—3. (*Crucifera.*) [From *eruo*, to draw, from its power of producing blisters.]

amphib"ium, (water-radish, y. J. ♃.) silique or rather silicle, oblong-ovate, declined; leaves lance-oblong, pinnatifid or serrate; petals longer than the calyx. Wet. 1-2 f.

palus"tre, (y. Ju. ⊙.) leaves lyrate pinnatifid; lobes confluent, unequally dentate, smooth; petals as long as the calyx; siliques short-turgid; root spindle-form. 18 i.

cheiranthoi'des, (g.) leaves lanceolate, sub-dentate, somewhat scabrous, green; siliques erect, spreading, twice as long as the pedicels; stigma small, sub-sessile. Flowers small. *S.*

as"perum, leaves linear-oblong: lower ones tooth-runcinate, all scabrous, pubescent; siliques spreading; style short and thick. *S.*

ERYTHRI'NA. 16—10. (*Leguminosa.*) [From *eruthros*, blushing.]

herba'cea, (r. M. ♃.) small leaves ternate; leafets rhomboidal, glabrous; spikes long, stem herbaceous, prickly. 2-4 f.

cris"ta-galli. (coxcomb evergreen, r. M. ♄.) leaves ternate; petioles prickly, glandular; stem unarmed.

ERYTHRO'NIUM. 6—1. (*Liliacea.*) [From *eruthros*, red, on account of the color of its juice.]

america'num, *dens-canis*, (dog-tooth violet, adder's tongue, y. Ap. ♃.) leaves lance-oval, punctate; petals oblong-lanceolate, obtuse at the point; inner ones 2-dentate near the base; style clavate; stigma entire; stigmas 3. 6-8 i.

albid"um, leaves elliptical-lanceolate, not punctate; segments of the petals linear-lanceolate obtuse, inner ones without dentures, sub-unguiculate; style filiform; stigma 3-cleft, lobes reflexed; flowers white. Wet meadows. Ap. May. Scape 6 inches high.

EUCHRO'MA. 13—3. (*Scrophularia.*) [From *eu*, fine, *chroma*, color.]

cocci'nea, (painted cup, y. and r. J. ♂.) leaves alternate, linear, gash-pinnatifid; divisions linear; bracts dilated, generally 3-cleft. longer than the flowers; calyx 2-cleft, about equal to the corolla, divisions retuse, emarginate; flowers yellow, with scarlet bracts. One variety, *pallens*, has yellow bracts. 10-16 i.

grandiflo'ra, (g. w. M. ♃.) leaves and bracts mostly 3-cleft; segments divaricate; calyx 4-cleft, partly oblique; corol longer than the calyx, divisions of the lower lip acuminate. Bracts not colored.

EUON"YMUS. 5—1. (*Rhamni.*) [From *eu*, good, *nomos*, name]

america'nus, (burning bush, spindle-tree, r-y. J. ♄.) branches opposite, smooth, square; leaves opposite, sub-sessile, elliptic-lanceolate, serrate; peduncles mostly 3-flowered, terete; calyx small; corolla 5-petalled; fruit warty, scarlet. Shady woods. 4-6 f.

atropurpu'reus, flowers dark purple; fruit bright red.

obova'tus, flowers green, tinged with purple.

EUPATO'RIUM. 17—1. (*Corymbifera.*) [From its discoverer, Eupator, king of Pontus.]

1. *Involucres not more than 5-flowered.*

hyssopifo'lium, (hyssop thorough-wort, hemp-weed, w. Au. ♃.) stem erect; lowest leaves opposite, lance-linear, sub-dentate; corymb sub-fastigiate; style much longer than the corol. 2 f.

sessilifo'lium, (w. Au. ♃.) leaves sessile, clasping, distinct, lance-ovate, rounded at the base, serrate, very glabrous; stem somewhat glabrous. 2 f.

trunca'tum, (w. Ju. ♄.) leaves sessile, clasping, distinct, lanceolate, truncate at the base, serrate, somewhat glabrous; stem pubescent.

verbenæfo'lium, (w. Au. ♃.) leaves sessile, distinct, lance-ovate, rugose, scabrous, upper ones with coarse teeth at the base, and with the summit entire. 2 f.

al"bum, (Au. ♃.) leaves sub-sessile, lance-oblong, roughish, serrate; inner scales of the calyx long, lanceolate, scarious-colored. Seashore. 18-24 i.

pubes"cens, (E. w. Au. ♃.) leaves sessile, distinct, ovate, sub-scabrous, veiny. lower ones doubly serrate; upper ones sub-serrate; stem panicled, pubescent; branches fastigiate. 18-24 i. *S. W.*

2. *Involucres more than 5-flowered.*

purpu'reum, (purple thoroughwort, joe-pye, p. Au. ♃.) leaves in fours or fives, petioled, lance-ovate, serrate, rugose-veined, roughish, stem hollow. 4-6 f.

perfolia'tum, (boneset, thoroughwort, w. Au. ♃.) leaves connate-perfoliate, oblong-serrate, rugose, downy beneath; stem villose. 2 f.

puncta'tum, (O. p. Au. ♃.) leaves in fours or fives, petioled, ovate, acuminate, serrate, scabrous both sides; stem solid, terete. 3-5 f. N. W. States.

verticilla'tum, (joe-pye's weed, p. Au. ♃.) leaves petioled, in threes or fours, lance-ovate, acuminate at each end, unequally serrate, somewhat glabrous; stem solid, smooth. 6-7 f.

cœlesti'num, (Au. ♃.) leaves petioled, heart ovate, obtusish, 3-nerved, obtusely serrate, slightly scabrous; involucre many-leaved; many-flowered, receptacle conic. 2-3 f.

aromat"icum, (w. Au. ♃.) leaves petioled, cordate-ovate, acute, 3-nerved, obtusely serrate, somewhat scabrous; flowers corymbed; scales of the involucres sub-equal. 2 f.

ageratoi'des, (w. Au. ♃.) leaves petioled, ovate-acuminate, 3-nerved, unequally and coarsely-toothed, serrate, glabrous; corymb many-flowered, spreading; involucre simple. 2-4 f.

fœnicula'ceum, (y-w. S. ♃.) stem panicled; leaves glabrous; lower ones pinnate. upper ones clustered; all filiform. 3-10 f. *S.*

cuneifo'lium, (E. w.) leaves petioled, obovate-lanceolate, slightly serrate at the summit; 3-nerved, pubescent on both sides. 1 f. *S.*

pinnati'fidum, (w. S. ♃.) leaves pinnatifid; lower ones verticillate; upper ones alternate; divisions linear; pubescent; flowers corymbed. 3-4 f. *S.*

parviflo'rum, (w. 8.) leaves sessile; narrow-lanceolate; very acutely serrate; pubescent on both sides; flowers corymbed; small; seeds angled. 2 f. *S.*

EUPHOR"BIA. 19—1. (*Euphorbia.*) [In honor of Euphorbus, physician to Juba, king of Mauritania.]

hypericifo'lia, (spurge, Ju. ☉.) smooth, branching, erect, spreading; branches divaricate; leaves opposite, oval-oblong, serrate; corymbs terminal; flowers small.

corolla'ta, the 5-rayed umbel dichotomous; floral leaves and those of the stem oblong, obtuse; inner segments of the involucre petaloid, obovate; flowers conspicuous. 1-2 f.

lathy'rus, the caper-tree; umbel dichotomous. Ex.

officina'rum, stem naked, many-angled. Affords the gum-resin imported from Africa, under the name of euphorbium. Ex.

macula'ta, (Ju. O. ☉.) stem procumbent, branching, hairy; leaves opposite, oval or oblong, serrulate, oblique at the base, short petioled, smooth above, hairy and pale beneath; flowers solitary and axillary, much shorter than the leaves.

ipecacuan"hae,(Ju.♃.) procumbent, small, glabrous; leaves opposite, oboval or lanceolate; peduncles axillary, elongated, 1-flowered.

lathy'rus, (spurge caper, J. ♂.) umbel 4-cleft, dichotomous; leaves opposite, entire, lanceolate, pointing four ways. Ex.

EUPHRA'SIA. 13—2. (*Pedicularee.*) [From *euphron,* delightful, pleasant to behold.]

officina'lis, (eye-bright, w-p. Ju. ☉.) leaves ovate, obtusely toothed; lower divisions of the lip emarginate.

EVOL"VULUS. 5—2. (*Convolvuli.*) [From *evolvo,* to roll outward.]

argente'us, (p. M. ♄.) stem simple, erect; leaves oblong, acute, silky-tomentose on both sides; peduncles flowered, short. *S.*

nummula'ris, (☉.) leaves roundish; stem creeping; flowers sub-sessile. *S.*

serice'us, (☉.) leaves lanceolate, sessile, silky beneath; peduncles short, 1-flowered. *S.*

EX"ACUM. 4—1. (*Gentianae.*)

pulchel"lum, (r. Au. ☉.) corolla 4-cleft; calyx 4-parted, divisions subulate; panicle corymbed; peduncles filiform.

FA'GUS. 19—13. (*Amentaceae.*) [From *phage,* to eat, its nuts being among the first fruits eaten by man.]

ferrugin"ea, (red-beech, y-w. M. ♄.) leaves ovate-oblong, acuminate, pubescent beneath, coarsely-toothed, at the base obtuse, sub-cordate, oblique; nuts ovate, acutely 3-sided.

sylvat"ica, leaves of a brighter green, and wood of a lighter color, than the preceding species. White beech.

FE'DIA. 3—1. (*Dipsaceae.*) [From *phede,* clemency, from its harmless properties.]

radia'ta, (wild lamb lettuce, w. J. ☉.) stem dichotomous; leaves spatulate-oblong, sub-entire; fruit pubescent, about 4-sided, naked at the apex. 8-18 i.

olito'ria, (lamb lettuce,) stem dichotomous; leaves lance-linear. Ex. See VALERIANELLA.

FERU'LA. 5—2. (*Umbelliferae.*) [From *ferio,* to whip.]

villo'sa, (giant fennel, w. Ju. ♃.) leaves on long petioles, ternate, the partitions quinate; leaflets ovate, serrate, rigid, veiny; stem villose.

assafo'tida, leaves alternate, sinuate, obtuse. A plant of Persia, which affords from its roots a gum known as the assafœtida of commerce.

FESTU'CA. 3—2.

ela'tior, (fescue-grass, O. J. ♃.) panicled, spreading, very branching, lax; spikelets ovate-lanceolate, 4-5-flowered; florets slightly armed; leaves flat; root creeping. 3-4 f.

tenel"la, (E. J. ☉.) panicle spiked, very simple, one-sided; spikelets about 9-flowered; bristles shorter than the subulate florets; culm filiform, angular above; leaves setaceous. 8-15 i.

praten"sis, (J. ♃.) panicle spreading, branched; spikelets linear, many-flowered, acute; leaves linear; root fibrous. 1-2 f.

spica'ta, (w. J.) spikelets alternate, sessile, erect; somewhat 5-flowered; florets subulate, sub-glabrous, with a long scabrous awn; linear leaves and culm glabrous.

grandiflo'ra, (E.) panicle simple, erect; spikelets very few; generally 7-flowered; florets acute, distant.

nu'tans, (nodding festuca, J. ♃.) panicle slender, diffuse, at length nodding; branches long, in pairs, naked below; spikelets lance-ovate, about 3-flowered; florets smooth, awnless, and nearly nerveless. 3 f.

FI'CUS. 20—3. (*Urticae.*)

ca'rica, (fig-tree, g. Ju. ♄.) leaves cordate, 3 or 5-lobed, repand-toothed; lobes obtuse, scabrous above, pubescent beneath. 5-8 f. Ex.

FLŒR"KIA. 6—1. (*Ranunculaceae.* From a German by the name of Flœrke.]

palus"tris, (false mermaid, w-y. M-y. ♃.) stem decumbent, terete, slender, smooth; leaves alternate, trifid and pinnatifid, with a long petiole. Marshes.

FOTHERGIL"LA. 11—2. (*Amentaceae.*)

alnifo'lia, (witch-alder, W. Ap. ♄.) leaves wedge-obovate, crenate-toothed above. *S.*

FRAGA'RIA. 11—12. (*Rosaceae.*) [From *fragro,* to smell sweet.]

virginia'na, (wild strawberry, w. M. ♃.) calyx of the fruit spreading; hairs on the petioles erect, on the peduncles close-pressed; leaves somewhat glabrous above.

grandiflo'ra, (pine-apple strawberry,) calyx of the fruit erect; hairs erect; leaves somewhat glabrous above. Ex.

ves"ca, (English strawberry, w. M. ♃.) calyx of the fruit reflexed; hairs on the petioles spreading, on the peduncles close-pressed. Ex.

canaden"sis, (mountain strawberry, M. ♃.) large; leaflets broad-oval, lateral ones manifestly petioled; pedicels long, recurved-pendulous; receptacle of the seeds globose, favose-scrobiculate, villose.

ela'tior, (hautboy strawberry, w. ♃.) calyx of the fruit reflexed; hairs on the peduncle and petiole spreading. Ex.

FRANKE'NIA. 7.--3. (*Caryophyllea*.) [From Prof. John Frankenius, of Upsal, Sweden.]

grandiflo'ra, (see heath.) leaves obovate-cuneiform, mucronate, with revolute margins, rather coriaceous, very minutely hairy and ciliate, particularly at the base; stem prostrate; branches and calyx minutely hairy. A dwarf-perennial. S.

FRASE'RA. 4—1. (*Gentianea*.)

verticil''lata, (American columbo, g-y. Ju. ♂.) leaves oblong-lanceolate, whorled or opposite, smooth; flowers on whorled peduncles. Medicinal. Swamps. 3-6 f.

FRAX''INUS. 20—2. (*Jasmina*.) [From *phraxis*, a hedge: used in making hedges.]

acumina'ta, (white ash, w-g. M. ♄.) leafets petioled, oblong, shining, acuminate, very entire, or slightly toothed, glaucous beneath; flowers calycled.

pen''dula, weeping ash.

or''nus, leaves pinnate. Flowering ash.

sambucifo'lia, (black ash, M. ♄.) leafets sessile, ovate-lanceolate, serrate, the lateral ones somewhat rounded and unequal at the base.

juglandifo'lia, (swamp ash, M. ♄.) leaves pinnate; leafets petiolate, ovate, opaque, serrate, glaucous beneath; axils of the veins pubescent; branches smooth; flowers calyculate.

FRITILLA'RIA. - 6—1. (*Liliacea*.) [From *fritillus*, a chess-board, in reference to the variegated petals of one of its species.]

imperia'lis, (crown imperial, r. and y. M. ♃.) flowers under a leafy crown, nodding; leaves lance-linear, entire. From Persia.

lanceola'ta, (p. Ju. ♃.) stem leafy, 1-2-flow leaves lance-linear, lower ones whorls petals lanceolate. S.

molea gris, (fritillary, Guinea-hen flower, p. and y. M. ♃.) leaves alternate, linear, channelled; stem 1-flowered; nectary linear; flower checkered.

al''ba, (w. Ap. ♃.) glaucous; leaves remotish, alternate, sessile, oblong-linear, flattish, oblique, obtuse, substriate beneath; flowers 1-3, axillary and terminal. 1 f.

FU'SCHSIA. 8—1. (*Onagra*.) [From a German botanist, Leonard Fuchs.]

magella'nica, (ear-drop, r.) peduncles axillary, 1-flowered; leaves opposite or in threes, very entire; flowers pendulous. Ex.

FU'CUS. 21—4. (*Alga*) [*Phucus*, the Greek for sea-weed.]

lo'rens, stem very short, dilated into a cup, sending out a fusiform, dichotomous receptacle. In the ocean.

FUMA'RIA. 16—6. (*Papaveracea*.) [From *fumus*, smoke]

officina'lis, (fumitory, r. J. ☉.) stem branching, spread; leaves more than decompound; leafets wedge-lanceolate, gashed. Naturalized. 6-10 i.

FUNA'RIA. 21—2. (*Musci*.) [From *funis*, a rope in allusion to its long pedicels.]

hygromet''ica, (hygrometer moss,) leaves ovate, acute, concave, entire, inflected; capsules swelling, drooping, pear-form; pedicels very long, twisting spirally when dry.

GALAC''TIA. 16—10. (*Leguminosa*.) [From *gala*, milk.]

mollis, (Ju. ♃.) stem twining, soft-pubescent; leaves ternate; leafets ovate-oblong, obtuse, pale beneath; racemes axillary, a little longer than the leaves, pedunculate; flowers pedicelled; calyx acuminate villose; legume compressed, villose; flowers small, purple. Milk plant. Pine barrens.

glabel''la, leafets shining above; stem smooth.

pilo'sa, stem twining, minutely and retrorsely hirsute; leafets oblong-ovate, finely hirsute on both surfaces, pale beneath; cemes much longer than the leaves; flowers on short pedicels, scattered and remote; legume villous. S.

GALAN''THUS. 6—1 (*Narcissi*.) [from *gala*, milk, *anthos*, flower, in allusion to its whiteness]

nival''is, (snow-drop, w. Ap. ♃.) leaves linear, keeled, acute, radical; scape 1-flowered. Ex.

plican''thus, (Russian snow-drop,) flowers smaller than the preceding.

GALAR''DIA. 17—3. (*Corymbifera*.)

pinnatifi'da, (y. p.) leaves pinnatifid divisions lance-linear, somewhat entire.

GA'LAX. 5—1. (*Sempervivea*.) [From milky, because of the whiteness flowers]

rotundifo'lia, (w. J. ♃.) very glab leaves round-reniform, toothed; spike long. S.

GALE'GA. 16—10. (*Leguminosa*.) [From *gala*, milk, because it increases the milk of animals who eat it.]

virginia'na, (goat's-rue, r-y. w. Ju. ♃.) erect; leafets 8-12 pairs, oval-oblong, mucronate, white-villose beneath; raceme terminal; legumes falcate, villose. 1 f.

GALEOP''SIS. 13—1. (*Labiata*) [From *gale*, a weasel, *opsis*, appearance]

lada'num, (red hemp-nettle, r-w. Ju. ☉) stem hairy, not swollen below the joints; leaves on short petioles, lanceolate, serrate, hairy; flowers whorled; upper lip of the corolla slightly crenate. 1 f. Waste grounds. Introduced.

tetra'hit, stem hispid, swollen between the joints; flowers rose-colored, with a white tube, lower lip dotted with purple.

GA'LIUM. 4—1. (*Rubicea*.) [From *gala*, milk, some species having the property of coagulating milk.]

trifi'dum, (bed-straw, w. Ju. ♃.) stem procumbent, scabrous backwards; cauline leaves in fives, branch leaves in fours, linear, obtuse, scabrous at the margin and on the nerves; terminal, few-flowered; pedicel short; corollas mostly 3 cleft.

asprel''lum, (rough bed-straw, w. Ju. ♃.) stem diffuse, very branching, prickly backwards; leaves in fives and sixes, lanceolate, acuminate: margins and nerves prickly; pedicels short. 18-24 i.

tincto'rium, (dyer's cleavers, w. Ju.) stem diffuse, smoothish; leaves linear, cauline leaves in sixes, branch leaves in fours; peduncles terminal, elongated, mostly 3-flowered. Wet woods. Stem weak and branching; leaves very narrow; corolla mostly 4-cleft. Used as a red dye.

obtu'sum, (E. w. J. ♃.) stem smooth; procumbent leaves in fours, oblanceolate, obtuse, rough on the margin and midrib.

brachia'tum, (bed-straw, E. w. Ju. ♃.) stem limber, long, brachiate-ramose, hispid; branches short; leaves in sixes, lance-oblong, acuminate, glabrous, margin and keel ciliate; branches whorled, the longest dichotomous; pedicels 2-flowered; fruit with hooks.

asgari'ne,(W.w. J. ♂.) stem limber, scabrous backwards; leaves in about eights, linear, and linear-oblanceolate, mucronate, hispid above, margin and keel prickly; fruit hook-bristled. 3-4 i.

tere'trum, (O. w. Ju. ♃.) stem procumbent, smoothish; leaves in fives or sixes, lance-obovate, mucronate, glabrous, scarcely ciliate at the margin; branchlets 3-flowered at the end; flowers pedicelled; fruit small.

borea'le, (O. w. Ju. ♃.) stem stiffly erect, smoothish, branching; branches short-erect; leaves in fours, linear-lanceolate, obtuse, revolved, with involute scabrous margins; panicled, terminal; fruit minutely bristled. 12-24 i.

..zens, (w-y. J. ♃.) stem erect, or slightly pubescent on the angles; leaves in fours, oval, obtuse, smooth, margin and nerves ciliate; peduncles short, divaricate, few-flowered; flowers remote, sub-sessile, alternate; fruit nodding, with hooked bristles. 6-12 i.

lanceola'tum, (p.Ju. ♃.) stem erect, very smooth, with remote joints; leaves in fours, lanceolate, generally acute, smooth, 3-nerved, margin sub-ciliate; peduncles long, divaricate; fruit sub-sessile,nodding,covered with hooked bristles. 1 f.

latifo'lium, (p. Ju.) stem erect, smooth; leaves by fours, oval, acute, membranous, the margins somewhat hispid; peduncles divaricate, loosely many-flowered. *S.*

uniflo'rum, (p. J. ♃.) stem assurgent, smooth; leaves generally by fours, linear, acute, revolute; peduncles generally solitary, 1-flowered. 10-12 i. *S.*

hispidu'lum, stem procumbent, pubescent, much branched; leaves by fours, lanceolate, dotted, scabrous. *S.*

GAULTHE'RIA. 10—1. (*Ericæ.*) [From *Gaulthier*, a physician and naturalist.]

procum"bens, (spicy wintergreen, w. J. ♃. or ♄.) stem procumbent; branches erect; leaves obovate, acute at the base; flowers few, nodding. Berries red, consisting in part of the permanent calyx; a little mealy; pleasant tasted.

hispidu'la, (creeping wintergreen, w. M. ♄.) stem creeping, hispid; leaves oval, acute; flowers solitary, axillary, sub-sessile, having but 8 stamens, short-bellform.

shal"lon, (w. J. ♄.) erect, fruticose; leaves ovate, sub-cordate, serrulate; raceme 1-sided, bracted; pedicels 2-bracted in the middle. *S.*

GAU'RA. 8—1. (*Onagra.*)

bien"nis, (r-y. Au. ♂.) stem having leaves purplish, sessile, lanceolate, toothed; flowers in terminal spikes. Banks of streams.

angustifo'lia, (w. Ju. ♃.) leaves clustered, linear, repand, undulate; fruit oblong, 4-angled, acute at each end. 3 f.

mol"lis, leaves lanceolate, entire, clothed with soft hairs.

GELSEMI'NUM. 5—2. (*Bignoneæ.*)

sempervi'rens, (y. March, ♃.) stem twining, smooth, glabrous; leaves opposite, perennial, lanceolate, entire, dark green above, paler beneath; petioles short. *S.* Nearly allied to Bignonia.

GENIS"TA. 16—10. (*Leguminosæ.*) [From *genu*, a knee, on account of its joints.]

tincto'rea, (dyer's broom, y. Ju. ♄.) root creeping; stem sub-erect, suffruticose; branches terete, striate, erect; leaves lanceolate, smooth; flowers in spiked racemes; legumes smooth. Hills. Introduced. Affords a yellow dye. Ex.

GENTIA'NA. 5—2. (*Gentianæ.*) [From *Gentius*, king of Illyria.]

quinqueflo'ra, stem square, branched; leaves ovate-lanceolate, sub-clasping, acute, 3-nerved, flowers somewhat in fives, axillary and terminal; corolla sub-campanulate, 5-cleft, segments lanceolate, mucronate; calyx very short. Woods. Aug. Flowers small, pale blue.

ochroleu'ca, large flowers. yellowish-white, striped inside with blue and purple.

crini'ta, (fringed gentian, b. Sept. ♃.) stem terete; branches long, 1-flowered; leaves lanceolate, acute; corolla 4-cleft, divisions obovate, gash ciliate. 18 i.

sapona'ria, (b. Oct. ♃.) leaves ovate, lanceolate, acute, 3-nerved; flowers whorl-capitate, sessile; corolla ventricose, closed, 10-cleft, interior segments unequally 3-cleft, as long as the exterior ones; segments of the calyx ovate, shorter than the tube. 18 i. Soap gentian.

lu'tea, (yellow gentian, y.) leaves broad-ovate, nerved; corollas about 5-cleft, wheel-form, whorled.

cates"baei, (Oc.) rough; leaves narrow-lanceolate; segments of the calyx linear-lanceolate, twice as long as the tube; corolla with the border erect, the interior segments short, 2-cleft, fimbriate. *S.*

GERA'NIUM. 15—10. (*Gentianæ.*) [From *geranos*, a crane, because its pistil is long, like a crane's bill.]

macula'tum, (crow-foot geranium, r. and b. J. ♃.) erect; pubescence reversed: stem dichotomous; leaves opposite, 3.5 parted, gashed, upper ones sessile; peduncles 2-flowered; petals obovate. 1.2 f.

sanguin"eum, (bloody geranium, ♃.) peduncle 1-flowered; leaves 5-parted, 3-cleft, orbicular; capsule bristly at the top. Ex A

robertia'num, (herb-robert, p. Sept ⊕.) leaves ternate or quinate, pinnatifid; peduncles long, 2-flowered; calyx angular, hairy; carpels small, wrinkled; stem long. Plant fetid.

carolinia'num, diffuse, pubescent; leaves opposite, 5-lobed, crowded toward the top; flowers small, white. *S.*

cæspito'sum, radical leaves reniform, deeply cleft; flowers red. *S.*

pusil'lum, (small crane's bill, b. M. ☉.) leaves sub-reniform, 7-lobed; lobes 3-cleft; peduncles short, 2-flowered; petals emarginate, scarcely longer than the awnless calyx; carpels keeled, pubescent; seeds smooth. Probably synonomous with *dissectum*.

dissec'tum, (wood geranium, r. Ju. ☉.) leaves 5-parted; lobes opposite, petiolate, 3-cleft, linear; peduncles short, 2-flowered; petals emarginate, rather shorter than the awned calyx; carpels hairy, not rugose; seeds reticulate. 12 i.

columbi'num, (long-stalked geranium, C. M. ⚳.) peduncles 2-flowered, longer than the leaves; leaves 5-parted; lobes many-cleft, linear; petals emarginate, of the length of the awned calyx; carpels glabrous.

GERAR"DIA. 13—2. (*Scrophularia*.) [From *Gerarde*, a writer on plants in 1597.]

tenuifo'lia, (p. Au. Sept. ♂.) very branching; leaves linear, acute, scabrous; peduncles axillary, longer than the flowers; teeth of the calyx acute. 6-10 i.

fla'va, (false foxglove, y. Ju. ⚳.) pubescent; stem nearly simple; leaves sub-sessile, lanceolate, entire, or toothed, lower ones sub-pinnatifid, gashed; flowers axillary, opposite, sub-sessile. 2-3 f.

glau'ca, (oak-leaf foxglove, y. Ju. ⚳.) smooth; stem panicled; leaves petioled, pinnatifid, paler beneath, the upper ones lanceolate; flowers axillary, opposite, on pedicels. 3-5 f.

pedicula'ria, (lousewort foxglove,y. Sept. ♂.) pubescent, brachiate-panicled; leaves oblong, doubly gash-serrate and pinnatifid; flowers axillary, opposite, pedicelled; divisions of the calyx leafy, gash-toothed. Var. *pectina'ta*, stem and branches densely pilose; leaves ovate, pectinately sub-bipinnatifid, soft pubescent; calyx hirsute. 2 f.

purpu'rea, (p. Au. ⚳.) stems with opposite branches; leaves linear, slender; flowers axillary, opposite, sub-sessile; segments of the calyx subulate. 12-18 i.

aphyl'la, (p. Ju. ⚳.) stem naked, nearly simple, with small, deciduous, opposite, ovate scales; corolla longer than the peduncle. 3 f. *S.*

fascicula'ta, (p. S. ⚳.) stem rigid, erect, branching near the summit; leaves opposite, and in threes, sometimes alternate, linear, clustered, very scabrous; peduncles much shorter than the leaves. *S.*

GE'UM. 11—12. (*Rosacea*.)

riva'le, (purple avens, p. J. ⚳.) pubescent; stem simple; radical leaves interruptedly pinnate, cauline ones 3-cleft; flowers nodding; petals as long as the calyx; awns plumose, nearly naked at the top, minutely uncinate. 10 i.

virginia'num, (avens, w. Ju. ⚳.) pubescent; radical and lower cauline leaves ternate, upper ones lanceolate; stipules ovate,

sub-entire; flowers erect; petals shorter than the calyx; awns hooked, naked; at the apex twisted, hairy. Var. *trilobum*, has the radical leaves 3-lobed, or ternate. 2 f.

stric"tum, (upright avens, O. y. J. ⚳.) hirsute; leaves all interruptedly pinnate, the odd one largest; leafets ovate, toothed, stipules gashed; divisions of the calyx 5, alternately linear, short; flowers erect; petals roundish, longer than the calyx; awn naked, hooked. 2 f.

al'bum, (w. Ju. ⚳.) pubescent; radical leaves pinnate, cauline ones ternate, upper ones simple, 3-cleft; lower stipules gashed; flowers erect; petals of the length of the calyx; awns hooked, naked, hairy at end. 2 f.

triflo'rum, (W. w. ⚳.) pilose; stem simple, somewhat 3-flowered; radical leaves interruptedly pinnate; leafets wedge, gash-toothed; petals oblong, as long as calyx; awns very long, villose. *S.*

coccin"eum, a splendid plant, a native of Chili, with large orange-scarlet flowers.

urba'num, (y. M. ⚳.) flowers erect; awns hooked, naked; stem erect, branching, hairy; radical leaves pinnatifid in fives, cauline ones palm-ternate, upper ones ovate, 1-lobed; stipules large, sub-orbicular. 18 i. Ex.

GILLE'NIA. 11—5. (*Rosacea*.)

trifolia'ta, (Indian physic, w.) leaves ternate; leafets lanceolate, serrate; stipules linear; flowers in loose, terminal panicles, large, medicinal, resembling ipecac.

stipula'cea, (w. J. ⚳.) radical leaves pinnatifid, stem leaves ternate; leafets incisely serrate; stipules foliaceous, ovate, incised, toothed and clasping; flowers in loose, terminal panicles, large. Var. *inca'na*, has ternate leaves, with leafets gash-toothed. 2 f.

GLAU'CIUM. 12—1. (*Papaveraceæ*.) [From *glaukos*, sea-green, from its color.]

lu'teum, (horned poppy, y. Ju. ⚳.) stem glabrous; cauline leaves clasping, repand; peduncles 1-flowered; silique tuberculate, little scabrous. *S.*

...DIO'LUS. 3—1. (*Iridea*.) [Diminutive of *gladius*, a sword, from the shape of its leaves.]

ensiform"is, (p. r. b. Ju. ⚳.) leaves ensiform, glabrous, entire; flowers spiked, colors various; root bulbous.

GLAUX. 5—1. (*Lysimachia*.) [From *glaukos*, sea-green.]

marati'ma, (black salt-wort, r-w. ⚳.) leaves roundish, entire, fleshy; stem leafy. 4-5 i. Marshes on the sea-coast.

GLECHO'MA. 12—1. (*Labiatæ*.) [From *glechos*, sweet.]

hedera'cea, (ground ivy, gill-overground, b. and r. M. ⚳.) leaves reniform, crenate; stem rooting. Var. *cordata*, leaves cordate.

GLEDITSCH"IA. 20—6. (*Leguminosæ*.) [From *Gleditsch*, professor of botany at Frankfort.]

triacan"tha, (honey-locust, w. J. ♄.) thorn strong, cross branched; a large tree, with oval and oblong leaves pinnate; legumes large, not caducous.

monosper"ma, pods small, 1-seeded. Water locust.

GLYCIRRHI'ZA. 16—10. (*Leguminosæ*.) [From *glukos*, sweet, and *riza*, root.]
gla'bra, legume glabrous; leaves pinnate; root tuberous, sweet. Liquorice. Ex.
lepido'ta, (w. Ju. ♃.) leafets oblong, acute, silky, villose; legumes racemed, oblong, hispid. 3-5 f.

GNAPHA'LIUM. 17—2. (*Corymbiferæ*.) [From *gnaphalon*, cotton.]
margaritace'um, (large-flowered life-everlasting, y. and w. Ju. ♃.) leaves linear, lanceolate, gradually narrowing, acute; stem branching above; corymb fastigiate; flowers pedicelled; flowers with white, pearly rays, and yellow disks. 1-2 f.
...yceph"alum, (sweet-scented life everlasting, y.w. Ju. ⊙.) leaves lance-linear, ...ous above, downy beneath; ...d, downy; corymbs terminal.
...eum, (early life-everlasting, o. ...hoots procumbent; stem sim... leaves spatulate, ovate, and ...ved; corymb close pressed; ...ous; inner scales of the calyx ...utish, colored. 6-10 i.
...um, (y.-w. J. ⊙.) herbaceous, ...; leaves obovate-spatulate, ...ath; flowers axillary and ...nerate spikes. 6-8 f.
... (Ju. ♃.) stem erect, simple, ...rs in a leafy spike, axillary; l...es lance-linear, downy.
...ns, (neglected life-everlasting, y. ...m erect, much branched; leaves ...eolate, very acute, decurrent, ...d woolly beneath, naked above; ...in dense, terminal, roundish clus...
...reum, (p. Ju. Oc. ♃.) herbaceous; ...m erect, simple; leaves linear-spatulate, tomentose beneath; flowers sessile, clus...ed, axillary, and terminal. 8-12 i.
...iligino'sum, (marsh cudweed, Au. ⊙.) ...m herbaceous, branched, diffuse, woolly; leaves linear-lanceolate; flowers in terminal crowded clusters, which are shorter than the leaves. 4-6 i.
germa"icum, (common cudweed, Au. ⊙.) stem herbaceous, erect, proliferous at the summit; leaves lanceolate, acute, downy; flowers capitate in the axils of the branches, and terminal. 6-8 i.
dio'ica, (w. M. ♃.) stoloniferous, creeping; leaves tomentose beneath chiefly, radical ones spatulate, obscurely 3-nerved at the base, cauline ones lance-linear; stem simple; flowers corymbose, capitate, diœcious.

GOMPHRE'NA. 5—1. (*Amaranthi*.) [From the surname of Pliny, the naturalist.]
globo'sa, (globe amaranth, bachelor's button, r. Au. ⊙.) stem erect; leaves lance-ovate; heads solitary; peduncles 2-leaved. Ex.

GONOLO'BUS. 18—5. (*Apocyneæ*.) [From *gonia*, angle, *lobus*, a pod.]

obliqu'us, (false choak-dog, p. J. ♃.) stem climbing, hairy; leaves ovate-cordate, villose, acute; corymbs axillary; segments of the corolla ovate, acuminate, oblique, revolute; calyx small. 4-5 f.
hirsu'tus, (p. Ju. ♃.) stem twining; younger branches very hairy; leaves cordate-oval, acuminate, pubescent on both sides; segments of the corolla linear-oblong; follicles oblong, muricate; umbels 3-4 flowered. 3-4 f.
macrophyl"lus, (y. Ju.) leaves broad, cordate, with the sinus closed; abruptly acuminate; follicles muricate; lobes of the crown divided. *S.*
prostra'tus, (p.) stem prostrate, herbaceous; leaves reniform-cordate, acute, tomentose underneath. 6-12 i. *S.*
viridiflo'rus, (g.) smooth, twining; leaves sub-reniform-cordate, auricled at the base, acuminate, somewhat long-peduncled; divisions of the corolla oblong-linear, oblique; obtuse follicles ribbed. *S.*

GOODYE'RA. 18—1. (*Orchideæ*.) [John Goodyer.]
pubes"cens, (rattlesnake leaf, scrophula weed, y. w. Ju. ♃.) leaves radical, ovate, petioled, veins colored, reticulate; scape sheathed; scape and flower pubescent; lip ovate, acuminate; petals ovate. 10-15 i.
re'pens, (w. Ju. ♃.) radical leaves ovate, petioled, reticulate; scape sheathed; scape and flowers pubescent; flowers one-sided; lip and petals lanceolate. 8 i.

GORDO'NIA. 15—12. (*Malvaceæ*.) [In honor of James Gordon.]
lasian"thus, (w. Ju. ♄.) leaves lance-oblong, shining, glabrous; flowers long, peduncled; capsules conical, acuminate. Evergreen. *S.*
pubes"cens, flowers large, white, with gold-colored stamens. Shrub. 5-6 f. Ex.

GOSSYP"IUM. 15—12. (*Malvaceæ*.) [From an Egyptian word, *gottipium*.]
herbace'um, (cotton, Au. ♂.) leaves 5-lobed, mucronate, one gland beneath; stem herbaceous, smooth. 5 f. Ex.

GRATIO'LA. 2—1. (*Scrophularia*.) [Diminutive of *gratia*, so called on account of its supposed admirable qualities.]
virgin"ica, (creeping hedge-hyssop, w. and y. ♃.) stem pubescent, assurgent, terete; leaves smooth, lanceolate, sparingly dentate, serrate, alternate, and connate at the base; leafets of the calyx equal; sterile filaments none. 6-8 i.
aure'a, flowers bright yellow, on axillary peduncles; stem 4-angled, rooting at the base.
carolin"ensis, (w. Ju. ♃.) stem smooth, somewhat branched, procumbent at the base, 4-sided above, terete below; leaves sessile, lance-oblong, obtusish, dentate, 3-nerved; peduncles pubescent, short; divisions of the calyx lance-linear, equal, entire; bracts broader, expanding; corolla pubescent within; sterile filaments none; capsule globose.

anagalloi'dea, (water hedge-hyssop, w-b. Ju. ♃.) sub-erect, very smooth; stem 4-sided; leaves oblong-oval, sparingly denticulate, shorter than the flowers; calyx without bracts, subulate, pubescent; corol smooth within; divisions generally obtuse. 3-6 i.

missouria'na, (J. y. ♃.) erect, terete, nearly simple; leaves narrow, lanceolate, connate, opposite, toothed at the apex; peduncles longer than the leaves; segments of the calyx linear-lanceolate, more than half as long as the tube of the corol, bracts longer than the calyx; whole plant viscid-pubescent. 4-6 i.

florida'na, (y. Mar. ♃.) glabrous, erect; leaves lanceolate, obsoletely denticulate, acutish; peduncle longer than the leaf; flowers largish, divisions emarginate. 9 i.

visco'sa, (w-p. Ap. ♃.) stem assurgent, viscid-pubescent, sub-terete; leaves smooth, sessile, lanceolate, acutish, dentate, 3-nerved; peduncles long; divisions of the calyx equal, lance-linear; bracts broader, expanding, shorter than the calyx; corolla pubescent within; sterile filaments two; capsules ovate, as long as the calyx.

pilo'sa, (w. Ju. ♃.) erect, branching, very hairy; stem 4-sided; leaves sessile, oval, dentate; flowers sub-peduncled; divisions of the calyx unequal; two intermediate ones small, setaceous; corolla smooth within; sterile filaments 2, very minute. 1-2 f.

GYMNOCLA'DUS. 20—10. (*Leguminosa*.)
canaden"sis, (coffee-tree, w. J. ♄.) leaves bipinnate; leafets oval, acuminate, pubescent; flowers in racemes.

GYMNOSTY'LES. 17—4. (*Corymbifera*.)
[From *gumnos*, naked, and *stulos*, style.]
stolonif"era, (M. ♃.) herbaceous, procumbent, creeping, glabrous; leaves pinnatifid; flowers sessile at the root.

GYNANDROP"SIS. 6—1. *(Capparides.)*
pentaphyl"la, (w. Ju. ♃.) smooth; leaves quinate and ternate; leafets entire, sub-serrulate; stamens inserted on the pedicel of the germ. 2 f.

GYRO'MIA. See MEDEO'TA.

GYROPHO'RA. 21—5. (*Alga*) [From *gyros*, a circle, and *sphero*, spherical.]
pennsylva'nica, frond tawny olive, under side rough granulate; receptacles marginated. On rocks and mountains.

HABENA'RIA. 18—1. (*Orchideæ*.) [From *habena*, a thong.]
psyco'des. (g-w. Ju. ♃.) lip 3-parted; segments finely divided; petals obtuse; horn filiform, clavate, ascending, longer than the germ.

cilia'ris, (orchis, y. Ju. ♃.) lip lance-oblong, pinnate-ciliate, twice as long as the petals; spur longer than the germ. 1-2 f.

dilata'ia, (giant orchis, w. or g. J. ♃.) spur shorter than the germ; lip entire, linear, with the base dilated, of the length of the spur; bracts of the length of the flower; stem leafy; in the mountain woods the flowers are green, in the meadows white. 1-4 f.

bractea'ta. (vegetable satyr, g-w. M. ♃.) lip linear, emarginate, obsoletely 3-toothed;

spur short, sub-inflated, somewhat 2-lobed; bracts twice as long as the flowers, leaf-like, spreading; roots palmate. 6-10 i.

macrophyl"la, (g-y. J. ♃.) lip lanceolate, entire, acuminate; spur longer than the germ, serrate, nearly straight; upper petals ovate, acute; scape with 2 broad-oval sub-erect leaves at the base. 1 f.

quin"queseta, (w.) lip 3-parted; lateral segments setaceous; inner petals 2-parted; lower segment setaceous, nearly as long as the outer petal; spur twice as long as the germ; leaves lance-oval; bracts acuminate. 2 f.

inte'gra, (y. Ju. ♃.) lip oblong, entire, longer than the inner petals; spur longer than the germ, acute at the point; stem leafy; bracts shorter than the flowers.

fusces"cens.(p-y. Ju. ♃.) lip ovate, toothed at the base; petals spreading; spur subulate, of the length of the germ; bracts longer than the flowers.

herbio'la, (y. J. ♃.) lip oblong, obtuse, toothed at the base; palate 1-toothed; spur filiform, shorter than the germ; bracts longer than the flowers. 1 f.

grandiflo'ra; (p, J.) lip dependent, twice as long as the petals, 3-parted; divisions wedge-form, fringed; middle one largest, with convivent fimbria; lateral petals fimbriate; spur ascending, clavate, longer than the germ; leaves oval oblong. 2 f.

inci'sa, (w-p. Ju. ♃.) lip 3-parted; divisions wedge-form, gash-toothed, middle one emarginate; lateral petals obtuse, subdentate; spur subulate, ascending, of the length of the germ. 2.4 f.

fis"sa, (p. Ju. ♃.) lip 3-parted; divisions wedge-form, toothed; intermediate one 2-lobed; spur filiform, clavate, ascending, longer than the germ.

re'pens, (y-g. Au. ♃.) lip 3-parted; lateral segment setaceous; inner petals 2-parted; lower segment setaceous, scarcely longer than the outer petals; horn as long as the germ; leaves narrow, lanceolate; bracts acute. *S*. See ORCHIS.

HALE'SIA. 15—12. (*Malvaceæ*.)
tetrapte'ra, (snow-drop tree, w. Ap. ♄.) leaves lance-oval, acuminate, serrulate; corolla 4-cleft; fruit 4-winged.

dipte'ra, (w. Ap. ♄.) leaves lance-oval and ovate, acuminate, serrulate; petals 4; flowers octandrous; fruit compressed, with two large wings. *S*.

parviflo'ra, (♄.) fruit unequally and somewhat 4-winged, clavate, small; flowers small. *S*.

HALYME'NIA. 21—4. (*Algæ*.)
palma'ta, frond flat, sub-palmate; divisions oblong, sub-simple; color reddish purple; substance at first thin and membranaceous, at length passing into a soft leathery substance. In the sea.

HAMAME'LIS. 4—2. (*Berberides*.)
virgin"ica, (witch hazel, y. Oct. ♄.) leaves obovate, acute, toothed, cordate, with a small sinus. Var. *parvifo'lia*, leaves oblong-ovate, upper part undulate, coarse crenate, pubescent, and somewhat hirsute beneath; divisions of the calyx oblong.

Blossoms in the fall, and perfects the fruit the next summer. 5-15 f.

HAMILTO'NIA. 20—5. (*Thymeleæ.*)
oleife'ra, (oil-nut, g-y. J. ♇.) pubescent; leaves oblong, entire, acuminate; flowers in terminal racemes, small. Whole plant oily.

HEDEO'MA. 13—1. (*Labiatæ.*)
pulegioï des, (pennyroyal, b. J. ☉.) pubescent; leaves oblong, serrate; peduncles axillary, whorled. 6-8 i.
hispi'da, (Ju. ☉.) branching, pubescent; leaves linear, acutish at both ends, very entire, veined, revolute at the margin; whorls many-flowered; calyx strigose. 3-6 i. *S.*
bracteo'la, pubescent; stem simple, slender; leaves linear, sub-lanceolate, acute at each end, entire; pedicels 3-5-flowered; bracts setaceous. *S.*

HEDE'RA. 5—1. (*Caprifolia.*) [From *hædus*, a kid]
he'lix, (English ivy, g-w. S. ♄.) leaves 3-5-lobed; floral ones ovate; umbel erect. Evergreen. Ex.

HEDYO'TIS. 4—1. (*Rubiaceæ.*)
glomera'ta, (w. g. M. ☉.) stem procumbent; leaves opposite, lanceolate, attenuate at the base, pubescent; flowers in clusters, forming whorls.
lanceola'ta, glabrous; stem erect, 4-sided, angles somewhat winged; leaves sessile, lanceolate, acute; stipules lanceolate, membranaceous; corymbs trichotomous, terminal; corol funnel-form, with exsert anthers.

HEDYSA'RUM. 16—10. (*Leguminosa.*) [From *edus*, sweet, aroma, smell.]
panicula'tum, (p. Ju. ♃.) erect; leaves ternate, lance-linear, smoothish, revolute at the margin; stipules subulate; panicle terminal; loment hispid; joints somewhat triangular. 2-3 f.
stric'tum, (p. Ju. ♃.) stiffly erect, glabrous, simple; leaves ternate, sub-linear, net-veined; stipules subulate; racemes axillary and terminal; loments about 2-jointed; joints ternate-triangular, hispid. 2-3 f.
nudiflo'rum, (p. Ju. ♃.) leaves ternate, broad-oval, acuminate, sub-glaucous beneath; scape panicled, glabrous, radical, taller than the stem; joints of the loment round-triangular. 1-2 f.
viridiflo'rum, (g. and p. Au. ♃.) stem erect, branched, scabrous; leaves ternate, ovate, obtuse, scabrous above, villose, and very soft beneath; panicle terminal, very long, naked; joints of the loment triangular. 3 f.
rotundifo'lium, (p. Au. ♃.) stem prostrate, hairy; leaves ternate; joints of the loment sub-rhomboidal. 2-4 f.
acumina'tum, (p. Ju. ♃.) erect, simple, pubescent; leaves ternate, ovate, conspicuously acuminate, a little hairy; panicle terminal, on a very long, naked peduncle; joints of the loment roundish. 1-2 f.
canaden"se, (bush trefoil, r. Ju. ♃.) erect, smoothish; leaves ternate, lance oblong; stipules filiform; flowers racemed; bracts lance-ovate, acuminate, ciliate; joints of the loment obtusely triangled, hispid. 3 f.
borea'le, leaves pinnate, leafets oblong-

ovate, hairy; stipules sheathing, subulate; racemes on long peduncles; loments with smooth, roundish joints. (p. Ju. ♃.) Mountains.
obtu'sum, (p. and g. Au. ♃.) erect, slender, sub-pubescent; leaves ternate, ovate, obtuse, sub-cordate at the base; stipules subulate; panicle terminal; joints of the loment sub-orbiculate, reticulate, hispid. 1-2 f.
linea'tum, (♃.) stem creeping, striped with green; leaves ternate, roundish, sub-sessile; racemes long, with small scattered flowers; joints of the loment lenticular. *S.* See DESMO'DIUM.

HELE'NIUM. 17—2. (*Corymbiferæ.*) [From Helena, wife of Menelaus, king of Sparta.]
autumna'le, (false sun-flower, y. Au. ♃.) leaves lanceolate, serrate, sub-decurrent; stem corymbed above; disk florets 5-cleft; rays flat, reflexed. Var. *pubes"cens*, leaves pubescent. 3-5 f.

HELIAN"THEMUM. See CIS"TUS.

HE'LIAN"THUS. 17—3. (*Corymbiferæ.*) [From *elios*, the sun, *anthos*, flower, on account of its broad yellow disk and rays : and not, as is often supposed, from its turning with the sun, which is not the fact with respect to this flower.]
angustifo'lius, (y. and p. O. ♃.) stem slender, slightly scabrous; leaves narrow-lanceolate, revolute at the margin, scabrous, entire, glaucous beneath; scales of the calyx lance-linear, ciliate, expanding; chaff 3-toothed. 3-5 f.
mol'lis, (y. Ju. ♃.) stem smooth below, scabrous above; leaves lance-ovate, acute, serrate, scabrous above, pubescent and hoary beneath; flowers few, terminal. 3-6 f.
pauciflo'rus, (y.) leaves lance-linear, acuminate, serrate, smoothish; stem naked, trichotomous, few-flowered; calyx close-imbricate; divisions ovate. 4-5 f. *S.*
trachelifo'lius, (y. Au. ♃.) leaves ovate-lanceolate, opposite acuminate, serrate, triply-nerved, very scabrous on both sides; scales of the calyx lance-linear, ciliate; outer ones longest. 3-4 f.
decapeta'lus, (y. Sept. ♃.) leaves ovate, acuminate, remotely serrate, 3-nerved, scabrous; scales of the calyx lanceolate, sub equal, sub-ciliate; rays 10 or 12. Flowers in large terminal panicles.
gigan"teus, (y. Sept. ♃.) leaves alternate, lanceolate, serrate, scabrous, paler beneath, nearly sessile, ciliate at the base, scales of the calyx lanceolate, ciliate; flowers in a loose, terminal panicle; rays 12-14, not large. 5-6 f.
atroru'bens, hispid, stem naked toward the summit, loosely paniculate; leaves opposite, spatulate, oblong-ovate, crenate. 3-nerved, scabrous on the upper side; scales of the calyx ovate-lanceolate, as long as the disk; rays yellow; disk dark purple.
corona'rium, French honeysuckle, a native of Italy. 4 f. Flowers scarlet. Ex.
tubero'sus, (Jerusalem artichoke, y. S. ♃.) leaves 3-nerved, scabrous; lower ones heart-ovate, upper ones ovate, acuminate; petioles ciliate; root tuberous. Naturalized. 4-8 f.

an''nuus, (common sun-flower, y. and w
Ju. ♃.) leaves all cordate, 3-nerved ; peduncles thickening upward ; flowers nodding. 6-10 f. Naturalized.

pubes''cens, (y. Au. ♃.) hoary-pubescent ;
stem villose ; leaves sessile, heart-ovate,
clasping, 3-nerved, crenulate, very soft ;
scales of the involucre lanceolate, villose.
2-3 f. S.

longifo′lius, (y. S. ♃.) very glabrous ;
stem panicled ; branches few-flowered at
the summit ; leaves sub-sessile, very long-
lanceolate, 3-nerved, very entire ; lower
ones serrate ; scales of the involucre ovate,
acute ; outer ones linear, divaricate. 4-7 f.
S.

tomento′sus, (y. S. ♃.) stem rough ; leaves
lance-ovate, tapering to the summit, acute,
serrulate, scabrous above, tomentose beneath, generally alternate ; scales of the
involucre leafy, squarrose-lanceolate ; chaff
3-cleft. 4-6 f. S.

HELIOP''SIS. 17—2. (Corymbifera.) [From
elios, the sun, opsis, appearing like.]

la′vis, (ox-eye, Ju. ♃.) stem glabrous ;
leaves opposite, ovate, serrate, 3-nerved,
smooth. 3-5 f.

sca′bra, (W.) leaves shortly petioled,
ovate, 3-nerved, deeply serrate, apex very
entire, both sides scabrous ; involucre pubescent.

HELIOTRO′PIUM. 5—1. (Boragina.) [From
elios, the sun, trope, turning ; a name given
by Dioscorides, because, as he says, the
flower turns with the sun.]

in''dicum, (turnsole, b. Ju. ☉.) leaves
heart-ovate, acute, roughish ; spikes solitary ; fruit bifid. 8-12 i. S.

curassavi′cum, (y. w. J. ☉.) leaves lance-
narrow-lanceolate, succulent, glabrous,
without veins ; spikes conjugate. 6-12 i. S.

europe′um, (w. ☉.) leaves ovate, very
entire, tomentose, rugose spikes conjugate.
Inodorous. S.

HELLEBO′RUS. 12—12. (Ranunculacea.)
[From ellein, destructive of life, bora, food,
from its poisonous qualities.]

fæ′tidus, (hellebore,) stem many-flowered,
leafy ; leaves pedate, remotely serrate, coriaceous ; corolla somewhat converging.

HELO′NIAS. 6—3. (Junci.)

angustifo′lia, (J. ♃.) scape leafy ; leaves
linear, subulate ; raceme simple, terminal ;
capsules oblong, covering at the summit ;
seeds linear. 2 f.

latifo′lia,(p-b. M. ♃.) scales leafless; spike
ovate, crowded ; bracts linear-lanceolate ;
leaves lanceolate, mucronate, nerved.

dic′cia, scape leafy ; leaves lanceolate,
broader near the root ; racemes diœcious,
spiked ; pedicels very short, without bracts ;
segments of the perianth linear ; stamens
exserted ; flowers white, in a terminal,
spiked raceme. Unicorn plant. Blazing
star. 2 f.

du′bia, leaves very long and narrow,
grass-like ; scape naked ; spike slender ;
flowers small, sessile. 2-3 f. S.

HEMEROCAL′LIS. 6—1. (Asphodeli.) [From
emera, day, and kallos, beauty, beauty of the
day.]

fla′va, (yellow day-lily, y. Ju. ♃.) leaves
broad-linear, keeled ; petals flat, acute ;
nerves of the petals undivided. Ex.

ful′va, leaves very long, linear, carinate ;
three inner petals obtuse, undulate ; nerves
of the outer petals branching ; flowers large,
fulvous ; scape 3-4 f. Introduced. Tawny
day-lily.

japon''ica, (w. Au.) leaves cordate, acuminate ; corol funnel-shaped. Japan.

HEMIAN''THUS. 2—1. (Scrophularia.)

micran''tha, (w. Au. ☉.) leaves opposite, crowded, sessile, obscurely 3-nerved,
glabrous ; succulent stem creeping, dichotomous ; flowers axillary, solitary, minute.
Banks of rivers.

HEPAT''ICA. 12—12. (Ranunculacea.) [From
epar, the liver : probably from the belief that
it was of use in complaints of this organ.]

acutil''oba, or triloba, (heart liverleaf, w.
and p. Ap. ♃.) leaves cordate, 3-5 lobed ;
lobes entire, acute ; leaves of the calyx
acute. Grows in woods, preferring the
north side of hills and mountains. 5 i.

america′na. (kidney liverleaf, w. and p.
Ap. ♃.) leaves heart-reniform, 3-lobed ;
lobes entire, round-obtuse ; leaves of the
calyx obtuse. Grows chiefly in woods,
preferring the south side of hills and mountains. 5 i.

HERAC′LEUM. 5—2. (Umbellifera.) [Named
either from Hercules, or the city of Heraclea,
near which it grew.]

lana′tum, (cow-parsnip, w. Ju. ♃.) leaves
ternate, petioled, tomentose beneath ; leaflets
round-cordate, lobed ; partial involucres 5-6
leaved ; fruit orbicular. One of our largest
umbelliferous plants, with a white, woolly
aspect. Flowers white, in very large, terminal umbels. Poisonous.

spondyli′um, leaves pinnate ; leaflets 5,
oblong. S.

HERPES''TIS. 12—2. (Scrophularia.) [From
erpo, creeping.]

cuneifo′lia, (b. Au. ♃.) very smooth ;
leaves opposite, cuneate-obovate ; peduncles as long as the leaves ; corolla 5-cleft ;
stem creeping.

rotundifo′lia, (b. Au.) finely pubescent ;
leaves oval, roundish, many-nerved ; peduncles opposite, as long as the leaves ; corolla 4-cleft. S.

amplexicau′lis, (Au.) stem woolly ; leaves
cordate, clasping, entire, obtuse ; peduncles
shorter than the leaves ; corolla 4-cleft. S.

HES''PERIS. 14—2. (Crucifera.) [From
esperos, evening.]

pinnatifi′da, (p. J. ♂.) lower leaves lyrate, pinnatifid ; upper ones lanceolate, unequally serrate ; borders of the petals obovate, entire ; pedicel becomes longer than
the calyx ; stem smooth. 1 f.

matrona′lis, (dame's violet, sweet rocket,
p-w.) pedicels of the length of the calyx ;
petals obovate ; leaves ovate lanceolate,
toothed. Ex.

pygmæ′a, (p. ☉.) leaves lance-linear, attenuate at the base, entire, or dentate ; pubescence 2-parted, appressed ; siliques numerous, erect, compressed, pubescent ; stem
erect, simple,

menzie"sii, (♃.) leaves spatulate, fleshy; pubescence 2-parted, appressed; siliques spreading (when young); stem very short, erect, simple.

tris"tis, (yellow rocket, ♂.) stem hispid; branches spreading. Ex.

HETERANTHE'RA. **3—1.** (*Narcissi.*) [From *eteros*, other or different, and *aner*, anther, because the anthers are of different sizes in the same flower.]

renifor"mis, (Ju. Aug. w. ♃.) leaves orbicular, reniform; spatha oblong, acuminate, 3-5 flowered.

HEUCHE'RA. **5—2.** (*Saxifragæ.*) [Heucher.]

america"na, (alum-root, r. Ju. ♃.) viscidly-pubescent; scape and leaves somewhat scabrous; leaves radical, on long, pubescent petioles; flowers in a long, terminal panicle; stamens exserted; calyx short, obtuse; petals lanceolate, as long as the calyx. 2-3 f.

pubes"cens, dusty-pubescent; calyx large, bell-form; stamens scarcely exserted; flowers large, red and yellow.

his"pida, (p. J. ♃.) hispid, scabrous; scape, petioles, and leaves, glabrous beneath; leaves hispid-pilose above, acute-lobed, toothed; teeth very short, sub-retuse, mucronate; peduncles of the panicle few-flowered; calyx shortish, sub-acute; petals spatulate, as long as the calyx; stamens exsert.

caules"cens, (w. J. ♃.) suffruticose at the base; scape at the base, and petioles pilose; leaves glabrous above, pilose at the nerves beneath, acutely lobed, ciliate, dentate; teeth acute, mucronate; calyx short, villose; petals linear, twice as long as the calyx; stamens exsert.

acerifo"lia, (♃.) petioles hirsute; leaves smooth, glaucous beneath, acutely 5-lobed, unequally toothed; teeth mucronate; scape smooth; panicle elongated, lax-flowered; petals short; stamens exsert.

HIBIS"CUS. **15—12.** (*Malvaceæ.*) [From *ibis*, the stork, who is said to be fond of it.]

phæni"ceus, (phœnicean mallows, r. Ju. ♃.) leaves ovate, acuminate, serrate, and crenate, lower ones 3-cuspidate; peduncles jointed; seeds woolly. 6.8 f. Ex.

milita"ris, (w-r. Au. ♃.) very glabrous; leaves 3-lobed, hastate, acuminate, serrate; corolla tubular, campanulate; capsule ovate, acuminate, glabrous; seeds silky.. 3-4 f.

virgin"icus, (sweat-weed, r. Au. ♃.) downy, rough; leaves acuminate, unequally toothed, lower ones cordate, undivided, upper ones cordate-oblong, 3-lobed; peduncles axillary, and in terminal racemes; flowers nodding; pistils nodding. 2-4 f.

syr"iacus, (althea frutex, w. and p. Au. ♄.) leaves wedge-ovate, 3-lobed, toothed; outer calyx about 8-leaved, of the length of the inner. 5-10 f. Ex.

esculen"tus, (okra, y. Ju. ☉.) leaves heart-5-lobed, obtusish, toothed; petiole longer than the flower; outer calyx about 5-leaved, caducous, bursting lengthwise. 3-5 f. Ex.

trio"num, (beautiful ketmia, flower of an hour, ☉.) flowers yellowish-white, with the lower part purple; calyx inflated; leaves toothed. Ex.

moscheu"tus, (marsh mallows, w-p. Au. ♃.) leaves tomentose beneath; petioles bearing the peduncles; calyx tomentose. Swamps. Flowers large, white, with a purple centre.

grandiflo"rus, leaves large, coriaceous, 3-lobed, tomentose on both sides, hoary beneath; flowers large, red. 5-7 f. S. Ex.

inca"nus, (y-w. S. ♃.) leaves ovate, acuminate, obtusely serrate, hoary-tomentose on both sides; peduncles axillary; calyx tomentose, nearly equal.

coccin"eus, (r. Au. ♃.) very glabrous; leaves palmate-5-parted; divisions lance-linear, acuminate, remotely serrate at the apex; capsule ovate, glabrous. 4-8 f. S.

sca'ber, (y. and p. Au. ♃.) stem scabrous; lower leaves cordate, angled, upper ones palmate, 3-5 lobed; lobes irregular, dentate, angled; calyx very hispid, twice as long as the bracts. 3 f. S.

carolin"ia"nus, (p. Au. ♃.) leaves cordate, ovate, acuminate, serrate, smooth on both sides, sometimes sub-3-lobed; seeds hispid. 4-6 f. S.

vesica'rius, (African hibiscus, y-p. ☉.) 2 f.

HIERA'CIUM. **17—1.** (*Cichoraceæ.*) [From *hierax*, a hawk.]

marian"um, (O. y. Ju. ♃.) stem erect, villose; leaves oval-obovate, strigose, villose on the keel, lower ones sub-dentate; peduncles and involucre downy. 1-2 f.

runcina'tum, hirsute; leaves radical, oval-oblong, runcinate; scape few-flowered, angular; involucre glandular-pilose. 1 f. S.

grono'vii, (y. Ju. ♃.) scape leafy, naked above, corymb-panicled; calyx and peduncles glandular-pilose; radical leaves obovate and lanceolate, ciliate, very pubescent. Var. *nudicau'le*, stem about 1-leaved; panicle somewhat fastigiate. 2.3 f.

panicula'tum, (y. Ju. ♃.) somewhat glabrous; stem erect, leafy, panicled, white-woolly below; pedicels capillary; leaves lanceolate, naked, toothed, membranaceous. 2-4 f.

fascicula'tum, somewhat glabrous; stem erect, leafy, simple, glabrous; leaves sessile, oblong, acute, sharply toothed; teeth elongated; branches of the panicle divaricate, short; pedicels pubescent, somewhat fascicled.

macrophyl'lum, (y. ♃.) very tall; stem erect, leafy, hispid, sulcate; leaves cordate, half-clasping, ovate-oblong, remotely coarse-toothed, nearly naked; nerves and veins pubescent beneath; panicle divaricate-corymbed; peduncles elongated, naked, glabrous; calyx glabrous.

veno'sum, (vein-leaf hawkweed, y. Ju. ♃.) scape naked, corymb-panicled, glabrous; leaves lance-obovate, with thin hairs above, and naked beneath, margin ciliate, glandular-toothed, veins colored; calyx glabrous. 1-2 f.

auranti'acum, (orange hawkweed, y. ♃.) scape leafy, hispid; flowers corymbed; peduncles glomerate; leaves oblong, acutish, pilose-hispid. Ex.

kal"mii, (y. Au. ♃.) stem erect, sub-villose; leaves sessile, lanceolate, acuminate, sharply and divaricately toothed; panicle sub-corymbose; pedicels downy. 2 f.

HIPPO'PHÆ. 20—8. (*Æleagni.*) [From *ippos*, a horse, *phao*, to destroy.]

canaden"sis, (sea buckthorn, M. ♄.) leaves ovate, nearly smooth above, argenteus beneath. 6-8 f.

argen"tea, both sides of the leaves covered with silver scales. 12-18 f.

HIPPU'RIS. 1—1. (*Naides.*) [From *ippos*, a horse, *oura*. tail.]

vulga'ris. (mare's-tail, y-g. M. ♃.) leaves linear, and lance-linear, verticillate.

monta'na, slender in sixes, linear, acute.

maritI'ma, leaves in fours or sixes, lance-olate, obtuse, scarcely gangrenous. 9-18 i.

HOL"CUS. 3—2. (*Gramineæ.*) [From *olkos*, the Greek name of a plant with awns like barley.]

lana'tus, (soft-grass, w. Ju. ♃.) perfect floret inferior and awnless, sterile one with a curved awn included in the glume; root fibrous; culm 18 inches high; panicle oblong, contracted, whitish.

HO'PEA 15—12. (*Malvaceæ.*) [Dr. John Hope.]

tincto'ria, (sweet leaf, y. Ap. ♄.) leaves lance-oblong, glaucous, pubescent beneath; flowers sessile, axillary, in clusters. 15-18 f.

HOR"DEUM. 3—2. (*Gramineæ.*)

juba'tum, (J. ♂.) lateral florets abortive; awns of the calyx and corolla 6 times as long as the flowers. 2 f.

vulga're, (barley, Ju. ☉.) florets all perfect, awned, in two erect rows. Ex.

dis"tichon, (J. ☉.) lateral florets imperfect, awnless; seeds angular, imbricate. Ex.

pusil"lum, lateral florets staminate or neuter, awnless, acute; four inner glumes coriaceous, dilated, all short-awned; awns scabrous, decumbent. 4-6 i.

HORTEN"SIA 10—3. (*Caprifolia.*)

specio'sa, (changeable hydrangea, r. and w. J. ♄.) leaves broadly ovate, serrate, acuminate; flowers corymbed. From the East Indies. This is the common flower-pot shrub called hyderindia, and by corruption of this word hyderangea.

HOTTO'NIA. 5—1. (*Lysimachia.*) [John Hotton.]

infla'ta, (water-feather, Ju. ♃.) stem thick, generally submersed; scape jointed; flowers whorled, on peduncles; leaves long, pectinate. Stagnant waters.

HOUSTO'NIA. 4—1. (*Gentiana.*) [Dr. Houston.]

serpyl"lifolia, (b. M. ♃.) procumbent, cespitose; leaves spatulate, obtuse; peduncles terminal, 1-flowered, very long.

tenel"la, (b. J. ♃.) stem creeping, filiform; leaves round, acute, nerved; peduncles terminal, 1-flowered, very long, smaller than the last.

rotundifo'lia, (w. Ap. ♃.) creeping; leaves ovate, roundish, abruptly narrowed at the base; peduncles axillary, solitary, 1-flowered; leaves evergreen.

ciliola'ta, (p.) radical leaves ovate, obtuse, narrow at the base, ciliate at the margin, cauline ones ovate-spatulate, sessile; co-

rymbs terminal, pedicelled; peduncles trichotomous; divisions of the calyx lance-linear; stem smooth, branched above.

pubes"cens, leaves wedge-form, acute, pubescent, lower ones sub-petioled, lance-olate, upper ones sub-oval, sessile; panicle trichotomous, terminal.

pa'tens, (p. Mar. ☉.) small; stem branching, dichotomous, with scabrous angles; flowers solitary, terminal, and axillary. 1-2 i.

cæru'lea, (innocence, Venus'-pride, b. and w. M. ♃.) stem erect, setaceous, dichotomous; radical leaves spatulate, cauline ones oblanceolate, opposite; peduncles 1-flowered, elongated. 4-6 i.

longifo'lia, (b-w.) leaves narrow; flowers terminal, nearly sessile.

purpu'rea, purple flowers in terminal corymbs.

HOY'A. 18—5. (*Apocyneæ.*)

carno'sa, (w-r.) leaves ovate; flowers bearded, wax-like, distilling a honey-like fluid. A vine.

HUDSO'NIA. 12—1. (*Cesti.*)

ericoi'des, (false heath, y. J. ♄.) pubescent; stem suffruticose, sub-erect; branches elongated; leaves filiform, subulate; peduncles lateral, elongated; calyx cylindrical, obtuse; capsule pubescent; 1-seeded. 4-6 i. Pine barrens.

tomento'sa, hoary-pubescent. Sea-shore.

monta'na, decumbent, smoothish, cespitose; leaves long, filiform-subulate, sub-imbricate; peduncles terminal, solitary; calyx bell-form, woolly; capsule villose. 3-5 i.

HU'MULUS. 20—5. (*Urtica.*) [From *humus*, the ground, because, without support, it trails on the ground.]

lu'pulus, (hop, gr-y. Au. ♃.) stem twining with the sun; leaves lobed. One of the best of tonics.

HYACIN"THUS. 6—1. (*Asphodeli.*) [Said to have been named from the friend of Apollo, who, according to the poets, was changed into this flower.]

orienta'lis, (garden hyacinth, r. Ap. ♃.) corolla funnel-form, half 6 cleft, ventricose at the base. Ex.

mus"cari. (musk hyacinth, r. Ap. ♃.) corollas ovate, all equal. Ex.

botryoi'des, (grape hyacinth, b. Ap. ♃.) corollas globose, uniform; leaves cylindric, channelled, straight. Ex.

racemo'sus, (hare-bell hyacinth,) flowers thick, ovate, those at the top sessile; leaves lax-pendent, linear.

como'sus, (purple grape-hyacinth,) corollas angular-cylindric; upper flowers long-peduncled.

HYDRAN"GEA. 10—2. (*Saxifrago.*) [From *udor*, water, and *aggeion*, a vessel, in allusion to the shape of the seed-vessel.]

vulga'ris, (hydrangea, w. Au. ♄.) leaves oblong-ovate, obtuse at the base, acuminate, glabrous beneath; cymes naked. 4 f.

radia'ta, leaves cordate, serrate, tomentose, and white beneath; cymes terminal, radiate; flowers white, very ornamental. Shrub. 6 f. For the cultivated hydranges, see HORTEN"SIA.

corda'ta, (M. J. ♃.) leaves broad ovate,

slightly cordate at base, acuminate, coarsely toothed, glabrous underneath; cymes generally radiate.

quercifo'lia, (w-r. M. J.) leaves oblong, sinuate and lobed, dentate, tomentose underneath; cymes radiate, paniculate. 4-5 f.

HYDRAS''TIS. 12—12. (*Ranunculaceæ*.) [From *udor*, water.]

canaden''sis, (yellow pucoon, w-r. Ap. 2f.) stem with two opposite leaves above; leaves petioled, emarginate at the base, palmate, serrate, gashed; peduncle terminal, solitary, 1-flowered; roots yellow. Used by the Indians as a die.

HY'DROCHA'RIS. 20—9. (*Hydrocharides*.) [From *udor*, water, and *charis*, grace, this little plant being considered as ornamental to placid waters.]

spongio'sa, monœcious, leaves floating, round, cordate, reticulate underneath, with vesicles at base.

HYDROCHLO'A. 19—6. (*Gramineæ*.) [From *udor*, water, and *eloa*, oil, because the leaves have a very oily appearance.]

flui'tans. (Ju. 2f.) floating in water, culm long, slender, branching; leaves linear, flat; spike solitary, axillary, setaceous; about 4-flowered.

HYDROCOT''YLE. 5—2. (*Umbelliferæ*.) [From *udor*, water, *kotule*, a cavity.]

umbella'ta. (w. M. 2f.) leaves peltate, crenate, emarginate at base; umbels many-flowered, on long peduncles.

vulga'ris, (g-w. J. 2f.) leaves orbicular, peltate, slightly crenate; scape interruptedly spiked, few-flowered.

ranunculoi'des, (Ju. 2f.) creeping; glabrous; leaves orbicular-reniform, somewhat 5-lobed; lobes obtuse, crenate, the middle one smaller and more distinct; umbels subcapitate, few-flowered; pedunculate, 5 to 10-flowered.

america'na, root tuberous; stem filiform, with creeping suckers; leaves reniform, slightly 7-lobed, crenate; umbels 4-6 flowered, axillary; petals greenish-white. Wet places.

interrup'ta, stem creeping at the joints; leaves peltate; flowers pinnate, white, in small umbels, much shorter than the petioles. Marsh penny-wort.

HYDRO'LEA. 5—2. (*Convolvuli*.) [From *udor*. water, *elaia* oil.]

quadrival''vis, (b. Ju. 2f.) spinose, pilose; leaves long-lanceolate; flowers nearly sessile, axillary.

corymbo'sa, without spines, flowers terminal.

spino'sa, leaves lanceolate, hirsute; flowers terminal. S.

HYDROPEL''TIS. 12—12. (*Ranunculacea*.) [From *udor*, water, *pelte*, a shield.]

purpu'rea, (water-shield, p. Ju. 2f.) leaves peltate, tinged with purple; peduncles solitary, 1-flowered. Whole plant covered with a viscid gelatine; stem long, floating.

HYDROPHYL''LUM. 5—1. (*Boragineæ*.) [From *udor*, water, *phyllum*, a leaf.]

virgini'cum, (water-leaf, w. J. 2f.) smoothish; leaves pinnatifid and pinnate; segments with deep serratures; clusters of flowers crowded; peduncles larger than the petioles. 18 i.

canaden''se, somewhat hairy; leaves large, about 5-7-lobed; flowers blue and white, in clusters.

linea're. (Ap. 2f.) pilose; leaves linear; racemes elongated. S.

HYOSCYA'MUS. 5—1. (*Solana*.) [From *sus*, a swine, and *kuamos*, a bean, because the plant is unsightly.]

ni'ger, (hen-bane, y-p. Ju. ♂.) leaves clasping, sinuate; flowers veiny, sessile. Introduced.

HYPER''ICUM. 12—5. (*Hypericæ*.) [From *uper*, over, *eikon* evil spirits, because it was thought to have power over such.]

corymbo'sum, (y. Ju. 2f.) erect, glabrous, darkly-punctate; stem terete, branching; leaves clasping, oblong-oval, obtuse; corymbs terminal, brachiate, dense-flowered; divisions of the calyx lanceolate, acute. 18-24 i.

parviflo'rum, (y. J. 2f.) erect, small, glabrous; dichotomous-ramose, somewhat 4-sided; leaves ovate oblong, sub-cordate, obtuse, nerved. sessile; panicles terminal, dichotomous-corymbed; petals shorter than the lanceolate calyx. 6-12 i.

perfora'tum, (y. J. 2f.) erect, branching; stem 2-edged; leaves oblong, obtuse, transparently punctate; panicle terminal, brachiate, leafy; petals twice as long as the acute, lanceolate calyx; 3 styles. St. John's wort.

virgin''icum, (p. Au. 2f.) flowers with 9-12-stamens, distinctly arranged in 3 parcels, and separated by nectaries; 3 styles; leaves oval, obtuse, clasping; stem compressed. 1-2 f.

ascyroi'des, smooth; stem square, winged at the base; leaves sessile, acute; styles free, as long as the stamens; flowers and leaves large; capsules nearly as large as nutmegs, yellow. River banks.

puncta'tum, stem terete, leaves subclasping; flowers in dense corymbs; styles 3, longer than the stamens. Whole plant dotted with black.

canaden''se, erect. small, few-flowered, stem 4-sided, dichotomous above; leaves sessile, linear; capsules red. 9-19 i.

kalmia'num, (laurel-leaved hypericum,) shrubby, very branching corymbs terminal. 3-4 f. Cultivated as ornamental.

prolif'cum, leaves more narrow than the preceding; flowers smaller, numerous. Cultivated.

HYPOPEL''TIS. 21—1. (*Filices*.) [From *upo*, under, and *pelte*, a shield.]

obtu'sa, (Au. 2f.) fronds 8-14 inches, high, bi-pinnate; divisions sub-remote; segments oblong, the lower ones crenate-dentate; rachis pubescent above. Rocky woods.

HYPOX''IS. 6—1. (*Narcissi*.)

erec'ta, (star-grass, y. Ju. 2f.) pilose; scape 2-3-flowered; leaves lance-linear; divisions of the corolla lance-oblong. Var. *gramin'ea*, has longer and narrower leaves; more flowers, longer lance-linear divisions to the corolla, and altogether a more grassy appearance.

filifo'lia, (Ap.) leaves filiform, somewhat angled, hairy scape generally 2-flowered. S.

junce'a, (J. 2{.) pilose; leaves filiform, channeled, very entire; scapes 1-flowered. S.

HYP"TIS. 13—1. (*Labiatæ.*)

radia'ta, (w. Au. 2{.) heads of flowers opposite; peduncles as long as the internodes; bracts lanceolate; leaves oblong, serrate.

HYSSO'PUS. 13—1. (*Labiatæ.*) [A Hebrew name.]

nepetoi'des, (giant hyssop, g-y. Ju. 2{.) stem acutely 4-angled; leaves opposite; calyx small; bracts dilated. Woods. 3 6 f.

officina'lis, (garden hyssop,) flowers whorled; leaves lance-linear.

IBE'RIS. 14—1. (*Cruciferæ.*) [From *Iberis*, the ancient name of Spain.]

umbella'ta, (purple candy-tuft,) leaves lanceolate, acuminate; lower ones serrate, upper ones entire. Ex.

ama'ra, (white candy-tuft,) leaves irregularly dentate, narrow towards the base, somewhat spatulate, fleshy. 1 f. Ex.

ICTO'DES. 4—1. (*Aroideæ.*) [From *iktis*, a skunk.]

fœ'tida, (skunk cabbage, fetid hellebore, p. Ap. 2{.) stemless; leaves radical, heart-ovate; very large spadix supporting the flowers in a sub-globose head. Odor resembles that of a skunk.

I'LEX. 4—4. (*Rhamni.*)

opa'ca, (evergreen holly, g-w. M. ♄.) leaves evergreen, ovate, acute, spinose, glabrous, flat; flowers scattered at the base of the shoots of the preceding year. A middle-sized tree.

canaden"sis, (mountain holly, g-y. M. ♄.) leaves deciduous, ovate, glabrous. 3-5 f.

vomito'ria, leaves oval-obtuse, obtuse at each end, glabrous; umbels lateral, sub-sessile. 6-8 f. S.

laxiflo'ra, (J. ♄.) leaves ovate, sinuate-toothed, spinose, shining, flat; peduncles super-axillary, in the young branches aggregate, lax-ramose. S.

cas"sena, (Mar. Ap. ♄.) leaves oval, obtuse at each end, crenately serrate. 6-15 f. S.

laurifo'lia, (leaves large, elliptic, acutish, very entire, sempervirent; pedicels elongated, sub-3-flowered. S.

myrtifo'lia, (M. ♄.) leaves linear-lanceolate, mucronate, rigid, very glabrous; fertile flowers solitary. 6-10 f. S.

ligus"trina, (J. ♄.) leaves linear-lanceolate, cuneate at the base, generally entire; fertile flowers solitary. 6-10 f. S.

ILLI'CIUM. 12—12. (*Lauri.*)

parviflo'ra, (y. M. ♄.) leaves alternate, lanceolate, entire, perennial; petals and sepals round, concave. 6-10 f. S.

florada'num, flowers purple; leaves acuminate; petals numerous, oblong, and linear. S.

IMPA'TIENS.* 5—1. (*Geraneæ.*)

---

*The capsules are remarkable for bursting open with an elastic spring, at the slightest touch, hence the generic name Impatiens.

pal"lida, (jewel-weed, touch-me-not, y. Ju. ☉.) peduncles solitary, 2-4-flowered; calcarate petals conic, dilated, shorter than the rest; spur recurved, very short; flowers sparingly punctate; leaves rhomb-ovate, mucronate-toothed. 2-4 f.

ful'va, (speckled jewel-weed, y-r.) peduncles solitary, 2-4-flowered; leaves rhombic-ovate; mucronate-dentate; calcarate petal longer than the rest; flowers with crowded spots.

balsami'na, (garden ladies'-slipper,) peduncles aggregate, 1-flowered; leaves lanceolate, upper ones alternate; calcarate petal (or nectary) shorter than the other petals. Of various colors. 1-3 f.

INDIGOFE'RA. 16—10. (*Leguminosæ.*) [From *fero*, to bear, added to indigo.]

tincto'ria, (indigo, ♄.) leaves pinnate, oblong, glabrous, in 4 pairs; racemes shorter than the leaves; legume terete, somewhat arched. Ex.

carolin"ia'na, (b. Au. 2{.) leaves pinnate; leaflets oval and obovate; spikes longer than the leaves; pods 2-seeded, reticulate, veiny. 3-7 f. S.

INU'LA. 17—2. (*Corymbiferæ.*) [Fabled to have sprung from the tears of Helen.]

hele'nium, (elecampane, Au. ♄.) leaves clasping, ovate, rugose, tomentose beneath; scales of the calyx ovate. Naturalized. 3-5 f.

IONI'DIUM. 5—1. (*Cisti.*) [*Ion*, violet, from the resemblance.]

con"co'lor, (green violet, w-g. M. 2{.) straight, erect; leaves broad-lanceolate, sub-entire; stipules subulate entire; peduncles short; petals connivant; emarginate; spur 0. 20 i.

IPO'MEA. 5—1. (*Convolvuli.*) [From two Greek words, signifying like a vine.]

nil, (morning-glory, b. Ju. ☉.) hirsute; leaves cordate, 3-lobed; peduncles short, 1-3-flowered; calyx very villose, long, acuminate.

bona'nox, (w. Ju. ☉.) very glabrous; leaves cordate, entire or angled; peduncles 1-3-flowered; calyx awned; corolla undivided, tube long. S.

coccin"ea, (scarlet morning-glory, y-r. ☉.) pubescent; leaves cordate, acuminate; peduncles about 5-flowered; corolla tubular. West Indies.

quam"oclit, (crimson cypress-vine, r-w.) leaves pinnatifid, linear; flowers sub-solitary, corolla tubular; dark red. East Indies.

lacuno'sa, (starry ipomea, w-p.) glabrous; leaves cordate, acuminate; peduncles short, about 1-flowered; calyx hairy.

dissec"ta, (Ju. 2{.) stem, petioles, and peduncles very pilose; leaves glabrous, 7-lobed; divisions sinuate; peduncles 1-flowered; divisions of the calyx oval; corol bell-form. S.

trichocar"pa, (p. Ju. ☉.) leaves entire, cordate, or 3-lobed, villose; calyx ciliate; capsules hirsute; peduncles about 2-flowered. S.

caroli'na, (b.) leaves digitate; leaflets petioled; peduncles 1-flowered. S.

I'RIS. 3—1. (*Iridæ.*) (From *iris*, the rainbow.)

crista'ta, (b-y. Ap. ♃.) bearded ; beard crested ; scape generally 1-flowered, as long as the leaves. 2-4 f. *S.*

tri'pet''ala,(E. M. ♃.) bearded ; stem terete, longer than the leaves ; rudiments of the inner petals 3-toothed, middle tooth acuminate. 2 f. *S.*

cupre'a, (r-y.) beardless ; stem terete, flexuous, equalling the leaves ; capsules large, 6-angled. 3 f. *S.*

ver''na, (b. M.) without beard or stem ; 1-flowered ; leaves grass-like ; tube very long. On the earth. *S.*

prismat''ica, (b. y. J. ♃.) flowers beardless ; leaves linear ; stem round, many-flowered ; germs triangular, twice grooved on the sides. 1-2 f.

plica'ta, (garden iris, p. w. M. ♃.) bearded ; stem many-flowered, higher than the leaves ; petals undulate-plicate, erect ones broadest. 18-24 j. Ex.

pu'mila, (dwarf flower-de-luce, b. M. ♃.) bearded ; scape 1-flowered ; leaves ensiform, glabrous ; tube of the corolla exsert ; petals oblong, obtuse. 6-10 j. Ex.

ochroleu'ea, (yellow iris, y. M.) beardless ; leaves ensiform, depressed, striate ; scape sub-terete ; germ 6-cornered. Ex.

versic''olor, (snake-lily, blue-flag, b. J. ♃.) leaves ensiform ; stem acute on one side ; capsules oblong, 3-sided, with obtuse angles. Var. *sulca'ta*, inner petals longer than the stigmas ; germ with sulcate angles and concave sides ; capsule oblong, ventricose ; angles somewhat furrowed. Var. *commu'nis*, stem erect, flexuous ; leaves narrow-ensiform ; inner petals a little shorter than the stigmas ; angles of the germ not grooved when young, sides deep-concave ; capsule cylindric, oblong. 2-3 f.

lacus'tris, (b. ♃.) beardless ; leaves short-ensiform ; scape much shorter than the leaf, 1-flowered ; petals attenuated on the tube ; capsule turbinate, 3-sided, margined ; seed roundish, smooth ; root tuberous.

missou'riensis, (y. b. ♃.) beardless ; stem terete, higher than the leaves, somewhat 3-flowered ; leaves narrow, ensiform ; capsules oblong-linear ; flowers bicolored. 12-16 i.

sam''buci'na, (elder-scented iris, garden-iris, b. p. w. M. ♃.) stem many-flowered, higher than the leaves ; divisions of the corolla emarginate, outer ones flat ; leaves inflex-falcate at the apex ; spaths membranaceous at the apex ; lower flowers peduncled ; stigmas with acute, serrate divisions. 18-24 i. Ex.

sibiri'ca, (b. Ju. ♃.) beardless ; stem hollow, terete, higher than the leaves, somewhat 3-flowered ; leaves linear ; capsules short, 3-angled, obtuse at each end. *S.*

ISAN''THUS. 13—1. (*Labiatæ.*) [From *isos*, equal, *anthos*, flower.]

cæru'leus, (blue gentian, false pennyroyal, b. Ju. ♋.) viscid, hairy ; leaves lance-oval, acute at both ends, 3-nerved ; peduncles 1-2 flowered.

ISA'TIS. 14—1. (*Cruciferæ.*) [Name given by Dioscorides, origin unknown.]

tincto'ria, (woad, J. ♂.) radical leaves crenate, cauline ones sagittate, oblong.

ISNAR''DIA. (See LUDWIGIA.)

ISOME'RIS. 6—1. (*Cappar''ideæ.*) [From *isos*, equal, and *meris*, divisions.]

arbo'rea, (y, ♄.) leaves crowded, trifoliate ; leafets lanceolate, somewhat mucronate, glabrous ; calyx campanulate ; segments triangular-ovate, acuminate.

I'TEA. 5—1. (*Saxifrage.*)

virgin''ica, (w. J. ♄.) leaves alternate, lanceolate, acuminate, serrulate, pubescent beneath ; flowers in terminal racemes. 4-8 f.

I'VA. 17—4. (*Corymbiferæ.*)

frutes''cens, shrubby ; leaves opposite, lanceolate, deeply serrate ; heads globular, depressed. Seacoast. 3-8 f. High-water shrub. Flowers green.

imbrica'ta, (Au. ♄.) perennial, glabrous ; leaves linear-lanceolate, cuneate, succulent, the upper alternate and very entire ; involucrum imbricate ; chaff of the receptacle spatulate. *S.*

xanthifo'lia, (Au. ♋.) leaves opposite, petioled, heart-ovate, acuminate, doubly-serrate, soft-villous, hoary beneath ; spikes naked, panicled. 5-6 f. *S.*

cilia'ta, (Ju. ♋.) herbaceous ; leaves lance ovate, sub-serrate ; spike somewhat crowded ; bracts lanceolate, acuminate ; bracts and petioles long-ciliate. 2 f. *S.*

IX''IA. 3—1. (*Iridæ.*) [From *iksos*, glue, from the gummy juice of some plants which first bore the name.]

chinen''sis, (blackberry lily, y. r. J. ♃.) corolla about 6-petalled ; stem flexuous ; leaves ensiform. Ex.

cœlesti'na, (b. M. ♃.) leaves linear-subulate, much shorter than the 1-flowered scape. *S.*

JASMI'NUM. 2—1. (*Jasminæ.*) [From *ion*, a violet, and *asme*, odor]

fru'ticans, (jasmine, y. ♄.) leaves alternate, ternate, simple ; leafets obovate, wedge-form, obtuse ; branches angled. Ex.

officina'le, (jasmine, w. ♄.) leaves pinnate, opposite ; leafets acuminate. Ex.

JATRO'PHA. 19—15. (*Euphorbiæ.*) [From *Jatros*, an ancient physician.]

stimulo'sa, (w. Ju. ♃.) hispid, with prickles ; leaves palmate-lobed ; lobes toothed ; cymes short peduncled. 6-8 i. *S.*

elas''tica, the juice affords the elastic gum called caoutchouc, or Indian-rubber.

mani'hot affords the cassada root. *S.*

JEFFERSO'NIA. 8—1. (*Papaveraceæ.*) [In honor of Thomas Jefferson, named by Barton]

diphyl''la, (twin-leaf, w. M. ♃.) stemless ; peduncles naked, 1-flowered ; leaves in pairs.

JU'GLANS. 19—12. (*Terebint'aceæ*)

cine'rea, (butternut, M. ♄.) leafets numerous, lanceolate, serrate, rounded at the base, soft-pubescent beneath ; petioles villose ; fruit oblong-ovate, viscid, long-peduncled.

ni'gra, (black walnut, M. ♄.) leafets numerous, lance-ovate, serrate, sub-cordate, narrowed above ; petioles and under side of the leaves sub-pubescent ; fruit globose, with scabrous punctures, nut wrinkled.

re'gia, (madeira nut, M. ♄.) leafets about 9, oval, glabrous, sub-serrate, numerous, sub-equal; fruit globose. Ex. Var. *frax-inifo'lia*, has 9 or 10 leafets, oblong, serrate, smooth, lateral lower one adnate on the common petiole.

JUN"CUS. 6—1. (*Junci.*)

effu'sus, (2̸.) scape minute-striate (soft); panicle loose, very branching, spreading: leafets of the calyx lanceolate, acuminate, rather longer than the obovate, obtuse capsule. 2-3 f.

ten"uis, (Ju. 2̸.) stem erect, filiform, somewhat dichotomous at the summit, nearly terete, leaves setaceous, channeled; flowers solitary, approximate, sub-sessile; calyx longer than the obtuse capsule. 1 f.

nodo'sus, (Ju. 2̸.) stem somewhat leafy; leaves nodose-articulate; heads about 2, globose; one of them lateral and peduncled, the other sessile; sepals mucronate, shorter than the acuminate capsule; leaves few, and very slender. 8-10 i.

bufo'nius, (toad-rush, Ju. ☉.) stem dichotomous above, panicled; leaves filiform, setaceous, channeled; flowers sub-solitary, sessile, 1-sided; sepals very acuminate, much longer than the oval-ovate capsule. 3-6 i.

acumina'tus, (Au. 2̸.) stem leafy, erect; leaves terete, nodose-articulate; panicle terminal, compound; heads 3-6 flowered, peduncled, and sessile; sepals lance-linear, somewhat awned, shorter than the acute capsule. 18 i.

polyceph"alus, (Au. 2̸.) stem leafy, erect; leaves compressed, nodose, articulate; panicle decompound; heads globose, many-flowered; flowers triandrous; sepals somewhat awned, rather shorter than the triangular, acute capsule. 18.24 i.

re'pens, (J. 2̸.) creeping; stem geniculate, branching; leaves linear, flat; fascicles lateral and terminal; flowers triandrous; leaves of the calyx subulate, carinate, very acute. 6-10 i. S.

biflo'rus, (Ju. 2̸.) stem terete; leaves linear, flat; panicle decompound, long; fascicles 2-flowered. 3 f. S.

JUNGERMAN"NIA. 21—3. (*Hepaticæ.*) [From John Gotlob Juncker, a learned German of the last century.]

complana'ta, stem branched, creeping; leaves roundish, very entire; ears sub-ovate, flattish. On smooth bark; very rarely on rocks.

palma'ta, frond short, somewhat ascending, digitate-palmate, nerveless. Dark green. Rotten-wood in wet places. Most of the jungermannia are in fruit late in the spring; some, however, in the winter.

JUNIPE'RUS. 20—15. (*Coniferæ.*) [From *juvenis*, young, *pario*, to bring forth, because it produces its young berries while the old are ripening.]

commu'nis, (juniper-tree,) leaves in threes, spreading, mucronate, longer than the berry.

virginia'na, (red cedar,) leaves adnate at the base, in threes. Small tree. Berries covered with a blue powder.

sabi'na, (savin,) leaves opposite, obtuse, glandular in the middle. Small shrub.

JUSSI'ÆU. 10—1. (*Onagræ.*) [In honor of the elder Jussieu.]

grandiflo'ra, (y. Ju. 2̸.) creeping, stem erect and ascending; leaves lanceolate, entire; peduncles and calyx villose.

erec"ta, (y. Au. 2̸.) erect, glabrous; leaves lanceolate; flowers octandrous, sessile. S.

subacau'lis, (y. J. 2̸.) creeping, glabrous; leaves lance-linear, repand toothed; flowers solitary, octandrous, peduncled; alternate filaments shortest; petals obovate. S.

leptocar"pa, (y. ☉.) erect; stem and calyx somewhat hirsute; leaves lanceolate, attenuate at each end; flowers sessile; capsule slender, cylindric. S.

JUSTI'CIA. 2—1. (*Acanthi.*) [In honor of Justice, author of the "British Gardener's Director."]

pedunculo'sa, spikes axillary; flowers crowded, leaves lanceolate; peduncles elongated, alternate. Water willow.

adhato'da, (malabar nut, p. ♄.) leaves lance-ovate; helmet of the corolla concave. Ex.

hu'milis, (W. Ju. 2̸.) spikes axillary and terminal, few-flowered; flowers distant; bracts linear; leaves oblong-lanceolate. S.

brachia'ta, (r. Ju. 2̸.) peduncles axillary, in whorled racemes; pedicels in pairs; bracts obovate, 3-flowered; leaves ovate, acute; petioles very long; stem six-angled, very branching. S.

KAL"MIA. 10—1. (*Rhododendra.*) [In honor of its discoverer, Kalm.]

latifo'lia, (laurel, w. and r. Ju. ♄.) leaves long-petioled, scattered, and in threes, oval, smooth both sides; corymbs terminal, with viscid hairs. 3-20 f.

angustifo'lia, (sheep-laurel, J. ♄.) leaves in threes, petioled, oblong, obtuse, sometimes rusty beneath; corymbs lateral; bracts linear; peduncles and calyx with glandular hairs. Var. *ova'ta*, taller; leaves broader, sub-ovate. 2-3 f.

glau'ca, (swamp-laurel,) branches ancipital; leaves glaucous beneath.

cunea'ta, (w. r. ♄.) leaves scattered, sessile, wedge-shaped, oblong, pubescent underneath, at the summit slightly awned; corymbs lateral, few-flowered. S.

hirsu'ta, (r. ♄.) branches, leaves, and calyx hairy; leaves opposite and alternate, nearly sessile, lanceolate; peduncles axillary, solitary, 1-flowered, longer than the leaves. S.

KO'CHIA. 5—2. (*Atriplices.*)

denta'ta, (J. ☉.) leaves lanceolate, sinuate, toothed; stem erect, very branching. Resembling Chenopodium.

KOELE'RIA. 3—2. (*Gramineæ.*) [In honor of M. Kohler, a German botanist.]

pennsylva'nica, (2̸.) panicle long, very slender, rather loose; spikelets shining, the terminal ones 3-flowered. 2 f.

trunca'ta, (J. 2̸.) leaves flat, smooth; panicle oblong, racemose; glumes 2-flowered, with a third abortive floret, unequal; lower glume a little scabrous, obtuse. paleas smooth. 2-3 f.

panicula'ta, panicle oblong, smooth; spikelets 2-3 flowered, shining; glumes awnless, very unequal, largest one obtusely truncate. *S.*

crista'ta, (Ju. ♃.) spike somewhat lax; spikelets diverging, 3-4 flowered, somewhat awned and wrinkled, sub-ciliate on the keel. *S.*

KRAME'RIA. 4—1. (*Leguminosa.*) [In honor of two German botanists, Kramer, father and son.]

lanceola'ta, (y. ♄.) hoary-pubescent; leaves lanceolate, acute, villose; pedicels axillary, about twice as long as the leaves.

KRI'GIA. 17—1. (*Cichoracea.*)

virgini'ca, flowers small. orange-yellow; primary leaves roundish, entire, the rest lyrate, nearly smooth; scape 1-flowered. Dwarf dandelion.

amplexicau'lis, (y. Ju. ♃.) glaucous; leaves of the root spatulate, lanceolate, and oval, toothed; scapes somewhat leafy and branched.

carolin"ia'na. leaves runcinate, nearly glabrous; scapes very long, and with the base of the involucrum glandularly hairy. 6-12 i. *S.*

dandeli'on, (y. ♃.) glabrous, slightly glaucous; leaves linear-lanceolate, entire, scape 1-flowered. 8-18 i. *S.*

, (y. ♃.) very glabrous, stemless; ab-lyrate, oblong, acute; scape d, naked; divisions of the involucr-acute. Arkansas.

. 17—1. (*Corymbifera*) [Adam

rioi'des, (W. Au. ♃.) smooth; etioled, broad-lanceolate, serrate; terminal, few-flowered, crowded. hady woods. False boneset.

ia, pubescent; leaves narrower, and glandular beneath; flowers low. Mountains.

o'sa, (w. S.) pubescent, glutinous; nceolate, tapering toward the sumhed and toothed, crowded; flowers panicled. 2 f.

I"GA. 3—1. (*Cyperoidea.*) [From C, a Danish botanist.]

ceph"ala, (false bog-rush, Oc. ♃.) form, 3-angled; head globose, scsrolucrum 3 leaved, very long. 3-12 i. ila, (dwarf kyllinga, J. ♃.) head , sessile, solitary; involucre short, staceous; flowers diandrous. 3-6 i. *S.* 'ulata, (three-headed kyllinga, Ju. umes 3, unequal, imbricate; heads generally 3, ovate, sub-acute, sessile. 3-5 f. *S.*

LACTU'CA. 17—1. (*Cichoracea*) [From *lac*, milk, on account of the juice from the stalk.]

elonga'ta, (wild lettuce, y. Ju. ♂. or ♄.) leaves smooth, lower ones runcinate, amplexicaul, upper ones lanceolate, sessile; flowers panicled. 4-6 f.

sati'va, (lettuce, y. Ju. ⊕.) leaves roundish, cauline ones cordate; stem corymbed. Var. *roma'na,* has oblong, straight leaves, narrowed at the base. Var.*cris"pa,*has sinuate-crenate leaves, toothed, undulated, crisped, radical ones hairy on the keel. Var. *lacinia'ta,* has the lower leaves pinnatifid, and the upper ones runcinate. Ex.

hirsu'ta, (y. p. ♃.) lower part of the stem and leaves hairy; radical leaves lyrate; segments truncate, sub-dentate, upper ones partly runcinate, pinnatifid; flowers in racemes.

integrifo'lia, (y. Ju. ♂.) leaves sagittate, entire, unarmed, and clasping; flowers panicled. 3-4 f.

sanguin"ea, (wood-lettuce, r. Au. ♂.) leaves amplexicaul, runcinate, glaucous beneath, with the midrib filamentous; flowers panicled. 2-3 f.

gramin:fo'lia, (p. Ju.) leaves unarmed, generally undivided, simple at the base, long-linear; panicle leafless, loose, branched, few-flowered; stem erect, simple, flowers all peduncled. 3 f. *S.*

sagittifo'lia, (y. r. Au.) stem erect, glabrous; leaves lance-oblong, acute, entire, glabrous, pale beneath, close-sessile, sagittate at the base; flowers panicled. *S.*

LA'MIUM. 13—1. (*Laviata.*) [From *Lamium,* a mountain of Ionia, where it grew.]

amplexicau'le, (dead-nettle, r. Nov. ⊙.) floral leaves broadly cordate, sessile, amplexicaul, crenate, radical leaves petioled. 6-10 i.

purpu'reum, (p. ⊙.) leaves cordate, crenate-serrate, petiolate, upper ones crowded; stem nakedish downwards. 4-8 i.

garga'nicum, (dead-nettle, ♃.) leaves cordate, concave; throat of the corolla inflated; tube short. Ex.

hispidu'lum, (w.) leaves long-petioled, broad-cordate, pubescent; axils 1-flowered; stem hispid. *S.*

LANTA'NA. 13—2. (*Pedicularea.*)

cama'ra, (y. Au. ♄.) leaves opposite, lance-ovate, crenate and serrate, scabrous; stem rough, not prickly; flowers in umbellate heads, leafless. 2-4 f. *S.* Hot-house plants, nearly allied to the verbenas.

LATHY'RUS. 16—10. (*Leguminosa.*) [From *lathyros,* leguminous.]

odora'tus, (sweet pea, J. ⊙.) peduncles 2-flowered; tendril with ovate oblong leafets; legumes hirsute. Ex.

latifo'lius, (everlasting-pea, Au. ♃) peduncles many-flowered; tendril with 2 lance-ovate leaves; membranaceous between joints. Ex.

palus"tris, (w-p. Ju. ♃.) stem smooth, winged, weak; leafets in 3 pairs, oblong, mucronate; stipules acute, semi-sagittate; peduncles 3-5 flowered, a little longer than the leaves; legume compressed. Low grounds.

myrtifo'lius, flowers smaller than the preceding, purple and rose-colored; leafets 4, reticulate, scabrous on the margin; peduncles longer than the leaves, 3-4-flowered. Salt marshes.

veno'sus, numerous leafets, veiny; peduncles shorter than the leaves, 4-5-flowered.

mariti'mus, (beach pea, p. Ju. ♃.) stem compressed, 4-angled; stipules sagittate; leafets numerous, sub-alternate, obovate; peduncles shorter than the leaves, about 7-flowered.

sati'vus, (chick vetch, ⊙.) peduncles 1-

flowered, tendrils with 2 and 4 leafets; legumes ovate, compressed, with two narrow wings on the back. Ex.

grandiflo'rus, (2[.) remarkable for the large size of its flowers. Ex.

tingita'nus, (tangier pea,) a tall plant, the flowers of which are dark purple. Ex.

magellan"icus, (2[.) the foliage very beautiful with blue flowers. Ex.

pusil'lus, (p. M.) cirrhi 2-leaved, simple; leaves linear lanceolate, peduncles 1-flowered, long; stipules falcate; stem angled and winged. S.

decaphyl"lus, (p. 2[.) leaves in 5 pairs; leafets oval-oblong, mucronate; peduncles 3-4-flowered; stem 4-angled. S.

LAU'RUS. 9—1. (*Laur.*) [From *laus*, praise, because it was used to crown the heads of distinguished persons.]

ben"zoin, (spice bush, fever bush, g. y. Ap. ♄.) leaves wedge-obovate, whitish, subpubescent beneath; flowers in clustered umbels; buds and pedicels glabrous. 4-10 f.

sas"safras, (sassafras-tree, y. M. ♄.) leaves entire and lobed on the same plant; flowers mostly diœcious. 10-25 f.

carolin"ensis. leaves perennial, oval, lanceolate, coriaceous, glaucous beneath, peduncles simple, terminated with a few-flowered fascicle; outer segments of the calyx half as long as the inner. A large shrub. Flowers polygamous, in small clusters, pale yellow; drupe dark blue. From Georgia to Delaware.

perse'a, alligator pear of the West Indies, an eatable fruit.

cinnamo'num, the inner bark affords the cinnamon of commerce. Indies.

no'bilis, leaves veined, lanceolate and perennial; flowers 4-cleft. This is the poet's laurel, the fabled favorite of Apollo. It is a handsome evergreen shrub; berries and leaves fragrant. Native of Italy.

camphora'tus, (camphor-tree, ♄.) leaves about 3-nerved, lance-ovate; panicle spreading. From Japan.

æstiva'lis, leaves veined, oblong, acuminate, rugose underneath; branches axillary above. S.

catesbya'na, (w. M. ♄.) panicles on short peduncles; segments of the corolla oblong, obtuse, nearly equal, deciduous, leaves perennial, broad-lanceolate. 6-9 f. S.

genicula'ta, (y. Ap. ♄.) stem dichotomous, flexuous flowers in umbels; leaves small, oval, smooth. 10-15 f. S.

melissæfo'lia, (y. Ap. 2[.) root creeping, leaves cordate-lanceolate, strongly veined, pubescent beneath; flowers in clustered umbels; buds and pedicels villose. 2-3 f. S.

LAVANDU'LA. 13—I. (*Labiatæ.*) [From *lavo.* to wash, so called, because, on account of its perfume, it was used in baths.]

spica'ta, (lavender, Au. 2[.) leaves sessile, lance-linear, with revolute margins; spike interruptedly naked. Ex.

LAVATE'RA. 15—12. (*Malvaceæ.*) [In honor of Lavater, a celebrated writer on physiognomy.]

trimen"sis, (red lavatera,) lower leaves

angled; upper ones 3-lobed, with the middle lobe longest; peduncles solitary. 2 f. Introduced.

arbo'rea, (tree-mallows, 8. ♂.) stem woody; leaves downy, plaited, 7-angled; flowers large, purplish, rose-color, darker on the base, on aggregated, axillary stalks. Ex.

thurin"gia'ca, (gay mallows, 2[.) peduncles solitary; lower leaves angled upper ones 3-lobed, the middle lobe longest.

LECHE'A. 3—3.

ma'jor, (pin-weed, g-p. Ju. 2[.) erect, hirsute; leaves lance-oblong, mucronate; panicle leafy; branches bearing flowers at their tops; flowers in fascicled racemes, one-way, on short pedicels. Dry woods and hills. 1-2 f.

mi'nor, stem assurgent, smoothish, branched; leaves linear-lanceolate, acute; panicle leafy; branches elongated; flowers racemose. 8 i.

thymifo'lia, (Ju. 2[.) whole plant whitish villose; stem erect; pedicels very short, leaves linear, acute; panicle leafy, elongated; branches very short; flowers minute, in lateral and terminal fascicles. 1 f.

LE'DUM. 10—1. (*Ericeæ.*) [From the *ledon* of the ancient Greeks, supposed to have been a species of Cistus.]

latifo'lium, (Labrador tea, w. r. J. 2[.) leaves oblong, replicate at the margin, ferruginous, tomentose beneath; stamens 5, as long as the corolla. Evergreen shrub, irregularly branched, woolly; flowers in long, terminal corymbs.

palus'tre, leaves linear, revolute on the margin; stamens 10, longer than the corolla. A shrub smaller than the preceding, with narrower leaves.

buxifo'lia, a small compact-growing plant with box-like leaves; clusters of white flowers, petals tinged with pink.

LEER"SIA. 3—2. (*Gramineæ.*) [In honor of Leers, who wrote on botany in 1775.]

virgin"ica, (white grass, Ju. y. 2[.) panicle simple; the lower branches diffuse; flowers appressed, monandrous, sparingly ciliate on the keel. 2-4 f.

oryzoi'des, (cut grass, Au. 2[.) panicle diffuse, sheathed at the base; flowers triandrous, spreading; keel of the glumes conspicuously ciliate. 3-5 f.

len"ticula'ris, (catch-fly grass, Ju. 2[.) panicle erect; flowers large, nearly orbicular, diandrous, imbricate; keel and nerves ciliate. 2-4 f.

LEIOPHYL"LUM. 10—1. (*Ericeæ*) [From *leios.* smooth, and *phullon,* leaf.]

buxifo'lium, (sand myrtle, w. ♄.) leaves small, lance-oval, entire, glabrous, lucid, revolute at the margin; corymbs terminal. 6-18 i.

LEM"NA. 19—2. (*Naides.*) [From *lemo,* deprived of bark.]

trisul"ca, (duck's meat, ⊕. f.) fronds thin, elliptic-lanceolate, caudate at one extremity, at the other serrate; root a single fibre. Young fronds produced from lateral clefts, of the same shape as the parent plant, and again proliferous before they are detached. Flowers very minute. Water.

polyrrhi'za, (water flax-seed, Ju. ☉.) fronds obovate rotundulate, compressed; roots numerous, fascicled. Stagnant waters.

LEON"TICE. 6—1. (*Berberides.*)

thalictroi'des, (poppoose-root, false cohosh, p-y. Ap. ♃.) leaves bi-triternate; leafets 2-3-lobed; flowers paniculate, from the centre of the leaves. 1 f.

LEON"TODON. 17—1. (*Cichoracea.*) [From *leon,* a lion, *odons,* tooth, from the shape of its leaves.]

tarax"acum, (dandelion, y. Ap. ♃.) outer calyx reflexed; scape 1-flowered; leaves runcinate, with toothed divisions. Introduced.

palus"tre, (marsh-dandelion, ♃.) leaves sinuate-toothed, somewhat glabrous; outer involucre scales short, erect, ovate.

LEONU'RUS. 13—1. (*Labiatæ.*) [From *leon,* a lion, and *oura,* tail.]

cardia'ca, (motherwort, w-r. Ju. ♃.) leaves 3-lobed, toothed, bases wedge-form; calyx prickly, less than the corolla. Naturalized. 2-4 f.

marrubias"trum, (r. Au.) leaves lanceolate, toothed; calyx somewhat prickly, as long as the corolla. Naturalized. 2-4 f.

LEPID'IUM. 14—1. (*Cruciferæ.*) [From *lepis,* a scale, from its supposed virtue in cleansing the skin.]

virgini'cum, (wild pepper-grass, w. J. ♃.) radical leaves pinnatifid; cauline leaves lance-linear; flowers with 4 petals; stamens 2-4; pouch orbicular, flat, emarginate, shorter than the pedicel. Sandy fields.

campes"tre, (field pepper-grass,) cauline leaves sagittate. Hills.

sati'vum, (pepper-grass, w. Ju. ☉.) leaves oblong, many-cleft.

rudera'le, flowers diandrous, apetalous; radical and cauline leaves pinnatifid or incised; branch leaves linear, very entire; silicles broad-oval, emarginate, spreading; cotyledons incumbent.

menzie'sia, flowers diandrous, apetalous; radical leaves bi-pinnatifid; cauline and branch-leaves many, pinnatifid, upper ones linear, very entire. One variety with its radical leaves hispid; another, pubescent.

monta'num, (♃.) nearly glabrous, decumbent; silicles elliptical, slightly emarginate, wingless; style conspicuous; leaves pinnatifid, and bi-pinnatifid; segments oblong; upper leaves trifid or entire. Oregon.

califor"nicum, (☉.) stem somewhat hirsutely pubescent, much branched; silicles nearly orbicular, emarginate, wingless; flowers diandrous (petals 4); leaves nearly glabrous, laciniately pinnatifid. California.

LEPTAN"DRA. 2—1. (*Scrophularia.*)

virgin"ica, (w. Ju. Aug. ♃.) leaves verticillate, in fours or fives, lanceolate-serrate, petioled. 3-4 f. Culver'sphysic.

LEP"TOPO'DA. 17—3. (*Corymbiferæ.*)

fimbria'ta, (y. Ap. ♃.) stem viscid pubescent, striate; leaves alternate, lance-linear, half clasping, glabrous, punctate; cauline

ones gash-toothed, chaff of the egret torn. 2 f. Florida.

decur"rens, (y. Ap. ♃.) stem very glabrous; leaves linear-lanceolate, toothed, glabrous, decurrent; chaff of the pappus fimbriate. 12-18 i. S.

LEPU'ROPE'TALON. 5—3. (*Saxifraga.*)

spat"ula'tum, (w. Ap. ☉.) glabrous; stem erect and procumbent, somewhat succulent, sub-angled; leaves alternate, sessile, lance-spatulate, obtuse, entire. 1.2 i.

LESPEDE'ZA. 16—10. (*Leguminosæ.*) [In honor of Lespedes.]

polysta'chia, (bush-clover, w. r. Aug. ♃.) stem erect, branched, very villose; leaves on very short petioles; leafets round-oval, obtuse; spikes oblong, axillary, pedunculate, twice as long as the leaves; corolla and legume as long as the calyx: flowers in dense racemes, on peduncles longer than the leaves. 2-4 f.

viola'cea, longer leaves and petioles than the preceding, is more branching, and has violet-colored flowers.

procum"bens, slender and procumbent, pubescent; racemes sub umbellate; flowers in pairs, purple with yellow spots. 2-3 f.

capita'ta, leaves on very short petioles; spikes capitate, on short peduncles, conglobate, terminal; calyx villose, as long as the corolla, legume much longer. Borders of woods. Aug. 2-3 f. Flowers purple.

angus"tifo'lia, (w-p. S. ♃.) leafets linear-lanceolate, hoary, pubescent; racemes capitate, longer than the leaves; corolla longer than the calyx; stem erect. 3-5 f.

stu'vei, (p. S.) stem erect, simple, silky-villose; leafets oval; racemes pedunculate, scarcely longer than the leaves, loose; lower loments naked, pubescent. 2-3 f.

sessiliflo'ra, (bush-clover, p. Ju. ♃.) stem erect, branching; leafets oblong; clusters of flowers numerous, sessile; pods acute, scarcely covered by the minute calyx. 1-3 f.

diver"gens, (p. Ju. ♃.) diffuse, branched; leafets oblong, obtuse, close-pressed, hairy underneath; racemes not as long as the petioles; flowers in pairs; legumes ovate-reticulate, smooth. Probably a variety of the *viola'cea.* 1-2 f.

re'pens, (p. Ju. ♃.) leaves ternate; leafets roundish elliptical; emarginate; racemes axillary; legume repand.

prostra'ta, (p. Au. ♃.) smooth, prostrate; leaves short petioled; leafets obovate-elliptic, obtuse; racemes axillary and terminal, sub-paniculate; peduncles very long; legumes oval, sub-pubescent.

longifo'lia, (♃.) erect, angled, pubescent; petioles short; leafets oblong, glabrous above, silk-silvery and close-pressed underneath; racemes fascicled-corymbed, many-flowered, axillary and sub-terminal; legume shorter than the acuminate lobes of the calyx. Louisiana.

frutes"cens, (♃.) stem erect; leafets elliptical, obtuse, silky-pubescent; flowers in sub-capitate fascicles, shorter than the leaves, conglomerate towards the summit of the stem; loments hairy, shorter than the villous calyx. 2-3 f. *S.*

LEU'CAS. 13—1. (*Labiatæ.*) [From *leukos*, white.]

martinien''sis, leaves entire; whorls many-flowered, capitate. Native of India.

LEWIS'IA. '12—1. (*Portulaceæ.*) [In honor of Gen. Lewis, the leader of the first expedition to the Rocky Mountains.]

redivi'va, (w. Ju. 2/.) leaves radical, linear, somewhat fleshy, obtusish; scape 1-2-flowered; pedicel geniculate at the base; capsule oblong. *S.*

LIA'TRIS. 17—1. (*Corymbiferæ.*)

spica'ta, (gay feather, Aug. 2/.) leaves linear, entire, smooth, cordate at the base, nerved and punctate; flowers in spikes; scales of the calyx linear-oblong, obtuse. Meadows. Flowers purple. 3-6 f.

pilo'sa, stem simple, pubescent; leaves long, linear, hairy, ciliate; flowers in loose racemes, bright purple, small.

ele'gans, (p. r. Oct. 2/.) stem simple, villose; leaves lance-linear, sub-scabrous beneath; raceme cylindrical; flowers crowded; inner scales of the calyx colored. *S.*

scario'sa, (blue blazing-star,) leaves tapering to both ends; calyx squarrose below, racemed; scales spatulate, with colored membranaceous margins. 3 f.

squarro'sa, (r. S. 2/.) stem simple, pubescent; leaves linear, very long; raceme few-flowered, leafy; calyx large; scales leafy, lanceolate, mucronate, rigid and spreading; segments of the florets linear, villous internally. 2-3 f.

cylin''drica, (p. Au. 2/.) slender, hirsute; leaves grass-like; spike few-flowered; involucre sub-sessile, cylindric, few-flowered; scales round at the summit, abruptly mucronate. 1-2 f.

pycnos''ta'chya, (S. 2/.) stem simple, hairy; leaves straight, narrow-linear, pubescent; spike long; flowers clustered, sessile; involucrum appressed, squarrose at the summit. *S.*

as''pera, (S. 2/.) stem somewhat branching, scabrous-pubescent; leaves linear-lanceolate, very rough; heads short, spiked, distinctly alternate, solitary, sessile; scales of the involucrum roundish, obtuse-connivent. *S.*

graminifo'lia, (p. S. 2/.) stem simple, glabrous; leaves linear, very long, glabrous, nerved, margin somewhat scabrous, midrib hairy above; flowers in spikes, rather distant, nearly sessile; scales of the involucrum oblong, obtuse, mucronate, ciliate, appressed, the interior colored. 2-4 f. *S.*

hetrophyl''la, (S. 2/.) stem simple, glabrous; leaves lanceolate, glabrous, smooth; upper ones lance-linear, much smaller; heads spiked, short peduncled, sub-squarrose; scales of the involucrum lanceolate, acute, naked. *S.*

tenuifo'lia, (p. S.) stem slender, glabrous; lower leaves crowded, linear, a little hairy at base, upper ones setaceous; racemes very long; pedicels leafy; scales of the involucrum oblong, mucronate. 2-4 f. *S.*

resino'sa, (p.) glabrous; leaves linear, crowded; heads spiked, oblong, 4-5-flowered; scales of the involucrum obtuse, appressed, resinous, finally hoary.

secun''da, (p. S. 2/.) stem reclining, pubescent; leaves linear, glabrous, sparingly fringed at the base; racemes secund; scales of the involucrum lanceolate, acute, appressed. 2-3 f. *S.*

sphæroi'dea, (p. S. 2/.) leaves smooth, lower ones broad, lanceolate, upper ones narrow; flowers racemed, large, solitary, alternate; involucrum nearly globular; the scales oval, erect. 2-4 f. *S.*

Flowers in corymbs; roots fibrous.

pauciflo'ra, (2/.) stem simple, glabrous; leaves linear; panicle virgate, leafy, with the branches short; few-flowered; involucrum sessile, secund, 3-5-flowered; scales erect, lanceolate, acute, glabrous.

panicula'ta, (p. S. 2/.) stem simple, hairy viscid; leaves lanceolate, nerved, nearly glabrous; panicle contracted; involucrum generally 5-flowered; scales lanceolate. 1-2 f.

odoratis''sima, (p. S. 2/.) very glabrous; stem simple; leaves ovate and lanceolate, nerved, toothed, slightly glaucous; panicle corymbose; involucrum 7-8-flowered; scales obovate, obtuse. 3-4 f.

tomento'sa, (p. S. 2/.) stem simple, with the cuneate-lanceolate leaves hairy; corymb few-flowered, depressed-divaricate; involucrum tomentose; scales ovate, acute. 2 f.

wal''teri, (p. S. 2/.) leaves lanceolate, acute, glabrous, dotted, attenuate at base; stem simple, hairy near the summit; involucrum many-flowered; scales acute; tomentose. 2 f.

frutico'sa, (p. ♄.) glabrous; stem fruticose; branches corymbed; leaves wedge-obovate, punctate; involucrum sub-5-flowered; divisions acuminate. Florida.

squamo'sa, (2/.) pulverulent-canescent, corymbose; little corymbs 3-5-flowered; radical leaves linear, long; cauline ones appressed, very short; involucrum sub-hemispheric; scales acute. *S.*

LIGUS''TICUM. 5—2. (*Umbelliferæ.*) [From *Liguria* in Italy, its native country.]

sco'ticum, (Scottish loveage, w. Ju. 2/.) lower leaves bi-ternate, upper ones ternate; leafets broad, smooth, serrate, entire at the base, dark green; flowers white with a reddish tinge; stem erect, smooth, striate, 12 inches high; umbels many-rayed; petals inflexed. The root is acrid, and is used by the people of the Hebrides as a substitute for tobacco. Very abundant on the sea-coast in Scotland; found in salt marshes in this country.

levisti'cum, (smellage,) leaves many, upper ones toothed. Medicinal. Ex.

LIGUS''TRUM. 2—1. (*Jasmineæ.*)

vulga're, (prim, w. J. ♄.) leaves lanceolate, acutish; panicle compact. Introduced. Sometimes called privet; very common in England.

lu'cidum, and *spica'tum*, sub-evergreen shrubs or low trees, natives of China.

LIL"IUM. 6—1. (*Liliaceæ*.) [From *leios*, graceful, on account of its beauty.]

philadel"phicum, (red lily, r. y. J. ♃.) leaves whorled, lance-linear; 3-nerved, nerves hairy beneath; corolla erect, bell-form, spreading; petals lanceolate, having claws.

canaden"se, (nodding lily, y. r. Ju. ♃.) leaves remotely whorled, lanceolate; peduncles terminal, elongated, mostly in threes; corolla nodding; petals spreading. 2-3 f.

super"bum, (superb lily, y. p. Ju. ♃.) leaves lance-linear, 3-nerved, glabrous; lower ones whorled; upper ones scattered; flowers in a pyramid raceme; petals revolute. 3-6 f. Wet meadows.

cates"bæi, (Southern lily,) leaves scattered, lance-linear, very acute; stem 1-flowered; corolla erect; segments with long claws, undulate on the margin, reflexed at the summit; flowers scarlet, spotted with yellow and brown. Stem 18 i.

pennsylva'nicum, leaves scattered, lance-linear, the upper ones whorled; stem about 1-flowered; peduncles woolly; corolla erect, woolly without; flowers red and yellow.

mar"tagon, (Turk's cap,) leaves narrow, peduncles terminal; petals reflexed so as to give the corolla the appearance of a turban; flowers scarlet, with varieties; stem 2-3 feet high. Ex.

tigri'num, (tiger lily,) leaves scattered; petals reflexed; flowers in whorls; dark orange, spotted with black; stem bulbiferous. A very showy plant, of easy culture. 4-5 f. Ex.

japon"icum, (Japan lily,) corolla elongated into a tube; flowers very large, pure white, with a streak of blue; stem 4-5 feet high, generally with 2 flowers. Ex.

pu'dicum, stem 1-flowered; corolla bell-form, nodding; petals erect, sessile, spatulate-obovate, flat within; yellow. S.

umbella'tum, flowers 1 to 5, terminal, erect; petals unguiculate, spreading, red. S.

can"didum, (white lily, w. J. ♃.) leaves lanceolate, scattered, tapering to the base; corolla bell-form, glabrous within. Ex.

bulbif"erum, (orange lily, y. J. ♃.) leaves scattered, 3-nerved; corolla campanulate, erect, scabrous within. Ex.

carolinia'num, (Au. ♃.) leaves verticillate and scattered, lanceolate, cuneate at base; flowers few, (1-3,) terminal; peduncles thick; corolla revolute, orange-colored, spotted with dark purple. Perhaps a variety of the superbum. 2 f. S.

pompo'nium, (J.) a splendid species with scarlet flowers.

LIMNAN"THES. 10—1. (*Geraniæ*.) [From *limnus*, a water nymph, *anthos*, flower.]

douglass"ii, (y. w.) leaves bipinnatifid; the divisions often alternate. Plant slightly succulent. California.

LIMNET"IS. 3—1. (*Gramineæ*.) [From *limnes*, a pool or marsh, alluding to its place of growth.]

cynosuroi'des, (many-spiked salt-grass, Au. ♃.) spikes numerous (10-40), peduncled, panicled, spreading; leaves broad, flat, at length convolute; a short awn on one of the glumes; styles 2-cleft at the summit. 4-9 f.

junce'a, (rush salt-grass, Ju. ♃.) leaves 2-ranked, convolute, spreading; spikes few (1-3) peduncled; peduncles smooth; paleas obtusish; styles 2. 18 i.

gla'bra, (Au. ♃.) leaves concave, erect; spikes alternate, sessile, erect, appressed; paleas smoothish on the keel; style deep-cleft. 3-5 f.

LIMO'NIA. 10—1. (*Aurantia*.)

acidis"sima, leaves pinnate; leafets roundish-oval, crenate; spines germinate. Florida.

LIMOSEL"LA. 13—2. (*Scrophularia*.) [From *limus*, slime or mud.]

subula'ta, (mudwort, Aug. ♃.) leaves linear, very narrow, scarcely dilated at the apex; scape 1-flowered, as long as the leaves. Muddy shores. Stem an inch high; flowers very small, bluish white.

LINDER"NIA. 2—1. (*Scrophularia*.) [In honor of Von Lindern.]

attenu'ata, (false hedge hyssop, w-p. Ju. ☉.) leaves lanceolate and obovate, narrowed at the base; peduncle shorter than the leaves, erect.

dilata'ta, leaves dilated at the base, clasping; peduncles longer than the leaves; flowers pale purple. Inundated banks. Stem 4-sided, 6 inches high, smooth.

montico'la, (June. ♃.) stem slender, dichotomous; radical leaves spatulate, punctate; cauline ones linear, small, remote; peduncles very long; flowers pale blue; stem erect. 4-6 inches high.

grandiflo'ra, (♃.) leaves roundish, entire, nerveless, half-clasping; peduncles very long, axillary and terminal; stem creeping. S.

refrac'ta, (b. J. ♃.) radical leaves spatulate-oval; upper ones subulate; flowers solitary, axillary and terminal; peduncles refracted, after flowering; stem slender, erect, branching, glabrous. 8-12 i. S.

LINNÆ'A. 4—1. (*Caprifolia*.) [In honor of Charles Von Linnæus.]

borea'lis, (twin-flower, w. r. J. ♃.) stem prostrate; branches erect, each bearing 2 flowers; leaves roundish, crenate. Woods and hills. Evergreen, creeping. Has been found at Green Island, Troy, N. Y.

LI'NUM. 5—5. (*Caryophylleæ*.) [From *leios*, smooth or soft, on account of its texture.]

usitatis"simum, (common flax, b. Ju. ☉.) leafets of the calyx ovate, acute, 3-nerved; petals crenate; leaves lanceolate, alternate; stem sub-solitary. Ex.

virgin"icum, (Virginia flax, y. ☉.) stem erect, slender, smooth; radical leaves oval and spatulate; cauline leaves long and narrow; panicle lax, corymbose.

ri'gidum, (y.) divisions of the calyx ovate, acuminate, 3-nerved, ciliate; petals oblong, narrow; leaves stiffly erect, linear, short. 6 i. Missouri.

selaginoi'des, (w-r.) glabrous; stems a span high, suffruticose, corymbosely attached at the summit, leaves crowded, alternate, very small, linear and very narrow, mucronate, piliferous; flowers terminal, sub-sessile, petals shorter than the calyx; ovary 10-celled. Texas.

LIPA'RIS. 18—1. (*Orchidea.*) [From *lipos*, fat, so called on account of its unctuous property.]
liliifo'lia, (y-w. Ju. 2/.) leaves 2, ovate-oblong; scape angular; flowers racemose; segments of the perianth linear; lower ones setaceous, reflexed; lip concave, obovate, mucronate. 6-8 i. Wet woods.

LIQUIDAM"BER. 19—12. (*Amentaceæ.*) [From *liquidum*, fluid, and *amber*, fragrant, alluding to the gum which distils from this tree.]
styraciflu'a, (sweet gum-tree, M. ♄.) leaves palmately-lobed; lobes acuminate, serrate, with sinuses at the base of veins, villose. A resinous juice called liquid *amber*, is obtained by wounding the bark of this tree. By boiling the leaves, a different gummy substance, called liquid *storax*, is obtained.

LIRIODEN"DRON. 12—13. (*Magnolia.*) [From *leiron*, a lily, and *dendron*, a tree.]
tulipif"era, (white wood, tulip-tree, y-r. J. ♄.) leaves truncate at the end, with 2 side-lobes. A beautiful flowering tree. 90.150 f.

LISIAN"THUS. 5—1. (*Gentianæ.*)
glaucifo'lius, (y.) stem herbaceous, terete; leaves oblong-ovate, sessile, glaucous, acute, 3-nerved; flowers terminal, corymbed; peduncles elongated.
russelia'nus, a ligneous plant with handsome purple flowers. Ex.

LISTE'RA. 18—1. (*Orchideæ.*) [Named from Martin Lister, physician to Queen Anne.]
corda'ta, stem with 2 opposite, roundish, cordate leaves; raceme loose; column without any appendage behind; lip elongate, 2-toothed at the base, deeply bifid, the segments divaricate and acute. Swamps. Stem 4-6 i. Flowers distant and minute.
convallarioi'des, (lily orchis,) column porrected; lip oblong, dilated, and obtusely 2-lobed at the extremity; stem 6 inches, very slender; root fibrous; flowers dark brown and green, larger than the preceding.
pubes"cens, (g-w. J. 2/.) leaves radical, ovate, acute; scape leafless, pubescent, loosely flowered; flowers on pedicels, lip 2-lobed, scarcely longer than the connivent petals; capsules clavate; root palmate. S.

LI'THOSPER"MUM. 5—1. (*Boragineæ.*) [From *lithos*, a stone, and *sperma*, seed, on account of the hardness of its seed.]
arven"se, (corn gromwell, w. M. ☉.) stem erect, branched; leaves sessile, lance-linear, rather acute, veinless, rough, hairy; calyx a little shorter than the corolla; segments spreading; nuts rugose; plant hispid, pilose; flowers solitary, axillary. Fields. Introduced.
officina'le, (common gromwell, y. M. 2/.) stem covered with rigid hairs; leaves broad-lanceolate, acute, rough on the upper surface, hairy on the lower; tube of the corolla as long as the calyx; nuts smooth. Fields. Flowers axillary, pale yellow.

mariti'mum, has blue flowers.
denticula'tum, has purple flowers.
pilo'sum, (y. 2/.) simple, pilose-hirsute; leaves linear, acuminate, sessile, approximate; flowers fascicled, sessile, smallish; divisions of the corolla oblong, entire.
torrey'i, (J. 2/.) strigose-hispid; leaves oblong-linear, obtusish, scattered; stem low, branching; fascicles terminal, few-flowered; lobes of the corolla oblong, entire. 9 i.
angustifo'lium, (w. Ju. ☉.) nut turgidly ovate, shining, with hollow punctures on every part; flowers mostly lateral; leaves linear, with close-pressed pubescence; stem procumbent. S.
apu'lum, (y. Ju. ☉.) nut muricate; spikes terminal, 1-sided; bracts lanceolate; leaves lance-linear, acute. S.

LOBE'LIA. 5—1. (*Campanulaceæ.*) [In honor of Mathias Lobellus.]
cardina'lis, (cardinal flower, r. Ju. 2/.) erect, simple, pubescent; leaves lance-ovate, acuminate, denticulate; racemes somewhat 1-sided, many-flowered; stamens longer than the corollas. Damp. 1-2 f.
infla'ta, (Indian tobacco, b. Ju. ☉.) erect, branching, very hirsute; leaves ovate, serrate; racemes leafy; capsules inflated. 12-18 i.
kal"mii, (b. Ju. ☉.) slender, erect, sub-simple; radical leaves spatulate; cauline ones linear, delicately toothed; flowers racemed, alternate, remote, pedicelled. 6-24 i.
dortman"na, (b. Ju. 2/.) leaves linear, 2-celled, fleshy, obtuse; scape nearly naked; flowers in a terminal raceme, remote, pedicelled, nodding; leaves growing in a tuft about the root, spreading, recurved. Water gladiole.
syphilit"ica, flowers on short pedicels, in a long, leafy raceme, large, blue. Bogs. 2-3 f.
claytonia'na, stem erect, simple, pubescent; cauline leaves oblong, obtuse, nearly entire; radical leaves spatulate; raceme virgate, naked; flowers pale blue. 1-2 f.
puberu'la, covered with silky down; lower leaves obovate, upper lanceolate; flowers spiked, alternate, sub-sessile, bright blue, smaller than the syphilitica.
ful'gens, (native of Mexico,) leaves very long, alternate, sub-entire; raceme many-flowered; stamens and pistils as long as the corolla.
aph"ylla, (2/.) very small; stem filiform, sub-simple, scaly; peduncles remote, elongated. 4-6 i. Florida.
ama'na, (b. Ju. 2/.) stem erect, pubescent; leaves broad-lanceolate, doubly toothed; spike secund; margin of the calyx erect. 2-4 f. S.
glandulo'sa, (b. S. 2/.) erect; leaves linear-lanceolate, rather thick, denticulate; flowers in racemes. S.
michauxii, (p. Ju. ☉.) glabrous, branching above; leaves petioled, ovate, crenate-dentate; lowest ones roundish; racemes lax; peduncles elongated. S.

LO'LIUM. 3—2. (*Gramineæ.*)
peren"ne, (M. 2/.) florets much longer

than the calyx, unarmed, linear-oblong, compressed. Introduced. 18 i.

temulen"tum, (Ju. ⊙.) florets shorter than the glumes, as long as the bristle at their extremity; culm scabrous above. 2 f.

LONICE'RA. 5—1. (*Caprifolie.*) [From Lonicer, a botanist of the 16th century.]

semper"virens, (r. y. M. ♄.) spikes with distant, nakedish whorls; corollas sub-equal; tube ventricose above; leaves ovate and obovate, glaucous beneath; upper ones connate-perfoliate; leaves perennial.

/ *caprifo'lium*, (honeysuckle, ♄.) corollas ringent-like, terminal; flowers crimson; sessile leaves connate-perfoliate at the top. Ex.

parviflo'ra, (r-y. J. ♄.) spikes verticillate, capitate; leaves deciduous, glaucous beneath, all connate-perfoliate; corolla ringent, gibbous at the base; filaments bearded.

periclyme'num, (woodbine, J. ♄.) flowers in ovate, imbricate, terminal heads; leaves all distinct. Var. *quercifo'lia*, leaves sinuate. Ex.

fla'va, (yellow honeysuckle, J. ♄.) spikes whorled, terminal; corolla ringent; flowers bright yellow.

hirsu'ta, (rough woodbine,) leaves pubescent and ciliate; flowers yellow pubescent; berries orange.

gra'ta, has scarlet flowers. Mountains.

cilio'sum, (J. ♄.) spikes with whorled heads, sub-sessile; corolla sub-equal; tube hirsute, ventricose in the middle; leaves somewhat clasping, sessile, and petioled, ovate, glaucous beneath, margin ciliate, upper ones connate-perfoliate; flowers yellow. *S.*

LOPHIO'LA. 6—1. (*Junci.*)

aure'a, (y. Ju. ♃.) leaves radical, ensiform, shorter than the scape; scape erect, with one or two short leaves; flowers in a crowded corymb; root creeping. Sandy swamps.

LUDWIG"IA. 4—1. (*Onagra.*) [From Professor Ludwig, of Leipsic.]

pilo'sa, (y. Ju. ♃.) stem erect, branched, hairy; leaves alternate, oblong, sessile; peduncles 1-flowered, axillary; capsule globose, quadrangular. Swamps.

alternifo'lia, stem nearly smooth; leaves alternate, lanceolate, somewhat scabrous on the margins and under side; segments of the calyx large, colored, persistent; flowers yellow, 4-petalled, on short peduncles.

palus"tris, petals 0; stem prostrate, creeping; leaves opposite, smooth, succulent Grows in stagnant waters.

uniflo'ra, stem straight, simple; leaves alternate, lanceolate, acute, glabrous; flower terminal; petals longer than the calyx. Perhaps synonymous with *alternifo'lia.*

mol"lis, (Au. ♃.) villose; stem erect, much branched; leaves lanceolate; flowers generally clustered; capsule globose, 2-leaved. *S.*

capita'ta, (y. J. ♃.) erect, virgate; leaves linear-lanceolate, glabrous; flowers mostly in terminal heads; bracts longer than the calyx. *S.*

pedunculo'sa, (y. J. ♃.) stem procumbent, radicant; leaves opposite, lanceolate; peduncles longer than the leaves. 3-6 i. *S.*

linea'ris, (y. Au. ♃.) erect, branching, angled near the summit; leaves linear, glabrous; flowers sessile. 2 f.

decur"rens, (y. Au. ♃.) stem erect; leaves ovate-lanceolate, decurrent; flowers octandrous. 2 f. *S.*

na'tans, (y. Ju.) swimming and creeping; leaves opposite, lance-spatulate; flowers axillary, sessile; petals and calyx equally long. *S.*

LUNA'RIA. 14—1. (*Crucifera.*) [From *luna* the moon, moon-form.]

an"nua, (honesty, p. ♂.) leaves obtusely toothed; silicles oval, obtuse at both ends. Naturalized.

redivi'va, (satin-flower. b-p. ♃.) leaves with mucronate teeth; silicles tapering to both ends; flowers odorous. Ex.

LUPINAS"TER. 16—10. (*Leguminosa.*)

macroceph"alus, (y. and p. M. ♃.) leafets nine, oblanceolate; petioles very long; stipules cuneate, gash-3 toothed; teeth of the calyx filiform, plumose. *S.*

LUPI'NUS. 16—10. (*Leguminosa.*) [From the Greek *lupe*, grief, on account of its acrid juices.]

peren"nis, (wild lupine, p. M. ♃.) stem and leaves smoothish; leaves digitate, with about 8-10 leafets, which are oblanceolate, obtusish; calyxes alternate, not appendaged; banner emarginate; keel entire. 12-18 i.

hirsu'tus, (garden lupine, p. ⊙.) calyxes appendaged, alternate; banner 2-parted; keel 3-toothed. Ex.

al"bus, (white lupine, w. Au. ⊙.) calyx not appendaged, alternate; banner entire; keel 3-toothed. Ex.

pilo'sus, (rose lupine, r. w. ⊙.) calyx whorled; banner 2-parted; keel entire. Ex.

lu'teus, (yellow lupine, y. ⊙.) keel 3-toothed. Ex.

mutab"ilis, herbaceous, very branching; attains the height of four or five feet.

nootkaten"sis, (Ju. ♃.) stem and leaves hirsute; leaves digitate; leafets (7-8) lanceolate, obtuse; calyxes whorled, without appendages; banner emarginate; keel entire.

decum"bens, (p.) suffruticose, sub-decumbent; flowers on pedicels, somewhat whorled, bracted; calyx silky-hirsute; banner and keel entire; leafets lance-oblong, acute and obtuse, silky underneath.

seri'ceus, (r. p. Ju. ♃.) stem and leaves silky-tomentose; leaves digitate; leafets (7-8) lanceolate, acute, silky both sides; calyxes somewhat whorled, without appendages; banner gashed; keel entire.

villo'sus, (hairy lupine, w. r. p. J. ♃.) very villose; leaves simple, oblong; calyxes not appendaged, alternate in a long spike; banner 2-cleft; keel entire. long. Florida.

diffu'sus, (Ap. ♃.) villose, silky; stems numerous, diffuse, decumbent; leaves

124   LUZULA—LYGODESMIA.

simple, oblong, obovate; petioles and stipules short, naked. *S.*

argen"teus, (y-w. Ju. ⚇.) leaves digitate; leafets (5.7) lance-linear, acute, glabrous above, silvery-silky underneath; calyxes alternate, not appendaged; banner obtuse; keel entire. *S.*

na'nus, (b. ⚇.) a native of California.

polyphil"lus, (⚇.) very vigorous exotics, with spikes of flowers from 1 f. to 18 i. in length.

latifo'lius, (⚇.) a native of California. Has very long spikes of blue flowers.

LUZU'LA. 6—1. (*Juncæ.*)

pilo'sa, (M. ⚇.) leaves hairy; panicle sub-cymose; peduncles 1-flowered, reflexed; leafets of the perianth acuminate, shorter than the capsule; radical leaves numerous, hirsute. Woods. 6-12 i.

melanocar"pa, culm leafy; leaves sublanceolate, smooth; panicles capillary, loose; capsule black. Mountains.

campes"tris, (M. ⚇.) leaves hairy; spikes sessile and pedunced, glume-like sepals acuminate, longer than the obtuse capsule. 1 f. Mich.

spica'ta, (Au.) leaves narrow, hairy at the throat; spike nodding, compound; glume-like sepals acuminate-awned, about as long as the roundish capsule. 8 i.

LYCH"NIS. 10—5. (*Caryophylla.*) [From *luchnos*, a torch.]

chalcedon"ica, (scarlet lichnis, r. J. ⚇.) flowers fascicled, level top, or convex. Ex.

floscu'culi, (ragged robin, ⚇.) petals torn; capsules 1-celled, roundish. Ex.

apet"ala, (⚇.) calyx inflated; corol shorter than the calyx; stem about 1-flowered. Canada.

alpi'na, (r. ⚇.) glabrous; flowers in dense umbelled heads; petals 2-cleft; styles 4. Labrador. Canada.

visca'ria, (clammy lichnis, ⚇.) stem geniculate, viscous, petals entire; capsule 5-celled. Ex.

LY'CIUM. 4—1. (*Polemonia.*) [From the country Lycia.]

carolin"ia'num, (p. Ju. ♄.) unarmed; leaves clustered, cuneate, fleshy; flowers 4-cleft. 3-5 f. *S.*

barba'rum, (matrimony vine, J. r. y. ♄.) stem angled; branches erect; leaves lanceolate, tapering to both ends; calyx mostly 3-cleft. Ex.

LYCOPER"DON. 21—6. (*Fungi.*) [From *lukos*, a wolf, and *perdo*, to explode, so named because it was supposed to be the excrements of this animal.]

bovis"ta, (common puff-ball,) at first white and oboconic, becoming black and spherical; outer coat downy, which peeling off, leaves the leathery inner coat; seeds black, lighter than air, and appearing like smoke. In meadows.

LYCOPO'DIUM. 21—1. (*Filices*) [From *lukos*, a wolf and *pous*, foot, so called from its supposed resemblance.]

Spikes pedunculate.

complana'tum, (ground pine, g-y. Ju. ⚇.) creeping, erectish; branches alternate, dichotomous; leaves bifareous, connate,

spreading at the tips; spikes in pairs, peduncled. Woods.

clava'tum, (club-moss, Ju.) stem creeping, branches ascending; leaves scattered, incurve-bristle-bearing, serrate; spikes in pairs or single, cylindrical, pedunculate; scales ovate-acuminate, dentate. Pine woods.

carolin"ia'num, (Ju. ⚇.) stem creeping; leaves somewhat distichus, spreading, lanceolate, very entire; peduncle erect, solitary, elongated, 1-spiked; bracts sub-lanceolate, entire. Sandy swamps.

2. *Spikes sessile; leaves surrounding the stem.*

dendroi'deum, (tree-weed, g. Ju. ⚇.) erect; branches erect; leaves in 6 equal rows; spikes numerous, solitary, sessile. Woods. About a span high.

rupes"tre, (festoon-pine, Ju. ⚇.) stem creeping; branches sub-divided, ascending; leaves scattered, imbricate, linear-lanceolate, ciliate, ending in hairs; spikes solitary, sessile. terminal. Rocks and side hills.

3 *Spikes sessile; leaves distichus.*

albid'ulum, leaves ovate, acute, denticulate, alternate, close-pressed; spikes terminal, long, 4-sided.

apo'dum, (Ju. ⚇.) leaves ovate, acute, denticulate, flat, superficial ones alternate, acuminate; spikes terminal, sub-solitary.

4. *Capsules axillary.*

lucid'ulum, (moonfruit pine, M. y. ⚇.) leaves in 8 rows, linear-lanceolate, denticulate, acute, spreading, reflexed, shining; stem ascending, bifid.

tristach"ymum, (Ju. ⚇) stems erect; branches alternate, dichotomous, sub compressed; leaves lanceolate, acute, appressed, pointing 4 ways; peduncles solitary, elongated, 3-spiked; spikes terete; scales roundish, acuminate. *S.*

LYCOP"SIS. 5—1. (*Boraginea.*) [From *lukos*, a wolf, and *opsis*, aspect, because it is a rough-looking plant.]

arven"sis, (b. Ju. ⚇.) leaves lanceolate, repand-toothed; racemes in pairs; flowers sessile; whole plant hispid.

virgin"ica, (w. J. ⚇.) small, hispid; under leaves spatulate. upper ones linear-oblong, entire; racemes solitary; flowers on peduncles.

LYCO'PUS. 2—1. (*Labiata.*) [From *lukos*, a wolf, and *pous*, foot, sometimes called wolf's-claw.]

europæ'us, (water horehound, w. Au. ⚇.) smooth; stem acutely 4-cornered; leaves narrow-lanceolate, with large acute teeth, lower ones somewhat pinnatifid; segments of the calyx acuminate, terminating in short spines. 1.2 f.

virgin"icus, (bugle-weed, w. J. ⚇.) leaves broad-lanceolate, serrate, tapering and entire at the base; calyx shorter than the seed, spineless; flowers in whorls. Wet places.

uniflo'rus, (w. J. ⚇.) small, root tuberous; stem simple; leaves oval, obtuse, obtusely toothed; axils 1-flowered.

LYGODES"MIA. (See PRENAN"THES.)

LYGO'DIUM. 21—1. *(Filices.)* [From *ly-godes*, pliant, and *ima*, one, or *meiou*, to diminish.]

palma'tum, (climbing fern, g-y. Au. ♃.) stem flexuous and climbing; fronds conjugate, cordate, palmate, 5-lobed; lobes entire, obtuse; spikelets oblong-linear, in a compound terminal spike. 3-4 f.

LYSIMA'CHIA. 5—]. *(Lysimachia.)* [From Lysimachus, its discoverer.]

stric"ta, (loose-strife, y. Ju. ♃.) raceme terminal, very long, lax; leaves opposite, lanceolate, sessile; petals lanceolate, spreading. 1-2 f.

cilia'ta, (y. ♃.) sub-pubescent; leaves opposite, long-petioled, sub-cordate, oval; petioles ciliate; pedicels somewhat in pairs; flowers nodding. 2-4 f.

quadrifo'lia, (y. J. ♃.) leaves verticillate in fours and fives. ovate-lanceolate, acuminate; peduncles axillary, 1-flowered, by fours; segments of the corolla oval, entire, often obtuse. 12-18 i.

thyrsiflo'ra, (y. J. ♃.) stem simple, smooth; leaves sessile, lanceolate, opposite, acute, paler underneath; racemes lateral, long-peduncled; flowers small. Appearance unlike the rest of the genus. 12-18 i.

hy'brida, stem smooth, somewhat branched; leaves mostly opposite, lanceolate, acute at each end, on short, ciliate petioles; flowers nodding; corolla about as long as the calyx.

revolu'ta (y. J. ♃.) stem quadrangular, branched; leaves opposite, sessile, long-linear, margin revolute; peduncles 1-flowered, sub-terminal, nodding. 12-18 i.

herbemon"ti, (♃.) flowers in terminal racemes, lower ones verticillate, upper ones scattered; leaves by fours, ovate-lanceolate, sessile, 3-nerved. 2 f. S.

lanceola'ta, (♃.) very smooth; leaves by fours, rather petiolate, lanceolate, prominently acuminate; peduncles by fours, many-flowered, upper flowers in racemes; segments of the corolla ovate and acute. S.

heterophyl"la, (Ju. ♃.) leaves opposite, lower ones roundish, upper ones linear, sessile; flowers nodding. 12-18 i. S.

angustifo'lia, (y. ♃.) very smooth, branching; leaves opposite and whorled, long-linear, punctate; racemes terminal, short; segments of the corolla oblong. S.

nummula'ria, (money-wort,) an evergreen trailer in a moist soil, producing shoots two and three feet long. Ex.

verticilla'tum, an upright plant, with a profusion of showy yellow flowers. Ex.

capita'ta, (y. J. ♃.) stem smooth, simple, punctate; leaves opposite, sessile, broad-lanceolate, punctate; peduncles axillary, elongated; flowers in dense heads, 6-7 parted. Swamps. Stem 1 f.

quadriflo'ra, branching; stem smooth; leaves sessile, opposite, long-linear; peduncles in fours, sub-terminal, 1-flowered. 2-3 f.

LYTH"RUM. 11—1. *(Salicaria.)* [From *luthron*, blood, so called from its color.]

salica'ria, (purple loose-strife, p. Ju. ♃.) pubescent; leaves opposite and ternate, sessile, lanceolate, cordate at the base; flowers with 12 stamens (sometimes 5 or 8), terminal, whorled-spiked; capsule oblong. Wet meadows. Stem 2 f.

ala'tum, (p. Ju. ♃.) very glabrous; stem winged; flowers hexandrous, axillary, solitary, sessile. 2-3 f. S.

verticilla'tum, (swamp willow-herb, p. Au. ♃.) pubescent; leaves opposite, or in threes, lanceolate, petioled; flowers axillary, somewhat in whorls; fruit globose; stamens 10. Wet grounds. 2 f.

hyssopifo'lium, (dwarf grass-poley, w. p. ♃.) leaves alternate and opposite, lance-linear, sub-oval; flowers solitary, axillary. Hexandrous. 6-10 i.

virga'tum, (p. Ju. ♃.) leaves opposite, lanceolate, glabrous; stem panicled; flowers axillary in threes, on pedicels; stamens 12. S.

linea're, (w. Ju. ♃.) smooth, virgate; leaves generally opposite, linear, acute; flowers axillary, solitary, hexandrous. 3-4 f. S.

diffu'sum, (p. Au.) 1 foot in height.

MACBRI'DEA. 13—1. *(Labiatæ.)* [In honor of Dr. McBride.]

pul"chra, (p. and w. Au. ♃.) stem erect, simple; leaves opposite, acute, lanceolate, ciliate, serrulate, punctate. glabrous beneath, somewhat hairy above, upper ones sessile, lower ones attenuated at the base as if petioled. 12-18 i. S.

MACRO'TRYS 12—1. *(Ranunculaceæ.)* [From *makros*, large, and *botrus*, a raceme.]

racemo'sa, (bug-bane, blacksnake root, cohosh, w. Jd. ♃.) leaves decompound; leafets oblong-ovate, gash-toothed; racemes in wand-like spikes; capsules ovate. Woods. 3-9 f.

MAGNO'LIA. 12—12. [From Magnol, who wrote on botany in 1720.]

glau'ca, (sweet-bay, swamp-laurel, w. J. ♄.) leaves glaucous beneath, perennial, obtuse, elliptical; flowers 9-12 petalled; petals obovate, concave. A large shrub, with whitish bark; flowers solitary, odorous. Var. *latifo'lia*, has deciduous leaves. Var. *longifo'lia*, has leaves acute at both ends, perennial. N. J. to Car.

acumina'ta, (cucumber-tree, b-y. J. ♄.) leaves deciduous, oval, acuminate, pubescent beneath; flowers 6-9 petalled; petals obovate. Mountains. Penn. to Car. A tree, sometimes 70 feet high.

tripe'tala, (umbrella tree, w. J. ♄.) leaves large, deciduous, cuneate-lanceolate. acute, silky when young; petals 9, oval-lanceolate, acute, the outer ones reflexed. Mountains, woods. Penn. to Geo. A small tree, with very large leaves and flowers.

grandiflo'ra, (big laurel magnolia, w. M. ♄.) leaves evergreen, oval, thick, leathery; petals broad, obovate, abruptly narrowed into a claw. 60-80 feet. S.

macrophyl"la, (w. J. ♄.) leaves very large, oblong, cuneate-obovate, sinuate and auriculate at base, glaucous beneath; petals 6, ovate-obtuse. 30-35 f. S.

corda'ta, (y. M. ♁.) leaves broad, oval or ovate-lanceolate, at base slightly cordate, somewhat tomentose beneath; petals oblong-lanceolate, acute. 40-50 f. *S.*

pyramida'ta, (Ap. ♄.) leaves rhomb-oboval, abruptly acute, both sides colored alike, sub-cordate and auricled at the base; lobes divaricate; petals lanceolate, gradually acute. *S.*

MALAX"IS. 18—1. (*Orchideæ.*) [From *malakia,* softness, from the delicacy of the plants.]

lilifo'lia, (twayblade, w. y. p. J. ♃.) scape 3-cornered; inner petals filiform, reflexed, 2-colored; lip concave, obovate, mucronate; leaves 2, lanceovate, or oval. 4-8 i.

longifo'lia, (y-g. J. ♃.) leaves broad-lanceolate, longer than the scape; spike oblong; lip cordate, concave, channeled, shorter than the petals; bulb roundish; scape 2-leaved. 3-7 i.

MALACHODEN"DRON. 15—5. (*Durantia.*) [From *malake,* soft, *dendron,* tree.]

ova'tum, (w. M. ♄.) leaves ovate, acute; flower solitary, sub-sessile. 6-12 f. *S.*

MAL"OPE. 15—12. (*Malvaceæ.*)

malacoi'des, (y. ☉.) leaves oblong, acute, entire, crenate, glabrous on the upper surface; peduncles solitary, axillary. 12-18 i. *S.*

MAL"VA. 15—13. (*Malvaceæ.*) [From *mollis,* soft.]

rotundifo'lia, (low mallows, r. w. J. ♃.) leaves heart-orbicular, obsoletely 5-lobed; peduncles bearing the fruit declined; stem prostrate. Probably introduced.

sylves"tris, (mallows, r-b. J. ♂. and ♃.) stem erect; leaves about 7-lobed, acutish; peduncles and petioles hairy. Ex.

cris"pa, (curled mallows, Au. ☉.) stem erect; leaves angular, crisped; flowers axillary. glomerate. Ex.

coccin"ea, (r. Au. ♃.) hoary-tomentose, covered with stellate hairs; racemes terminal; stem diffuse. *S.*

moscha'ta, (musk mallows, ♃.) erect; radical leaves reniform, gashed; cauline leaves 5-parted, pinnate, many-cleft; leafets of the involucre linear. Naturalized.

virga'ta, (whip-stalk mallows, r. ♃.) leaves deeply 3-lobed, toothed, cuneate at the base; peduncles in pairs, longer than the petioles. Ex.

abutiloi'des, leaves with 5 angular lobes, tomentose; peduncles 2-cleft, generally 4-flowered; axillary capsules many-seeded. *S.*

carolin"iana, (r. Au. ☉.) leaves 5-lobed or palmate, notched and toothed; peduncles longer than the petioles; petals entire; fruit villose; stem prostrate, branching. *S.*

triangula'ta, (p. Ju.) hirsute, sub-decumbent; lower leaves triangular cordate; upper ones 3.5-lobed, irregularly toothed; flowers racemed. 12-18 i. *S.*

peda'ta, (♃. p.) somewhat scabrous with stellate hairs; leafets pedately 5-7-parted; segments laciniately toothed; flowers on elongated peduncles in a loose panicle; calyx naked, slightly hirsute. 2-4 f. *S.*

MAL'VAVIS"CUS. 15—12. (*Malvaceæ.*)

florida'nus, (r. ♃.) pilose, hirsute, herbaceous, leaves cordate-ovate, crenate-obtusish, small, short-petioled; peduncles axillary, nodding, towards the end of the branches. Florida.

MARCHAN"TIA. 21—2. (*Hepaticæ.*) [From Marchant, a naturalist.]

polymor"pha, (brook liverwort, g-y. Ju. ♃.) pistillate receptacles radiated; staminate ones peduncled, peltate; fronds crowded together, lobed, nerved, and covered with small decussate veins; pistillate peduncles very long; nerves of the frond generally brown. On earth and stones, in wet or damp places.

MARRU'BIUM. 13—1. (*Labiatæ.*) [From a Hebrew word, *marrob,* a bitter juice.]

vulga're, (horehound, w. Ju. ♄.) leaves round-ovate, toothed, rugose, veined; calyx toothed, setaceous, uncinate. Introduced.

MARSHAL"LIA. 17—1. (*Corymbiferæ.*)

lanceola'ta, (p. M. ♃.) stem simple, le[ ] below, naked near the summit; leave[ ] the root obovate, of the stem lon[ ], late; scale[ ] the involucrum spatulate. [ ]-24 i. *S.*

latifo'lia, (Au. ♃.) stem simp[ ] oblong-lanceolate, acuminate, [ ] lowest ones sheathing; scales of the [ ] ucrum acute; chaff of the receptacle l[ ] *S.*

angustifo'lia, (p. Ju. ♃.) stem br[ ] ing; lower leaves narrow-lanceolate, [ ] ones linear; scales of the involucrum [ ] subulate; chaff linear. Var.*cyanan"the[ ]* corolla pale purple; anthers sky blue; ste[ ] simple, angular, very pubescent near the top; leaves lance-linear, 3-nerved; scales of the calyx lanceolate, acuminate. 2 f. *S.*

MARTYN"IA. 12—2. (*Bignonæ.*) [In honor of the botanist, Martyn.]

probosci'dea, (martinoe, w. p. y. Ju. ☉.) stem short, branching; leaves alternate, cordate, entire, villose; pericarp terminating in a long proboscis. 1-2 f. *S.*

MATRICA'RIA. 17—2. (*Corymbiferæ.*) [Named from its efficacy in the diseases of females.]

chamomil"la, (wild chamomile, w. M. ☉.) leaves bi-pinnate; scales of the involucre obtusish. Ex.

MECONOP"SIS. 12—1. (*Papaveraceæ.*) [From *mekon,* a poppy, *opsis,* aspect, resembling a poppy.]

diphyl'la, (y. m. ♄.) leaves 2, glaucous, sessile, hairy; lobes rounded and obtuse; capsules 4-valved-echinate. 1 f.

petiola'tum, stem 4-sided; leaves very broad, long-petioled, pinnatifid-lobed. *S.*

heterophyl'la, (r. ♃.) leaves few and remote, pinnately divided; segments of the lower ones ovate, incised and petioled; of the upper linear, entire, somewhat confluent.

MEDEO'LA. 6—3. (*Asparagi.*)
virgin"ica, (Indian cucumber, g·y. m. ♄.) leaves in whorls, lance-oval, acuminate; pedicels aggregated, terminal; root white. 12-18 i.

MEDICA'GO. 16—10. (*Leguminosa.*) [Called *medike*, by Dioscorides, on account of its supposed medicinal virtues.]
lupuli'na, (hop medick, y. J. ☉.) spikes oval; legumes reniform, 1-seeded; stipules entire; leaves obovate; stem procumbent.
intertex"ta, (y. Au.) stem procumbent; leafets obovate, toothed; stipules ciliate, toothed; peduncles somewhat 2-flowered; legume pilose, spiral, oval; spines straight, thick, rigid, and acute. Sandy fields. Conn. to Car. Introduced.
sati'va, (p. Ju. ♃.) peduncles racemed; legume smooth, cochleate; stipules entire; leaves oblong, toothed. Naturalized.
tribuloi'des, (hedge-hog. ☉.) peduncles 2-flowered; legume cochleate, cylindric, flat both sides, aculeate, conic 2-ways, reflexed; stipules toothed; leaves toothed, obovate. Ex.
scutella'ta, (snail-shell, bee-hive, Ju. ☉.) peduncles about 2-flowered; legumes unarmed, cochleate in an orbicular form, with a convex base and a flat top; st●●●s toothed; leaves oblong, toothed. Ex●●
macula'ta, (p.) stem prostrate; leafets obcordate, toothed, spotted; stipules toothed; peduncles 3-5-flowered; legumes compactly spiral, furrowed on the margin, and fringed with a double row of long-curved spines; seeds reniform, ●●●ish. S.
denticula'ta, (p.) nearly ●●abrous; stem prostrate; leafets obcordate; stipules laciniate; peduncles 2 to 5-flowered; legumes broad. loosely-spiral and flat, with 1-3 convolutions, reticulated; the margin thin, keeled, with a double compact row of subulate-curved prickles. 1-2 f. S.

MELAMPY'RUM. 13—2. (*Pedicula·es.*) [From *melas*, black, and *puros*, wheat.]
america'num, (cow-wheat, w. Ju. ☉.) slender; lower leaves linear, entire; floral ones lanceolate, toothed behind; flowers axillary, distinct. Var. *latifo'lium*, has very broad leaves. Woods. S.

MELAN"THIUM. 6—3. (*Junca.*) [From *melas*, black, *anthos*, flower.]
virgin"icum, (g·y. black flower.) panicle pyramid-form, very large; petals ovate; leaves long, linear-lanceolate, flat, smooth; flowers become black. 3-4 f.
hybrid"um, (bunch-flower, w. J. ♃.) panicles racemose; petals sub-orbicular, plaited with long claws; glands connate. 2 f.
glau'cum, (g-w. Ju. ♃.) root a tunicated bulb; leaves glaucous, gramineous, margined; racemes mostly simple, few-flowered; segments of the perianth roundish, clawed, with two peculiar spots; seeds subulately-winged. 1-3 f. Northern lakes. Canada. Mich.
monoi'cum, (Ju. ♃.) panicle with the lower flowers sterile; upper ones fertile, racemed; petals oblong, flat with short claws; styles half the length of the germ. S.

MELANANTHE'RA. 17—1. (*Corymbifera.*) [From *melas*, black, and *anthos*, flower.]
hasta'ta, (w. S. ♃.) leaves hastate, 3-lobed; chaff of the receptacle lanceolate, acuminate. Var. *loba'ta*, leaves deeply 3-lobed. Var. *pandura'ta*, leaves slightly 3-lobed, panduriform. 4-6 f. S.

ME'LIA. 10—1. (*Melia.*) [From *meh*, honey.]
azed"arach, (pride of China, ♄.) leaves doubly pinnate; leafets smooth, ovate, toothed. 30-40 f. S.

MELI'CA. 3—2. (*Graminea.*)
specio'sa, (melic grass, J. ♃.) smooth; panicle loose, erect, few-flowered; branches simple; leaves flat, pubescent beneath; florets obtuse. 3-4 f. Charleston, S. C.
diffu'sa, (J. ♃.) panicle diffuse, very branching; stem erect, pubescent; flowers acute, beardless.
gla'bra, (large flowered melica, Ap.) stem glabrous; leaves narrow, scabrous; panicle erect, loose branches simple, few-flowered; flowers with the glumes unbearded. 2-3 f. S.

MELILO'TUS. 16—10. (*Leguminosa.*) [From *meli*, honey, and *lotus*, a plant.]
officina'lis, (yellow melilot-clover, y. J. ☉.) stem erect, branching; leafets lanceolate, oblong; spikes axillary, paniculate; legume 2-seeded, rugose; flowers in long yellow racemes. 2-4 f.
al"ba, (white melilot-clover, w. J. ☉.) stem erect; leafets variable, (oval, ovate, obovate, and oblanceolate,) mucronately serrulate; banner longer than the wings; racemes axillary, panicled; the longest raceme 6 to 10 times as long as the longest leafet at its base; legumes oval. 3-6 f. Probably introduced, but now very common, and growing wild.
occidenta'lis, (y.) erect; leafets linear-oblong or obovate, serrate, truncate at the extremity; flowers minute; teeth of the calyx unequal, as long as the tube; legume 1.2 seeded, ovate-orbiculate, slightly wrinkled. California.

MELIS"SA. 13—1. (*Labiata.*) [From *melissa*, a bee, because it affords honey.]
officina'lis, (balm, w. b. Ju. ♃.) flowers whorled half-way round, sub-sessile; bracts oblong,. pedicelled; leaves ovate, acute, serrate. Naturalized.

MELO'THRIA. 19—15. (*Cucurbitacea.*) [From *melon*, fruit, and *thrion*, food.]
pendu'la, (small creeping cucumber, y. J. ☉.) leaves sub-reniform, lobed, and angled, slightly hispid: fruit oval, smooth, pendulous. A slender vine, running over small shrubs and herbs on the banks of streams; stem hairy; leaves petioled; tendrils 5-6 inches high; flowers axillary; the steril in small racemes, the fertile solitary.

MENISPER'MUM. 20—12. [From *mena*, the moon, and *sperma*, seed; seed crescent-form.]
canaden"se, (moon-seed, y. Ju. ♃.) leaves peltate, cordate, round-angular; racemes compound; petals 8.
smilaci'num, (y. Ju. ♃.) racemes generally simple; petals 4-leaves peltate, somewhat glabrous, cordate, nearly round, obtusely angled, glaucous beneath. S.
lyo'ni, (Ju. ♃.) racemes simple; petals

6; stamens 12; leaves palmate-lobed, cordate, very long petioled. S.

MEN"THA. 12—1. (*Labiatæ*.) [From Minthe, the daughter of Cocytus, who is said to have been changed into this herb.]

canaden"se, (w. p. Au. ⚥.) flowers whorled; leaves lance-ovate, serrate, petioled, hairy; stamens as long as the .corolla. Sandy soils. Stem 1 f.

borea'lis, (w. p. J. ⚥.) ascending, pubescent; leaves petioled, ovate-lanceolate, acute at both ends; flowers in whorls, stamens exsert, twice as long as the corolla. Horse-mint.

piperi'ta, (peppermint, p. Au. ⚥.) spikes obtuse, interrupted below; leaves subovate, somewhat glabrous, petioled; stem glabrous at the base. Naturalized. 1-2 f. Ex.

vir"idis, (spearmint, p. Au.) leaves lanceolate, sessile; spikes elongated, interrupted; stamens long. 1-2 f. Ex.

ten"uis, (America spearmint, w. J. ⚥.) glabrous; leaves opposite, ovate-lanceolate, serrulate, petioled; spike slender, terminal, with verticils very small, distant at base; stamens shorter than the corolla. 1-2 f.

arven"sis, (field-mint. p. Ju.) hairy, branching; leaves ovate; flowers whorled; calyx bell-form. Naturalized. 1 f.

MENTZE'LIA. 11—1. (*Onagræ*.) [In honor of Dr. Mentzel.]

au'rea, (y.) stem dichotomous; leaves lance-ovate, deeply angular-crenate; flowers sessile; petals oval, acuminate, entire; plant rough. 12 i. S.

oligosper"ma, decumbent; flowers large. S.

MENYAN"THES. 5—1. (*Gentianæ*.) [From *mene*, mouth, and *anthos*, flower.]

trifo'liata, (buck-bean, r. J. ⚥.) leaves ternate, petioled, sheathing, smooth; flowers pale, in a terminal raceme. Marshes.

MENZIE'SIA. 8—1. (*Ericæ*.) [Named by Smith, in honor of Menzies.]

cæru'lea, (mountain-heath, Ju. ♄.) stem branched, woody below; leaves scattered, crowded, linear, toothed; peduncles terminal, aggregate, 1-flowered, flowers bell-shaped, 5-cleft, decandrous; calyx very acute. An evergreen shrub, resembling the heath. White hills, N. H., and other cold, elevated regions. Flowers large, purple, on long, red peduncles.

globula'ris, leaves lanceolate, glaucous beneath, nerves pubescent; calyx 4-cleft; flowers globose, octandrous. Mountains. Penn. to Car. Shrub. 4 f. Flowers yellowish brown.

ferrugin"ea, leaves lance-obovate; flowers urceolate, octandrous. S.

empetrifor"mis, (r. Ju. ♄.) leaves linear, serrulate, concave beneath; peduncles terminal, aggregate; flowers bell-form; calyx obtuse, decandrous. S.

polifo'lia, (St. Daboec's heath,) flowers larger, and more globular than those of the common heaths. Found wild in Ireland.

MESEMBRYAN"THEMUM. 11—5. (*Ficoideæ*.) [From *mesembria* mid-day, and *anthos* flow-

er, so called because its flowers expand at noon.]

crystali'num, (ice plant, w. Au. ⚥.) branching; leaves alternate, ovate, papillose; flowers sessile; calyx broad-ovate, acute, retuse. Ex.

pomeridia'num, (⚥.) flowers of a brilliant yellow.

MES"PILUS. 11—5. (*Rosaceæ*.)

germani'ca, (medlar, ♄.) leaves lance-ovate, downy beneath; flowers sessile, solitary. Ex.

oxycan"tha, (English hawthorn.)

MICRAN"THEMUM. 2—1. (*Lysimachiæ*.) [From *mikros*, small, and *anthos*, flower.]

orbicula'tum, (w. Au. ⚥.) stem prostrate, terete; orbicular, abruptly narrowed at the base; flowers peduncled.

emargina'tum, (w. Au. ⚥.) leaves oval and obovate, emarginate, sessile, flowers sessile; stem prostrate or creeping. Wet places.

MICROPE'TALON. 10—4. (*Caryophylleæ*.) [Named from the diminutive size of the petals.]

lanceola'tum, (blind starwort, Ju. ⚥.) glabrous; leaves lanceolate, narrow at both ends; flowers panicled; petals ovate, very short or wanting. Damp. 6-8 i.

longi..., (long-leaf starwort, w. J. ⚥.) stem decumbent or sub-decumbent; leaves lance-linear, opposite, entire.

lanugino'sum, (Ju. ⚥.) closely pubescent; leaves lanceolate, tapering to a petiole; peduncles generally solitary, long, finally reflect... owers without petals. S.

MICROS"TYL... 18—1. (*Orchideæ*.) [From *mikros*, small, and *stulos*, style.]

ophioglossoi'des, (g-w. J. ⚥.) scape 1-leafed; leaf ovate, amplexicaule; lip truncate, emarginate. Roots of trees.

brachypo'da, (Ju.) stem 1-leaved; racemes sub-spiked, lateral petals refracted; lip triangular-hastate, cucullate, acuminate.

MIE'GIA. 3—1. (*Gramineæ*.)

macrosper"ma, (cane, Ap. ⚥.) stem terete-glabrous, hollow, rigid; leaves distichous, lanceolate, flat, sub-acuminate, pubescent underneath. 3-15 f. Var. *gigant"ea*, much taller. 30-40 f.

MIKA'NIA. 17—1. (*Corymbiferæ*.) [In honor of Professor Mikan of Prague.]

pubes"cens, (w-p. S. ⚥.) stem climbing, pubescent; leaves cordate, acuminate, angularly dentate, pubescent on both sides; divaricate, equal.

scan"dens, (climbing thoroughwort, w. Au. ⚥.) stem glabrous, climbing; leaves cordate, toothed, acuminate.

MIL"LIUM. 3—2. (*Gramineæ*)

effu'sum, (millet, Ju. ⚥.) panicle diffuse, compound; branches horizontal; glumes ovate, very obtuse; paleas awnless, smooth and shining; leaves broad-linear. 5-8 f.

amphicar"pon, (Au. ⚥.) leaves linear-lanceolate, hairy, ciliate; panicle simple, contracted, bearing perfect flowers; fertile flowers in solitary, elongated radical scapes, at length subterraneous; glumes acuminate. 1-2 f. Sandy swamps.

pun"gens   dwarf millet grass M ⚥.

calm erect; leaves lanceolate, very short, pungent, at length involute; panicle contracted; branches generally in pairs, 2-flowered; flowers awnless, ovate; paleas hairy. 12-18 i. Rocky hills.

ni'gricans, (African millet,) flowers in panicles, crowded; glumes shining, becoming black; leaves ensiform, very long. Ex.

MIMU'LUS. 13—2. (*Scrophularia*.) [From *mimus*, a mimic.]

rin"gens, (monkey-flower, b. Ju. ♃.) erect, glabrous; leaves sessile, lanceolate-acuminate, serrate; peduncles axillary, opposite, longer than the flower; teeth of the calyx acuminate. 1-2 f.

ala'tus, (b. Ju. ♃.) erect, smooth; leaves petioled, ovate, acuminate, serrate; stem square, winged. 2 f.

lute'us, (yellow monkey-flower,) erect, stoloniferous; leaves roundish-ovate, lower ones petioled-obtuse, upper ones sessile, acute.

lewis"ii, (p. Au. ♃.) erect, small, pubescent; leaves sessile, lance-oblong, nerved, mucronate-denticulate; flowers few, terminal, very long peduncled; teeth of the calyx acuminate. 6-8 i. S.

glutino'sus, a shrubby species with orange-colored flowers. Ex.

MIRAB"ILIS. 5—1. (*Nyctagines*.) [From the Latin *mirabilis*, wonderful.]

jal"apa, (four-o'clock, r. y. Ju. ♃.) flowers heaped, peduncled; leaves glabrous.

dichoty'ma, (Mexican four-o'clock, ♃.) flowers sessile, erect, axillary, solitary. Ex.

longiflo'ra, (w. Au. ♃.) flowers crowded, very long, nodding; leaves sub-villose. Ex.

MITCHEL"LA. 4—1. (*Rubiaceæ*.) [In honor of the late Dr. Mitchell of New York.]

re'pens, (w. Ju. ♃.) stem creeping, branched; leaves smooth, roundish, opposite. Woods.

MITEL"LA. 10—2. (*Saxifraga*.)

dyphyl"la, (w. M. ♃.) leaves somewhat lobed; lobes acute-dentate; stem erect, with two opposite leaves above the middle. 12-18 i.

cordifo'lia, (w. M. ♃.) radical leaves cordate, sub-3-lobed, doubly crenate; scape naked, or with a single leaf, scaly at the base; petals fimbriate-pinnatifid. 6-8 i.

prostra'ta, (Ju. ♃.) root creeping; stem prostrate; leaves alternate, round-cordate, sub-acute, obtusely sub-lobed. Canada.

MOLUCEL"LA. 13—1. (*Labiatæ*.) [From *moluca*, to bite, on account of its sharp taste]

læ'vis, (shell-flower, w-g. Ju. ♀.) calyx campanulate, 5-toothed; teeth equal, awnless; leaves petioled, round-ovate, toothed.

MOLLU'GO 3—3. (*Caryophylleæ*.) [From *mollis*, soft.]

verticilla'ta, (carpet-weed, w. Ju. ♀.) leaves verticillate, wedge-form, acute; stem branched, depressed; peduncles 1-flowered.

MOMOR"DICA. 19—15. (*Cucurbitaceæ*.)

echina'ta, (w. Au. ♀.) pomaceous; berry 4-seeded, roundish, setose, echinate; leaves cordate, 5-lobed, angled, acuminate, entire; calyx 6-cleft; corolla 6-parted.

balsami'na, (balsam apple, S. ♀.) poma-

ceous; berry angled, tubercled; leaves glabrous. spreading, palmate. Ex.

MONAR"DA. 2—1. (*Labiatæ*.) [So called from Monardes, a Spanish botanist.]

did"yma, (mountain-mint, r. J. ♃.) leaves ovate, acuminate, sub-cordate, somewhat hairy; flowers in simple or proliferous heads; outer bracts large, colored, lanceolate. Var. *angustifo'lia*, leaves lance ovate, acuminate, pubescent; stem pubescent. 18-24 i.

fistulo'sa, (y. Ju. ♃.) stem obtuse-angled, nearly smooth, hollow, leaves oblong-lance-olate, acuminate, coarsely serrate; calyx 5-toothed, long, curved, bearded; corolla rough, pale.

puncta'ta, (y-b. S. ♃.) nearly smooth; stem white, downy; leaves smooth; flowers whorled; bracts lanceolate, colored, longer than the whorl; corolla yellow, dotted with brown; calyx 5-toothed, unequal.

hirsu'ta, (b-p. Au. ♃.) whole plant hairy; leaves on long petioles; flowers small; bracts short; calyx 2-lipped; lower lip 3-toothed. 2-3 f.

oblongata, (wild burgamot, b. r. Ju. ♃.) pubescent; heads simple; exterior bracts ovate; calyx short, with the throat bearded, teeth divaricate; stem obtuse-angled, hairy above; leaves oblong, lanceolate, rounded at the base. Rocky situations. 2-3 f.

clinopo'dia, (y. p. Ju. ♃.) smooth; heads simple, terminal; exterior bracts ovate, wide, acute, entire; corolla pubescent, slender; leaves ovate-oblong, acuminate, serrate, hairy; stem obtuse-angled, glabrous.

cilia'ta, (p. Au. ♃.) hirsute; flowers verticillate; bracts ovate, glabrous, ciliate, as long as the calyx; leaves ovate-oblong, tapering, serrate.

rugo'sa, (w. Ju. ♃.) smoothish; heads simple, middling size; outer bracts ovate-undulate; calyx smoothish; leaves ovate, sub-cordate, acute, rugose; nerves beneath and petiole pilose; stem acute-angled, smoothish. 4 f. Canada.

gra'cilis, (y-w.) very glabrous; heads lateral and terminal; exterior bracts linear, ciliate; corollas short; leaves linear, lance-olate-serrate; stem obtuse-angled, broad-ovate, acuminate. S.

purpu'rea, (r. Ju. ♃.) somewhat glabrous; heads large, leafy; outer bracts large, colored, serrate; calyx colored; throat bearded; corolla long, nearly glabrous; leaves oblong-ovate, coarsely serrate; stem acutely angled. S.

MONOCE'RA. 3—2. (*Gramineæ*.) [From *monos*, one, and *keras*, horn.]

aromat"ica, (J. ♃.) spikes solitary; spikelets about 6-flowered; flowers awned, bearded at the margin; outer paleas roughened by glands awned on the back. S.

MONOTRO"PA. 10—1. (*Erica*.)

uniflo'ra, (bird's nest, Indian-pipe, w. J. ♃.) stem 1-flowered; flower nodding at first, at length erect; scales of the stem approximate. Whole plant ivory white at first. 4-8 i.

morisonia'na, (J. ♃.) scape long, straight, 1-flowered; scales distant; flowers erect; stamens 10.12. Shady woods.

lanugino'sa, (y-w. Ju. ♃.) scape bearing flowers in a spike; bracts and flowers hairy on all sides. Parasitic on roots.

europe'a, (y. J. ♃.) scape bearing flowers in a spike; flowers and scales on the stem glabrous outside; lateral flowers octandrous. Canada.

MONOTROP'SIS. 10—1. (*Erica.*)
odora'ta, (r-w. Mar.) flowers bell-form. in aggregate heads. 3-4 i. *S.*

MON''TIA. 2—3. (*Portulaccea.*)
fonta'na, (false spring-beauty,) leaves opposite; flowers axillary, small.

MO'RUS. 19—4. (*Urtica.*) From *mauros*, black, so called from the color of the fruit of one of its species.]
nigra, (black mulberry, Ju. ♄.) leaves heart-form, ovate, or sub-5-lobed; unequally toothed, scabrous. Ex.

al''ba, (white mulberry, M. ♄.) leaves heart-form, with oblique bases, ovate or lobed, unequally serrate, smoothish. From China and Persia. Naturalized. 15-20 f.

ru'bra. (red mulberry, M. ♄.) dioecious; leaves cordate, ovate-acuminate, often 8-lobed, equally serrate, scabrous, pubescent beneath; fertile aments cylindric. 15.30 f.

multicau'lis, (many-stemmed mulberry, ♄.) leaves cordate, ovate, acuminate, crenate, serrate, net-veined; sub-scabrous and pimpled beneath, sub-scabrous above; sprouts proliferous. 8-16 f. Leaves sometimes 12-14 inches long.

MU'COR. 21—6. (*Fungi*)
aspergil''lus, (mould,) stipe filiform, dichotomous; little beads terminal, sub-conjugate, oblong when mature. On putrid fungi in autumn.

MUH'LENBERG''IA. 3—2. (*Graminea.*) [In honor of Henry Muhlenberg, D.D., a distinguished botanist of Lancaster, Penn.]
diffu'sa, (dropseed grass, S. ♃.) culm decumbent, branching, diffuse; leaves lance-linear; panicles terminal and lateral, slender; branches appressed; awns about as long as the palea. 12-18 i.

erec''ta, (Au. ♃.) culm erect, simple, leaves lanceolate; panicle terminal, simple, racemed; awn twice as long as the palea; upper palea awned at the base. 2-3 f.

MYLOCA'RIUM. 10—1. (*Erica.*) [From *mule*, a mill, and *karua*, a kernel.]
ligustri'num, (buckwheat-tree, w. M. ♄.) leaves perennial, alternate, sessile, entire, glabrous; racemes simple, terminal. 6-15 f. *S.*

MYOSO'TIS. 5—1. (*Boraginea.*) [From *mus*, a mouse, *ous* (*otos*), an ear, the leaves being hairy like a mouse's ear.]
arven''sis, (forget-me-not, w-b. J. ☉.) seeds smooth; calyx-leaves oval, acuminate, very hirsute, longer than the tube of the corolla; stem very branching; racemes conjugate; leaves lance-oblong, hirsute. 4-8 i.

palustris, (scorpion-grass, b. M. ♃.) leaves lance-oval, rough; border of the corolla longer than the tube; flowers very small, bright blue. · Wet grounds.

suffrutico'sa, stem very branching, suffruticose, hirsute; leaves lance-linear, hispid-pillose; spikes terminal, many flowers on pedicels; calyx closed; nut smooth, ovate. 8 i.

na'na, (b. and y. ♃.) leaves oblong, villose, racemes few-flowered; nut smoothish; margin serrulate. *S.*

MYOSU'ROS. 5—12. (*Ranunculacea.*) [From *mus*, mouse, and *oura*, tail.]
min'mus, (Ap. ☉.) leaves linear, entire; seed 1-flowered; stamens 5-8; petals anther-form. 2-4 i. *S.*

MYRI'CA. 20—4. (*Amentacea.*) [The name is derived from the Greek; its original meaning is uncertain.]
ga'le, (Dutch-myrtle, sweet-gale, M. ♄.) leaves wedge-lanceolate, serrate at the apex, obtuse; steril aments imbricated; scales acuminate, ciliate; fruit in scaly heads, with a strong aromatic odor. 4-5 f. Bogs, mountains, and lakes.

cerife'ra, (bayberry, wax-myrtle, g-p. M. ♄.) leaves acute; steril aments loose; scales acute; fruit globular, naked. On boiling, a pleasant-flavored wax is obtained, which is used either alone or with tallow, in making candles. 5-18 f.

carolinien''sis, (Ap. ♄.) leaves cuneate-oblong, coarsely-toothed; staminate aments loose; scales acute; fruit globular, large. 3-5 f.

penn''sylva'nica, (M. ♄.) leaves oblong, acutish at each end, very entire or sparingly sub-serrate at the apex, revolute at the margin; staminate aments loose; scales acute; fruit globular, large. 3 f.

MY'RIOPHYL'LUM. 19—12. (*Onagra.*) [From *murios*, innumerable, and *phullon*, leaf, from the great number.]
verticilla'tum, (water milfoil, Ju. ♃.) leaves pinnate, capillary, upper ones pectinate-pinnatifid; flowers axillary, verticilate, upper ones staminate-octandrous.

tenel''lum, (Ju. ♃.) erect, nearly leafless; bracts entire, obtuse; petals linear, conduplicate and revolute; flowers mostly perfect, tetrandrous. 4-12 i.

scabra'tum, (J. p. ♃.) leaves pinnatifid; flowers verticillate-axillary; upper ones staminate, tetrandrous; lower ones pistillate; fruit 8-angled. 12 i. *S.*

MYR''TUS. 11—1. (*Labiata.*) [From *muros*, perfume.]
commu'nis, (myrtle, w. Ju. ♄.) flowers solitary; involucrum 2-leaved; leaves ovate. Ex.

NA'JAS. 19—1. (*Aroides*) [From *Nais*, a water nymph.]
canaden''sis, (water nymph,) small, filiform; leaves narrow-linear. Canada.

NARCIS''SUS. 6—1. (*Narcissi.*) [From *Narcissos*, a beautiful youth, according to mythology, changed into this flower.]
pseudo-narcis'sus,(daffodil, M. ♃.) spatha 1-flowered; nectary bell-form, erect, crisped, equalling the ovate petals. Ex.

tazet''ta, (polyanthos, M. ♃.) spatha many-flowered; nectary bell-form, plicate, truncate, thrice as short as the petals; petals alternately broader; leaves flat. Ex.

jonquil''la, (jonquil, M. ♃.) spatha many-flowered; nectary bell-form, short; leaves subulate. Ex.

poet"icus, (poet's narcissus, 2f.) spatha 1-flowered; nectary wheel-form, very short, scarious, crenulate; leaves inflexed at the margin. Ex.

NAR"DUS. 2—1. (*Cyperoideæ*.) [From *nardos*, spikenard, a term applied to aromatic grasses.]

stric"ta, (mat grass, 2f.) spike setaceous, straight; flowers 1-sided.

NARTHE'CIUM. 6—1. (*Junca.*) [From *narthex*, fennel.]

america'num, (y. Ju. 2f.) racemes lax, sometimes interruptedly spiked; pedicels with a setaceous bract below the flower, and another embracing the base; filaments with very short hair; leaves narrow-ensiform; flowers in a terminal spike or raceme; scape 1 f. Sandy swamps.

NASTUR"TIUM. 14—2. (*Cruciferæ.*)

officina'le, (w. Ju. 2f.) leaves pinnate; leafets ovate, sub-cordate repand; stem decumbent; petals longer than the calyx.

palus"tre, (J. Au. 2f.) leaves lyrate-pinnatifid; lobes confluent, unequally dentate, smooth; petals as long as the calyx; siliques short, turgid; root ensiform. 1-2 f.

amphib"ium, (y. Ju. 2f.) leaves oblong-lanceolate, pinnatifid or serrate; root fibrous; petals longer than the calyx; siliques elliptical. 1-3.

his"pidum, stem tomentose-villose; leaves somewhat villose, runcinate-pinnatifid; lobes rather obtusely-toothed; siliques ovate, turgid; petals not quite as long as the calyx.

sylves"tre, (y.) leaves pinnately divided; segments lanceolate, serrate, or incised; petals longer than the calyx; siliques oblong, somewhat torulose; style very short. Introduced.

tanacetifo'lium, (M. ♂. y.) leaves pinnately divided; segments sinuate-pinnatifid or toothed; siliques oblong-linear, nearly erect, acute; style short. S.

sessiliflo'rum. (y.) leaves cuneate-obovate, obtuse, repand-toothed or nearly entire; siliques sub-sessile, linear-oblong, obtuse, tipped with the nearly sessile stigma. S.

NEC"TRIS. 6—2. (*Ranunculaceæ.*) [From Gr *nekton*, swimming or floating.]

aquat"ica, (g-w. M. 2f.) submersed leaves opposite, many-parted, capillary; floating ones alternate, elliptic, peltate; flowers in racemes. S.

NEGUN"DO. 20—5. (*Acerineæ.*)

califor"nicum, young leaves villose, 3-foliate; leafets 3-lobed; lobes incised or toothed. California. S.

NELUM"BIUM. 12—12. (*Ranunculaceæ.*)

lute'um, (water chinquepin, sacred bean, Indian lotus, w. y. Ju. 2f.) corolla many-petalled; anthers produced in a linear appendage of the extremity; leaves peltate-orbicular, very entire. Lakes. Flowers larger than those of any other plant in North America, except one species of magnolia.

penta'pet"alum, (w. Ju. 2f.) leaves pel-

tate, orbicular, entire; calyx 5-sepalled; petals 5. S.

NEMOPAN"THES. 4—4. (*Rhamni.*) [From *nemos*, grove, and *panios* (from *pas*), all, common in all groves.]

canaden"sis, (wild holly, Canadian holly, g-y. M. ♄.) leaves deciduous, ovate-oblong, very entire, smooth, mucronate; peduncles sub-solitary, very long, 1-flowered; fruit obtusely quadrangular. Berries deep red. 3-6 f.

NEMOPHI'LA. 5—1. (*Boragineæ.*) [From *nemos*, a grove, and *phileo*, to love; so called from its habit.]

panicula'ta, (b. M. ♂.) very hairy; radical leaves sub-pinnatifid; cauline ones angularly lobed; divisions of the calyx with minute, oval appendages; flowers on short peduncles, somewhat paniculate. Moist woods.

phaceloi'des, (b. M. ♂.) succulent; stem 3-sided; leaves alternate, pinnatifid; peduncles very long, 1-flowered, opposite the leaves, and terminal.

NEOT"TIA. 18—1. (*Orchideæ.*) [The name is from the Greek, and signifies bird's nest.]

torti"lis, (summer ladies'-tresses. w. Ju. 2f.) radical leaves linear; scape sheathed; flowers spirally secund; lip somewhat 3-lobed; middle lobe larger, crenulate. 12 i.

gra'cilis, (ladies'-tresses, w. Ju.) radical leaves ovate; scape sheathing; flowers in a spiral row; lip obovate, curled; scape 8-12 inches, with a few sheathing leafets or scales; leaves on short petioles, sometimes falling off before the plant blossoms; flowers in a twisted spike. Var. *secun"da*,spike scarcely twisted, flowers more slender. Dry woods.

cer"nua, (nodding ladies'-tresses, w. Au. 2f.) leaves lanceolate, nerved; flowers in a dense spike, nodding; lip oblong, entire, acute.

odora'ta, (w.) stem leafy, glabrous; leaves lanceolate, acuminate; radical ones very long: spike somewhat loose; flowers recurved; lip ovate, entire; margin undulate, sub-fimbriate. S.

NEPE'TA. 13—1. (*Labiata.*) [Name is said to have been derived from Nepet, a town in Tuscany.]

cata'ria, (catmint, catnep, b-w. 2f.) hoary-pubescent; flowers in whorled spikes; leaves petioled, cordate, tooth-serrate.

NICOTIA'NA. 5—1. (*Solaneæ.*) [From Nicot, who first introduced it into Europe.]

taba'cum, (Virginian tobacco, w-r. Ju. ☉.) leaves lance-ovate, sessile, decurrent; flowers acute. Naturalized at the north.

rus"tica, (common tobacco, g-y. Au. ☉.) viscid-pubescent; stem terete; leaves petioled, ovate, very entire; tube of the corolla cylindrical, longer than the calyx; segments round, 12-18 i. Flowers in a terminal panicle or raceme. Introduced.

panicula'ta, (small-flowered tobacco, w-r. Ju. ☉.) leaves petioled, cordate, entire; flowers on pedicels, obtuse, clavate. Ex.

quadrival"vis, (b-w. Ju. ☉.) leaves oblong-ovate, petioled; flowers scattered, solitary, near the summit of the branches; corolla funnel-form; divisions oblong, acutish; capsule sub-globose, 4-valved. *S.*

NIGEL''LA. 12—4. (*Ranunculaceæ.*) [From *niger*, black, on account of its black seed.]

♂ damasce'na, (fennel-flower, lady-in-the-green, b. M. ☉.) flowers surrounded with a leafy involucrum, composed of linear bracts.

sati'va, (nutmeg-flower,) pistils 5; capsules muricate; roundish leaves sub-pilose, pinnatifid.

NOLI'NA. 6—3. (*Junci.*)

georgia'na, (W. M. ♃.) leaves long-linear, coriaceous, dry; scape with small subulate scales near the base; panicle racemose, spreading. 2-3 f.

NOS''TOC. 21—4. (*Algæ.*)

commu'ne, on the earth; frond ventricose, gelatinous. On the earth after a storm; an inch or two in extent; olive green.

NU'PHAR. 12—1. (*Papaveraceæ.*) [From the Greek, signifying water-lily.]

kalmia'na, (water-lily, Kalm's water-lily, Ju. ♃.) leaves cordate, lobes near each other; calyx 5-leaved; stigma gashed, with 8-12 radiated lines; leaves and flowers small.

lute'a, (yellow water-lily, y. Ju. ♃.) calyx with 5 obtuse sepals; stigma entire, 16-20 rayed; leaves cordate-oval; petals much smaller than the sepals, truncate. Water.

adve'na, calyx with 6 sepals; petals numerous, small; petioles semi-cylindrical.

sagittæfo'lia, (y. Ju. ♃.) leaves long, cordate-sagittate, obtuse; calyx 6-sepalled; petals none; anthers sub-sessile; stigma entire. Water. *S.*

NUTTAL''LIA. 15—12. (*Malvaceæ.*) [In honor of Thomas Nuttall.]

digita'ta, (r. M. ♃.) glaucous; lower leaves obsoletely digitate, sub-peltate; divisions linear; segments glabrous; upper leaves 3-parted and simple; peduncles somewhat racemed, very long. Poppy-like. 3-4 f.

NYMPHÆ'A. 12—1. (*Papaveraceæ.*) [From *nympha*, water-nymph.]

odora'ta, (pond-lily, w. Ju. ♃.) leaves round-cordate, entire, sub-emarginate; lobes spreading asunder, acuminate, obtuse; petals equalling the 4-leaved calyx; stigma 16-20-rayed; flowers large, odorous. The Egyptian lotus belongs to this genus.

NYS''SA. 20—5. (*Æliagni.*)

multiflo'ra, (sour or black gum, y-g. M. ♄.) leaves lanceolate, very entire, acute at each end; the petiole margined, and midrib villose; fertile peduncles many-flowered; flowers in umbellate clusters; drupe nearly round, dark blue. Low woods. 30-50 f.

biflo'ra, (tupelo-tree, swamp horn-bean,) leaves ovate-oblong, very entire, acute at each end, smooth; fertile peduncles 2-flowered; drupe oval, compressed. Swamps. 30-50 f.

aquat''ica, (M. ♄.) leaves oblong-lanceolate, entire, acute at each end, glabrous; fruit bearing peduncles 2-flowered.

capita'ta, (M. ♄.) leaves on short petioles, oblong-lanceolate and oval, nearly entire, pubescent and somewhat hoary beneath; staminate florets capitate; pistillate 1-flowered. *S.*

tomento'sa, (M. ♄.) leaves on long pet ioles, oblong, acuminate, acutely toothed, tomentose beneath; fruit bearing peduncles 1-flowered. *S.*

uniflo'ra, (Ap. ♃.) leaves on long petioles, oblong, acuminate, sparingly and angularly toothed, slightly pubescent beneath, lower ones sometimes cordate; fruit bearing peduncles 1-flowered. Swamps. *S.*

OBOLA'RIA. 13—2. (*Pediculares.*)

virgin''ica, (penny-wort, r. Ap. ♃.) stem simple; leaves oblong, truncate fleshy, purple beneath; flowers axillary, solitary, sessile. 3-4 i.

OCY'MUM. 13—1. (*Labiatæ*) [From *okus*, swift, on account of its rapid growth.]

basil''icum, (basil, ☉.) leaves ovate, glabrous; calyx ciliate. 6-12 i.

ŒNAN''THE. 5—2. (*Umbelliferæ.*)

sarmanto'sa, stem branching, weak, somewhat climbing; leaves gash-bipinnate; segments ovate, dentate, terminal one somewhat 3-lobed; umbels opposite the leaves, many-rayed; involucre 0; fruit oblong; style very long.

filifor''mis, (w. Au. ☉.) leaves simple, terete, jointed, acute; nut winged. Harper's Ferry.

ŒNO'THERA. 8—1. (*Onagra.*)

Capsules elongated, sessile.

bien''nis, (scabish, tree-primrose, y. J. ♂.) stem villose, scabrous; leaves lance-ovate, flat-toothed; flowers sub-spiked, sessile; stamens shorter than the corolla. 3-5 f.

parviflo'ra, (y. Ju. ♂.) stem smooth, sub-villose; leaves lance-ovate, flat; stamens longer than the corolla.

grandiflo'ra, (y. Ju. ♂.) stem nearly smooth, branched; leaves ovate-lanceolate, glabrous; flowers axillary, sessile, large; petals obcordate; stamens declining, shorter than the corolla. 2-3 f. Introduced.

murica'ta, (y. Ju. ♂.) stem purplish, muricate; leaves lanceolate, flat; stamens as long as the corolla. 1.2 f.

sinua'ta, (y. J. ☉.) stem diffuse, pubescent; leaves oval-oblong, toothed and sinuate; flowers axillary, villose; capsules prismatic. Var. *min''ima*, stem low, simple, 1-flowered; leaves entire.

Capsules obovate, clavate, angular, mostly pedicelled.

frutico'sa, (shrubby œnothera, sun-drop, y. Ju. ♃.) pubescent; stem branching from the base, divaricate; leaves sessile, lanceolate, acute, slightly toothed, pilose; flowers in a terminal raceme; petals broad-obcordate. Shady woods. Stem 12-18 inches high, purple. Var. *ambig''ua*, has smaller flowers.

hybri'da, stem erect, villose; leaves pubescent on both sides, lanceolate, remotely toothed, undulate; capsules somewhat spiked; flowers pale yellow. 9-18 i.

chrysan"tha, (dwarf-scabish) stem slender, minutely pubescent; leaves lanceolate, rather obtuse, flat, entire; segments of the calyx twice as long as the tube; capsule sessile; flowers small, bright yellow. -Mountains.

inca'na, (y.) stem slender, erect; leaves flat, hoary and tomentose, very entire, elliptic-ovate, acute; raceme few-flowered, naked; capsules sub-sessile, oblong, quadrangular. 6-9 i.

pu'mila, (y. Ju. ♃.) smooth; stem ascending; leaves lanceolate, entire, obtuse; capsules obovate, angled, sub-sessile. 8-12 i.

longifo'lia, (y. Au. ♂.) leaves lance-ovate denticulate; stem simple, very hairy; petals not in contact at the base, 2-lobed. Ex.

glau'ca, (y. ♃.) very glabrous; leaves broad oval, repand-denticulate, smooth, glaucous; capsule ovate-quadrangular, pedicelled. S.

linea'ris, (y. Ju. ♃.) pubescent, slender; leaves linear, entire; capsules long-peduncled, roundish, 4-angled, villose; corolla large. S. ♦

cæspito'sa, (w. r. Ju. ♃.) stemless; leaves lanceolate, gash-toothed; capsules oblong, sessile; tube of the calyx long; petals 2-lobed, distant. S.

pinnatifi'da, (w. J. ♂.) minutely pubescent; stem decumbent; radical leaves nearly entire; cauline ones pinnatifid; divisions linear, acute; capsules prismatic, grooved. 6-24 i. S.

frase'ri, (Ju. ♃.) nearly glabrous; stem near the base simple; leaves ovate, petiolate, denticulate; racemes leafy; capsules pedicelled, obovate, quadrangular. S.

macrocar"pa, (y. Au. ♂.) nearly glabrous; stem branching; leaves lanceolate, petioled, distantly glandular-denticulate, white-silky on the margin and nerves; petals obcordate, pointed; capsules elliptic, 4-winged, very short pedicels. S.

trilo'ba, (y. M. ☉. ♃.) stemless; leaves interruptedly pinnatifid, toothed, glabrous; petals 3-lobed at the apex; capsules 4-winged, large. S.

specio'sa, (w. J. ♃.) finely pubescent; leaves lance-oblong, toothed, sub-pinnatifid; raceme naked, at first nodding; capsules obovate, angled; stem suffruticose. S.

linifo'lia, (y. J. ♂.) leaves entire; radical ones lanceolate; cauline ones linear, crowded; raceme naked, terminal; capsule obovate, angled, pubescent; petals obcordate, longer than the stamens; stigma 4-lobed. S.

serrula'ta, (y. J.) leaves linear, somewhat spinose-serrate, acute; flowers axillary; sepals keeled; stigma 4-lobed; capsules cylindric, erect; stem suffruticose. S.

humifu'sa, (w.) stem prostrate, branching, villose; leaves lance-linear, sub-dentate or entire, silky-villose; flowers axillary; calyx villose; tube longer than the germ; capsule prismatic. S.

suaveo'lens, (y.) stem, calyx, and capsule, sub-pillose; leaves lance-ovate, obsoletely-toothed; petals large, emarginate; capsules elongated, furrowed, nearly uniform in thickness. Cultivated.

noctur'na, and *villo'sa*, both biennial plants, natives of Good Hope.

accau'lis, (♃.) a native of Chili.

OLE'A. 2—1. (*Jasminea.*) [Name from the Celtic word *olea*, signifying oil.]

america'na, (American olive, w. M. ♄.) leaves lanceolate-elliptic, entire; racemes compressed; bracts all persistent, connate, small. S.

europæ'a, leaves lanceolate, entire; racemes axillary, crowded. The drupes when green are used for pickles, when ripe they afford the oil called olive oil. Ex.

ONO'CLEA. 21—1. (*Felices*) [From *onos*, a vessel, and *kleid*, to close.]

sensib"ilis, (sensitive fern, J. ♃.) barren frond pinnate; fertile one doubly pinnate; stem glabrous. The leaflets slowly approach each other on pressing the stem in the hand.

obtusilo'ba, barren frond pinnate; fertile one doubly-pinnate; stem scaly.

ONOPOR"DON. 17—1. (*Cinarocephalæ.*)

acan"thium, (cotton thistle, p. Ju. ♂.) calyx scaly, scales spreading; leaves ovate-oblong, sinuate. Naturalized. Ex.

ONOSMO'DIUM. 5—1. (*Boraginea.*)

his"pidum, (y.-w. J. ♃.) very hispid; leaves lance-oval, acute, papillose, punctate; segments of the corolla subulate. 1-2 f.

mol"le, (w. Ju. ♃.) whitish villose; leaves oblong-oval, somewhat 3-nerved; segments of the corolla sub-oval.

OPHIOGLOS"SUM. 21—1. (*Filices*.) [From *Ophis*, serpent, and *glossa*, tongue.]

vulga'tum, (adder-tongue fern, ♃.) spike cauline; frond oblong-ovate, obtuse, closely reticulate.

bulbo'sum, (♃.) root bulbous; frond heart-ovate, obtuse; spike cauline. 6 i.

pusil"lum, spike cauline; frond cordate. 1 f. S.

OPHIORHI'ZA. 5—1. (*Gentiana.*) [From *ophis*, serpent, and *riza*, root, because the root is used as a specific against the bite of the viper in Judea.]

mitreo'la, (w. Ju. ☉.) leaves ovate, sessile with the margins scarious; stem erect. 12-18 i. S.

lanceola'ta, (w.-p. Au. ☉.) leaves long, lanceolate, finely serrulate, tapering at base. 18 i. S.

OPLOTHE'CA. 15—5. (*Gerania*)

florida'na, (w. Ju.) stem erect, pubescent, with tumid joints; leaves sessile, lance-linear, scabrous above, lanuginous beneath. 3-4 f.

OR"CHIS. 16—1. (*Orchidea.*) [A name derived from the Greek.]

spectab"ilis, (r. M. ♃.) lip obovate, undivided, crenate, retuse; petals straight; lateral ones longest; spur clavate, shorter than the germ; bracts longer than the flowers; stem leafless. 3-6 i.

tridenta'ta, (w. Ju. ♃.) lip ovate-lanceo

late, obtuse, 3-toothed ; petals obtuse ; spur filiform, longer than the germ. 6-12 i.

fla'va, (y. Ju. ♃.) lip 3-cleft, entire ; middle division larger ; spike compact ; bracts longer than the flower. 1-2 f.

fimbria'ta, (p. Ju. ♃.) lip 3-parted ; lobes all incisely fimbriate and wedge-form ; segments of the perianth oval, spreading, fimbriate-toothed ; spur filiform, clavate, longer than the germ ; leaves broad-lanceolate ; purple flowers, in a large spike. 2 f. Meadows.

obsole'ta, (J. ♃.) lip oblong, lanceolate, undivided ; petals erect ; horn obtuse, as long as the germ ; bracts very short ; root oval or palmate.

vires'cens, (g. Ju. ♃.) lip lanceolate, crenate ; petals connivent ; bracts longer than the flowers ; spur sub-inflated, obtuse ; root fascicled. 12-18 i.

hyperbo'rea, (g. y. Ju. ♃.) lip lanceolate, as large as the spreading petals ; spur subulate, shorter than the germ ; spike ovate ; bracts as long as the flowers ; roots fascicled.

obtusa'ta, (♃.) lip linear, very entire, longer than the horn which is the length of the germ ; leaf solitary, radical, sub-wedge-form, obtuse ; roots fascicled.

rotundifo'lia, (♃.) lip 3-cleft, middle segment 2-cleft ; spur shorter than the germ ; leaves roundish-oval ; root fascicled.

nive'a, (w.) lip linear, oblong, entire ; petals spreading ; horn filiform, longer than the germ ; lower leaves linear, very long, upper ones subulate. *S.*

vir''idis, (g w. Ju.) lip linear, 3 toothed at the apex ; petals connivent ; horn obtuse, sub-inflated ; bracts much longer than the flowers ; root fascicled. 3 i. *S.*

bidenta'ta, lip oval, oblong, 2-toothed at the base ; petals ovate, expanding ; horn shorter than the thickened germ ; leaves narrow-lanceolate ; stem nearly naked. 12-18 i. *S.*

See HABENARIA.

ORIGA'NUM. 13—1. (*Labiatæ*) [From *oros*, a mountain, and *gano*, to rejoice, so called because it grows upon the mountain sides.]

vulga're, (wild marjoram, r. Ju. ♃.) spikes round, panicled, heaped ; bracts ovate, longer than the calyx. 1-2 f.

majora'na, (sweet marjoram, ♄.) spikes roundish, ternate, compact, peduncled ; leaves petioled, oval, obtuse, smoothish. 6-12 i. Ex. A native of Portugal.

ORNITHOG''ALUM. 6—1. (*Asphodeli*.) [From ornis, a bird, and gala, milk, from the color of its flowers.]

umbella'tum, (star of Bethlehem, M. ♃.) flowers corymbed ; peduncles longer than the bracts ; filaments subulate. Naturalized. 6-8 i.

bractea'tum, (p. ♃.) scape bracted, 1-flowered, terete ; petals lance-oblong, obtusish ; filaments linear ; leaves channeled, filiform. 4 i. *S.*

pyramida'le, (prussian asparagus,) a native of Spain.

latifo'lium, a native of Egypt.

cauda'tum, Cape of Good Hope.

OR''NUS. 2—1. (*Jasmineæ*.) [From the Hebrew orn, an ash.]

america'na, (M. ♄.) leafets broad-ovate, serrate, terminal one obcordate. Shady woods. Resembles the genus fraxinus.

europe'a, affords the manna of commerce. The American ornus is thought by some to be but a variety of this. Ex.

OROBAN''CHE. 13—2 (*Pediculares*.) [From *orobos*, the wild pea, and *agchein*, to suffocate, so called because it twines around the orobos and destroys it.]

uniflo'ra, (cancer-root. b-w. M. ♃.) stem very short ; peduncles 2, elongated, scape-form, 1-flowered. naked ; scales smooth, concave ; lobes of the corolla oblong-oval, with a pubescent, colored margin. 4-6 f Parasitic. Woods.

america'na, (Ju. ♃.) stem simple, covered with ovate-lanceolate, imbricate scales ; spike terminal, smooth ; corolla recurved ; stamens exserted ; flowers brownish yellow, the spike covered by the scales of the stem. 6-8 i. Parasitic. Woods.

ludovicia'na, (p.) pulverulent, pubescent ; stem low, simple ; flowers and ovate-acute scales sub-imbricate ; calyx 2-bracted ; corolla recurved ; stamens enclosed, smooth. 3-4 i.

fascicula'ta, (p. Ju.) stem short, simple ; peduncles numerous, naked, nearly terminal, about the length of the stem ; scales few, ovate, concave, pubescent. 4-5 i.

ma'jor, stem erect, somewhat scaly and bulbous at the base, terminating in a spike of rather large purplish or brownish flowers. Parasitic. Ex.

ru'bra, very fragrant, similar in appearance to the preceding species. Ex.

ORO'BUS. 16—10. (*Leguminosæ*.) [From *erepto*, to eat, the root being considered nutritious.]

dis'par, (ervum, w-y. J. ♃.) leaves unequally pinnate ; leafets linear, obtuse ; stipules ovate, acute ; racemes sessile.

tubero'sus, the heath-pea. The Scotch islanders chew the root ; they hold the plant in high esteem. Ex.

ORON''TIUM. 6—1. (*Aroïdæ*.) [From *Orontes*.]

aquat''icum, (golden club, y. M. ♃.) leaves all radical, lance-ovate ; scape cylindrical, spiked ; flowers with a peculiar smell. Water. 1-2 f

ORTHOCAR''PUS. 13—2. (*Pediculares*.) [From orthos, erect, and carpos, fruit.]

lute'us, (y. Ju.) stem simple, terete, hirsute ; leaves alternate, sessile, acute, entire ; calyx, bracts and leaves viscid-pubescent. 12-14 i. *S.*

ORY'ZA. 6—2. (*Gramineæ*.) [From ores, Arabian.]

sati'va, (rice, ⊙.) culm jointed ; leaves clasping ; panicle terminal. Ex.

ORYZOP''SIS. 3—1. (*Gramineæ*.) [From oruza, rice, and opsis, resemblance.]

asperifo'lia, (mountain rice, M. ♃.) culm nakedish ; leaves rigid, erect, pungent at the point ; flowers in a racemose panicle. 18 i.

OSMORHI'ZA. (See URASPER''MUM.)

OSMUN''DA. 21—1. (*Filices*.) [From Osmund, who first used it as a medicine.]

cinnamo'mea, (flowering-fern, y. J. ♃.)

barren frond doubly pinnatifid ; segments oval, entire ; fertile fronds with opposite racemes, woolly. ·3-6 f.

clayto'nia'na, (2f.) fronds pinnate, bearing fruit at the summit, pinnate-pinnatifid, having small dense masses of fulvous down at their origin. 20-30 i.

rega'lis, (r-y. J. 2f.) frond bipinnate, terminal in several racemes, very branching, without hairs. Wet.

OS"TRYA. 19—12. (*Amentacea.*) [From *osteon*, a bone, on account of its hardness.]

virgin"ica, (iron-wood, hop-hornbeam, g. M. ♂.) leaves alternate, ovate-oblong, subcordate, acuminate, unequally serrate ; strobilums oblong-ovate. A small tree, with very hard and heavy wood. Fertile flowers enlarged into a sort of oblong cone, resembling the common hop. Woods. Can. to Car.

OXA'LIS. 10—5. (*Geraniæ.*) [From *oxus*, sour, on account of the juice.]

acetosel"la, (wood-sorrel, w. r. M. 2f.) stemless ; scape 1-flowered, longer than the leaves ; leaves ternate, broad-obcordate, with rounded lobes ; styles as long as the inner stamens ; root dentate.

viola'cea, (violet wood-sorrel, p. J. 2f.) stemless ; scape umbelliferous, 3-9 flowered ; flowers nodding ; leaves ternate, obcordate, smooth ; styles shorter than the outer stamens. Scape 4-6 i. Rocky woods.

stric"ta, (upright wood sorrel, y. J. ⊕.) hairy ; stem erect, sometimes procumbent, branched ; umbels about as long as the leaves ; leaves ternate, obcordate ; petals obovate, entire ; styles as long as the inner stamens. 4-10 i. Sandy fields. Flowers small, 4-6 in an umbel.

corniculata, (lady's wood-sorrel, y. M. ⊕.) pubescent ; stem prostrate ; umbels as long as the petioles ; petals obovate, slightly emarginate ; styles as long as the interior stamens ; leaves ternate, obcordate.

OXYCOC"CUS. 8—1. (*Ericea.*) [From *oxus*, sour, and *coccus*, a berry.]

macrocar"pus, (cranberry, r. J. ♄.) creeping ; stem ascending ; leaves oblong, flattish, obtuse, becoming white beneath ; pedicels elongated ; divisions of the corolla lance-linear ; berry large, bright scarlet. Wet grounds.

palus"tris, (J. ♄.) divisions of the corolla ovate ; berries purple, smaller than the preceding. Alpine bogs.

erythrocar"pus, (J. ♄.) erect ; leaves oval, acuminate, serrulate, and ciliate ; pedicels axillary ; corolla long, at last revolute, very bright scarlet, transparent, of exquisite flavor. A small shrub. *S.*

OXY'RIA. 6—2. (*Polygoneæ.*) [From *oxus*, acid, alluding to the leaves.]

renifor"mis, (J. 2f.) outer sepals oblong, half the length of the inner obovate valves ; radical leaves reniform, long-petioled ; stamens 2 ; styles 2. 2-3 i.

OXYTRO'PIS. 16—10. (*Leguminosæ.*)

lamber"ti, (p. Ju. 2f.) stemless ; silkypilose ; leafets numerous, oblong, acute at each end ; scape about equal to the

leaves ; spikes oblong, capitate ; bracts lance-linear, about equal to the calyx.

uralen"sis, (r-w.) stemless, villose, or silky ; leafets lance-oblong ; scapes longer than the leaves, scape and calyx hirsutewoolly, small heads many-flowered, ovate ; flowers spreading ; lower bracts longer than the calyx.

campes"tris, (y.) stemless ; leafets manypaired, lanceolate, silky ; scape often decumbent, sub-equal to the leaves ; spikes capitate or elongated ; bracts a little shorter than the calyx ; legumes erect, ovate, inflated, pubescent ; flowers erect.

PACHYSAN"DRA. 19—4. (*Euphorbia.*) [From *pachus*, thick, clumsy, and *aner* (andros), stamen.]

procum"bens, (g-w. J. 2f.) stem procumbent ; leaves short, oval, crenate-toothed above ; calyx minute-ciliate ; capsule finely pubescent. *S.*

PÆO'NIA. 12—3. (*Papaveraceæ.*) [From *Pæon*, who is said to have first applied it to medicinal purposes.]

officina'lis, (peony, r. J. 2f.) leaves decompound ; leafets lobed, lobes broad-lanceolate ; capsules downy. Ex.

brown"ii, (r-p. J. Ju.) carpels 5, oblong, very glabrous, erect ; leaves smooth on both sides, somewhat glaucous, biternate ; leafets ternately divided or pinnatifid, laciniate ; lacinia oblong, those of the lower leaves obtuse.

califor"nica, (Mar. A. r.) carpels 3, glabrous ; leaves smooth on both sides, ternate ; leafets broadly cuneate, nearly twice 3-cleft ; lacinia oblong-lanceolate, acute. Upper California.

PA'NAX. 5—2. (*Araliæ.*) [From *pan*, all, and *akos*, medicine, on account of its great virtues.]

quinquefo'lia, (ginseng, w. M. 2f.) root fusiform ; leaves ternate, or quinate ; leafets oval, acuminate, petioled-serrate. 1-2 f.

trifo'lium, (dwarf ginseng,) root tuberous, roundish ; stem simple, smooth ; leaves ternate ; leafets sub-sessile, lance-oblong, serrate ; styles often 3 ; berry 3-seeded. Woods. 4-6 i.

hor"ridum, fruticose, somewhat creeping, aculeate ; leaves palmate-lobed, gash-serrate ; umbels capitate, racemed ; 2 styles, 2 seeds. 10-12 f.

PANCRA'TIUM. 6—1. (*Narcissi.*) [From *pan*, all, and *krateo*, to conquer ; supposed by the ancients to have been a powerful medicine.]

mexica'num, (w. M. 2f.) spatha about 2-flowered ; leaves lance-oblong ; 6 teeth of the nectary bearing stamens, 6 simple. 18-24 i. *S.*

mariti'mum, (w. Ja. 2f.) spatha many-flowered ; leaves linear-lanceolate ; nectary funnel-form, with 12 teeth not bearing stamens. 12-18 i. *S.*

rota'tum, (w. Ju. 2f.) spatha many-flowered ; leaves linear-lanceolate ; nectary salver-form, tubular below, with 6 stamens bearing teeth, intermediate one gash-toothed ; stamens twice as long as the nectary. Splendid lily-like bulbous plants. *S.*

PAN"ICUM. 3—2. (*Gramineæ.*)

crus-gal"li, (barn-grass, Au. ⊙.) racemes alternate and in pairs; compound rachis 5-angled; glumes terminating in hispid bristles; sheath glabrous. 2-4 f.

his"pidum, (S. Oc. ⊙.) panicle compound; nodding racemes alternate; glumes terminating in hispid bristles; sheaths hispid. Salt marshes. 3-4 f.

clandesti'num, (Au. ⅛.) culm with short axillary branches; leaves broad-lanceolate, cordate at base; sheaths hispid, enclosing the short panicle; abortive floret neutral, 2-valved, upper valve obtuse. Moist woods. 2-3 f.

latifo'lium, (Ju. ⅛.) culm mostly simple, bearded at the joints; leaves oblong-lanceolate, smooth, or with the sheaths somewhat pubescent; panicle terminal, a little exsert, simple, pubescent; spikelets oblong-ovate; abortive floret antheriferous, 2 valved. 1 f.

pubes"cens, (J. ⅛.) erect, much branched, leafy, softly pubescent; leaves lanceolate, ciliate; panicle small, few-flowered, free; spikelets sub-globose, ovate, pubescent. 18 i.

dichot"omum, (Au. ⅛.) culm much branched, dichotomous above; branches fasciculate; leaves numerous, lanceolate, smooth; panicle simple, capillary, lax; abortive floret neutral; upper valve minute, bifid. Var. *curva'tum,* culm very tall; branches few, somewhat curved. Var. *fascicula'tum,* culm low, erect, or decumbent; branches and leaves dense-fascicled; panicles small, concealed among the leaves. Var. *gra'cili,* culm tall, slender; leaves membranaceous. 8-16 i.

ni'tidum, (panic grass, Ju. ⅛.) culm slender, simple, smooth; sheaths bearded at the throat; leaves few, broad-linear; panicle capillary, rather crowded, compound, remote, smooth; spikelets minute, obtuse, ovate, slightly pubescent; lower glume very small. Var. *cilia'tum,* culm hairy; leaves linear-lanceolate (lower ones broader), sparingly hirsute, ciliate on the margin; panicle with the branches and flowers pubescent. Var. *ramulo'sum,* culm more branched; panicle contracted; branches smooth. Var. *pilo'sum,* culm very hairy; lower leaves approximate and broad-lanceolate, upper ones linear, rather rigid, subpilose above, ciliate at base; sheaths villose, minutely papillose; panicle sub-contracted; branches virgate, flowers pubescent. Var. *gla'brum,* smooth except the base of the leaves, nearly simple; lower leaves short, approximate, sub-cartilaginous; panicle branched, almost verticillate; spikelets large; upper valve of the abortive floret entire. Var. *barba'tum,* culm simple, smooth; nodes hairy; leaves linear-lanceolate; sheaths smooth, except on the margin; flowers minutely pubescent. 1-2 f.

agrost"oi'des, (Au. ⅛.) culm compressed, smooth, erect; leaves very long; panicles lateral and terminal, pyramidal, spreading; branches bearing racemes; spikelets appressed; abortive floret neutral; valves nearly equal. 2-3 f.

virga'tum, (Au. ⅛.) whole plant very smooth; panicle diffuse, very large; flowers acuminate; abortive-floret nearly equal. 3-4 f.

capilla're, (S. ⊙.) culm nearly simple; sheaths very hairy; panicle large, capillary, expanding, loose; spikelets long-peduncled, acuminate, smooth; abortive floret 1-valved. Var. *sylvat"icum,* culm branched at the base, very slender; leaves linear. 1-2 f.

fus"co-ru'bens, (Au.) racemes linear, virgate; glumes clavate, colored, hairy under the divisions of the panicle. S.

- *ama'rum,* (Oc. ⅛.) very glabrous; leaves thick, glaucous, very bitter; panicle appressed; glumes acuminate. 2-3 f. S.

angustifo'linm, panicle few-flowered, expanding; leaves scattered, linear-lanceolate, glabrous beneath, sparingly ciliate. 1-3 f. S.

PAPA'VER. 12—1. (*Papaveraceæ.*) [From *pappa,* pap, so called because nurses mixed this plant in children's food to make them sleep.]

somnif'erum, (opium poppy, J. ⊙.) calyx and capsule glabrous; leaves clasping, gashed, glaucous. Ex.

rhé'as, (red corn-poppy, r. J. ⊙.) capsules glabrous, sub-globose; stems many-flowered, pilose; leaves gash-pinnatifid. Ex.

nudicau'le, (y. ♂.) capsule hispid; scape 1-flowered, naked, hispid; leaves sub-pinnate; leaflets lanceolate, lower ones somewhat gashed.

bractea'ta, (r. J.) capsules smooth; stem 1-flowered, rough; leaves scabrous, pinnate, serrate; flowers subtended by leafy bracts. 3 f. Ex.

du'bium, (r. w. Au. ⊙.) leaves pseudo-pinnate; segments lance-oblong, pinnatifid, incised, sessile, decurrent; stem with spreading hairs; peduncles with appressed bristly hairs; capsules obovoid-oblong, smooth. Naturalized in Chester co., Pa. 1-2 f.

PARIETA'RIA. 19—4. (*Urticeæ.*) [From a Latin word signifying wall, from its growing on old ruins.]

penn'sylva'nica, (pallitory J. ⊙.) leaves lance-oblong, veiny, opake, dotted; involucrum 3-leaved, longer than the flower. 12-15 i.

florida'na, (J.) leaves ovate, nearly round, obtuse, opake, dotted; flowers clustering as long as the involucrum; stem assurgent. 12-18 i. S.

PARME'LIA. 21—5. (*Algæ.*) [From *parma* shield, and *eilo,* to enclose.]

capera'ta, (shield lichen,) frond orbicular, pale yellow, becoming green, rugose, at length granulated, dark and hispid beneath; lobes plicate, sinuate-laciniate, roundish, somewhat entire; receptacles scattered; margin incurved, entire, at length pulverulent. On old timber, &c.

PARNAS'SIA. 5—4. (*Saxifrageæ.*) [From Mount Parnassus, the seat of the Muses.]

america'na, (flowering plantain, w. y. p. Ju. ⅛.) leaves radical, (often a leaf on the scape,) heart-orbicular, 5-9-nerved; nectaries 5, each divided into 3 filaments termi-

nated by little spherical heads. Damp or
wet. 6-18 i.

palus"tris, leaves all cordate, cauline
ones sessile; scale smooth, many-bristled;
flowers white, with veins of green or purple. Bog meadows.

parvifo'ra, very slender, with slender,
bristly scales, about 5; radical leaves ovate,
tapering into the petiole; cauline ones linear, oblong-sessile.

fimbria'ta, scales broad, wedge-form,
fleshy, crenate in the middle, within keeled, naked; radical leaves long-petioled, ear
kidney-form; cauline ones cordate, sessile,
inserted much above the middle; petals
fringed at the base.

asarifo'lia, (Ju. Au. ♃.) radical leaves
reniform; petals clawed; nectaries 3-cleft.
S.

PARTHE'NIUM. 17—4. (*Corymbiferæ.*)
integrifo'lium, (w. S. ♃.) leaves oblong,
unequally-toothed, rough; upper ones clasping. 1-2 f.

PAS"PALUM. 3—3. (*Gramineæ.*) [From *pas-
palos*, millet, which this plant resembles in
its seeds.]

ciliatifo'lium, (S. ♃.) stem decumbent;
leaves hairy, ciliate; spikes 1-2, indistinctly
3-rowed. 18 i.

seta'ceum, (paspalon-grass, ♃.) culm
erect; leaves and sheaths villose; spike
generally solitary; flowers in 2 rows. Sandy
fields. 1-2 f.

de'bile, stem weak, leaves and stem hairy;
spike generally one, slender, flowers alternate; 1-rowed. *S.*

florida'num, (J. S.) erect; lower leaves
very villose; upper ones scabrous, hairy;
sheaths long; spikes few; flowers in 2
rows, glabrous, large. 3-4 f.

præ'cox, (J.) erect; leaves lanceolate-
linear, glabrous; spikes many, alternate;
rachis narrow, hairy at base; flowers 3-
rowed. Damp soil. *S.*

dis"tichum, (creeping paspalum, joint-
grass, ♃.) stem creeping; leaves short,
somewhat glaucous, shining; spikes 2, one
sessile; glumes lanceolate. *S.*

PASSIFLO'RA. 15—5. (*Cucurbitaceæ.*) [The
term *flos passionis*, or passion-flower, was
before the time of Linnæus, applied to this
beautiful genus, because the instruments
of Christ's passion were thought to be represented by the parts of the flower.]

cæru'lea, (blue passion-flower, b. Ju. ♄.)
leaves palmate, 5-parted, entire; petioles
glandular; involucrum 5-leaved, entire;
threads of the crown shorter than the corolla. Ex.

lute'a, (yellow passion-flower, y. S. ♃.)
leaves cordate, 3-lobed, obtuse, nearly
smooth; petioles without glands; peduncles axillary, in pairs; petals much longer
than the calyx; stem climbing and slender.
Banks of streams.

incarna'ta, (w. p. Sept. ♃.) leaves
smooth; petioles with 2 glands; involucrum 3-leaved; leafets lanceolate, glandu-
lar-serrate; stem long, climbing; petals
white; nectary purple, longer than the
corolla; fruit sub-acid and spongy, eatable.

ala'ta, (winged passion-flower, Oct. ♄.)
leaves oblong-ovate, sub-cordate, entire,
veiny; petioles with 4 glands; stipules
lance-falcate; stem 4-cornered. Ex.

pelta'ta, (♄.) leaves peltate, deeply 3-
lobed, glabrous; lobes lance-linear; petioles
with 2 glands; peduncles solitary, axillary;
petals 0. *S.*

va'rei, (♃.) lower leaves 3-lobed, acute;
upper ones undivided, ovate; petioles with
2 glands; peduncles somewhat in pairs.
4-6 i. *S.* Cer.*lou'doni*, flowers of a most
brilliant crimson. Ex.

PASTINA'CA. 5—2. (*Umbelliferæ.*) [From
pasco, to feed]

sati'va, (parsnip, y. Au. ♂.) leaves simply pinnate; leafets glabrous. Var.*arven"sis*,
leafets sub-pubescent. This variety is often
found in situations which almost prove it to
be indigenous.

PEC"TIS. 17—2. (*Corymbiferæ.*)
angustifo'lius, (y. ⊙.) stem branching at
the base, diffuse; leaves narrow-linear,
mucronate, denticulate at the base; teeth
terminating in hairs; flowers terminal, solitary, short peduncled; involucrum 8-leaved;
chaff short, 5-toothed. - 3-2 i.

PEDICULA'RIS. 13—2. (*Pediculares.*) [From
pediculus, a louse.]

canaden"sis, (louse-wort, y-p. M. ♃.)
stem simple; leaves pinnatifid, gash-toothed; heads leafy at the base, hirsute; corolla with a setaceous, 2 toothed upper lip;
calyx obliquely truncate. 6-12 i.

pal"lida, (y. Ju. ♃.) stem smooth, branched; leaves sub-opposite, lanceolate, pinnatifid, toothed and crenate, scabrous on
the margin; helmet of the corolla truncated
at the apex; calyx bifid, with roundish
segments. 1-2 f.

resupina'ta, (p. Ju. ♃.) stem simple,
glabrous; leaves lanceolate, toothed, crenate; calyx bifid-truncate; upper lip of the
corol acute. Canada.

hirsu'ta, (r. ♃.) stem simple; leaves pinnate; leafets lanceolate, obtusely-toothed;
calyx hirsute, 5-cleft; upper lip of the corolla very obtuse.

ela'ta, (p. Ju. ♃.) stem simple; leaves
deeply pinnatifid; divisions lance-linear,
crenate; spike lax, somewhat leafy; calyx
glabrous, 5-toothed; upper lip of the corolla obtuse, truncate. 2 f.

gladia'ta, (y-p. J. ♃.) stem simple; leaves
lanceolate, pinnatifid, toothed; spikes leafy,
hairy; flowers alternate; capsule terminating in a long, flat point. 1 f.

PELARGO'NIUM. 15—7. (*Gerania.*) [From
pelargos, a stork, on account of the shape
of the pericarp, which resembles a stork's
bill.]

A. *Nearly stemless; roots tuberous.*

tris"te, (mourning geranium,) umbel
simple; leaves rough-haired, pinnate; leafets bi-pinnatifid; divisions oblong-acute,
flowers dark green.

daucifo'lium, (carrot geranium, ♃.) scape
very simple; leaves thrice pinnate, hirsute;
leafets lance-linear.

B. *Leaves simple, not angled.*

odoratis"simum, (sweet-scented geranium, ♄.) peduncles sub-5-flowered ; leaves round-cordate, very soft.

C. *Leaves simple, more or less angled, or lobed.*

zona'le, (horse-shoe geranium, ♄.) umbels many-flowered ; leaves heart-orbicular, obsoletely-lobed, toothed, with a colored zone around or near the margin.

ın' quinans, (scarlet geranium, ♄.) umbels many-flowered ; leaves round-reniform, hardly divided, crenate, viscid-downy.

acerifo'lium, (lemon or maple-leaf geranium, ♄.) umbels about 5-flowered ; leaves 5-lobe-palmate, serrate ; below wedge-form, undivided.

capita'tum, (rose-scented geranium, ♄.) flowers capitate ; leaves cordate, lobed, waved, soft ; stem diffuse.

quercifo'lium, (oak-leaf geranium, ♄.) umbels sub-many-flowered ; leaves cordate, pinnatifid, crenate ; sinuses rounded ; filaments ascending at the apex.

grave'olens, (sweet-rose geranium, ♄.) umbels many-flowered, sub-capitate ; leaves palmate, 5-lobed ; divisions oblong, obtuse ; margins revolute.

PENNISE'TUM. **3—2.** (*Graminea.*) [*Penna*, a feather, and *seta*, a bristle, from the feathery appearance of the involucre]

glau'cum, (fox-tail panic, J. ⊕.) perfect floret transversely rugose ; involucre of many fascicled bristles, scabrous upwards ; spike cylindrical. Var. *purpuras"cens*, sheaths hairy ; glumes and bristles of the involucre hairy. Introduced. 2 f.

pun"gens, (♃.) spike terete, strict ; involucre many-parted, 1-flowered ; segments terete, subulate, rigid, scabrous, a little longer than the florets. *S.*

PENTHO'RUM. **10—5.** (*Sempervivæ.*)

sedoi'des, (Virginian orpine, g-y. Ju. ♃.) stem branching, angled ; leaves lanceolate, sub-sessile, unequally and densely serrate ; spikes secund, terminal, panicled, alternate and cymed ; seeds pitted. 12-18 i. *S.*

PENTSTE'MON. **13—2.** (*Bignonea.*) [From *pente*, five, and *stema*, stamen. This plant, though it is placed in the class Didynamia, has the rudiment of a fifth stamen ; from hence its name.]

pubes"cens, (beard-tongue, w-p. J. ♃.) stem hairy ; leaves serrulate, lance-oblong, sessile ; flowers panicled ; the barren filament bearded from the apex to below the middle. Var. *latifo'lia*, has broad, smooth leaves. Var. *angustifo'lia*, has narrow, hairy, obscurely denticulate leaves. 1-2 f. Hill-sides.

læviga'tum, (p. J. ♃.) smooth ; leaves ovate-oblong, clasping at the base, slightly toothed, the lower ones entire ; flowers paniculate ; steril filament bearded near the top. 1-2 f. Low grounds.

frutes"cens, (p. ♄.) stem fruticose ; branches angled, pubescent above ; leaves lanceolate, obsoletely denticulate, sessile, nearly glabrous ; racemo terminal, subcorymbed ; sterile filament bearded. 12-18 i.

grandiflo'rum, very glabrous ; leaves half-clasping, ovate-oblong, entire ; upper ones roundish ; flowers solitary and axillary ; sterile filament partly pubescent at the summit ; segments of the calyx oblong, acute.

gra'cile, (p. J.) stem smooth, slender ; leaves smooth, linear, acute, half-clasping, sharp-serrulate ; sterile filament bearded longitudinally ; segments of the calyx linear-oblong. 12-24 i.

cæru'leum, (b. ♃.) smooth ; radical leaves linear, entire ; cauline ones lance-linear, entire ; all sessile ; sterile filament short, bearded above ; segments of the calyx lanceolate, acute, glabrous. *S.*

erian"thera, (p. J. ♃.) stem and leaves glabrous ; leaves sessile, lance-ovate, entire, sub-undulate at the margin ; peduncles many-flowered, secund ; segments of the calyx round-oval, acuminate ; sterile filament slightly bearded under the retuse point ; anthers pubescent. *S.*

al"bidum, (w.) low ; leaves lance-ovate, sub-serrulate, smooth, sessile ; flowers sub-fascicled, axillary and terminal ; sterile filament slenderly and interruptedly bearded ; segment of the calyx lance-linear, pubescent. 6-8 i. *S.*

dissec"tum, (p.) leaves opposite, sessile, glabrous, compoundly dissected ; segments linear and generally obtuse ; flowers in panicles ; stigma simple. 2 f. *S.*

campanula'tus, produces light purple flowers, from Mar. to Oc. 18 i. Ex.

ro'seus. has red flowers. Ex.

murraya'nus, the most beautiful species ; grows about two feet high, and produces brilliant scarlet flowers in August. Ex.

PERIPLO'CA. **18—5.** (*Apocynea*) [From *peri*, about, and *ploke*, twining.]

græ'ca, (milk-vine, p. m. ♄.) climbing, leaves opposite, ovate, acuminate ; flowers hairy within, and terminal. A native of the Canary Isles.

PETALOS'TEMON. **16—8.** (*Leguminosæ.*) [From *petalon*, a petal, and *stema*, a stamen, the petals and stamens united form a tube]

can"didum, (w. Ju. ♃.) spike cylindric, pedunculed ; bracts longer than the flower ; calyx glabrous ; leaves lanceolate, in 3 pairs.

viola'ceum, (r-p. Au. ♃.) bracts about equal to the calyx ; little bracts spatulate, caducous ; calyx silky ; leaves linear, in 2 pairs.

villo'sum, (r. Au. ♃.) villous ; stem decumbent ; spike large, cylindric, sub-sessile ; bracts shorter than the woolly, 5-toothed calyx ; leafets linear-oblong, about 7 pairs.

carne'um, (w. Ju. ♃.) spike cylindric, pedunculate ; bracts subulate, as long as the glabrous calyx ; leafets lanceolate, entire, small. 2-3 f. *S.*

corymbo'sum, (w. S. ♃.) peduncles in panicled corymbs ; calyx plumose ; leafets linear, unawned, glabrous. 2 f. Dry sandy pine barrens. *S.*

PEUCEDA'NUM. **5—2.** (*Umbelliferæ.*) [From *peuke*, fir, which its leaves resemble, and *dense*, dry or burning, alluding to the qualities of the root]

terna'tum, (sulphur-wort, 2f.) leaves ternate, long-petioled ; leaflets entire, long-linear, acute, alternated below ; involucrum nearly wanting ; involucel very short, 5-6 leaved ; fruit oblong-oval. 3 f. *S.*

PHA'CA. 16—10. (*Leguminosa.*) [A Greek word signifying lentil.]

villo'sa, (y. Ju. 2f.) nearly stemless, villose ; leaflets oval, glabrous above ; peduncles as long as the leaves ; legumes hoary ; villose, oblong.

tri-phyl''la, (2f.) stemless, assurgent ; leaves ternate ; leaflets lanceolate, sessile ; scape none ; legumes sessile, oblong. Leaves of silvery hue. *S.*

PHACE'LIA. 5—1. (*Boraginea.*) [From *phakelos*, bundle, alluding to the fascicled spikes.]

bipinnat''ifi'da, (b. M.) erect ; leaves pinnatifid ; segments incisely lobed ; racemes mostly bifid, oblong, many-flowered ; divisions of the corolla entire. 6 f.

fimbria'ta, (b. M. ☉.) leaves pinnatifid, the lobes undivided ; segments of the corolla fimbriate. 6.12 i.

parviflo'ra, (b. M. ☉.) stem diffuse, pubescent ; leaves sub-sessile, pinnatifid ; segments oblong, rather obtuse, entire ; racemes solitary ; pedicels short ; segments of the corolla round, very entire. 6-8 i.

integrifo'lia, erect, pubescent ; leaves petioled, ovate, crenate-serrate ; racemes numerous, dense-flowered.

heterophyl''la, (b. Ju. ♂.) erect, hispid ; leaves petioled, pinnatifid ; divisions lanceolate, entire, nerved, terminal one elongated ; floral leaves simple, lanceolate, petioled ; racemes 2-cleft, dense-flowered ; divisions of the corolla entire. *S.*

PHALAN''GIUM. 6—1. (*Asphodeli.*) [From *phalaga*, tarantula, whose bite it was supposed to cure.]

esculen''tum, (b. M. 2f.) root bulbous ; leaves all radical, linear, connate ; stigma minutely 3 cleft. 12-18 i.

cro'ceum, (y. 2f.) root bulbous ; scape much longer than the grass-like leaves ; spike pyramidal ; bracts short ; seed sub-globose, smooth. *S.*

PHALA'RIS. 3—2. (*Graminea.*) [From *phalos*, shining, so named from the appearance of its seed.]

america'na, (riband-grass, wild canary-grass, Ju. 2f.) panicle oblong, spiked ; glumes of the calyx boat-shaped, serrulate ; corolla unequal ; rudiments hairy. Var. *pic''ta* leaves variously striped. This variety is the riband-grass of the gardens. 2-5 f.

canarien''sis, (canary-grass, Ju. ☉.) panicle sub-spiked, ovate ; glumes boat-form, entire at the apex ; rudiments smooth. Introduced. 18 i.

PHASEO'LUS. 16—10. (*Leguminosa*) [From *phaselos*, a little boat, which its pods were thought to resemble.]

peren''nis, (wild kidney-bean, p. Ju. 2f.) twining, pubescent ; leaflets ovate, acuminate, 3-nerved ; racemes 1-3, axillary, paniculate, longer than the leaves ; bracts mi-

nute ; legumes pendulous, broad, falcate, mucronate ; flowers large. Dry woods.

luna'tus, (Carolina bean, Lima bean, g-w. Ju. ☉.) twining ; legumes cimeter-form, sub-lunate, smooth ; seeds compressed. Ex.

vulga'ris, (common pole-bean, p. w. Ju. ☉.) stem twining ; racemes solitary, shorter than the leaves ; peduncles in pairs ; bracts smaller than the calyx, spreading legumes pendulous. From the East Indies.

na'nus, (bush-bean, six-weeks-bean, ☉.) stem erect, smooth ; bracts larger than the calyx ; legumes pendulous, compressed, rugose ; seeds variously colored. Ex.

multiflo'rus, (scarlet runner, r. w. Ju. ☉.) twining, sub-glabrous ; leaflets ovate, acuminate ; racemes peduncled, longer than the leaves ; peduncles in pairs ; bracts close-pressed, shorter than the calyx ; legumes sub-scabrous. South America.

PHILADEL''PHUS. 11—1. (*Myrti*) [From *phileo*, to love, *adelphos*, a brother. This name was first given to the Galium or bed-straw, because by its roughness it attached itself to what was near.]

inodo'rus, (scentless syringa, w. J. ♄.) leaves acuminate, oval, entire ; divisions of the calyx acute ; style undivided, longer than the stamens ; stigmas 4, oblong ; flowers large. *S.*

corona'rius, (mock-orange, false syringa, w. J. ♄.) styles distinct ; leaves ovate, sub-dentate. Ex.

grandiflo'rus, (w. M. ♄.) leaves short-petioled, opposite, ovate, acuminate, denticulate, a little hairy ; segments of the calyx acuminate ; style undivided, longer than the stamens ; stigmas 4, linear. Cultivated. *S.*

hirsu'tus, (w. ♄.) leaves oblong-ovate, acute, sharply angular-denticulate, hirsute above, whitish-villose beneath ; style and stigma undivided ; peduncles 2-bracted near the summit. *S.*

PHILOX'ERUS. 15—5. (*Amaranthi*) [From *philos*, love, and *xeros*, dry or burnt ; alluding to the kind of soil in which it is found.]

vernic''ularis, (J. 2f.) glabrous ; stem creeping ; leaves sub-terete, fleshy ; flowers in terminal, solitary, oblong heads. *S.*

PHLE'UM. 3—2. (*Graminea.*)

praten''se, (timothy grass, J. 2f. and ♂.) spike cylindric, calyx mucronate-awned ; keel ciliate ; awn shorter than the calyx ; culm erect. Introduced. 2-3 f.

alpi'num, (Au.) spike ovate or cylindric-ovate, hirsute or villose ; spike often blackish or dark purple.

PHLOX. 5—1. (*Polemonea.*) [A Greek word signifying flame, from the bright color of the flowers of some of its species.]

panicula'ta, (smooth stem lichnidia, r. w. J. 2f.) glabrous, erect ; leaves lanceolate, narrowing gradually, flat ; margins rough ; corymbs panicled ; divisions of the corolla rounded ; calyx awned. Cultivated. 2-3 f.

macula'ta, (spotted lichnidia, r. w. Ju 2f.) stem erect, scabrous and spotted ; leaves oblong-lanceolate, smooth ; panicle oblong, many-flowered ; segments of the corolla rounded; teeth of the calyx acute, recurved. Var. *suave'olens*, stem without spots; corolla white. 2 f.

arista'ta, (r. w. J. ♃.) weak, erect, viscid-pubescent; leaves lance-linear; panicle lax, fastigiate; pedicels somewhat in pairs; divisions of the corollas somewhat obovate; tube curved, pubescent; teeth of the calyx long, subulate. Var. *divarica'ta*, corolla purplish blue. Var. *vi'rens*, corolla reddish purple. Var. *canes'cens*, corolla whitish rose-color. 1.2 f.

pilo'sa, (creeping lichnidia, p. w. J. ♃.) small, decumbent, pubescent; leaves linear-lanceolate, downy with the margins revolute; corymbs nearly fastigiate; teeth of the calyx subulate, acute. 12-18 i.

rep"tans, (b. p. J. ♃.) pubescent with creeping suckers; radical leaves obovate spatulate, cauline ones lance-oval; corymb spreading, few-flowered; segments of the corolla obovate; teeth of the calyx subulate, reflexed. 8.18 i.

seta'cea, (r. J. ♃.) cæspitose, pubescent; leaves fascicled, subulate, pungent, ciliate; flowers few, terminal, somewhat umbelled; segments of the corolla cuneate, emarginate; teeth of the calyx subulate, much shorter than the tube of the corolla. Rocks and sandy hills.

revolu'ta, (w-p. J. ♃.) glabrous; stems erect, sub-simple; leaves coriaceous, paler beneath, sub-sessile; lower ones lance-linear, acute at each end; upper ones lanceolate, rounded at the base; corymb sub-fastigiate, few-flowered; pedicels sub-scabrous; segments of the corolla obovate, slightly crenulate; calyx glabrous; segments lanceolate, acute, unawned, half as long as the corolla. 12-18 i. Damp woods. Md.

undula'ta, (b. Au. ♃.) erect, glabrous; leaves oblong-lanceolate. slightly waved, margins scabrous; corymbs paniculate; segments of the corolla somewhat retuse; calyx awned. 2 f. *S.*

caroli'na, (p. Au. ♃.) stem pubescent; leaves ovate-lanceolate, glabrous; corymbs sub-fastigiate, branches generally 3-flowered; teeth of the calyx linear-lanceolate. *S.*

glaber"rima, (p. Ju. ♃.) erect; leaves linear-lanceolate, glabrous; corymb terminal, nearly fastigiate; teeth of the calyx linear-lanceolate, acute. 1.2 f. *S.*

specio'sa, (w. p. J. ♄.) erect, glabrous. frutescent, very branching; leaves linear; upper ones alternate, dilated at the base; racemes panicle-corymbed; segments of the corolla wedge-oblong, emarginate; teeth of the calyx subulate, equalling the tube. *S.*

ova'ta, (r-p. J. ♄.) erect, glabrous; radical leaves ovate, acute, somewhat fleshy; cauline ones lanceolate; corymbs sub-fastigiate; segments of the corolla undulate, retuse; teeth of the calyx linear, acute. *S.*

ni'tida, (p. J. ♃.) erect, glabrous; stem scabrous; leaves ovate-oblong, sub-coriaceous; corymb fastigiate; segments of the corolla obovate, sub-retuse; teeth of the calyx lanceolate, mucronate. 18.24 i. *S.*

corda'ta, (Au.) erect; leaves oblong-cordate, sub-acuminate; margin scabrous; corymbs paniculate; teeth of the calyx long, awned. 1-2 f. *S.*

niva'lis, a low training perennial, producing white flowers in April and May. Ex.

canaden"sis, (Ap. M.) produces blue flowers, grows nearly a foot in height. Ex.

drummond".ii, the only annual species, and has many varieties.

subula'ta, (mountain-pink, r. M. ♃.) cespitose, white-pubescent; leaves linear-ciliate; corymbs 5-flowered; pedicels 3-cleft; divisions of the corolla wedge-form, emarginate; teeth of the calyx subulate, scarcely shorter than the tube of the corolla. Cultivated. 3-6 i.

pyr"ami'dalis, (p. Au. ♃.) erect, smooth; stem scabrous; leaves cordate-acute; panicle fastigiate, pyramidal; segments of the corolla wedge-form, truncate; leaves opposite, sessile, very entire. Mountain meadows. 2-3 f.

PHŒNICAU'LIS. 14—2. (*Cruciferæ*.)
cheiranthoi'des, (p. ♃.) scape slender; leaves entire, densely and stellately tomentose; flowers in simple corymbose racemes; siliques diverging horizontally; scape 4 6 i. with a few small sessile and partly clasping leaves. Oregon.

PHRAGMI'TES. 3—2. (*Gramineæ*.)
commu'nis, (Au. ♃.) calyx about 5-flowered; florets longer than the calyx. 6-12 f.

PHRY'MA. 13—2. (*Labiatæ*.)
leptosta'chya, (p. w. ♃.) leaves large, ovate, toothed, petioled; spikes terminal, slender; flowers opposite, small. Shady woods. 2-3 f.

PHYLLAC"TIS. 3—1. (*Dipsaceæ*.) [From *phullon*, leaf, and *ago*, to carry, from being stemless.]
obova'ta, (Oc.) stemless; root fusiform; leaves radiating. linear-spatulate, obtuse, hirsute-pilose. *S.*

PHYLLAN"THUS. 19—15.(*Euphorbiæ*.) [From *phullon*, a leaf, and *anthos*, flower, because the flowers in one of the original species (since placed in another genus) grow out of the leaves.]
obova'tus, (S. ⊕.) leaves alternate, oval-obtuse, glabrous; flowers few, axillary, pedicelled, nodding; stem erect; branches distichus.

PHYSA'LIS. 5—1. (*Solaneæ*.) [From *phusa*, to inflate, so called because its seed is contained in a kind of bladder.]
visco'sa, (yellow henbane, y. Ju. ♃.) leaves in pairs, heart-oval, repand, obtuse, sub-tomentose, a little viscous; stem herbaceous, paniculate above; fruit-bearing calyx pubescent. 2-3 f. Road-sides.

obscu'ra, (y. p. Au.) pubescent; stem prostrate, divaricate; leaves broad-cordate, sub-solitary, toothed; flower solitary, nodding; calyx hairy; flower pale yellow, with 5 purple spots at the base; anthers bluish. Hills.

pennsylvan'ica, (y. S. ♃.) stem branched; leaves ovate, obtuse; peduncles axillary, solitary, a little longer than the petioles. 1 f. Road-sides.

alkeken"gi, (winter-cherry,) leaves in pairs, entire, acute, sub-ramose below. Ex.

lanceola'ta, (y. J. 2⟨.) erect, densely pubescent; leaves mostly in pairs, oval-lanceolate, entire, narrowed at the base into a petiole; flower solitary, nodding; calyx villose. 1-2 f.

loba'ta, (g.) leaves oblong, somewhat fleshy, lyrate-lobed, narrowed into the petiole at the base, glabrous, revolute at the margin; stem herbaceous, branching; peduncles solitary.

pubes"cens, (y. Ju. ⊕.) leaves villose, viscous, slightly cordate; stem much branched; flowers solitary, pendulous; fruit-bearing calyx nearly globose, slightly angled. S.

somnife'ra, (y. Ju. 2⟨.) tomentose; leaves ovate, very entire; flowers crowded, shortpedicelled; corolla bell-form. S.

wal"teri, (2⟨.) pulverulent, sub-tomentose, very branching, dichotomous; leaves in pairs, broad-ovate, obtuse, long-petioled, entire; peduncles solitary, nodding; calyx fructiferous, somewhat glabrous. S. C.

angustifo'lia, (2⟨.) very glabrous, prostrate, dwarfish; leaves very long, linear, fleshy, solitary; peduncles nodding, filiform, solitary. West Florida.

PHYTOLAC"CA. 10—10. *(Atriplices.)* [From *phuton*, a plant, and *lakka*, gum-lac, on account of the color of its fruit.]

decan"dra, (poke-weed. w. Ju. 2⟨.) leaves ovate, acute at both ends; flowers racemed; berries flattened at the ends. 3-6 f.

PICKERIN"GA. 5—1. *(Ericæ.)*

panicula'ta, (♄.) evergreen; leaves entire, alternate, wedge-oblong, obtuse; flowers panicled. S.

PINCKNEY'A. 5—1. *(Rubiaceæ.)* [In honor of Gen. C. C. Pinckney, of S. C.]

pu'bens, (p. J. ♄.) leaves opposite, lanceolate, entire, thinly pilose and shining above; tomentose below, sub-acuminate; calyx superior, persistent, colored. 15-20 f. S.

PINGUIC"ULA. 2—1. *(Scrophulariæ.)* [From *pinguis*, fat, so called because its leaves are greasy to the touch.]

vulga'ris, (butter-wort, M. 2⟨.) spur cylindrical, acute, as long as the veinless petal; upper lip 2-lobed, lower one in 3 obtuse segments; leaves radical, spatulate ovate, fleshy; flowers solitary, nodding; tube of the corolla villose, purple. Wet rocks. Rochester, N.Y. Canada.

lute'a, border of the corolla 5-cleft; spur subulate, a little shorter than the tube. 6-8 i. Flowers yellow. S.

pu'mila, (b. Ap. 2⟨.) border of the corolla 5-cleft; segments emarginate; lobes entire; spur subulate, a little obtuse, as long as the tube. 3.5 i. S.

acutifo'lia, (Ju. 2⟨.) very glabrous; leaves erect, oval, very acute. S.

austra'lis, (r.) glabrous; nectary very short, incurved; flowers rather large. West Florida.

PI'NUS. 19—15.

A. *Leaves solitary, with separate bases.*

canaden"sis, (hemlock-tree, M. ♄.) leaves flat, denticulate, 2-ranked; strobiles ovate, terminal, scarcely longer than the leaves. The bark is used in tanning leather.

balsa'mea, (American silver-fir, balsam-fir, M. ♄.) leaves solitary, flat, glaucous beneath, somewhat pectinate at the summit; strobile cylindrical, erect. 40.50 f.

frase'ri, (J. ♄.) leaves short, emarginate, subsecund, erect above; cones ovate-oblong; bracts elongated, incisely denticulate.

taxifo'lia, (♄.) leaves solitary, flat, subdistichous; cones oblong; anthers didymous.

ni'gra, (M. ♄.) leaves solitary, 4-angled, scattered on all sides, erect, straight; cones ovate, scales elliptic, undulate along the margin, the summit denticulate.

al"ba, (M. ♄.) leaves 4-sided, incurved; strobiles sub-cylindric, lax; scales obovate, entire.

ru'bra, (M. ♄.) leaves solitary, subulate; strobiles oblong, obtuse; scales rounded, somewhat 2-lobed, entire on the margin.

B. *Leaves many, sheathed at the base.*
[*Leaves in pairs.*]

resino'sa, (yellow-pine, Norway-pine, red-pine, M. ♄.) leaves and sheath elongated; strobiles ovate-conic, rounded at the base, sub-solitary, about half as long as the leaves; scales dilated in the middle, unarmed. Bark of a reddish color, and much smoother than the pitch-pine, or white-pine. Often grows very tall and straight.

in"ops, (M. ♄.) leaves short, strobile recurved, oblong-conic, as long as the leaves; spines of the scales subulate, straight.

banksia'na, (scrub-pine, M. ♄.) leaves short, in pairs, rigid, divaricate, oblique; strobiles recurved, twisted; scales unarmed. Rocky grounds.

[*Leaves in threes.*]

rig"ida, (pitch-pine, M. ♄.) leaves with abbreviated sheaths; staminate aments erect-incumbent; strobiles ovate, scattered or aggregated; spines of the scale reflexed. Though very common, it grows most plentifully on barren, sandy plains.

varu'bilus, (yellow-pine, M. ♄.) leaves elongated, in pairs and threes, channelled; strobile ovate-conic, mostly solitary; spines of the scales incurved.

[*Leaves in fives.*]

stro'bus, (white-pine, M. ♄.) leaves in fives, slender; sheaths very short; strobile pendulous, cylindrical, longer than the leaves; scales loose. Timber soft-fine-grained and light. Extensively used. Sometimes 140 f.

C. *Leaves many, in a fascicle.*

pen"dula, (black larch, tamarack, hackmatack, M. ♄.) leaves deciduous; strobiles oblong; margins of the scale inflexed; bract guitar-form, with a slender point.

flex"ilis, leaves in fives, short, and rather rigid; sheaths short and lacerate; strobile erect; scales large, unarmed; branches very flexible.

la'rix, (common larch, ♄.) leaves deciduous; strobiles ovate-oblong; margins of the scales reflexed, lacerate; bracts guitar-form. Ex.

pun"gens, (table mountain pine, ♄.) leaves by pairs, short, acute; cones ovate-conical, spines of the scales long, subulate, incurved, lower ones reflexed. 40.50 f. *S.*

tæ'da, (M. ♄.) leaves long, by threes; sheaths long; strobiles oblong-conic, deflexed, shorter than the leaves; spines inflexed. Var. *heterophyl"la,*leaves in pairs and threes; bark smooth.

palus"tris, (long-leaved, yellow, or pitch-pine, M. ♄.) leaves by threes, very long; stipules pinnatifid, ramentaceous, persistent; strobiles subcylindric, muricate. Timber, extensively used in the Southern States.

PI'PER. 2—3. (*Urticeæ.*) [Originally *pippul,* in the Bengalese tongue.]

leposta'chyon, (Florida pepper, ⊙.) herbaceous, small, leaves obovate, obtuse, sub-3-nerved, pubescent; spikes axillary, filiform, erect, much longer than the leaves. 6-12 i. *S.*

PIPTATHE'RUM. 3—2. (*Gramineæ.*) [From *pipto,* to fall, and *theios,* harvest, summer.]

racemo'sum, or *nigrum,* (clustered or black seed millet grass, Au. ♃.) panicle simple; flowers racemose, ovate-lanceolate; corolla black, hairy; awn as long again as the glume.

PIS"TIA. 15—8. (*Geranis.*)

spathula'ta, (w. Ju.) leaves abruptly narrowed into the petiole, dilated, round and obtuse toward the summit. *S.*

PI'SUM. 16—10. (*Leguminosæ.*)

sati'vum, (pea, p. w. J. ⊙.) petioles terete; stipules round and crenate at the base; peduncles many-flowered. Var. *umbella'tum,* (bouquet-pea,) has the stipules 4-cleft, acute. Var. *quadra'tum,* (quadrate pea,) fruit ash-color, 4-sided. Var.*hu'mile,* (dwarf pea,) stem erect, not climbing; leafets roundish. Ex.

PIT'H'ERIA. 16—10. (*Leguminosæ.*) [In honor of Dr. Pitcher of U. S. A.]

galactoï'des, (r. ♃.) stem erect, rigid, branched, smooth; leaves trifoliate, oval, obtuse, glandular-dotted beneath. *S.* Florida.

PLANE'RA. 5—2. (*Amentaceæ.*)

aquat"ica, (M. ♄.) leaves ovate, acute, serrate, equal at the base, slightly scabrous, short-petioled. 25-30 f. *S.*

PLANTA'GO. 4—1. (*Plantaginea.*) [From *planta,* the sole of the foot, so called because its leaves are trodden under foot.]

ma'jor, (plantain, w. J. ♃.) leaves ovate, sub-dentate, sub-glabrous; scape terete; spike oblong, imbricate. 6-24 i.

lanceola'ta, (English plantain, ripple grass, J. ♃.) leaves lanceolate; spike short, ovate-cylindrical; scape angular; capsule 2-seeded. 1-2 f.

virgin"ica, (dwarf plantain, r-y. J. ⊙.) hoary-pubescent; leaves lanceolate-ovate, sub-denticulate; spikes cylindric, with remote flowers; scape angular; cap 2-seeded.

corda'ta, (w. J. ♃.) leaves ovate, cordate, broad, sub-dentate, smooth; spike very long; flowers sub-imbricate, lower ones

scattered; bracts ovate, obtuse; cells of the capsule 2-seeded. 12-18 i.

me'dia, (w. J. ♃.) leaves ovate, pubescent, short-petioled; scape terete; spike short, cylindric; cells 1-seeded. One variety has the leaves hirsute and the spikes branching.

mariti'ma, (sea plantain, Au. ♃.) leaves linear, grooved, fleshy, hairy near the base; scape round, terete; spike cylindric; bracts acutish. 6-10 i.

pusil"la, (Au. ⊙.) minutely pubescent; leaves linear-subulate, flat, entire, acute, scape terete, longer than the leaves; spike cylindric, loose; lower flowers distant; bracts ovate, acute. as long as the calyx.

cuculla'ta, (Ju. ♃.) leaves ovate-cucullate, sub-denticulate, 9-nerved, pubescent beneath; spike cylindric, imbricate; scape terete.

eriopo'da, (♃.) stemless petioles covered with long wool at the base; leaves broad-lanceolate, alternate at each end, long-petioled, glabrous, entire, 5-nerved; scape terete, glabrous, spike cylindric; flowers remote; stamens and styles long; bracts broad-ovate, obtusish; capsules 2-seeded. 9-12 i.

interrup"ta, (Ju. ♃.) leaves lanceolate, entire, hairy; spike long, slender, interrupted; flowers scattered, glabrous. *S.*

gla'bra, leaves ovate, denticulate, smooth; scape slender, sub-compressed, nearly equal to the leaves; flowers scattered; bracts ovate, acuminate. *S.*

gnaphaloï'des, (Ju. ⊙.) silky-villose; leaves lance-linear. very entire, scape terete, scarcely longer than the leaves; spike cylindric, intricate; bracts linear, with long, villose cilia. *S.*

PLATAN"THERA. 18—1. (*Orchideæ.*) [From *platys,* broad, *anthe'ra,* anther, from the width of that organ.]

orbicula'ta, (g-w. J. ♃.) leaves 2, radical, orbicular; scape with 2 or 3 bract-like leaves, many-flowered; bracts shorter than the flowers; lip lance-linear, obtuse; spur longer than the ovary. 12-18 i.

dilata'ta, (giant orchis, w. or g. J. ♃.) spur shorter than the germ; lip entire. linear, with the base dilated of the length of the spur; bracts of the length of the flower. On mountains the flowers are green, in the meadows white. 1-4 f.

PLATA'NUS. 19—13. (*Amentaceæ*) [From *plata,* broad, alluding to the size of the tree]

occident"alis, (buttonwood, American plane-tree, false sycamore, J. ♄.) leaves 5-angular, obsoletely lobed, toothed, pubescent beneath; stem and branches becoming white. One of the largest trees in North America.

PLATYS"TEMON. 12—12. (*Papaveraceæ*) [From *platus,* broad, *stemon,* strand or filament.]

califor"nicum, (y-w. ⊙. leaves half clasping. oblong, linear, obtuse, entire, alternate, 3-5-nerved; peduncles axillary, elongated, 1-flowered; plant sparsely covered with shaggy, spreading hairs. California.

PLATYSPET"ALUM. 14+1. (*Crucifera.*)
[From *platus*, broad, *petalon*, leaf.]
purpuras"cens, stigma 2-lobed, spreading; style manifest; scape naked, 1-leaved, and pubescent; silicles sub-glabrous.
du'bium, stigma undivided, sub-sessile; silicles and scapes pubescent.

PLATYSPER"MUM. 14—1. (*Crucifera.*)
[From *platus*, broad, *sperma*, seed.]
scapige'rum, (w. Mar. Ap. ⊙.) root sub-fusiform, scarcely fibrous; leaves all radical, spreading, sub-runcinate-pinnatifid; lobes mostly acute; lower ones gradually smaller, attenuated into a petiole; scapes digitate, erect, simple; very glabrous, 1-flowered; flowers small, erect.

PLECTRI'TIS. 3—1. (*Dipsaceæ.*) [From *plektron*, a spur, alluding to the form of the corolla.]
conges"ta, (r. ⊙.) glabrous; flowers in a dense whorl; bracts many-cleft, in subulate divisions. Var. *mi'nor*, leaves very narrow.

PLEE'A. 9—2. (*Junci.*) [From a Greek word signifying abundance, from the number of stamens.]
tenuifo'lia, (y. r. ♃.) very glabrous; leaves very narrow-ensiform; sheaths of the spike 1-flowered. 1-2 f.

PO'A. 3—2. (*Gramineæ.*) [From a Greek word, signifying grass.]
annu'a, (Ap. ⊙.) panicle sub-secund, divaricate; spikelets ovate-oblong, 5-flowered; florets free; culm oblique, compressed; root fibrous. 6-8 i.
praten"sis, (J. ♃.) panicle diffuse; upper leaves much shorter than the smooth sheaths; florets acute, 5-nerved, webbed at the base; stipule short-truncate, root creeping. 2-3 f.
aquat"ica, var. *america'na*, (Au. ♃.) panicle erect, semi-verticillate, diffuse; branches flexuous, smooth; spikelets linear, 6-8-flowered; florets ovate-obtuse, free; leaves broad-linear, smooth; sheaths smooth. 4-5 f.
trivia'lis, (Ju. ♃.) panicle equal, diffuse; spikelets oblong-ovate, about 3-flowered; florets webbed at the base, 5-nerved; culm and sheaths roughish; stipules oblong; root creeping. 2-3 f.
compres"sa, (blue-grass, Ju. ♃.) panicle contracted, somewhat secund; spikelets oblong, 3-6-flowered; florets webbed; glumes nearly equal; culm oblique, compressed; root creeping. Var. *sylves"tris*, panicle loose, spreading; spikelets 2-3-flowered; culm slender, nearly erect. 12-18 i.
seroti'na, (J. ♃.) panicle elongated, diffuse, at length somewhat secund; spikelets lance-ovate, 2-3-flowered; florets a little webbed at the base, yellow at the tip, obscurely 5-nerved; root creeping. 2 3 f.
nemora'lis, (♃.) panicle attenuated, weak; branches flexuous; spikelets ovate, about 3-flowered; florets loose, slightly webbed, acute, obsoletely nerved; stipule almost wanting. 2 f.
nerva'ta, (J. ♃.) panicle equal, diffuse; branches weak, at length pendulous; spikelets 5-flowered; florets free, conspicuously 7-nerved, obtuse. 3-4 f.
obtu'sa, (Au. ♃.) panicle ovate, contract-

ed; spikelets ovate, tumid, 5-7-flowered; florets free; glumes scarious; palea ovate, smooth, obtuse; lower one indistinctly 7-nerved; leaves as long as the culm, with the sheaths smooth. 2-4 f.
canaden"sis, (Ju. ♃.) panicle large, effuse; branches semi-verticillate, flexuous, at length pendulous; spikelets ovate, tumid, 5-8-flowered; florets free; lower palea acutish, 7-nerved; upper one very obtuse; stamens 2. 3-4 f.
capilla'ris, (Au. ⊙.) panicle very large, loose spreading, capillary; spikelets 3-flowered, ovate, acute; florets free; culm branched at the base; leaves hairy. 12 i.
pectina'cea, (Ju. ⊙.) culm cespitose, oblique; leaves hairy at the base; panicle capillary, expanding, pyramidal, hairy in the axils; spikelets linear, 5-9-flowered; florets free, acute, upper palea persistent. 8-12 i.
rep"tans, (Au. ⊙.) diœcious; culm branched, creeping; panicle fascicled; spikelets lance-linear, 12-20-flowered; florets acuminate. Var. *cæspito'sa*, culm very short, cespitose; spikelets much crowded, oblong. 8 i.
eragros"tis, (Ju. ⊙.) panicle equal, spreading; lower branches hairy in the axils; spikelets linear-lanceolate, 9-15-flowered; florets obtuse; root fibrous. 12-18 i.
ten"uis, (Au.) panicle branching, expanding, capillary; spikes 3-flowered, glabrous, long peduncled; leaves linear, very long. 12-18 i. *S.*
parviflo'ra, (J. ♃.) panicle diffuse, capillary; spikelets small, generally 3-flowered; flowers obtusish, striate, caducous; leaves distichous, flat. 12-18 i. *S.*
confer"ta, panicles terminal and axillary, erect, compressed, with clustered flowers; spikelets 8-flowered, glabrous. 2-3 f. *S.*
angustifo'lia, (M. ♃.) leaves linear, involute; panicle somewhat crowded; spikes lanceolate, acute, 4-flowered; flowers villose at the base. 1-2 f. *S.*
ni'tida, (Ju.) stem erect, very glabrous; panicle large, diffuse, capillary, sub-verticillate; peduncles long; spikes lanceolate, 8-flowered. 1 f. *S.*
rigi'da, (M. ♃.) panicle lanceolate, a little branched, secund; branches alternate, secund. 2-4 i. *S.*
airoi'dea, panicle attenuated, erect; branches capillary, loose, semi-verticillate; leaves with very long sheaths, short and acute; spikelets oblong, obtuse, sub-sessile, 4-6-flowered; glumes unequal, shorter than the palea. 4-5 f *S.*

PODOPHYL"LUM. 12—1. (*Ranunculaceæ.*)
[From *pous*, foot, and *phullon*, leaf, on account of the shape of its leaf.]
pelta'tum, (wild mandrake, may-apple, w. M. ♃.) stem terminated with 2 peltate, palmate leaves; flower single, inserted in the fork formed by the petioles of the leaves. Sometimes the plant is 3-leaved, and the flower inserted on the side of one of the petioles. 1-2 f.

PODOS"TEMUM. 19—2. (*Aroideæ.*)
ceratophyl"lum, (thread-foot, Ju. ♃.) stem

filiform, floating; leaves pinnate; flowers axillary. Attached to rocks and large stones in shallow waters.

abrotanoi'des, divisions next to frond very branching; the terminal ones capillary, dichotomous, many-cleft floral spathe elongated. No root distinct from the stem. *S.*

PODOSTIG"MA. 18—5. (*Apocyneœ.*) [From *pous*, foot, and *stigma*, stigma.]

pubes"cens, (y.-g. M. 2f.) erect; leaves linear; umbels terminal and axillary; petals erect, longer than the calyx; corpuscle pedicelled. 12-18 i. *S.*

vir"idis, (g. and p. M. 2f.) erect, leaves oblong, obtuse, petiolate; petals large, erect; umbels generally terminal. *S.*

POGO'NIA. 18—1. (*Orchideœ.*) [From a Greek word signifying beard.]

ophioglossoi'des, (snake-mouth arethusa. r. Ju. 2f.) root fibrous; scape with 2 distant leaves, 1-2-flowered; leaves lance-oval; lip fringed. 8-12 i.

verticilla'ta, (y-r. J. 2f.) leaves 5-verticillate; flower solitary; 3 outer segments of the perianth long and linear; the inner ones lanceolate, obtuse; lip 3-lobed, dilated, the middle lobe undulated; root fasciculate. Swamps.

divarica'ta, (p. J. 2f.) root fibrous; scape 1-flowered, with two distant, lance-oblong leaves; outer petals long-linear, expanding; lip sub-3-lobed, crenulate. 18.24 i. *S.*

POLANIS"IA. 12—1. (*Capparides.*)

graveo'lens, (false mustard, r. w. Ju. ⊙.) viscid-pubescent; leaves ternate; leaflets elliptical-oblong; flowers generally dodecandrous. 1 f.

tenuifo'lia, viscid-glandular; leaves 3-foliate, nearly glabrous; leaflets filiform-linear, longer than the petiole; petals very unequal, sub-orbicular, entire, on short claws; stamens 9.11; style longer than the ovary; pods linear, terete, minutely reticulated. glabrous. Georgia.

POLEMO'NIUM. 5—1. (*Polemonia.*) [An ancient name derived from *polemos*, war, because, according to Pliny, kings contended for the honor of its discovery.]

rep"tans, (Greek valerian, b. M. 2f.) leaves pinnate, leaflets 5-13; flowers terminal, nodding.

POLYCAR"PON. 3—3. (*Amaranthi.*) [From *polus*, many, and *karpos*, seed or fruit.]

tetraphyl"lum, (w. J.) leaves opposite and in fours, obovate, obtuse, entire, glabrous, narrowed at the base; calyx persistent; capsule ovate, 1-celled; stem branching, glabrous, striate, knotted. 3-6 i. *S.*

POLYCNE'MUM. 3—1. (*Atriplices.*) [From *polus*, and *kneme*, a leg or knee, from the number of jointed branches, or joints of the stem.]

america'num, (2f.) cespitose; leaves connate, crowded, subulate, 3-angled, rather pungent; flowers terminal. *S.*

POLYAN"THES. 6—1. ((*Narcissi.*) [From *polus*, many, and *anthos*, flower, because it bears many flowers.

tubero'sa, (tuberose, 2f.) flowers alternate, in pairs, rootlets tuberous; scape scaly; leaves linear, long; sweet-scented. Ex.

The polyanthus of the gardens belongs to the genus Primula.

POLYG"ALA. 16—6. (*Leguminosœ.*) [From *polus*, much, and *gala*, milk, from its milky juice.]

paucifo'lia, (flowering wintergreen, r. M. 2f.) small, large-flowered; stem simple, erect, naked below; leaves ovate, acute, glabrous, near the top of the stem; flowers crested, terminal, about in threes. 3-4 i.

sen"ega, (seneca snake-root, mountain-flax, r. or w. J. 2f.) stem erect, simple, leafy; leaves alternate, lanceolate; spike terminal, filiform; flowers alternate, not crested. Var. *d"bida*, leaves lanceolate or oval, spike somewhat crowded; flowers white, sub-sessile. 8-14 i.

polyg"ama, (ground-flower, p. J. 2f.) stems numerous; leaves linear-oblong, alternate downwards; racemes terminal and lateral, elongated; flowers sessile; radical racemes procumbent, with apterous flowers. 4-8 i.

purpu'rea, (r. Ju. ⊙.) stem fastigiately branched; leaves alternate, oblong-linear; flowers beardless, imbricated in obtuse cylindrical spikes; rachis squarrose; wings of the calyx cordate, ovate, erect, twice as long as the capsule. 12-18 i. Woods and hill-sides.

lu'tea, (yellow milkwort, y. S. ♂.) stem simple or branched; lower leaves spatulate, upper ones lanceolate; flowers in globular heads; wings of the calyx ovate, mucronate; bracts shorter than the flowers. 8-16 i. Pine barrens.

incarna'ta, (r. J. ⊙.) stem nearly simple, erect, glaucous; leaves scattered, subulate; spikes oval, oblong; tube of the corolla long, slender. *S.*

sanguin"ea, (r. Ju. ⊙.) erect; branches fastigiate; leaves linear; spikes crowded; flowers not fimbriated; rachis squarrose. 12-18 i.

verticilla'ta, (dwarf snake-root, w. J. ⊙.) erect, branching; leaves whorled and scattered; spike filiform, peduncled; flowers distinctly alternate, approximate, crested; calycine wings shorter than the fruit. 6.8 i.

ambig"ua, (p. ⊙.) erect; leaves linear, lower ones verticillate, the rest scattered; spikes acute, long-peduncled; flowers cristate; calycine wings round and veined, as long as the fruit; bracts deciduous.

crucia'ta, (r. g. Ju. ⊙.) stem fastigiate, wing-angled; leaves verticillate in fours linear-oblong; flowers in spiked sessile heads. 8-12 i.

corymbo'sa, (g-y. Ju. 2f.) stem erect, terete, nearly naked; lower leaves long, linear-lanceolate, stem leaves subulate, minute near the summit; racemes corymbed; rachis squarrose. 2-4 f.

seta'cea, (Ju. ⊙.) stem setaceous, nearly leafless, simple, sparingly branched near the summit; leaves small, setaceous, scattered; flowers minute, in a compact spike. *S.*

viridis"cens, (g-y. Ju.) stem simple; leaves cuneate, obovate, obtuse; head cylindric, squarrose; calycine wings conspicuously acuminate. 1-4 i. *S.*

baldwin"ia, (y-w. Ju.) stem slightly angied, branching near the summit; radical leaves spatulate, obtuse, cauline ones lanceolate, small; flowers capitate, heads squarrose, corymbed; calycine wings setaceous, acuminate. 2-3 f. *S.*

boykin"ia, (g-w.) flowers cristate; stem simple; leaves 4-5 verticillate, oblong-oval, lanceolate or acute; upper ones scattered; spike solitary, long-peduncled, lax-flowered. *S.*

cymo'sa, (J. Au. y.) cyme simple; spikelets ovate; wings elliptical-oblong, rather obtuse, mucronate; superior sepal half as large as the wings, rather obtuse; lateral petals distinct nearly to the base; crest minute; seed sub-globose, glabrous; stem simple, terete, attenuated upward; radical leaves linear-spatulate, cauline ones linear-subulate, minute. 2-5 f.

POL"YGO'NUM. 8—3. (*Polygoneæ.*) [From *polus,* many, and *gone,* a joint, on account of the many joints in its stem.]

1. *Flowers axillary.*

avicula're, (knot-grass, w. M. ⚥.) leaves lanceolate, scabrous at the margin; stipules short, laciniate; stem procumbent; flowers sub-sessile, axillary, minute. 6-12 i.

fagop"yrum, (buckwheat r-w. Ju. ⚥.) racemes panicled; leaves heart-sagittate; stem erectish, unarmed; angles of the seeds equal. 1-2 f. Ex.

orien"tale, (prince's feather, r. Au. ⚥.) stem erect; leaves very large, petioled, ovate, acuminate, minutely pubescent; stipules hairy, somewhat sabre-form: flowers in crowded, terminal spikes. 4-5 f. Old fields and road-sides. Flowers in large, pendulous, crimson spikes. Naturalized.

erec"tum, (w. J. ⚥.) stem branched; leaves broad, oval, petiolate; flowers pentandrous. 1-3 f.

mariti'mum, (w-r. Au. ♄.) stamens 8; leaves lanceolate, thick and glaucous, margin revolute; stipules lacerate; stem diffuse, prostrate, suffruticose. 1-2 f. *S.*

ten"ue, (w. Ju. ⚥.) stem erect, slender, branched, acute-angled; leaves long-linear, straight, acuminate; stipules tubular; apex villose; flowers alternate, sub-solitary. 9-18 i.

2. *Flowers in slender spikes.*

lapathifo'lium, (r. w. Au. ⚥) stipules awnless; stamens 6; styles 2; peduncles scabrous; spikes numerous, rather crowded; leaves lance-ovate, short-petioled, pubescent above. 2-4 f.

puncta'tum, (water-pepper, w. Au. ⚥.) flowers octandrous, glandular; styles 3; stipules ciliate, spotted; leaves lanceolate, glabrous; spike filiform, at first cernuous; bracts remotely alternate. 1-2 f.

mi'te, (tasteless knotweed, J. ⚥.) flowers octandrous, in crowded spikes; styles 3; leaves narrow-lanceolate, sub-hirsute; stipules hirsute, long-ciliate; bracts ciliate, subimbricate. 12-18 i.

virginia'num, (w. Ju. ⚥.) stamens 5; styles 2, unequal; stem simple, angular; leaves broad-oval; spikes virgate; flowers remote. 2-4 f.

bistortoi'des, (w-r. J. ⚥.) bracts 1-flow-

ered, 2-3 valved; leaves oval, flat, petioled; stem simple, 1-spiked.

3. *Flowers in thick crowded spikes.*

vivipa'rum, (r. Au. ⚥.) stem simple; spike linear, solitary; leaves lance-linear, margins revolute; bracts ovate, acuminate. 6 i.

barba'tum, (r. w. Ju.) stamens 6; styles 3; spike virgate, truncate; bristle ciliate; leaves oblong, acute, smoothish. 18-24 i.

persica'ria, (r. Ju. ⚥.) stamens 6; styles 2; spikes ovate oblong, erect; peduncles smooth; leaves lanceolate; stipules smoothish, ciliate. 1-2 f.

pennsylva'nicum, (Ju. ⚥.) flowers octandrous; spikes oblong; leaves lanceolate; stipules smooth and naked; stem geniculate. 2-4 f.

amphib"ium, (mud knotweed, Ju. r. ⚥.) leaves petiolate, oblong-lanceolate, sometimes cordate at base; flowers in dense, terminal spikes, pentandrous; styles bifid; stem nearly erect. Var. *terres"tre,* leaves smooth above, slightly pubescent beneath; spike ovate, oblong. Var. *aquat"icum,* leaves floating, ovate lanceolate; spike cylindric-oblong.

4. *Flowers in spiked panicled racemes.*

articula'tum, (joint weed, r. Ju. ⚥.) leaves linear, obtuse; flowers octandrous; styles 3; spikes paniculate, filiform, erect; pedicels solitary, articulate near the base. Sandy plains.

5. *Flowers sub-racemed; leaves cordate, sagittate, or hastate.*

sagitta'tum, (w. J. ⚥.) stem prostrate, square; the angles awned with reversed prickles; leaves sagittate; flowers octandrous, in small peduncled heads. Wet grounds.

arifo'lium, (r-w. Ju. ⚥.) stem prostrate, square, the angles with reversed prickles; leaves long-petioled, hastate; spikes few-flowered; flowers hexandrous; styles 2 or 1. 2-4 f.

convol"vulus, (w. r. Ju. ⚥.) stamens 8; styles 3; leaves petioled, oblong, hastate-cordate; stem long, twining; segments of the perianth bluntly keeled.

scan"dens, (climbing buckwheat, w. r. Au. ⚥.) stamens 8; styles 3; leaves broad-cordate; stipules truncate, naked; stem twining, glabrous; calyx bearing the fruit 3-winged.

cilino'de, (Ju. ⚥.) stamens 8; styles 3; leaves cordate; stipules sub-acute, surrounded at the base with an outer ciliate series; stem angled, prostrate or climbing; divisions of the calyx obtusely keeled.

hirsu'tum, (w. Ju. ⚥.) stamens 7; styles 3-cleft; spikes filiform; stem and stipules very hirsute; leaves lanceolate, hirsute, punctate. 2 f. *S.*

fimbria'tum, (w. Au.) spikes panicled; flowers solitary, fimbriate; stipule truncate, fringed; leaves linear, acute at each end. 2 f. *S.*

seta'ceum, (w. Ju.) flowers octandrous, styles 3-cleft; peduncles long, 2-spiked; spikes interrupted, hirsute; leaves broad-lanceolate, acuminate. hirsute; stipules hirsute, ciliate; stem erect, glabrous. 1-2 f. *S.*

gra'cile, (g-w. ☿.) diœcious, glaucous; racemes slender, filiform; flowers deflected, longer than the peduncles; peduncle articulated to the calyx; leaves spatulate-linear, obtuse; fruit lobger than the calyx. 1-4 f. *S.*

POLYM"NIA. 17—4. (*Corymbiferæ.*) [Named from *Polyhymnia,* the muse of eloquence.]

canaden"sis, (y. J. ♃.) viscid-villose; leaves denticulate, acuminate, lower ones pinnatifid, upper 3-lobed or entire. 2-4 f. Flowers in a loose terminal panicle. Shady hills.

uveda'lia, (yellow leaf-cup, y. Ju. ♃.) leaves opposite, 3-lobed, acute, decurrent into the petiole; lobes sinuate-angled; rays elongated. 3-5 f.

POLYPO'DIUM. 21—1. (*Filices*) [From *polus,* many, and *pous,* foot, because it has many roots.]

vulga're, (polypod, Ju. ♃.) frond deeply pinnatifid; divisions lance linear, obtuse, crenulate, approximate, upper ones gradually smaller; fruit dots solitary; root chaffy. 8-12 i.

hexagonop"terum, (Ju. ♃.) fronds bipinnatifid, rather smooth. circumference triangular, lower divisions deflexed; segments lanceolate, obtuse. ciliate, upper ones entire, lower ones adnate-decurrent; sori minute, solitary; stipe smooth. 12-16 i.

connec"tile, (Ju. ♃.) fronds bipinnatifid, ciliate, triangular; divisions opposite, contiguous, adnate; segments sub-elliptical; stipe chaffy; sori minute. 12 i.

virginia'num, (Ju. ♃.) fronds deeply pinnatifid; divisions lanceolate, obtuse, very entire, approximate, upper ones gradually smaller; sori and root solitary. *S.*

inca'num, (Ju. ♃.) fronds deeply pinnatifid; divisions alternate, linear, very entire, obtuse, upper ones gradually smaller, scaly beneath; stipe scaly, fruit-bearing at the apex; sori solitary. *S.*

POLYPRE'MUM. 4—1. (*Gentianeæ.*) [From *polus,* many, and *premnon,* stalk or shoot]

procum"bens, (w. Ju. ☿.) stem herbaceous, procumbent, furrowed, margins of the furrows sharply serrulate, dichotomous above; leaves opposite, linear, sessile, finely serrulate, sub-decurrent. 6-12 i.

POLYP"TERIS. 17—1. (*Corymbiferæ.*) [From *polus,* many, and *pteris,* a wing, alluding to the many-valved (or winged) egret]

integrifo'lia, (♃.) erect; sub-scabrous, branching above; leaves alternate, scabrous, entire, linear-lanceolate; style 2-cleft, longer than the stamens. 3-4 f. *S.*

POLYTRI'CHUM. 21—2. (*Musci.*) [From *polus,* many, and *thrix,* hair, so called from its resemblance to hair.]

juniperi'num, (hair-cap moss, M. ♃.) stem generally simple; leaves lance-linear, entire, flattish, somewhat spreading; the apophysis depressed. In dry woods, &c.

POMA'RIA. 10—1. (*Leguminosa.*)

glandulo'sa, (y. ♄.) branching; glandular-punctate; branches slender, sub-pubescent; leaves abruptly bipinnate; leaflets ovate, oblique at the base, entire, sessile, sub-pilose, smooth and pale-green above. *S.*

PONTEDE'RIA. 6—1. (*Narcissi.*) [Name from an ancient botanist, Pontidera.]

corda'ta, (pickerel-weed, b. Ju. ♃.) leaves heart-oblong, obtuse; spike many-flowered, compact; divisions of the corolla oblong. Var. *angustifo'lia,* leaves elongated, triangular, truncate, and sub-cordate at the base. 1-2 f.

POP"ULUS. 20—8. (*Amentaceæ.*) [The origin of the name is doubtful.]

tremuloi'des, (white poplar, American aspen, Ap. ♄.) leaves heart-roundish, abruptly acuminate; tooth-serrulate, glabrous, a little pubescent at the margin, with two glands at the base, on the upper side; petioles compressed, in the young state silky. 20 30 f.

balsamife'ra, (balsam poplar, Ap. ♄.) leaves ovate, acuminate, white, and net-veined beneath; buds resinous. 70-80 f.

angula'ta, (balm of Gilead, Ap. ♄.) leaves ovate-deltoid, acuminate, glabrous, branches wing-angled. 80 f.

diluta'ta, (Lombardy poplar, Italian poplar, Ap. ♄.) leaves glabrous both sides, acuminate, serrate, deltoid, the breadth equal to, or exceeding the length; branches erect, close to the stem. It is said no pistillate plant of this species has been brought to America; consequently no seeds are obtained from it. 40.80 f. Ex.

grandiden"tata, (tree poplar, Ap. ♄.) leaves round-ovate, acute, unequally and coarsely sinuate-toothed, glabrous, when young, villose; petioles compressed. Var. *pen"dula,* branches pendulous. 40-50 f.

betulifo'lia, (birch-leaf poplar, Ap. ♄.) leaves rhomboidal, long-acuminate, dentate, glabrous; young branches pilose. 30-40 f.

can"dicans, (Ap. ♄.) leaves cordate, ovate, acuminate, obtusely and unequally serrate, white beneath, sub-3-nerved, reticular veined; petioles hairy; buds resinous. 40-50 f.

læviga'ta, (cotton-tree, Ap. ♄.) leaves round-ovate, deltoid, acuminate, sub-cordate, unequally serrate, glabrous, glandular at base; petioles compressed; younger branches angled. 70-80 f.

heterophyl"la, (various leaved poplar, M. ♄.) leaves round-ovate, cordate; the sinus small, cordate and somewhat auricled; when young, tomentose. 70-80 f.

monolife'ra, (Ap. ♄.) leaves sub-cordate-deltoid, glabrous, glandular at the base, with cartilaginous, sub-pillose, hooked serratures; nerves spreading; petioles compressed above; older branches terete. 60-70 f.

græ'ca, (Athenian poplar, Ap. ♄.) leaves cordate-ovate, acuminate, obsoletely serrate; petioles compressed. 20-40 f. Ex.

PORCEL"IA. 12—12. (*Annona.*) [In honor of Porcel, a distinguished Spanish botanist.]

trilo'ba, (custard apple, paw-paw, p. Ap. ♄.) leaves smoothish, oblong-wedge-obovate; outer petals orbicular; fruit large, fleshy. 30 40 f.

parviflo'ra, (g-p. M. ♄.) leaves wedge-obovate, mucronate, under surface and branches rufous-pubescent; outer petals scarcely twice as long as the calyx. 2 f. *S.*

pyg'na'sa, (Ap. ♁.) leaves long-linear, wedge-form, obtuse, coriaceous, with the branches glabrous; outer petals obovate-oblong, much larger than the calyx. 6-18 i. *S.*

grandiflo'ra, (y-w. Ap. ♄.) leaves wedge-obovate, obtuse, under surface and branches rufous-pubescent; outer petals obovate, much larger than the calyx. 18-24 i. *S.*

PORTULAC"CA. 12—1. *(Portulacceæ.)* [From *porto,* to carry, *lac,* milk.]
olera'cea, (purslane, y. J. ☉.) leaves wedge-form; flowers sessile.
pilo'sa, (☉.) leaves subulate, alternate; axils pilose; flowers sessile, terminal. *S.*

POTAMOGE'TON. 4—4. *(Junci.)* [From *potamos,* a river, and *geiton,* adjacent, so called because it grows about rivers.]
natans, (pond-weed, g. J. ♃.) leaves long-petioled, floating, lance-oval; at first some are sub-cordate. On water.
flui'tans, (g. Ju. ♃.) lower leaves long, linear, upper ones lanceolate, nerved, coriaceous; all petioled. In water.
heterophyl"lum, (variegated pond-weed, g. Ju. ♃.) upper leaves floating, coriaceous, elliptical, petiolate, lower ones membranous, linear-lanceolate, sessile.
diversifo'lium, (g. Ju. ♃.) upper leaves floating, elliptical, petiolate, 5-nerved, lower ones filiform; spike axillary, almost sessile, few-flowered. Water.
perfolia'tum, (g. Ju. ♃.) leaves amplexicaul, cordate, ovate; spike few-flowered, on a short peduncle. Water.
- lu'cens, (g. Au. ♃.) leaves ovate-lanceolate, petiolate, pellucid, and finely veined; spike long, cylindrical.
cris"pum, (r-g. J. ♃.) leaves lanceolate, tapering, sessile, undulate, serrate; spike 8-10 flowered. Lakes.
pectina'tum, (g. J. ♃.) leaves setaceous, distichus, alternate, sheathing; spikes terminal, interrupted.
gramin"eum, (grass pond-weed, g. Ju. ♃.) leaves linear, grass-like, alternate, sessile; stipules broad; stem terete, sub-dichotomous. In July, some of these plants begin to raise their spikes of unopened flower-buds to the surface of the water; as soon as the stigmas are fertilized by the pollen, the spikes are again withdrawn, to ripen the fruit under water; others succeed them, and the process continues.
compres"sum, (g. Ju. ♃.) leaves linear, obtuse, sessile; stem compressed; spike 4-6 flowered.
rosterifo'lium, (g-y. Au. ♃.) leaves alternate, linear, closely sessile; stem flexuous, compressed, sub-alate; branches axillary; stipules lance-linear, acute; spikes many (20-40) flowered. 2 3 f.

POTENTIL"LA. 11—12. *(Rosaceæ.)* [From *potentia,* power, so named on account of its supposed power to heal diseases]

A. *Leaves digitate.*
canaden"sis, (common five-finger, y. M. ♃.) procumbent, sub-ramose, whitish-silky; stipules ovate, gashed; leaves wedge-ovate, gash-toothed; stem ascending and creeping, hirsute; peduncles solitary, elongated;

divisions of the calyx lance-linear; petals orbicular, sub-entire, of the length of the calyx. 2-18 i.
argen"tea, (silver five-finger, w-y. Ju. ♃.) stem prostrate and ascending, rarely sub-erect, branching, white-downy; stipules ovate, acute; leaves wedge-form, gash-toothed, silvery white beneath; petals retuse, scarcely longer than the calyx. 4-10 i.
sim"plex, (y. Ap. ♃.) erect, simple, hirsute; leaves oblong-oval, coarsely toothed; peduncles axillary, solitary, long, 1-flowered; petals nearly round, obcordate, longer than the calyx.
sarmento'sa, (y. M. ♃.) stem sarmentose; leafets obovate, obtuse, serrate, glabrous above, hirsute beneath; petals roundish, longer than the calyx.
rec"ta, (y. J. ♃.) erect; leaves in fives and sevens; leafets lanceolate, coarsely toothed; petals obcordate, larger than the calyx; corolla large, pale.

B. *Leaves pinnate.*
anseri'na, (tansey cinquefoil. y. J. ♃.) creeping; leaves interruptedly pinnate, numerous, gash-serrate, silky, white-downy beneath; peduncles solitary, 1-flowered.
frutico'sa, (shrubby cinquefoil. y. J. ♃.) stem fruticose, oblong, lanceolate, entire, approximate; stipules lanceolate, membranous, acute; flowers in corymbs, large; petals longer than the calyx. A shrub 2 feet high, much branched, hairy. Margin of swamps.
pennsylva'nica. (y. Ju. ♃.) erect, very soft, somewhat whitish-villose; leafets oblong, obtuse, sub-pinnatifid, woolly; panicle straight, many-flowered; segments of the calyx semi-oval.
supi'na, (y. J. ☉.) stem decumbent, dichotomous; leaves oblong, incisely serrate; peduncles axillary, solitary, 1-flowered.
argu'ta, (w. J. ♃.) stem erect, pubescent, viscous above; leaves unequally pinnate; leafets somewhat round-ovate, oblique at the base, doubly gash-toothed; stipules sub-entire; calyx acute, somewhat shorter than the corolla. 1-3 f.
humifu'sa, (y. M.) leaves digitate, quinate; leafets wedge-oblong, obtuse, gash-toothed, white-tomentose beneath; peduncles short, filiform, procumbent. 4-5 i. *S.*
russellia'na, (♄.) the most beautiful species produces rich, dark, scarlet flowers. A low shrubby plant. Ex.
formo'sa, deep red or purple flowers, blossoming from May to August. Ex.

C. *Leaves ternate.*
tridenta'ta, (mountain cinquefoil. w. Ju. ♃.) smoothish; stem ascending, dichotomous; leaves ternate-palmate; leafets wedge-oblong, coriaceous, 3-toothed at the summit, pubescent beneath; stipules lanceolate, acuminate; corymb loose, few-flowered; petals oblong-ovate, longer than the calyx; stem 3-6 inches high. Mountains. Frozen regions to Car.
norwe'gi'ca, (Norway cinquefoil. y. J. ☉.) hirsute; stem erect, dichotomous above; leaves ternate, palmate; leafets lance-

rhombic, simply and doubly serrate; flowers numerous, sub-corymbed, and axillary; petals obcordate, shorter than the calyx. 8-10 i. Old fields. Can. to Car.

villo'sa, (hairy five-finger, ♃.) assurgent, silky-villose; stipules broad, membranaceous, entire; leafets sessile, approximate, with shining, close-pressed hairs above, hoary-tomentose beneath; peduncles short, aggregate; petals obcordate, longer than the calyx.

hirsu'ta, (w. Ju. ♃.) erect, simple, very hirsute; leafets roundish, deeply dentate; stipules lanceolate, sub-entire; flowers axillary, sub-corymbed; petals oblong-linear, shorter than the calyx.

POTE'RIUM. 19—12. (*Rosaceæ.*) [From *poterion*, a cup, so called from the shape of the flowers.]

sanguisor''ba, (burnet, J. ♃.) stem somewhat angled, unarmed; leaves pinnate; leafets serrate; flowers in heads. Ex.

PRENAN''THES. 17—1. (*Cichoraceæ.*) [From *prenes*, drooping, and *anthos*, flower.]

al''ba, (white lettuce, w. p. Au. ♃.) radical leaves angled, hastate, toothed, somewhat lobed, cauline ones round-ovate, toothed, petioled, upper ones mostly lanceolate; panicle lax; the terminal fascicle nodding; calyx 8-cleft, 8-10 flowered. Var. *na'na*, leaves 3-parted, hastate, ovate, and lanceolate, sometimes all simple; racemes panicled or simple. 1-3 f.

altis''sima, (p. y. Au. ♃.) stem branching; leaves petioled, 3-lobed, angled, denticulate; margin scabrous; racemes axillary; flowers nodding; calyx about 5-flowered.

corda'ta, (w. y. Au. ♃.) stem panicled above; leaves petioled, cordate, toothed, ciliate; floral ones sessile, oblong, entire; panicle lax; raceme flowered. 4-6 f.

virga'ta, (w-p. Au. ♃.) glabrous; stem very simple; leaves all lyrate-sinuate; branches somewhat 1-sided; flowers pendent; involucre glabrous, 8-cleft, 10-flowered. 3-6 f.

crepidin''ea, (S.) leaves broad lanceolate, attenuated at the base, unequally tooth-angled; panicle fascicled, terminal, few-flowered, nodding; involucrum hirsute, 10-12 cleft, about 20-flowered. 4 6 f.

deltoi'dea, (p. S.) stem simple, glabrous; leaves deltoid, acuminate, acutely denticulate, sub-glaucous beneath; racemes axillary, few-flowered; involucrum 5-flowered. 2 f.

pauciflo'ra, stem branching, flexuous, panicled above; branchlets 1-flowered; flowers erect; leaves lance-linear, runcinate, glabrous; involucrum about 5-flowered.

illinoien''sis, stem simple, and with the leaves, very rough; leaves all undivided, lance-oval; raceme long; fascicles sub-sessile, erect, hirsute. S.

PRIMU'LA. 5—1. (*Primulaceæ.*) [From *primulus*, the beginning, so called because it blossoms in the beginning of spring. The natural family, Primulaceæ, is a division of Jussieu's order, Lysimachiæ.]

farino'sa, (bird's eye primrose, p. ♃.)

leaves obovate-spatulate, mealy beneath; umbel many-flowered; peduncles spreading; border of the corolla flat, as long as the tube, with obtuse, obcordate segments; scape 6-10 i. Leaves all radical.

mistasin''ica, (♃.) small, glabrous; leaves oval-spatulate, sub-dentate; scape elongated; umbel few-flowered; limb of the corolla reflexed; segments wedge-oblong, obtusely 2-cleft; capsule oblong, exsert.

angustifo'lia. (p.) leaves lance-oval, very entire, glabrous; scape 1-flowered; segments of the corolla ovate, very entire. 1-2 i.

cortusoi'des, a very ornamental species, producing red flowers from May to July. Ex.

præni'tens, (Chinese primrose,) a beautiful species, of which there are many varieties with pink, with white, and with semi-double flowers. Ex.

acau'lis, (primrose, ♃.) leaves rugose, toothed, hirsute beneath; scape 1-flowered. Ex.

auric''ula, (auricula primrose, ♃.) leaves serrate, fleshy, obovate; scape many-flowered; calyx mealy. Ex.

ve'ris, (cowslip, r-y. ♃.) leaves rugose, toothed; limb of the corolla concave; neck of the tube oblong; calyx inflated.

ela'tior, (oxlip primrose, w. y. ♃.) stalk many-flowered; limb of the corolla flat; flowers in an umbel, pale yellow, the centre deeper yellow; this is supposed to be a hybrid, between the primrose and cowslip.

vulga'ris, the English botanists describe the *acaulis* under this name; it is the polyanthus of the florist.

PRI'NOS. 6—1. (*Rhamni.*)

verticilla'tus, (winter berry, w. J. ♄.) stem much branched; leaves deciduous, oval, serrate, acuminate, pubescent beneath; flowers diœcious, 6 cleft; sterile ones axillary, sub umbellate; fertile ones aggregated, berries globose. Berries bright scarlet. 6-8 f. Swamps.

gla'ber, leaves evergreen, wedge-form, coriaceous, shining; pedicels axillary, mostly 3-flowered; berries black and shining, globose. 3-4 f. Ink-berry.

lævig'atus, (Ju. ♄.) leaves deciduous, lanceolate with appressed serratures, glabrous both sides, shining above; nerves beneath scarcely pubescent; flowers 6-cleft; pistillate flowers axillary, solitary, sub-sessile; staminate flowers scattered. 6-8 f.

ambig''uus, (w. J. ♄.) leaves deciduous, oval, entire, acuminate at each end; flowers 4-cleft; staminate ones crowded on the lower branches, pistillate ones solitary, on long peduncles. A small tree with whitish bark. 3-5 f.

integrifo'lia, (♄.) leaves deciduous, oval, entire, mucronate, on long petioles, glabrous on both sides; pistillate flowers solitary, long-peduncled. S.

lanceola'tus, (J. ♄.) leaves deciduous, lanceolate, finely and remotely serrulate, acute at each end, glabrous on both sides; pistillate flowers scattered generally in pairs, peduncled, 6 cleft; staminate ones aggregate, triandrous. *S.*

coria'ceous, (M. ♄.) leaves perennial, broad oval, acute, serrate near the apex, lucid above, minute-punctate beneath; pistillate flowers solitary, generally 8-parted; staminate ones aggregate-octandrous. Var. *latifo'lia,* leaves lance-obovate, acuminate. Var. *angustifo'lia,* leaves lanceolate, acute. 5-6 f. *S.*

PROSERPINA'CA. **3—3.** (*Hydrocharides.*) [From Proserpina, fabled as queen of the lower regions.]

palus"tris, (mermaid-weed, Au. ☉) upper leaves lance-linear, serrate: lower ones often pinnatifid; fruit angular, acute, stem procumbent. Wet places.

pectina'ta, distinguished from the former, by having the leaves all finely pectinate, and the fruit with rather obtuse angles.

PROSO'PIS. **10—1.** (*Leguminosæ.*) [From *Prosopon,* face, from the appearance of the frucification.]

glandulo'sa, (♄.) spines thick, cylindric-conic; leaves conjugate-pinnate, or pinnate one pair; leafets distant, 6.7 pairs, linear, sub-falcate, obtuse, glabrous, sub-coriaceous; petiole between the leaves and leafets glandular; legumes straight; spikes cylindric.

PRUNEL"LA. **13—1.** (*Labiatæ.*) [From *pruna,* a burn, because it heals burns.]

vulga'ris, var. *pennsylva'nica,* (heal-all, self-heal, J. ♃.) leaves petioled, oblong-ovate, toothed at the base; lips of the calyx unequal; upper one truncate, awned; ascending. 6-12 i.

NUS. 11—1. (*Rosaceæ.*) [*Prunus,* the Latin name for plum.]

A. *Flowers in racemes.*

virginia'na, (wild-cherry, rum-cherry, cabinet-cherry, w. M. ♄.) racemes erect, elongated; leaves oval-oblong, acuminate, unequally serrate, glabrous both sides; petioles generally bearing 4 glands. In open fields, the limbs of this tree spread out into an elegant oval top; but in dense forests, it grows to a very great height, with a few contracted branches.

america'na, (yellow or meadow plum, w. m. ♄.) leaves oblong-oval, acuminate, sharply serrate, veined; pedicels smooth; stipules mostly 3-parted; drupe oval or sub-globose, reddish yellow, with a coriaceous skin. Banks of streams; meadows.

mariti'ma, (w. M. ♄.) peduncles sub-solitary; leaves ovate-oblong, acuminate, doubly serrate.

seroti'na, (choke-cherry, w. J. ♄.) flowers in lax racemes; leaves oval, short-acuminate, opake, doubly and acutely serrate; midrib bearded on each side towards the base; petiole with 2 glands.

canaden"sis, (w. ♄.) flowers in racemes; leaves glandless, broad-lanceolate, rugose, sharply serrate, pubescent both sides, tapering into the petiole.

spino'sa, (English sloe, ♄.) peduncles solitary; leaves lance-oval, pubescent beneath; fruit straight; branches thorny. Ex.

cera'sus, (garden cherry, w. r. ♄.) umbel sub-peduncled; leaves lance-ovate, glabrous, conduplicate. Ex.

domes"tica, (plum, w. M. ♄.) peduncles sub-solitary; leaves lance-ovate, convolute; branches thornless. Var. *julia'na,* (damson plum,) fruit oblong, blue. Var. *claudia'na,* (sweet plum, horse-plum,) fruit round, at first green, becoming yellowish. Var. *enuclea'ta,* (stoneless plum,) the putamen obsolete. Ex.

can"dicans, has long clusters of white flowers, leaves woolly. Very ornamental. Ex.

cacomil'la, a native of Italy.

divarica'ta, has white flowers and yellow fruit. Ex.

PSORA'LEA. **16—10.** (*Leguminosæ.*) [From *psoraleus,* scabby; the plant being more or less glandular, which gives it a scurfy appearance.]

esculen"ta, (bread-root, b. J. ♃.) villose, leaves quinate-digitate, leafets lanceolate, unequal, flat, entire; spikes axillary, dense-flowered; divisions of the calyx lanceolate, scarcely as long as the corolla; legume ensiform, beaked; root fusiform. The root is used for food by the Indians.

canes"cens, (y. J. ♃.) hoary; leaves trifoliate, short-petioled, broad-lanceolate; spikes lax-flowered; flowers pedicelled; calyx hairy, not as long as the corolla. *S.*

tenuifo'lia, (b. S. ♃.) pubescent, branching; leaves trifoliate; leafets oval, rugose-punctate on both sides; peduncles axillary, about 3-flowered, longer than the leaves. 2 f. *S.*

lupinel"la, (p. J.) stem sparingly branched; leaves digitate, long-petioled; leafets filiform; racemes many-flowered, longer than the leaves; legumes rugose. 2 f. *S.*

longifo'lia, (♃.) wholly silky-villose; leaves trifoliate; leafets long-linear; spikes axillary, peduncled, lax-flowered, shorter than the leaves; teeth of the calyx and bracts subulate. *S.*

onobry'chis, stem smooth; leaves trifoliate; leafets lance-ovate, sub-pubescent; racemes axillary, long-peduncled; flowers 1-sided; legume sub-ovate, muricate, smooth. 3-5 f. *S.*

virga'ta, (b. ♃.) stem virgate, sub-pubescent; radical leaves oblong, ovate; cauline ones very narrow, glabrous; spikes axillary, shorter than the leaves. 2 f. *S.*

melilotoi'des, (p. J.) sub-pubescent; leaves trifoliate; leafets lance-oblong; spikes oblong; bracts broad-cordate, long-acuminate; pods round, nerved, very rugose. 1-2 f. *S.*

eglandulo'sa, (p. J. ♃.) pubescent, without glands; leaves trifoliate, oblong-lanceolate; spikes oblong; bracts broad, lanceolate, long-acuminate, and with the calyx villose. *S.*

multiju'ga, (p. J.) stem branching; leaves pinnate; leafets numerous (9.10 pairs) lance-oblong, obtuse, pubescent; spikes oblong; bracts small, membranaceous, glandless. 1-2 f. *S.*

PSYCHO'TRIA. **5—1.** (*Rubiaceæ.*) [From

pouches, cool, otruno, to excite, alluding to its properties.]

lanceola'ta, (♄.) branches and leaves reddish, hairy beneath; leaves lanceolate, acuminate at both ends; stipules clasping. roundish, caducous; corymb terminal, 3-forked at the base. Florida.

PTE'LEA. 4—1. (*Terebintaceæ.*) [From *ptelea,* elm, the fruit of this genus resembling that of the elm.]

trifolia'ta, (g-w. J. ♄.) leaves trifoliate; flowers panicled, diœcious. Var. *pentaphyl"la,* leaves quinate. Var. *pubes"cens,* leaves pubescent. 6.8 f.

baldwin'ii, leaves very small, glabrous; leafets sessile, oval, obtuse, the terminal ones cuneiform at the base; flowers tetandrous; styles none. Florida.

monophyl"la, leaves simple, lanceolate-ovate, nearly sessile; flowers racemed; fruit 3-winged. *S.*

PTE'RIS. 21—1. (*Filices.*) [From *pteron,* a wing, so called from the likeness of its leaves to wings.]

aquili'na, (common brake, Ju. ♃.) frond pinnate, 3-parted; barren branches doubly pinnate, with leafets lance-linear, obtuse pinnatifid, toothed; fertile branches pinnate, with leafets pinnatifid; divisions acutish, all ciliate.

atropurpu'rea, (rock brake, Ju. ♃.) frond pinnate; lower leafets lanceolate, obtuse, ternate or pinnate; at the base obtusely truncate or sub-cordate. Var. *veno'sa,* leafets veined beneath; stipe angled. Var. *puncta'ta,* leafets punctate beneath; stipe terete, dark purple. 3-10 i.

cauda'ta, (Au. ♃.) frond 3-parted, pinnate; barren divisions bi-pinnate; leafets linear, elongated, obtuse, entire; lower ones bi-pinnatifid; fertile branches pinnate; leafets remotish below; at the base pinnatifid, dentate.

peda'ta, (Ju. ♃.) frond deeply 5-lobed; palmate; lobes pinnatifid; segments lance-linear, acute. 6 i. *S.*

PTEROCAU'LON. 17—2. (*Corymbiferæ*) [From *pteron,* a wing, and *kaulos,* a stem.]

pycnostach"ya, (black-root, w. Au. ♃.) stem erect, simple, winged; leaves lanceolate, slightly undulate, dentate, tomentose and white beneath; spike cylindric; flowers clustered. *S.*

PTEROSPO'RA. 10—1. (*Ericæ.*) [From *pteron,* a wing, *spora,* seed.]

androm"eda, (Albany beech-drops, r-y. Ju. ⊕.) scape purple, very tall, bearing a many-flowered raceme; flowers lateral and terminal, nodding; peduncles filiform, longer than the flowers; lanceolate scales below, none above. 1-2 f.

PULMONA'RIA. 5—1. (*Boragineæ.*) [From *pulmo,* the lung, so called on account of its efficacy in diseases of the lungs.]

virgin"ica, (b. M. J. ♃.) smooth; stem erect; radical leaves obovate, oblong, obtuse leaves of the stem narrower; flowers in terminal racemes or fascicles; calyx much shorter than the tube of the corolla; segments lanceolate, acute; leaves somewhat glaucous; flowers large, bright blue. Plant becomes black by drying.

officina'lis, (spotted lung-wort, b. M. ♃.) leaves ovate, hairy, generally speckled with white on the upper side; the lower leaves on long petioles, the upper ones sessile; flowers violet-blue. 12 i. Ex.

alpi'na, (b. ♃.) nearly glabrous; stem simple, assurgent; leaves spatulate-ovate; flowers in terminal fascicles, sub-sessile; segments of the calyx oblong, obtusish, ciliate, about half the length of the corolla. 6 i.

lanceola'ta, (b. w. ♃.) glabrous, erect, radical leaves very long-petioled, lanceolate; cauline ones linear-oblong; flowers sub-peduncled; calyx short. *S.*

cilia'ta, (b.) glabrous; leaves lance-ovate, attenuate at each end, ciliate on the margin; flowers fascicle-panicled, pedicelled; corolla tubular-bell-form; calyx short, 5-parted; segments ovate, obtuse. 1 f. *S.*

PU'NICA. 11—1. (*Rosaceæ.*) [From *punicus,* Carthaginian.]

grana'tum, (pomegranate, ♄.) leaves lanceolate; stem woody. Ex.

PURSH"IA. 11—1. (*Rosaceæ.*) [In honor of Frederic Pursh, author of the North American Flora.]

tridenta'ta, (♄.) branches erect; branchlets numerous, short; leaves in fascicles, simple, 3-toothed, white beneath; flowers terminal or solitary. A North American shrub, with small yellow flowers, quite hardy.

PYCNAN"THEMUM. 13—1. (*Labiatæ.*) [From *puknos,* dense, *anthos,* flower, on account of its crowded inflorescence.]

A. *Stamens exsert.*

in"canum, (wild basil, mountain-mint, w. r. Ju. ♃.) leaves oblong-ovate, acute, sub-serrate, white-downy; flowers in compound heads, lateral ones peduncled; bracts setaceous. 1.5 f.

arista'tum, (w. Au. ♃.) leaves lance-ovate, sub-serrate, on very short petioles, whitish; heads sessile; bracts awned; flowers very small, in one or two sessile whorls and a terminal head; bracts and calyx terminated by long awns.

linifo'lium, (Virginian thyme, w. Ju. ♃.) stem straight, much branched, somewhat scabrous; leaves linear, 3-nerved, very entire, smooth; heads terminal, in a fasciculate corymb, stem 12-18 inches high, with trichotomous, fastigiate branches; flowers minute, shorter within. Woods.

virgin"icum, (narrow-leaf Virginian thyme, w. J. ♃.) pubescent; leaves sessile, lance-linear, entire, punctate; heads terminal, corymbed; bracts acuminate. 12-18 i. Mich.

B. *Stamens included.*

verticilla'tum, (w. Au. ♃.) leaves lance-ovate, sometimes toothed; whorls sessile, compact; bracts acuminate. 2 f. Mountains.

lanceola'tum, leaves linear-lanceolate, entire, veined; heads terminally sessile, in fascicled corymbs.

mu'ticum, (w. Ju. ♃.) leaves lance-ovate, sub-dentate, ribbed, sub-glabrous; heads terminal; bracts lanceolate, acutish. 18-24 i.

PYRO'LA. 10—1. (*Ericæ.*) [From *pyrus*, a pear, so called on account of the shape of the leaf.]

rotundifo'lia, (shin-leaf, pear-leaf wintergreen, w. J. ♃.) style declined; leaves rounded, or broad-oval, obsoletely serrulate, sub-coriaceous, shining; petiole about as long as the lamina; scape many-flowered. 6-12 i.

ellip"tica, (g-w. Ju. ♃.) leaves membranaceous, elliptical-ovate, serrulate, rather acute, lamina longer than the petiole; scape nearly naked; bracts subulate; calyx 5-toothed; style declined; scape 10 i.

asarifo'lia, (g-w. Ju. ♃.) leaves reniform, coriaceous, half as long as the dilated petiole; raceme many-flowered; stigma clavate; the disk elongated and 5-lobed. Dry woods.

secun"da, (one-sided wintergreen, g-w. Ju. ♃.) stamens erect; style straight; leaves ovate, acute; secund. 2-3 i. Sandy woods.

uniflo'ra, (J. ♃.) flower solitary; leaves orbicular, serrate; stigma acute; style straight. 5-toothed; flower terminal, large, white, fragrant, nodding. Chiefly in northern latitudes; rare.

aphyl"la, style declined; scape and stalk leafless, scaly; scales lanceolate, membranaceous; scape angular.

mi'nor, (w-r. Ju. ♃.) style straight; leaves round-oval, serrulate; scape sub-naked; spike with flowers reversed.

PY'RUS. 11—5. (*Rosaceæ.*) [Origin of the name doubtful.]

corona'ria, (crt b-apple, w-r. M. ♄.) leaves broad-oval, at the base rounded, sub-angled or sub lobed, serrate, smooth; peduncles corymbed. Flowers sweet-scented.

commu'nis, (pear, w. r. M. ♄.) leaves ovate, serrate, (rarely entire); peduncles corymbed. Ex.

ma'lus, (apple, w. r. M. ♄.) flowers in sessile umbels; leaves ovate-oblong, acuminate, serrate, glabrous; claws of the petals shorter than the calyx; styles glabrous. Var. *sylves"tris*, (wild-apple,) leaves ovate, serrate; fruit small, rough to the taste. The various kinds of apples are but varieties of the same species.

cydo'nia, (quince, w. J. ♂.) flowers solitary; fruit tomentose; leaves ovate, entire. Ex.

angustifo'lia, (M. ♄.) leaves lance-oblong, at the base acute, slightly crenate-toothed, shining; peduncles corymbed. Fruit very small. Florida.

prunifo'lia, (Siberian crab, w. r. M. ♄.) umbels sessile; pedicels pubescent; styles woolly at the base; leaves ovate, acuminate. 12-15 f.

spectab"ilis, (Chinese crab or garland flowering wild apple,) produces very showy flowers in May. Quite hardy. Ex.

corona'ria, (sweet-scented crab,) large and beautiful pink blossoms, very fragrant. Ex.

astracan"ica, (moscow or transparent crab,) fruit very large, wax-colored, almost transparent when ripe. Ex.

salvifo'lia, (w.) leaves woolly. Ex.

amyg"dalæfor"mis, leaves silvery-white; fruit shaped like that of the almond. Ex.

floribun"da, grows about four feet high, and sends down weeping branches, which are covered with a profusion of white flowers. Ex.

QUER"CUS. 19—12. (*Amentaceæ.*) [From *quero*, to inquire, because the Druids gave their divinations from this tree.]

1. *Fructification biennial; leaves setaciously mucronate.*

Leaves entire.

phel"los, (willow-oak, M. ♄.) leaves deciduous, linear-lanceolate, tapering at each end, very entire, glabrous, mucronate; acorn roundish. Var. *humil"is*, low and straggling; leaves shorter. 30-60 f.

imbrica'ria, (shingle-oak, M. ♄.) leaves deciduous, oblong, acute at each end, mucronate, very entire, shining-pubescent beneath; cup shallow; scales broad-ovate; acorn sub-globose. 40-50 f.

2. *Leaves dentate or lobed.*

trilo'ba, (downy black-oak, M. ♄.) leaves oblong-cuneiform, acute at the base, sub-3-lobed at the apex; lobes equal and mucronate, tomentose beneath; cup flat; acorn depressed-globose.

aquat"ica, (water-oak, M. ♄.) leaves obovate, cuneiform, glabrous, very entire; apex obscurely 3-lobed, middle lobe longest; cup hemispheric; acorn sub-globose; leaves very variable. 30-40 f.

ni'gra, (barren oak, blackjack, M. ♄.) leaves coriaceous, cuneiform, sub-cordate at the base, dilated, and retusely 3-lobed at the apex; when young, mucronate, glabrous above, rusty and pulverulent beneath; cup turbinate; scales obtuse and scarious; acorn short, ovate. Small.

cates"baei, (barren scrub-oak, M. ♄.) leaves short-petioled, cuneate at the base, oblong, deeply sinuate, glabrous; lobes 3-5, divaricate, dentate, acute; cup turbinate, large; scales obtuse, marginal ones inflexed; acorn ovate. 15—30 f. Bark used by tanners.

palus"tris, (pin-oak, M. ♄.) leaves long-petioled, oblong, deeply sinuate, glabrous; axils of the veins villose beneath; lobes divaricate, dentate, acute; cup flat, smooth; acorn sub-globose.

tincto'ria. (black-oak, M. ♄.) leaves obovate-oblong, slightly sinuate, pubescent beneath; lobes oblong, obtuse, obscurely toothed, mucronate; cup flat; acorn depressed, globose; bark dark-colored.

banniste'ri, (scrub-oak,) leaves on long petioles, wedge-obovate, 3-5 lobed, entire on the margin, grayish-tomentose beneath; lobes setaceously mucronate; cup sub-turbinate; acorn sub-globose. Dry hills and barrens. 4-6 f.

ru'bra, (red-oak,) leaves large, bright green; sinuses rounded; cup of the corolla shallow, base flat.

coccin"ea, (scarlet-oak,) distinguished by

the brilliant red of its leaves toward the close of autumn ; acorn short, ovate ; cup turbinate, scaly. The wood is used for cooper's staves.

3. *Fructification annual ; fruit peduncled ; leaves awnless, lobed.*

obtusilo'ba, (iron-oak, post-oak, M. ♄.) leaves oblong, sinuate, cuneate at the base. pubescent beneath ; lobes obtuse, the upper dilated ; cup hemispherical ; acorn oval. 30-50 f.

al"ba, (white-oak, M. ♄.) leaves oblong, sinuate-pinnatifid, pubescent beneath ; lobes obtuse, entire, narrowed at their bases, particularly on full-grown trees ; fruit peduncled ; calyx somewhat bowl-form, tubercled, flattened at the base ; acorn ovate. Fertile forests throughout the U. S. Timber firm and durable, of great use in shipbuilding, and in many other arts. 70-100 feet high.

macrocar"pa, (over-cup oak, M. ♄.) leaves downy beneath. deeply lyrate, sinuate-lobed ; lobes obtuse, repand, upper ones dilated ; cup deep, upper scales setose ; acorn short-ovate. A large tree.

olivæfor"mis, mossy-cup oak, M. ♄.) leaves oblong, smooth, glaucous beneath, deeply and unequally sinuate-pinnatifid ; cup very deep, crenate above ; acorn elliptic-oval. Hills. A large tree.

4. *Leaves entire, dentate.*

pri"nus, (swamp chestnut-oak, M. ♄.) leaves long petioled, obovate, acute, pubescent beneath, coarsely toothed ; teeth dilated, callous at the point ; cup deep, attenuate at the base ; acorn ovate.

chin"quapin, (dwarf chestnut-oak, chinquapin, M. ♄.) leaves obovate, obtuse, glabrous, short-petioled, coarsely toothed, glaucous beneath ; teeth nearly equal, dilated, callous at the apex ; cup hemispheric ; acorn ovate. A low shrub. 3-4 f.

monta'na, (rock chestnut oak, M. ♄.) leaves petioled, broad-obovate, oblong, white-tomentose beneath, shining above, coarsely toothed, obtuse and unequal at the base ; teeth nearly equal, very obtuse ; fruit in pairs, short-peduncled ; cup hemispheric, scales tuberculate, rugose ; acorn ovate. 30-50 f.

casta'nea, (yellow-oak, M. ♃.) leaves long-petioled, lance-oblong, obtuse at base, acuminate, tomentose beneath, coarsely toothed ; teeth unequal, dilated, acute, callous at the apex ; cup hemispheric ; acorn ovate, sub-globose. Mountains. 60-70 f.

bi'color, (swamp white-oak, M. ♄.) leaves short petioled, oblong, obovate, white tomentose beneath, coarsely toothed, entire at the base ; teeth unequal, spread, acutish, callous at the apex ; fruit in pairs, long-peduncled ; cup hemispheric ; acorn oblong-ovate. Var. *mol"lis.*leaves toothed, sub-ferruginous and soft-pubescent beneath. 60-70 f. S.

vir"ens, (live-oak, M. ♄.) leaves perennial, coriaceous, oblong-oval, entire, margins revolute, obtuse at base, acute at the apex, stellate-pubescent beneath ; fruit pedicelled ; cup turbinate ; acorn oblong. 40-60 f. Florida.

pu'mila, (Ap. ♄.) leaves deciduous oblong-lanceolate, sub-undulate, acute and mucronate at the apex, glabrous above tomentose beneath ; acorn nearly spherical. 2 f. S.

mariti'ma, (Ap. ♄.) leaves perennial, coriaceous, lanceolate, entire, glabrous, tapering at the base, acute at the apex, mucronate ; acorn oval. 4-10 f. S.

hemispher"ica, (M. ♄.) leaves perennial, lance oblong, undivided, 3-lobed, and sinuate ; lobes mucronate, glabrous on both sides. Resembles the *aquat"ica*. S.

laurifo'lia, (Ap. ♄.) leaves nearly perennial, sessile, lance-oblong, sub-acute, tapering at the base, entire, glabrous on both sides ; acorn sub-ovate. Var. *obtu'sa*, leaves obtuse at the apex. 40-50 f. S.

cine'rea, (Ap. ♄.) leaves perennial, coriaceous, oblong-lanceolate, entire, margins sub-revolute, mucronate at the apex, stellate. tomentose beneath ; fruit sessile ; acorn sub-globose. 20 f. S.

myrtifo'lia, (♄.) leaves perennial, coriaceous, small, oblong-ovate, acute at each end, glabrous, shining and reticulate above, margins revolute. S.

na'na, (♄.) leaves cuneate, glabrous, 3-lobed at the summit, sub-sinuate at the base ; lobes divaricate, mucronate ; middle one largest ; axils of the veins beneath pubescent ; acorn ovate, sub-globose. S.

michaux"ii, (Ap. ♄.) leaves petioled, obovate, obtuse at the base, unequally dentate, sinuate, tomentose beneath ; fruit generally in pairs ; acorn very large, ovate. 50-60 f. S.

lyra'ta, (Ap. ♄.) leaves glabrous, sinuate, oblong ; lobes oblong, sub-acute, upper ones broad, angled ; cup as long as the globose nut ; acorn nearly covered. 60-70 f. S.

i'lex, (evergreen oak,) a very ornamental shrub. Ex.

lucumbea'na, (turkey-oak,) grows rapidly, and forms a very handsome pyramidal tree. Ex.

RANUN"CULUS.   12—12.   (*Ranunculaceæ.*)
[Diminutive of *rana*, a frog, because it is found mostly in places where frogs abound.]

A. *Leaves divided.*

abor"tivus, (y. M. ♃.) glabrous ; stem striate, naked below ; radical leaves heart-reniform, obtusely crenate, cauline ones petioled, ternate, angled, upper ones sessile, branches about 3-flowered. 9-15 l.

re'pens, (y. M. ♃.) pubescent ; leaves ternate, 3-cleft, gashed ; creeping shoots sent off in the summer ; peduncles furrowed ; calyx spreading. Damp.

a'cris, (crowfoot, buttercup, y. M. ♃.) hairs close-pressed ; leaves 3-parted, many-cleft, upper ones linear ; peduncles terete ; calyx spreading. 1-2 f.

scelera'tus, (celery crowfoot, y. Au. ♃.) radical leaves petioled, 3-parted, the segments lobed, cauline ones sessile, 3-lobed, carpels small, numerous, forming an oblong head ; stem 1 f., succulent, branched.

his"pidus, (hairy crowfoot, w-y.) stem and petioles with stiff, spreading hairs ; calyx hairy ; styles short. Wet ground.

recurva'tus, calyx and corolla recurved; carpels uncinate; stem erect; petioles covered with stiff, spreading hairs. Shady woods.

fluvia'tilis, (river-crowfoot, w. y. M. 2/.) stem submersed; leaves dichotomous, capillary.

bulbo'sus, (y. M. 2/.) very hirsute; leaves ternate, 3 cleft, gashed and toothed; stem erect, many-flowered; petals obcordate, shorter than the reflexed sepals; root bulbous.

pennsylva'nicus, (y. Au. 2/.) stem pilose, erect, branching; leaves ternate, villose; segments sub-petiolate, acutely 3-lobed, incisely serrate; calyx reflexed; petals about equalling the calyx; styles of the fruit straight. 1-2 f.

hirsu'tus, (pale buttercup, y. Ju. 2/.) hirsute; leaves ternate; stem erect, many-flowered; peduncles sulcate; calyx reflexed; fruit globose; carpels tubercled; root ... Wet fields.

... M. 2/.) leaves ternate, ...bothed and incised, cuneate, ...etioled, floral leaves incised ...ncle 1-3 flowered; petals ... spreading; carpels mar...ort, ...style. 6-8 i.

...(y. J... ; leaves 3-...ky; ped-...al ...nding.

... 2/.) ...le ...ided ...is ... ...ate-...tha ...er-...ersed ... r. ...ves p ...d, dissected ...s, all immer ...e ...eathing and au... r. ...es sessile, all immersed, fili-...d, circinate; segments short; ... acute, nearly smooth.

...icus, (w.y. M. 2/.) pubescent; ...ple, sub-naked; radical leaves ter-nate; leafets 3-lobed; lobes acute, gashed; calyx reflexed.

hedera'ceus, (Ju. 2/.) stem creeping; leaves sub-reniform, about 3-5 lobed; lobes broad, entire, very obtuse; petals oblong, scarcely longer than the calyx; stamens 5-12; carpels glabrous. S.

echina'tus, (y.) simple, rather glabrous; leaves roundish, 3-lobed; petals twice as long as the calyx. S.

tomento'sus, (y. 2/.) stem ascending, very villose, 1-2 flowered; leaves petioled, tomentose, 3-cleft, upper ones sessile, ovate, entire; calyx very villose, sub-reflexed. S.

carolinia'nus, (y.) stem erect, branched, and with the petioles appressed, pubescent; leaves glabrous, 3-cleft or 5-lobed; lobes ovate, somewhat gashed, toothed; calyx glabrous, reflected, a little shorter than the petals. S.

trachysper''mus, (y. M.) stem, petiole, and leaves, villose, with the hair spreading; leaves 3-cleft; lobes acutely gashed; peduncles short, opposite the leaves; carpels tubercled, with the point hooked. 12-15 i. S.

murica'tus, (y. Ap. ⊙.) leaves petioled, glabrous, roundish, 3-lobed, coarsely toothed; stem erect or diffuse; peduncles opposite the leaves; calyx spreading; carpels rough-tubercled on both sides, with a straight-acuminate point. 12-18 i. S.

B. Leaves undivided.

lin''gua, (great spearwort, y. Au. 2/.) leaves long, lanceolate, serrate, semi-amplexicaulis; stem erect, smooth, many-flowered; flowers large. Banks of streams. 2-3 f.

flammu'la, (spearwort, y. Ju.) leaves glabrous, lance-linear, lower ones petioled, stem decumbent, rooting; peduncles opposite the leaves; flowers smaller than the preceding. Swamps. 12-18 i.

pusil''lus, (y. Ju. 2/.) erect; leaves petioled, lower ones ovate, upper ones lance-oblong; petals about as long as the calyx. 6-12 i.

rep''tans, (w-y. Ju. 2/.) leaves linear-sub-ulate; stems filiform, creeping, geniculate; joints 1-flowered. 6-10 i.

nemoro'sus, produces yellow flowers from May to August. Ex.

illyr''icus, remarkable for its silky, white leaves. Ex.

plantagin''eus, (Ap. w.) leaves glaucous, lanceolate.

RAPHA'NUS. 14—2. (Cruciferæ.) [From radios, root, phainesthai, to grow quickly.]

sati'vus, (garden radish, w. J. ⊙.) leaves lyrate; silique terete, torose, 2-celled. There are several varieties of this species— one has a fusiform, another a globose, another a black root. Ex.

raphanis''trum, (wild radish, y. Au.) leaves simple, lyrate; pod jointed, 1-celled, striate, 3-8 seeded. 1-2 f. Stem hispid. Fields.

RENSSELAE'RIA. 19—12. (Aroideæ.) [In honor of Gen. Stephen Van Rensselaer, of Albany, N Y.]

virgin''ica, (g. J. 2/.) scapes several from one root; leaves on long petioles, oblong, hastate-cordate, with the lobes obtuse, a foot or more long; spatha lanceolate, involute, border undulate, closely embracing the spadix, which is long and slender; berries 1-seeded. 12-18 i.

RESE'DA. 12—5. (Capparides.) [From reseda, to appease, so called from its supposed virtues in allaying inflammation.]

odora'ta, (mignonette, w-y. Ju. ⊙.) leaves entire and 3-lobed; calyx equalling the corolla. Ex.

luteo'la, (dyer's weed, y.) leaves lanceolate, undulate, entire, each side of the base toothed; calyx 4-cleft; flowers in a spike. Introduced.

micran''thus, (y.) hairy; leaves petiolate, somewhat rhombic-ovate, crenate, some 3-parted or 3-cleft, cauline ones sub-sessile, with 3-5 linear-oblong segments; sepals with a broad, membranaceous border, as long as the corolla. Ex.

RHAM''NUS. 5—1. (Rhamni.) [From rao, to destroy, on account of the many thorns of some of its species.]

alnifo'lius, (dwarf-alder, w-g. M. ♄.) unarmed; leaves oval, acuminate, serrulate, pubescent on the nerves beneath; flowers diœcious; peduncles 1-flowered, aggregate; calyx acute; fruit turbinate; berries black. Rocky hills.

cathar"ticus, (buckthorn, y-g. ♄.) branches spiny; leaves opposite, ovate; flowers 4-cleft, diœcious. Mountain woods.

frangulo'dens, (w-g. M. ♄.) unarmed; leaves oval, acuminate, serrulate, pubescent at the nerves beneath; peduncles aggregate, 1-flowered; calyx acute; fruit turbinate; berries black.

carolin"ia'nus, (w. J. ♄.) unarmed; leaves alternate, oval-oblong, sub-entire, ribbed, glabrous; umbels peduncled; flowers all fertile; berry black, globose. 4-6 f.

lanceola'tus, (♄.) unarmed; leaves nearly opposite, oval, serrulate; flowers very minute, divisions spiked, alternately sessile on the rachis; style 3-cleft; berries 3-seeded.

parvifo'lius, unarmed; leaves ovate, serrulate, when young, pubescent, acute, or emarginate; flowers solitary, or 2 to 3 together, axillary, short-pedicelled, tetandrous; petals minute, 2-lobed, partly surrounding the very short stamens; styles 2, united below, very short and conical. *S.*

minutiflo'rus, (Oc. ♄.) unarmed; leaves nearly opposite, oval, serrulate; flowers very minute, divisions spiked, alternately sessile on the rachis; style 3-cleft; berries 3-seeded.

RHE'UM. 9—3. (*Polygonæ.*) [From *Rha,* an ancient name of the Wolga, on whose banks it was discovered.]

palma'ta, (rhubarb, J. 2f.) leaves palmate, acuminate. Ex.

rhapon"ticum, (pie rhubarb, w. J. 2f.) leaves heart-ovate, obtuse and acute, smooth; veins sub-pilose beneath, the sinuses at the base dilated; petioles furrowed on the upper side, rounded at the edge; radical leaves very large. 2-4 f. Ex.

RHEX"IA. 8—1. (*Melastomiæ.*)

maria'na, (w-r. Ju. 2f.) very hairy; leaves lanceolate, acute at each end, 3-nerved, sub-petiolate; calyx tubular, nearly smooth. Var. *purpu'rea,* has purple flowers; petals obovate, hairy on the outer surface.

virgin"ica, (deer-grass, meadow-beauty, p. Ju. 2f.) stem with winged angles, square, somewhat hairy; leaves sessile, ovate-lanceolate, ciliate, serrate, 3-7 nerved, sprinkled with hairs on both sides; corymbs dichotomous. Wet meadows. 1 f.

cilio'sa, (p. Ju. 2f.) stem nearly square, smooth; leaves sub-petioled, oval, serrulate, ciliate, 3-nerved, glabrous beneath, slightly hispid above; flowers involucred. 12-18 l.

glabel"la, (deer-grass, p. Ju.) glabrous, stem terete; leaves lanceolate and ovate, 3-nerved, denticulate, slightly glaucous; calyx glutinous. 2-3 f. *S.*

serrula'ta, (p. 2f.) stem nearly square, glabrous; leaves small, sub-petioled, roundish-oval, acute, smooth on both sides, margin serrulate, base sub-ciliate; flowers peduncled, about in threes; calyx glandular-hirsute. 6-10 i. *S.*

lu'tea. (y. Ju. ⦹.) hirsute; leaves linear-lanceolate, sometimes wedge-form at the base, 3-nerved; panicle pyramidal; anthers erect. terminal. 18 i. *S.*

angustifo'lia, (w. Ju. 2f.) anthers incumbent; leaves linear and lance-linear, somewhat clustered; plant hirsute. *S.*

stric"ta, (p. J. 2f.) stem 4 angled, straight, winged, glabrous, bearded at the joints; leaves sessile, narrow-lanceolate, acuminate, 3-nerved, glabrous on both sides; corymb dichotomous. *S.*

linearifo'lia, (y.) stem cylindrical. sub-pubescent; leaves alternate, linear. oblong, obtuse, sessile, pubescent on both sides; flowers generally solitary. *S.*

RHINAN"THUS. 13—2. (*Pediculares.*) [From *rin,* nose, and *anthos,* flower.]

cristagal"li, (yellow-rattle, y. J. ⦹.) upper lip of the corolla arched; calyx smooth; leaves lanceolate, serrate, opposite: flowers axillary, somewhat spiked, yellow Meadows.

RHIZOPHO'RA. 12—5. (*Salicariæ.*) [From *rhizo,* root, and *phero,* to bear, on account of its peculiar root.]

man"gle, (mangrove, ♄.) leaves acute, ovate, opposite; peduncles axillary; fruit clavate, subulate.

RHO'DODEN"DRON. 10—1. (*Rhododendra.*) [From *rodon,* a rose, *dendron,* tree; so called because it resembles the rose.]

max"imum, (wild rosebay, E. r. Ju. ♄.) leaves oblong, glabrous, paler beneath; umbels terminal, dense; corollas somewhat bell-form. Var. *rose'um,* corolla pale rose-color; segments roundish; leaves obtuse at the base. Var. *al"bum,* corolla smaller, white, segments oblong; leaves acute at the base. Var. *purpu'reum,* corolla purple; segments oblong; leaves obtuse at the base, green on both sides. 4-20 f.

pon"ticum, (rosebay, p. ♄.) leaves oblong, glabrous, both sides colored alike; corymbs terminal; corolla bell-wheel-form; petals lanceolate. A native of Asia Minor.

lappon"icum, (p. Ju. ♄.) flowers in terminal, leafy clusters, campanulate; stamens mostly 8; leaves elliptical, punctured, coriaceous, evergreen; shrub 8-10 i. White hills.

albiflo'rum, (w. ♄.) erect; leaves deciduous, lance-oval, very entire, membranaceous, glabrous, fasciculate in the apex of the branches; peduncles fasciculate, lateral and terminal; calyx sub-foliaceous, hispid; corolla rotate-campanulate; stamens 10, erect, equal. 2-3 f.

puncta'tum, (r. Ju. ♄.) leaves oval, lanceolate, glabrous, with resinous dots beneath; umbels terminal; corolla funnel-form; capsules long. 4-6 f. *S.*

catawbien"se, (r. J. ♄.) leaves short-oval, glabrous, roundish-obtuse at each end; umbels terminal; segments of the calyx narrow-oblong; corolla campanulate. 3-4 f. *S.*

arbo'reum, grows about 20 feet high, with immense bunches of dark scarlet or crimson velvet-like flowers. These flowers secrete honey in such abundance, that when the tree is shaken, the drops of honey fall from it like rain. The leaves are large and silvery beneath.

chrysan"thum, a dwarf species, with yellow flowers. Ex.

RHODO'RA. 10—1. (*Rhododendra.*)

canaden"sis, (false honeysuckle. p. M. ♄.) leaves alternate, oval, entire, pubescent-glaucous beneath; flowers in terminal umbels or clusters, appearing before the leaves. Mountain bogs. 2 f.

RHUS. 5—3. (*Terebintaceæ.*) [From *reo*, to flow, so called because it was supposed to be useful in stopping hæmorrhages.]

gla'brum, (sleek-sumach, g. r. Ju. ♄.) branches, petioles, and leaves, glabrous; leaves pinnate, many-paired; leafets lance-oblong, serrate, whitish beneath; fruit silky. The leaves are used for tanning morocco leather. Berries red and sour. 6 12 f.

ver"nix, (poison sumach, y-g. J-Ju. ♄.) very smooth; leaves pinnate; leafets in many pairs, oval, abruptly acuminate, entire; panicles loose; flowers diœcious. A small tree.

toxicoden"dron, (g-y, J–Ju. ♄.) stem erect; leaves ternate; leafets broad, oval, entire or sinuate, dentate, sub-pubescent beneath; flowers diœcious, in sessile, axillary racemes. 1 ½ f. Var. *rad"icans* (poison-ivy), stem climbing.

typhi'na, (stag's-horn sumach, y-g. J. ♄.) branches and petioles very villose; leafets in many pairs, lance oblong, acuminate, acutely serrate, pubescent beneath; flowers in oblong, dense panicles, diœcious; clusters of fruit covered with a purple, velvety down; berries red, and very sour. Rocky hills.

copalli'num, (gum-copal tree, mountain sumach, y-g. Jn. ♄.) petioles winged, appearing as if jointed; leafets many-paired, oval-lanceolate, very entire, shining on the upper surface; panicle sessile; flowers diœcious. Fruit red, hairy, small.

aromat"icum, (y. M. ♄.) leafets sessile, ovate-rhomboid, dentate, pubescent beneath; flowers amentaceous, diœcious. 2-6 f. Mountains.

pu'milus, (Ju. ♄.) low; branches and petioles pubescent; leafets oval, sharply toothed, tomentose beneath; fruit silky and downy. Poisonous. 1 f. *S.*

lauri'num, very glabrous; leaves elliptical or elliptic-ovate, obtuse or emarginate, often mucronate; panicles crowded; stamens 5; filaments very short. California.

co'tinus, (purple fringe-tree, p-g. Ju. ♄.) leaves simple, obovate and ovate; panicled racemes plumose. A small tree, with very minute flowers supported on capillary, downy, or hairy peduncles. Indigenous in Siberia, Austria, and Lombardy, often called the periwig-tree from the curious appearance of the seed-vessels which look like a powdered wig. Ex.

vernicife'ra, (varnish or Japan sumach,) a native of India and Japan, where it is much esteemed on account of its gum, which forms the best varnish.

RHYNCHOS"PORA. 3—1. (*Cyperoideæ.*) [From *runchos*, the beak of a bird, and *spora*, a

34*

seed, the permanent style forming a beak to the seed.]

al"ba, (Ju. ♃.) spike corymb-fascicled; culm triangular above; leaves setaceous; pericarp somewhat lenticular; bristles about 10. 12-18 i.

glomera'ta, (false bog-rush, J. ♃.) spikes clustered in corymbs, distant, by pairs; stem obtusely angled; pericarp obovate, wedge-form, very glabrous. 12-18 i.

rariflo'rus, (M. ♃.) stem and leaves setaceous; panicle loose, few-flowered; seed obovate, rugose; bristles as long as the seed. 1 f. *S.*

inexpan"sa, (Ju.) stem obscurely 3-angled; panicles remote, pendulous; seed oblong, compressed, rugose; bristles scabrous, twice as long as the seed. 2 f. *S.*

dis"tans, (Ju.) stem 3-angled; flowers in distant clusters; seed lenticular, slightly furrowed; bristles setaceous. 12-20 i. *S.*

puncta'ta, fascicles lateral and terminal, clustered near the summit of the stem; seeds rugose, dotted, shorter than the bristles. 1-2 f. *S.*

RI'BES. 5—1. (*Cacti.*) [Origin of the name doubtful]

flo'ridum, (wild black-currant, M. ♄.) unarmed; leaves punctate both sides; racemes pendent; calyx cylindric; bracts longer than the pedicels. 3-4 f.

triflo'rum, (wild gooseberry, g. M. ♄.) spine sub-axillary; leaves glabrous, 3-5-lobed, gash-toothed; peduncles sub-3-flowered; pedicels elongated; bracts very short; petals spatulate, undulate; style hirsute, half 2 or 3-cleft, exsert, berry glabrous, pale red. 3-4 f.

ru'brum, (currant, g. M. ♄.) unarmed; racemes glabrous, nodding; corolla flat; petals obcordate; leaves obtusely 5-lobed; stem erect; berries red. 2-4 feet. Ex.

ni'grum, (black currant, g. M. ♄.) unarmed; leaves punctate beneath; racemes lax; flowers bell-form; bracts shorter than the pedicels; berries black. 5-3 f. Ex.

grossula'ria, (English gooseberry, g. M. ♄.) branches prickly; petioles hairy; bracts 2-leaved; berry glabrous or hirsute. 2-4 f. Ex.

albiner'vium, (g-y. M. ♄.) leaves short, acutely lobed, smoothish; nerves white; racemes recurved; berries red, smooth.

tri'fidum, (y-g. m. ♄.) leaves moderately lobed, smooth above, pubescent beneath; racemes lax, pubescent; flowers rather flat; segments of the calyx about 3-cleft; petals spatulate, obtuse; berries hairy, red.

ri'gens, (mountain currant, m. ♄.) unarmed; branches straight; leaves long-petioled, acutely lobed and dentate, reticulate-rugose, pubescent beneath; racemes lax; becoming stiffly erect; segments of the calyx obovate, obtuse; berries red, hispid.

glandulo'sum, (r-y.) branches prostrate; leaves lobed, smoothish; younger ones pubescent; racemes sub-erect; petals deltoid; bracts minute; berry hispid, most of the plant, particularly the calyx, covered with glandular hairs. 2-3 f.

gra'cile, (M. ♄.) spines sub-axillary; leaves on slender petioles, pubescent on both sides; lobes acute, dentate, incised; peduncles slender, erect, about 2-flowered; calyx tubular-campanulate; berries glabrous. 2-3 f.

oxycanthoi'des, (smooth gooseberry, M. ♄.) larger spines sub-axillary; smaller ones scattered; leaves glabrous; lobes dentate; peduncles short, about 2-flowered; berries purple, glabrous. 3 f.

cynos"bati, (prickly gooseberry, g. M. ♄.) sub-axillary spines by pairs; leaves short-lobed, gash-toothed, soft, pubescent; racemes nodding, few-flowered; calyx erect, campanulate; berries aculeate, dark brown.

resino'sum, (g. Ap. ♄.) unarmed, covered with resinous, glandular hairs; leaves 3-5-lobed, roundish; racemes erect; calyx flattish; petals obtuse-rhomboid; bracts linear, longer than the pedicels; berries hirsute. *S.*

rotundifo'lium, (♄.) spines sub-axillary; leaves roundish, lobes obtuse; peduncles 1-flowered; limb of the calyx tubular; berries glabrous. *S.*

ni'veum, (snowy-flowered gooseberry,) has pendulous white flowers, and dark purple fruit. Ex.

specio'sum, (fuschia-flowered gooseberry,) flowers scarlet, stamens very long; leaves sub-evergreen.

puncta'tum, an evergreen species, a native of Chili; flowers bright yellow; leaves shining.

RICI'NUS. 19—15. (*Euphorbia.*) [From *rin*, nose, and *kunos*, a dog, because the capsules stick to the noses of dogs.]

commu'nis, (castor-oil plant, palma-christi, ☉.) leaves peltate, palmate; lobes lanceolate, serrate; stem with hoary mealiness. 4-6 f. Ex.

RIVI'NA. 4—1. (*Atriplices.*) [In honor of *Rivinus*, the great German botanist.]

læ'vis, leaves ovate, acuminate, glabrous, flat; stem terete; racemes simple.

hu'milis, (♄.) racemes simple; leaves tetandrous; leaves pubescent. *S.*

ROBIN"IA. 16—10. (*Leguminosæ.*)

pseudo-aca'cia, (locust-tree, false acacia, w. M. ♄.) leaves pinnate, with a terminal leafet; stipules thorny, or a thorn; racemes pendent; teeth of the calyx unawned; legumes smooth. 30.40 f.

visco'sa, (clammy locust, Ju. ♃.) racemes of one-flowered pedicels; pinnate leaves with a terminal leafet; branches and legumes viscid; racemes axillary, dense-flowered, erect; flowers varying from red to white. *S.* Cultivated.

hispi"da, (rose-locust, Au. r. ♄.) racemes axillary; calyx acuminate; most of the plant hispid; leaves pinnate with a terminal leafet; leafets round-oval, mucronate, sometimes alternate. 3-6 f. *S.* Cultivated.

ROCHEL"IA. 5—1. (*Boragineæ.*)

virginia'na, (w. b. J. ☉.) pilose, leaves oblong-lanceolate, acuminate, large, scabrous above; racemes divaricate; fruit densely covered with hooked bristles. 2 f. Rocky hills.

lap"pula, (b. Ju. ☉.) leaves linear-oblong; stem branched above; corolla longer than the calyx; border erect-spreading. 12-18 i.

RO'SA. 11—12. (*Rosaceæ.*) [The Latin name rosa, is from the Greek *rodon*, red.]

parviflo'ra, (wild-rose, r. w. ♄.) germs depressed, globose; germs and peduncles hispid; petioles pubescent, sub-aculeate; stem glabrous; prickles stipular, straight; leafets lance-oval, simply serrate, glabrous; flowers somewhat in pairs; very variable. 1-3 f.

rubig- ...... with the original laws on file in this office, and do certify that the same are correct transcripts therefrom and of the whole of the said originals.
　　　　　N. S. BENTON, Sec'y of State.

THE CONVENTION.
Monday, June 15.

Communications were received from several county surrogates and clerks of the Supreme Court, in answer to the resolution relative to the judiciary. Referred to a committee of five to revise and digest.

RESOLUTIONS OFFERED.

By Mr. Talmadge to make no imprisonment for debt a constitutional provision, and in relation to the fraudulent contraction of debts.

Mr. T. presented a plan or a system of judiciary. A long debate took place upon the printing of the documents. In the journal, a rule adopted, that the substance of the subject only of this and similar documents be entered on the journal.

By Mr. Nellis, in relation to the construc-

...... germ depressed-globose, with the peduncles glabrous. Dry hills.

sabifo'lia, (climbing rose, r. Ju. ♄.) tube of the calyx sub-globose; with the peduncles glandular-hispid; stem smooth; prickles short, solitary, uncinate; leaves petioled, ternate; leafets ovate, acute, serrate, glabrous above, white, downy beneath, segments of the calyx viscid-pilose; flowers corymbed. 6-8 f.

micran"tha, (r.w. J. ♄.) tube of the calyx ovate, with the peduncles somewhat hispid; prickles hooked; leafets ovate, acute, with reddish glands beneath. 4-8 f.

pimpinel'lifo'lia, (burnet rose, r. ♄.) leaves obtuse, petioles scabrous; peduncles glabrous; stem with straight prickles scattered. Very small. Ex.

parvifo'lia, (small-leaf rose, ♄.) small; tube of the calyx ovate, sub-glabrous; peduncles glandular; stem and petioles with slender prickles; leafets rugose; a little villose beneath, ovate, glandular, serrate.

setige'ra, (J. ♄.) fruit globose, with the petioles and veins prickly; branches glabrous; prickles by pairs and scattered; leafets 3-5, acuminate, glabrous; leafets of the calyx feathered with bristles. 5 8 f. *S.*

læviga'ta, (Cherokee rose. w. Ap. ♄.) fruit oblong, hispid; leaves perennial. ternate; leafets lanceolate, serrate. lucid, coriaceous; flowers solitary, terminal. 5-20 f. *S.*

lutes"cens, (y-w. J. ♄.) fruit globose, and with the peduncles glabrous; branches hispid-spiny; leafets (7) glabrous, oval; petioles unarmed; flowers solitary; segments of the calyx lanceolate, cuspidate; petals oval, very obtuse. *S.*

musco'sa, (moss-rose. r. Au. ♄.) germs ovate; calyx, peduncles, petioles, and branches, hispid, glandular-viscid, (moss-like); spines of the branches scattered, straight. Ex.

moscha'ta, (musk-root, ♄.) germs ovate; germs and peduncles villose; stem and petioles prickly; leafets oblong, acuminate, glabrous; panicle many flowered. Ex.

burgundia'ca, (Burgundy-rose, ♄.) germs sub-globose; germ and peduncles hispid; leafets ovate, pubescent beneath; corolla small, full, fleshy, white; disk obscure. Var. *provincia'lis*, has scattered, reflexed prickles on the branches, and glandular serratures. Ex.

semperflo'rens, (monthly-rose, ♄.) germs ovate-oblong, tapering to both ends; germs and peduncles hispid; stem prickly; flowers in erect corymbs. Resembles damascena. Ex.

al"ba, (white-rose, w. J. ♄.) germs ovate. glabrous or hispid; stem and petioles prickly; leafets ovate, villose beneath. Ex.

centifo'lia, (hundred-leaved rose, r. ♄.) germs ovate; germs and peduncles hispid; stem hispid, prickly; leaves pubescent beneath; petioles unarmed. Ex.

cinnamo'mea, (cinnamon-rose, ♄.) germs globose; germs and peduncles glabrous; stem with stipular prickles; petioles somewhat unarmed; leafets oblong. Stem brown, cinnamon-color. Ex.

multiflo'ra, (Japan-rose, ♄.) germs ovate; germs and peduncles unarmed, villose; stem and petioles prickly. Branches generally purple; leafets ovate; flower small, panicled. Ex.

spinosis"sima, (Scotch-rose, ♄.) germs globose, glabrous; peduncles hispid; stem and petioles very hispid. Var. *scot"ica*, is smaller. Loudon says that there are 300 varieties of this rose in a nursery at Glasgow; and that florists enumerate upwards of 900 sorts of roses. Ex.

ROSMARI'NUS. 2—1. (*Labiatæ.*) [From ros. dew, and *marinus*, of the sea.]

officina'lis, (rosemary, ♄.) some leaves are green both sides; others whitish beneath, linear; margins revolute. Ex.

ROTBOL'LIA. 3—2. (*Gramineæ.*) [In honor of Rotboll, professor of botany at Copenhagen.]

dimidia'ta, (hard grass, 2⟁.) spike compressed, linear; flowers secund; glumes 2-flowered; outer floret staminate; inner one perfect. *S.*

cilia'ta, culm erect, tall; spikes terete,

long-peduncled; flowers pedicelled, secund; margins and pedicels of the rachis villose; glumes and paleas each 2. 3-4 f. *S.*

RU'BIA. 4—1. (*Rubiaceæ.*) [From *ruber*, red; on account of the color of its roots.]

tincto'ria, (madder,) leaves lanceolate, about in sixes; stem prickly, climbing. Var. *sylves"tris*, lower leaves in sixes, upper ones in fours, or in pairs. Ex.

brown"ii, (y. 2⟁.) hispid; leaves by fours oval; peduncles solitary, single-flowered; stem decumbent. Berries purple, smooth. *S.*

RU'BUS. 11—12. (*Rosaceæ.*) [From *ruber*, red, on account of the color of its fruit.]

ide'us, (garden raspberry, w. M. ♄.) leaves quinate-pinnate and ternate; leafets rhomb-ovate, acuminate, downy beneath; petioles channeled; stem prickly, hispid, flowers sub-panicled. Var. *america'nus*, branchlets nearly glabrous; stem and petioles terete; leaves all ternate; pedicels somewhat prickly. 4 6 f.

villo'sus, (high blackberry, w. J. ♄.) pubescent, hispid, and prickly; leaves digitate, in threes or fives; leafets ovate, acuminate, serrate, hairy both sides; stem and petioles prickly; calyx short, acuminate; racemes naked; petals lance-ovate. 4-6 f.

strigo'sus, (red raspberry, w. J. ♄.) unarmed, rigidly hispid; leafets 3, or pinnate-quinate, oval, at the base obtuse, acuminate, marked with lines, and white-downy beneath; terminal one often sub-cordate; fruit red, sweet.

occidenta'lis, (black raspberry, w. g. ♄.) branches and petioles glaucous and prickly; leaves ternate, oval, acuminate, sub-lobate and doubly serrate, white-downy beneath; petioles terete; prickles recurved. 4-8 f.

trivia'lis, (creeping blackberry, dewberry, w. J. ♄.) sarmentose-procumbent; petioles and peduncles aculeate, hispid, with the prickles recurved; stipules subulate; leaves ternate or quinate, oblong-oval, acute, unequally serrate, sub-pubescent; pedicels solitary, elongated. Var. *flagella'ris*, has orbicular petals, and small, smooth leaves.

odora'tus, (flowering raspberry, r. J. ♄.) unarmed, erect, viscid; hispid leaves simple, acutely 3-5-lobed; corymbs terminal, spreading; flowers large; berries rather dry and thin. 3-6 f.

frondo'sus, (leafy raspberry, J. 2⟁.) stem erect, prickly; leaves ternate or quinate, pubescent, simple; racemes leafy; upper flowers opening first; petals orbicular. 3-6 f. Road-sides.

seto'sus, (bristly raspberry, w-r. J. 2⟁.) stem erect, reclining, rigidly hispid; leaves ternate or quinate, smooth and green on both sides.

his"pidus, (w. J. ♄.) sarmentose-procumbent; stem. petioles, and peduncles, strongly hispid; leaves ternate, gash-serrate, naked, middle one pedicellate. Berries black, large.

canaden"sis, (J. ♄.) stem purple, smoothish; leaves digitate, in tens, fives, and threes; leafets lanceolate, acutely serrate, naked on both sides; stem unarmed; bracts lanceolate; pedicels elongated, 1-3-flowered; calyx 5-7-cleft.

obova'lis, (M. ♃.) stem becoming a little woody, hispid with stiff hairs; leaves ternate; leafets round-obovate, serrate, naked; stipules setaceous; racemes sub-corymbed, few-flowered; bracts ovate; pedicels elongated. 2-4 f.

cuneifo'lius, (w. J. ♄.) branches, petioles, and peduncles, pubescent; prickles few, recurved; leaves ternate and quinate, palmate; leafets cuneate-obovate, entire at the base, sub-plicate, tomentose beneath; racemes loose; pedicels solitary, 1-flowered. 2-3 f.

stella'tus, (p. ♃.) herbaceous, small; stem unarmed, erect, 1-flowered; leaves simple, cordate, 3-lobed, rugose-veined; petals lanceolate.

chamæmo'rus, (cloud-berry, w. J. ♃.) herbaceous, small; stem unarmed, 1-flowered, erect; leaves simple, sub-reniform, with rounded lobes; petals oblong. Canada.

peda'tus, (♃.) small, herbaceous, creeping; leaves pedate-quinate, gashed; peduncles filiform, bracted in the middle; calyx nearly glabrous, reflexed.

RUDBECK"IA. 17—3. *(Corymbosa.)* [In honor of two botanists of the name of Rudbeck, who lived in the 17th century.]

purpu'rea, (p. Ja. ♃.) very rough; lower leaves broad-ovate, alternate at the base, remotely toothed, cauline ones lance-ovate, acuminate at each end, nearly entire; ray-florets very long, deflected, bifid. High grounds. Stem 3-4 f. Ray purple; disk brown; involucrum imbricate.

ful"gida, (y. Oct. ♃.) stem hispid, branches long, virgate, and 1-flowered; leaves lance-oblong, denticulate, hispid; scale of the involucrum as long as the ray; ray florets 12-14, 2-cleft at the summit; stem 2-3 feet high, branched.

pinna'ta, stem furrowed, hispid; leaves all pinnate; flowers very large, yellow; rays long, reflexed; disk ovate, purple.

lacinia'ta, (cone-flower, cone-disk sunflower, y. Au. ♃.) lower leaves pinnate; leafets 3-lobed; upper ones ovate; egret crenate; stem glabrous. Damp. 6-10 f.

dis"color, (y. and p. Au. ♃.) branches corymbose, 1-flowered; peduncles naked, elongated; leaves lanceolate, hairy, strigose; scales of the involucrum ovate, acute; petals lanceolate, entire, two-colored, as long as the involucrum. 2 f.

trilo'ba, (y. and p. Au. ♃.) stem paniculate, branches divaricate, leafy; leaves lanceolate, acuminate at each end, serrate; lower ones 3-lobed; scales of the involucrum linear, deflexed. 4-5 f.

hir"ta, (y. and p. Ja. ♃.) very hirsute; stem virgate, sparingly branched, 1-flowered; leaves alternate, sessile, lower ones spatulate-lanceolate, hirsute; scales of the involucrum imbricate in a triple series,

shorter than the ray; chaff obovate, acute, 2-3 f.

digita'ta, (y. Au. ♃.) stem branching, glabrous; lower leaves pinnate; leafets pinnatifid; upper ones simply pinnate; highest 3-cleft; egret crenate. 4-8 f.

læviga'ta, (y. ♃.) very glabrous; leaves lance-ovate, acuminate at each end, triplinerved, sparingly toothed; scales of the involucrum lanceolate, as long as the ray. S.

mol"lis, (p. S. ♃.) stem hispid, villose, branching; leaves sessile, lance-ovate, dentate, soft-tomentose; florets of the ray numerous, three times as long as the involucrum. 2-3 f. S.

rad"ula, (♂.) stem hispid below, glabrous above, nearly naked; peduncles very long, 1-flowered; leaves ovate, attenuate, tuberculate-hispid; involucrum imbricate; scales ovate, acuminate, ciliate. S.

apet"ala, (♃.) scabrous; stem elongated, 1-flowered, very pilose at the base; rays mostly wanting; leaves radical, sub-sessile, very broadly ovate, sub-rotund. Ala. Geo.

spatula'ta, (Au. ♂.) slender, minutely pubescent; stem 1-flowered; leaves obovate-spatulate, entire; involucrum expanding, imbricate; florets of the ray 3-toothed. Mountains of Carolina.

bi'color, (y. b-r. ⊙.) pilose, sub-scabrous; stem somewhat 1-flowered; leaves oblong, sessile, rarely sub-serrate, obtusish; lower ones sub-ovate, petioled; segments of the involucrum oblong; scales lanceolate, hirsute; rays short, bi-colored. 18 i. Ark.

RUEL"LIA. 13—2. *(Pediculares.)*

stre'pens, (b. Ju. ♃.) erect, hairy; leaves on petioles, opposite, lance-ovate, entire; peduncles 3-4-flowered; segments of the calyx linear-lanceolate, acute, hispid, shorter than the tube of the corolla; flowers axillary; stem 8-12 i. Shady woods. Penn. to Geo.

cilio'sa, (w. p. J. ♃.) erect, branching; leaves nearly sessile, ovate-oblong; margins, nerves, and veins, fringed with long white hair; bracts lanceolate, short; segments of the calyx linear, hispid, ciliate, with whitish hairs; corolla sub-equal. S.

hirsu'ta, (b. Oct.) hirsute, branching; leaves oval-lanceolate, nearly acute, sessile; segments of the calyx subulate, hispid, a little longer than the tube of the corolla; style very long. 12-18 i. S.

RU'MEX. 6—3. *(Polygoneæ.)* [From *rumex*, a spear, which the leaves of some of the species resemble.]

cris"pus, (dock, Ju. ♃.) valves of the calyx ovate, entire, all bearing grain-like appendages on their backs; leaves lanceolate, undulate, acute. 2-3 f.

ascetosel"lus, (field-sorrel, g. p. M. ♃.) valves without grains; leaves lance-hastate; flowers diœcious. 6-12 i.

aceto'sus, (garden sorrel, ♃.) stem elongated; leaves oblong, clasping, sagittate, acute. Ex.

patien"tia, (garden-dock, patience, ♃.) valves entire, one of them bearing a grain-like appendage; leaves lance-ovate. Naturalized.

obtusifo'lius, (J. ♃.) valves ovate, toothed, one chiefly graniferous; radical leaves heart-oblong, obtuse; stem a little scabrous. Introduced. 2-3 f.

alpi'nus, polygamous; valves veined, very entire, naked; leaves cordate, obtuse, wrinkled,. large, rhubarb-like. New Haven, Conn.

palli'dus. (white dock, J. ♃.) valves ovate, entire, hardly larger than the grain; spikes slender; stems numerous; leaves lance-linear, acute. Salt marshes.

verticilla'tus, valves entire, graniferous; flowers semiverticillate; racemes leafless; leaves lanceolate; sheaths cylindrical. 2 f.

sanguin"eus, valves oblong, small, one graniferous; leaves heart-lanceolate, mostly variegated with red. 2-3 f.

acu'tus, (M. ♃.) valves oblong, somewhat toothed, all graniferous; leaves cordate, oblong, acuminate. large; whorls leafy. Introduced. 2-3 f.

aquat"icus, (water dock, Ju. ♃.) valves ovate, entire, graniferous; leaves lanceolate, acute; flowers whorled.

britan"nicus, (yellow-rooted water dock, J. ♃.) valves entire and graniferous; leaves broad-lanceolate, flat, smooth; whorls of flowers leafless; sheaths obsolete. 2-3 f. Swamps.

pul"cher, (Ju. ♃.) valves toothed, one conspicuously graniferous; radical leaves panduriform. Naturalized.

veno'sus, (Ap. ♃.) valves large, heart-reniform, entire, net-veined; leaves small, lance-oval, entire, veined. 12 f. S.

hastatu'lus, (Ap. ♃.) valves round-cordate, entire, graniferous; leaves petioled, oblong, hastate; auricles entire. Diœcious. 1-3 f. S.

persic"aroi'des, (Ju. ⊙.) valves toothed, graniferous; leaves lanceolate, petioled, undulate, entire, smooth. 6-12 i. S.

crispatu'lus, valves obtusely cordate, crested, 3-toothed; one naked, two unequally graniferous; spikes leafless; lower leaves oval; upper ones lanceolate, all undulate. S.

RUP"PIA. 4—4. (*Aroideæ*.)
mariti'ma, (sea teasel-grass, J. ♃.) floating; leaves pectinate, obtuse; flowers spiked.

RU'TA. 10—1. (*Rutaceæ*.) [From *ruo*, to preserve, because it was supposed to preserve health.
grave'olens, (rue,) leaves more than decompound; leafets oblong, terminal ones obovate; petals entire. Ex.

SA'BAL. 6—3. (*Palmæ*.)
pu'mila, (Ju. ♄.) leaves fan-shape; scape panicled; flowers sub-sessile, small; berry dark colored. 4-6 f. Florida.
min"ima, root creeping; fronds palmate, plicate; fruit brownish. 8 i. S.

SABBA'TIA. 5—1. (*Gentianeæ*.) [In honor of *Liberatus Sabbati*, author of a work called "*Hortus Romanus*."]
campanula'ta, (p. Au. ♂.) stem terete; leaves lanceolate-linear, smooth; calyx as long as the corolla. 1 f. Flowers termi-

nal, sub-solitary, on long branches. Wet grounds.

stella'ris, segments of the calyx half as long as the corolla; leaves somewhat fleshy, obscurely 3-nerved; flowers solitary, at the extremity of the branches, forming a small corymb; rose-colored. 12-18 i. Salt marshes.

angula'ris, (American centaury, r. Au. ⊙. and ♂.) stem square, somewhat winged; leaves clasping; branches opposite. 1-2 f.

calyco'sa, flowers 7-9-parted; calyx leafy; leaves sessile.

corymbo'sa, (w. S.) flowers corymbed, corymbs few-flowered; leaves somewhat clasping; corolla 4-6-parted. Swamps.

chloroi'des, (r. Au. ♂.) weak; leaves lanceolate, erect; branches few, 1-flowered; flowers 7-12 parted; segments of the calyx linear, shorter than the corolla. Var. *erec'ta*, stem erect, rigid; leaves linear; corolla generally 10-parted; segments lanceolate. Var. *coria'cea*, stem sparingly branched, erect; lower leaves sub-oval; corolla thick, coriaceous, 18-parted. Var. *flexuo'sa*, stem flexuous; leaves lance-linear; corolla 12-parted; segments long, lanceolate. 2-3 f.

panicula'ta, (w. Au. ♃.) much branched; panicle diffuse; leaves lineo-lanceolate; stem sub-terete; branches alternate. 1-2 f.

brachia'ta, (r. Ju.) leaves lanceolate; panicle long; branches brachiate, about 3-flowered; corolla twice as long as the calyx; stem erect, slightly angled. S.

gentianoi'des. (r. Au.) erect; leaves long, linear, acute; flowers axillary and terminal, sessile, upper ones crowded; corolla about 10-parted. S.

SAC"CHARUM. 3—2. (*Gramineæ*.) [The name is said to be of Arabic origin, derived from *soukar*, sugar.]
officina'rum, (sugar-cane,) flowers panicled; in pairs, one sessile and one pedicelled; corolla 1-valved, awnless. From the East Indies.

SAGI'NA. 4—4. (*Caryophylleæ*.)
procum"bens, (pearl-wort, w. Ju. ♃.) stems procumbent, smooth, branched; leaves linear-mucronate; petals very short. 2.4 i. Borders of streams. Peduncles larger than the leaves.

ape'tala, (⊙.) stems somewhat erect, sub-pubescent; flowers alternate; petals nearly obsolete, pale green.

erec'ta, (⊙.) glabrous; stem about 1-flowered; leaves linear, acute; peduncles strict; sepals, petals, and stamens 4. 2 i. Introduced.

fontina'lis, (Ap. ⊙.) apetalous, stem procumbent, branching, dichotomous above; leaves opposite, linear-spatulate, en'ire; pedicels solitary, alternate, longer than the leaves. 8-15 i.

SAGITTA'RIA. 19—12. (*Junca*.) [From *sagitta*, an arrow; so called from the shape of the leaves of some of the species.]
sagittifo'lia, (arrow-head, w. Ju. ♃.) leaves lanceolate, acute, sagittate; lobes lanceolate, acute, straight. Var. *latifo'lia*, leaves ovate, sub-acute, sagittate; lobes

o**te**, slightly acuminate, straight. Var. *ma'jor*, leaves large, abruptly acute; scape sub-ramose. Var. *gra'cilis*, leaves linear; lobes much spreading, linear, long, acute. Var. *hasta'ta*, leaves oblong-lanceolate, sagittate; lobes expanding, long, very narrow. Var. *pubes"cens*, leaves, stems, bracts, and calyx, very pubescent. 1-2 f.

heterophyl'la, (w. Au. ♃.) leaves simple, linear, and lanceolate, acute at each end, or elliptical and sagittate, with the lobes linear and divaricate; scape simple, few-flowered; fertile flowers sub-sessile; bracts short, sub-orbiculate. 1 f.

obtu'sa, (w. J. ♃.) leaves sagittate, dilated-ovate, rounded at the extremity, mucronate; lobes approximate, oblong, obliquely acuminate, straight; scape simple; bracts ovate, acute. Diœcious. Ponds.

rig"ida, (w. Ju. ♃.) leaves narrow-lanceolate, carinate below, rigid, very acute at each end; scape ramose. Monœcious. Deep water.

acutifo'lia, (w. Ju. ♃.) leaves subulate, sheathed at the base, convex on the back; scape simple, few-flowered; bracts dilated, acuminate. 6 i.

na'tans, (w. Ju. ♃.) leaves floating; lance-oval, obtuse, 3-nerved, attenuated at base, lower ones sub-cordate; scape simple, few-flowered; lower peduncles elongated. 3-6 i.

gramin"ea, (w. Ju. ♃.) leaves lance-linear, glabrous, long, 3-nerved, somewhat perennial; bracts ovate, acuminate. Monœcious.

lancifo'lia, (w. J. ♃.) leaves broad, lanceolate, acute at each end, glabrous, coriaceous, entire, somewhat perennial; scape simple; seed compressed, sub-falcate. 2-3 f. Marshes.

pusil"la, (Au. ☉.) leaves linear, obtuse and short; summits foliaceous; scape simple, shorter than the leaves; flowers monœcious, few; fertile one solitary, deflexed; stamens mostly ♀. Muddy banks. 2.4 i.

SALICOR"NIA. 1—1. (*Atriplices*.) [From *sal*, salt, and *cornu*, a horn.]

herba'cea, (samphire, glasswort. Au. ☉.) herbaceous, spreading; joints compressed at the apex, emarginate-bifid. Var. *virgin"ica*, has the branches undivided, and the jointed spikes long. The fructification is very obscure, but it may be known by its leafless, nearly cylindric, jointed branches. It grows in salt marshes along the seaboard. Onondaga salt springs. 12-18 i.

ambig"ua, (shrubby samphire, Ju. ♃.) perennial, procumbent, branching; joints crescent-shaped, small; spikes alternate and opposite; calyx truncate.

mucrona'ta, (dwarf samphire, Au. ☉.) low, herbaceous; joints 4-angled at base, compressed, and truncate at the top; spikes oblong, with mucronate scales.

SA'LiX. 20—2. (*Amentaceæ*.) [From *sal*, near, and *lis*, water.]

vimina'lis, (osier, basket-willow. Ap. ♄.) branches slender and flexible; filaments yellow; anthers orange; aments appear before the leaves; leaves white, silky be-

neath. Banks of streams. Middle-sized tree. Introduced.

babylo'nica, (weeping-willow, M. ♄.) branchlets pendent; leaves lanceolate, acuminate, serrate, glabrous, upper and lower sides of different colors; stipules roundish, contracted; aments flower as soon as the leaves appear; germs sessile, ovate, glabrous. Supposed to be the willow on which the Israelites hung their harps when captive in Babylon. Introduced.

can"dida, (white willow, Ap. ♄.) leaves lance-linear, very long, obscurely denticulate at the extremity, pubescent above, white-downy beneath; margins revolute; stipules lanceolate, as long as the petioles; aments cylindric; scales lance-obovate, very long, villose. 3-4 f. Shady woods.

muh'lenberg"ia'na, (dwarf or speckled willow, Ap. ♄.) leaves lanceolate, acutish, sub-entire, white-hairy, rugose-veined beneath; margin revolute; stipules lanceolate, deciduous; aments precede the leafing; scales oblong; margins villose; germs lance-ovate, silk-villose, long pedicelled; styles short; stigmas bifid.

tris"tis, (mourning willow, Ap. ♄.) leaves lance-linear, acute at each end; margins revolute, smoothish above, rugose-veined and downy beneath; stipules none. 3-4 f

re'pens, (creeping willow, J. ♄.) creeping; leaves lance-oval, entire, acute, glabrous, somewhat silky beneath; stipules none; aments appearing before the leaves, ovate, diandrous; scales obovate, obtuse, hairy, fuscous at the point; germs ovate-oblong, pedicelled, pubescent; style very short; stigmas 2-lobed; capsule smooth. Very small.

obova'ta, (♄.) diffuse; leaves obovate, obtuse, very entire, glabrous above, silkyvillose beneath; stipules none; aments flower at leafing-time, sessile, oblong. diandrous; scales obovate; apex black, pilose.

lambertia'na, (Ap. ♄.) leaves lance-obovate, acute, glabrous, sub-serrate at the apex, discolored; scales round, black; filament 1; anthers 2; germs sessile, ovalovate, silky; style short; stigmas ovate, emarginate. Introduced.

fusca'ta, (sooty willow, Ap. ♄.) leaves lance-obovate, acute, glabrous, sub-serrate, glaucous beneath, when young, pubescent; stipules very narrow; aments nodding; scales obtuse; germs short-pedicelled, ovate, silky; stigma sessile, 2-lobed.

pedicel"la'ris, (stem-berried willow, Ap. ♄.) branchlets smooth; leaves lance-obovate, acute, entire, both sines glabrous, and colored alike; stipules none; aments pedunculate, glabrous; scales oblong, half the length of the pedicels, scarcely pilose; germs ovate-oblong, glabrous, long-pedicelled; stigmas sessile, 2-cleft. Catskill Mountains.

rosmarin"ifo'lia, (rosemary willow, Ap. ♄.) leaves straight, lance-linear, acute at each end, entire, pubescent above, silky beneath; stipules lanceolate, erect; aments precede the leafing; scales oblong, obtuse, ciliate; germs pedicelled, lanceolate, villose; stigmas sub-sessile, bifid. 3 f.

conife'ra, (rose willow, Ap. ♄.) leaves
lance-oblong, remotely serrate, acute, glabrous above, flat and downy beneath; stipules
lunate, sub-dentate; aments precede the
leafing; scales lanceolate, obtuse, villose;
germs pedicelled, lanceolate, silky; style
bifid; stigmas 2-lobed; cone-like excres-
cences at the end of the branches. 4-8 f.

myricoi'des, (gale-leaf willow, Ap. ♄.)
leaves lance-oblong, acute, biglandular at
the base, obtusely serrate, smooth, glaucous
beneath; stipules ovate. acute, glandular-
serrate; aments villose, leafy at the base;
scales lanceolate, obtuse, villose, black;
germs long-pedicelled, style bifid; stigmas
bifid.

prinoi'des, (Ap. ♄.) leaves oval-oblong,
acute, remotely undulate-serrate, glabrous,
glaucous beneath; stipules semicordate,
incisely-toothed; aments precede the leafing; germs pedicelled, ovate, acuminate,
silky; style long; stigmas bifid. 6 8 f.

dis"color, (bog willow, Ap. ♄.) leaves
oblong, rather obtuse, glabrous, remotely
serrate, entire near the summit, glaucous
beneath; stipules deciduous, lanceolate,
serrate; aments flower near leafing time,
diandrous, oblong, tomentose; scales oblong, acute, hairy, black; germs sub-sessile, lanceolate, tomentose;style of middling
length; stigmas 2-parted.

angusta'ta, (Ap. ♄.) leaves lanceolate,
acute, very long, gradually attenuated at
the base, serrulate, glabrous; stipules semicordate; aments precede the leafing; erect,
smoothish; germs pedicelled, ovate, smooth;
style bifid; stigmas 2-lobed. *

longifo'lia, (long-leaf willow, M. ♄.)
leaves linear, acuminate at each end, elongated, remotely toothed, smooth; stipules
lanceolate, toothed; aments peduncled, tomentose; scales flat, retuse; filaments
bearded at the base; twice the length of
the scales. 2 f.

purshia'na, (♄.) leaves long, lance-linear, gradually attenuate above, sub-falcate.
acute at base, close-serrate, glabrous on
both sides, silky when young; stipules lunate, toothed, reflexed. 8-15 f.

ni'gra, (M. ♄.) leaves lanceolate, acute at
each end, serrulate, green on both sides;
petiole and midrib tomentose above; stipules dentate; aments cylindric; scales oblong, very villose; filaments 3-6, bearded
at the base; germs pedicelled, ovate,
smooth; style very short; stigmas bifid.
15.20 f. Banks of streams.

lu'cida, (M. ♄.) leaves ovate-oblong, cus-
pidate-acuminate, rounded at the base, serrate, glabrous both sides, shining; stipules
oblong, serrate, aments triandrous; scales
lanceolate, obtuse, pilose at base, serrate,
smooth at the apex; germs lanceolate-subulate, smooth, style bifid; stigmas obtuse.
A small tree. ♂..

corda'ta, (heart-leaf willow, Ap. ♄.)
leaves lance-oblong, acuminate, sub-cordate at base, rigid, smooth, acutely serrate,
paler beneath; stipules large, cordate, obtuse; stamens 3; scales lanceolate, black,

woolly; germs pedicelled, smooth; style
very short; stigmas bifid. 6-8 f.

gri'sea, (gray willow, Ap. ♄.) leaves lanceolate, acuminate, serrulate, glabrous
above, silky or naked beneath; stipules
linear, deflexed, deciduous; scales oblong,
hairy, black at the apex; germs oblong,
pedicelled, silky; stigma sessile, obtuse;
branches purple. very brittle at the base.
6 8 f.

al"ba, (M. ♄.) leaves lanceolate, acuminate, silky on both sides; lower serratures
glandular; stipules obsolete; aments elongated; scales lance-oval, pubescent; germs
sub-sessile, ovate-oblong, at length smooth;
style short; stigma 2-parted, thick. Introduced.

vitelli'na, (yellow willow, M. ♄.) leaves
lanceolate, acuminate, thickly serrate, glabrous above, paler and somewhat silky beneath; stipules none; aments cylindrical;
scales ovate-lanceolate, pubescent externally, germs sessile, ovate-lanceolate;
stigmas sub-sessile, 2-lobed. Introduced.

russelia'na, (♄.) leaves lanceolate, acuminate, serrate, glabrous; florets generally
triandrous; germs pedicelled, subulate,
smooth; styles elongated. Tall tree. Introduced.

herba'cea, (Ju. ♄.) leaves round; stipules
none; scales obovate, villose; germs sub-
sessile, glabrous. Forms a kind of turf
rising not more than an inch from the
ground, yet forming a perfect miniature
tree.

houston"ia'na, (♄.) leaves lance-linear,
acute, finely serrate, glabrous, shining, 1-
colored; stipules none; aments appearing
with the leaves, cylindric, villose; scales
ovate, acute; filaments 3-5, bearded at the
middle. *S.*

SALSO'LA. 5—2. (*Atriplices.*) [From *sal*,
salt; so called on account of its saline
properties.]

ka'li, (prickly salt-wort, Ju. ⊕.) decumbent; leaves subulate, rough; stem bushy,
flowers solitary. Sea-shore. Burnt for the
alkaline salts which it contains.

so'da, (salt-wort) smooth, ascending.

tra'gus, (Ju. ⊕.) herbaceous, smooth,
spreading; leaves subulate, fleshy, mucro-
nate-spinous; flowers sub-solitary; calyx
sub-ovate; margin flattened, discolored.

SAL"VIA. 2—1. (*Labiate.*) [From *salvo*, to
save; so called in reference to its qualities.]

lyra'ta, (wild sage, b. M. ♃.) stem nearly covered with reflexed hairs; radical
leaves lyrate-dentate; upper lip of the corolla very short; flowers about 6 in a whorl.
Woods. 1 f.

clayto'ni, leaves cordate, ovate, sinuate,
toothed, rugose; flowers violet, in whorls.
Woods. 8-12 i.

urticifo'lia, viscous and villose; leaves
ovate-oblong, very pubescent; flowers blue,
in remote whorls. Mountains.

officina'lis, (sage, b. J. ♃ or ♄.) leaves
lance-ovate, crenulate; whorls few-flowered; calyx mucronate. Ex.

scla'ra, (clarry, ♂.) leaves rugose, cordate, oblong, villose, serrate; floral bracts

longer than the calyx, concave, acuminate.
Ex.

splen"dens, (scarlet sage, r. ♃.) leaves ovate and lance-ovate, flat, smooth beneath; flower long; calyx and corolla scarlet, downy; style exsert. Ex.

azu'rea, (narrow-leaved sage, b. w. Au. ♃.) leaves lance-linear, smooth; calyx pubescent, 3-cleft; segments short. 4 6 f. *S.*

trichos"temmoi'des, (b. ☉.) leaves lanceolate, serrate; racemes terminal; flowers opposite; corolla equal to the 3-cleft calyx; stem brachiate-branched. *S.*

obova'ta, (downy-leaved sage, Ju.) leaves large, obovate, toothed, pubescent; stem slightly angled; whorls 6-flowered. 18 i. *S.*

coccin"ea, (r. Ju. ♃.) leaves cordate, acute, tomentose, serrate; corolla twice as long as the calyx, and narrower. 1 f.

au'rea, flowers golden-yellow.

formo'sa, a shrubby plant with dark scarlet flowers. Ex.

pa'tens, flowers of the richest blue.

denta'ta, flowers white.

purpu'rea, flowers purple.

SALVIN"IA. 21—1. (*Filices.*)

na'tans, (♃.) leaves elliptic, sub-cordate, obtuse, with fascicled bristles above; fruit sub-sessile, aggregated. Lakes and still waters.

SAMBU'CUS. 5—3. (*Caprifoliæ.*) [From *Sabucca*, (Hebrew,) the name of an ancient musical instrument, made from the wood of this shrub.]

canaden"sis, (black-berried elder, w. J. ♄.) branchlets and petioles glabrous; leafets about in 4 pairs, oblong-oval, glabrous, shining, acuminate; cyme lax, divided into about 5 parts. 8-15 f.

pubes"cens, (red-berried elder. w. M. ♄.) bark warty; leafets in 2 pairs, lance-oval, pubescent beneath; flowers raceme-panicled, or in a crowded bunch. 6-12 f.

SAMO'LUS. 5—1. (*Lysimachia.*) [Supposed to be named from the island of Samos.]

valeran"di, (water pimpernell, brook-weed, w. Ju. ♃.) erect; leaves obovate, entire; racemes many-flowered; pedicels with a minute bract. Wet grounds. 8-12 i.

ebractea'tus, (w. ♃.) stem short, robust, smooth, divided at the base; leaves obovate, obtuse, somewhat fleshy, attenuate at the base; racemes elongated, sub-pubescent; pedicels filiform, without bracts. *S.*

SANGUINA'RIA. 12—1. (*Papaveraceæ.*) [From *sanguis*, blood; so named either from the color of its root, or its use in stopping hemorrhages.]

canaden"sis. (blood-root, w. Ap. ♃.) leaves sub-reniform, sinuate-lobed; scape 1-flowered. A variety, *stenopet'ala*, has linear petals. 6-10 i.

SANGUISOR"BA. 4—1. (*Rosaceæ.*) [From *sanguis*, blood, and *sorbeo*, to absorb; so named from its medicinal qualities.]

canaden"sis, (burnet saxifrage, w. Ju. ♃.) flowers in a long, cylindric spike; stamens several times longer than the corolla. The leaves resemble the burnet. 3-5 f.

me'dia, stipes shorter than the preceding, and tinged with red. Wet meadows, chiefly on mountains.

SANIC"ULA. 5—2. (*Umbellifera.*) [From *sano*, to heal; so called from its virtues in healing.]

● *maryland"ica*, (w. June—Au. ♃.) leaves all digitate; leafets oblong, deeply serrate; staminate flowers numerous, pedicelled. 2 f.

canaden"sis, (♃.) leaves palmate; segments petioled; divisions gash-serrate, lateral ones 2-parted; flowers polygamous, staminate ones short-pedicelled; lobes of the calyx entire. Canada.

SANTOLI'NA. 17—1. (*Cosymbiferæ.*) [From *santalum*, saunders, because it swells like the saunders-wood.]

suaveo'lens, (y. Ju. ☉.) smooth; stem fastigiate; leaves sub-bipinnatifid; divisions acute, linear; peduncles terminal, 1-flowered.

SAPIN"DUS. 8—3. (*Sapindi.*) [From two words, *sapo indus*, Indian soap, the rind of the fruit being used as a substitute for soap.]

sapona'ria, (w. ♄.) leaves glabrous, abruptly pinnate; leafets lance-oval, fruit glabrous. *S.*

SAPONA'RIA. 10—2. (*Caryophylleæ.*) [From *sapo*, soap, the juice being found to have saponaceous properties.]

officina'lis, (soap-wort, bouncing bet, w. J. ♃.) calyx cylindric; leaves lance-ovate, opposite, sub-connate, entire. Naturalized. 10-18 i. Ex.

vacca'ria, (field soap-wort, r. Ju. ☉.) calyx pyramidal, 5-angled, smooth; bracts membranaceous, acute; leaves ovate-lanceolate, sessile. Introduced.

SARRACE'NIA. 12—1. (*Papaveraceæ.*) [This name is said, by some, to have been given in honor of Dr. Sarrazin, by others, it is thought to have originated in the resemblance of the peculiar flower of the plant to the head of a Saracen enveloped in his crimson turban; thus the plant is sometimes called Turk's-head.]

purpu'rea, (side-saddle flower, p. J. ♃.) leaves radical, short, gibbose-inflated, or cup-form, contracted at the mouth, having a broad, arched, lateral wing; the contracted part of the base hardly as long as the inflated part. Scape with a single, large, nodding flower. In marshes. 1-2 f.

heterophyl'la, has palish yellow flowers, and is more slender than the preceding.

ru'bra, (r-p. ♃.) leaves slender; lateral wing linear; appendage ovate, erect, obtuse, mucronate, contracted at the base. 6-10 i. *S.*

fla'va, (y. J. ♃.) leaves large, funnel-form, throat expanding; lateral wing nearly wanting; appendage erect, contracted at base; reflexed at the sides. 18-24 i. *S.*

variola'ris, (y. J. ♃.) leaves slightly ventricose, with the tube near the summit spotted on the back; appendage arched, incurved; lateral wing slightly dilated; stigma acute at the angles. 12-18 i. *S.*

drummon"dii, (p.) leaves erect, very long; tube dilated above, with very narrow wing; tube and lamina whitish and strongly reticulated with purplish veins. Florida.

psittaci'na, (p. Mar.) leaves short, reclined, marked with white spots; tube inflated, with a broad semi-obovate wing; lamina ventricose, recurved, so as nearly to close the tube. *S.*

SATURE'J A. 13—1. (*Labiatæ.*) [From *satyri, satyrs.*]
horten"sis. (summer savory, b-w. Ju. ☉.) peduncles axillary, somewhat in a cyme; leaves lanceolate, entire; stem brachiate.
monta'na, (winter savory, ♄.) peduncles somewhat 1-sided; segments of the calyx acuminate, mucronate; leaves mucronate.

SAURU'RUS. 7—4. (*Naiades.*) [From *saura,* a lizard, and *oura,* tail.]
cer"nuus, (lizard's-tail, swamp-lily, Au. ♃.) stem angular, sulcate; leaves alternate, heart-oblong, acuminate. 1-2 f. Swamps.

SAUSSU'REA. 17—1. (*Ericeæ.*) [From *saura,* lizard, and *oura,* tail, alluding to the shape and scaly appearance of the long spike of flowers.]
montico'la, sparingly woolly; leaves linear, entire; leafets of the involucrum oblong-cylindric, villose, lanceolate, acute.

(*Saxifraga.*) [From *go,* to break, because a remedy against the

axifrage, w. M. ♃.) eaves oval, obtuse, the petiole; flowers

er saxifrage, y-g. M. es oblong-lanceolate, ty, obsoletely toothicle oblong, flowers ar, longer than the for. 18-28 i. Root

ak geranium, creep-♃.) leaves roundish, off creeping shoots; elongated. Ex.

aggregate, spatulate, acaton, simple, with cartilaginous teeth; stem simple, pilose, leafy; calyx smooth. 3 i.
serpyl"lifo'lia, (♃.) erect; leaves small, oval, glabrous; stem 1-flowered, few-leaved; petals obovate.
androsa'cea, (w. ♃.) pubescent; leaves petioled, linear, spatulate; stem leafy, 1-2-flowered.
bronchia'lis, (♃.) stoloniferous; leaves imbricate, subulate, flat, mucronate, spinose, ciliate; stem panicled.
niva'lis, (alpine saxifrage, w. J. ♃.) leaves roundish, wedge-form, crenate-before, decurrent into the petiole; stem naked, simple; racemes crowded. 2 i.
ge'um, (♃.) leaves reniform, toothed, veinless and pilose; stem naked, panicled.
leucan"themifo'lia, (w. r. y. Ju. ♃.) very hirsute; leaves spatulate-oval, with acute and large teeth; panicles long, diffuse; calyx reflexed, persistent; petals unequal. 18-24 i.
ero'sa, (y-g. Ju. ♃.) nearly glabrous; leaves oblong-lanceolate, acute, erose, dentate; panicles oblong; branches divaricate; stem naked.
semi-pubes"cens, (y. ♃.) leaves not petioled, oblong-oval, obtuse, very glabrous.

denticulate; flowers pedicelled, disposed in dense corymbs; calyx pilose-glandulose; sepals triangular-ovate, acute; petals ovate, obscurely 3-nerved, somewhat equalling the calyx. Cultivated.

SCABIO'SA. 4—1. (*Dipsaceæ.*) [From *scaber,* rough; so called from its rough surface.]
stella'ta, (star scabious, y-w. ☉.) corolla 5-cleft, radiate; leaves irregularly lobed, and toothed; outer crown of the seeds orbicular, large, many-nerved.
atropurpu'rea, (sweet scabious, r. ♃.) outer crown of the seed short, lobed, and crenate; receptacle cylindric.

SCHEUCHZE'RIA. 6—3. (*Junci.*) [Named from Scheuchzer]
palus'tris, (flowering rush, g-y. J. ♃.) leaves sheathing at the base, linear; flowers in a small, terminal raceme. Swamps.

SCHIZÆ'A. 21—1. (*Felices.*)
pusil'la, (one sided fern, Ju. ♃.) frond simple, linear, compressed, tortuous; spikes conglomerate, inflexed one way. 3-6 i.

SCHIZAN"DRA. 19—5. (*Menisperma.*) [From *schiro,* to split, and *anei,* a stamen, the stamens being nearly separated by fissures in the receptacle.]
coccin"ea, (r. and y. M. ♄.) glabrous; leaves alternate, lanceolate, sub-denticulate, petioled, sometimes sub-cordate, climbing. 10-15 f. *S.*

SCHŒ'NUS. 3—1. (*Cyperoideæ.*)
mariscoi'des, (water-bog rush. Ju. ♃.) culm terete or sub-sulcate, leafy; leaves channeled, semi-terete; umbel terminal; fascicles on spikes, 3 on each peduncle; seed naked, rounded at the base. 2 f.
hispidu'lus, peduncles axillary and terminal, 3-spiked; spikes globose, pedicelled; leaves filiform, hispid. *S.*
effu'sus, (saw grass, Au.) stem leafy, obtusely 3-angled; leaves aculeate; panicle terminal, very long, diffuse; pericarp ovate, longitudinally wrinkled. 6-10 f. *S.*
seta'ceus, peduncles axillary and terminal, generally 3-flowered; stem 3-angled; leaves setaceous. *S.*

SCHOL"LERA. 3—1. (*Narcissi.*) [Named from a German teacher.]
gramin"ifolia, (yellow-eyed water grass, y. Ju. ♃.) leaves all linear, grass-like; stem slender, floating. 6-18 i.

SCHRANK"IA. 15—10. (*Leguminosa.*) [From *Schrank,* a German.]
sensiti'va, prickly; leaves pinnate; leafets in pairs, under ones very small. Sensitive plant. known by some botanists as the Mimosa sensitiva.
uncina'ta, (p. Ju. ♃.) stem prostrate, angled, prickly; leaves alternate, abruptly bi-pinnate; leafets small, thinly pilose, irritable. 2-3 f. *S.*

SCHWAL"BEA. 13—2. (*Scrophulariæ.*) [Named in honor of Schwalbe.]
america'na, (p-y. J. ♃.) simple, pubescent; leaves lanceolate; racemes terminal; flowers alternate, sub-sessile. 2 f. Chaff-seed. Pine barrens.

SCIL"LA. 6—1. [From *skillo,* to dry; so

35

called from its property of drying up humors.]

mariti'ma, (squill, w.) scape long, naked, many-flowered; bracts bent back; root bulbous. Ex.

SCIR"PUS. 3—1. (*Cyperoideæ.*) [An ancient Latin name for the Bull-rush.]

1. *Seed surrounded with bristles at the base. (Style articulated to the seed; base dilated and persistent. Seed often lenticular. Spikes terminal, solitary.)*

capita'tus, (Ju. ♃.) stem terete or subcompressed; spike ovate, obtuse; seed oval, compressed, smooth. Wet places. 8-18 i.

ten"uis, (Ju. ♃.) culm slender, quadrangular; spike elliptical, acute at each end; glumes ovate, obtuse; stamens 3; styles 3-cleft; seed rugose. 8-12 i.

pusil'lus, (J. ♃.) culm compressed, subangular; spike ovate, compressed; seed obovate; stamens 3; style 2-3-cleft. Salt marshes. 1 i.

acicula'ris, (Ju.) culm setaceous, quadrangular; spike ovate, acute, 3-6-flowered; glumes somewhat obtuse; stamens 3; styles bifid; seed obovate. 3-6 i.

planifo'lius, (J.) culm triquetrous; radical leaves linear, flat, nearly equalling the culm; spike oblong, compressed, shorter than the cuspidate bracts at the base. 8 i. Swamps.

(Style filiform, not bearded, deciduous.)

lacus"tris, (J. ♃.) culm terete, attenuated above, naked; panicle sub-terminal; spikes peduncled, ovate. 4-8 f.

america'nus, (Au. ♃.) culm nearly naked, triquetrous; sides concave; spikes lateral, 1-5, ovate, conglomerate, sessile; glumes round-ovate, mucronate; seed triquetrous, acuminate. 3-5 f.

debi'lis, (Au. ♃.) culms cespitose, deeply striate; spikes about 3, ovate, sessile; glumes ovate, obtuse, mucronate; margins of ponds. 8-12 i.

brun"neus, (Au. ♃.) culm leafy, obtusely triangular; cyme decompound; involucrum 3-4-leaved; spike round-ovate, clustered in about sixes; glumes ovate, obtuse. 2-3 f.

atrovi'rens, (Ju. ♃.) culm triangular, leafy; cyme terminal, compound, proliferous; involucrum 3 leaved; spikes conglomerate, ovate, acute; glumes ovate, mucronate, pubescent. Wet meadows. 2 f.

macrosta'chyus, (Au. ♃.) culm triquetrous, leafy; corymb clustered; involucrum about 3-leaved, very long; spikes oblong; glumes ovate, 3-cleft; middle segment subulate and reflexed; style 3-cleft. 3-4 f.

Style filiform, deciduous. Bristles much longer than the seed.

eriopho'rum, (red cotton grass, Au. ♃.) stem obtusely triquetrous, leafy; panicle decompound, proliferous, nodding; spikes peduncled; bristles surrounding the pericarp exsert. 4-5 f. Swamps.

linea'tus, (leafy scirpus, Ju.) culm triquetrous, leafy; panicles terminal and lateral, decompound, at length nodding; involu-

crum 1-2-leaved; spikes ovate; glumes lanceolate, somewhat carinate; bristles longer than the seed. 2-3 f.

2. *Seed naked at the base. (Style simple at the base, not articulated to the seed, deciduous.)*

autumna'lis, (flat stemmed scirpus, S. ♃.) culm compressed, ancipitous; umbel compound; involucrum 2-leaved; spikes lanceolate, acute, a little rough; glumes mucronate, carinate. 8-12 i. Low woods.

, *sim"plex*, (J. ♃.) culm columnar; spike somewhat ovate; glumes sub-ovate, obtuse; pericarp obovate. 3-angled. 8-13 i. S.

filifor'mis, (Au.) spike cylindric, oblong, obtuse; scales roundish; pericarp naked at the summit; culm filiform, terete. S.

tubercula'tus, (Au.) culm columnar, striate; glumes very obtuse, loose, appressed; seed somewhat 3-angled; tubercle sagittate, larger than the seed. 12 i. S.

equiseto'ides, (J.) culm erect, terete, doubly jointed; spike cylindric, terminal; scales very obtuse. 18-24 i. S.

genicula'tus, (Ju.) culm terete, growing in distinct clusters; spike ovate-oblong; scales round-ovate. S.

quadrangula'tus, (M. ♃.) culm erect, glabrous, acutely 4-angled; 3 sides concave; one wider, flat; spike cylindric; glumes very obtuse. 1-2 f. Swamps. S.

sylvat"icus, (wood rush, ♃.) spikes oblong, crowded; corymb leafy more than decompound; culm leafy, triquetrous; scales oblong, obtusish. green. S.

SCLERAN"THUS. 10—2. (*Portulacceæ.*) [From *skleros*, hard, and *anthos*, flower, alluding to its hard calyx.]

an"nuus, (knawel. ☉.) stems slightly pubescent; calyx of the fruit spreading, acute. Stems numerous, procumbent. Flowers very small, green, in axillary fascicles. Dry fields.

peren"nis, (♃.) calyx of the fruit with obtuse, spreading segments. England.

SCLE'RIA. 19—3. (*Cyperoideæ.*) [Named from its hard and polished fruit.]

tri'glomera'ta, (whip grass, J. ♃.) culm acutely triangular, scabrous; leaves lance-linear, channeled, a little scabrous, sparingly pilose; spikes fascicled, lateral and terminal; glumes ciliate; nut smooth. 2 f.

pauciflo'ra, (Au. ♃.) culm triquetrous, glabrous; leaves linear, glabrous; spikes lateral and terminal, few-flowered, the lateral ones pendulous, fasciculate; glumes smooth, nut rugose. Wet meadows. 12-18 i.

verticilla'ta, (Au. ♃.) stem simple, triquetrous, smooth; leaves glabrous; spike glomerate, naked, clusters alternate; glumes glabrous, nut globose, mucronate, transversely corrugate. 1 f.

oligan"tha, (M.) stem slender, triquetrous, glabrous; leaves narrow, nerved, slightly scabrous; spikes 2-3, sub-terminal, sessile, one lateral, one remote, long peduncled; nut very smooth, shining. 12-18 i. S.

gra'cilis, stem filiform, triquetrous, and

with the leaves glabrous; spikes few-flowered, fascicled, sub-terminal; glumes glabrous; nut smooth, shining. 1 f. *S.*

SCOLOPEN"DRIUM. 21—1. (*Filices.*) [From *skolopendra*, centipede; so called from the numerous roots and branches, or from little marks upon the frond resembling this insect.]
officina'rium, (caterpillar fern, Ju. ♃.) frond simple, ligulate, entire, cordate at base, sub erect; stipe chaffy. 8-15 i.

SCROPHULA'RIA. 13—2. (*Scrophularia.*) [From *scrofula*, the king's evil; so called because the leaves were formerly considered a remedy for scrofulous tumors.]
marylan"dica, (fig-wort, g-p. Ju. ♃.) leaves cordate, serrate, acute, rounded at the base; petioles ciliate below; panicle fasciculate, loose, few-flowered; stem obtusely angled. 2-4 f.
lanceola'ta, leaves lanceolate, unequally serrate; petioles naked; fascicles corymbed. 2-3 f. Wet meadows. Flowers greenish yellow.

SCUTELLA'RIA. 13—1. (*Labiate.*)
lateriflo'ra, (scullcap, b. Ju. ♃.) branching, glabrous; leaves long-petioled, ovate, toothed; cauline ones sub-cordate; racemes long, lateral, leafy. Damp. 1-2 f. At one time in repute as a remedy for hydrophobia.
galericula'ta, (common scullcap, b. J. ♃.) branching; leaves sub-sessile, lance-ovate, sub-cordate at the base, crenate, white-downy beneath; flowers axillary, solitary, or in pairs; flowers large. Damp. 12-18 i.
integrifo'lia, (b. Ju. ♄.) stem nearly simple, densely pubescent; leaves sub-sessile, oblong, obtuse, wedge-form at the base, obscurely toothed; racemes loose, leafy; flowers opposite, often in panicles. Var. *hys"sopifo'lia*, has the leaves all linear. 18-24 i. Swamps.
gra'cilis, (b. J. ♃.) stem sub-simple; leaves opposite, remote, broad-ovate, toothed, veined, smooth, sessile, margins scabrous; upper ones smaller, entire; flowers axillary. 12-18 i.
amllg"ua, (b. Ju. ♃.) stem sub-decumbent, branched divaricately from the base; leaves sessile, ovate; flowers small, axillary. 3-6 i.
pilo'sa, (b. J. ♃.) erect, pubescent; leaves distant, ovate, obtuse, crenate, rugose, petioled, lower ones sub-cordate; racemes panicled; flowers crowded; bracts lanceolate, entire; calyx hispid. 18-24 i.
canes"cens, (b. ♃.) branched; leaves ovate, acute, petiolate, acutely toothed, hoary-villous beneath; lower ones sub-cordate; racemes pedicelled, sub-panicled; axillary and terminal; bracts lance-ovate, longer than the calyx. 2-3 f.
leviga'ta, (b. M. ♃.) simple, smooth, slender; leaves petioled, opposite, ovate, coarse, serrate, veined, sub-acuminate, tapering to the base, entire at the base and apex, glabrous, paler beneath; raceme simple, terminal; flowers sub-pubescent, erect, upper bracts smaller, entire. Open woods. 12-18 i.

nervo'sa, (b. Au. ♃.) nearly simple, glabrous; leaves sessile, ovate, dentate, nerved; raceme terminal, loose, leafy.
angustifo'lia, (J. ♃.) simple, finely pubescent; leaves linear; flowers axillary, opposite; stamens sub-exsert. *S.*
serra'ta, (b. Ju. ♃.) erect, branching, pubescent; leaves short petioled, acuminate, ovate, serrate, dotted beneath; racemes terminal, loose, often panicled; bracts lanceolate, short; stamens shorter than the corolla. *S.*

SECA'LE. 3—2. (*Graminea.*) [From *seco*, to cut or mow.]
cerea'le, (rye, J. ♂.) glumes and bristles scabrous-ciliate; corolla smooth. Introduced.

SE'DUM. 10—5. (*Sempervivæ.*) [From *sedo*, to assuage, because it allays inflammation.]
terna'tum, (false ice-plant, w. J. ♃.) small, creeping; leaves flat, round-spatulate, ternate; flowers somewhat 3-spiked, sometimes octandrous. Cultivated.
tele'phium, (orphine, live-forever, r. w. Ju. ♃.) leaves flattish, tooth-serrate, thickly scattered; corymb leafy; stem erect. Ex.
anacamp"seros, (stone-crop, ♃.) leaves wedge-form, entire, sub-sessile; stem decumbent; flowers corymbed. Ex.
telephio'des, (p. Ju.) leaves broad, flat, ovate, acute at each end; corymbs many-flowered. 1 f. Harper's ferry.
nuttal"ii, leaves roundish, flat, entire, scattered; cymes terminal, 3-forked.
lanceola'tum, leaves sub-alternate; lower ones crowded, lance-oblong, acutish, glabrous; stem branched, assurgent; flowers cyme-corymbed; petals spreading, lanceolate.
pulchel"lvm, (p.) glabrous; stems assurgent; leaves scattered, obtuse, linear; lower ones oblong-oval; cyme many-spiked; flowers sessile, octandrous. *S.*
rhodio'la, (g. y. ♃.) erect, simple; leaves glaucous, fleshy, sessile, imbricate, toothed above; cymes terminal, branching. 8 i. *S.*
pusil"lum, (w. Ju. ♃.) glabrous; leaves nearly terete, oblong, alternate; flowers sub-terminal, few, sub-pedicelled, alternate. 2-4 i. *S.*

SELI'NUM. 5—2. (*Cruciferæ.*)
aure'a, (y. ☉.) stem glabrous, sub-divided at the base, acute-triangular; leaves somewhat succulent, smooth; peduncles axillary, angular. 4-6 i. *S.*

SEMPERVI'VUM. 12—12. (*Sempervivæ.*) [From *semper*, always, and *vivo*, to live.]
tecto'rum, (houseleek, Au. ♃.) leaves ciliate; bulbs spreading; nectaries wedge-form, crenulate. Ex.
arbo'reum, (tree houseleek,) stem woody, smooth, branching; leaves wedge-form, glabrous, with soft spreading hairs. Ex.

SENE'CIO. 17—2. (*Corymbosæ.*) [From *senesco*, to grow old; so called because some of its species are covered with a grayish pubescens, like the hair of an aged person.]

A. *Florets tubular; those of the ray wanting.*

hieracifo'lia, (fire-weed, w. J. ☉.) stem virgate, paniculate; leaves clasping, oblong, acute, unequally, acutely, and deeply toothed; involucre smooth; seeds pubescent; stem 2-6 f. high, succulent. branching toward the summit; flowers in a compound, terminal panicle. Road-sides.

vulga'ris, (groundsel, y. ♃.) flowers in crowded corymbs. Stem 18 i. Cultivated grounds. Introduced.

B. *Flowers with ray florets.*

aure'a, (y. ♃.) radical leaves ovate, cordate, serrate, petiolate; cauline ones pinnatifid, toothed, the terminal segments lanceolate; peduncles thickened; flowers somewhat umbelled. Shady woods. 2 f.

obova'ta, (y. J. ♃.) stem smoothish; radical leaves obovate, crenate-serrate, .petiolate; cauline ones pinnatifid, toothed; flowers somewhat umbelled, on long peduncles; rays 10-12. 1 f. Rocky hills.

paupercu'lua, (♃.) simple, erect, nearly naked; leaves lanceolate, radical ones sub-entire and gash-toothed; corymb few-flowered; involucrum smooth; rays small.

gra'cilis, (y. J. ♃.) slender; radical leaves very long, petioled, orbicular, sub-cordate, crenate; cauline ones few, very remote, linear-oblong, dilated at the base, incisely toothed; peduncles very short, hairy, sub-umbelled; involucrum smooth; rays few, very short. 1 f.

balsami'ta, (y. J. ♃.) stem and peduncles villose at the base; radical leaves oblong, serrate, petioled; cauline ones lyrate or pinnatifid; flowers sub-umbelled. Meadows. 1-2 f.

cilia'ta, (w.) pilose; leaves lance-linear, ciliate. *S.*

tomento'sa, (M. ♃.) stem simple, hoary and woolly; leaves petioled, oval, lanceolate, serrulate; corymb sub-umbelled; seed pubescent. 2 f. *S.*

fastigia'ta, (y.) leaves of the root oblong, cordate-ovate, crenate-toothed, glabrous, cauline ones pinnatifid; segments-gash-toothed; flowers sub-umbelled; peduncles and involucrum glabrous. 2 3 f. *S.*

loba'ta, (butter-weed, y. M. ☉.) glabrous; leaves pinnatifid. lyrate, lobes round, sub-repand; corymb compound; highest peduncles sub-umbelled; seed oblong, striate. 1-3 f. *S.*

SERPIC"ULA. 20—9. (*Hydrocharides.*) [From *serpo*, to creep.]

canaden"sis, (little-snake weed, w. Ju. ♃.) perfect flowers triandrous; stigmas reflexed, bifid; leaves linear, acute, somewhat whorled, glabrous, denticulate; pistillate corollas tubular.

SESA'MUM. 13—2. (*Bignonia.*) [An Egyptian name.]

in"dicum, (oily grain, bene-benni, r-w. Au. ☉.) leaves lance-ovate; outer ones 3-lobed; upper ones undivided, serrate. 2-4 f.

SESBA'NIA. 16—10. (*Leguminosæ.*)

vesica'ria, (y. Au.) leaves pinnate; leaflets oblong, obtuse, glabrous; racemes shorter than the leaves. 5-7 f. *S.*

macrocar"pa. (y. and p. S. ☉.) glabrous; leaves pinnate; leaflets elliptic, glabrous, entire, sub-glaucous beneath; racemes axillary, few flowered; legumes slender, nearly terete. 4-12 f. *S.*

SES"ELI. 5—2. (*Umbelliferæ.*) [Origin of the name doubtful.]

triter"na'tum, (M. y. ♃.) leaves triternate; leaflets long, linear; umbels hemispheric; involucrum leafy, linear; leaflets equal length with the umbels.

SESLE'RIA. 3—2. (*Gramineæ.*)

dactyloi'des, (moorgrass, g.) culm leafy setaceous; leaves short, flat, subulate, sub-pilose; spikes 2-3, few-flowered; calyx entire, acuminate; stipules bearded. 4.5 i. *S.*

SESU'VIUM. 11—5. (*Ficoideæ.*)

sessi'le, (r. Ju.) flowers sessile; leaves linear-oblong, flat. Stem succulent. Sea-coast.

pedun"cu'latum, (w. Au.) prostrate. terete; leaves linear-lanceolate, obtuse, entire, succulent; flowers solitary, axillary, short-ped uncled, polyandrous. *S.*

SEYME'RIA. 13—2. (*Scrophulariæ.*)

tenuifo'lia, (y-p. Au. ☉.) glabrous. very branching; leaves compound-pinnatifid, segments filiform, opposite, and alternate; corolla sub-rotate. 3-4 f. *S.*

pectina'ta, leaves pectinate-pinnatifid.

macro'phyt"la, (y. Ju.) branched; lower leaves sub-pinnatifid or deeply toothed, upper ones lanceolate, entire; corolla very woolly; stamens scarcely exserted. 4-5 f. *S.*

SIBBAL"DIA. 5—5. (*Rosaceæ.*) [After Sir Robert Sibbald, author of Scotia Illustratica.]

procum"bens, (y. Ap. ♃.) leaves ternate; leaflets wedge-form, 3-toothed, smooth above, hairy beneath.

erec"ta, var. *parviflo'ra*, (r-w. Ju. ♂.) erect, branching; radical leaves about twice 3-cleft; segments sub-divided; leaves of the stem sessile, alternate, sub-bipinnatifid. 4-6 i. *S.*

SIC"YOS. 19—15. (*Cucurbitaceæ.*) [From the Greek *sikuos*, a cucumber.]

angula'ta, (single-seed cucumber, w. ☉.) leaves cordate; back lobes obtuse, 5-angled, scabrous, denticulate; tendrils umbellate; sterile flowers corymbose-capitate, with the common peduncle long; fertile flowers sessile; fruit small, ovate, hispid.

SI'DA. 15—12. (*Malvaceæ.*) [Origin of the name doubtful.]

abu'tilon, (Indian mallows, y. Ju. ☉.) leaves round-cordate. acuminate, toothed, tomentose; peduncles solitary, shorter than the petioles; capsule 2-awned, truncate. 4-6 f.

spino'sa, (y. Ju. ☉.) leaves ovate-lanceolate, serrate, dentate, with a sub-spinose tubercle at the base of the petiole; stipules setaceous; pedicels axillary, sub-solitary, mostly shorter than the stipules and petiole; carpels 5, bi-rostrate; seeds triquetrous, ovoid. 1-2 f.

napœ'a, (w. Ju. ♃.) leaves palmately 5-lobed, glabrous; lobes oblong, acuminate, toothed; peduncles many-flowered; capsules awnless, acuminate. 2-4 f. Rocky places.

dioi'ca, (w. Oc. ♃.) leaves palmately 7-lobed, rough; lobes lanceolate, incisely dentate; peduncles many flowered, sub corymbed, bracted; flowers diœcious. 4-5 f.

cris"pa, (w. Au. ☉.) leaves oblong-cordate, acuminate, crenate, upper ones sessile; peduncles solitary, longer than the petiole, when in fruit, deflected; capsules inflated, awnless, crisp-undulate.

his"pida, (y. Au. ♃.) hispid; leaves lanceol~~ate~~ ~~peduncles~~ solitary, axillary; ~~form.~~ 1-2 f. S.

~~.)~~ slender, glabrous; alternate; peduncles ~~< angular.~~ 12-18 i. S.

~~ι. ♃.)~~ stem suffruti-~~s~~ oblong, lanceolate, entire at base; ped-~~n~~ the petioles; cap-~~.~~ S.

~~>~~wer leaves triangu-~~>~~per ones palmate, ~~d.~~ S.

~~a~~tely pubescent and ~~o~~led, deeply 3-part-~~>~~arted, intermediate ~~r~~minal; styles 12. ~~ι~~souri.

~~p~~id; leaves ovate, ~~a~~se, serrate; flow-~~p~~sules 5, 2-beaked.

(Corymbiferæ.) ~~.)~~ leaves dentate, ~~t~~a 3-toothed, trian-~~l~~ate, pinnatifid, up-~~e,~~ tuberculate; ex-~~.~~i ray florets very

~~t~~osacea) [After M. ical collector] . creeping, stolon-~~s~~ pinnate; pinnæ pex; stipules fili-~~n~~ the calyx; style

le, 3-flowered, na-pinnæ cuneate, pinnatifid; petals ~~a~~s long, silky-vil-

~~h~~yllœ) ~~s~~atchfly, p. M. J. ~~.~~ radical leaves lanceolate; pani-~~.~~ slightly emargi-nate. 8-12 i.

virgin"ica, (r. J. ♃.) erect, or decumbent; viscidly pubescent; leaves lance-oblong, scabrous on the margin; panicle dichotomous; petals bifid; stamens exsert. 12 i.

rotundifo'lia, (r. Ju.) decumbent; stem, calyx, and margin of the leaves very pilose; leaves broad-oval; flowers few, trichotomal; petals gashed, sub-4-cleft. S.

infra'ta, calyx bladder-like, and beautifully veined; flowers white, petals bifid. Bladder campion. Rocky hills. Ex.

arme'ria, (w-r. Au. ☉.) flowers fascicled, fastigiate; upper leaves cordate, glabrous; petals entire. 35*

co'nica, calyx of the fruit conic, striate. Ex.

dichot"oma, racemes in pairs, terminal, 1-sided; flowers intermediate, peduncled. Ex.

noctur"na, (w. J. ☉.) flowers spiked, alternate; sessile, secund; petals bifid.

stella'ta, (w. Au. ♃.) leaves verticillate in fours, oval-lanceolate, long acuminate; calyx inflated; petals lacerate, fimbriate. 2-4 f. Hill-sides.

noctiflo'ra, (w-r. Ju. ☉.) calyx veiny, 10-angled; teeth of the tube equal; petals 2-cleft; stem dichotomous. Ex.

quinquevul"nera, (r. Ju. ☉.) hirsute; leaves cuneate-oblong, upper ones linear; petals entire, roundish; fruit alternate, erect. 8-12 i. S.

ova'ta, (r-w. J. ♃.) leaves ovate, lanceolate, acuminate, nearly smoothish; raceme terminal, compound; calyx ovate; stamens and styles exsert; stem simple.

fimbria'ta, (M.) stem pubescent; leaves obovate, ciliate; petals large, fimbriate, white; flowers generally 3, in a terminal fascicle. 6-8 i. S.

antirrhi'na, (Ap. ☉.) stem pubescent near the base, sometimes spotted; leaves narrow, spatulate, lanceolate, ciliate; panicles dichotomous; petals small, bifid; stamens included. 1-2 f. S.

axilla'ris, (p. Au.) viscid-pubescent; stem branched; leaves ovate, oval, petioled, subdentate; flowers axillary, sessile, solitary. 8 i. S.

SILPH"IUM. 17—4. (Corymbiferæ.)

perfolia'tum, (ragged-cup, y. Au. ♃.) stem 4-angled, smooth; leaves opposite, connate, ovate, serrate. 6 f. Rays 24. Mountains.

trifolia'tum, leaves verticillate by threes; panicle trichotomous; stem 4-6 f. high, mostly purple; ray florets about 14, long, bright yellow.

integrifo'lium, (y. Au. ♃.) stem 4-angled, rough; leaves opposite, erect, sessile, oblong, entire, scabrous; flowers few, short-peduncled. 4 f.

terna'tum, (y. Ju. ♃.) stem terete, glabrous; leaves verticillate by threes, petioled, lanceolate, sub-denticulate, somewhat scabrous, ciliate at the base; upper ones scattered, sessile; panicle dichotomous; calyx ciliate. 4-6 f.

gummif"erum, (y. Ju.) erect, hispid, gumniferous; leaves sinuate, pinnatifid, sub hispid beneath; flowers large, axillary, subsessile; scales of the involucrum ovate, acuminate, outer ones fringed or hispid on the margins. 2-3 f.

terebin"thina'ceum, (y. Ju. ♃.) erect, glabrous; radical leaves large, round, or reniform, cordate, slightly lobed and toothed, cauline leaves alternate, ovate, serrate, scabrous; panicle compound, many-flowered. 4-5 f.

lacinia'tum, (y. Ju. ♃.) stem simple, hispid above; leaves pinnatifid, alternate, petioled; segments tooth-sinuate; flowers panicled; scales of the involucrum sub-cordate, acuminate. 8-12 i. S.

compos"itum, (y. Ju. ♃.) smooth; cauline

Out in every tempest,
Out in every gale,
Buffeting the weather,
Wind, and storm and hail
In the meadow mowing,
In the shadowy wood,
Letting in the sunlight
Where the tall oaks stood.
Every flitting moment
Each skilful hand employs,
Bless me! were there ever
Idle farmer's boys?

Though the palm be callous,
Holding fast the plow,
The rounded cheek is ruddy,
And the open brow
Has no lines and furrows
Wrought by evil hours,
For the heart keeps wholesome,
Trained in Nature's bowers;
Healthy, hearty pastime,
The spirit never cloys;
Heaven bless the manly,
Honest farmer's boys!

At the merry husking,
At the apple-bee,
How their hearts run over
With genial, harmless glee!
How the country maidens
Blush with conscious bliss
At the love-words whispered
With a parting kiss!
Then the winter evenings,
With their social joys!
Bless me! they are pleasant,
Spent with farmer's boys!

leaves sinuate, pinnatifid, radical ones ternate, sinuate, many-cleft; flowers small,
panicled. 2-4 f. *S.*

conna'tum, (y. Au. ♃.) erect. terete, hispid; leaves opposite, connate, scabrous,
remotely serrate; panicle terminal, dichotomous. 6 f. *S.*

pinnatif'dum, (y. Au.) stem somewhat
glabrous; leaves sinuate, pinnatifid, sub-
scabrous, a little hairy beneath; flowers
large; scales of the involucrum oval, outer
ones roundish. 4-6 f. *S.*

lœviga'tum, (y. Au.) stem simple, 4-an-
gled, furrowed, glabrous; leaves sessile,
ovate-acuminate, slightly serrate. sub-cordate at the base, glabrous; scales of the
involucrum ovate, ciliate. 2 f.

scaber"rimum, (y. Au.) stem sub-angled;
angles rough above; leaves short-petioled,
ovate, sub-acuminate, serrate, rigid, scabrous; flowers corymbed; scales of the
involucrum ovate, ciliate. 3-4 f. *S.*

atropur"pu'reum, (y. Au. ♃.) terete,
smooth; leaves verticillate by fours, lanceolate, scabrous, sub-entire, sub-sessile, ciliate
at base, upper ones scattered; panicle dichotomous. 4 f. *S.*

denta'tum, (y. Au.) erect, somewhat glabrous; lower leaves opposite, upper ones
alternate, all lanceolate, sinuate-toothed,
pilose, scabrous; flowers corymbed; scales
of the involucrum broad-ovate, ciliate.
2-3 f. *S.*

ela'tum, (y. ♃.) leaves petioled, alternate,
cordate, sinuate; scales of the involucrum
obtuse. *S.*

reticula'tum, (y. ♃.) leaves alternate,
ovate-lanceolate, cordate, serrate, rather obtuse, a little villose. *S.*

SINA'PIS. 14—2. (*Cruciferæ.*)

ni'gra, (common mustard, y. J. ♁.) silique
glabrous, 4-angled, close-pressed to the
stem; leaves at the top lance-linear, entire,
smooth. Naturalized.

al"ba, (white mustard,) pod mostly hispid, spreading; flowers corymbose. 1-2 f.
Introduced.

arven"sis, (y. Ju. ♁.) stem and leaves
hairy; siliques glabrous, many-angled. uneven, about three times the length of the
style; style slender, ancipital. Introduced.

SIPHONY'CHIA. 5—1. (*Amaranti.*) [From
siphon, tube, funnel, and *nuchios*, night.]

america'na, leaves oblanceolate, shorter
than the internodes, a little hairy below,
ciliate. rather obtuse; stem much branched,
minutely and retrosely pubescent; flowers
in small, glomerate cymes at the ends of
the branches.

SI'SON. 5—2. (*Umbelliferæ.*)

majus, glabrous; leaves cut-pinnate; lobes
with cartilaginous margins, sharply serrulate, those of the lower ones lanceolate, of
the upper ones many-cleft and linear.

rubricau'le, leaves semi-verticillate, cut
tri-pinnate; segments capillaceous; partial
involucres compound, longer than the umbellets.

SISYM"BRIUM. 14—2. (*Cruciferæ.*) [From
sisubos, fringe, so called from its fringed
roots.]

officin"ale, (y. Ju. ♁.) leaves runcinata,
hairy; flower in a long raceme; pod subulate. 1-2 f. Stem hairy, branched. Road-
sides.

canes"cens, (y. Ap. ♁.) leaves bi-pinnatifid, hoary; segments dentate, obtuse, sometimes obovate; petals as long as the calyx;
siliques sub-angled, ascending, shorter than
the peduncle; stigma capitate. 1-2 f.

cheiranthoi'des, (y. J. ♁.) siliques erect;
fruit-bearing pedicels spreading; leaves
nearly entire, lanceolate. Canada.

SISYRIN"CHIUM. 15—3. (*Iridæ.*)

an"ceps, (blue-eyed grass, b. ▓▓▓) scape
or culm simple, 2-edged or ▓▓▓▓
glume-like spatha of 2 unequal ▓▓▓
tending above the flower; p▓▓
nate. Hedge-mustard. 6-12 ▓
mucrona'tum, scape simple, ▓
tha colored, one of the valves ▓
long, rigid point; stem setace ▓
Flowers 3-4 in a spatha, blue. ▓

SI'UM. 5—2. (*Umbelliferæ.*) [▓
move, from its agitation in the ▓
latifo'lium, (water-parsnip, ▓
root creeping; stem erect, angul ▓
pinnate; leafets ovate, lanceolate ▓
smooth, serrate, sometimes pinnati ▓
bels terminal, large, rayed; invol ▓
many-leaved. 2-4 f. The leaves that gr▓
in water are bi-pinnatifid. Swamps.

linea're, leafets linear, lanceolate, acutely
and finely serrate; stem tall.

SMI'LAX. 20—6. (*Asparagi.*) [From *smileus*,
to cut; so called from the roughness of its
leaves and stalk.]

1. *Stems frutescent.*

sarsaparil"la, (Ju. ♃.) stem prickly,
slightly 4-angled; leaves unarmed, ovate-
lanceolate, cuspidate, sub-5-nerved, glaucous beneath; peduncles long.

quadran"gula'ris, (Ju. ♄.) leaves unarmed, ovate, sub-cordate, acute, 5-nerved;
stem prickly, 4-angled; berries black.

ca'du'ca, (J. ♄.) stem flexuous, aculeate;
leaves ovate, mucronate, membranaceous,
5-nerved; common peduncle scarcely longer
than the petiole.

pandura'ta, (Ju. ♄.) aculeate; leaves
ovate-panduriform, acuminate, 3-nerved;
peduncle twice as long as the petiole.
Sandy woods.

laurifo'lia, (Ju. ♄.) aculeate; branches
unarmed; leaves coriaceous, perennial,
oval-lanceolate, slightly acuminate. 3-nerv-
ed; umbels short, peduncled.

pseu'dochi'na, (J. ♄.) unarmed cauline
leaves cordate, ramose ones oblong-ovate,
5-nerved; peduncles very long.

rotundifo'lia, (green-brier, w-g. Ju ♃.)
stem prickly, sub-terete; leaves unarmed,
roundish-ovate, short-acuminate, cordate,
5-7 nerved; berries spherical.

2. *Stems herbaceous.*

pedun"cula'ris, (Jacob's ladder, w-g. M.
♃.) stem round, climbing; leaves round-
ovate, cordate, acuminate, 9-nerved; umbels
long-pedicelled. 3-5 f. Low grounds.

herba'cea, (bobea tea, g. J. ♃.) stem erect,
simple, slightly angled; leaves long-peti-

oled, oval, nerved, pubescent beneath; umbels with long, compressed peduncles; berries spherical.

tamnoi'des, (Ju. ♄.) stem round, aculeate; leaves ovate-oblong, acute, sub-panduriform, obsoletely cordate, 5-nerved; common peduncle longer than the petiole.

hasta'ta, (Ju. ♄.) stem angled, prickly; branches unarmed; leaves lanceolate, acuminate, hastate-auricled at the base, 3-nerved, prickly, ciliate on the margin. Var. *lanceola'ta*, leaves long, narrow, lanceolate. *S.*

bona'nox, (Ju. ♄.) stem unarmed, angled; leaves heart-ovate, smooth, 7-nerved, prickly, ciliate. *S.*

ova'ta, (Ju. ♄.) generally unarmed; leaves ovate, acute, cuspidate, uniformly colored; common peduncle shorter than the petiole. *S.*

cin"cidifo'lia, (♄.) prickly; leaves unarmed, round-cordate, acuminate, 5-nerved, glabrous, net-veined, short-petioled. *S.*

walte'ri, (Ju.♃.) aculeate; leaves cordate, ovate, smooth, 3-nerved; berries 3-seeded, acuminate. *S.*

al'ba, (J. ♄.) generally unarmed; stem obsoletely angled; leaves lance-elongated, coriaceous, glabrous, entire, 3-nerved, umbels short-peduncled, few-flowered. *S.*

pu'mila, (S. ♄.) unarmed; leaves cordate, ovate, entire, somewhat 5-nerved, soft-pubescent beneath; umbels short-peduncled; pedicels very short; berries oblong, acute; stem prostrate; corolla 0. 2-4 f. *S.*

lanceola'ta, (J. ♄.) unarmed; leaves lanceolate and ovate, acute or acuminate, 3-nerved, very glabrous, perennial; umbels many-flowered; peduncles short; berries red. *S.*

ru'bens, a very handsome species, the tendrils of which are of a bright red. Ex.

excel"sa, remarkable for the large size of the leaves. Ex.

SOLA'NUM. 5—1. (*Solaneæ.*) [From *solor*, comfort, because some species give ease by their narcotic quality.]

dulcama'ra, (bitter-sweet, p-b. Ju. ♄.) stem unarmed, woody, climbing; lower leaves mostly cordate, glabrous, upper ones mostly guitar-hastate, few-flowered; corymbs opposite to the leaves. This is the true bitter-sweet, though the celastrus scandens is called so by some. Damp.

ni'grum, (deadly night-shade, w-p-b. J. ⨀.) stem unarmed, erectish. or erect; branches angled, dentate; leaves ovate, repand, glabrous; racemes 2-ranked, nodding. 1-2 f. Ex.

tubero'sum, (potato, b-w. Ju. ♄.) stem wing-angled, unarmed; leaves interruptedly pinnate; leaflets entire; flowers subcorymbed; roots knobbed, tuberous. Cultivated.

lycoper"sicum, (love-apple, tomato, y. S. ⨀.) stem unarmed; leaved pinnatifid, gashed; racemes 2-parted, leafless; fruit glabrous, torulose. Ex.

pseudo-capsi'cum, (Jerusalem cherry, ♄.) stem woody; leaves lanceolate, repand; umbels sessile. Ex.

carolin"ense, (horse-nettle, b. J. ⨀.) stem aculeate; leaves ovate-oblong, tomentose, hastate-angled; racemes lax. 1-2 f.

flavid"um, suffruticose, densely tomentose; branchlets and calyx aculeate; leaves solitary, oblong, obtusish, lower ones repand-sinuate, upper ones obsoletely sinuate; racemes about 3-flowered.

melonge'na, (egg-plant. J. ⨀.) unarmed; leaves ovate, tomentose; peduncles pendent, incrassate; calyx unarmed. Ex.

mammo'sum, (y. Ju. ⨀.) stem aculeate, herbaceous; leaves cordate, angled, lobed, villose on both sides and prickly. *S.*

virginia'num, (b. Ju. ⨀.) stem erect, aculeate; leaves pinnatifid, prickly; segments sinuate, obtuse; margins ciliate; calyx prickly. *S.*

verbascifo'lium, (♄.) stem unarmed, frutescent; leaves ovate, tomentose, entire; corymbs bifid, terminal. *S.*

hirsu'tum, (p. ♃.) small, pilose, hirsute; leaves broad-obovate; raceme somewhat 3-flowered; peduncles filiform. *S.*

SOLE'A. 5—1. (*Cisti.*)

con'color, (Ap. w-y. ♃.) stem simple, erect; leaves wedge-form, lanceolate, sessile, irregularly toothed above; peduncles short, 2 3 flowered; calyx nearly as long as the petals; spur none. 2-4 f. Rocks. Green violet.

SOLIDA'GO. 17—2. (*Corymbiferæ*) [From *sohdo*, to make firm, from its supposed virtue in healing wounds.]

A. *Flowers one-sided; leaves with three combined nerves.*

canaden"sis, (Canadian golden-rod, y. Ju. ♃.) stem downy; leaves lanceolate, serrate, rough; racemes panicled, recurved; rays hardly longer than the disk; stem angular; leaves sessile, three inches long, sometimes nearly entire. 2-5 f.

pro'cera, (great golden rod, y. Ju. ♃.) erect, villose; leaves lanceolate, serrate, scabrous, villose beneath; racemes erect, spike-form, before flowering, nodding; rays short. 4-7 f. Low grounds.

cilia'ris, (fringed golden-rod, y. ♃.) stem erect, smooth, angular; leaves lanceolate, sub-3-nerved, smooth, scabrous on the margin; racemes panicled, secund; peduncles glabrous; bracts ciliate; rays short. 3 f.

reflex'a, (y. Ju. ♃.) erect, villose; leaves lanceolate, sub-serrate, scabrous, reflexed; branches panicled, sub-secund, reflexed. Pine woods.

gigan"tea, (giant golden-rod, y. Au. ♃.) stem erect, glabrous; leaves lanceolate, smooth, serrate, rough-edged, obscurely 3-nerved; racemes panicled; peduncles rough-haired; rays short. 4-7 f.

lateri'flora, (side-flowered golden-rod, y. Au. ♃.) stem erect, a little hairy; leaves lanceolate, slightly 3-nerved, glabrous, rough-edged, lower ones sub-serrate; racemes panicled, a little recurved, sub-secund; flowers large, the rays being much longer than the calyx; stem striated, often purplish, pinnatifid, with numerous lateral flowering branches. 2-3 f.

B. *Racemes or flowers 1-sided; leaves veiny.*

altis"sima, (variable golden-rod, y. Au.

2f.) stem erect, rough-haired; leaves lanceolate, lower ones deeply serrate, scabrous, rugose. The panicled racemes are very numerous, and spread every way, so as to bring the one-sided flowers upward; rays short; the serratures of the leaves irregular; it is hairy or villose, and sometime the racemes diverge but little. This species is variable. 3-6 f.   169

as"pera, (y. Au. 2f.) erect, terete, hairy; leaves ovate, somewhat elliptic, very scabrous, rugose, serrate, nerveless; racemes panicled, secund. 3-5 f.

nemora'lis, (woolly golden-rod, y. Au. 2f.) erect, tomentose; radical leaves somewhat cuneate, serrate, caulines ones lanceolate, hispid, entire; racemes panicled. Plant grayish. 1-3 f.

ulmifo'lia, (elm golden-rod, y. Au. 2f.) erect, smooth, striate; leaves elliptic, deeply serrate, acuminate, villose beneath, radical ones obovate; racemes panicled; peduncles villose; rays short. 3-4 f.

argu'ta, (y. Oc. 2f.) erect, smooth; leaves glabrous, acutely and unequally serrate, radical ones oblong-ovate, cauline ones elliptic; racemes panicled; rays elongated. 2-3 f.

jun"cea, (rush-stalk golden-rod, y. Au. 2f.) erect, smooth, slender; leaves lanceolate, glabrous, smooth, rough-edged, lower ones serrate; racemes panicled. 2-3 f.

ellip'tica, (oval-leaf golden-rod, y. Au. 2f.) erect, smooth; leaves oval, smooth, serrate; racemes panicled; rays middle-sized. 2-3 f.

recurva'ta, (y. S. 2f.) erect, pubescent; leaves lanceolate, serrate, rough-edged; racemes elongated, panicled, recurved. Shady woods.

sempervi'rens, (narrow-leaf golden-rod, y. S. 2f.) erect, smooth; leaves lanceolate, narrow, long, somewhat carnose, smooth, entire, rough-edged; peduncles hairy. 3-5 f. Swamps.

odo'ra, (sweet-scented golden-rod, y. Au. 2f.) pubescent; leaves lance-linear, entire, smooth, scabrous on the margin; racemes panicled. The flowers, when dried, form an excellent substitute for tea, and the leaves, when distilled, yield a fragrant volatile oil.

pat"ula, (spread golden-rod, y. S. 2f.) stem erect, glabrous; leaves oval, serrate, glabrous, radical ones oblong-spatulate; racemes panicled, spreading; peduncles pubescent; stem wand-like, angular, and striate; stem leaves sessile, about an inch long, pointed, the radical ones resemble those of the ox-eyed daisy; racemes about an inch long; flowers rather large. 2 f.

C. *Racemes erect.*

bi-color, (white golden-rod, w. Au. 2f.) stem hairy; leaves oval, hairy, lower ones serrate, those on the flower branches entire, numerous, and small; scale and calyx obtuse; racemes are short and compact; rays white, somewhat numerous and shortish; disk florets rather numerous. 2-4 f.

specio'sa, (y. S. 2f.) tall, smooth; branches virgate; leaves lanceolate, sub coriaceous; lower ones sparingly serrate; racemes terminal, erect, compound; peduncles short; rays about 5, elongated. 3-6 f.

virga'ta, (y. Au. 2f.) stem smooth, simple; leaves lanceolate, somewhat cuneate, obtuse, entire, glabrous, close-pressed; upper ones gradually smaller; branches of the panicle elongated, racemed at the summit; peduncles erect, smooth, slender. 2 f.

petiola'ris, (late golden-rod, y. Oc. 2f.) villose; leaves elliptic, roughish, petioled; racemes numerous, short; rays elongated. 2-3 f.

stric"ta, (willow-leaf golden-rod, y. Au. 2f.) erect, glabrous; radical leaves serrate, cauline ones lanceolate, entire, smooth, scabrous on the margin; racemes panicled, erect; peduncles smooth. 2 f. Sandy woods.

gramin"ifo'lia, (y. S. 2f.) stem angled, branching; leaves lanceolate-linear, entire, nearly erect, 3-5-nerved, a little scabrous; corymbs terminal, fastigiate; heads clustered; florets of the ray as long as the disk.

tenuifo'lia, (pigmy golden-rod, y. S. 2f.) stem angled, scabrous; branches fastigiate; leaves linear, narrow, expanding, slightly 3-nerved, scabrous, axils leafy; corymbs terminal, fastigiate; heads clustered; ray florets about 10, scarcely exceeding the disk. 1-2 f.

cæ'sia, (blue-stem golden-rod, y. Au. 2f.) stem smooth, tinged with purple, sub-glaucous; leaves lanceolate, smooth, serrate, sometimes rough-edged; racemes erect; rays middle-sized. 2-3 f.

liv"ida, (purple stem golden-rod, y. S. 2f.) stem smooth, panicled, dark purple; leaves lanceolate, serrate, smooth, margins scabrous; branches racemed at the extremity; rays elongated.

lithosper"mifo'lia, (y. S. 2f.) stem pubescent, branched; leaves lanceolate, scabrous, tapering, 3-nerved, entire; ray-florets elongated.

puber"ula, (y. 2f.) stem brownish, simple, sub-pubescent, terete; leaves lanceolate, entire, sub-pubescent, tapering; radical ones sub-terete; racemes spiked, axillary; peduncles pubescent; scales of the involucrum lance-linear, acute; ray-florets elongated, about 10.

læviga'ta, (y. S. 2f.) erect, smooth; leaves lanceolate, fleshy, entire, very smooth, radical leaves sub-ovate; racemes panicled, erect; peduncles scaly, villose; rays elongated, about 10. 4-5 f.

limonifo'lia, (y. Oc. 2f.) stem oblique, smooth, generally purple; leaves lanceolate, somewhat carnose, entire, smooth; racemes panicled, erect; peduncles scaly, smooth; rays long. 3-5 f. Salt marshes.

flexicau'lis, (zigzag golden-rod, y. S. 2f.) stem flexuous, smooth, angled; leaves ovate, acuminate, serrate, glabrous; racemes axillary, erect, short, scattered; rays middle-sized. 2-3 f. Woods.

rigid"a, (y. S. 2f.) stem corymbed, hairy,

scabrous; leaves ovate oblong, rough, with small, rigid hairs; those of the stem very entire, lower ones serrate; flowering branches panicled; racemes compact; rays elongated; scales of the involucrum obtuse. 3-4 f.

latifo'lia, (y. S. Oc. ♃.) stem somewhat flexuous, angular, smooth; leaves broad-ovate, acuminate, deeply serrate, glabrous; petioles winged; racemes axillary. 18 i. Dry woods.

vimin"ea, (twig golden-rod, y. Au. Oc. ♃.) erect, sub pubescent; leaves lance-linear, membranaceous, attenuate at base, glabrous; margins scabrous; lower ones sub-serrate; racemes erect; rays elongated. Banks of streams.

virgau'rea, (European golden-rod, y. ♃.) stem terete, pubescent, flexuous; leaves serrate, roughish, attenuate at the base; racemes panicled, erect; rays elongated; flowers large. 1-3 f. Var. *alpi'na*, small; leaves obovate or lanceolate. 3-6 i. The only species common to both continents.

novebo'racen"sis, (star golden-rod, y. Oc. ♃.) stem nearly leafless; branches fastigiate; leaves rough, radical ones ovate-oblong, petioled; flowers large. 2-3 f. Sandy fields.

Southern species.
1. Racemes one-sided.

cineras"cens, (y. S. ♃.) stem slender, pubescent; leaves long, linear-lanceolate, attenuate at base, serrate, sub-scabrous, pubescent; racemes recurved; peduncles and ray-florets elongated; seeds pubescent. 3 f.

tortifo'lia, (y. S. ♃.) stem pubescent; leaves linear-lanceolate, sub-serrate, expanding, twisted, the upper surface and midrib scabrous, nearly glabrous beneath; panicle pyramidal; racemes recurved. 3 f.

corymbo'sa, (y. S. ♃.) stem robust and virgately erect, glabrous; branches hispid; lower leaves lance-oblong; upper ones ovate, all fleshy, glabrous, rigid, margins scabrous and ciliate; racemes corymbed; lower ones recurved; ray-florets elongated. 4-6 f.

pitch"eri, racemes glabrous; leaves glabrous, thickly set, lance-oblong, acuminate at each end, sharply serrate; panicle pyramidal, few-flowered; pedicels pubescent; liguli abbreviated. Ark.

pyramida'ta, (y. S. ♃.) stem terete, hispid; leaves oblong, acute, somewhat amplexicaul, sessile, glabrous, margins scabrous, rarely and obsoletely toothed; panicle naked, pyramidal; branches reflexed; peduncles squamose. 4-6 f.

retror"sa, (y S. ♃.) stem terete, glabrous, somewhat amplexicaul, pubescent towards the summit; leaves closely sessile, linear, tapering above, glabrous, pellucid-punctate, reflexed margins rough; branches of the panicle recurved.

2. Racemes erect.

pulverulen"ta, (y. ♃.) stem simple; stem and leaves pulverulent-pubescent; leaves sessile; lower ones elliptic, serrate; upper ones obovate, entire, margins scabrous;

racemes erect, spike-form; ray-florets elongated. 3-4 f.

pubes"cens, (y. Oc. ♃.) stem branching, pubescent, slightly scabrous, generally colored, with numerous branches rigidly erect; leaves long, lanceolate, tapering at base, pubescent; lower ones serrate; racemes erect, panicled; ray-florets middle sized. 3-4 f.

paucifos"culo'sa, (y. S. ♃.) smooth, suffruticose; leaves lanceolate, obtuse, nerveless; panicle compound, many-flowered; the clusters erect; involucrum oblong, 5-flowered; floret of the ray, one.

glomera'ta, (y. ♃.) stem simple, low; leaves glabrous, lance-oblong, serrate; lower ones broad-oval, acuminate; racemes simple, composed of axillary heads, upper ones clustered; involucrum turgid, many-flowered.

angustifo'lia, (y. S. ♃.) stem glabrous, generally colored, with many slender, erect branches above; leaves subulate-linear, entire, glabrous; racemes erect, panicled; ray-florets middle sized. 2-3 f.

ela'ta, (y. S. ♃.) stem terete, hairy, tomentose above; leaves lance-oval, acute, sub-entire, veiny, tomentose beneath; racemes erect, panicled; ray-florets elongated. 2-3 f.

salici'na, (y. S. ♃.) stem tall, slender, pubescent above, somewhat scabrous; branches virgate, long, erect; leaves lanceolate, sessile, scabrous above, glabrous beneath; lower ones serrate; racemes sub-secund; branches short, sometimes recurved. 4-5 f.

hirsu'ta, sub-pilose; stem simple; racemes erect; flowers sub-racemose-glomerate; leaves elliptic-ovate, scattered; lower ones spatulate, finely crenate.

squarro'sa, (y. S.) stem branching, pubescent; leaves lanceolate, acute, serrate, softly pubescent beneath; lower ones tapering at base; racemes compound, erect; flowers large; involucrum squarrose; ray-florets about 10, scarcely longer than the involucrum. 3-5 f.

SON"CHUS. 17—1. (*Corymbiferæ*)

olera'ceus, (sow-thistle, y. Ju. ⊙.) leaves lance-oblong, clasping, slightly toothed and serrate; peduncles axillary and terminal, covered with cotton-like down. Waste grounds. 2-4 f. Introduced.

arven"sis, root creeping; leaves runcinate, denticulate, cordate at the base; involucre hispid; flowers large, deep yellow; stem 2 f.

macrophyl"lus, (b. Au. ♃.) leaves lyrate, cordate at base, hairy beneath; peduncles hairy, naked; flowers panicled. 4-7 f.

spinulo'sus. (y. Au. ⊙.) leaves clasping, undulate, spinose, oblong; flowers somewhat umbelled. 2 f. Salt marshes.

leucopha'us, (b-w. Ju. ♂.) peduncles squamose; flowers racemed; leaves runcinate, acuminate; stem virgate and panicled. 2-5 f. Swamps.

florida'nus, (b. Ju. ♂.) peduncles sub-squamose; flowers panicled; leaves lyrate-runcinate, denticulate, petioled. 3-6 f.

acumina'tus, (b. Au. ♂.) peduncles sub-squamose; flowers panicled; radical leaves sub-runcinate; cauline ones ovate, acuminate, petioled, denticulate in the middle. 3-5 f. Woods.

pallid"us, (y. J. ♃.) raceme compound, terminal; leaves lance-ensiform, amplexicaul, dentate. 2-3 f.

corolinia'nus. (y. Au. ☉.) erect, glabrous, fistulous; leaves lanceolate, acute, undulate, sub-spinose, toothed, auricled at the base, semi-amplexicaul; flowers somewhat umbelled. 1-3 f. S.

SOPHO'RA. 10—1. (*Leguminosa*.)
serice'a, leaves pinnate; leafets wedge-oval, smooth above, silky-villose beneath; spikes many-flowered, sub-sessile; flowers white. ♃. 1 f.

japon"ica, a tree which produces large bunches of cream-colored flowers in August and September. The drooping sophora, a variety of the japonica, is very different in appearance, being a trailing shrub, which sends out shoots six or eight feet long, in a single season. Ex.

SOR"BUS. 11—5. (*Rosaceæ*.) [From *sorbeo*, to suck up, because its fruit stops hemorrhages.]
america'na, (mountain-ash, w. M. ♄.) leaves pinnate; leafets lance-oblong, acute, serrate, very smooth; flowers in terminal corymbs. The yellowish berries remain on the tree during winter. 13-20 f.
microcar'pa, fruit small, scarlet.

SOR"GHUM. 3—2. (*Graminea*) [An Indian name.]
sacchara'tum, (broom-corn, y-g. Au. ☉.) panicle somewhat whorled, spreading; seeds oval; glumes covered with permanent, softish hairs; leaves linear. From the East Indies. 6-8 f.
vulga're, (Indian millet,) panicle compact, oval, nodding when mature; seed naked.

SPARGA'NIUM. 19—3. (*Typha*.) [From *sparganon*, a band or fillet, from the long linear form and pliant texture of the leaves.]
ramo'sum, (bur-reed, w. Ju. ☉.) the 3-sided bases of the leaves concave on the two outsides; the general fruit stem branched; stigmas linear. In water generally. Flowers in round heads; the staminate heads above the pistillate, and considerably the smallest.
angus"tifo'lium, (floating bur-reed, w. Au. ♃.) leaves flat, long linear, very narrow, much longer than the stem, weak; the part above water floating on its surface. Grows in great abundance in the little lake on Catskill Mountain, near the Mountain House.

SPARGANOPH"ORUS. 17—1. (*Corymbiferæ*.) [From *sparganon*, a crown, and *phero*, to bear.]
verticilla'tus, (water-crown-cup, p. Au. ♃.) leaves linear, verticillate; pods few, terminal; egret 5-toothed, submersed.

SPAR"TIUM. 16—10. (*Leguminosa*.) [From *sparto*, a rope; so called because the tough branches and bark are used in making cordage.]
junce'um, (Spanish broom, g. ♄.) branch-

es opposite, virgate, with terminal flowers; leaves lanceolate, glabrous.
scopa'rium, (Scotch broom, g. ♄.) leaves ternate, solitary, and oblong; flowers axillary; legumes pilose at the margin; branches angular.

SPER"GULA. 10—5. (*Caryophylleæ*.) [From *spergos*, to scatter.]
arven"sis, (spurry, w. Ju. ☉.) leaves whorled; panicles dichotomous; peduncles of the fruit becoming reflexed.
saginoi'des, (pearl-wort spurry, w. J. ☉.) glabrous; leaves opposite, subulate, awnless; peduncles solitary, very long, smooth. 2-3 i.
ru'bra, (red sand-wort, r. J. ☉.) stem prostrate, glabrous; leaves filiform, fleshy, larger than the joints; stipules cuneate-membranaceous, sheathing; stamens 5; capsule angular or globose. 8 i.

SPERMACO'CE. 4—1. (*Rubiaceæ*.) [From *sperma*, seed, and *akoke*, a sharp point; the seeds being pointed.]
ten"uior, (w. Ju. ☉.) lanceolate; flowers verticillate, stamens included; seeds hirsute. S.
diodi'na, (Ju. ☉.) stem terete; leaves linear-lanceolate, sessile; flowers axillary, sessile; stamens shorter than the corolla. Dry soils. S.
involucra'ta, (w.) stem very hispid; leaves ovate, lanceolate, acuminate; stipules many-bristled; heads terminal, involucred; stamens longer than the corolla. 1 f. S.
gla'bro, (w. J. ☉.) stem procumbent, glabrous; leaves ovate-lanceolate, glabrous; flowers verticillate; seeds glabrous. S.

SPIGE'LIA. 5—1. (*Gentianæ*.) [Named by Linnæus, in honor of Adrian Spigelias, a botanist who wrote in 1606.]
maryland"ica, (Indian pink-root, p. J. ♃.) stem 4-sided; leaves all opposite, sessile, lance-ovate, entire. 9-18 i. Sometimes called worm-grass, on account of its efficacy in cases of disease arising from worms.

SPINA'CIA. 20—5. (*Polygoneæ*.) [From *Ispania*, Spain, whence it originated.]
olera'cea, (spinach, J. ☉.) fruit sessile, prickly or unarmed; leaves hastate-sagittate; stem branched. 1-2 f. Ex.

SPIRÆ'A. 11—5. (*Rosaceæ*.) [From *spira*, a pillar; so named from its spiral stalk.]

Stem more or less woody.

salicifo'lia, (meadow-sweet, willow hard-hack, r. w. J. ♄.) leaves lance-ovate or obovate, serrate, glabrous; flowers in panicled, spreading racemes. Var. *al"ba*, has white petals, and often the twigs are reddish. The small branches are generally killed by frost in the winter, as also of the next species. 2-4 f.
tomento'sa, (steeple-bush, purple hard-hack, meadow-sweet, r. Ju. ♄.) leaves lanceolate, unequally serrate, downy beneath; racemes in a crowded, sub-panicled spike. 2-3 f.
hypericifo'lia, (John's-wort, hard-hack, w. M. ♄.) leaves obovate, entire or toothed at the apex; umbels sessile. Cultivated. 3 f.
opulifo'lia, (nine-bark, snow-ball, hard

hack. w. J. ♄.) leaves sub-ovate. lobed, doubly toothed or crenate. glabrous: corymbs terminal, crowded; capsules inflated; flowers trigynous. Wet. 3-5 f.

crena'ta, (♄.) leaves obovate, crenulate at the apex, acute, 3-nerved; corymbs crowded. peduncled.

capita'ta, (J. ♄.) leaves ovate, somewhat lobed, doubly toothed, reticulate beneath, tomentose; corymbs terminal, crowded, sub-capitate, long-peduncled; calyx tomentose.

sorbifo'lia, (w. Au. ♄.) flowers panicled: leaves pinnate; leafets uniform, serrate. A native of Siberia.

mo'nogy'na, (♄.) leaves glabrous, broad-ovate. sub-3-lobed, gash serrate; corymbs umbelled; pedicels glabrous; segments of the calyx erect, spreading.

2. *Stem herbaceous. Leaves pinnate.*

arun"cus, (goat's beard, w. J. ♃.) leaves 2-3 pinnate, shining; spikes in panicles; styles 3-5. Var. *america'na*, very long, slender spikes. 4-6 f. Mountains.

loba'ta, (r. Ju. ♃.) leaves glabrous, terminal one large, 7-lobed, lateral ones 3-lobed; corymbs proliferous.

ulma'ria, (queen of the meadow, w. Au. ♃.) leaves pinnate, downy beneath; the terminal leafets larger, 3-lobed; the lateral ones undivided; flowers in a proliferous corymb; stem herbaceous. Ex.

betulifo'lia, (r. J. ♄.) leaves glabrous, broad-ovate, gash-toothed; corymbs terminal, compound, fastigiate, leafy. 1 f.

ulmifo'lia, (w.) corymbs fastigiate; leaves large. Ex.

bel"la, (J.) corymbs of beautiful rose-colored flowers. Ex.

aricefo'lia, (Ju.) a beautiful species, producing loose panicles of feathery, whitish flowers, A native of California.

STA'CHYS. 13—1. (*Labiata.*) [From *stachus*, a spike.]

as"pera, (hedge-nettle, clown-heal, w-p. Ju ♃.) stem erect, hispid backward; leaves sub petioled, lanceolate, acutely serrate, very glabrous; whorls about 6 flowered; calyx with spreading spines. Var. *tenuifo'lia*, leaves very thin and slender. Fields.

hyssopifo'lia, scarcely pubescent, slender, erect; leaves sessile, lance linear; whorls about 4-flowered; flowers sessile, purple; corolla little hairy. Meadows.

sylvat"ica, leaves cordate, ovate-acuminate, serrate, hairy; floral ones nearly linear; whorls of 6 flowers; calyx hairy, with 5 acute teeth; flowers purple; lower lip of the corolla whitish with dark spots; fetid. Woods.

veluti'na, (b.) stem simple, quadrangular, villose or sub-hispid; leaves lance-ovate, crenate, serrate, opposite and pointing four ways, clasping, close-sessile; nerves silky-tomentose; whorls about 6-flowered; corolla sub-pil)se. 1 f.

pilo'sa, (r. ♃.) hirsutely pilose; leaves sub-sessile, serrate, acute, oblong-ovate; calyx very pilose; whorls somewhat 6-flowered.

latifo'lia, (p. Ju. ♃.) whorls many-flow-

ered, spiked; upper lip 2-cleft with acute segments; leaves broad, cordate, rugose, hairy. Ex.

his"pida, (y-p. Ju. ♃.) stem and leaves hispid; leaves petioled, nearly sessile, ovate-oblong, acute, obtusely serrate; whorls about 4-flowered; calyx glabrous; corolla large, rather longer than the stamens. 2 f. S.

tenuifo'lia, stem erect, angled, smoothish; leaves petioled, oval-lanceolate, serrate, acuminate; whorls 6-flowered; calyx very pubescent. 18-24 i. S.

interme'dia, (♃.) leaves oblong, sub-cordate, crenate; stem somewhat woolly; whorls many-flowered. S.

STAPHYLE'A. 5—3. [From *staphule*, a tumor.]

trifo'lia, (bladder-nut, y-w. M. ♄.) leaves in threes; racemes pendent; petals ciliate below. When the fruit is ripe, it consists of 2 or 3 inflated, adnate, sub-membranous capsules, each containing from 1 to 3 hard, small nuts. 6-12 f.

STAT"ICE. 5—5. (*Plumbagines.*)

limoni'um, (marsh-rosemary, sea-lavender, Au. ♃.) scape paniculate, terete; leaves radical, linear, flat, smooth; flowers sessile, secund, in a very large and much-branched panicle. Salt marshes.

arme'ria, leaves all radical, linear, flat; scape bearing a round head of rose-colored flowers, which are intermixed with scales, and have a 3-leaved, general involucre. Rocks near the seashore. Striped.

STELLA'RIA. 10—3. (*Caryophylleæ.*) [From *stella*, a star; so called from the starlike appearance of its flowers.]

me'dia, (chickweed, w. M. to Nov. ☉.) stem procumbent, with pubescent leaves on opposite sides; peduncles axillary and terminal, 1-flowered; petals white, deeply cleft; stamens 5-10. 9-13 i. Road-sides.

lanceola'ta, (♃.) leaves lanceolate, acute at each end; petals about as long as the calyx; stigmas mostly 4, or wanting; flowers solitary, axillary, and terminal, on slender peduncles. 6-18 i.

longifo'lia, (long-leafed starwort,) leaves linear, acute, spreading, with the margins often scabrous; panicle very long; petals 2-parted, broad-obovate. 12-15 i. Moist woods.

pu'bera. (w. M. ♃.) pubescent; leaves sessile, ovate, ciliate; pedicels dichotomous, recurved; petals longer than the calyx. 6-12 i.

borea'lis, (w. Ju.) stem angular, dichotomous; leaves lance-oval; peduncles axillary, elongated, flowered; petals deeply cleft, about equal to the calyx. White Mountains.

lon"gipes, (w.) weak, very glabrous, glaucous; leaves linear, subulate, spreading; peduncles terminal, dichotomously branched; bracts membranaceous; pedicels much elongated; petals broad-ovate, deeply bifid, a little longer than the obscurely 3-nerved calyx. Woods near Lake Ontario.

prostra'ta, (Ap. ☉.) stem slightly channelled, prostrate, hollow, forked, sub-pubescent; peduncles solitary, long; flowers small, heptandrous; calyx erect. 1-4 f. *S.*

jamesia'na, viscid-pubescent; leaves lanceolate, sub-falcate, sessile, acute; stem somewhat branched, weak; panicles lax, divaricate; petals 2-lobed, about twice the length of the oblong-acute divisions of the calyx. *S.*

gla'bra, (w. M.) stem slender, glabrous; leaves subulate-linear, expanding; peduncles erect, axillary, 1-flowered; petals emarginate, much longer than the calyx.

STE'VIA. 17—1. (*Corymbiferæ.*) [After an eminent Spanish botanist.]

callo'sa, (r. ☉.) leaves linear, crowded, somewhat succulent, callous at the apex; upper ones alternate; flowers divaricate, sub corymbed; egret about 8 leaved, erose, short. *S.*

STILLIN"GIA. 19—15. (*Euphorbia.*) [From Stillingfleet, who wrote on gardening in 1759.]

sylvat"ica, (y. J. ♃.) herbaceous; leaves sessile, oblong-lanceolate, serrulate; scaly bracts nearly as long as the staminate flowers. *S.*

sebif"era, (Ju. ♄.) leaves rhomboid, acuminate, entire, with a gland below the base on the petiole; staminate flowers pedicelled. Introduced. 20-40 f. *S.*

ligustri'na, (Ju) fruticose; leaves lanceolate, tapering at each end, glabrous, entire, petioled; staminate florets short-pedicelled. 6-12 f. *S.*

STI'PA. 3—2. (*Gramineæ.*)

avena'cea, (feather grass, M. ♃.) stem terete, glabrous; leaves striate, glabrous; panicle spreading; branches whorled with branchlets; awns naked, twisting. Var. *bi'color*, fruit bearded at the base, obovate. *S.*

stric"ta, panicle long, narrow; peduncles very straight, jointed; awns naked; somewhat flexuous. *S.*

STI'PULICI'DA. 3—1, (*Amaranti*) [From *stipula*, the stipule, and *cædo*, to cut, the stipule being divided into many segments.]

seta'cea, (w. M.) erect, smooth, branched; lower leaves small, opposite, spatulate; on the branches none; at each fork 2 fimbriate stipules. 6-10 i. *S.*

STOKE'SIA. 17—1. (*Corymbiferæ.*) [After John Stokes, an eminent botanist.]

cya'nea, (b. ♃.) stem leafy; leaves lanceolate; peduncles axillary, 1-flowered. *S.*

STREPTAN"THUS. 14—2. (*Cruciferæ.*)

sagitta'tus, (r. ☉.) leaves sagittate, acute, clasping, entire; petal oblong-oval, not maculate.

ovalifo'lius, (Arkansas cabbage,) leaves oval. Grows in Arkansas.

STREPTO'PUS. 6—1. (*Liliaceæ.*) [From *streptos*, twisted, *pous*, foot.]

ro'seus, (r. M. ♃.) smooth and shining; stem dichotomous, terete; leaves clasping, serrulate, ciliate; anthers short, 2-horned. 12-18 i. Mountains.

distor"tus, (g y. M. ♃.) pedicels distorted or twisted, and geniculate in the mid-

dle; anthers much longer than the filaments. 2 f. Shady, alpine woods.

laniguno'sus, hoary-pubescent; flowers greenish, larger than the preceding. Mountains.

STROPHOS"TYLES. 16—10. (*Leguminosæ.*)

angu'losa, (p. Au. ☉.) leaves ternate; leafets angular, 2-3-lobed; peduncles longer than the leaves; flowers capitate.

helvo'la, flowers red, prostrate, sometimes twining; leaves ternate, deltoid-oblong; flowers capitate; banner short; wings large, expanded.

STUAR"TIA. 15—12. (*Malvaceæ.*)

pentagy'na, (w-y. Ju) sepals lanceolate; styles distinct; capsules 5-angled; leaves oval or ovate, acuminate, entire or mucronately serrulate, somewhat pubescent beneath. N. C. to Geor.

virgin"ica, (w. M. ♄.) leaves ovate, acuminate; flowers axillary; calyx ovate; petals entire. 6-12 f. *S.*

STYL"IPUS. 11—12. (*Rosaceæ.*) [From *stulos*, column, from the receptacle being columnar.]

ver"na, (y. J. ♃.) sparingly pubescent; radical leaves interruptedly pinnate; cauline ones pinnate and pinnatifid; leafets gash-toothed; stem procumbent at the base, branching above; stipules large, roundish, gash-toothed; petals longer than the calyx; awns naked; flowers small.

STYLOSAN"THES. 16—10. (*Leguminosæ.*) [From *stulos*, a column, and *anthos*, flower.]

ela'tior, (pencil-flower, y. Au. ♃.) stem pubescent on one side; leaves glabrous, lanceolate; bracts ciliate; heads 2-3-flowered. 9-15 i.

STY'RAX. 15—12. (*Malvaceæ.*) [Name from the Greek.]

grandifo'lium, (w. Ap. ♄.) leaves obovate, acuminate, tomentose beneath; racemes simple, axillary, leafy near the base. 4-12 f. *S.*

ben"zoin, a tree producing a balsam, the preparations of which are much used for medicinal purposes.

læ've, (w. Ap. ♄.) branches virgate, slightly geniculate; leaves lanceolate, acuminate at each end, serrate, glabrous; racemes lateral, leafy; flowers axillary and terminal; corolla tomentose. 4-6 f. *S.*

pulverulea"tum, (w. Ap. ♄.) leaves oval, acute, tomentose beneath; racemes lateral, leafy, few-flowered; corolla very fragrant. 18 i. *S.*

gla'brum, (w. Ap. ♄.) branches diffuse, spreading; leaves oval-lanceolate, acute at each end, finely serrulate, membranaceous, glabrous, thin; racemes lateral, leafy; corolla large. 6-8 f. *S.*

SUBULA'RIA. 14—1. (*Cruciferæ.*) [From *subula*, an awl.]

aquat"ica, (w. Ju. ☉.) scape 1-2 inches high; radical leaves entire, subulate. Water.

alpi'na, (♃.) stem branching; leaves obovate. *S.*

SWER"TIA. 4—1. (*Gentianeæ.*) [Named from Emanuel Swert.]

deflex"a, (g. y. Au. ♂.) stem 4-sided;

branches short; leaves opposite, sessile, ovate; corolla bell-form, with horns. 18 i. Swamps.

pusil"la, (false gentian, b. J. ☉.) corolla rotate twice as long as the calyx; stem simple, 1-flowered; leaves oblong. 1 i. High mountains.

fastigia'ta, (Ju. ♃.) stem branching; corolla bell-wheel-form; flowers fastigiate, clustered; pedicels in pairs; leaves spatulate-obovate, nerved. *S.*

SYE'NA. 3—1. (*Narcissi.*) [In honor of Syen, superintendent of the garden at Leyden.]

fluvia'tilis, (J. ♃.) leaves crowded, subulate; flowers axillary, solitary, long-peduncled; peduncle recurved after flowering. 2-3 i. *S.*

SYM"PHITUM. 5—1. (*Boraginea.*) [From *sumphio*, to unite, because it was supposed to heal wounds.]

officina'le, (comfrey, y-w. J. ♃.) leaves ovate-sub-lanceolate, decurrent, rugose. Naturalized. 2-4 f.

SWIETE'NIA. 10—1. (*Meliæ.*) [So named from Van Swieten, to whom a statue was erected by the Empress Maria Theresa.]

mahogan"ii, leaves lanceolate-ovate, acuminate; racemes axillary, pubescent. Mahogany-tree. *S.*

SYMPHO'RIA. 5—1. (*Caprifolia.*) [From the Greek, signifying a cluster.]

glomera'ta, (r-y. Au ♄.) racemes axillary, capitate, glomerate; leaves opposite, ovate, on short petioles; flowers small, numerous; berries purple. 3-4 f. Sandy fields. Penn. to Car.

racemo'sa, (r. Ju. ♄.) racemes terminal; corolla bearded within; leaves elliptical, ovate, opposite; corolla pale red; berries white. 2-3 f. Snow-berry.

occiden"ta'lis, leaves very large; racemes drooping.

SYNAN"DRA. 13—1. (*Labiata.*) [From *sun*, together, and *aner*, stamens; so called because the anthers cohere.]

grandiflo'ra, (y-w. J. ♃.) leaves cordate, ovate, acuminate, upper ones sessile, clasping; lower ones sessile, sub-petioled; flowers solitary, sessile. 1 f. *S.*

SYRIN"GA. 2—1. (*Jasminea.*) [From a Turkish word, signifying pipe, because pipes were made from its branches.]

vulga'ris, (lilac, b-p. w. M. ♄.) leaves cordate; flowers in a thyrse. Ex.

per"sica, (Persian lilac, b. M. ♄.) leaves lanceolate, entire, and pinnatifid. Ex.

chinen"sis, (Chinese lilac, b. M. ♄.) branches rigid, mottled; leaves lanceolate. Ex.

TAGE'TES. 17—2. (*Corymbifera.*)

erec"ta, (African marygold, y. Ju. ☉.) leaves pinnate; leafets lanceolate, ciliate, serrate; peduncles 1-flowered, incrassate, sub-inflated; calyx angled. Ex.

pat"ula, (French marygold, y. Ju. ☉.) stem spreading; leaves pinnate; leafets lanceolate, ciliate-serrate; peduncles 1-flowered, sub-incrassate; calyx smooth. Ex.

TALI'NUM. 12—1. (*Portulaceæ.*)

teretifo'lium, (p. Ju. ♃.) leaves terete,

subulate, fleshy; cyme terminal, dichotomous, corymbose; flowers pedunculate, polyandrous. 4-10 i. Rocks. Penn. to Va.

parviflo'rum, small; leaves slender; stamens 5-10. Ark.

TAMARIN"DUS. 15—3. (*Leguminosa.*) [From the Arabic *tamarhindi*, or Indian date.]

in"dica, (tamarind,) leaves abruptly pinnate; leafets 16-18 pairs, downy, obtuse, entire; flowers lateral, yellow; pods brown. Ex.

TANACE'TUM. 17—2. (*Corymbifera.*) [A corruption of *athanasia*, an ancient name for tansey.]

vulga're, (tansey, y. Ju. ♃.) leaves doubly-pinnate, gash-serrate. Naturalized. Var. *cris"pum*,(double tansey,) leaves crisped and dense.

huronen"sis, (y. ♃.) flowers large, corymbed; ray-florets irregular, 4-5-cleft; leaves pseudo-bi-pinnate, gash-serrate, sub-tomentose beneath; pedicels thickened.

TAX"US. 20—15. (*Coniferæ.*)

canaden"sis, (yew, Ap. ♄.) leaves linear, distichus, revolute on the margin; receptacle of the staminate flowers globose. 4-8 f.

bacca'ta, (the common English yew,) leaves flat, dark green, smooth and shining above; flowers imbricated; berries scarlet.

TEPHRO'SIA. 16—10. (*Leguminosa.*) [From *tephros*, ash colored, alluding to the foliage.]

virginia'na, (goat's-rue, r. Ju. ♃.) erect, villose; leafets numerous, oblong-lanceolate, acuminate; raceme terminal, sub-sessile; legumes falcate, villose. 1 f. Dry woods.

hispid"ula, (r. M. ♃.) stem slender, very much divided, pubescent; leaves pinnate; leafets (11-15) elliptic, sub-retuse, mucronate, hairy beneath; racemes as long as the leaves, few-flowered; pods mucronate, slightly hispid. 2 f. *S.*

paucifo'lia, (r. Ju. ♃.) stem generally decumbent, very villose; leaves scattered, pinnate; leafets oval, cuneate at base, villose beneath; peduncles much longer than the leaves; few-flowered. *S.*

chrysophyl"la, (Ju.) prostrate, pubescent; leaves pinnate by fives, sub-sessile; leafets cuneate, obovate, obtuse, coriaceous, glabrous above, silky beneath; peduncles opposite the leaves, long, about 3-flowered; pods nearly straight. *S.*

el'egans, (r-p. ♃.) decumbent, sparingly pubescent; leaves sub-sessile; leafets (15-17) oblong-oval; peduncles filiform, few-flowered; segments of the calyx acuminate. Ala.

TEU'CRIUM. 13—1. (*Labiata.*) [From Teucer, who is said to have been its discoverer.]

canaden"se, (wood-sage, germander, r. Ju. ♃.) pubescent; leaves lance-ovate, serrate, petioled; stem erect; spikes whorled, crowded; bracts longer than the calyx. Var. *virgin"icum*, upper leaves sub-sessile; bracts about the length of the calyx. 1-3 f.

lancinia'tum, somewhat pubescent; leaves pinnately 5 parted; upper ones 3-parted; segments linear; flowers axillary, solitary, pedicelled; pedicels much shorter than the leaves.

36

beton"icum, has loose spikes of fragrant crimson flowers. *Ex.*

THA'LIA. 1—1. (*Orchidea.*) [In honor of John Thalius.]

dealba'ta, (p. Au. ♃.) spatha 2-flowered; leaves ovate, revolute at the summit; panicle white-pulverulent. *S.*

THALIC"TRUM. 12—12. (*Ranunculacea.*) [From *thallo*, to flourish.]

dio'icum, (meadow rue, w-r. M. ♃.) flowers diœcious; filaments filiform; leaves about 3-ternate; leafets roundish, cordate, obtusely lobed, glabrous; peduncles axillary, shorter than the leaves. 1-2 f.

pubes"cens, (w. Au. ♃.) leafets woolly, lobed, margin revolute, finely pubescent beneath.

cornu'ti, (g-y. Ju. ♃.) leaves decompound; leafets ovate, obtusely 3-lobed, glaucous beneath, with the nerves scarcely prominent; flowers mostly diœcious; filaments subclavate; fruit sessile, striate. 2-5 f. Wet grounds.

clava'tum, (♃.) leaves glabrous, without stipes; flowers monœcious; filaments clavate; pericarp compressed, with a very short style. *S.*

alpi'num, a dwarf species.

THAS"PIUM. 5—2. (*Umbelliferæ.*) [From the isle of Thaspia.]

actæifo'lium, (Ju. ♃.) leaves gash-biternate; segments oval, equally dentate; umbels sub-verticillate; lateral ones sterile. 3 f. Canada.

atropurpu'reum, (p. J. ♃.) radical leaves petioled, cordate, undivided; cauline ones gash-pinnate; segments 3 to 7, short petioled, ovate, oblong, all cartilaginous-dentate. 2-3 f.

THE'A. 12—1. (*Meliæ.*) [A Chinese name.]

bohe'a, (bohea tea, M. ♄.) flowers 6-petalled; leaves oblong-oval, rugose. From China and Japan.

vir"idis, (green tea, ♄.) flowers 9-petalled; leaves very long-oval. *Ex.*

THER"MIA. 10—1. (*Leguminosæ.*) [From *thermes*, temperature; a plant of warm climates.]

rhombifo'lia, (y. ♃.) leaves ovate-rhomboid, silky-pubescent beneath; stipules leaf-like, round, ovate, oblique, shorter than the petiole; flowers racemed. *S.*

THE'SIUM. 5—1. (*Eleagni.*) [From a Greek word signifying garland.]

umbella'tum, (false toad-flax, w. g. J. ♃.) erect; leaves oblong; umbels axillary, 3-5-flowered; peduncles longer than the leaves. 9-15 i.

THLAS"PI. 14—1. (*Cruciferæ.*) [From *thlao*, to break, so called because it appears broken.]

bursa-pasto'ris, (shepherd's-purse, w. M. ⊕.) hirsute; silicles deltoid, obcordate; radical leaves pinnatifid.

arven"se, (penny-cress, w. J. ⊕.) leaves oblong, sagittate, coarsely toothed, smooth; pouch sub-orbicular, shorter than the pedicel; its wings dilated longitudinally; flowers in a raceme. 1 f.

tubero'sum, (Ap. ⊕.) flowers large, rosaceous; stem 1-5 inches high, simple, pubes-

cent; upper leaves sessile, radical leaves long-petioled; root tuberous, pouch orbicular.

allia'ceum, (⊙.) leaves oblong, obtuse, dentate, glabrous; silicle sub-ovate, ventricose. Introduced.

THU'JA. 19—15. (*Coniferæ.*) [From *thuon*, odor, so called from its fragrant smell.]

occidenta'lis, (American arbor-vitæ, M. ♄.) branches ancipetal; leaves imbricated, in 4 rows, ovate-rhomboidal; strobiles obovate. Mountains. A small tree with very tough branches. Leaves resembling scales.

gigan"tea, leaves imbricate 4-ways, ovate, obtusish, closely incumbent, sub-equal; strobiles loose; scales obval, 200 feet high, and 12 feet in diameter.

articula'ta, produces the gum Sandarach; the wood is said to resist fire, and is also supposed to be the sandal-wood of the ancients.

THYM'US. 12—1. (*Labiatæ.*) [From *thum*, odor.]

vulga'ris, (thyme, b-p. J. ♃.) erect; leaves ovate and linear, revolute; flowers in a whorled spike. *Ex.*

serpyl'lum, (wild thyme, b-p. J. ♃.) stems branched, creeping; leaves elliptic-ovate, obtuse, flat, petioled, ciliate at base; flowers capitate. 4-8 i. Naturalized.

lanugino'sus, (lemon thyme, ♃. ♄.) stem creeping, hirsute; leaves obtuse, villose; flowers capitate. *Ex.*

grandiflo'ra, very ornamental. *Ex.*

THY'SANOCAR"PUS. 14—1. (*Cruciferæ.*) [From *thasanos*, fringe, and *karpos*, fruit, the pods having fringe on the edge.]

curvi'pes, flowers racemed, small; leaves mostly radical, pinnatifid; silicle pendulous; stem solitary, erect. West of Rocky Mountains.

oblongifo'lius, silicles nearly orbicular, wingless, hispid, with uncinate hairs; petals about half as long as the calyx; leaves oblong, toothed, densely and stellately hirsute. Oregon.

TIAREL'LA. 10—2. (*Saxifraga.*) [From *tiara*, an ornament for the head.]

cordifo'lia, (mitre-wort, w. M. ♃.) leaves cordate, acutely lobed, dentate; teeth mucronate; scape racemed; petals with long claws; flowers in a simple terminal raceme. Shady woods. 8-10 i.

menzie'sii, (♃.) leaves ovate, heart-shaped, acute, lobes short, dentate; cauline ones alternate, distant; raceme filiform, somewhat spiked; calyx tubular. 1 f.

trifolia'ta, (♃.) leaves ternate; leafets sub-rhomboid, serrate, pilose; racemes terminal; small corymbs of flowers alternate; calyx campanulate.

TIGA'REA. 11—1. (*Rosaceæ.*)

tridenta'ta, (y. Ju. ♄.) leaves crowded towards the ends of the branches, 3-toothed, villose above, hoary-tomentose beneath; flowers terminal, solitary. *S.*

TI'GRIDIA. 15—3. (*Iridea.*) [So called from its spotted appearance, resembling a tiger.]

ensifor"mis, (tiger flower,) spatha 2-leaved; two outer petals longer than the other four; leaves ensiform, nerved. Mexico.

TIL'IA. 12—1. (*Tiliaceæ*.) [From *ptelea*, the Greek name.]

gla'bra, (bass-wood, lime-tree, y-w. Ju. ♄.) leaves round-cordate, abruptly acuminate, sharply serrate, sub-coriaceous, glabrous: petals truncate at the apex, crenate; style about equalling the petals; nut ovate. Large tree. Wood soft and white. Leaves often truncate at the base.

pubes'cens, (y-w. Ju. ♄.) leaves truncate at the base, sub-cordate, oblique, denticulate-serrate, pubescent beneath; petals emarginate; nut globose, smooth. Var. *leptophyl'la*, leaves lax, serrate, very thin.

laxiflo'ra, (M. ♃.) leaves cordate, gradually acuminate, serrate, membranaceous, smooth; panicles loose; petals emarginate; styles longer than the petals; fruit globose. Near the seacoast.

heterophyl'la, (J ♄.) leaves ovate, at base oblique or equally truncate and cordate, serrate, white-tomentose beneath; fruit globose. *S.*

TILLAND"SIA. 6—1. (*Narcissi*.) [Named from Tillandsius, professor of Medicine at Albq.]

utricula'ta, (wild pine, bladder tillandsia, w.) leaves concave, broad, their base enlarged; panicle branching; flowers sessile; stamens longer than the corolla. 3 f. The leaves are often found containing nearly a pint of water. *S.*

usneoi'des, stem gray, diffuse, filiform, pendulous, branching. Parasitic. From its peculiar appearance, suspended from trees to which it has fastened itself, it is called *old man's beard*.

recurva'ta, (p.) leaves subulate, recurved; scape setaceous, longer than the leaves, generally 2-flowered at the summit. *S.*

TIPULA'RIA. 18—1. (*Orchideæ*.)

disco'lor, (w. Au.) leaf solitary, plaited, and longitudinally-nerved; flower in nodding racemes.

TOFIEL"DIA. 6—3.

pubes"cens, (p-w. Ju. ♃.) leaves sub-radical, ensiform, narrow, smooth, rachis and pedicels scabrous; spike oblong, interrupted; scape 18 i. Swamps.

glutino'sa, (♃.) scape and pedicels glutinous, scabrous; spike with a few alternate fascicles; capsule ovate, twice the length of the calyx.

glaber"rima, (w. Oc. ♃.) very glabrous; leaves linear, gladiate; flowers racemed; buds approximate, nearly whorled, 1-flowered. *S.*

gla'bra, (g-w.) scape terete; leaves linear, ensiform; spike oblong, short, dense; peduncled, solitary, angular; capsules membranaceous. 8-10 i. *S.*

TRADESCAN"TIA. 6—1. (*Junci*.) [From John Tradescant.]

virgin"ica, (spider-wort, b-p. M. ♃.) erect, branching; leaves lanceolate, elongated, glabrous; flowers sessile; umbel compact, pubescent. Cultivated. 1-2 f.

rosa, flowers smaller than the prece-

ding; inner segments rose-colored, longer than the outer.

TRA'GIA. 19—3. (*Euphorbeæ*.) [Named after a famous German herbalist.]

ramo'sa, stem herbaceous, pilose, very branching; leaves petioled, lance-ovate, sharply serrate, hirsute beneath, sub-cordate at the base; racemes axillary, filiform, few-flowered. 8 i.

u'rens, (Ju. ♃.) erect; leaves lanceolate, sessile, obtuse, sub-dentate at the apex; stem and branches pubescent. Var. *suboval lis*, leaves oblong-oval, sometimes wedge-form. Var. *lanceola'ta*, leaves lanceolate, sub-dentate, and entire. *S.*

urticifo'lia, (Ju. ☉.) stem erect, hirsute; leaves cordate, ovate, serrate, alternate, short-petioled. 12.18 i. *S.* Dry soils.

macrocar"pa, (Ju. ☉.) climbing, hispid; leaves deeply cordate, ovate, dentate. *S.*

TRAGOPO'GON. 17—1. (*Cichoraceæ*.) [From *tragos*, a goat, and *pogon*, beard, so called from its downy seed.]

porrifo'lium, (vegetable-oyster, goat-beard, salsify, p. Ju. ♂.) calyx longer than the rays of the corolla; the florets very narrow, truncate; peduncles incrassate. Ex.

pra'ten"sis, (go-to-bed-at-noon, y. ♂.) has large flowers, which close in the middle of the day, and a curious, feathery head of seeds. Ex.

TRE'POCAR"PUS. 5—2. (*Umbelliferæ*.)

æthu'sæ, (w.) umbels 5-rayed; fruits four times as long as broad; leaves many-cleft, with linear lobes. Arkansas.

TRIB'ULUS. 10—1. (*Butaceæ*.)

max"imus, (y. Ju.) leaves pinnate; leafets about 4-pairs, outer ones largest; pericarps 10-seeded, not spiny. 1-2 f.

triju'ga'tus, (y. ☉.) leafets in 3 pairs, terminal ones largest, pubescent beneath; capsules 5, 1-seeded, muricate, spineless.

TRI'CHO'PHYL"LUM. 17—2. (*Corymbiferæ*.) [From *thrix*, hair, and *phullon*, a leaf.]

lana'tum, (y. Ju. ♃.) woolly in all parts; leaves linear, pinnatifid above; peduncles elongated, 1-flowered; rays 2-toothed; akenes glabrous, 5-angled.

oppositifo'lium, (Ju. ♃.) decumbent, branching, short, hoary-pubescent; leaves opposite, palmate, 3-cleft; segments ligulate, simple, or divided; peduncle filiform, mostly dichotomous, scarcely longer than the leaves. 6-12 i. *S.*

TRI'CHOSTE'MA. 13—1. (*Labiatæ*.) [From *trichos*, hair, and *stema*, stamens.]

dichot"oma, (blue curls, b. Au. ☉.) leaves lance-ovate; branches flower-bearing, 2-forked; stamens very long, blue, curved. Var. *linea'ris*, somewhat pubescent; leaves linear. 6-12 i.

TRIENTA'LIS. 7—1. (*Lysimachiæ*.)

america'na, (chick wintergreen, w. Ju. ♃.) leaves lanceolate, serrulate, acuminate; petals acuminate. 3-6 i.

TRIFO'LIUM. 16—10. (*Leguminosæ*.) [From *tres*, three, *folium*, leaf.]

re'pens, (white-clover, w. M. ♃.) creeping; leafets ovate oblong, emarginate, serrulate; flowers in umbelled heads; teeth of the calyx sub-equal; legumes 4-seeded.

praten"se, (red-clover, r. M. ⚄.) ascending, smooth; leafets ovate, sub-entire; stipules awned; spikes dense-ovate; lower tooth of the calyx shorter than the tube of the corolla, and longer than the other teeth. 2-3 f.

arven"se, (rabbit-foot, w. J. ⚄.) heads very hairy, oblong-cylindrical; teeth of the calyx setaceous, longer than the corolla; leafets villose. narrow, obovate; banner deciduous. 6-12 i.

reflex"um, (r. J. ⚄.) pilose; stem ascending; leafets obovate; stipules oblique, cordate; heads globose; flowers pedicelled; at length reflexed. 12—18 i. Dry hills.

agra'rium, (y. J. ⚄.) stem ascending, with erect branches; leafets lanceolate-cuneate, 'obtuse, intermediate one sessile; stipules lanceolate, acute; heads oval, imbricate; banner deflexed. persistent; teeth of the calyx subulate, glabrous, unequal. 6-14 i. Sandy soils.

campes"tre, (y. J. ⚄.) stem sub-diffuse; branches decumbent; spike ovate. imbricate; banner deflexed, persistent; leafets lanceolate-ovate, intermediate one petioled.

stolo'nif'erum, (running buffalo-clover, w. J. ⚄.) stoloniferous, glabrous; lower leaves long-petioled; leafets obovate or cuneate, serrulate. retuse or emarginate at the apex; stipules membranaceous, broad-lanceolate; flowers in globose heads, pedicelled, erect, at length reflexed; segments of the calyx nearly equal, narrow, smooth, longer than the tube. 4-8 i.

procum"bens, (yellow clover. y. J. ⚄.) procumbent, pubescent; leafets oval; peduncles long, setaceous; racemes short; loments sub-orbicular. 2-3 f.

carolin"ia'num, (p-w. Ap. ⚄.) small, procumbent; leafets obcordate (the upper one only emarginate), ternate, hairy, dentate; stipules 2-cleft; heads capitate, peduncled, reflexed, few-flowered; corolla scarcely exserted; legumes 3-4 seeded. 3-10 i. S.

TRIGLO'CHIN. 6—3. (Junci.)

palus"tre, (arrow-grass, g. Ju. ⚄.) fruit 3 united capsules, nearly linear, attenuated at the base; scape very slender, 1 foot long; leaves fleshy, nearly as long as the scape; flowers, small, greenish, in a terminal spike. Marshes.

marati'mum, fruit of 6 united capsules, ovate-oblong. Salt marshes.

triand"rum, (Ju.) triandrous; flowers 3-cleft, short-pedicelled; leaves terete, linear. 6-9 i.

TRI'GONEL"LA. 16—10, (Leguminosæ.) [Alluding to its little triangular flower.]

fœ'num-grœ'cum, (fenu-greek, ⚄.) stem erect; leaves wedge-oblong; legumes sessile, solitary. straight, erectish, sub-falcate, acuminate.

seri'cea, (y. Ju. ⚄.) leaves ternate, sessile, oblong, acute. silky villose; peduncles axillary, 1-flowered, longer than the leaf; flowers 1-bracted; divisions of the calyx linear; legume glabrous, very long. S.

TRIL"LIUM. 6—3. (Asparagi.) [From triles, triple.]

pen"dulum, (nodding wake-robin, w. M.

⚄.) peduncles erect, with the flower a little nodding; petals ovate, shortly acuminate, spreading, flat, longer than the calyx; leaves rhomboid, acuminate, sessile.

erec"tum, (false wake-robin, p. w-y. M. ⚄.) peduncles erect or erectish, with the flowers a little nodding; petals ovate, acuminate, spreading, equalling the calyx; leaves rhomboid, acuminate. Var. atropurpu'reum, petals large, dark-purple. Var.a'bum, petals smaller, white; germ red. Var. fla'vum, petals yellow; both petals and calyx leaves longer and narrower. 12.18 inches high; leaves often 3-4 inches broad; peduncles about 3 inches long. 9-16 i.

ses"sile, (p. Ap. ⚄.) leaves sessile, broad-ovate, acute; flowers closely sessile; petals lanceolate-ovate. very acute, alternate at base, erect, as long as the recurved calyx; stem smooth. 8-10 i. Leaves clouded with dark-green. Shady woods.

viri'de, leaves solitary, with whitish spots on the upper surface; petals dark-green.

pic"tum, peduncle somewhat erect; leaves ovate, acuminate, rounded at the base. abruptly contracted into a short petiole; flowers white, with purple veins near the base.

cer"nuum, (w. M. ⚄.) peduncle recurved; petals lanceolate, acuminate, flat, recurved, as long as the calyx; leaves rhomboid, on short petioles; flowers small, berries red. 12-18 i.

grandiflo'rum, peduncle a little inclined, nearly erect; flower solitary; petals spatulate, connivent at the base, much longer than the calyx; leaves broadly rhomboid, ovate, sessile. abruptly acuminate. Rocky banks of streams. Flower much larger than in any of the preceding species, varying from white to rose-color; stem 8-12 i.

petiola'tum, (p. J. ⚄.) leaves long-petioled, lance-oval, acute; flowers sessile, erect; petals lance-linear, erect, a little longer than the calyx.

pusil"lum, (dwarf wake-robbin, r. M. ⚄.) leaves oval-oblong, obtuse, sessile, peduncle erect; petals scarcely longer than the calyx.

obova'tum, (r. w. ⚄.) leaves rhomb-ovate, acuminate, close sessile; peduncles erect; petals obovate, obtusish, flat, spreading.

stylo'sum, (w-r.) slender; leaves sub-petioled, lance-oval, acute at both ends; peduncle much shorter than the flower, recurved; petals undulate, expanding, oblong-obtuse, larger than the calyx; germ styliferous; style 1. 8-10 i. S.

nervo'sum, (r-w. M.) leaves lanceolate and ovate, acute at each end, .membranaceous, nerved; peduncle recurved; petals lance-oblong, larger than the calyx. 6-8 i. S.

ova'tum, (p. Ap. ⚄.) leaves ovate, gradually acute, closely sessile; peduncle erect; petals oblong, acute, expanding, a little longer than the linear sepals. S.

TRIOS"TEUM. 5—1. (Caprifolia.)

perfolia'tum, (fever-root, p. J. ⚄.) leaves connate, spatulate, lanceolate, acuminate, pubescent beneath, margin undulate; flowers 1-3, in the axils of the leaves, sessile; berries purple, or yellow; the root is medicinal. Rocky woods. 2-3 feet high. N.Y. to Car.

angustifo'lium, (y. Ju. ♃.) stem hairy; leaves sub-connate, lanceolate, acuminate; peduncles opposite, 1-flowered. 2-3 f. *S.*

TRIPHO'RA. 18—1. (*Orchideæ.*) [From the Greek, signifying to bear three flowers.]

pen"dula, (p. 8. ♃.) root tuberous; stem leafy, about 3-flowered at the summit; leaves ovate, alternate; flowers pedunculate; stems often in clusters. 4-6 i. Roots of trees.

TRIP"SACUM. 19—3. (*Gramineæ.*)

dactyloi'des, (sesame grass, J. ♃.) spikes numerous (3-4), aggregate; florets staminate near the summit, pistillate below; spike large. Var. *monosta'chyon*, spike solitary.

cylin"dricum, spike solitary, cylindrical, separating into short joints; flowers all perfect.

TRIT"ICUM. 3—2. (*Gramineæ.*) [From *tero*, to thresh.]

hyber"num, (winter-wheat, J. ♂.) calyx glume 4-flowered, tumid, even, imbricate, abrupt, with a short compressed point; stipule jagged; corollas of the upper florets somewhat bearded. There are several varieties of this species which were introduced by culture. Ex.

æsti'vum, (summer wheat, J. ☉.) glumes 4-flowered, tumid, smooth, imbricated, awned. Considered a variety of the *hyber"num*.

compos"itum, (Egyptian wheat,) spike compound; spikelets crowded, awned. Few species of wheat, but many varieties.

tri'colo'rum, stem slender, weak; flowers red. black, and yellow. Ex.

re'pens, spikelet oblong, 5-flowered; glumes subulate, many-nerved; florets acuminate; leaves flat; root creeping. Fields. A troublesome weed.

pauciflo'rum, spike erect, simple; spikelets about 2-flowered: culm terete, simple, leafy, striate, smooth; leaves somewhat glaucous, ribs and margin scabrous. 2 f.

TROL"LIUS. 12—12 (*Ranunculceæ.*) [From the German, signifying to roll; so called from the roundness of the flower.]

america'nus, (globe-flower, y. M. ♃.) leaves palmate; sepals 5-10, spreading; petals 5-10, shorter than the stamens; flowers large, terminal; resembles a ranunculus. Wet grounds.

TROPÆO'LUM. 8—1. (*Gerania.*) [From *tropæon*, a warlike trophy.]

ma'jus, (nasturtion. Indian cress, y. and r. Ju. ☉. and ♃.) leaves peltate. sub-repand; petals obtuse, some of them fringed. Ex.

peregri'num, (canary-bird flower,) flowers numerous, pale yellow. Ex.

TROXI'MON. 17—1. (*Cichoraceæ.*) [A Greek word, signifying eatable.]

glau'cum, (y. Ju. ♂.) scape 1-flowered; leaves lance-linear, flat, entire, glaucous; divisions of the calyx imbricate, acute, pubescent.

TU'LIPA. 6—1. (*Liliaceæ.*) [The name is said to be of Persian origin, and to signify a turban.]

suave'olens, (sweet tulip, M. ♃.) small; stem 1-flowered, pubescent; flowers erect; petals obtuse, glabrous; leaves lance-ovate. Ex.

gesnria'na, (common tulip, M. ♃.) stem 1-flowered, glabrous; flower various-colored, erect; petals obtuse, glabrous; leaves lance-ovate. Ex. The various kinds of tulips which are cultivated, are only varieties of the *gesneria'na*.

sylves"tris, (y. wild French tulip,) flowers very fragrant.

præ'cox, (Van Thol's tulip,) a dwarf species, flowers generally in March or April.

TUL'LIA. 13—1. (*Labiatæ.*) [In honor of Prof. Tully, of New Haven, Conn.]

pycnan"themoïdes, (false mountain mint, r-p. Au. ♃.) leaves tapering to the base, remotely toothed, ovate, acuminate, hoary above and glaucous beneath; bracts of the striate calyx subulate. 2-3 f. Tennessee.

TURRI'TIS. 14—2. (*Cruciferæ.*) [From *turris*, a tower.]

ova'ta, (w. M. ♂.) leaves rough, radical ones ovate, toothed, cauline ones clasping.

gla'bra, erect; radical leaves petioled, dentate, upper ones broad-lanceolate, sagittate, glabrous, semi-amplexicaul, glaucous; legume narrow-linear, stiffly erect; petals scarcely longer than the calyx. Naturalized about New Haven. Hudson's Bay.

TUSSILA'GO. 17—2. (*Corymbiferæ.*) [From *tussis*, a cough, and *ago*, to drive away; so called on account of its medicinal properties.]

farfa'ra, (colt's foot, y. Ap. ♃.) scape single-flowered, scaly; leaves cordate, angular, toothed, downy beneath. The flower appears long before the leaves. 4-6 i.

frig'ida, (y. M. ♃.) thyrse fastigiate, many-flowered, bracteate; leaves roundish cordate, unequally toothed, tomentose beneath. 5-10 i. Mountains.

sagitta'ta, (♃.) thyrse ovate, fastigiate; leaves radical, oblong, acute, sagittate, entire; lobes obtuse.

TY'PHA. 19—3. (*Typhæ.*)

latifo'lia, (cat-tail, reed-mace, Ju. ♃.) leaves linear, flat, slightly convex beneath, staminate and pistillate aments close together. Wet. 4-6 i.

UDO'RA. 20—9. (*Hydrocharides.*)

canaden"sis, (w. Au. ♃.) leaves whorled, in threes and fours, lanceolate, oblong or linear, serrulate; tube of the perianth filiform; stem submersed, dichotomous. Still waters. Ditch-moss. Can. to Vir.

U'LEX. 16—10. (*Leguminosæ.*)

europe'us, (furze M. ♄.) leaves lance-linear, villose; bracts ovate; branchlets erect; *stric'ta*, (Irish furze, y.) without spines. 8-10 f. Ex.

na'na, (y.) seldom exceeds two feet in height. Gravelly soils. Ex.

UL'MUS. 5—2. (*Amentaceæ.*)

america'na, (elm, white-elm, g-p. Ap. ♄.) branches smooth; leaves oblique at the base, having acuminate serratures a little hooking; flowers pedicelled; fruit fringed with dense down. Var. *pen'dula*, has hanging branches and smoothish leaves. 40-70

36*

f. Flowers appear before the leaves, a magnificent tree.

ful"va, (slippery-elm, M. Ap. ♄.) branches scabrous, white; leaves ovate-oblong; very acuminate, pubescent on both sides; buds tomentose, with a thick tawny wool; flowers sessile, smaller than the white-elm; leaves larger; stamens often 7. The mucilage of the inner bark medicinal.

nemoral"is, (river-elm, Ap. ♄.) leaves oblong, somewhat glabrous, equally serrate, nearly equal at base; flowers sessile.

racemo'sa, (♄.) flowers in racemes; pedicels in distinct fascicles united at their bases; leaves ovate, acuminate, auriculate on one side, doubly serrate, glabrous above, minutely pubescent beneath; stamens 7-10; stigmas 2, recurved.

ala'ta, (whahoo, Mar. ♄.) branches on each side winged with a cork-like bark; leaves nearly sessile, oblong-oval, acute, doubly serrate, nearly equal at base; fruit pubescent, ciliate. 30 f. *S.*

UL"VA. 21—4. (*Algæ.*)
lin"za, frond lance-linear; margin undulate-crisped; about an inch broad, tapering at the base, green. Seashore.

URASPER"MUM. 5—2. (*Umbelliferæ.*) [From *ours*, a tail, and *sperma*, *seed*.]
clayto'ni, (sweet cicely, J. ♃.) leaves compound, hairy; leafets gash-toothed; umbels axillary and terminal, about 5-rayed; style as long as the villose germ, filiform, reflexed. 2 f.

URE'DO. 21—6. (*Fungi.*) [From *uro*, to burn, on account of its burnt color.]
linea'ris, (yellow grain-rust, J. ⊙.) linear, very long, stained yellow, at length but obscurely colored. On the culms and leaves of barley, oats, rye, wheat, &c.

UR"TICA. 19—4. (*Urticeæ.*) [From *urendo*, burning; on account of the sensation it causes.]
dio'ica, (common nettle, J. ♃.) leaves opposite, cordate, lance-ovate, coarsely serrate; flowers diœcious; spikes panicled, glomerate in pairs, longer than the petioles. 2-3 f.

pu'mila, (rich-weed, Ju. ⊙.) leaves opposite, ovate, acuminate, 3-nerved, serrate; lower petioles as long as the leaves; flowers monœcious, triandrous, in clustered corymbs, shorter than the petioles; stem succulent, almost transparent. 6-12 i. Wet grounds.

u'rens, (stinging nettle, J. ⊙.) stem hispid; leaves opposite, elliptic, about 5-nerved, acutely serrate; spikes glomerate, in pairs. 12-14 i.

canaden"sis, (Canada nettle, Ju. ♃.) leaves cordate-ovate, acuminate, hispid on both sides; panicles axillary, mostly in pairs, divaricately branched; the lower staminate ones longer than the petioles; upper pistillate ones elongated; stem hispid, stinging. 5-6 f. Var. *divarica'ta*, leaves smooth; panicles solitary, spreading. 4-6 f.

chamædroi'des, (Mar. ⊙.) stem glabrous; leaves opposite, sub-sessile, ovate, serrate, strigose beneath; clusters of flowers axil-

lary, sessile, sub-globose, reflexed; prickles stimulant, white. 4-6 i. *S.*

reticula'ta, (r-y.) leaves deep green. A native of Jamaica.

US"NEA. 21—5. (*Filices.*)
plica'ta, frond pendulous, smooth, pale; branches lax, very branching, sub-fibrous; the extreme ones capillary; receptacles flat, broad, ciliate; the hairs very slender and long. On trunks and branches of trees; most common on dry, dead limbs of evergreens, from which it often hangs in long, green locks.

UTRICULA'RIA. 2—1. (*Scrophularia.*) [From *utriculus*, a little bladder.]
vulga'ris, (bladder-wort, y. Au. ♃.) floating; stem submerged, dichotomous; leaves many-parted, margins bristly; scape 5-9-flowered; upper lip of the corolla entire, broad, ovate; spur conical, incurved; flowers in racemes. Ponds.

stria'ta, floating; scape 2-6-flowered; root furnished with air-vessels; corolla large, yellow striate with red; spur much shorter than the lower lip.

purpu'rea, scapes axillary, generally 2 or 3 inches long; flowers purple. Ponds on mountains. Mass. to Flor.

infla'ta, (y. Au. ♃.) radical leaves verticillate, inflated, pinnatifid at their extremities; lower lip of the corolla 3-lobed; spur deeply emarginate. Ponds.

stria'ta, (y. J. ♃.) floating; scape 2-6-flowered; upper lip of the corolla ovate-round, sub-emarginate, margin waved; lower lip 3-lobed, sides reflected; spur straight, obtuse, shorter than the lower lip. Swamps.

gib"ba, (y. Ju. ♃.) floating; scape mostly 2-flowered; spur shorter than the lower lip of the corolla, obtuse, gibbous in the middle. 1-3 i. Ponds.

cornu'ta, (y. Au. ♃.) scape rooting, erect, rigid; flowers 2-3, sub-sessile; inferior lip of the corolla very wide 3-lobed; spur very acute, lengthened out longer than the corolla. 10-12 i. Wet rocks.

persona'ta, (y. ⊙.) scape rooting, many-flowered; upper lip of the corolla emarginate, reclined; lower one small, entire, palate large; spur linear-subulate, acutish. Bogs.

seta'cea, (y. J. ♃.) scape rooting, filiform; upper lip of the corolla ovate, lower one deeply 3-lobed; spur subulate, entire. 3-6 i. Swamps.

inte'gra, (y. ⊙.) floating; scape 1-2-flowered; upper lip of the corolla sub-3-lobed; lateral lobes sub-involute; lower lip entire; spur nearly equalling the lower lip. *S.*

biflo'ra, (y. Ju.) spur subulate, obtuse, about as long as the lower lip; scape about 2-flowered; leaves setaceous. *S.*

UVULA'RIA. 6—1. (*Liliaceæ.*) [From *uvula*, a membrane of the throat, the soreness of which this is supposed to heal.]
perfolia'ta, (bell-wort, y. M. ♃.) leaves perfoliate, oval-obtuse (lance-linear or oval-oblong in the young state); corolla bell-liliaceous, scabrous or granular within; anther cuspidate. 8-12 i.

sessilifo'lia, (γ. M. ♃.) stem smooth; leaves sessile, oval-lanceolate, glaucous beneath; petals flat, smooth within; capsules stiped. 6-12 i.

grandiflo'ra, leaves perfoliate, oblong, acute; perianth smooth within, anthers without awns; nectaries nearly round; pistil shorter than the stamens; whole plant larger than the preceding species.

puber"ula, leaves colored alike on both sides, oval, rounded at the base, somewhat amplexicaul; capsule sessile, ovate. 8-12 i. S.

VACCIN"IUM. 10—1. (Ericeæ.) [A corruption of baccinium, a berry.]

resino'sum, (whortleberry, a. p. M. ♃.) leaves slender, petioled, oblong-oval, mostly obtuse, entire, bedewed with resinous beneath; racemes lateral, 1-sided; short, somewhat bracted; corolla nic, 5-cornered; berries black. riety has a yellowish green, and has a reddish yellow corolla. 1-4 f.

tbo'sum, (high whortleberry, w. M. wer-bearing branches almost leaves oblong-oval, acute at each end; ng leaves pubescent; racemes short, bracted; corolla cylindrical-ovate. s and wet woods; 4 to 8 feet high. large, black, sub-acid.

o'sum, whortleberry; leaves ovate-sprinkled with resinous dots, glaumeath; racemes lateral, loose, bracdicels long, filiform; corolla ovate, ulate; berries large, black, sweet, later than the other species.

tylea'nica, low blue-berry; branch, angular; leaves sessile, shining; 2-18 i. high, much branched; flow. red, 6 to 8 in a fascicle; berries lue, somewhat glaucous. Dry hills. Geo.

n"eum, (J. ♃. squaw whortleberry,) anching; leaves glaucous beneath; corolla campanulate, spreading; anthers exserted. 2-3 f. Berries large, greenish white. Dry woods. Car. to Flor.

vitis"idea, (bilberry, w-r. M. ♃.) evergreen; low, leaves punctate beneath, obovate, emarginate, revolute, sub-serrulate; racemes terminal, nodding.

dumo'sum, (bush-whortleberry, w. M. ♃.) branchlets, leaves and racemes sprinkled with resinous dots; leaves obovate, cuneate at base, mucronate, entire; racemes bracted; pedicels short, axillary, sub solitary; corolla campanulate; segments rounded; anthers included. Var. hirtel"lum, racemes and calyx pilose; berries hispid. 12-18 i. Pine woods.

ligustri'num, (p. r. J. ♃.) branches angular; leaves sub-sessile, erect, lanceolate, mucronate, serrulate; fascicles gemmaceous, sessile; flowers nearly sessile; corolla urceolate. Dry woods.

uligino'sum, (b. r-w. Ap. ♃.) leaves obovate, obtuse, entire, smooth above, veined and glaucous beneath; flowers sub-solitary, octandrous; corolla short, ovate, 4-cleft. Var. alpi'num, (winter-green whortleberry,) leaves entire, obovate; flowers sub-solita-

ry; berries oblong, crowned with the style. 1-2 f. High mountains.

tenel"lum, (dwarf whortleberry, r-w. M. ♃.) racemes bracted, sessile; corolla ovate cylindric; leaves oblong-elliptic, sub-cuneiform, serrulate, nearly smooth. White hills.

obtu'sum, (♃.) evergreen, creeping; leaves elliptic, round-obtuse at each end, mucronate, entire, glabrous, coriaceous, small; peduncles axillary, solitary, 1-flowered.

ova'tum, (M. ♃.) evergreen; leaves ovate, acute, revolute, serrate, smooth, coriaceous, petioled; racemes axillary and terminal, bracted, short; corolla cylindric; calyx acute.

ni'tidum, (r. M. ♃.) erect; branches distichous; leaves nitid, oval-obovate, acute at each end, glabrous, serrate; racemes terminal, corymbed, bracted, nodding; corolla cylindric; leaves perennial. S.

myrtifo'lium, (♃.) creeping, very glabrous; leaves petioled, oval, lucid, revolute, denticulate; clusters axillary, nearly sessile; corolla campanulate with 5 short teeth; anthers unawned at the back. Berries small, pedicelled, black; leaves perennial. S.

arbo'reum, (farkleberry, w. M. ♃.) leaves broad-lanceolate, oval, serrulate, mucronate, shining above; pubescent beneath; racemes leafy; corolla campanulate; anthers awned.

crassifo'lium, (r. J. ♃.) diffuse; branches ascending; leaves oblong-lanceolate, acute at each end, serrate, rigid, glabrous; racemes terminal, corymbed, bracted, few-flowered; flowers nodding; calyx appressed; corolla campanulate, deeply 5-parted. Leaves evergreen.

myrsini'tis, (p. m. ♃.) erect, branching; leaves small, sessile, ovate, mucronate, serrulate, lucid above, glandular punctures beneath; racemes short, bracted, axillary, and terminal; corolla urceolate. Berries black; leaves perennial. Var. lanceola'tum, leaves lanceolate, acute at each end. Var. obtu'sum, leaves roundish, obovate. S.

gale'zans, (r-w. M. ♃.) leaves sessile, cuneato-lanceolate, serrulate, pubescent; fascicles sessile; corolla urceolate; stamens included, awnless; style exserted. S.

VALERIA'NA. 2—1. (Dipsaceæ.) [From Valerias, who first described it.]

dio'ica, (r. J. ♃.) glabrous, radical leaves sub-spatulate, ovate, entire, very long, petioled; cauline ones few, pinnatifid; divisions lanceolate, entire.

phu, cauline leaves pinnate, radical ones undivided; stem smooth. The Valerian of medicine. Ex.

VALERIANEL"LA. 2—1. (Dipsaceæ.) [A diminutive of Valeriana; from which this genus was separated.]

rhombicar"pa, (b-w. J. ☉.) stem dichotomous above, ciliate-angled; radical leaves obovate; cauline ones spatulate-oblong, ciliate; upper leaves toothed at the base; involucrum ciliate, scarious at the apex;

fruit compressed, rhomboidal. 4-6 i. Meadows. Md.

VALLISNE'RIA. 20—2. (*Hydrocharides.*) [From Anthony Vallisneri.]

spira'lis, (tape-grass, w. Au. ♃.) leaves floating, linear, obtuse, serrulate at the summit, tapering at the base, radical; peduncle of the pistillate flower long; of staminate short, erect. Grows in still water.

VERA'TRUM. 6—3. (*Junci.*)

vir''ide, (Indian poke, white hellebore, g. J. y. ♃.) racemes paniculate; bracts of the branches oblong-lanceolate, partial ones longer than the sub-pubescent peduncles; leaves broad-ovate, plaited. 3-5 f. Meadows and swamps. Abundant in the valleys of the Green Mountains.

angustifo'lium, (g-y. J. ♃.) flowers diœcious; panicle simple; petals linear; leaves very long, linear-keeled. Mountains.

parviflo'rum, (g. Ju.) leaves oval, lanceolate, flat, glabrous; panicle slender; spreading; petals acute at each end, staminiferous. *S.*

VERBAS''CUM. 5—1. (*Solanea.*) [From *barbascum*, on account of its being bearded.]

thap'sus, (mullein, y. J. ♂.) leaves decurrent, downy both sides; stem generally simple, though sometimes branched above; flowers in a cylindric spike. 3-6 f.

blatta'ria, (moth mullein, sleek mullein, w-y. J. ♃.) leaves glabrous, tooth serrate; lower ones oblong-obovate; upper ones heart-ovate, clasping; pedicels 1-flowered, in a terminal, panicled raceme. Var.*al''ba*, leaves toothed; flowers white. Var.*lu'tea*, leaves doubly serrate; flowers yellow. 2-3 f.

lychni'tis, (y. J. ♂.) stem angular; leaves oblong, cuneate, white-downy beneath; spikes lax, lateral and terminal.

phœ'nicum, (♃.) a very handsome species.

VERBE'NA. 12—1. (*Labiata.*) From *herba'na*, a name of distinction for herbs used in sacred rites. The vervain in former times was held sacred, and employed in celebrating sacrificial rites.]

hasta'ta, (vervain, simpler's joy, p-w. Ju. ♃.) erect, tall; leaves lanceolate, acuminate, gash-serrate; lower ones sometimes gash-hastate; spikes linear, panicled, subimbricate. Var. *pinnatifi'da*, has the leaves gash-pinnatifid, coarsely toothed. Var. *oblongifo'lia*, leaves lance-oblong, deeply serrate, acute; spikes filiform, panicled. 2-5 f.

urticifo'lia, (nettle-leaf vervain, w. Ju. ♃.) erect, sub-pubescent; leaves oval, acute, serrate, petioled; spikes filiform, loose, axillary, terminal; flowers tetrandrous. 2-3 f.

spu'ria, (b. Au. ☉.) stem decumbent, branched, divaricate; leaves laciniate, much divided; spikes filiform; bracts exceeding the calyx. 1-2 f. Sandy fields.

angustifo'lia, (b. J. ♃.) erect, mostly simple; leaves lance-linear, attenuate at the base, remotely toothed, with elevated veins; spikes filiform, solitary, axillary and terminal.

stric''ta, (b. Ju. ♃.) stems rigidly erect; leaves sessile, obovate, serrate, sub-tomentose, very hirsute; spikes straight, imbricate, fascicled. Var. *mol''lis*, (p.) stem simple, terete, villose; leaves ovate, acutish, unequally gash-toothed; teeth acute, hirsute above, soft villose beneath; spikes dense-flowered, terminal, somewhat in threes. Perhaps a distinct species. *S.*

panicula'ta, (p. Ju. ♃.) scabrous; leaves lanceolate, coarsely serrate, undivided; spikes filiform, imbricate, corymb-panicled. 4-6 f. *S.*

bipin''nûtifi'da, (b. J. ♃.) hirsute; leaves 3-cleft, bipinnatifid; divisions linear; nuts deeply punctate. *S.*

carolin''ia'na, (p. J. ♃.) scabrous; leaves oblong-obovate, obtuse, unequally serrate, tapering at base, sub-sessile; spikes very long, filiform; flowers distinct. 2 f. *S.*

melin''dres, (common scarlet verbena,) stem prostrate. Introduced from S. America.

tweedia'na, stem erect; flowers crimson.

sabin''ii, stem prostrate; flowers lilac; one variety has white flowers. Ex.

sulphu'rea, stem prostrate; flowers yellow. Ex.

VERBESI'NA. 17—2. (*Corymbiferæ.*)

siegesbeck''ia, (y. Au. ♃.) stem winged; leaves opposite, ovate-lanceolate, acuminate at each end, acutely serrate; corymbs brachiate; branches irregularly many-flowered at the summit; root creeping; stem erect, 4-6 f., 4-winged; ray-florets 3-toothed. Shady woods. Penn. to Car. Crownbeard.

virgin''ica, (w. Ju. ♃.) stem narrowwinged; leaves alternate, broad, lanceolate, sub-serrate; corymb compound; involucrum oblong, pubescent, imbricate; ray-florets 3 or 4; seeds four-angled. 3-6 f.

sinua'ta, (w. Oc. ♃.) stem pubescent, striate; leaves alternate, sessile, sinuate, attenuate at base; flowers corymbed; involucrum imbricate; ray-florets 3-5. 4-6 f. *S.*

VERNO'NIA. 17—1. (*Corymbiferæ.*)

noveboracen'sis, (flat-top, p. Au. ♃.) leaves numerous, lanceolate, scabrous, serrulate; corymbs fastigiate; scales of the involucre filiform at the summit; flowers in a large terminal corymb; stem 4-6 f. Branching towards the top. Wet grounds. Can. to Car.

tomento'sa, (p. Au. ♃.) stem tomentose above; leaves long, narrow, lanceolate, acutely serrate, slightly scabrous above, hoary tomentose beneath; corymb fastigiate; scales of the involucrum filiform at the apex. 3-5 f.

angustifo'lia, (p. Ju. ♃.) stem simple, somewhat scabrous; leaves numerous, long, linear, nearly entire; corymb sub-umbelled, scales of the involucrum rigid, mucronate. 3 f. *S.*

altis''sima, (p. Au. ♃.) stem glabrous; leaves lanceolate, serrate, somewhat scabrous; involucrum small, hemispheric; scales ovate, acute, ciliate, unawned, closely appressed. Var. *margina'ta*, (p.) leaves narrow-lanceolate; glabrous, very entire; corymb fastigiate; involucrum hemispheric-turbinate; scales arachnoid-ciliate, a little mucronate. Perhaps a distinct species. *S.*

scaberri'ma, (p. Ju. ♃.) stem simple; leaves lance-linear, denticulate, scabrous, hairy; corymb sub-umbelled; scales of the involucrum lanceolate, mucronate. 2 f. *S.*

fascicula'ta, (Au. ♃.) leaves long, linear, sparingly serrate; flowers corymbed, approximate; involucrum ovoid, smooth; scales unarmed. *S.*

VERON"ICA. 2—1. *(Pediculares.)*

officina'lis, (speedwell, b. M. ♃.) spikes lateral, peduncled; leaves opposite, obovate, hairy; stem procumbent, rough-haired. 9.12 i.

anagal'lis, (brook pimpernel, b. J. ♃.) racemes opposite, long, loose; leaves lanceolate, serrate; stem erect. 12-18 i.

beccabun"ga, (brook-lime, b. J. ♃.) racemes opposite; leaves oval-obtuse, subserrate, glabrous; stem procumbent, rooting at the base. Probably a variety of the last. 9-18 i.

serpyllifo'lia, (b. M. to Au. ♃.) racemes spiked, many-flowered; leaves ovate, slightly crenate; capsules broad-obcordate; stems procumbent, 3-5 inches long, sometimes creeping; flowers pale, in a long terminal spike, or raceme. Meadows. Introduced.

scutella'ta, racemes axillary, alternate; pedicels divaricate; leaves linear, dentate-serrate; stem erect, weak. 6.12 i. Flowers flesh-colored, racemed. Moist places.

agres"tis, flower peduncled; leaves on short petioles, cordate-ovate, deeply serrate; segments of the calyx ovate-lanceolate; stem procumbent; flowers small, pale blue, axillary, solitary. Sandy fields. Can. to Car.

alpi'na, leaves opposite, lance-oblong, acute, toothed; corymb terminal; calyx hispid.

arven"sis, (field veronica, w-b. M. ☉.) stem procumbent; flowers solitary; lower leaves opposite, petioled, cordate-ovate, serrate; floral leaves alternate, lanceolate, sessile, longer than the peduncles; segments of the calyx unequal. Var. *renifor"mis,* leaves reniform, entire, sessile. 3-8 i.

hederifo'lia, (ivy speedwell, w-b. M. ☉.) flowers solitary; leaves as long as the petioles, round-cordate, 5-lobed; upper ones 3-lobed; segments of the calyx cordate, ciliate, acute; stem procumbent.

peregri'na, (Maryland veronica, w. Mar. ☉.) flowers solitary, sessile; leaves oblong, obtuse, toothed and entire; lower ones opposite, upper ones alternate, linear-lanceolate. 4-8 i.

renifor"mis, (b. J. ♃.) stem creeping; spikes peduncled; peduncles lateral, axillary, 1-bracted; leaves opposite, long-petioled, heart-reniform, gash-crenate. *S.*

VESICA'RIA. 14—1. *(Cruciferæ.)*

didymocar"pa, white-downy, down-stellated; calyx equal; silicles large, inflated, in pairs; radical leaves broad-ovate-spatulate; the rest lanceolate, sub-entire. Rocky Mountains, and West.

VEXILLA'RIA. 16–-10. *(Leguminosæ.)* [From *vexillum,* a banner.]

virgin"ia'na, (butterfly weed. p. Ju. ♃.) stem twining, and with the ovate leafets glabrous or sub-pubescent; peduncle 1-4-flowered; calyx 5 parted, about as long as the lanceolate-bracts; legume linear. compressed; flowers larger than those of any other North American papilionaceous plant. Hedges. Penn. to Car.

maria'na, stem climbing, glabrous; leaves ternate; leafets lance-oval; peduncles solitary, 1-3-flowered; calyx tubular-campanulate, glabrous, much longer than the bracts; legume torulose. Banks of streams. Flowers large, pale blue.

plumi'era, climbing; leaves ternate, ovate-oblong, acuminate; calyx campanulate, shorter than the ovate bracts; corolla large, silky. *S.*

VIBUR"NUM. 5—3. *(Caprifolia.)*

oxycoc"cus, (high cranberry, r-w. J. ♄.) leaves 3-lobed, acute at the base, 3-nerved; lobes divaricate, acuminate, remotely and obtusely toothed; petioles glandular; cymes radiate; flowers of the ray large, abortive. Small shrub with spreading branches; fruit large, red, acid. 5-8 f. Mountain woods.

lantanoi'des, (hobble-bush. w. M. ♄.) branches flexuose, often procumbent; leaves orbicular-ovate, abruptly acuminate, unequally serrate; nerves and petioles pulverulent-tomentose; cymes closely sessile; fruit ovate. 4-8 f. Fruit red, black when fully ripe. Mountains.

pyrifo'lium, (w. J. ♄.) smooth; leaves ovate-oblong, acute. crenate, serrate; petiole naked; cymes sub-pedunculate; fruit oblong-ovate. 5-10 f.

lenta'go, (sheep-berry, w. J. ♄.) glabrous; leaves broad-ovate, acuminate, hook-serrate; petioles margined, undulate; cymes sessile. The branches, when full grown, often form a fastigiate top. Berries black, oval, and pleasant tasted; somewhat mucilaginous. 8-15 f.

acerifo'lium, (maple guelder-rose, dock-mackie, w, J. ♄.) leaves heart-ovate, or 3-lobed, acuminate, sharp serrate, pubescent beneath; cymes long peduncled; stem very flexible; leaves broad and sub-membranaceous. 4-5 f. Leaves applied to inflamed tumors by the Indians.

nu'dum, w. M. ♄.) glabrous; leaves oval, sub-entire; margins revolute; petioles naked; cymes peduncled; flowers small, crowded. Berries black. 8-12 f.

pubes"cens, (w. J. ♄.) pubescent; leaves short-petioled, ovate, acuminate, dentate-serrate, villose beneath; cymes peduncled; fruit oblong. 6 f. High grounds.

cassinoi'des, (J. ♄.) glabrous; leaves lanceolate, acute at each end, crenate; margins slightly revolute; petioles keeled, without glands. Swamps.

denta'tum, (arrow-wood, w. M. ♄.) smoothish; leaves long-petioled, orbicular-ovate, dentate-serrate, plicate, glabrous both sides; cyme peduncled; fruit sub-globose. Fruit blue. 8 f.

obova'tum, (M. ♄.) glabrous; branches virgate; leaves obovate, crenate, dentate or entire, obtuse; cymes sessile, fruit ovate, roundish. Var. *punicifolium*, leaves obovate, entire or slightly crenate at the apex, obtuse. 4-8 f. *S.*

ni'tidum, (♄.) very glabrous; leaves lance-linear, shining above, obscurely serrate or entire, small; branches quadrangular. *S.*

mol'le, (J. ♃.) leaves roundish-cordate, plicate, toothed, pubescent beneath; petioles sub-glandular; cymes with rays. Berries red. *S.*

lævigatum, (w. J. ♄.) stem much branched; leaves lanceolate, smooth, remotely serrate, entire at base; branchlets 2-edged. *S.*

Exotic.

op"ulus, (guelder-rose, snow-ball, w. J. ♄.) leaves 3-lobed, sharp-toothed; petioles; glandular, smooth; flowers in compact cymes, surrounded with radiating florets. Var. *ro'seum*, has the whole cyme made up of radiating florets.

li'nus, (laurestine, r-w. ♄.) leaves ovate, entire, with tufts of hair in the axils of the veins beneath; flowers in smooth cymes.

VI'CIA. 16—10. (*Leguminosæ.*) [From *vincio*, to bind together, as the tendrils of this plant twine around other plants.]

caroli'nia'na, (M. ♃.) smoothish; leafets 8-10; stipules lance-oval, entire; peduncles many-flowered; flowers distant; teeth of the calyx short; style villose at the top; legume smooth, obliquely veined; stem long and climbing; flowers small, white, the standard tipped with black. Mountains. Penn. to Car.

sati'va, (common vetch-tare, b. J. ⊕.) leafets 10-12; stipules with a dark spot beneath; style bearded at top; flowers small. 1-2 f.

crac'ca, (tufted vetch, P. Au. ♃.) stem sub-pubescent; leaves pinnate; flowers small, pale, numerous, drooping, imbricated. Meadows. New E.

america'na, (p. J. ♃.) peduncles many-flowered, shorter than the leaves; stipules semi-sagittate, dentate; leafets numerous, elliptical-lanceolate, smooth, obtuse, mucronate. Shady woods. Niagara. Genesee Falls.

acutifo'lia, (w. Ap.) peduncles few-flowered; stipules lanceolate, entire; leafets (6) linear, acute at each end; stem glabrous, somewhat angled; legume glabrous, many-seeded. 2-3 f. *S.*

fa'ba, (garden-bean, windsor-bean, w. and black, J. ⊕.) stem many-flowered, erect, strong, legumes ascending, tumid, coriaceous; leafets oval-acute, entire; stipules sagittate, toothed at the base. From Persia.

VILLAR"SIA. 5—1. (*Gentianæ.*)

lacuno'sa, (w. Au. ♃.) leaves reniform, sub-peltate, slightly crenate, lacunose beneath; petioles long, bearing the flowers; corolla smooth; stem long, filiform, floating; flowers somewhat umbelled. Ponds and Lakes.

corda'ta, (Ju. ♃.) leaves cordate, variegated; petioles glabrous, bearing the flow-

ers; corolla campanulate. Shallow streams. *S.*

VIN"CA. 5—1. (*Apocyneæ.*) [From *vincio*, to bind, on account of its usefulness in making bands, or its creeping stem.]

mi'nor, (periwinkle, b. Ap. ♄.) stem procumbent; leaves lance-oval, smooth at the edges; flowers peduncled; teeth of the calyx lanceolate. Ex.

VI'OLA. 5—1. (*Cisti*, or, according to the divisions of Lindley, *Violaceæ.*) [From *ion*, because first described in Ionia.]

A. *Stemless, or with a subterranean stem.* [*Leaves more or less reniform, always cordate, younger cucullate; proper color of the corolla violet.*]

cuculla'ta, (b. p. M. ♃.) glabrous; leaves cordate, somewhat acuminate, orbenate-dentate; autumnal ones largest, very exactly reniform; peduncle somewhat 4-sided, longer than the leaves; divisions of the calyx subulate, acuminate, emarginate behind, or very entire; petals (as in many American species) oblique, veiny, very entire, white at the base, upper one generally naked, glabrous, lateral ones bearded, and with the upper one marked with a few blue lines. Var. *papiliona'cea*, petioles and peduncles longer; leaves sub-lance-ovate; beards of the lateral petals often yellow. Var. *tetrago'na*, peduncle strong, exactly 4-sided; petals azure-color, veinless. Var. *villo'sa*, leaves, petals, and peduncles villose. 4-8 i.

palma'ta, (b-p. M. ♃.) mostly villose; leaves heart-reniform, palmate, 5-7-lobed; lobes often narrow, and gashed, middle one always larger; sometimes villose both sides, sometimes only beneath; often glabrous, all of them very often purple beneath, the first spring ones are ovate, entire; petioles sub-emarginate; peduncle somewhat 4-sided, longer than the leaves; divisions of the calyx lance-ovate, ciliate, very entire behind; petals all very entire, veiny, and white at the base; upper ones narrow, smaller, sometimes villose at the base, yet often naked, glabrous; lateral ones densely bearded, and with the upper one marked with a few blue lines. One variety has white flowers. 3-6 i.

soro'ria, (b-p. M. ♃.) leaves orbicular or roundish-cordate; the sinus often closed, crenate-serrate, mostly pilose, thickish, purple beneath, flat, appressed to the ground; petioles short, somewhat margined; stipules small, lanceolate; segments of the calyx short, glabrous, entire behind; petals obovate, entire; lateral ones densely bearded; stigma depressed, with a deflexed beak; capsule smooth.

[*Leaves oblong or ovate, never reniform; younger ones cucullate.*]

sagitta'ta, (E. b-p. ♃.) glabrous; leaves ciliate, oblong, not acute, sagittate-cordate, dentate, gashed at the base (or furnished with elongated divaricate teeth); peduncle somewhat 4-sided, longer than the leaves; divisions of the calyx lanceolate, acuminate, emarginate behind; petals all very entire, veiny, white at the base;

upper one generally naked, glabrous; lateral ones densely bearded, and with the upper one marked with a few blue lines; spur elongated behind. A variety has the leaves more or less villose. Dry.

[*Leaves ovate or lanceolate; corolla white, with the lateral petals narrower.*]

amœ'na, (E. w. Ap. 2{.) glabrous; leaves ovate, sub-acuminate, crenate, sometimes sub-villose above; petioles long, spotted with red; peduncle somewhat 4-sided, equalling or exceeding the length of the leaves, spotted; divisions of the calyx lanceolate; petals all very entire, green at the base; lateral ones sometimes with the base pubescent, and with the upper one marked with a few blue lines. Moist woods. Flowers odorous.

primulifo'lia, (primrose-leaved violet. w. J. 2{.) stoloniferous; leaves oblong, sub-cordate, abruptly decurrent into the petiole; nerves beneath and scape somewhat pubescent; sepals lanceolate; petals obtuse; the two lateral ones a little bearded and striate; stigma capitate, rostrate. Var. *villo'sa*, leaves very green; petioles densely villose, becoming hoary. Flowers odorous.

blan''da, (smooth violet, w. Ap. 2{.) glabrous; leaves round, sometimes sub-ovate, crenate, appressed to the ground, sometimes sprinkled with a few short hairs above; petioles pubescent; peduncles somewhat 4-sided, longer than the leaves; segments of the calyx lanceolate, obtuse; petals all very entire, green at the base; lateral ones slightly bearded and striate, the lower one distinctly striate and somewhat bearded; stigma depressed, rostrate; corolla small, odorous. 2-4 i. Wet, low grounds.

[*Stemless, not belonging to the preceding divisions.*]

rotundifo'lia, (O. M. y. 2{.) glabrous; leaves thickish, appressed to the earth, broad-ovate or orbicular, cordate, crenate; nerves pubescent beneath; sinus closed; peduncle somewhat 4-sided, as long as the leaves; divisions of the calyx oblong, obtuse; petals sometimes emarginate, upper ones small; lateral ones somewhat bearded, and with the upper one marked with a few yellowish brown lines; spur very short. Woods. 1-3 i.

peda'ta, (M. p-b. 2{.) glabrous; leaves sometimes ciliate, variously divided, very open, pedately 9-parted; divisions linear, and obtusely lanceolate, generally 3-lobed at the apex, often simply lanceolate, with the apex 5-7-lobed; peduncle somewhat 4-sided; divisions of the calyx lanceolate, acute, ciliate, emarginate behind; petals all white at the base, veinless, very entire, very glabrous, naked; upper one truncate, and marked with a few very blue lines, sometimes obsolete. Var. *veluti'na*, has the two lower petals of a very deep violet-color, and appears like velvet. Var. *al'ba*, has white flowers. Dry. 3-4 i. *S.*

B. *Caulescent.*

pubes''cens, (y. 2{.) villose-pubescent; stem simple, erect, terete, leafless below; leaves broad-ovate, cordate, dentate; petioles short; stipules large, ovate, dentate; peduncles 4-sided, shorter than the leaves; bracts subulate; minute divisions of the calyx lanceolate; petals all very entire, veinless; upper one naked, glabrous; lateral ones bearded, and with the upper one marked with a few blue lines; lower ones often becoming reddish outside; spur short, gibbous, acutish; stigma pubescent, scarcely beaked. Varies in pubescence, leaves are sometimes glabrous; the capsules are either glabrous or woolly. 4-12 i., rarely 4 f.

rostra'ta, (beaked violet, b-p. M. 2{.) smooth; stem diffuse, erect, terete; leaves orbicular and ovate, cordate, crenate-dentate, younger ones cucullate; stipules linear, acuminate, furnished with elongated, linear teeth; peduncles filiform, axillary, very long; segments of the calyx lanceolate, acute, entire behind; petals all very entire, veinless, naked, beardless, upper and lateral ones marked with a few blue lines; spur straight, linear, compressed, obtuse, double the length of the petals; stigma sub-clavate; root woody, perpendicular, fibrous. Var. *barba'ta*, lateral petals bearded. 6-10 i.

stria'ta, (striped violet, y-w. J. 2{.) smooth; stem oblique, branching, angular; leaves roundish, ovate, sub-acuminate, crenate-dentate, sometimes sub-pubescent; petioles long; stipules large, oblong-lanceolate, dentate-ciliate; peduncles quadrangular; bracts linear, rather large; segments of the calyx lanceolate, acuminate, ciliate, emarginate behind; petals entire, upper one marked with a few blue lines, naked, smooth, sometimes a little villose, lateral ones bearded, lower one occasionally a little villose; spur sub-porrected; stigma pubescent behind.

muh'lenberg''ia'na, (slender violet, b-p. M. 2{.) smooth; stem weak, sub-prostrate; leaves reniform-cordate, upper ones ovate, crenate; stipules lanceolate, serrate-ciliate, sub-pinnate; peduncles somewhat quadrangular, axillary, longer than the leaves; bracts minute, subulate; segments of the calyx linear, acute, sub-ciliate; petals all entire, veinless; upper one naked, glabrous, lateral ones bearded, the upper one marked with a few blue lines; spur porrected, compressed, obtuse; stigma ciliate behind; beak ascending. 3-6 i.

hasta'ta, (halbert-violet. y. M. 2{.) smooth; stem erect, simple, terete, leafy above; leaves long-petioled, cordate-lanceolate or hastate, acuminate; lobes obtuse, dentate; stipules minute, ciliate-dentate; petals all very entire, lower ones dilated, sub-3-lobed, lateral ones slightly bearded; spur short, gibbose, acutish; stigma truncate, or pubescent; capsule glabrous, or pubescent on all sides. Var. *gibba*, has no hastate or deltoid leaves. 8-12 i. Mountains.

canaden''sis. (r. w-y. M. 2{.) smooth; stem sub-simple, erect, terete; stipules entire, membranaceous, oblong, sub-ovate or lance-ovate; leaves broad-cordate, acuminate, serrate, slightly pubescent on the nerves, lower ones long-petioled; peduncle somewhat 4-sided; bracts subulate; flowers regular, large; segments of the calyx subulate, acute, entire behind; petals white, very entire, veiny, becoming. yellow at the base, lower ones pale violet, upper ones broad, spreading, lateral ones bearded, with the upper one marked with a few blue lines; stigma short, pubescent; spur very short; flowers odorous. 6-24 i. Moist. Rocky woods.

C. Exotic.

tri'color, (garden-violet, heart's-ease, pansy, p. y. b-p. M. ♃.) stem angular, diffuse, divided; leaves oblong, deeply crenate; stipules lyrate-pinnatifid.

odora'ta, (sweet-violet, b. M. ♄.) stemless; scions creeping; leaves cordate, crenate, smoothish; calyx obtuse; two lateral petals with a bearded or hairy line.

VIRGIL"IA. 10—1. (Leguminosæ.) [In honor of the poet Virgil.]

lute'a, (y. J. ♄.) leaves pinnate; leafets alternate, ovate, short, acuminate, glabrous; racemes elongated, pendulous; legumes petioled, flat. The bark is used in dying yellow. S.

VIS"CUM. 20—4. (Caprifolia.) [From the Greek ixos, altered by the Æolians into biskos. The Greeks had a great veneration for this plant on account of its supposed medical virtues, and the Druids ascribed to it many miraculous powers.]

ru'brum, (♄.) leaves lance-obovate, obtuse; spikes axillary, whorled. S.

purpu'reum, (♄.) leaves obovate, obtuse, obsoletely 3-nerved; spikes axillary; flowers opposite. S.

verticilla'tum, (mistletoe, g. w. J. ♃.) branches opposite and whorled; leaves wedge-obovate, 3-nerved; spikes axillary, a little shorter than the leaves; berries yellowish white. On the branches of old trees.

VI'TIS. 5—1. (Vitices.)

labrus"ca, (plum-grape, w-g. J. ♄.) leaves broad-cordate, lobe angled, white-downy beneath; fertile racemes small; berries (blue, flesh-color, and green) large. Var. labruscoi'des, (fox-grape), has smaller fruit, approaching a tart taste.

vulpi'na, (frost-grape, g-w. J. ♄.) leaves cordate, acuminate, gash-toothed, glabrous both sides; racemes lax, many-flowered; berries small; leaves very variable, but the uppermost mature leaves will agree with the description.

æsti'valis, (summer-grape, J. ♄.) leaves 3-5 lobed, younger ones rust-downy beneath, when old nearly smooth; sinuses rounded; racemes opposite the leaves, crowded, oblong; berries deep-blue or purple. Woods, on banks of streams.

ripa'ria, (w-g. M. ♄.) leaves unequally incisely toothed, shortly 3-lobed, pubescent on the petioles, margins, and nerves; flowers sweet-scented.

bipinna'ta, (g-w. Ju.) leaves bipinnate, glabrous; leafets incisely serrate; flowers

pentandrous; berry 2-celled; cells 1-2 seeded. S.

indi'visa, (J. ♃.) leaves simple, cordate or truncate at base, somewhat 3-nerved, pubescent on the nerves beneath; flowers pentandrous and pentapetalous; berry 1-celled, 1-2 seeded. Swamps.

vinif'era, (wine-grape, J. ♄.) leaves sinuate-lobed, naked or downy. Ex.

WARE'A. 14—2. (Cruciferæ.)

cuneifo'lia, (w.) leaves nearly sessile, rather thick, oblong, obtuse, attenuate at the base; siliques with the valves somewhat convex. 1-2 f. Georgia and Florida.

amplexifo'lia, (p. ⊕.) silique two-edged, pendulous; leaves oblong-ovate, half-clasping. 1 f. S.

XAN"THIUM. 19—5. (Corymbiferæ.) [From xanthos, yellow, a color said to be produced by the plant.]

struma'rium, (cockle-burr, sea-burdock, Au. ⊕.) stem unarmed; leaves ovate, angulate-dentate, sub-cordate, and strongly 3 nerved at base; fruit oval, pubescent, armed with rigid, hooked bristles. 3-6 f.

spino'sum, (thorny clot-weed, S. ⊕.) spines ternate; leaves 3-lobed; flowers axillary, solitary. 2-4 f.

XANTHOX"YLUM. 20—5. (Terebintaceæ.) [From xanthos, yellow, and xulon, wood, alluding to its color.]

fraxin'eum, (prickly ash, tooth-ache bush, g-w. M. ♄.) prickly; leaves pinnate; leafets lance oval, sub-entire, equal at base; petioles terete, unarmed; umbels axillary. 8-12 f. The bark is pungent, and is used for medicinal purposes.

tricar"pum, (J. ♄.) leaves glabrous. pinnate; leafets petioled, falcate-lanceolate, crenate-serrate; petioles unarmed; flowers bearing petals; capsules mostly in threes; leaves very aromatic and pungent.

XANTHORHI'ZA. 5—12. (Ranunculaceæ.) [From xanthos, yellow, and riza, root]

apifo'lia, (parsley yellow-root, Ap. ♄.) leaves 3-ternate; petioles dilated and clasping at the base; flowers racemed. 1-3 f. Banks of streams.

XEROPHYL'LUM. 6—3. (Junci.)

asphodeloi'des, (w. J. ♃.) filaments dilated toward the base, and equalling the corolla; racemes oblong, crowded; bracts setaceous; scape leafy; leaves subulate, 3-5 f.

te'nax, (w. J. ♃.) scape leafy; racemes lax; bracts membranaceous; petals elliptic; filaments filiform, exceeding the corolla; leaves subulate-setaceous, very long. S.

XYLOS"TEUM 5—1. (Caprifolia.)

cilia'tum, (fly-honeysuckle, twin-berry, w-y. M. ♄.) berries distinct; leaves ovate and sub-cordate, margin ciliate, in the young state villose beneath; corolla a little calcarate at the base, tube ventricose above, divisions short, acute; style exsert. 3-4 f.

solo'nis, (swamp twin-berry, y. M. ♄.) berries united in one, bi-umbilicate (never distinct), two flowers situated on one germ; leaves oblong-ovate, villose. Berries dark-purple. 2 f.

XYRIS. 3—1. (*Junci.*) [From a Greek word, signifying pointed.]
carolin"ia'na, (yellow-eyed grass, E. y. Au. 2⁄.) leaves linear, grass-like; stem or scape two-edged; head ovate, acute; scales obtuse. 9-18 i.
brevifo'lia, (y. Au. 2⁄.) leaves subulate-ensiform, short; interior valves of the calyx shorter than the exterior, somewhat gash-toothed. 12 i.
jun"cea, (M. 2⁄.) leaves terete, hollow, acute; scape terete, sheathed at the base; calyx about as long as the roundish bracts; head oval. 6-12 i. S
in"dica, (y. J. 2⁄.) leaves long, grass-like, tortuous; scape tortuous; heads globose; scales nearly round, obtuse. S.
fimbria'ta, (feathered xyris, J. 2⁄.) heads lax-imbricate; calyx much longer than the ˙····· fimbriate; leaves long, sword-shaped.

6—1. (*Liliaceæ.*) [From *Juca*, the ˙me. S.]
o'sa, (silk-grass, w. Au. 2⁄.) leaves lanceolate, broad, entire, ˙ on the margin; stigmas re-reading. 2-5 f.
˙, (w. Au. 2⁄.) caulescent, branch-s broad-lanceolate, plaited, entire; ˙eolate. 2-4 f.
˙, leaves lance-linear. with callous ˙, rigid. 10-12 f.

20—12. (*Coniferæ.*) [From *semia*, ˙r loss, in allusion to the fact that inate aments produce no seed.]
o'lia, (♄.) frond pinnate; leaflets ˙, roundish-obtuse, attenuate at ˙utely serrate toward the apex; ˙rous, somewhat 4-cornered. S.
˙, flowers diœcious; leaves rigid, ˙s, erect; nut oblong, erect, scaly, ˙d. A native of New South

HEL"LIA. 19—1. (*Naides.*)
˙ris, anthers 4-celled; stigmas en-˙icarps toothed on the back; stem ˙orm; flowers small, axillary. Horn˙ed. Ditches.
˙'dia, (false pond-weed, Ju. ⊙.) ˙-celled; stigmas dentate-crenate; ˙oth, entire on the back; stem fili-˙aves entire. Salt-marsh ditches.

IA. 13—2. (*Scrophulariæ*)
o'ra, (b-w. Ju. 2⁄.) leaves ovate ˙orm, serrate above; spikes solitary, filiform peduncles, forming conical ˙stem herbaceous, creeping, 6-8 ˙ˑmes long, procumbent.
lanceola'ta, leaves linear-lanceolate; spikes solitary. Banks of streams.

ZE'A. 19—3. (*Gramineæ.*) [An ancient Greek name.]
mays, (Indian-corn, y-g. Ju. ⊙.) leaves lance-linear, entire, keeled. S.
ZIGADE'NUS. 6—3. (*Junci.*)
glaberri'mus, (w. J. 2⁄.) scape leafy; bracts ovate, acuminate; petals acuminate; leaves long, recurved, channelled. 2-4 f.
el"egans, (w. J. 2⁄.) scape nearly naked; bracts linear; petals ovate, acute. S.
ZIN"NIA. 17—2. (*Corymbiferæ.*)
viola'cea, (r-p. Ju.) leaves ovate-acute, sessile, sub-crenate; chaff imbricate-serrate.
el"egans, (p. J. ⊙.) heads stalked; leaves amplexicaul, cordate, ovate, sessile, opposite; stem hairy; scales serrate. 2 f Mexico.
multiflo'ra, (r. S. ⊙.) flowers peduncled; leaves opposite, sub-petioled, lance-ovate. S.
ZIZA'NIA. 19—6. (*Gramineæ.*)
aquat"ica, (wild rice, Au. 2⁄.) panicle pyramidal, divaricate and sterile at the base, spiked and fertile above; pedicels clavate; awns long; seed linear. In water.
milia'cea, (Au. 2⁄.) panicle effuse, pyramidal; glumes short-awned; staminate and pistillate flowers intermingled; style 1; seed ovate, smooth; leaves glaucous. 6-10 f. In water.
ZI'ZIA. 5—2. (*Umbelliferæ.*)
corda'ta, (y. J. 2⁄.) radical leaves undivided, cordate, crenate, petiolate, cauline ones sub-sessile, ternate; segments petiolate, ovate, cordate, serrate; partial involucre 1-leaved. 12-18 i. Fruit black. Canada to Florida.
au'rea, (golden alexanders, y. J. 2⁄.) leaves biternate, shining; leaflets lance-oval; umbels with short peduncles. 1-2 f.
integerri'ma, (y. J. 2⁄.) very glabrous; leaves biternate, sub-glaucous, lower ones thrice ternate, upper ones twice; leaflets oblique, oval, entire; umbels with elongated peduncles. 12-18 i. Mountains.
ZI'ZIPHUS. 5—1. (*Rhamni.*)
volu'biles, (g-y. Ju. 2⁄.) unarmed; leaves ovate, ribbed, entire; umbels axillary, peduncled; stem twining. S.
ZOR"NIA. 16—10. (*Leguminosæ.*)
tetraphyl"la, (y. J. 2⁄.) leaves digitate; leaflets 4, lanceolate, glabrous; spikes axillary, peduncled; flowers alternate, 2-bracted; bracts roundish. S.
ZOS"TERA. 19—1. (*Naides.*) [From *zoster*, a girdle.]
mari'na, (sea-eel grass, Au. 2⁄.) leaves entire; stem terete; flowers very small; leaves long. In salt water.

SECTION V.

VOCABULARY,

OR

EXPLANATION OF BOTANICAL TERMS

A, in composition, signifies privation, or destitute of; as *acaulis*, referring to a plant without a caulis or stem.

Abor'tive flower. Falling off without producing fruit.

—— stamens, not furnished with anthers.

—— pistil. Defective in some essential part.

—— seed. Not becoming perfect, through want of the fertilizing influence of the pollen.

Abrupt leaf. A pinnate leaf with an old or terminal leafet.

Acal'yces. (From *a*, signifying without, and *calyx*, a flower cup.) A class in an ancient method of arrangement, consisting of plants without a calyx.

Acau'les. (From *a* wanting, and *caulis*, a stem.) The 20th class in Magnolius's method, including plants without stems.

Acer'ose leaf. Linear and permanent, as in the pine.

Ache'nium, one of Mirbel's genera of fruits.

Acic'ular. Needle-shaped.

Acina'ciform. Cimeter-shaped.

A'cinus. A small berry which, with many others, composes the fruits of the mulberry and raspberry; the plural is *acini*.

Acotyled"onous. (From *a* without, and *cotyledon*, a seed lobe.) Plants destitute of seed lobes, and which consequently put forth no seminal or seed-leaves, as mosses and ferns.

Acu'leus. (From *acus*, a needle.) A prickle, or sharp point; common to the rose and raspberry.

Accum'bent. The corcle lying against the back of the cotyledons.

Acu'minate. Taper-pointed, the point mostly curved towards one edge of the leaf, like an awl.

Acute. Less gradually sharp-pointed than acuminate. An obtuse angle, or any other mathematical angle, is a:ute in botanical language.

Adel'phous. (From the Greek *adelphos*, a brother or an equal) Applied to plants whose stamens are united by their filaments, whether in one or two sets.

Adnate'. Growing together.

Adversifo'liæ (From *adversus*, opposite, and *folium*, a leaf.) Plants whose leaves stand opposite to each other, on the same stem or branch. Name of the 5th class in Sauvage's *Methodus foliorum*; as exemplified in the labiate flowers.

Æsti'vales. (From *æstas*, summer.) Plants which blossom in summer. The second division of Du Pas's method, with reference to the four seasons of the year, consisting of herbs which blossom in summer.

A'fora. (From *a*, without, and *fores*, a door.) Having no doors or valves. The name of a class in Camerius's method, consisting of plants whose pericarp or seed-vessel is not furnished with internal valves.

Aga'mous. (From *a*, without, and *gamos*, marriage.) A term derived from the indelicate notions of the last century, respecting the sexual distinctions of plants; and which, whatever analogies may actually exist between the vegetable and animal kingdoms, should as far as possible be excluded from the science. Were it to be otherwise, the study of Botany ought to be limited to the medical profession. Of all studies, that of Botany should not be accompanied by aught that might pain or disgust a delicate mind. Plants without any visible stamens or pistils are by French botanists called *agamous*.

Ages of plants. *Ephemeral* are such as spring up, blossom and ripen their seed in a few hours or days; *annual* live a few months or one summer; *biennial*, spring up one summer and die the following; *perennial*, live an indefinite period.

Agglom'erated. Bunched, crowded together.

Ag'gregate. (From *aggregare*, to assemble.) Many springing from the same point; this term was at first applied to compound flowers, but there is at present a sevenfold division of aggregate flowers:

. *aggregate*, properly so called;
compound, *amentaceous*,
umbellate, *glumose*,
cymose, *spadiceous*.

Ag'gregate flower is erected on peduncles or footstalks, which all have one common receptacle on the stem; they sometimes have one common calyx, and are sometimes separately furnished with a calyx.

Ai'grette. See Egret.

A'la. A Latin word signifying a wing. It is sometimes used to express the angle formed by the stem with the branch or leaf. Linnæus and some others use the term *ala*, as the name of a membrane, affixed to some species of seed which serves as a wing to raise them into the air, and thus promote their dispersion.'

A'læ. The two lateral or side petals of a papilionaceous flower.

Albu'men. The farinaceous, fleshy or horny substance which constitutes the chief bulk of monocotyledonous seeds; as wheat, rye, &c.

Albur'num. (From *albus*, white.) The soft white substance which in trees is found between the liber, or inner bark, and the wood, and becoming solid, in progress of time is converted into wood. From its colour and comparative softness, it has been styled the fat of trees. It is called the *sap-wood*, and is formed by a deposite of the cambium, or descending sap; in one year it becomes wood; and a new layer of alburnum is again formed by the descént of the cambium.

Al'gæ. Flags; these, by Linnæus, comprise the plants of the order *Hepaticæ* and *Lichenes.*

Al'pine. Growing naturally on high mountains.

Alter'nate. Branches, leaves, flowers, &c. are alternate, when beginning at different distances on the stem; opposite, is when they commence at the same distances, and base stands against base.

Alter'nately-pinnate leaf; when the leafets are arranged alternately on each s'de of the common footstalk or 'petio1e.

Alve'c'ate. Having cells which resemble a honeycomb.

Am'bitus. The outer rim of a frond receptacle, &c.

A'ment. Flowers collected on chaffy scales, and arranged on a thread or slender stalk; their scales mixed with the flowers resemble the chaff in an ear of corn; in the willow and poplar, an ament supports both staminate and pistillate flowers on distinct roots. Flowers supported by an ament are generally destitute of a corolla.

Amplexicau'lis. Clasping the base of the stems.

Anal'ysis. To analyze a plant *botanically*, is fo ascertain its name, by observing its organs, and comparing them with scientific descriptions of plants.

Ancip'etal. Having two sharp edges like a sword.

An'dria. Signifies stamen.

Androg'ynous plants. Such as bear staminate and pistillate flowers on the same root; as the oak and Indian corn: such plants belong to the class Monœcia.

Angiocar'pus. Fungi bearing seeds internally.

Angiosper'mal. (From *angio*, a vessel, and *sperma*, seed.) Plants whose seeds are enclosed or covered.

An'gular. Forming angles; when the stems, calyxes, capsules, &c. have ridges running lengthwise.

Angustifo'lius. Narrow-leaved.

An'nual. A plant which lives but one year. The herbage is often annual, while the root is perennial; in this case the plant is said to be perennial.

An'nulated. Having a ring round the capsules; as in ferns; or in mush-rooms having a ringed stipe.

An'nulus. A ring.

Anom'alous. (From *a*, without, and *nomos*, law.) Irregular, or whatever forms an exception to a general rule. The 11th class in Tournefort's method is called *anomalæ*, including plants whose corollas are composed of irregular and dissimilar parts; as the columbine, monk's-hood, violet, larkspur, &c.

An'ther. (From *anthos*, a flower; so called as indicating its importance.) That part of the stamen which contains the pollen; it is of various forms, as linear, awl-shaped, heart-shaped, round, &c.; it is one-celled, two-celled, &c.

Antherid'ium. A mass of pollen.

Antherif'erous. Flowers bearing anthers without filaments.

An'thus. (From the Greek *anthos*.) A

flower, generally referring to the petals only.

Antiscorbu'tics. Substances which cure eruptions.

Apet'alous. (From *a*, without, and *petalum*, a petal.) Having no petals or corollas; such flowers are termed *incomplete;* such as are destitute of either stamens or pistils are called *imperfect*

Apet'alæ. A class formed by some of the ancient botanists, including plants destitute of corollas.

A'pex. The top or summit.

Aphyl'lous. (From *a*, without, and *phyllon*, a leaf.) Destitute of leaves.

Aphyl'læ is the name given by an ancient botanist to a class of plants without leaves, comprising garlic, rush, mushrooms, &c.

Apothe'cia. The fructifications of the lichens.

Appen'daged. Having bracts, thorns, prickles, &c.

Appres'sed. Closely pressed; as leaves against the stem, &c.

Approx'imate. Growing near each other.

Ap'terous. Without wings.

Aquat'ic. (From *aqua*, water.) Growing in, or near water. *Aquaticæ* was an ancient name for a class including all plants which grow in water.

Ar'bor. A tree; a perennial plant, which rises to a great height. Most trees spring from seeds having two cotyledons; they are therefore called dicotyledonous plants. The ancient botanists divided plants into trees and herbs; but this distinction is too vague to form the basis of classification.

Arbo'reus. Like a tree.

Arbusti'vus. (From *arbustum*, a shrub.) An ancient class of plants containing shrubs; as the myrtle, mock-orange (*philadelphus*,) &c.

Arch'ed. Curving above, vaulted.

Ar'cuate. (From *arcus*, a bow.) Bent like a bow.

Arena'rius. Growing in sand.

Argen'teus. Silver-coloured.

Ar'id. Dry.

A'ril, (*arillus*.) The external coat or covering of seeds which, drying, falls off spontaneously.

Aris'tate. (From *areo*, to be dried.) Awned, ending a bristle.

Aro'ides. So called from *arum.*

Arms, (*arma.*) Offensive weapons. Plants are said to be armed, when they are furnished with prickles, thorns, &c.

Aromat'ic. Sweet-scented.

Aromat ícæ. The name of a class of Dioscorides, Clusius, Bauhin, and

some other botanists, who arranged plants according to their virtues and sensible qualities.

Ar'row-form. Shaped like an arrow-head, the hind lobes acute.

Artic'ulated. Jointed; as in the culm or stem of the grasses.

Arundina'ceous. (From *Arundo*, a reed.) Resembling reeds.

Arven'sis. Growing in cultivated fields.

Ascend'ing. Rising from the ground obliquely.

Ascid'isate. Pitcher-form. From the Greek *askidion*, a bottle or pitcher.

Asperifo'lius. Rough-leaved.

Astrin'gents. Substances which condense the fibres.

Atten'uated. Gradually diminished or tapering.

Auric'ulate. Having appendages resembling ears.

Awl-form. Sharp at the point, and curved to one side.

Awn. A short stiff bristle

Ax'il. The angle between a leaf and stem on the upper side.

Ax'illary. Growing out of the axils; leaves are said to be axillary when they proceed from the angle formed by the stem and branch.

Ax'is. The elongated part of a petiole, upon which are attached many flowers. A centre. A line, real or imaginary, through any body.

B

Ba'ca. A berry. It is a pulpy pericarp, enclosing seeds without capsules. A berry is said to be proper, when it is formed of the pericarp or seed vessel; improper or singular, when it is formed of any other parts. In the mulberry and rose, a large, fleshy and succulent calyx becomes a berry. In the strawberry, a berry is formed of the common receptacle; in the raspberry, of a seed.

Baccif'erus. Bearing berries.

Ban'ner. The upper petal in a papilionaceous flower.

Barb. A straight process, armed with teeth pointing backwards.

Barba'tus. Bearded.

Bark. The covering of vegetables, consisting of several parts; as cuticle, cellular integument, &c. The bark consists of as many layers as the tree on which it grows has years: a new layer being formed from the cambium, or from the alburnum, every year. The newest layer of bark is called liber.

Bar'ren. Producing no fruit; containing stamens only.

Beak'ed. Terminating by a process shaped like the beak of a bird.

Ber'ry. A pulpy pericarp enclosing seeds without capsules. See Bacca.

Bi, derived from bis, signifying two.

Bicor'nis. Anthers with two horns.

Bi'dens. Having two teeth.

Bien'nial. Living two years, in the second of which the flower and fruit are produced; as in wheat.

Bi'fid. Two-parted.

Bila'biate. Corolla with two lips.

Bi'nate. Two growing together.

Bipin'nate. Twice pinnate.

Biter'nate. Twice ternate. The petiole supporting three terante leaves.

Bi'valve. Two-valved.

Blas'teme. From the Greek *blastema*, a bud.

Bor'der. The brim or spreading part of a corolla.

Bot'rus. A cluster, like grapes.

Brach'ilate. Branches opposite, and each pair at right angles with the preceding.

Bract. Floral leaf; a leaf near the flower which is different from the other leaves of the plant. In the crown-imperial the Bracts are at the termination of the flower stem; from their resemblance to a hair, they are called coma.

Branch. A division of the main stem or mainroot.

Branch'let. Subdivision of a branch, a twig.

Bre'vis. Short.

Bruma'les. (From *bruma*, winter.) Plants which blossom in winter.

Bud. The residence of the infant leaf and flower.

Bulbs. Called roots; sometimes found growing on the stem; strictly speaking, bulbs are buds, or the winter residence of the future plants. Annual plants do not have bulbs; they are only preserved by seeds.

Bun'dle. See Fascicle.

C

Cadu'cous. (From *cado*, to fall.) Falling early; as the calyx of the poppy.

Cæs'pitose. Forming turfs, several roots growing together.

Cal'amus. Reed-like.

Calca'reous. Containing lime; as shells of oysters, &c.

Cal'carate. Resembling or being furnished with a spur.

Calli. Small callosities or protuberances.

Calyb'ion. (From *kalubion*, a little cabin.) A genus in Mirbel's second class of fruit.

Calyc'ulated. Having an additional calyx.

Calyp'tra. The cap or hood of pistillate mosses, resembling an extinguisher set on a candle.

Calyx. From the Greek, signifying a *flower-cup;* in most plants it encloses and supports the corolla. It is defined by Linnæus to be the termination of the outer bark.

Cam'bium. The descending sap, which every year forms a new layer of bark and one of wood. It descends between the bark and the wood, so that the new wood is formed externally and the new bark internally.

Campan'ulate. Bell-form.

Campes'tris. Growing in uncultivated fields.

Can'cellated. Appearing like latticework.

Canes'cent. White or hoary.

Cap'illary. Hair-like.

Cap'itate. Growing in heads.

Cap'sule. A little chest; that kind of hollow seed vessel which becomes dry and opens when ripe; a capsule that never opens is called a *samara*.

Carcer'ular. (From *carcer*, prison.) A seed contained in a covering, whose sides are compressed. One of Mirbel's genera of fruits, in the order Carcerulares.

Cari'na. The keel or lower folded petal of a papilionaceous flower.

Car'inated. Keeled, having a sharp back like the keel of a vessel.

Carmin'ative. A medicine used to dispel wind.

Carno'se. Of a fleshy consistence.

Car'pel. A term used for the divisions of the fruit. Each carpel generally forms a distinct cell.

Car'pos. From the Greek *karpos*, fruit.

Caryophyl'leous. Pink-like corolla, having five petals with long claws, all regular and set in a tubular calyx.

Cat'kin. See Ament.

Cau'date. Having a tail; as in some seeds.

Cau'dex. The main body of a tree, or root.

Caules'cent. Having a stem exclusive of the peduncle or scape.

Cau'line. Growing on the main stem.

Cau'lis. The main, herbage-bearing stem of all plants, called in French *la tige.*

Cell. The hollow part of a pericarp or anther: each cavity in a pericarp that contains one or more seeds, is called a cell. According to the number of these cells, the pericarp is one-celled, two-celled, three-celled, &c.

Cel'lular. Made up of little cells or cavities.

Ceno'bion. From the Greek, signifying a community; one of Mirbel's genera of fruits.

Cerion. A carcerulate fruit, forming one of Mirbel's genera of fruits.

Cerea'lis. Any grain from which bread is made. (From *Ceres*, goddess of corn.)

Cer'nuus. When the top only droops.

Chaf'fy. Made up of short membranous portions like chaff.

Cha'mepy'this. From the Greek *kamia*, on the ground, *pithus*, the pine-tree. This is the specific name of some plants.

Chan'nelled. Hollowed out longitudinally with a rounded groove.

Cho'rion. A clear limpid liquor contained in a seed at the time of flowering. After the pollen is received, this liquor becomes a perfect embryo of a new plant.

Cic'atrice. The mark or natural scar from whence the leaf has fallen.

Cil'iate. Fringed with parallel hairs.

Cine'reous. Ash-coloured.

Cin'gens. Surrounding, girding around.

Cir'rose. Bearing tendrils. From Cirrus, a tendril or climber.

Clasp'ing. Surrounding a stem with the base of the leaf.

Class. The highest division of plants in the system of Botany. Linnæus divided all plants into 24 classes; 3 of these are now rejected, and the plants which they included placed in the remaining 21 classes. The ancient botanists knew neither methods, systems, nor classes: they described under chapters, or sections, those plants which appeared to them to resemble each other in the greatest number of relations.

Cla'vate. Club-shaped, larger at the top than the bottom.

Clau'sus. Closed, shut up.

Claw. The narrow part by which a petal is inserted.

Cleft. Split, or divided less than half way.

Climb'ing. Ascending by means of tendrils, as grapes; by leaf-stalks, as the Clematis; by cauline radicles or little fibrous roots, as the creeping American ivy.

Clinanthe. The dilated summit of a peduncle, bearing flowers. The receptacle.

Club-shaped. See Clavate.

Clus'tered. See Racemed.

Cly'peate. Form of a buckler. See Peltate.

Coad'nate. United at the base.

Coarc'tate. Crowded.

Coated. With surrounding coats or layers.

Coccineous. Scarlet-coloured.

Cochleate. Coiled spirally, like a snail-shell.

Coc'cum. A grain or seed; tricoccous, 3-seeded, &c.

Cæru'leus. Blue.

Coleop'tile. From *koleos*, an envelope, and *ptilon*, a bud.

Co'leorrhize. From *koleos*, an envelope, and *riza*, a root.

Colli'nus. Growing on hills.

Col'oured. Different from green; in the language of botany, green is not called a colour. White, which in reality is not a colour, is so called in botany. The primitive colours and their intermediate shades and gradations, are by botanists arranged as follows:

Water-colour,	*hyalinus*.
White,	*albus*.
Lead-colour,	*cinereus*.
BLACK,	*niger*.
Brown,	*fuscus*.
Pitch-black,	*ater*.
YELLOW,	*luteus*.
Straw-colour,	*flavus*.
Flame-colour,	*fulvus*.
RED,	*rubex*.
Flesh-colour,	*incarnatus*.
Scarlet,	*coccineus*.
PURPLE,	*purpureus*.
Violet-colour,	*cæruleo-purpureus*.
BLUE,	*cæruleus*.
Green,	*viridis*.

White is most common in roots, sweet berries, and the petals of spring flowers. Black, in roots and seeds. Yellow, in anthers, and the petals of compound flowers. Red, in the petals of summer flowers and acid fruits. Blue and violet-colour, in the petals. Green, in the leaves and calyx.

Columella. The central pillar in a capsule or fruit of any kind.

Column. The filaments in gynandrous plants united with the style; the whole is termed a column.

Co'ma. A tuft of bracts on the top of a spike of flowers.

Comose. Sessile bracts.

Common. Any part is common, which includes or sustains several parts similar among themselves.

Compound. Made up of similar simple parts.

——— *flowers*. Such a are in the class Syngenesia, having florets with united anthers.

Compound leaf. When several leafets grow on one petiole.

—— *- raceme.* When several racemes grow along the side of a peduncle.

——— *umbel.* Having the peduncles subdivided into peduncles of lesser umbels.

——— *petiole.* A divided leaf stalk.

——— *peduncle.* A divided flower stalk.

Compressed. Flattened.

Concave. Hollowed on one side.

Conceptacle. Single-valved capsule.

Conchology. The science which treats of shells.

Cone. A scaly fruit like that of the pine. See Strobilum.

Conglom'erate. Crowded together.

Con'ic. With a broad base, gradually narrowing to the top like a sugar-loaf.

Conif'erous. Bearing cones.

Con'jugate. In pairs.

Con'nate. Opposite, with the bases united or growing into one, forming the appearance of one leaf. Anthers are sometimes connate.

Conni'vent. Converging, the ends inclining towards each other.

Contin'uous. Uninterrupted.

Contor'ted. Twisted.

Contrac'ted. Close, narrow.

Conver'ging. Approaching or bending towards each other.

Con'vex. Swelling out in a roundish form.

Con'volute. Rolled into a cylindric form, as leaves in the bud.

Cor'culum, or *Corcle.* The embryo or miniature of the future plant, which is found in seeds often between the cotyledons.

Cor'date. Heart-shaped, side lobes rounded.

Coria'ceous. Resembling leather; thick and parchment-like.

Cor'nu. A horn or spur.

Cornic'ulate. Horn-shaped.

Corol'la, or *corol.* (A word derived from *corona,* a crown.) Usually encloses the stamens.

Corona'tus. Crowned; as the thistle-seed is crowned with down.

Cor'tex. (From *corium,* leather, or hide, and *tego,* to cover.) The rind or coarse outer bark of plants; the organization of the outer and inner barks differs chiefly in the firmness of their textures.

Cor'tical. Belonging to the bark.

corydalis. Helmet-like.

Co'rymb. Inflorescence, in which the flower stalks spring from different heights on the common stem, forming a flat top.

Costate. Ribbed.

Cotyl'edons. (From *kotule,* a cavity.) Seed lobes. The fleshy part of seeds which in most plants rises out of the ground and forms the first leaves, called seminal or seed leaves. These lobes in the greatest proportion of plants, are two in number; they are very conspicuous in the leguminous seeds; as beans, peas, &c. The cotyledons are externally convex, internally flat, and enclose the embryo or principle of life, which it is their office to protect and nourish.

Cre'mocarpe. (From *kremao,* to suspend, and *karpos,* fruit.) A name given by Mirbel to a genus of fruits.

Creeping. Running horizontally; stems are sometimes creeping, as also roots.

Cre'nate. Scalloped, notches on the margin of a leaf which do not point towards either the apex or base.

Cre'nulate. Finely crenate.

Cres'cent-form. Resembling a half-moon.

Crest'ed. Having an appearance like a cock's-comb.

Crini'tus. Long-haired.

Crowded. Clustered together.

Crowned. See Coronatus.

Cru'ciform. (From *crux, crucis,* a cross.) Four petals placed like a cross.

Crusta'ceous. Small crusty substances lying one upon another.

Cryptog'amous. Plants which have stamens and pistils concealed.

Cu'bit. A measure from the elbow to the end of the middle finger.

Cucul'late. Hooded or cowled, rolled or folded in, as in the spatha of the Arum, or wild turnip.

Cucurbita'ceous. Resembling gourds or melons.

Cu'linary. Suitable for preparations of food.

Culm or *straw.* (From the Greek *kalama,* stubble or straw; in Latin *culmus.*) The stem of grasses, Indian corn, sugar-cane, &c.

Culmif'erous. Having culms; as wheat, grasses, &c.

Cune'iform. Wedge-form, with the stalk attached to the point.

Cup'ule. A cup, as in the acorn.

Curv'ed. Bent inwards. See Incurved.

Cus'pidate. Having a sharp straight point. (The eye-tooth is cuspidate.)

Cuticle. The outside skin of a plant, commonly thin, resembling the scarf or outer skin of animals. It is considered as forming a part of the bark.

Cya'nous. Blue.

Cy'athiform. Shaped like a common wine-glass.

Cylin'drical. A circular shaft of nearly equal dimensions throughout its extent.

Cyme. Flower stalks arising from a common centre, afterward variously subdivided.

Cymose. Inflorescence in cymes.

Cypse'le. (From the Greek, *kupselion.*) A little chest.

D

De'bilis. Weak, feeble.

Decan'drous. Plants with ten stamens in each flower.

Decaphyl'lous. Ten-leaved.

Decid'uous. Falling off in the usual season; opposed to *persistent* and *evergreen,* more durable than *caducous.*

Decli'ned. Curved downwards.

Decomposi'tion. Separation of the chemical elements of bodies.

Decompound'. Twice compound, composed of compound parts.

Decomposi'ta. Name of an ancient class of plants, having leaves twice compound; that is, a common footstalk supporting a number of lesser leaves, each of which is compounded.

Decum'bent. Leaning upon the ground, the base being erect. This term is applied to stems, stamens, &c.

Decur'rent. When the edges of a leaf run down the stem, or stalk.

Decur'sive. Decurrently.

Decus'sated. In pairs, crossing each other.

Deflec'ted. Bending down.

Defolia'tion. Shedding leaves in the proper season.

Dehis'cent. Gaping or opening. Most capsules when ripe are dehiscent.

Del'toid. Nearly triangular, or diamond-form, as in the leaves of the Lombardy poplar.

Demer'sus. Under water.

Dense. Close, compact.

Den'tate. Toothed; edged with sharp projections; larger than serrate.

Dentic'ulate. Minutely toothed.

Denu'date. Plants whose flowers appear before the leaves; appearing naked.

Deor'sum. Downwards.

Depres'sed. Flattened, or pressed in at the top.

Descrip'tions. In giving a complete description of a plant, the order of nature is to begin with the root, proceed to the stem, branches, leaves, appendages, and lastly to the organs which compose the flower, and the manner of inflorescence. Colour and size are circumstances least to be regarded in description; but stipules, bracts, and glandular hairs, are all of importance.

Dextror'sum. Twining from left to right, as the hop-vine.

Diadel'phous. (From *dis,* two, and *adelphia,* brotherhood.) Two brotherhoods. Stamens united in two parcels or sets; flowers mostly papilionaceous; fruit leguminous.

Diagno'sis. The characters which distinguish one species of plants from another.

Di'amond-form. See Deltoid.

Dianthe'ria. (From *dis,* two, and *anther.*) A class of plants including all such as have two anthers.

Dichot'omous. Forked, divided into two equal branches.

Dicli'nia. Stamens in one flower, and pistils in another; whether on the same plant or on different plants.

Dicoc'cous. Containing two grains or seeds.

Dicotyled'onous. With two cotyledons or seed lobes.

Didy'mous. Twined, or double.

Didyna'mia. (From *dis,* twice, and *dunamis,* power.) Two powers. A name appropriated to one of the Linnæan classes.

Dieresil'lia. (From *diairesis,* division.) One of Jussieu's orders of fruits.

Difform. A monopetalous corolla whose tube widens above gradually, and is divided into unequal parts; any distorted part of the plant.

Diffrac'ted. Twice bent.

Diffu'sed. Spreading.

Digitate. Like fingers. When one petiole sends off several leafets from a single point at its extremity. ↾

Digyn'ia. Having two pistils.

Dimid'iate. Halved.

Diœ'cious. Having staminate and pistillate flowers on different plants.

Dis'coid. Resembling a disk, without rays.

Disk. The whole surface of a leaf, or of the top of a compound flower, as opposed to its rays.

Disper'mus. Containing two seeds.

Dissep'iment. The partition of a capsule.

Dissil'iens. A pericarp, bursting with elasticity; as the Impatiens.

Di'stichus. Growing in two opposite ranks or rows.

Divar'icate. Diverging so as to turn backwards.

Diver'ging. Spreading; separating widely.

Diur'nus. Enduring but a day.

Dor'sal. Belonging to the back.

Dotted. See Punctate and Perforated.

Droop'ing. Inclining downward, more than nodding.

Drupe. A fleshy pericarp, enclosing a stone or nut.

Dru'peole. A little drupe.

Drupa'ceous. Resembling, or bearing drupes.

Dul'cis. Sweet.

Dumo'sus. Bushy.

Du'plex. Double.

E

Eared. Applied to the lobes of a heart-form leaf, to the side lobes near the base of some leaves, and to twisted parts in plants which are supposed to resemble the passage into the ear.

Ebur'neus. Ivory-white.

Echi'nate. Beset with prickles, as a hedge-hog.

Ecos'tate. Without nerves or ribs.

Edible. Good for food, esculent.

Efflorescen'tia. (From *effloresco,* to bloom.) A term expressive of the precise time of the year, and the month in which every plant blossoms. The term *efflorescence* is applied to the powder substance found on Lichens.

Effolia'tion. Premature falling off of leaves, by means of diseases or some accidental causes.

Effuse. Having an opening by which seeds or liquids may be poured out.

Egg-form. See ovate.

E'gret or *Ai'grette.* The feathery or hairy crown of seeds, as the down of thistles and dandelions. It includes whatever remains on the top of the seed after the corolla is removed. The egret is stiped, when it is supported on a foot-stem; it is simple, when it consists of a bundle of simple hairs; it is plumose, when each hair composing the crown has other little hairs arranged along its sides.

Ellip'tic. Oval.

Elon'gated. Exceeding a common length.

Emar'ginate. Having a notch at the end, retuse.

Em'bryo. (From *embrao,* to bud forth.) The germ of a plant; called by Linnæus the corculum.

Emol'lient. A medicine which softens and relaxes the animal fibre.

En'docarp. The inside skin of a pericarp.

Endog'enous. Applied to stems which grow from the centre outwardly, as in monocotyledons.

Eno'dis. Without joints or knots.

En'siform. Sword-form, two-edged, as in the flag and iris.

Entire. Even and whole at the edge.

Entomol'ogy. The science which treats of insects.

Epi. A Greek word, signifying upon; often used in composition.

E'picarp. (From *epi,* upon, and *karpos,* fruit.) The outer skin of the pericarp.

Epider'mis. (From *epi,* upon, and *derma,* skin.) See Cuticle.

Epig'ynous. (From *epi,* upon, and *gynia,* pistil.)

Ep'isperm. (From *epi,* upon, and *sperma,* seed.)

Equinoc'tial flowers. Opening at stated hours each day.

E'quitant. Opposite leaves alternately enclosing the edges of each other.

Erect'. Straight; less unbending than strictus.

Ero'ded. Appearing as if gnawed at the edge.

Es'culent. Eatable.

Ev'ergreen. Remaining green through the year, not deciduous.

Excava'tus. Hollowed out.

Exog'enous. A term applied to stems which grow externally.

Exotic. Plants that are brought from foreign countries.

Expan'ded. Spread.

Expec'torant. (From *expectoro,* to discharge from the breast.) Medicines which promote a discharge from the lungs.

Ex'serted. Projecting out of the flower or sheath.

Eye. See Hilum.

F

Facti'tious. (From *facio,* to make.) Not natural, produced by art.

Fam'ilies. A term in Botany implying a natural union of several genera in to groups; sometimes used as synonymous with Natural Orders.

Fal'cate. Sickle-shaped; linear and crooked.

Fari'na. (From *far,* corn.) Meal or flour. A term given to the glutinous parts of wheat and other seeds, which is obtained by grinding and sifting. It consists of gluten, starch, and mucilage. The pollen is also called farina.

Fas'cicle. A bundle.

Fascic'ulate. Collected in bundles.

Fastig'iate. Flat-topped. Branches are said to be fastigiate when they keep in a similar direction to the main stem, and their boughs point upwards.

Fuvo'sus. Resembling a honeycomb.

Faux. Jaws. The throat of the corolla.

Feb'rifuge. (From *febris*, a fever, and *fugo*, to drive away.) That which possesses the property of abating fever.

Ferns. Cryptogamous plants, with the fruit on the back of the leaves, or in spikes made up of minute capsules opening transversely.

Fer'tile. Pistillate, yielding fruit.

Ferruginous. Iron, rust-like.

Fi'bre. Any thread-like part.

Fil'ament. The slender thread-like part of the stamen.

Fil'ices. (From *filum*, a thread.) Ferns.

Fil'iform. Very slender. ·

Fim'briate. Divided at the edge like fringe.

Fis'tulous. Hollow or tubular, as the leaf of the onion.

Flabel'liform. Fan-shaped.

Flac'cid. Too limber to support its own weight.

Flagel'liform. Like a whip-lash.

Flam'meus. Flame-coloured.

Fla'vus. Yellow.

Flesh'y. Thick and pulpy.

Flex'uous. Serpentine, or bending in a ziz-zag form.

Flo'ra. Considered by the heathens as the goddess of flowers; descriptions of flowers are often called Floras.

Flo'ral leaf. See Bract.

Flo'ret. Little flower; part of a compound flower.

Flo'rist. One who cultivates flowers.

Flos'cular. A tubular floret.

Flow'er, (*Flos.*) A term which was formerly applied almost exclusively to the petals. At present a stamen and pistil only are considered as forming a perfect flower.

Flow'er-stalk. See Peduncle.

Folia'ceous. Leafy.

Fol'ioles. Leafets; a diminutive of *folium*, a leaf. The smaller leaves which constitute a compound leaf.

Fol'ium. Leaf. Leaves are fibrous and cellular processes of the plants, of different figures, but generally extended into a membranous or skinny substance.

Fol'licle. A seed-vessel which opens lengthwise, or on one side only.

Foot'-stalk. Sometimes used instead of peduncle and petiole.

Fork'ed. See Dichotomous.

Frag'ilis. Breaking easily.

Frond. The leaf of cryptogamous plants; formerly applied to palms.

Fronder'cence. (From *frons*, a leaf.) The time in which each species of plants unfolds its first leaves. See Frondose.

Frondo'se, (*Frondosus.*) Leafy, or leaf-like.

Fructifica'tion. The flower and fruit, with their parts.

Fructif'erous. Bearing or becoming fruit.

Fruc'tus. The fruit is an annual part of the plant, which adheres to the flower and succeeds it; and after attaining maturity, detaches itself from the parent plant, and on being placed in the bosom of the earth, gives birth to a new vegetable. In common language, the fruit includes both the pericarp and the seed, but strictly speaking, the latter only is the fruit, while the former is but the case or vessel which contains it.

Fru'tescent. Becoming shrubby.

Fru'tex. A shrub.

Fu'gax. Fugaceous, flying off.

Ful'cra. Props, supports: as the petiole, peduncle, &c.

Ful'vous. Yellowish.

Fun'gi. The plural of *fungus*, a mushroom.

Fun'gous. Growing rapidly with a soft texture like the fungi.

Fu'nicle. The stalk which connects the ovale to the ovary.

Funnel-form. Tubular at the bottom, and gradually expanding at the top.

Fu'siform. Spindle-shaped; a root thick at the top and tapering downwards.

G

Gal'ea. A helmet.

Gem'ma. A bud seated upon the stem and branches, and covered with scales, in order to defend it from injury. The bud resembles the seed in containing the future plant in embryo; but this embryo is destitute of a radicle, though if the bud is planted in the earth, a radicle is developed.

Gemma'ceous. Belonging to a bud; made of the scales of a bud.

Gener'ic name. The name of a genus.

Genic'ulate. Bent like a knee.

Ge'nus. (The plural of genus is *genera.*) A family of plants agreeing in their flower and fruit. Plants of the same genus are thought to possess similar medicinal powers.

Germ. The lower part of the pistil which afterward becomes the fruit.

Germination. The swelling of a seed and the unfolding of its embryo.

Gib'bous. Swelled out commonly on one side.

Glabel'lous. Bald, without covering.

Gla'brous. Sleek, without hairiness.

Gland. A small appendage, which

seems to perform some office of secretion or exhalation.

Glan'dular. Having hairs tipped with little heads or glands.

Glau'cous. Sea-green, mealy, and easily rubbed off.

Glome. A roundish head of flowers.

Glom'erate. Many branchlets terminated by little heads.

Glume. The scales or chaff of grasses, composing the calyx and corolla; the lower ones are called the calyx, all others the corolla; each scale, chaff, or husk, is called a valve; if there is but one, the flower is called *univalve,* if two, *bivalve.*

Glu'tinous. Viscid, adhesive.

Gon. (From *gonu,* a knee or angle;) as pentagon, five-angled; hexagon, six-angled; polygon, many-angled.

Graft'ing, is the process of uniting the branches or buds of two or more separate trees. The bud or branch of one tree, is inserted into the bark of another, and the tree which is thus engrafted upon is called the stock.

Gram'ina. Grasses and grass-like plants. Mostly found in the class Triandria.

Gramin'eous. Grass-like; such plants are also called culmiferous.

Grandiflo'rus. Having large flowers.

Gran'ular. Formed of grains, or covered with grains.

Grave'olens. Having a strong odour.

Grega'rious. In flocks, plants growing together in groups.

Groov'ed. Marked with deep lines.

Gru'mose. Thick, crowded.

Gymnocarp'es. (From *gumnos,* naked, and *karpos,* fruit.) Mirbel's first class of fruits, containing such as have fruit without being covered or concealed.

Gymnosper'mia. (From *gumnos,* naked, and *sperma,* seed.) Having naked seeds.

Gynan'drous. Stamens growing upon the pistil.

Gyn'ia, From the Greek, signifying pistil.

H

Habita'tio or *Habitat.* The native situation of plants.

Habit. The external appearance of a plant, by which it is known at first sight.

Hair. See Pilus.

Hair-like. See Capillary.

Hal'berd-form. See Hastate.

Hand-form. See Palmate.

Hang'ing. See Pendent.

Has'tate. Shaped like a halberd; it differs from arrow-shaped in having the side processes more distinct and divergent.

Head. A dense collection of flowers, nearly sessile.

Heart. See Corculum and Corcle.

Heart-form. See Cordate.

Hel'met. The concave upper lip of a labiate flower.

Helminthol'ogy. The science which treats of worms.

Hepat'ic. Liver-like.

Herb. A plant which has not a woody stem.

Herba'ceous. Not woody.

Her'bage. Every part of a plant except the root and fructification.

Herba'rium. A collection of dried plants.

Herb'ist. One who collects and sells plants.

Hexag'onal. Six-cornered.

Hi'ans. Gaping.

Hi'lum. The scar or mark on a seed at the place of attachment of the seed to the seed-vessel.

Hir'sute. Rough with hairs.

His'pid. Bristly, more than hirsute.

Hoary. Whitish-coloured, having a scaly mealiness, not unlike glaucous.

Holera'ceous. Suitable for culinary purposes. The term is derived from *holus,* signifying pot-herbs. One of the natural orders of Linnæus, called *holeraceæ,* includes such plants as are used for the table, or in the economy of domestic affairs.

Hon'eycup. See Nectary.

Hood'ed. See Cucullate, or cowled.

Hora'rius. Continuing but an hour.

Horizon'tal. Parallel to the horizon.

Horn. See Spur.

Hum'ilis. Low, humble.

Husk. The larger kind of glume, as the husks of Indian corn.

Hyber'nalis. Growing in winter.

Hy'brid. A vegetable produced by the mixture of two species: the seeds of hybrids are not fertile.

Hy'po. (From *upo,* under.) Much used in the composition of scientific terms.

Hypocrater'iform. Salver-shaped, with a tube abruptly expanded into a flat border.

Hypog'ynous. Under the style

I

Ichthyol'ogy. The science of fishes.

Icosan'drous. Having about twenty stamens growing on the calyx.

Im'bricate. Lying over, like scales, or the shingles of a roof.

Imper'fect. Wanting the stamen or pistil.

Incarna'tus. Flesh-coloured.

Inci'sor. Front tooth.

Inclu'ded. Wholly received, or contained in a cavity; the opposite of Exsert.

Incomplete. Flowers destitute of a calyx or corolla are said to be incomplete.

Incum'bent. When the corcle is at the edges of the cotyledon.

Incras'sate. Thickened upward, larger towards the end.

In'crement. The quantity of increase.

Incum'bent. Leaning upon or against.

Incurv'ed. Bent inwards.

Indi'genous. Native, growing wild in a country. (Some exotics, after a time, spread and appear as if indigenous.)

In'durated. Becoming hard.

Indu'sium. A covering; plural, *indusia.*

Infe'rior. Below; a calyx or corolla is inferior when it comes out below the germ.

Infla'ted. . Appearing as if blown out with wind, hollow.

Inflex'ed. The same as incurved.

Inflores'cence. (From *infloresco,* to flourish.) The manner in which flowers are connected to the plant by the peduncle, as in the whorl, raceme, &c.

Infrac'tus. Bent in, with such an acute angle as to appear broken.

Infundibulifor'mis. Funnel-form.

Insert'ed. Growing out of or fixed upon.

Insi'dens. Sitting upon.

Insigni'tus. Marked.

Inte'ger. Entire.

Interno'de. The space between joints; as in grasses.

Interrup'tedly-pinnate. When smaller leafets are interposed among the principal ones.

Intor'tus. Twisted inwards.

Introdu'ced. Not originally native.— Brought from some other country.

Involucrum. A kind of general calyx serving for many flowers, generally situated at the base of an umbel or head.

Involu'cel. A partial involucrum.

In'volute. Rolled inwards.

Irides'cent. (From *Iris,* the rainbow.) Reflecting light.

Irreg'ular. Differing in figure, size, or proportion of parts among themselves.

Irritabil'ity. The power of being excited so as to produce contraction; this power belongs to vegetables as well as animals.

J

Jag'ged. Irregularly divided and sub divided.

Jaws. See Faux.

Joints. Knots or rings in culms, pods, leaves, &c.

Jugum. A yoke; growing in pairs.

Juxta-position. (From *juxta,* near, and *pono,* to place.) Nearness of place

K

Keel. The under lip of a papilionaceous flower.

Keel'ed. Shaped like the keel of a boat or ship.

Ker'nel. See Nucleus.

Kid'ney-shaped. Heart-shaped without the point, and broader than long.

Knee. A joint.

Knob'bed. In thick lumps, as the potato.

Knot. See Joints.

L

La'biate. Having lips, as in the class Didynamia.

Lacin'iate. Jagged, irregularly torn, lacerated.

Lactes'cent. Yielding a juice, usually white like milk, sometimes red, as in the blood-root.

Lac'teus. Milk-white.

Lacu'nose. Lowered with little pits or depressions.

Lacus'tris. Growing about lakes.

Læ'vis. Smooth, even.

Lam'ellated. In thin plates.

Lam'ina. The broad or flat end of a petal, in distinction from its claw. The expanded part of a leaf. In a more general sense, any thin plate or membrane.

La'nate. Woolly.

Lance'olate. Spear-shaped, narrow, with both ends acute.

Lance-o'vate. A compound of lanceolate and ovate, intermediate.

Lanu'ginous. Woolly.

Lat'eral. (From *latus.*) On one side

La'tent. (From *lateo,* to hide.) Hidden, concealed.

Lar'va. The caterpillar state of an insect.

Lax. Limber, flaccid.

Leaf'et. A partial leaf, part of a compound leaf.

Leaf'-stalk. See Petiole.

Leg'ume. A pod or pericarp, having its seeds attached to one side or suture; as the pea and bean.

Legu'minous. Bearing legumes.

Lepan'thium. A term used for a petal-like nectary; like that of the larkspur and monk's-hood.

Li'ber. The inner bark of plants.

Lig'neous. Woody.

Lig'num. Wood.

Lig'ulate. Strap or riband-like, flat, as the florets of the dandelion.

Lilia'ceous. A corolla with 6 petals gradually spreading from the base.

Limb. The border or spreading part of a monopetalous corolla.

Lin'ear. Long and narrow, with parallel sides, as the leaves of grasses.

Lip. The under petal in a labiate corolla.

Littori'bus. Growing on coasts, or shores.

Li'vidous. Dark purple.

Lobe. A large division, or distinct portion of a leaf or petal.

Loc'ulus. (From *locus*, a place.) A little space.

Lo'ment. A pod resembling a legume, but divided by transverse partitions.

Longifo'lius. Long-leaved.

Longis'simus. Very long.

Lu'cidus. Bright and shining.

Lunate or *Lunulate.* Shaped like a half moon.

Lu'rid. Of a pale dull colour.

Lu'teus. Yellow.

Ly'rate. Pinnatifid, with a large roundish leafet at the end.

M

Macula'tus. Spotted.

Mares'cent. Withering.

Mar'gin. The edge or border.

Mar'itime. Growing near the sea.

Medul'la. The pith or pulp of vegetables. The centre or heart of a vegetable.

Mellif'erous. (From *mel,* honey.) Producing or containing honey.

Mem'branous or *Membrana'ceous.* Very thin and delicate.

Mes'ocarp. The middle substance of the pericarp, having the *epicarp* on the outer, and the *endocarp* on the inner side.

Mes'osperm. That part of the seed which corresponds to the mesocarp of the pericarp.

Midrib. The main or middle rib of a leaf, running from the stem to the apex.

Minia'tus. Scarlet vermilion colour.

Mola'res. Back teeth, grinders.

Mol'lis. Soft.

Mollus'cous. Such animals as have a soft body without bones; as the oyster.

Monadel'phous. Having the stamens united in a tube at the base.

Monil'iform. Granulate, strung together like beads.

Monoceph'alous. (From *mono,* one, and *kephale,* head.) The term is applied to pericarps, which have but one summit, as the wheat, while the geum and anemone have as many as they have styles; they are polycephalous.

Monocotyled'onous. Having but one cotyledon.

Monœ'cious. Having pistillate and staminate flowers on the same plant.

Monopet'alous. The corolla all in one piece.

Monophyl'lous. Consisting of one leaf.

Monosep'alous. A calyx of one leaf or sepal.

Monosper'mus. One seed to a flower.

Monta'nus. Growing on mountains.

Moon-form. See Crescent-form.

Mosses. The second order of the class Cryptogamia.

Mucronate. Having a small point or prickle at the end of an obtuse leaf.

Multiflo'rus. Many-flowered.

Mul'tiplex. Many-fold, petals lying over each other in two rows.

Mul'tus. Many.

Mu'ricate. Covered with prickles.

N

Na'ked. Destitute of parts usually found.

Na'nus. Dwarfish, very small.

Nap. Downy, or like fur, tomentose.

Napifor'mis. Resembling a turnip.

Narcot'ic. (From *narco,* to stupify.) A substance which has the power of procuring sleep, as Opium.

Na'tant. Floating.

Natural character. That which is apparent, having no reference to any particular method of classification.

Natural history. The science which treats of nature.

Nec'tary. (From *nectar,* the fabled drink of the gods.) The part of a flower which produces honey; this term is applied to any appendage of the flower which has no other name.

Nemoro'sus. Growing in groves, often given as a specific name, as *Anemone nemorosa;* the ending in *a* denotes the adjective as being in the feminine gender; the adjective in Latin varying its termination to conform to the gender of the substantive.

Nerves. Parallel veins.

Nerved. Marked with nerves, so called, though not organs of sensibility like the nerves in the animal system.

Nic'titans. (From a word which signifies to twinkle or wink.) Applied as a specific name to some plants which appear sensitive; as the *Cassia nictitans.*

Ni'ger. Black.
Nit'idus. Glossy, glittering.
Niv'eus. Snow-white.
Nod'ding. Partly drooping.
Node, Nodus. Knot.
No'men. A name.
Notch'ed. See Crenate.
Nu'cleus. Nut, or kernel.
Nu'dus. See Naked.
Nut, Nux. See Nucleus.
Nu'tant. See Nodding, Pendulous.

O

Ob. A word which, prefixed to other terms, denotes the inversion of the usual position; as, obcordate, which signifies inversely cordate.
Obcon'ic. Conic with the point downwards.
Obcor'date. Heart-shaped with the point downwards.
Oblance'olate. Lanceolate with the base the narrowest.
Oblique. A position between horizontal and vertical.
Oblong. Longer than oval, with the sides parallel.
Obo'vate. Ovate with the narrower end towards the stem, or place of insertion
Ob'solete. Indistinct, appearing as if worn out.
Obtuse. Blunt, rounded, not acute.
Ochraceous. Colour of yellow ochre.
Odora'tus. Scented, odorous.
Officina'lis. Such plants as are kept for sale as medicinal, or of use in the arts.
Oid, Oi'des. This termination imports resemblance, as *petaloid*, like a petal; *thalictroides*, resembling a thalictrum, &c.
Opaque. Not transparent.
Oper'culum. The lid which covers the capsules of mosses.
Opposite. Standing against each other on opposite sides of the stem.
Orbic'ular. Circular
Orchid'eous. Petals like the orchis, four arched, the fifth longer.
Ornithol'ogy. That department of zoology which treats of birds.
Os. A bone. A mouth.
Os'seous. Bony, hard.
O'vary. A name sometimes given to the outer covering of the germ, before it ripens.
O'vate. Egg-shaped, oval with the lower end largest.
Ovip'arous. Animals produced from eggs, as birds, &c.
Ovules Little eggs: the rudiments of seeds which the germ contains before its fertilization; after which the ovules ripen into seeds.
O'vum. An egg.

P

Pal'ate. A prominence in the lower lip of a labiate corolla, closing or nearly closing the throat.
Palca'ceous. See Chaffy.
Pal'mate. Hand-shaped; divided so as to resemble the hand with the fingers spread.
Palus'tris. Growing in swamps and marshes.
Pan'icle. A loose, irregular bunch of flowers with subdivided branches, as the oat
Pan'icled. Bearing panicles.
Panex'tern. The outer covering of the pericarp.
Panin'tern. The inner covering of the pericarp.
Papil'io. A butterfly.
Papiliona'ceous. Butterfly-shaped,—an irregular corolla consisting of four petals; the upper one is called the banner, the two side ones wings, and the lower one the keel, as the pea. Mostly found in the class Diadelphia.
Papil'lose. Covered with protuberances.
Pappus. The down of seed, as the dandelion; a feathery appendage. See Egret.
Parisit'ic. Growing on another plant and deriving nourishment from it.
Paren'chyma. A succulent vegetable substance; the cellular substance; the thick part of leaves between the opposite surfaces; the pulpy part of fruits, as in the apple, &c.
Partial. Used in distinction to general.
Parti'tion. The membrane which divides pericarps into cells, called the dissepiment. It is said to be parallel when it unites with the valves where they unite with each other. It is contrary or transverse when it meets a valve in the middle or in any part not at its suture.
Parted. Deeply divided; more than cleft.
Patens. Spreading, forming less than a right angle.
Pau'ci. Few in number.
Pec'tinate. Like the teeth of a comb, intermediate between fimbriate and pinnatifid.
Pedate. Having a central leaf or segment and the two side ones which are compound, like a bird's foot.
Ped'icel. A little flower-stalk, or partial peduncle.
Pedun'cle. A stem bearing the flower and fruit.
Pel'licle. A thin membranous coat.
Pellu'cid. Transparent or limpid.

Pel'tate. Having the petiole attached to some part of the under side of the leaf.

Pendent. Hanging down, pendulous.

Pen'cilled. Shaped like a painter's pencil or brush.

Peregri'nus. Foreign, wandering.

Peren'nial. Lasting more than two years.

Perfo'liate. Having a stem running through the leaf; differs from connate in not consisting of two leaves.

Per'forate. Having holes as if pricked through; differs from punctate, which has dots resembling holes.

Pe'ri. Around.

Per'ianth. (From *peri,* around, *anthos,* flower.) A sort of calyx.

Pericarp. (From *peri,* around, and *karpos,* fruit.) A seed-vessel or whatever contains the seed.

Perid'ium. The round membranous case which contains the seeds of some mushrooms.

Perig'ynous. From *peri,* around, and *gynia,* pistil.

Periph'ery. The outer edge of the frond of a lichen; the circumference of a circle.

Pe'risperm. (From *peri,* around, and *sperma,* seed.) Around the seed. Skin of the seed.

Peristo'mium. The fringe or teeth around the mouth of the capsule of mosses, under the lid.

Permanent. Any part of a plant is m··'d to be permanent when it remains ..·nger than is usual for similar parts in most plants.

Persis'tent. Not falling off. See Permanent.

Per'sonate. Masked or closed.

Pe'tal. The leaf of a corolla, usually coloured.

Pe'tiole. The stalk which supports the leaf.

Phenog'amous. Such flowers as have stamens and pistils visible, including all plants except the cryptogamous.

Physiol'ogy. Derived from the Greek, a knowledge of nature.

Phytol'ogy. The science which treats of the organization of vegetables, nearly synonymous with the physiology of vegetables.

Pi'leole. The outer covering of the germinating leaves of monocotyledonous plants; that which formed the primordial leaf.

Pi'leus. The hat of a fungus.

Pillu. See Columella and Column.

Pilose. Hairy, with distinct, straightish hairs.

Pilus. A hair

Pimpled. See Papillose.

Pinna. A wing feather, applied to leafets.

Pinnate. A leaf is pinnate when the leafets are arranged in two rows on the side of a common petiole, as in the rose.

Pinnat'ifid. Cut in a pinnate manner. It differs from pinnate, in being a simple leaf deeply parted, while pinnate is a compound of distinct leafets.

Pistil. The central organ of most flowers, consisting of the germ, style, and stigma.

Pis'tillate. Having pistils but no stamens.

Pith. The spongy substance in the centre of the stems and roots of most plants. See Medulla.

Placenta. The internal part of the germ or ovary to which every ovule is attached, either immediately or by the funicle.

Plaited. Folded like a fan.

Plane. Flat with an even surface.

Pli'cate. See Plaited.

Plumo'se. Feather-like.

Plu'mula or *Plume.* The ascending part of a plant at its first germination.

Plu'rimus. Very many.

Pod. A dry seed-vessel, not pulpy, most commonly applied to legumes and siliques.

Podetia. The pedicels which support the frond of a lichen.

Po'dosperm. (From *podos,* a part, and *sperma,* seed.) Pedicel of the seed. The same as the funicle.

Pointal. A name sometimes used for pistil.

Pollen. Properly fine flower, or the dust that flies in a mill. The dust which is contained within the anthers.

Pollin'ia. Masses of pollen, as seen in the class Gynandria.

Po'lus. Many.

Polyan'drous. Having many stamens inserted upon the receptacle.

Polyceph'alous. See Monocephalous.

Polyg'amous. Having some flowers which are perfect, and others with stamens only, or pistils only.

Polymor'phous. Changeable, assuming many forms.

Polypet'alous. Having many petals.

Polyphyl'lous. Having many leaves.

Polysep'alous. A calyx of more than one leaf, or sepal.

Pome. A pulpy fruit, containing a capsule, as the apple.

Porous. Full of holes.

Por'rected. Extended forward.

Præmorse. Ending bluntly, as if bitten off; the same as abrupt.

Pras'inus. Green, like a leek.
Praten'sis. Growing in meadow land.
Prickle. Differs from the thorn in being fixed to the bark, the thorn is fixed to the wood.
Prismat'ic. Having several parallel flat sides.
Probos'cis. An elongated nose or snout, applied to projecting parts of vegetables.
Process. A projecting part.
Procum'bent. Lying on the ground.
Prolif'erous. A flower is said to be proliferous when it has smaller ones growing out of it.
Prop. Tendrils and other climbers.
Prox'imus. Near.
Pseudo. When prefixed to a word, it implies obsolete or false.
Pubes'cent. Hairy, downy, or woolly.
Pulp. The juicy cellular substance of berries and other fruits.
Pulver'ulent. Turning to dust.
Pu'milus. Small, low.
Punctate. Appearing dotted. See Perforated.
Pungent. Sharp, acrid, piercing.
Purpu'reus. Purple.
Pusil'lus. Diminutive, low.
Puta'men. A hard shell.
Pyriform. Pear-shaped.
Pyx'ide. (From *puxis*, a box.) Name of one of Mirbel's genera of fruits.

Q

Quadran'gular. Having four corners or angles.
Quater'nate. Four together.
Quinate. Five together.

R

Raceme. (From *rax*, a bunch of grapes, a cluster.) That kind of inflorescence in which the flowers are arranged by simple pedicels on the sides of a common peduncle; as the currant.
Ra'chis. The common stalk to which the florets and spikelets of grasses are attached; as in wheat heads. Also the midrib of some leaves and fronds.
Radiate. The ligulate florets around the margin of a compound flower.
Ra'dix. A root; the lower part of the plant which performs the office of attracting moisture from the soil, and communicating it to the other parts of the plant.
Rad'ical. Growing from the root.
Radicle. The part of the corculum which afterward forms the root; also the minute fibres of a root.
Ra'meus. Proceeding from the branches.
35*

Ramif'erous. Producing branches.
Ramose. Branching.
Ramus. A branch.
Ray. The outer margin of compound flowers.
Recep'tacle. The end of a flower-stalk: the base to which the different parts of fructification are usually attached.
Recli'ned. Bending over with the end inclining towards the ground.
Rectus. Straight.
Recurv'ed. Curved backwards.
Reflex'ed. Bent backwards, more than recurved.
Reg'mate. (From *regma*, to break with an explosion.) Name of one of Mirbel's genera of fruits.
Refrig'erant. (From *refrigero*, to cool.) Cooling medicines.
Re'niform. Kidney-shaped, heart-shaped without the point.
Repand. Slightly serpentine, or waving on the edge.
Repens. Creeping.
Resu'pinate. Upside down.
Retic'ulate. Veins crossing each other like net-work.
Retuse. Having a slight notch in the end, less than emarginate.
Rever'sed. Bent back towards the base.
Rev'olute. Rolled backward or outward.
Rhomboid. Diamond-form.
Rib. A nerve-like support to a leaf.
Riband-like. Broader than linear.
Rigid. Stiff, not pliable.
Ring. The band around the capsules of ferns.
Ringent. Gaping or grinning; a term applied to some labiate corollas.
Root. The descending part of a vegetable.
Rootlet. A fibre of a root, a little root.
Rosa'ceous. A corolla formed of roundish spreading petals, without claws or with very short ones.
Rose'us. Rose-coloured.
Rostel. That pointed part of the embryo, which tends downward at the first germination of the seed.
Rostrate. Having a protuberance like a bird's beak.
Rotate. Wheel-form.
Rotun'dus. Round.
Rubra. Red.
Rufous. Reddish yellow.
Rugose. Wrinkled.
Run'cinate. Having large teeth pointing backward, as the dandelion.
Rupes'tris. Growing among rocks.

S

Sagit'tate. Arrow-form.
Salif'erous. Bearing or producing salt.

Sulsus. Salt-tasted.

Salver-form. Corolla with a flat spreading border proceeding from the top of a tube: flower monopetalous.

Sam'ara. A winged pericarp not opening by valves, as the maple.

Sap. The watery fluid contained in the tubes and little cells of vegetables.

Supor. Having taste.

Sarmen'tose. Running on the ground, and striking root from the joints only, as the strawberry.

Sar'cocarp. (From *sarz*, flesh, and *karpos*, fruit.) The fleshy part of fruit.

Scaber, or *Sca'brous.* Rough.

Scandens. Climbing.

Scape. A stalk which springs from the root, and supports flowers and fruit but no leaves, as the dandelion.

Sca'rious. Having a thin membranous margin.

Scattered. Standing without any regular order.

Scions. Shoots proceeding laterally from the roots or bulb of a root.

Secernant stimulants, are medicines which promote the internal secretions.

Secund. Unilateral, arranged on one side only.

Segment. A part or principal division of a leaf, calyx, or corolla.

Sempervi'vens. Living through the winter, and retaining its leaves.

Sepal. Leaves or divisions of the calyx.

Septa. Partitions that divide the interior of the fruit.

Septiferous. Bearing septa.

Serrate. Notched like the teeth of a saw.

Ser'rulate. Minutely serrate.

Sessile. Sitting down; placed immediately on the main stem without a foot-stalk.

Seta. A bristle.

Seta'ceous. Bristle-form.

Setose. Covered with bristles.

Shaft. A pillar, sometimes applied to the style.

Sheath. A tubular or folded leafy portion including within it the stem.

Shoot. Each tree and shrub sends forth annually a large shoot in the spring and a smaller one from the end of that in June.

Shrub. A plant with a woody stem, branching out nearer the ground than a tree, usually smaller.

Sic'cus. Dry.

Sil'icle. A seed-vessel constructed like a silique, but not much longer than it is broad.

Silique. A long pod or seed-vessel of two valves, having the seed attached to the two edges alternately.

Simple. Not divided, branched or compounded.

Sin'uate. The margin hollowed out resembling a bay.

Si'nus. A bay; applied to the plant, a roundish cavity in the edge of the leaf or petal.

So'ri. Plural of sorus; fruit-dots on ferns.

Sorose. A genus of fruits in Mirbel's classes.

Spa'dix. An elongated receptacle of flowers, commonly proceeding from a spatha.

Spa'tha. A sheathing calyx opening lengthwise on one side, and consisting of one or more valves.

Spat'ulate. Large, obtuse at the end, gradually tapering into a stalk at the base.

Spe'cies. The lowest division of vegetables.

Specif'ic. Belonging to a species only

Sper'ma. Seed.

Spike. A kind of inflorescence in which the flowers are sessile, or nearly so, as in the mullein, or wheat.

Spike'let. A small spike.

Spin'dle-shaped. Thick at top, gradually tapering, fusiform.

Spine. A thorn or sharp process growing from the wood

Spinescent. Bearing spines or thorns.

Spino'sus. Thorny.

Spi'ral. Twisted like a screw.

Sporules. That part in cryptogamous plants which answers to seeds.

Spur. A sharp hollow projection from a flower, commonly the nectary.

Spur'red-rye. A morbid swelling of the seed, of a black or dark colour sometimes called ergot; the black kind is called the malignant ergot Grain growing in low, moist ground, or new land, is most subject to it.

Squamo'se. Scaly.

Squarro'se. Ragged, having divergent scales.

Stamen. That part of the flower on which the artificial classes are founded.

Stam'inate. Having stamens withou pistils.

Standard. See Banner.

Stel'late. Like a star.

Stem. A general supporter of leaves flowers, and fruit.

Stemless. Having no stem.

Ster'il. Barren.

Stigma. The summit, or top of the pistil.

Stipe. The stem of a fern, or fungus

also the stem of the down of seeds, as in the dandelion.

Stip'itate. Supported by a stipe.

Stip'ule. A leafy appendage, situated at the base or petioles, or leaves.

Stolonif'erous. Putting forth scions, or running shoots.

Stramin'eous. Staw-like, straw-coloured.

Strap-form. Ligulate.

Stratum. A layer; plural, strata.

Striate. Marked with fine parallel lines.

Strictus. Stiff and straight, erect.

Strigose. Armed with close thick bristles.

Strobilum. A cone, an ament with woody scales.

Style. That part of the pistil which is between the stigma and the germ.

Styli'des. Plants with a very long style.

Sua'vis. Sweet, agreeable.

Sub. Used as a diminutive, prefixed to different terms to imply the existence of a quality in an inferior degree; in English, may be rendered by somewhat; it also signifies under, or less than.

Sub'acute. Somewhat acute.

Subero'se. Corky.

Submersed. Growing under water.

Sub'sessile. Almost sessile.

Subterra'neous. Growing and flowering under ground.

Subtus. Beneath.

Sub'ulate. Awl-shaped, narrow and sharp pointed. See Awl-form.

Suc'culent. Juicy; it is also applied to a pulpy leaf, whether juicy or not.

Sucker. A shoot from the root by which the plant may be propagated.

Suffru'ticose. Somewhat shrubby, shrubby at the base; an under shrub.

Sulcate. Furrowed, marked with deep lines.

Super. Above.

Supradecom'pound. More than decompound; many times subdivided.

Superior. A calyx or corolla is superior, when it proceeds from the upper part of the germ.

Supi'nus. Face upwards. See Resupinate.

Suture. The line or seam formed by the junction of two valves of a seed-vessel.

Syco'ne. (From *sucon*, a fig.) A name given to one of Mirbel's genera of fruits.

Sylves'tris. Growing in woods.

Syn'carpe. (From *sun*, with, and *karpos*, fruit.) A union of fruits.

Syngene'sious. Anthers growing together, forming a tube; such plants as constitute the class Syngenesia, being also compound flowers.

Syn'onyms. Synonymous, different names for the same plant.

Synop'sis. A condensed view of a subject, or science.

T

Taxon'omy. (From *taxis*, order, and *nomos*, law.) Method of classification.

Teeth of Mosses. The outer fringe of the peristomium is generally in 4, 8, 16, 32, or 64 divisions: these are called teeth.

Tegens. Covering.

Teg'ument. The skin or covering of seeds; often burst off on boiling, as in the pea.

Tem'perature. The degree of heat and cold to which any place is subject, is wholly dependant upon latitude, being affected by elevation; the mountains of the torrid zone produce the plants of the frigid zone. In cold regions white and blue petals are more common, in warm regions red and other vivid colours; in the spring we have more white petals, in the autumn more yellow ones.

Ten'dril. A filiform or thread-like appendage of some climbing plants, by which they are supported by twining round other objects.

Tenel'lus. Tender, fragile.

Tenuifo'lius. Slender-leaved.

Tenuis. Thin and slender.

Ter'ete. Round, cylindrical, tapering.

Ter'minal. Extreme, situated at the end.

Ter'nate. Three together, as the leaves of the clover.

Tetradyn'amous. With four long and two short stamens.

Tetran'drous. Having four stamens.

Thorn. A sharp process from the woody part of the plant; considered as an imperfect, indurated bud.

Thyrse. A panicle which is dense.

Tige. See Caulis.

Tincto'rious. Plants containing colouring matter.

Tomen'tose. Downy; covered with fine matted pubescence.

Tonic. (From *tono*, to strengthen.) Medicines which increase the tone of the muscular fibre.

Toothed. See Dentate.

Torose. Uneven, alternately elevated and depressed.

Torulose. Slightly torose.

Trachea. Names given to vessels supposed to be designed for receiving and distributing air.

Transverse. Crosswise.

Trichot'omous. Three-forked.

Trifid. Three-cleft.

Trifo'liate. Three-leaved.

Trilo'bate. Three-lobed.

Triloc'ular. Three-celled.

Trun'cate. Having a square termination, as if cut off.

Trunk. The stem or bole of a tree.

Tube. The lower hollow cylinder of a monopetalous corolla.

Tuber. A solid fleshy knob.

Tuberous. Thick and fleshy, containing tubers, as the potato.

Tubular. Shaped like a tube, hollow.

Tu'nicate. Coated with surrounding layers, as in the onion.

Turgid. Swelled, inflated.

Turbinate. Shaped like a top, or pear.

Twining. Ascending spirally.

Twisted. Coiled.

U

Uligino'sus. Growing in damp places.

Umbilicate. Marked with a central depression.

Umbel. A kind of inflorescence in which the flower-stalks diverge from one centre, like the sticks of an umbrella.

Umbellif'erous. Bearing umbels.

Umbo. The knob in the centre of the hat or pileus of the fungi tribe, originally the top of a buckler.

Unarmed. Without thorns or prickles.

Uncinate. Hooked.

Unctuo'sus. Greasy, oily.

Un'dulate. Waving, serpentine, gently rising and falling.

Unguis. A claw.

Unguic'ulate. Inserted by a claw.

Uniflo'rus. One-flowered.

U'nicus. Single.

Unilat'eral. Growing on one side.

Urce'olate. Swelling in the middle, and contracted at the top in the form of a pitcher.

Utricle. A little bladder, a term applied to capsules of a peculiar kind.

V

Valves. The parts of a seed-vessel into which it finally separates; also the leaves which make up a glume, or spatha.

Variety. A subdivision of a species, distinguished by characters which are not permanent.

Vaulted. Arched; with a concave covering.

Veined. Having the divisions of the petiole irregularly branched on the under side of the leaf.

Ven'tricose. Swelled out. See Inflated.

Vermifuge. A medicine for the cure of worms.

Vernal. Appearing in the spring.

Verrucose. Warty, covered with little protuberances.

Vertical. Perpendicular.

Verticil'late. Whorled, having leaves or flowers in a circle round the stem.

Vesic'ular. Made up of cellular substance.

Vespertine. Flowers opening in the evening.

Vex'illum. See Banner.

Vil'lose. Hairy, the hairs long and soft.

Viola'ceous. Violet-coloured.

Villus. Soft hairs.

Vires'cens. Inclining to green.

Virgate. Long and slender. Wand-like

Vir'idis. Green.

Virgultum. A small twig.

Virose. Nauseous to the smell, poisonous.

Viscid. Thick, glutinous, covered with adhesive moisture.

Vitel'lus. Called also the yolk of the seed; it is between the albumen and embryo.

Vit'reus. Glassy.

Vivip'arous. Producing others by means of bulbs or seeds, germinating while yet on the old plant.

Vul'nerary. (From *vulnus*, a wound.) Medicines which heal wounds.

W

Wedge-form. Shaped like a wedge, rounded at the large end, obovate with straightish sides.

Wheel-shaped. See Rotate.

Wings. The two side petals of a papilionaceous flower.

Wood. The most solid parts of trunks of trees and shrubs.

Z

Zool'ogy. The science of animals.

Zo'ophytes. The lowest order of animals, sometimes called animal plants though considered as wholly belonging to the animal kingdom. Many of them resemble plants in their form, and exhibit very faint marks of sensation.

SECTION VI.

SYMBOLICAL LANGUAGE OF FLOWERS.

BESIDES the scientific relations which are to be observed in plants, flowers may also be regarded as emblematical of the affections of the heart and qualities of the intellect. In all ages of the world, history and fable have attached to flowers particular associations; consecrating them to melancholy remembrances, to glory, friendship, or love. In oriental countries, a *selam*, or boquet of flowers, is often made the interesting medium of communicating sentiments, to which words are inadequate.

The authorities for the emblems here adopted, are, "Flora's Dictionary," "Garland of Flora," "*Les Vegeteux Curieux*," and "*Emblems des Fleurs*." In a few cases, alterations have been made, in order to introduce sentiments of a more refined and elevated character, than such as relate to mere personal attractions.

A

Acacia. Friendship.

Acanthus. Indissoluble ties.

Aconitum. (*Monk's hood.*) Deceit. Poisonous words.

Adonis autumnalis. Sorrowful remembrances.

Agrostemma. (*Cockle.*) Charms please the eye, but merit wins the soul.

Althea. I would not act contrary to reason.

Aloe. Religious superstition. Think not the Almighty wills one idle pang, one needless tear.

Amaranthus. Immortality. Unchangeable.

A. melancholicus. Love lies bleeding.

Amaryllis. Splendid beauty. Coquetry.

Anemone. Anticipation. Frailty.

Apocynum. Falsehood.

Arbor Vitæ. (*Thuja occidentalis.*) Friendship unchanging.

Arum. Deceit. Ferocity. Treachery.

Asclepias. (*Milk-weed.*) Cure for the heartache. The miserable have no medicine but hope.

Aster. Beauty in retirement.

Auricula. Elegance. Pride.

B

Bachelor's button. Hope, even in misery.

Balm. Sweets of social intercourse.

Balsam. (*Impatiens.*) Impatience. Do not approach me.

Bay. (*Laurus.*) I change but with death.

Box. Constancy.

Broom. Humility.

Broom-corn. Industry.

C

Calla ethiopica. (*Egyptian lily.*) Feminine delicacy.

Camellia japonica. (*Japan rose.*) Pity is easily changed to love.

Campanula. (*Bell-flower.*) Gratitude.

Cape jasmine. (*Gardenia florida.*) My heart is joyful.

Cardinal flower. (*Lobelia cardinalis.*) High station does not secure happiness.

Carnation. (*Dianthus*) Disdain. Pride.

Catch-fly. (*Lychnis.*) I am a willing prisoner.

Cedar. (*Juniperus*) You are entitled to my love.

Chamomile. Bloom in sorrow. Energy to act in adversity.

China-aster, double. (*Aster chinensis*) Your sentiments meet with a return.

China-aster, single. You have no cause for discouragement.

Chrysanthemum, red. Love.

Chrysanthemum, white. Truth needs no protestations.

Chrysanthemum, yellow. A heart left to desolation.

Citron. Beautiful, but ill-humoured.

Clematis. (*Virgin's-bower.*) Mental excellence.

Cock's-comb. (*Amaranthus.*) Foppery. Affectation.

Columbine, purple. (*Aquilegia canadensis.*) I cannot give thee up.

Columbine, red. Hope and fear alternately prevail.

Convolvulus. Uncertainty.

Cornus. Indifference. A changed heart.

Cowslip. (*Primula.*) Native grace.

Crocus. Cheerfulness.

Crown-imperial. (*Fritillaria imperialis.*) Power without goodness.

Cypress. Disappointed hopes. Despair.

D

Dahlia. For ever thine.

Daisy. (*Bellis perennis.*) Unconscious beauty.

Dandelion. Smiling on all. Coquetry.

E

Eglantine. (*Rosa rubiginosa.*) I wound to heal.

Elder. (*Sambucus.*) Compassion yielding to love.

Everlasting. (*Gnaphalium.*) Never-ceasing remembrance.

F

Fox-glove. (*Digitalis.*) I am not ambitious for myself, but for you.

Fuschsia. (*Ladies' ear-drop.*)

It were all one,
That I should love a bright particular star,
And think to wed it.

G

Geranium, fish. Thou art changed.

Geranium, oak. Give me one look to cheer my absence.

Geranium, rose. Many are lovely, but you exceed all.

H

Hawthorn. (*Cratægus.*) Hope! I thee invoke!

Heart's-ease. (*Viola tricolor.*) Forget me not.

Hibiscus. Beauty is vain.

Holly. (*Ilex.*) Think upon your vows.

Hollyhock. (*Althæa rosea.*) Ambition.

Honeysuckle. (*Lonicera.*) I strive with grief. Fidelity.

Houstonia cerulea. Meek and quiet happiness. Innocence.

Hyacinth. Love is full of jealousy.

Hydrangea. A boaster. Superior merit, when assumed, is lost.

Hypericum. (*St. John's wort.*) Animosity

I

Ipomœa. Busybody. Busybodies are a dangerous sort of people.

Iris. I have a message for you.

Ivy. (*Vitis hedera.*) Female affection. I have found one true heart.

J

Jasmine. You bear a gentle mind. Amiability.

Jonquil. (*Narcissus.*) Affection returned.

L

Laburnum. (*Cytisus laburnum.*) Pensive beauty.

Ladies'-slipper. (*Cypripedium.*) Capricious beauty.

Larkspur. (*Delphinium.*) Inconstancy. Inconstant as the changing wind.

Laurel. (*Kalmia.*) Oh what a goodly outside falsehood hath!

Lavender. Words though sweet may be deceptive.

Lemon. (*Citrus lemonium.*) Discretion. Prudence.

Lilac. (*Syringa.*) First love.

Lily, white. (*Lilium candidum.*) Purity. With looks too pure for earth

Lily, yellow. False. Light as air.

Lily of the valley. (*Convallaria.*) Delicacy. The heart withering in secret.

Locust, the green leaves. Affection beyond the grave. Sorrow ends not when it seemeth done!

Lupine. Indignation.

M

Magnolia. Perseverance.

Marigold. Cruelty. Contempt.

Mirabilis. (*Four-o'clock.*) Timidity.

Mignonette. (*Reseda odorata.*) Moral and intellectual beauty.

Mimosa. (*Sensitive plant.*) My heart is a broken lute!

Mock orange, or *Syringa.* (*Philadelphus.*) Counterfeit. I cannot believe one who has once deceived me.

Myrtle. (*Myrtus.*) Love.

Myrtle, withered. Love betrayed.

N

Narcissus. Egotism. The selfish heart deserves the pain it feels.

Nasturtion. (*Tropæolum.*) Honour to the brave. Wit.

Nettle. (*Urtica.*) Scandal.

Nightshade. Suspicion. Artifice. Skepticism.

O

Oleander. Beware. Shun the coming evil. In vain is the net spread in the sight of any bird.

Olive. Peace. After a storm comes a calm.

Orange flowers. Bridal festivity.

P

Parsley. (*Apium.*) Useful knowledge.

Passion-flower. (*Passiflora.*) Devotion.

Peach blossom. Here I fix my choice.

Periwinkle. (*Vinca.*) Recollection of the past.

Phlox. Our souls are united.

Pine. (*Pinus resinosa.*) Time and philosophy.

Pine. Spruce. Farewell! for I must leave thee.

Pink, single white. (*Dianthus.*) Ingenuousness. Stranger to art.

Pink, single red.

> A token of all the heart can keep
> Of holy love, in its fountain deep.

Pink, China. (*Dianthus chinensis.*) Aversion. Though repulsed, not in despair.

Pink, variegated. Refusal. You have my friendship, ask not for more.

Pæony. (*Pæonia.*) Anger. Ostentation.

Polyanthus. Thou knowest my confidence in thee.

Pomegranate flower. (*Punica.*) Mature and beautiful.

Poppy, red. Consolation. Let the darkness of the past be forgotten in the light of hope.

Poppy, white.

> Doom'd to heal, or doom'd to kill—
> Fraught with good, or fraught with ill.

Poppy, variegated. Beauty without loveliness.

Primrose. (*Primula.*) Be mine the delight of bringing modest worth from obscurity.

Primrose, evening. (*Œnothera.*) Inconstancy. Be not beguiled with smooth words. Man's love is like the changing moon.

R

Ranunculus. Thou art fair to look upon, but not worthy of affection.

Rosemary. Keep this for my sake. I'll remember thee.

Rose-bud. Confession. Thou hast stolen my affections.

Rose, Burgundy. Modesty and innocence united to beauty.

Rose, damask. Sweeter than the opening rose.

Rose, red. The blush of modesty is lovely.

Rose, moss. Superior merit.

Rose, white.

> I would be,
> In maiden meditation, fancy free.

Rose, white, withered. Emblem of my heart. Withered like your love.

Rose, wild. Simplicity. Let not your unsophisticated heart be corrupted by intercourse with the world.

Rose, cinnamon. Without pretension. Such as I am, receive me. Would I were of more worth for your sake.

Rue. (*Ruta.*) Disdain. This trifling may be mirth to you, but 'tis death to me.

S

Sage. (*Salvia.*) Domestic virtues. Woman's province is home.

Scarlet lychnis. (*Lychnis chalcedonica.*) I see my danger without power to shun.

Snapdragon. (*Antirrhinum.*) I have been flattered with false hopes.

Snow-ball. (*Viburnum.*) Virtues cluster around thee. A union.

Snow-drop. (*Galanthus.*) Though chilled with adversity, I will be true to thee. I am not a summer friend.

Solidago. (*Golden rod.*) Encouragement.

Sorrel. (*Rumex.*) Wit ill-timed. He makes a foe who makes a jest.

Speedwell. (*Veronica.*)

> True love 's a holy flame,
> And when 'tis kindled, ne'er can die.

Spider-wort. (*Tradescantia.*) The pledge of *friendship*, 'tis all my heart can give. Wouldst thou then counsel me to fall in love?

Star of Bethlehem. (*Ornithogalum.*) Reconciliation. Light is brightest when it shines in darkness.

Stock july-flower. You are too lavish of your smiles.

Strawberry. (*Fragaria.*) A pledge of future happiness.

Sumach. (*Rhus.*) Splendour. Wealth cannot purchase love. Have you never seen *splendid misery?*

Sun-flower. (*Helianthus.*) You are too aspiring.

Sweet-pea. Departure. Must you go!

Sweet-william. (*Dianthus barbatus*) Finesse. One may smile and be a villain. I cannot smile when discontent sits heavy at my heart.

T

Thistle. (*Carduus.*) Misanthropy. O that the desert were my dwelling place!

Thorn-apple. (*Stramonium.*) Alas! that falsehood should appear in such a lovely form!

Thyme. Less lovely than some, not more estimable.

Tuberose. (*Polyanthus tuberosa.*) Blessings brighten as they take their flight!

Tulip. Vanity. Thou hast metamorphosed me! This love has been like a blight upon my opening prospects.

Tulip-tree. (*Liriodendrum.*) Rural life favourable to health and virtue.

V

Verbena. Sensibility.

The heart that is soonest awake to the flowers, Is always the first to be touch'd by the thorns.

Violet, blue. Faithfulness. I shall never forget.

Violet, white. Modest virtue.

W

Wall-flower. (*Cheiranthus.*) Misfortune is a blessing when it proves the truth of friendship.

Water-lily. The American lotus. (*Nymphæa.*) An emblem of silence

Weeping-willow. (*Salix.*) Forsaken. Ask not one to join in mirth whose heart is desolate.

Wood sorrel. (*Oxalis.*) Tenderness and affection.

Woodbine. (*Lonicera.*) Fraternal love.

Y

Yarrow. (*Achillea.*) To heal a wounded heart.

GENERAL INDEX

TO THE

LECTURES AND APPENDIX.

The figures refer to the pages—Ap. stands for Appendix.

INDEX

INDEX·

TO THE

NATURAL ORDERS.

The numbers refer to the pages of the Appendix.

COMMON NAMES OF PLANTS.

In the following index, either the whole name of the genus, or one or two of its first syllables, are annexed to the common name. By a reference to the alphabetical arrangement of genera, the Specific description, the Artificial Order and Class, and the Natural Order, are ascertained.

Acacia. Robi-
Adam's needle. Yuc-
Adder-tongue. Erythro-
Adder-tongue fern. Ophi-
Agrimony. Agri-
Albany beech-drops. Pte-
Alder. Alnus.
Alum-root. Heuch-
American laurel. Kal-
American cowslip. Caltha.
American oil-nut. Hamil-
American water-cress. Carda-
American papaw-tree. Asi-
Anemone. Anem-
Angelica. Angel-
Anise-tree. Illi-
Apple. Pyrus.
Apricot. Armeni-
Arbor vitæ. Thuja.
Arrow-grass. Triglo-
Artichoke. Cyna- Helian-
Arrow-head. Sagit-
Ash. Fraxi-
Asparagus. Aspar-
Asphodel. Aspho-
Atamasko-lily. Amaryl-
Avens. Geum.

w

Bachelor's-button. Gomphre-
Balm. Melis-
Balsamine. Impa-
Balsam-apple. Momor-
Balm of Gilead. Popu- Amyr-
Barley. Horde-
Barberry. Berber-
Bass-wood. Tilia.
Bay-berry. Myrica.
Beard-grass. Androp-
Beard-tongue. Pentste-
Bean. Phase-
Bear-berry. Arbu-
Bed-straw. Galium.
Beech. Fagus.
Beech-drops. Epiphe-
Beet. Beta.
Bell-wort. Uvula-
Bell-flower. Campan-
Billberry. Vac-
Birch. Betula.
Bitter-vetch. Orobus.
Bird's-nest. Monotropa.
Bird-wort. Aristo-
Bind-weed. Convol-
Blackberry. Rubus-
Blackberry-lily. Ixia.
Black-flower. Melanthium.
Black-hoarhound. Ballo-
Black-walnut. Juglans.
Bladder-campion. Cucubalus.
Bladder-nut. Staph-
Bladder-senna. Colut.
Bladder-wort. Utricu-

Blazing-star. Helo-
Blessed-thistle. Centau-
Blind-starwort. Mie-
Blite. Blitum.
Blood-marigold. Zinnia.
Blood-root. Sanguin-
Blue-bell. Campan-
Blue-curls. Trichos-
Blue-hearts. Buchne-
Blue-eyed grass. Sisy-
Blue-bottle. Centau-
Blue-gentian. Isan-
Bog-rush. Kyl-
Boneset. Eupa-
Borage. Bora-
Bouncing-bet. Sapo-
Box. Buxus.
Box-wood. Cornus.
Brake. Pteris.
Bread-grass. Ares-
Bristled-panic. Pen-
Broom-corn. Sorgh-
Buck-eye. Æsc-
Buckwheat. Polygo-
Buck-bean. Menyan-
Buckthorn. Rhamnus.
Bugloss. Anchu-
Bulrush. Juncus.
Burnet. Poteri-
Burdock. Arcti-
Burnet-saxifrage. Sanguisor-
Burr-reed. Sparga-
Bush-clover. Hedys- Lespe-
Bush-honeysuckle. Dierv-
Butternut. Juglans.
Butterfly-weed. Vexil-
Butter-wort. Pinguic-
Button-bush. Cephal-
Button-wood. Plata-

Cabbage. Bras-
Cahlops. Trib-
Campion. Lych-
Cancer root. Epiph-
Cane. Mie-
Canna. Canna.
Caraway. Carum.
Cardinal-flower. Lobel-
Carolina allspice. Calycan-
Carpet-weed. Mollug.
Carrot. Daucus.
Castor-oil plant. Rici-
Catalpa. Catal-
Catch-fly. Silene.
Catnep. Nepeta.
Cat-tail. Typha.
Caterpillar-fern. Scol-
Celery. Apium.
Centaury. Sabba-
Chamomile. Aneth-
Chara. Cha-
Cherry. Prunus. Ceras-

Chestnut. Casta-
Chess. Broom-grass. Bro-
Chick-wintergreen. Trien.
Chick-pea. Cicer.
Choke-berry. Aronia.
Cinque-foil. Poten-
Cives. Allium.
Clarkia. Clar-
Clover. Trifo-
Club-rush. Seir-
Cockle. Agros-
Cock-foot grass. Panicum.
Cockscomb. Amar-
Coffee-bean. Gymno-
Cohosh. Macro-
Colic-weed. Cory-
Colt's foot. Tussil-
Columbine. Aqui.
Comb tooth thistle. Cardu-
Comfrey. Symph-
C?e-flower. Rud-
?al-tree. Erythrythrina.
Coral-root. Coral-
Coreopsis. Coreop-
Coriander. Corian-
?. Coro-
?-thistle. Onop-
Cotton? Gossyp-
?rsley. Herac-
?heat. Melampy-
?. Doli-
?ry. Oxyc-
?g-cucumber. Meloth
?ng-vetch. Ervum.
?wberry. Empe-
Cro??eard. Verbes-
??perial. Friti-
??oot. Ranun-
?cumber. Cucum-
Culv??physic. Leptan-
??Ribes.
??-leaf. Mitel-
?t-grass. Leer-
??-vine. Ipo-

Daffodil. Narci-
??lum. Diosp-
??ion. Leon-
Darnel-grass. Loli-
?y-flower. Comme-
?y-lily. Hemero-
Dead-nettle. Lami-
Deadly nightshade. Arop-
Deer-grass. Rhex-
Dew-berry. Rubus.
Dill. Aneth-
Ditch-moss. Udo-
Dittany. Cuni-
Dock. Rumex.
Dodder. Cuscu-
Dog-tooth violet. Erythro-
Dog-bane. Apoc-
Dog-wood. Cornus.
Dragon-head. Dracoceph-
Dry-strawberry. Dali-
Duck's-meat. Lemna.
Dwarf-dandelion. Krig-
Dyer's-broom. Genis-

Ear-drop. Fuschia.
Elder. Sambu.

Elecampane. Inu-
Elephant's-foot. Eleph-
Elm. Ulmus.
Enchanter's nightshade. Cir-
Endive. Cicho-
English cowslip. Primu-
English primrose. Prima-
English water-cress. Erysim-
European ivy. Hedera.
Evening primrose. Œnoth-
Eye-bright. Euphr-

False papaw-tree. Cari-
False rush-grass. Leer-
False bog-rush. Pehyr-
False saffron. Cartha-
False spiked-alder. Elliot-
False syringa. Philad.
False toad-flax. Thesi-
False wake-robin. Trill-
Fan-palm. Chamæ-
Feather-leaf. Hydro-
Feather-grass. Sti-
Fennel. Aneth-
Fescue grass. Festu-
Fenu-greek. Trig-
Fever-few. Chrysan-
Fever-root. Trios-
Field-sorrel. Rumex.
Field-thyme. Clini-
Fig-tree. Ficus.
Fire-weed. Sene-
Flag. Iris.
Flax. Linum.
Flower-de-luce. Iris.
Flowering almond. Amyg.
Flowering arum. Oron-
Flowering ash. Ornus.
Flowering fern. Osmun-
Flowering nettle. Galeop-
Flowering raspberry. Rubus.
Fork-fern. Heros-
Fool's-parsley. Areth-
Four-o'clock. Mirab-
Fringe-tree. Chion-
Fringe-tree, purple. Rhus-
Frost-plant. Cistus.
Fumitory. Fuma-

Garden artichoke. Cynara.
Garden daisy. Chrysan-
Garden ladies-slipper. Impa-
Gayfeather. Liatris.
Gentian. Gentia-
Geranium. Pelarg-
Gill-over-ground. Glech-
Ginseng. Panax.
Globe-flower. Troll-
Globe-thistle. Echi-
Goat's-rue. Gale-
Gold-basket. Alyss-
Gold-of-pleasure. Alyss.
Gold-thread. Coptis.
Golden-rod. Solid-
Golden-saxifrage. Chrys-
Gooseberry. Ribes.
Gourd. Cucur-
Grape-fern. Botrych-
Grape-vine. Vitis.
Grass-pink. Cymbid-
Grass-wrack. Zos-
Greek valerian. Polemo-

Green-brier. Smilax.
Gromwell. Lithos-
Ground-ivy. Glech-
Ground-nut. Api-
Ground-pine. Lycopo-
Groundsel-tree. Baccha-

Hardhack. Spirea.
Hawk-weed. Hiera-
Hawthorn. Cratægus.
Hazel-nut. Corylus.
Heath. Eri-
Hedge-hyssop. Grati-
Hedge-mustard. Sisym-
Hedge-nettle. Stach-
Hellebore. Helleb-
Hemp. Cannab-
Henbane. Hyoscy-
Hickory. Carya.
High Cranberry. Vibur-
High healall. Pedic-
High-water shrub. Iva.
Hog-weed. Ambro-
Hoarhound. Marr-
Hollyhock. Alth-
Honey-locust. Gledit-
Hop. Humu-
Horn-beam. Ostr-
Horn-wort. Cera-
Horned poppy. Arge-
Horse-chesnut. Æscu-
Horse-radish. Coch-
Horse-balm. Collinson-
Hound-tongue. Cynog-
House-leek. Semper-
Hydrangea. Hydran-

Ice-plant. Mesem-
Indian corn. Zea.
Indian cucumber. Mede-
Indian mallows. Sida.
Indian physic. Gille-
Indian reed. Canna.
Indigo. Indi-
Innocence. Hous-
Iron-wood. Ostr-

Jasmine. Jas-
Jerusalem artichoke. Heliar
Jewel-weed. Impa-
Job's-tear. Coix.
Jonquil. Narcis-
Judas'-tree. Cercis.
Juniper-berry. Juni-

Knawell. Scleran-
Knot-grass. Polyg-

Labrador tea. Ledum.
Ladies'-mantle. Alche-
Ladies'-tresses. Neot-
Ladies'-slipper. Cypri-
Lady-in-the-green. Nigel-
Lamb-lettuce. Fe-
Larkspur. Delph-
Lavender. Lavan-
Lavatera. Lava-
Leaf-flower. Phyll-
Leather-leaf. Androm-
Leather-wood. Dir-
Leek. Allium.
Lemon. Citrus.

Leopard's-bane. Arni
Lettuce. Lact-
Lichnidia. Phlox.
Lilac. Syr-
Lily-of-the-valley. Con-
Lily. Lilium.
Limodore. Tipu.
Liquorice. Glycyrr-
Live-forever. Sedum.
Liver-leaf. Hepat-
Lizard-tail. Sauru-
Locust-tree. Robin-
Loose-strife. Lysim-
Lop-seed. Phry-
Lucerne clover. Medic-
Lung-wort. Pulmo-

Madder. Rub-
Magnolia. Magn-
Maiden-hair. Adian-
Malabar-nut. Justi-
Mangrove. Rhizo-
Maple. Acer.
  arjoram. Ori-
  arsh penny-wort. Hydroc-
  arsh rosemary. Stati-
  arigold. Tagetes. Calen-
Matrimony-vine. Lycium.
Mat-grass. Nar-
Mayweed. Anthe-
Meadow-rue. Thal-
Medlar. Mespi-
Meliot-clover. Meli-
Mermaid-weed. Proa.
Melic grass. Meli-
Mezereon. Daphne.
Mignonette. Rese-
Milk-weed. Ascle-
Milk-willow herb. Lytn-
Milk-vine. Periplo-
Milk-vetch. Astrag-
Mint. Mentha.
Mistletoe. Viscum.
Mitre-wort. Tiar-
Mock-orange. Philad-
Monkey-flower. Mimu-
Monk's-hood. Aconi-
Moon-seed. Menis-
Moor-grass. Sesle-
Morning-glory. Ipo-
Motherwort. Leonu-
Mountain-ash. Sorbus.
Mountain-daisy. Bellis.
Mountain-flax. Polyg-
Mountain-mint. Pycnan-
Mountain-rice. Ory-
Mouse-ear. Ceras-
Mud-purslane. Portu-
Mulberry. Morus.
Mullein. Verbas-
Mullein, pink. Agrostem-
Muskmelon. Cucumis.
Muskmallows. Hibis-
Mushroom. Agaricus.
Myrtle. Myrtus.

Nasturtion. Trop-
Necklace-weed. Acta-
Nettle. Urt-
Nettle-tree. Celtis.
Night-shade. Solan-

Oak. Quercus.
Oak of Jerusalem. Cheno-
Oat. Avena.
Oil-nut. Hamil-
Oily grain. Sesam-
Old man's beard. Tilland-
Olive. Olea.
Onion. Allium-
Orach. Atrip-
Orange. Citrus.
Orange-root. Hydras-
Orchard-grass. Dact-
Orchis. Orch-
Ox-eyed daisy. Chrysan-

Painted-cup. Bart-
Paper-mulberry. Brous-
Papoose-root. Leontice. Cl. 6. Or. 1.
Parnassus-grass. Parnas-
Parsley. Api-
Parsnip. Pasti-
Partridge-berry. Mitch-
Passion-flower. Passi-
Pea-nut. Arachis. Cl. 16. Or. 10.
Pea. Pisum.
Peach. Amyg-
Pear. Pyrus.
Pearl-wort. Sagina.
Pellitory. Parieta,
Penny-royal. Hede-
Penny-wort. Obo-
Peony. Pæo-
Pepper. Piper.
Pepper-grass. Lepid-
Peppermint. Menth-
Peperidge-tree. Nyssa.
Periwinkle. Vinca.
Persimmon. Diospy-
Pheasant-eye. Adonis.
Physic-nut. Jatro-
Pickerel-weed. Ponted-
Pig-weed. Cheno-
Pine. Pinus.
Pink. Dian-
Pink-root. Spig-
Pipe-wort. Eri-
Pipsissiwa. Chimaph-
Plantain. Plant-
Plum. Prunus.
Poke-weed. Phyto-
Poison-hemlock. Cicu.
Poison-ivy. Rhus.
Polyanthos. Narc-
Polypod. Polypo-
Pomegranate. Punica.
Pond-weed. Potam-
Poplar. Popu-
Poppy. Papav-
Potato. Sola-
Pot marigold. Calen-
Prim. Ligustrum.
Prickly-ash. Xanthor
Prickly-pear. Cactus.
Pride of China. Melia-
Prince's pine. Chimaph-
Puccon. Batsch-
Pumpkin. Cucur-
Purslane. Portu-

Quake-grass. Briza.
Queen of the meadow. Spir-
Quince. Pyrus.

Radish. Raph-
Raspberry. Rubus.
Rattle-box. Croto-
Red-cedar. Juni-
Red-pepper. Caps-
Red-top grass. Agros-
Red-root. Dila-
Reed. Arum.
Riband-grass. Phal-
Rice. Oryza.
River-nymph. Caulinia.
Rocket. Hesp-
Rock-rose. Cistus.
Rose. Rosa.
Rose-bay. Rhododen-
Rose-campion. Agrostem-
Rose-locust. Robin-
Rosemary. Rosni-
Rue. Ruta.
Ruel. Ruellia.
Rush-grass. Juncas.
Rye. Secale.

Sacred bean. Nelum-
Saffron of Europe. Crocus.
Sage. Salvia.
Salt-wort. Sals-
Salt-grass. Lim-
Salsify. Tragop
Samphire. Sali-
Sanicle. Sanic-
Sand-wort. Arenaria.
Sand-myrtle. Leioph-
Sarsaparilla. Aralia.
Sassafras. Laurus.
Satin-flower. Luna-
Savin. Juni-
Savory. Satureja.
Saxifrage. Saxif-
Scabish. Œnoth-
Scarlet pimpernel. Anagal-
Scorpion-grass. Myoso-
Scouring-rush. Equise-
Scrofula-weed. Goodye-
Scull-cap. Scu-
Sea-buckthorn. Hippo-
Sea-burdock. Xan-
Sea-kale. Brassica.
Sea-holly. Eryng-
Seasame-grass. Trip-
Self-heal. Prun-
Sensitive fern. Onoc-
Shad-flower. Aronia.
Shell-flower. Molu-
Shepherd's purse. Thlaspi.
Shield-fern. Aspid-
Shin-leaf. Pyro-
Side-saddle flower. Sarra-
Silk-weed. Ascle-
Single-seed cucumber. Sicyos.
Skunk's cabbage. Ictodes.
Sleek-leaf. Leioph-
Smellage. Ligusticum.
Snake-head. Chelone.
Snake-mouth. Pogo-
Snap-dragon. Antirr-
Snow-ball. Vibur.
Snow-berry. Sympho-
Snow-drop tree. Halesia.
Soap-wort. Sapin- Sapo-
Solomon's seal. Conval-
Southern wood. Arte-

Spanish-broom. Sparti-
Speedwell. Veron-
Spear-grass. Poa.
Spear-arum. Rens-
Spearmint. Mentha.
Spicy wintergreen. Gaultheria.
Spice-bush. Laureus.
Spider-wort. Trades-
Spikenard. Aralia.
Spindle-tree. Euon-
Spinage. Spina-
Spleen-wort. Asple-
Spring-beauty. Clay-
Spruce. Pinus.
Spurge. Euphor-
Spurry. Sper-
Squash. Cucur-
Squills. Scilla.
Star-of-Bethlehem. Ornith-
Star-flower. Aster.
Star-grass. Hyp-
Star-wort. Stel-
Stock july-flower. Cheir-
Stone-crop. Sedum.
Stork's-bill geranium. Erod-
St. John's-wort. Hyper-
St. Peter's-wort. Ascy-
Stramonium. Datu-
Strawberry. Fraga-
Succory. Cicho-
Sugar-cane. Saccha-
Sulphur-wort. Peuceda-
Sumach. Rhus.
Sun-flower. Helian-
Swamp-willow herb. Deco-
Sweet-basil. Ocy-
Sweet-brier. Rosa.
Sweet-cicely. Uras-
Sweet-flag. Acorus.
Sweet-fern. Comp-
Sweet gum-tree. Liquid-
Sweet pea. Lathy-
Sweet pepper-bush. Clethra.
Sweet vernal-grass. Anthox-
Sweet-william. Dianthus.
Swine-thistle. Sonchus.
Syringa. Phil-

Tallow-tree. Stillin-
Tamarind. Tam-
Tansey. Tana-
Tape-grass. Valis-
Tassel-flower. Cacal-
Tea. Thea.
Teasel. Dips-
Thistle. Cnicus.
Thorn-apple. Datu-
Thorn-bush. Cratæ-
Thoroughwort. Eupa-
Thread foot. Podos-
Three-bird orchis. Triph-
Three-seed mercury. Acaly-
Thyme. Thymus.
Tiger-flower. Tig-
Timothy grass. Phleum.
Tobacco. Nicotia-
Toothache-tree. Zanthox-
Tooth-cup. Amman-
Tooth-root. Dent-
Tower-mustard. Turri-
Trailing arbutus. Epig-

Trumpet-flower. Bign-
Tuberose. Polyan-
Tulip. Tulipa.
Turnip. Bras-
Tway-blade. Listera.
Twin-flower. Linnæa.

Valerian. Valer-
Vanilla-plant. Epid-
Vegetable oyster. Tragop-
Venus' fly-trap. Dionæa.
Vervain. Verbe-
Vetch. Vicia.
Violet. Viola.
Viper's bugloss. Echi-
Virginian loose-strife. Gaura.
Virginian orpine. Pentho-
Virginian snake-root. Aristo-
Virgin's-bower. Clem-

Wall cress. Arab-
Wall flower. Cheir-
Walnut. Carya.
Water-arum. Calla.
Water crown-cup. Sparg-
Water dropwort. Oznan-
Water hemp. Acni-
Water leaf. Hydro-
Water milfoil. My-
Watermelon. Cucur.
Water parsnip. Sium.
Water shield. Villar-
Water plantain. Alis-
Wax-bush. Cuph-
Wheat. Trit-
Whip grass. Sele-
White-cedar. Cupres- Thu-
White lettuce. Prenan-
White pond-lily. Nymph-
Whitlow grass. Draba.
Whortleberry. Vaccin-
Wild bean. Stropos-
Wild bean-vine. Amphi-
Wild cucumber. Momor-
Wild geranium. Gera-
Wild honeysuckle. Azal-
Wild indigo. Baptis-
Wild ladies'-slipper. Cypri-
Wild lamb-lettuce. Vale-
Wild mandrake. Podoph-
Wild oats. Dantho-
Wild pine. Tilland-
Wild rice. Ziza-
Wild tobacco. Lobel
Wild turnip. Arum.
Willow. Salix.
Willow-herb. Epil-
Winter cherry. Phys-
Witch alder. Fother-
Witch hazel. Hama-
Woad. Isatis.
Wood sorrel. Oxal-
Woodbine. Loni-

Yam root. Diosc-
Yarrow. Achill-
Yellow-eyed grass. Xyris.
Yellow-root. Zanth-
Yellow-rattle. Rhin-
Yew. Taxus.

The Greeks considered plants to
possess thoughts? and soul.
Pythagoras was the first who drew
up a catalogue of plants.

_____ Aschick ____ the first _____
the Caliphs the _____ attention the Botany —
Lord Bacon was the author of the
inductive systems of botany.

_____ followed Linnaeus.

" American Botanists "

_____ Nuttall. Torry. Eaton.

Grey. Paine _____ ____

_____ is derived from herb or grass
Lily _ flower of _____ — May 1846.

commence at page 2̶1̶8̶.
233.
201

Nag. .. 247..
Age .. 16
P. .. 32

The Greeks considered plants to
possess thoughts and soul.
Pythagorus was the first who drew
up a catalogue of plants.

??? Aschiol ??? the first ???
the Caliphs to ??? attention ??? Botany -
Lord Bacon was the author of the
inductive systems of Botany.

??? Linnaeus.
"American Botanists"

??? Nuttall . Torry . Eaton
Gray . Paine ——— ——

??? is derived from herb or grass
??? flower of ??? — Aug 1846.

wch

commence   J 233.   Nag. 247..
           201   Sage 16
                          Pap. 38

Martin J Skidmore

HW 24W9 9

CARNATIONS AND PICOTEES.—What is the difference between Carnations and Picotees? They are advertised separately in the nurserymen's catalogues, and I could never learn the difference.—Please inform me and oblige—A RURAL READER.

REMARKS.—The distinction is one made by Florists and relates merely to the coloring of the flower. The Carnation is divided into three classes, and the Picotee is one of these, as follows:—*Flakes*, are of only two colors, with wide stripes running from the center of the flower to the outer edge of the petal. *Bizarres* have at least three colors, in irregular stripes and blotches. *Picotees* are finely spotted or lined, with scarlet, purple, &c., generally on a white ground. The outer edge of the petals usually have a dark stripe. Florists have made a great many arbitrary rules as to what should be considered a good flower, which it will not, perhaps, be profitable for us to heed at present. The Carnation is a queen among the flowers, and should receive more homage from the lovers of nature's beauties.

THE TUBEROSE.—To get flowering bulbs the method is to plant out the little offsets that surround the old bulb, each Spring, in rows a foot or so apart. These will not flower the first Summer, but make plump flowering bulbs for the next. In the Fall take them up just before frost kills the top. Lay them in some place where it will not freeze, to dry off a little; then trim off the tops and keep the bulbs dry, and where the thermometer does not go below 40°, until they are wanted for planting out the ensuing Spring.

They are better started first in a hotbed or other warm place, so that their season of flowering is early enough to escape the Fall frost. The old bulbs will not flower again. The double only is worth growing; the flower is of a waxy white, and highly fragrant.

THE ART OF GROWING TREES FROM CUTTINGS.—

Professor Delacroix, of Besancon, in France, has discovered a mode of propagating from cuttings, which is not only successful in case of roses and other plants easy to live, but apples, pears, plums, apricots, &c. Out of an hundred cuttings put out in June, not one but was thriving in August in the open air, without shade or extra care, except watering a few times soon after they were planted. His method is to put the whole cutting in the ground, bent in the form of a bow, with the centre part up, and just on a level with the surface, at which point there must be a good bud or shoot, which is the only part exposed to the air; the other being protected by the earth from drying up supports and gives vigor to the bud, which starts directly into leaf, and in its turn helps the cutting to form roots and the whole them forms a thriving tree. The method of setting them is to form two drills about three inches apart, with a sharp ridge between, over which bend the cutting, and stick an end in each drill, and cover up and press the earth firmly, and water freely. Cuttings should be of the last year's growth, fresh and vigorous.

THE DAHLIA.

EDS. RURAL:—I desire some information concerning Dahlias. I had but one sprout last spring, which was, by the way, a good bloomer, having over one hundred and fifty flowers. I took it up last week and find the roots corresponding with the stock, several large ones weighing from one to two pounds, and measuring from four inches to twenty. Now I desire to save all that will grow; and 1st—How shall I keep them? 2d—How shall I divide from stock and from each other? 3d—To tell those that will grow and bloom from those tha will not. In answering the above you will greatly oblige—A READER, *Monongahela City, Pennsylvania, Nov.*, 1857.

REMARKS.—As soon as the roots are taken up in the fall, expose them for a few days to the sun and air until they become dry. Then prepare a sufficient quantity of clean sand by thoroughly drying it in the sun, or by a stove; place a layer of the sand in the bottom of a clean barrel and on this place a layer of roots; then another layer of sand. Keep adding alternate layers of sand and roots till full, and then place the barrel in a cool, dry place in the cellar, where they will remain in perfect order until spring. To propagate plants, take a heavy, sharp knife, and separate the roots by cing clean through the collar of the plant, or part of union between the stem and root. The e or buds of the future plant will be found clustd around the collar. Preserve on en root and plant in the border, and i c shoot presents itself when the nce growing it must be rubbed off the strong roots will flower if we season is favorable. While on th

THE DAHLIA AND ITS CULTURE.

EDS. RURAL:—A few remarks may not be uninteresting to some of your readers, on the culture of the Dahlia, the king of border plants, and the glory of the flower garden. My observations may ot be of any great value to practical men, but I ave no doubt they will prove so to many of your eaders. The Dahlia is becoming more popular it deserves to be, but its beauties are not fully eloped under common treatment. The first int is to select good varieties, and it is a well own fact among practical florists that the best rieties of any hybrid plants are the first to degenerate under improper treatment. I would advise all that have Dahlias out in the ground to cut eir tops to within six inches of the surface, ecuring their names to a stalk, and then covering e root with eight or ten inches of short manure, leaf mold. This secures them against injury all slight frosts, and allows the roots to ripen, that they will keep much better through the inter, and be stronger in the spring. The roots ould be taken up carefully and turned top downards until dry; then they may be stored away in a dry place, where they will not be subject to ither frost or heat. It will be necessary to examine them occasionally to see that no mold makes its appearance, and if any is discovered remove them to a dryer place until it is subdued.